FISHERIES TECHNIQUES
SECOND EDITION

Fisheries Techniques

is a special project of the

Education Section of the American Fisheries Society.

Funding support for the publication of this book

was provided by

Advanced Telemetry Systems, Inc.
John D. Arndt
Carolina Power & Light Company
Chevron USA, Inc.
Duke Power Company
Edison Electric Institute
New York Power Authority
Sonotronics
Virginia Power

FISHERIES TECHNIQUES
SECOND EDITION

Edited by

Brian R. Murphy

Department of Fisheries and Wildlife Sciences
Virginia Polytechnic Institute and State University

and

David W. Willis

College of Agriculture and Biological Sciences
South Dakota State University

American Fisheries Society
Bethesda, Maryland, USA
1996

Suggested Citation Formats

Entire Book

Murphy, B. R., and D. W. Willis, editors. 1996. Fisheries techniques, 2nd edition. American Fisheries Society, Bethesda, Maryland.

Chapter within the Book

Bettoli, P. W., and M. J. Maceina. 1996. Sampling with toxicants. Pages 303–333 *in* B. R. Murphy and D. W. Willis, editors. Fisheries techniques, 2nd edition. American Fisheries Society, Bethesda, Maryland.

Original drawings and modifications to figures by
Betsy A. Sturm
Fairbanks, Alaska

Library of Congress Catalog Number 96-85438

ISBN 1-888569-00-X

Printed in the United States of America on recycled, acid-free paper.

American Fisheries Society
5410 Grosvenor Lane, Suite 110
Bethesda, Maryland 20814-2199, USA

Dedication

The editors wish to dedicate this book to the fisheries scientists who were most instrumental in our development as educators and fisheries professionals: Gary J. Atchison (BRM), Larry A. Nielsen (BRM), Stephen A. Flickinger (DWW), Donald W. Gabelhouse, Jr. (DWW), and Richard O. Anderson (BRM, DWW).

In Memoriam

Special dedication to the memory of chapter author, fisheries professional, educator, and friend Richard Frie (1957–1995), who passed away during the final stages of preparing this edition.

Contents

Contributors

Richard O. Anderson (Chapter 15): 3618 Elms Court, Missouri City, Texas 77459, USA.

Douglas J. Austen (Chapter 2): Illinois Department of Natural Resources, Division of Fisheries, 600 North Grand Avenue West, Springfield, Illinois 62701, USA.

Charles R. Berry, Jr. (Chapter 3): National Biological Service, South Dakota Cooperative Fish and Wildlife Research Unit, South Dakota State University, Box 2140B, Brookings, South Dakota 57007, USA.

Phillip W. Bettoli (Chapter 10): Tennessee Tech University, Box 5114, Cookeville, Tennessee 38505, USA.

H. Lee Blankenship (Chapter 12): Washington Department of Fish and Wildlife, 600 Capitol Way North, Olympia, Washington 98501-1091, USA.

Stephen H. Bowen (Chapter 17): Michigan Tech University, 1400 Townsend Drive, Houghton, Michigan 49931, USA.

Stephen B. Brandt (Chapter 13): Buffalo State College, Great Lakes Center, C215 Classroom Building, 1300 Elmwood Avenue, Buffalo, New York 14222-1095, USA.

Michael L. Brown (Chapter 2): South Dakota State University, Department of Wildlife and Fisheries Science, Box 2140B, Brookings, South Dakota 57007, USA.

Dennis R. DeVries (Chapter 16): Auburn University, Department of Fisheries and Allied Aquaculture, Auburn, Alabama 36849, USA.

C. Andrew Dolloff (Chapter 18): U.S. Forest Service, Department of Fisheries and Wildlife Sciences, 138 Cheatham Hall, Blacksburg, Virginia 24061-0321, USA.

Mary C. Fabrizio (Chapter 21): Great Lakes Lab, 1451 Green Road, Ann Arbor, Michigan 48105, USA.

C. Paola Ferreri (Chapter 7): Pennsylvania State University, School of Forest Resources, 207 Ferguson Building, University Park, Pennsylvania 16802, USA.

Richard V. Frie (Chapter 16): University of Wisconsin, College of Natural Resources, Stevens Point, Wisconsin 54481, USA.

Christopher S. Guy (Chapter 12): Kansas Cooperative Fish and Wildlife Research Unit, Kansas State University, 205 Leasure Hall, Manhattan, Kansas 66506, USA.

Daniel B. Hayes (Chapter 7): Michigan State University, Department of Fisheries and Wildlife, 13 Natural Resources Building, East Lansing, Michigan 48824, USA.

Wayne A. Hubert (Chapter 6): National Biological Service, Wyoming Cooperative Fisheries Research Unit, Box 3166, Biological Sciences Building, Laramie, Wyoming 82071, USA.

Steven W. Kelsch (Chapter 5): University of North Dakota, Department of Biology, Box 9019, Grand Forks, North Dakota 58202, USA.

William E. Kelso (Chapter 9): Louisiana State University, School of Forestry, Wildlife and Fisheries, LSU Agricultural Center, Baton Rouge, Louisiana 70803, USA.

Jeffrey Kershner (Chapter 18): Utah State University, Department of Fish and Wildlife, Logan, Utah 84322-5210, USA.

Barbara A. Knuth (Chapter 22): Cornell University, Department of Natural Resources, 122A Fernow Hall, Ithaca, New York 14853, USA.

Michael J. Maceina (Chapter 10): Auburn University, Department of Fisheries and Allied Aquaculture, 203 Swingle Hall, Auburn, Alabama 36849, USA.

Stephen P. Malvestuto (Chapter 20): Fishery Information Management Systems, P.O. Box 3607, Auburn, Alabama 36831-3607, USA. *Present address*: 2032 Whelk Way, St. George Island, Florida 32328, USA.

Thomas E. McMahon (Chapter 4): Montana State University, Biology Department, Fish and Wildlife Program, Bozeman, Montana 59717, USA.

Steve L. McMullin (Chapter 22): Virginia Polytechnic Institute and State University, Department of Fisheries and Wildlife Sciences, Blacksburg, Virginia 24061-0321, USA.

Brian Murphy (Coeditor; Chapter 1): Virginia Polytechnic Institute and State University, Department of Fisheries and Wildlife Sciences, Blacksburg, Virginia 24061-0321, USA.

Robert M. Neumann (Chapter 15): University of Connecticut, Department of Natural Resources Management and Engineering, 1376 Storrs Road, U-87, Storrs, Connecticut 06269-4087, USA.

Larry A. Nielsen (Chapter 12): Pennsylvania State University, School of Forest Resources, 113 Ferguson Building, University Park, Pennsylvania 16802, USA.

Donald J. Orth (Chapter 4): Oak Ridge National Laboratory, Environmental Science Division, Building 1505 MS 6038, Oak Ridge, Tennessee 37831-6038, USA.

Charles F. Rabeni (Chapter 11): National Biological Service, Missouri Cooperative Fish and Wildlife Research Unit, 112 Stephens Hall, Columbia, Missouri 65211, USA.

James B. Reynolds (Chapter 8): National Biological Service, Alaska Cooperative Fish and Wildlife Research Unit, P.O. Box 757020, Fairbanks, Alaska 99775-7020, USA.

R. Anne Richards (Chapter 21): National Marine Fisheries Service, Northeast Fisheries Center, Woods Hole, Massachusetts 02543, USA.

D. Allen Rutherford (Chapter 9): Louisiana State University, School of Forestry, Wildlife and Fisheries, LSU Agricultural Center, Baton Rouge, Louisiana 70803, USA.

Barbara Shields (Chapter 5): University of Michigan, Fisheries Division, Museum of Zoology, Ann Arbor, Michigan 48109-1079, USA.

Richard J. Strange (Chapter 14): University of Tennessee, Department of Forestry, Wildlife, and Fisheries, Box 1071, Knoxville, Tennessee 37901, USA.

William W. Taylor (Chapter 7): Michigan State University, Department of Fish and Wildlife, East Lansing, Michigan 48824, USA.

Russell Thurow (Chapter 18): U.S. Forest Service, Intermountain Research Station, 316 East Myrtle Street, Boise, Idaho 83702, USA.

David W. Willis (Coeditor; Chapter 1): South Dakota State University, Department of Wildlife and Fisheries Sciences, NPBL 138, Box 2140B, Brookings, South Dakota 57007-1696, USA.

Jim Winter (Chapter 19): Texas Tech University, Department of Range, Wildlife and Fisheries Management, Lubbock, Texas 79409, USA.

Alexander V. Zale (Chapter 4): National Biological Service, Montana Cooperative Fish Research Unit, Department of Biology, Bozeman, Montana 59717, USA.

List of Species

The colloquial names of many fish species have been standardized in *Common and Scientific Names of Fishes from the United States and Canada* (5th edition, 1990) and *World Fishes Important to North Americans* (1990), published by the American Fisheries Society. The reference for vertebrate species is *Checklist of Vertebrates of the United States, the U.S. Territories, and Canada* (1987), published by the U.S. Fish and Wildlife Service; for insects the reference is *Common Names of Insects & Related Organisms* (1989), published by the Entomological Society of America. Throughout this book, species listed in those publications are cited only by common name. The respective scientific names of these species follow.

Fish

albacore .... *Thunnus alalunga*
alewife .... *Alosa pseudoharengus*
American brook lamprey .... *Lampetra appendix*
American plaice .... *Hippoglossoides platessoides*
American shad .... *Alosa sapidissima*
angel shark .... *Squatina californica*
Antarctic toothfish .... *Dissostichus mawsoni*
Atlantic cod .... *Gadus morhua*
Atlantic croaker .... *Micropogonias undulatus*
Atlantic halibut .... *Hippoglossus hippoglossus*
Atlantic herring .... *Clupea harengus*
Atlantic mackerel .... *Scomber scombrus*
Atlantic menhaden .... *Brevoortia tyrannus*
Atlantic salmon .... *Salmo salar*
Atlantic silverside .... *Menidia menidia*

basking shark .... *Cetorhinus maximus*
bigmouth buffalo .... *Ictiobus cyprinellus*
black bullhead .... *Ameiurus melas*
black crappie .... *Pomoxis nigromaculatus*
blue catfish .... *Ictalurus furcatus*
blue tilapia .... *Tilapia aurea*
blueback herring .... *Alosa aestivalis*
bluegill .... *Lepomis macrochirus*
brook trout .... *Salvelinus fontinalis*
brown trout .... *Salmo trutta*
bullheads .... *Ameiurus* spp.
burbot .... *Lota lota*

capelin .... *Mallotus villosus*
chain pickerel .... *Esox niger*
channel catfish .... *Ictalurus punctatus*
chinook salmon .... *Oncorhynchus tshawytscha*

coelocanth .... *Latimeria chalumnae*
coho salmon .... *Oncorhynchus kisutch*
Colorado squawfish .... *Ptychocheilus lucius*
common carp .... *Cyprinus carpio*
cutthroat trout .... *Oncorhynchus clarki*

delta smelt .... *Hypomesus transpacificus*

Eurasian perch .... *Perca fluviatilis*
European smelt .... *Osmerus eperlanus*

flathead catfish .... *Pylodictis olivaris*
freshwater drum .... *Aplodinotus grunniens*

gizzard shad .... *Dorosoma cepedianum*
golden shiner .... *Notemigonus crysoleucas*
goldfish .... *Carassius auratus*
goosefish .... *Lophius americanus*
grass carp .... *Ctenopharyngodon idella*
grayling .... *Thymallus thymallus*
green sunfish .... *Lepomis cyanellus*

haddock .... *Melanogrammus aeglefinus*
half-banded sea perch .... *Hypoplectrodes maccullochi*

inland silverside .... *Menidia beryllina*
Iowa darter .... *Etheostoma exile*

king mackerel .... *Scomberomorus cavalla*

lake herring .... *Coregonus artedi*
lake trout .... *Salvelinus namaycush*
lake whitefish .... *Coregonus clupeaformis*
largemouth bass .... *Micropterus salmoides*
lemon shark .... *Negaprion brevirostris*
longear sunfish .... *Lepomis megalotis*

mosquitofish .... *Gambusia* spp.
mottled sculpin .... *Cottus bairdi*
mountain whitefish .... *Prosopium williamsoni*
Mozambique tilapia .... *Tilapia mossambica*
mullets .... *Mugil* spp.
muskellunge .... *Esox masquinongy*

northern anchovy .... *Engraulis mordax*
northern pike .... *Esox lucius*

ohrid rifle minnow .... *Alburnoides bipunctatus ohridanus*

Pacific cod .... *Gadus macrocephalus*
Pacific hake .... *Merluccius productus*
Pacific halibut .... *Hippoglossus stenolepis*
Pacific herring .... *Clupea pallasi*
paddlefish .... *Polyodon spathula*
palmetto bass .... *Morone saxatilis* × *M. chrysops*
plaice .... *Pleuronectes platessa*

pumpkinseed .... *Lepomis gibbosus*

rainbow smelt .... *Osmerus mordax*
rainbow trout .... *Oncorhynchus mykiss*
razorback sucker .... *Xyrauchen texanus*
red drum .... *Sciaenops ocellatus*
red shiner .... *Cyprinella lutrensis*
redbreast sunfish .... *Lepomis auritis*
redear sunfish .... *Lepomis microlophus*
redfish .... *Sebastes* spp.
rock bass .... *Ambloplites rupestris*
rough silverside .... *Membras martinica*
roughtongue kapenta .... *Limnothrissa miodon*

sauger .... *Stizostedion canadense*
saugeye .... *Stizostedion vitreum* × *S. canadense*
sea lamprey .... *Petromyzon marinus*
shortnose sturgeon .... *Acipenser brevirostrum*
shad (freshwater) .... *Dorosoma* spp.
silver lamprey .... *Ichthyomyzon unicuspis*
smallmouth bass .... *Micropterus dolomieu*
smallmouth buffalo .... *Ictiobus bubalus*
sockeye salmon .... *Oncorhynchus nerka*
South American lungfish .... *Lepidosiren paradoxa*
southern flounder .... *Paralichthys lethostigma*
spot .... *Leiostomus xanthurus*
spotfin shiner .... *Cyprinella spiloptera*
spotted bass .... *Micropterus punctulatus*
spotted sucker .... *Minytrema melanops*
sprat .... *Sprattus sprattus*
squawfish .... *Ptychocheilus* spp.
steelhead .... *Oncorhynchus mykiss*
striped bass .... *Morone saxatilis*

threadfin shad .... *Dorosoma petenense*
topminnows .... *Fundulus* spp.

walleye .... *Stizostedion vitreum*
warmouth .... *Lepomis gulosus*
white bass .... *Morone chrysops*
white crappie .... *Pomoxis annularis*
white perch .... *Morone americana*
white sturgeon .... *Acipenser transmontanus*
white sucker .... *Catostomus commersoni*

yellow bass .... *Morone mississippiensis*
yellow bullhead .... *Ameiurus natalis*
yellow perch .... *Perca flavescens*
yellowfin tuna .... *Thunnus albacares*

Other Animals

American alligator .... *Alligator mississippiensis*
American lobster .... *Homarus americanus*

Antarctic krill .... *Euphasia superba*

common frog .... *Rana temporaria*

estuarine crocodile .... *Crocodylus porosus*

green sea turtle .... *Chelonia mydas*

harp seal .... *Phoca groenlandica*
hide beetle .... *Dermestes maculatus*
hooded seal .... *Cystophora cristata*

leatherback turtle .... *Dermochelys coriacea*
loggerhead turtle .... *Caretta caretta*

manatee .... *Trichechus manatus*

northern shrimp .... *Pandalus borealis*
northern water snake .... *Nerodia sipedon*

red king crab .... *Paralithodes camtschaticus*

sea scallop .... *Placopecten magellanicus*
sea urchin .... *Stongylocentrotus drobachiensis*
sperm whale .... *Physeter macrocephalus*
spiny lobster .... *Panulirus cygnus*

Weddell seal .... *Leptonychotes weddelli*

Preface

This text represents a complete revision of the popular first (1983) edition of *Fisheries Techniques*. Like the first, this edition was created under the auspices of the Education Section of the American Fisheries Society. Like the first, it exists because concerned fisheries scientists and educators contributed thousands of hours of volunteer effort to it.

This text is written as an introduction to the ways in which fisheries data are collected. The book is primarily intended for graduate and advanced undergraduate courses in fisheries curricula, but it is written so that students with little or no previous exposure to fisheries techniques can use it. Secondarily, the book is intended to serve as a basic reference for practicing fisheries professionals. Therefore, each chapter begins with a basic explanation of the technique(s) under discussion, and then elaborates in increasing detail the uses, advantages, and limitations of those technique(s).

In most cases, discussion is limited to application of the techniques themselves; interpretation of the data collected with these techniques is generally beyond the scope of this book, although some suggestions for further reading on that topic are given. As with any introductory book of this type, every facet of the techniques could not be described here. Readers are encouraged to consult the references in each chapter for in-depth discussions of various techniques. Likewise, it was impossible to include every fisheries data collection technique that a practicing professional may encounter, but we have endeavored to include all of the commonly used techniques.

Some changes from the first edition bear elaboration here. A glossary was added to this edition to improve its functionality as a teaching tool. Chapters also were added to cover statistical considerations in fisheries sampling, and the sampling of invertebrate organisms. Conversely, several chapters from the first edition were combined or eliminated to reduce redundancy with more recent publications by the American Fisheries Society (*Inland Fisheries Management in North America* and *Methods for Fish Biology*).

Each chapter in the book was reviewed in outline and draft, as listed below. In selecting reviewers, we attempted to represent both the academic and fisheries management communities to assure that this book remained a useful tool to both groups. For their volunteer efforts as reviewers we thank C. Adams, S. Ahlstedt, C. A. Annett, N. A. Auer, E. P. Bergersen (2), N. Billington, J. C. Boxrucker, F. J. Bulow, B. M. Burr, R. F. Carline, D. Christensen, C. E. Cichra, D. W. Coble, W. G. Duffy, K. D. Fausch, S. A. Flickinger, L. M. Gigliotti, W. Grabriel, D. M. Green, R. S. Hayward, T. Hillman, J. Hoxmeier, B. L. Johnson, B. M. Johnson, D. J. Jude, C. C. Kohler, A. L. Kolz, V. B. Kuechle, J. J. Magnuson, S. P. Malvestuto, S. L. McMullin, M. R. Meador, L. E. Miranda, L. R. Mitzner, T. D. Mosher, J. J. Ney, R. L. Noble, V. M. O'Connell, L. L. Olmsted, D. Partridge, E. P. Pister, J. M. Pitlo, Jr., G. R. Ploskey, C. W. Prophet, M. R. Ross, L. G. Rudstam, D. B. Sampson, C. G. Scalet, S. Schaffer, D. J. Schill, R. J. Sheehan, A. J. Temple, J. W. Terrell, D. H. Wahl, A. S. Weithman, B. G. Whiteside, and G. R. Wilde.

Many of the chapters in this publication are based on chapters in the first edition that were prepared by other authors. The editors and chapter authors gratefully acknowledge the contributions of those first-edition authors on whose work some of this book is based. Some illustrations by Susan Lampton in the first edition are reproduced here, including the cover art.

The editors appreciate the assistance with Chapter 1 provided by David Johnson and Larry Nielsen, the authors of a similar chapter in the first edition. Many ideas and sections of the original chapter were used in this book with their gracious permission. Steve Kelsch and Barbara Shields (Chapter 5) acknowledge assistance provided by Anne Kelsch and Donovan Verrill. Wayne Hubert (Chapter 6) thanks Christine Waters for editorial assistance. Jim Reynolds (Chapter 8) thanks Kathy Pearse for typing and editorial assistance. William Kelso and Allen Rutherford (Chapter 9) acknowledge C. Fred Bryan, William Herke, James Ditty, and Darrel Snyder for their valuable assistance. Charles Rabeni (Chapter 11) thanks the Wildlife Supply Co. for providing several of the photographs in his chapter. Richard Strange (Chapter 14) acknowledges Tom Hill and Victoria Clyde as internal reviewers. Richard Anderson and Robert Neumann (Chapter 15) acknowledge the work of Stephen Gutreuter, a coauthor of this chapter in the first edition. Dennis DeVries and Richard Frie (Chapter 16) appreciate Brett Johnson's contribution of the figure included in Box 16.1, Dave Willis' contribution of the table included in Box 16.2, and the Wisconsin Department of Natural Resources's contribution of the photograph used in Figure 16.4. Some of this chapter is based on A. Jearld's chapter on age determination that appeared in the first edition, including a number of the original figures. Partial support for the preparation of Chapter 16 came from National Science Foundation grants DEB-9108986 and DEB-9410323. Andy Dolloff, Jeff Kershner, and Russell Thurow (Chapter 18) thank Dave Beauchamp for assistance provided during this project and Linc Freeze for providing photographs. Jim Winter (Chapter 20) thanks V. B. "Larry" Kuechle, Richard Huempfner, and Donald Brumbaugh, who made numerous suggestions for improving the text. Mary Fabrizio and Anne Richards (Chapter 21 which also is contribution 888 of the National Biological Service, Great Lakes Fisheries Center) thank Ann Zimmerman for assisting with literature searches and retrieval of references. Finally, Barbara Knuth and Steve McMullin (Chapter 22) thank Nancy Connelly and Tommy Brown for helpful suggestions on content and for permission to include their survey instrument in the chapter.

The editors thank the Steering Committee members who were involved in various stages of planning this project: Doug Austen, Mike Brown, Cynthia Jones, Robert Kendall, Steve McMullin, John Ney, Terry Roelofs, Chuck Scalet, Jim Schooley, Beth Staehle, and Bruce Wilkins. We also thank volunteers who helped to construct, review, and improve the glossary: Mike Allen, Michael L. Brown, Eric M. Hallerman, L. Esteban Miranda, and Harold L. Schramm. Brian R. Murphy also thanks Kathryn Fabrycky, Vivian Gonzales, Diana Murphy, and John Murphy for editorial assistance throughout the preparation of this volume.

Finally, we thank Paul Brouha and John Fritts for their efforts in fund-raising for this project. The authors also wish to thank American Fisheries Society Education Section Past Presidents Don Orth, Joseph Margraf, Wayne Hubert, and Christopher Kohler for their support of this project; Janet Harry for editorial assistance, and Beth Staehle and Robert Kendall for their friendship and guidance during the entire production cycle.

The mention of specific products in this book does not constitute endorsement of these products by either the authors or the American Fisheries Society.

BRIAN R. MURPHY
DAVID W. WILLIS

Chapter 1

Planning for Sampling

DAVID W. WILLIS AND BRIAN R. MURPHY

1.1 PLANNING: AN ESSENTIAL PREREQUISITE FOR SUCCESSFUL MANAGEMENT AND RESEARCH

Planning is essential for successful fishery management and research (see Box 1.1 for definitions of fishery, fishery management, and fishery research). Planning is needed to decide goals for a fishery and to establish measurable objectives. Once goals and objectives for a fishery have been established, then a series of steps for conducting investigative work can be planned. An often overlooked aspect of planning is the logistics of the sampling. Not only do biologists need to decide what, how, when, where, and why to sample, but they also need to plan for such items as travel, equipment, supplies, personnel requirements, and so forth. Failure to plan adequately for sampling will result in less effective use of economic and worker resources.

1.1.1 Justification for Sampling

Before sampling can be undertaken and data collected, the justification for sampling must be clearly defined and a sampling plan must be created. Clear justification for sampling by a biologist is essential, especially given the high level of public interest and involvement that typically occurs in fishery management and research. To ensure public support, biologists must be prepared to explain the need for a research project or the basis for a particular management plan. Such explanation may occur at public meetings, through newspaper or magazine articles written by the biologist, or through interviews in newspapers or on television.

The planned research must address an identified problem that cannot be solved by use of existing information. If a particular problem is deemed important, the first effort toward addressing it should be an evaluation of historical records and scientific literature. Only if the answers are truly unavailable, and the question is of sufficiently high priority, should an investigation be undertaken. The clientele being served, perhaps the anglers fishing a particular water body or citizens concerned about water quality in a local stream, should also recognize the problem and understand the approach being taken to achieve a solution. Regular communication with client groups concerning the problem, research objectives, and progress made will help create a valuable partnership.

The fishery management process has five fundamental steps: definition of goals, selection of objectives, identification of questions to ask, implementation of actions, and evaluation of actions. The insight provided by the evaluation step sometimes

Box 1.1 Definitions of Fisheries Terms

Community: All populations within a defined geographic area at a given time.

Fishery: A system composed of three interacting components: *habitat*, the environment, including both living and nonliving components, in which an organism lives; *biota*, the living organisms in an ecosystem, including fishes, plankton, aquatic insects, birds, and mammals; and *humans*, who are both users of fishery resources (e.g., recreational anglers or commercial harvesters) and competitors for water.

Fishery management: Manipulation of the three interacting elements in a fishery to meet intended and desirable objectives.

Fishery research: Diligent and systematic inquiry to develop methods, facts, or principles.

Growth: The addition of biomass by individuals, and thus accrued to the population, and generally measured as change in weight; also, increase in length of individuals.

Mortality: The death of fishes, which can be subdivided into natural mortality and fishing mortality.

Population: All individuals of the same species within a defined geographic location at a given time.

Population dynamics: The study of changes in numbers or biomass of organisms in populations and the factors (recruitment, growth, and mortality) that influence those changes.

Population structure: The size structure, age structure, or sex ratio of a population. The three dynamic rate functions (recruitment, growth, and mortality) interact and result in population structure.

Recruitment: Number of fish hatched or born in any year that survive to reproductive size; also, the number of individuals that reach a harvestable size, a particular size or age, or a size captured by a particular sampling gear.

leads to new goals and objectives, and the process begins anew. Further explanation of the management process, as detailed by Krueger and Decker (1993), will not be repeated here.

Because there are economic, social, political, and ecological components that influence the management process, setting goals and objectives is not easy. Both external and internal forces can affect the management process. External forces include the viewpoints of the various user groups with respect to the fishery in question. Internal forces include the diversity of viewpoints and knowledge among the individuals within a conservation agency. Different user groups often have different goals for the same fishery, and conservation agency personnel often espouse different approaches to an optimal solution. Ultimately, administrators are responsible for establishing priorities for various projects.

For the purposes of this chapter, we will assume that goals and objectives have already been decided, and we will focus specifically on the investigation process that can be followed to draw management conclusions.

1.1.2 Top-Down Planning

The top-down planning process (Phenicie and Lyons 1973) allows tasks to be accomplished in a logical, explicit manner. In this process, the objective is attained if and only if questions can be answered at some level of analysis. The process continues down through successive levels until all questions can be answered with available techniques.

Assume that you are the management biologist in charge of "Pristine Lake." There had been local complaints about the quality of the fishery in this lake, a public meeting was held, and administrators in your agency decided that evaluation of the problem was of high priority. The top-down planning process can be applied to the Pristine Lake scenario (Figure 1.1). Note that the three components of a fishery (habitat, biota, and human users) are all included in level 2 questions. To answer the primary question of the top-down plan (level 1), a biologist will need to determine recruitment, growth, and mortality of largemouth bass and bluegills, the fish community composition, water quality characteristics throughout one entire year, vegetation characteristics, and angler harvest.

1.2 STEPS IN CONDUCTING AN INVESTIGATION

Fishery management is driven by the knowledge base provided by fishery research. Management biologists must be able to conduct applied research or be effective at interpreting and applying research completed by others.

Sampling is undertaken so that judgements about the entire situation can be made by looking at only a portion of the situation. Sampling is essential in fishery management because it is rarely possible to observe completely any of the three primary components of a fishery. Fishery scientists sample fish populations, the environments in which these populations live, and the human users of fish populations. Sampling is also undertaken to evaluate important interrelationships among the three fishery components, such as how human-induced changes in aquatic ecosystems affect fish populations and fisheries (Hayward and Margraf 1987). How samples are collected determines how well the sampling data will reflect reality and will help achieve an understanding of the three dynamics of a fishery. Sampling, as part of a particular investigation, is frequently done by fishery management and research personnel in a state or federal conservation agency, by undergraduate or graduate students and faculty at universities, and by private fishery consultants.

1.2.1 Review the Literature

The first step in any investigation is a careful review of previous work. At times, this may involve primarily a review of published literature; at other times, it may involve a review of historic sampling data.

Computer technology has greatly improved the process of literature review. Most university libraries have computer search capabilities wherein key words can be entered and a list of pertinent references can be obtained from a database. One such database is *Fish & Fisheries Worldwide* (National Information Services Corp., Baltimore, Maryland), which requires a CD–ROM drive and is available in many university libraries. The advent of interactive personal computer systems—the computer information "superhighway"—will soon allow computer literature searches at the most remote field offices. Substantial changes in the ease of information access and transfer are expected within the next decade.

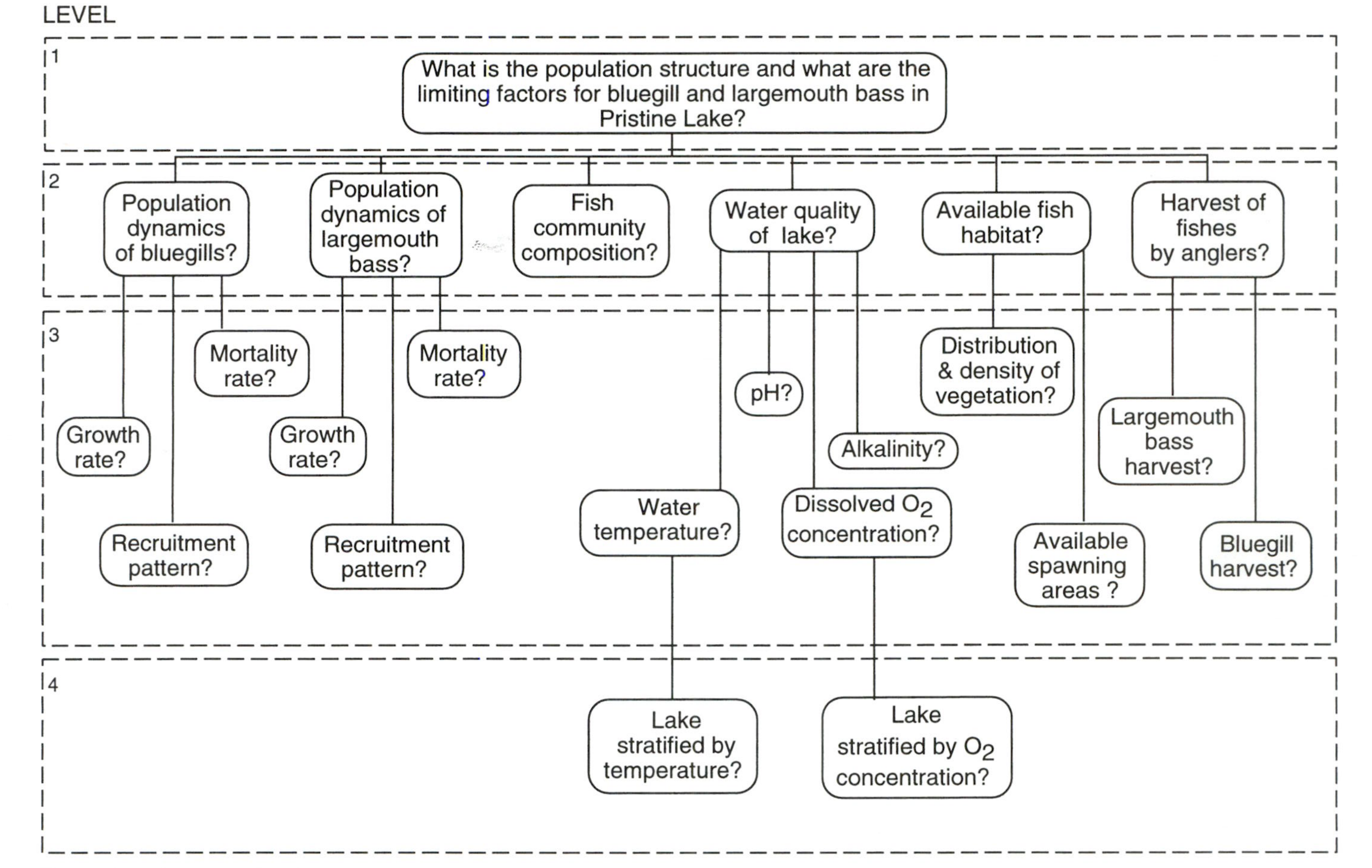

Figure 1.1 An example of top-down planning (Phenicie and Lyons 1973) for Pristine Lake. The question in the uppermost box (level 1) can be answered only after the questions in level 2 are answered. Similarly, questions in level 2 can be answered only after level 3 questions are answered, and so forth (modified from Johnson and Nielsen 1983).

1.2.2 Assess the Environment

Once the literature review has been completed, biologists should assess the biological, chemical, and physical aspects of the environment in which they are working and assess the influence of these aspects on the fish population or community under study. The environment—the first component of the fishery triumvirate—influences resident fish populations and, in turn, the fish populations influence the environment. Chemical aspects of the environment that might be of interest include nutrients (e.g., phosphorus) that influence productivity and oxygen levels; physical aspects might include temperature and basin morphometry. Temperature affects the fish community that can exist in a given body of water. Biological aspects of the environment that might be of interest include the type and amount of aquatic plant growth in a water body, as well as the diversity and abundance of aquatic invertebrates. Aquatic habitat measurements are discussed in Chapter 4 and invertebrate sampling in Chapter 11.

Perhaps the goal of a client group is to provide a rainbow trout fishery in a water body. A fishery biologist could determine whether the water temperature regime over the entire year is suitable for rainbow trout. If it is not suitable, much time and effort can be saved, and poor public relations can be avoided if an ill-conceived stocking program does not occur.

Perhaps a decrease in the abundance of a key prey fish species in a water body has been observed. Managers fear the consequences for popular, predatory sport fish, and they want to know why the prey have declined and what the decline means for the predators. A biologist might determine that the growth and abundance of predatory fishes in the water body will ultimately be reduced by the decline in prey abundance, and further, that the prey fish decline has been caused by pollution. Knowing this indirect environmental effect on a publicly valued resource, fishery managers can mobilize their principal clientele (fishers) and their counterparts in other agencies to change the (probably uninformed) public behavior that led to the water body's pollution.

1.2.3 Sample the Biota

The aquatic biota, including fishes, is the second component of a fishery. Sampling fish is very much a matter of selecting appropriate gear. Some gears are much more effective than others for collecting particular fish species. For example, largemouth bass cannot be collected effectively with trap or gill nets (Chapter 6), but they can be effectively sampled by electrofishing (Chapter 8). In addition, some gears may be more effective for particular sizes of a fish species. For many fish species, for example, trap nets more effectively capture larger individuals than smaller individuals (Laarman and Ryckman 1982).

Seasonal changes in vulnerability occur for various fish species and sizes of fish within a species. For example, largemouth bass are generally nearest shore and in shallower water during spring and fall. Thus, they are more effectively captured by electrofishing at those times of year, because electrofishing is more effective in shallow-water than in deepwater sampling.

Other components of the aquatic biota also are typically sampled, such as the aquatic invertebrates (Chapter 11). An understanding of the diversity and abundance of aquatic invertebrates on which fishes feed could provide much insight into the dynamics of a fish community.

The actual sites of data collection must be carefully selected. Because the biota of an entire water body can rarely be collected entirely, a sample must be collected according to a sampling design. Common sampling designs include simple random sampling, stratified random sampling, clustered sampling, and systematic sampling. These sampling designs are explained further in section 2.5.2.

1.2.4 Sample the Catch and Harvest

The human dimension is the third component of a fishery. Fishery biologists commonly use creel and port surveys to assess the total landings and to estimate the total catch represented by landings (Chapters 20 and 21). Very often, surveys are designed in a stratified random statistical design (Chapter 2). But preferences of human users can also be assessed. For example, many state conservation agencies complete angler use and preference studies. Agency biologists seek to determine which fish species and what types of water bodies anglers prefer. Similarly, they may assess the sizes of fish that anglers prefer to catch—for example, many small fish or a few large fish.

The economic value of fishery resources is another aspect of the human dimension. Knowledge of the economic value of a fishery (Chapter 22; Weithman 1993) can be valuable to a manager who must convince a variety of interest groups that sport or commercial fishing may be one of the better uses for a water body with limited water recharge rates and many competing uses. For example, a demonstration that sport or commercial fishing returns greater economic value should influence allocation decisions by policy makers.

1.2.5 Determine How Much to Sample

If field data are to be useful, they must be acquired according to established statistical procedures. Thus, samples from all three components of a fishery must be statistically reliable. Fishery field data tend to be quite variable and often require substantial sampling effort and a careful design. For example, assume that you are sampling white crappies in trap nets. You will find that some nets catch no white crappies and a few nets catch many. If only two nets were used, one catching 0 white crappies and the other catching 100, would you be comfortable stating that you averaged 50 crappies per net? Of course you wouldn't. A sufficient number of nets will need to be set to obtain a reliable average. Experimental design, sampling design, and data analysis are covered in depth in Chapter 2.

1.2.6 Determine What Data to Collect

A decision must be made as to what information to collect from sampled fishes. Common measurements include the number of fishes caught per unit operation of a gear, which at times serves as an index of relative abundance for a population (Box 6.1). Fish length data are often collected to assess size structure (Chapter 15), and weight–length information may be used to determine fish condition (relative plumpness, Chapter 15). Scales or other bony structures are sometimes removed from sampled fishes so that age and growth can be assessed (Chapter 16). Perhaps information needs to be collected on other components of the fishery, such as aquatic invertebrates (Chapter 11) or social and economic factors (Chapter 22).

The specific information collected depends primarily on the objectives established for an investigation. The more in-depth an investigation, the more expensive that evaluation becomes in terms of worker hours (Box 1.2). The challenge is to maximize the benefit : cost ratio in terms of effective fisheries management.

Box 1.2 A Cost–Benefit Hierarchy for Information about Fish Populations and Communities (from Johnson and Nielsen 1983)

The lowest-cost level of information available about fish communities is how many species are present and susceptible to a given sampling gear. That information is relatively easy to obtain but may be of limited value. At the other end of the cost scale are radio tracking and food habit information, which are expensive to collect but which may have high value. Between these extremes are activities that will provide more information at increasing cost. A list of these activities and an estimate of their relative cost follows. Each activity is compared with the first activity.

Activity	Information	Relative cost	Comments
Species enumeration	Number of species present	1	Useful in sampling
Numbers of fish caught of each species	Relative abundance of the species present	×2	Usually the minimal level of information needed
Length of fish	Relative year-class strength, growth, and mortality, especially for young fish; proportional stock density (see Chapter 15)	×4	A great deal of helpful information added, particularly for fast-growing fishes at a temperate latitude
Weight of fish	Weight–length curves, condition factors, relative weight (see Chapter 15)	×12	Most field scales are not accurate enough; must construct special shelter or move indoors to weigh; condition factors require both accurate lengths and weights
Age determination	More accurate than length distributions for calculating year-class strength, age distribution, growth history, and mortality (see Chapter 16)	×120	Must have accurate length measurements; requires extra handling of fish (see Chapter 5) and laboratory time for analysis
Radio or sonic tagging	Exact information about fish location and movements. May be combined with information about water depth and temperature (see Chapter 19)	×1200	Much information gained about movements of relatively few fish; equipment cost and maintenance can be quite high

If an investigator draws up a list of this sort, the options become much more clear. Optimization (i.e., a balance between the cost of an activity and the new information provided) is the key. The information gained must meet the study objectives, and costs must stay within the budget. If these two criteria are not met, then either the objectives or the budget must be modified.

1.2.7 Analyze the Data

Once data have been collected, they must be analyzed. The primary purpose of this book is to discuss sampling techniques; data analysis techniques are covered in Chapter 2 and, in much greater detail, in other publications (e.g., Ricker 1975; Ney 1993; Van Den Avyle 1993; Weithman 1993). Data analysis generally requires a substantial amount of training or experience. However, a modest knowledge of statistics allows one to interact effectively with analytical experts.

Common analysis techniques involve estimation of recruitment, growth, and mortality rates, assessment of population size and age structure, and estimation of population density and biomass. Habitat characteristics can also be assessed to determine whether quantity or quality of habitat is a limiting factor for a fish population or community. Socioeconomic data are analyzed to evaluate the desires of constituents and the economic value of a resource.

Recall that management is a dynamic and continuing process. Data analysis allows assessment of measurable objectives. If objectives are not being met, then management techniques can be modified, data collection and analysis can be continued under another frame of reference, and another evaluation can be completed.

1.2.8 Communicate Results

The final, essential, and very often overlooked aspect of the investigation process is the communication of results. Writing and speaking skills are essential for a fishery professional (Decker and Krueger 1993).

Where and how information from the investigation should be presented depends on the type of project completed. Generally, some type of technical report must be completed, especially if the biologist is an employee of a state, provincial, or federal agency. Some reports may also be developed into a manuscript that is submitted for publication in a peer-reviewed journal, depending on the quality and originality of the investigation. However, written reporting should not stop here. It is also important to relate the results of the project to the public, especially if client groups have been involved in the investigation since the identification of goals and objectives. Common means of transmitting popular information include newspaper press releases and articles written for conservation magazines.

Verbal communication of investigation results should also include technical and popular forums. Again depending on the quality and originality of the work, many professionals communicate the results of investigations at technical meetings. These meetings might be simply a gathering of an agency's fishery staff. However, presentations may also be made at a state, regional, or international meeting of a professional society such as the American Fisheries Society (AFS). In addition, it is important to present the results of the project to involved client groups. Such presentations might be made at a meeting of a lake association, a public meeting for all interested citizens, or a meeting of a special-interest group such as a fishing club.

1.3 SAMPLING CONSIDERATIONS

1.3.1 Altering the Sampling Design

It is inevitable that a predetermined sampling schedule will have to be altered at times. Vehicles or equipment can break down, storms can occur, and staff can become sick. There are no standard procedures used to determine what to do when sampling is impossible. Biologists must use their best judgement.

Three general options are available if a sample cannot be collected. First, the specific sampling unit may be omitted if many sampling points have been scheduled; indeed, this is an argument for including more sampling units than needed in the original design. Second, it may be possible to add another randomly selected site or time by the same process in which the original samples were chosen. Third, an adjacent site or time may be selected to replace the missing sample. The third alternative is most appropriate if the date of sampling needs to be changed, especially if the problem is related to an equipment failure. This alternative is less appropriate if changing the sampling site could bias the data.

Whether these or other options are used to adjust a sampling design, two rules must be followed. First, an explicit, written plan for dealing with sampling problems must be developed before sampling begins. This maintains the objectivity of sampling, assures consistent changes, and avoids poor decisions in the field. Second, every departure from the sampling design must be thoroughly described in a field notebook or on the data sheets. Why the change was made, how it was made, and differences in conditions between the original and alternate sampling unit all need to be recorded.

1.3.2 Standardized Sampling

Because of the numerous gear-related, season-related, and location-related biases in sampling data for many fishes, most conservation agencies design standardized sampling programs. Standardized sampling means that effective gears are chosen for the expected fish species present, used at an effective time of year, and set in standard locations from year to year. Thus, a long-term data set is established, and trends in the sample variables can be monitored over time.

As an example, consider the standardized electrofishing samples collected in reservoirs by the Texas Parks and Wildlife Department (Texas Parks and Wildlife Department 1993). The agency's objectives are to estimate relative abundance (Box 6.1) and population structure indices (Chapter 15) for target species and to collect target species for age determination (Chapter 16) and genetic analysis (Chapter 5). Target species include all black bass, important sunfishes (bluegill, redear sunfish, and redbreast sunfish), and gizzard and threadfin shad. The boats, generators, and electrofishing control units are standardized by size, capacity, and manufacturer. For safety reasons, all crews consist of at least three people, and all crew members must be trained in cardiopulmonary resuscitation and first aid, wear rubber boots and gloves, and be familiar with electrofishing safety procedures (Chapter 8). Electrofishing is conducted only in the spring and fall when surface water temperatures range from 16 to 23°C. All sampling is done at night, permanent sampling stations are established, and minimum sampling effort varies by size of reservoir. All sampling stations are electrofished for 15 min per station. Reservoirs less than 405 ha in surface area have a minimum of four standard stations, those from 405 to 4,050 ha have six stations, and those greater than 4,050 ha have eight stations.

We do not mean to imply that standardized sampling is the only sampling undertaken by conservation agencies with responsibilities for freshwater fisheries. Standardized sampling is routinely undertaken, but it is not the only sampling that is done. Commonly, additional or more specific sampling will be undertaken for specific research investigations.

Box 1.3 The Barkley Lake Rotenone Study (from Johnson and Nielsen 1983)

Cove rotenone sampling (rotenone is a fish toxicant) is one useful method for gathering information about fish populations (see Chapter 10). An ambitious volunteer sampling was conducted on Barkley Lake, Kentucky, in 1978 (Summers and Axon 1980). The organization for this particular study illustrates the need for careful planning in fishery studies.

The Reservoir Committee of the Southern Division, American Fisheries Society, decided in June 1976 to sponsor this large-scale evaluation of cove rotenone sampling. Based on prior experience with a smaller project 10 years earlier, the committee set the sampling date for September 1978, thus leaving 26 months for design and planning.

During the next 12 months, the committee defined the objectives of the project and chose the 85-ha Crooked Creek Bay of Barkley Lake as the sampling location. The project included 10 distinct objectives, addressing not only cove rotenone sampling but also fish kill counting guidelines, mark–recapture techniques, and effectiveness of fish attractors. The incorporation of these additional objectives allowed for even greater use of the data with a minimum of extra effort.

After selecting the Barkley Lake site, the Reservoir Committee sought legal authority and public support for the project. A public meeting, held in September 1977 at Barkley Lake State Park, was attended by 100 local citizens. Description of the sampling and its scientific value by officials of the Kentucky Department of Fish and Wildlife Resources converted the anticipated resistance to unanimous public support for the project. The Nashville office of the U.S. Army Corps of Engineers, a participant in the study, prepared the necessary environmental impact assessment, which was subsequently reviewed by appropriate groups and approved by the U.S. Environmental Protection Agency.

Field work at the site began in March 1978, when 15 coves and the sites for four fish attractors were chosen. Once the specific sites were identified, decisions could be made regarding the number of workers and amount of equipment needed for sampling. Commitments for personnel and equipment were secured from 14 state, 3 federal, and 2 private agencies, and from 15 universities. Ultimately, 400 people, 5,800 m of block nets, 2,500 L of rotenone, 120 boats, 46 rotenone pumps, 56 sorting tables, 79 weighing scales, and 350 kg of potassium permanganate were used for the project.

Sampling activities began on 18 September, when two main block nets were sewn into 700-m and 300-m lengths to enclose the entire Crooked Creek Bay. During the following 2 d, the nets were set, and sites for 13 other block nets, electrofishing zones, and fish-processing stations were marked. For three nights, fish were captured by electrofishing, tagged, and returned to the bay. A largemouth bass fishing tournament was held in the bay on 23–24 September to obtain the recapture sample for the mark–recapture population estimate.

Most of the volunteers arrived on Sunday, 24 September; equipment was assigned and moved to pre-marked staging areas. On Monday, block nets were set across individual coves, and crews received final instructions. Electrofishing

Box 1.3 Continued.

crews captured, tagged, and released fish in each blocked cove on Monday night. Rotenone was applied to all areas on Tuesday morning, and fish were collected throughout the next 2 d. In all, over 3 million fishes, weighing more than 84,000 kg, were recovered. After fish were processed, they were taken to prearranged waste disposal sites or were plowed into the ground to fertilize wildlife food plots in the area.

Data recorders, assigned to each sampling area, received blank data forms each morning and returned completed forms to project leaders each evening. Forms were proofread, checked for legibility, and coded for computer analysis. The project produced, in total, 3,178 data sheets. By midday on Thursday, 28 September, block nets had been removed, and most participants had headed for home.

Within 1 year, data were analyzed and completed reports for various aspects of the project were published in the *Proceedings of the Annual Conference of the Southeastern Association of Fish and Wildlife Agencies* (Summers and Axon 1980). From first commitment to final report, the project took 39 months; data collection took 2 d.

1.3.3 Logistics of Sampling

Once all decisions have been made about how and what to sample and how to record and analyze the data (Chapter 2), the next step is to execute the data collection activities. Proper preparation for the daily field activities is as necessary as the project planning that came before. A sampling trip that is well organized and well equipped will likely be successful.

Perhaps the most frustrating situation in sampling is to arrive at a field site only to find that the nets have holes, the battery for the boat motor is dead, an essential piece of equipment has been left behind, or the electrofishing boat does not work. Because fishery equipment usually is shared by many people, one cannot be sure that everything was returned in good condition after its last use.

The first step in avoiding equipment problems is to make a careful list of all needs by visualizing the sampling activities as you read through the written sampling plans. Write down everything. For any item that can fail or break, add a backup to the list. Include spare parts for items that wear out in the field. Assign specific people the responsibility of procuring and maintaining each piece of equipment needed.

The next step is to collect the equipment and make sure that each item works. Test batteries, start motors, stretch nets, and calibrate meters. Look for equipment needing preventive maintenance and perform that maintenance before sampling. Repair all items that need it well before the sampling date. What looks like a small hole in a net may take hours to reweave, and getting parts for a small engine may require weeks. Generally, it is best to have an equipment list that can be checked before leaving for field work.

Be prepared to make emergency repairs in the field. Carry supplies and tools in surplus ammunition boxes, which are nearly indestructible and waterproof. At the least, carry a slot and Phillips screwdriver, pliers, spark plug wrench, electrical tape,

Box 1.4 Code of Practices and Ethics: Standards of Professional Conduct for Members of the American Fisheries Society

Members of the American Fisheries Society (AFS) have an obligation to perform their duties in an ethical manner. First and foremost, members accept the responsibility to serve and manage aquatic resources for the benefit of those resources and of the public, based on the best available scientific data, as specified by the Society's North American Fisheries Policy. They act ethically in their relationships with the general public and with their employers, employees, and associates, and they follow the tenets of the Society's Equal Opportunity Policy. They strive to preserve and enhance the dignity of the fisheries profession. All members must adhere to the Standards of Professional Conduct as herein established.

Section I *Dignity and Integrity of the Profession*

Members of the AFS shall

1. Avoid actual or apparent dishonesty, misrepresentation, and unprofessional demeanor by using proper scientific methodology, by adhering to the Society's Guidelines for Use of Fishes in Field Research by fully documenting technical conclusions and interpretations, and by encouraging these practices in others;
2. Give just credit for professional work done by others;
3. Make the fisheries profession more effective by exchanging information and experiences with colleagues, students, and the public via formal publications, reports, and lectures; informal consultations; and constructive interactions with professional societies, journalists, and government bodies;
4. Approve only those plans, reports, and other documents they have helped prepare or have supervised and with which they agree;
5. Make professional recommendations and decisions to benefit fishery resources and the public, base them on the best available scientific data and judgments, and clearly give the consequences both of following and not following them;
6. Restrict, to the extent possible, criticisms of technical results and conclusions to professional forums such as meetings and technical journals;
7. Expose scientific or managerial misconduct or misinformation, including misrepresentation of fisheries information to the public, through established institutional procedures or by informing the president of the AFS;
8. Treat employees justly and fairly with respect to recruitment, supervision, job development, recognition, and compensation.

Section II *Relationships with Clients and Employers*

Members of the AFS shall

1. Serve each client or employer professionally without prejudice or conflict of interest;
2. Advertise professional qualifications truthfully, without exaggeration, and without denigration of others;

Box 1.4 Continued.

3. Maintain confidential relationships with employers and clients unless authorized by the employer or required by law or due process to disclose information, and refrain from using confidential information for personal gain or the advantage of others;
4. Reject all attempts by employers and others to coerce or manipulate professional judgment and advice, exercise professional judgment without regard to personal gain, and refuse compensation or other rewards that might be construed as an attempt to influence judgment.

Section III *Relation with the Public*

Members of the AFS shall

1. Communicate with the public honestly and forthrightly within constraints imposed by employer or legal confidentiality;
2. Oppose the release of selective, biased, or inaccurate fisheries information that might mislead the public or prevent it from gaining a balanced view of a subject;
3. Express opinions on a fisheries subject only if qualified by training and experience to do so and only if fully informed about the subject;
4. Clearly delineate professional opinion from accepted knowledge or fact in all public communications;
5. Base expert testimony to a court, commission, or other tribunal on adequate knowledge and honest conviction and give balanced judgments about the consequences of alternative actions.

duct tape, baling wire, electrical wire, and first aid supplies. Add additional supplies for special equipment. Carry operating and repair manuals for specialized equipment even if you are familiar with it.

Sampling requires people, and much of fishery sampling requires many people. Managing the sampling crew is critical for efficient and effective collection of data. The first consideration is having enough people. Although having too many people should be avoided in some situations, such as overcrowding a boat, the usual problem is having too few people during parts of the task. Plan for a crew that is sufficiently large to do the most labor-intensive task efficiently.

Organizing the workers at the site is necessary for successful sampling. There must be an understood hierarchy of authority and responsibility among the field workers. A rule of thumb is that one person can effectively supervise five others; therefore, it will be necessary to assign supervisory roles to more than one person when a large field crew is used. The leader must assign specific tasks to each individual and preferably keep each person doing the same tasks throughout the sampling period. In this way, the activity will be done consistently, and the responsible person can answer questions about that activity should they arise later. However, cross-training is also important. Perhaps during the next sampling trip, different individuals will be present and different responsibilities might be assigned.

Whenever possible, one person should record all data. If more than one person is needed, each should be told exactly how to record, including instructions about units of measurement and number of decimal places. Recorded data are the only link between the field and the final results, so an experienced and capable worker should do this job.

Suitable paper should be used for recording data. Indelible ink can be used on standard paper stock, or waterproof paper may be used. Copies of data should be made and stored in a separate location for protection of those data. If it is possible that legal questions may arise over the collected data, such as for a fish kill investigation (Chapter 14), use indelible ink in a bound notebook.

The person who functions as leader of the field activity should not have a specific job assignment. The leader assures that sampling occurs smoothly and according to design. The leader should oversee the setup of the field activity, assign and explain tasks, check to see that workers are following the design and instructions, make on-the-spot decisions, and assist where extra help is needed.

The last step before sampling is informing relevant people about sampling plans. The appropriate law enforcement offices should always be notified so that they can answer calls from casual observers. State fishery agencies and nearby universities should be notified so that other sampling or projects in the area will not be affected. Fishing groups and outdoor writers might be notified so that they can observe and inform their members or readers about the work. Good relations with the public are as important for fishery management as are good data.

Many activities need to be completed before and after the actual sampling. An excellent example of the range of such activities is evident in a large-scale study that was completed at Barkley Lake in Kentucky (Box 1.3). Although this project is of interest for its research objectives, the logistics for accomplishing this large-scale investigation are also worthy of study.

1.4 ETHICS

The reliability of collected samples is not determined solely by the sampling design and field techniques used. It is also determined by the reliability of the collector. The actions and ethics of biologists that undertake the sampling are of great importance. Ethics is defined as a set of moral principles or values and thus can vary among individuals in relation to their personal and professional backgrounds. Purposeful misreporting of sampling data would obviously be an example of an ethical compromise. However, ethical situations are not always so readily apparent. For example, consider a biologist that compromises a study design during field work because of fatigue. This situation can be further compromised by the failure of the biologist to report the deviation when the data are reported. This is but one of many ethical dilemmas that biologists constantly encounter (Brown et al. 1993).

The AFS has established a Code of Practices and Ethics, as have most professional societies. This code sets standards of professional conduct for AFS members (Box 1.4). We present this code to help readers develop personal ethics as a professional biologist. Ethical considerations also involve the treatment of organisms. The AFS has endorsed the ethical treatment of animals (Appendix 5.1).

1.5 REFERENCES

Brown, P. B., T. D. Beard, Jr., R. A. Jacobson, J. S. Mattice, P. J. Pristas, and M. G. Hinton. 1993. Should we eat these fish? A situational ethics survey of AFS members. Fisheries 18(2):19–24.

Decker, D. J., and C. C. Krueger. 1993. Communication: catalyst for effective fisheries management. Pages 55–75 *in* Kohler and Hubert (1993).

Hayward, R. S., and F. J. Margraf. 1987. Eutrophication effects on prey size and food available to yellow perch in Lake Erie. Transactions of the American Fisheries Society 116:210–223.

Johnson, D. L., and L. A. Nielsen. 1983. Sampling considerations. Pages 1–21 *in* L. A. Nielsen and D. L. Johnson, editors. Fisheries techniques. American Fisheries Society, Bethesda, Maryland.

Kohler, C. C., and W. A. Hubert, editors. 1993. Inland fisheries management in North America. American Fisheries Society, Bethesda, Maryland.

Krueger, C. C., and D. J. Decker. 1993. The process of fisheries management. Pages 33–54 *in* Kohler and Hubert (1993).

Laarman, P. W., and J. R. Ryckman. 1982. Relative size selectivity of trap nets for eight species of fish. North American Journal of Fisheries Management 2:33–37.

Ney, J. J. 1993. Practical use of biological statistics. Pages 137–158 *in* Kohler and Hubert (1993).

Phenicie, C. K., and J. R. Lyons. 1973. Tactical planning in fish and wildlife management and research. U.S. Fish and Wildlife Service, Bureau of Sport Fisheries and Wildlife, Resource Publication 123, Washington, DC.

Ricker, W. E. 1975. Computation and interpretation of biological statistics of fish populations. Fisheries Research Board of Canada Bulletin 191.

Summers, G. L., and J. R. Axon. 1980. History and organization of the Barkely (*sic*) Lake rotenone study. Proceedings of the Annual Conference Southeastern Association of Fish and Wildlife Agencies 33(1979):673–679.

Texas Parks and Wildlife Department. 1993. Fishery assessment procedures. Texas Parks and Wildlife Department, Inland Fisheries Division, Austin.

Van Den Avyle, M. J. 1993. Dynamics of exploited fish populations. Pages 105–135 *in* Kohler and Hubert (1993).

Weithman, A. S. 1993. Socioeconomic benefits of fisheries. Pages 159–177 *in* Kohler and Hubert (1993).

Chapter 2

Data Management and Statistical Techniques

MICHAEL L. BROWN AND DOUGLAS J. AUSTEN

2.1 INTRODUCTION

It is often said that fisheries management is an imprecise science and often an art. Biological systems and human exploitation of them vary, and this variation makes observation and management challenging. Managers are responsible for enumerating change, assessing management actions, and quantifying human influences on the fisheries they manage. Thus, managers and researchers use a diverse array of statistical and data management tools in formulating and evaluating management programs. This chapter is a guide to the role of statistics in fisheries assessment. If a management program is designed with data management and statistical analysis in mind, it will be more successful than one planned without quantitative controls.

2.1.1 Audience, Scope, and Limitations

This chapter is an overview of statistical applications to fisheries, not a resource for the mechanics of these methods. Numerous textbooks and user guides written at many levels of sophistication provide details of statistical procedures (Table 2.1). We also give little attention to statistical software, which is continually changing. We assume that readers have been exposed to basic statistics at the level of an introductory undergraduate statistics course, but we have tried to make the material understandable to all readers.

Statistical and quantitative thinking must be involved in all aspects of a management evaluation or research project. Many are the stories of frustrated graduate students who took their data to statistical consultants for analytical help, only to be told that the data were insufficient or the design was inappropriate for the problem being addressed, and that the statistician should have been consulted before data were collected. Similarly, public agency files are replete with reports of investigations that were inadequate in scope for honest evaluations of management programs. Before putting a net in the water or starting a generator, each biologist needs to ask, "Will the data I collect and the design of my study allow me to evaluate the problem I am supposed to address?" Given the constraints on agency resources (time and money), the urgency of preventing inappropriate or improper sampling or management practices seems clear.

Fisheries professionals have a scientific and societal obligation to provide well-designed and appropriate evaluations and assessments of resource management activities. In this light, we cover in this chapter (1) data collection in the field,

Table 2.1 Statistics books commonly referenced in the 1993 volumes of the *North American Journal of Fisheries Management* (volume 13) and the *Transactions of the American Fisheries Society* (volume 122).

Authors (year)	Principal subject matter
Cochran (1977)	Descriptions of sampling designs and sample size determinations
Cohen (1988)	Thorough coverage of power analyses
Conover (1980)	Applied nonparametric statistics
Draper and Smith (1981)	Applied simple and multiple linear regression analysis
Johnson and Wichern (1988)	Applied multivariate statistics
Neter et al. (1989)	Applied linear regression models
Siegel (1956)	Applied nonparametric statistics
Snedecor and Cochran (1989)	Broad coverage of basic statistical methods
Sokal and Rohlf (1981)	Applied statistics for biologists
Steele and Torrie (1980)	General statistical and sampling methods
Winer (1971)	Experimental design
Zar (1984)	Applied statistics for biologists

(2) computerized management of the data, (3) definition of problems in statistical terms, (4) visual representation of data, and (5) the role of statistics in collecting and interpreting data.

2.2 DATA HANDLING AND DATABASE MANAGEMENT

2.2.1 Data Recording

Fisheries surveys, management assessments, and laboratory experiments produce large volumes of data, some of which cover many years. These data are our observations of nature and provide the information upon which we make decisions concerning the management of our aquatic resources. Collecting the data often requires expensive equipment and many hours of labor, making the data very costly. Thus, it is vitally important that data be recorded accurately and quickly and kept safe from damage. Biologists employ a wide variety of methods to record information, including paper notebooks, waterproof field sheets, electronic measuring boards and balances, video equipment, and waterproof field computers. Each method has advantages and disadvantages. The key in selecting proper data-recording methods is to keep in mind the need to record data accurately, safely, and, if possible, quickly.

Standardized field data collection sheets have been developed for many fisheries management agencies. Generally, different field sheets are developed for different types of studies. For example, a form used for a fisheries survey might resemble Figure 2.1, whereas one used for a habitat assessment would have a totally different format designed specifically for the data being collected. It is often best to have fisheries data sheets printed on waterproof or heavy stock paper that is less susceptible to damage than typical photocopier paper. Pencils are always preferable to pens as writing implements because most inks are water soluble. It cannot be stressed enough that data should be written legibly and completely; illegible or omitted data cannot be reconstructed accurately later. Back in the office it is always best to copy data sheets and store the copies away from the original data.

Electronic field data loggers are steadily improving and diversifying. These devices can be as simple and single-task-oriented as electronic measuring boards with dynamic memory. These are preprogrammed to record fish lengths, which are later

FISHERIES DATA SYSTEM

Page ______ of ______

Date: ____________
Water body: ____________
Project: ____________
Gear: Type ____________
Size ____________
Number ____________

Location: ____________
Effort: ____________
Elapsed time: ____________
Distance/Area ____________
Collector: ____________

Water: Temp./depth ____________
Transparency ____________
Weather: Air temp. ____________
Wind direction ____________
Wind speed ____________

Species	Total length	Weight	Sex	Scale	Species	Total length	Weight	Sex	Scale	Species	Total length	Weight	Sex	Scale

Figure 2.1 Generalized example of a field survey form used to record biological data for fishes and other pertinent collection and environmental information. Variables such as species may be codified within an agency (e.g., abbreviations such as WAE for walleye and LMB for largemouth bass may be used). Total length and weight are continuous measures, sex is determined by observation, and "scale" indicates whether a scale was taken from an individual fish.

downloaded to a computer. The next level of sophistication is represented by field computers programmed with database entry screens. These computers have serial ports through which digital calipers and balances can transmit length and weight information. Field computers also interface with global positioning system receivers and digital sonars to produce spatial and depth information for lakes. The possibilities are endless, but caution must be taken when data are trusted to field electronic media. Data should be downloaded to laboratory computers at the first opportunity to ensure that they have been properly recorded. Portable printers can be used with the data loggers to provide real-time printouts of the data for checking. The key to data security with field computers is to make several backups and keep paper printouts of the data on file.

2.2.2 Data Management

For collected data to be truly useful, biologists must have a convenient and fast means of accessing and organizing this information. Natural resource agencies have understood the importance (and the requirement) of data management systems for many years (Clark et al. 1977). Virtually all state, provincial, and federal agencies now have database systems that run on machines ranging from personal computers to agencywide mainframes. New biologists must develop a facility with database systems to succeed at their jobs. In this section we discuss various types of database systems, issues in computerized storage of information, and the role of database management systems in fisheries management agencies.

A telephone book or baseball card collection is, in broad terms, a database. More specifically, a database is "a collection of data organized logically and managed by a unifying set of principles, procedures, and functionalities that help guarantee the consistent application and interpretation of those data across the organization" (Mattison 1993). The database package is the computer software that logically organizes data and manages the storage and retrieval of data. Database software is available in a wide variety of qualities and prices.

Agencies have developed many databases to organize and store their information (Table 2.2). One example of a database management system currently in use is the Fisheries Analysis System (FAS) that was developed by the Illinois Natural History Survey for the Illinois Department of Conservation. This system is used to store and analyze data collected during standard fish population surveys by fisheries management biologists. The FAS provides data input screens that mimic field data sheets, and it offers a wide variety of custom-programmed outputs such as tables of condition indices and graphs showing length–weight scatter plots (Bayley and Austen 1989). In the FAS, field biologists enter their own data and run their own analyses. Data also are shipped to a central database manager who loads the data into a larger computer that holds a comprehensive statewide database. Many other agencies have developed systems that provide similar capabilities customized for their needs.

2.2.3 Databases

The actual structure, or architecture, of a database determines the methodology for logically organizing data (Mattison 1993). Architectures vary widely, and many database systems are hybrids of several architectures. Currently popular commercial databases (in 1996) include Borland's dBase IV and Paradox, and NorthCon Technologies' Double Helix, all of which are primarily run on personal computers (PCs). Oracle (Oracle Corp.) is a relational database used on machines ranging from PCs to supercomputers. The user's manual for each software generally provides an

Table 2.2 Examples of aquatic-related databases.

Database name	Agency	Type of information
STORET	U.S. Environmental Protection Agency	Primarily water quality
National Water Information System II	U.S. Geological Survey	Major river basin hydrology, limnology, and biology
FISHNET	Ontario Ministry of Natural Resources	Fisheries survey data
Environmental Mapping and Assessment Program (EMAP), surface waters	U.S. Environmental Protection Agency	Assessment of trends in ecological status of lakes

understanding of the architecture and the mechanics of getting information into and out of the database. Each database has different strengths and weaknesses and is amenable to different types of data. Before developing a database, the user should carefully study the relationships in the data to be stored and either contact computer consultants or examine several database products. Computer magazines and books frequently carry reviews of database packages (Venditto 1992; Seiter 1994).

2.2.4 Media Life Expectancy and Data Backups

As with any magnetic device, removable computer "floppy" disks and tapes slowly degrade to a point of unreliability. The maximum physical lifetime for floppy disks may be 5–10 years, depending upon the environment in which they are stored. Even optical storage devices such as compact disks may degrade to uselessness after 30 years (Rothenberg 1995). If data files are to be kept on disk for long periods of time, regular recopying of the data onto new disks or tapes should be scheduled. Having data backed up on disk provides some security against loss, but the frequency of backups and the safety of their storage needs to be considered. For networked databases, a network administrator who manages the computer systems should be responsible for backups, which are often done daily or weekly. Many larger systems have different levels of backup; some backups are stored on site while others are kept in different buildings as a security precaution.

Possibly of more importance than media lifespan is the obsolescence of the technology used to write the digital file. Software and hardware used to write a disk often become obsolete before the integrity of the media declines (Rothenberg 1995). For example, many PCs presently sold do not include drives for 5.25-in floppy disks, yet many valuable data have been stored on such disks. Eight-inch disks and punch cards are even more outdated media. Software, too, evolves, and data that are not rewritten in new software versions may become unreadable once manufacturers stop supporting the old software. Because data collected now may be useful to biologists 50 years from now, regular translations of data to new physical and software media are essential elements of data management. Field personnel, no less than system managers, must stay aware of the electronic future as they plan their programs.

2.2.5 Data Integrity

Insuring that data are accurate becomes increasingly important as the size of the anticipated database increases; the larger the database, the greater the cost of finding and correcting errors. Because of this, most database systems have developed some formalized method of insuring that erroneous data are kept to a minimum (Figure 2.2). In most large databases, errors will never be completely eliminated.

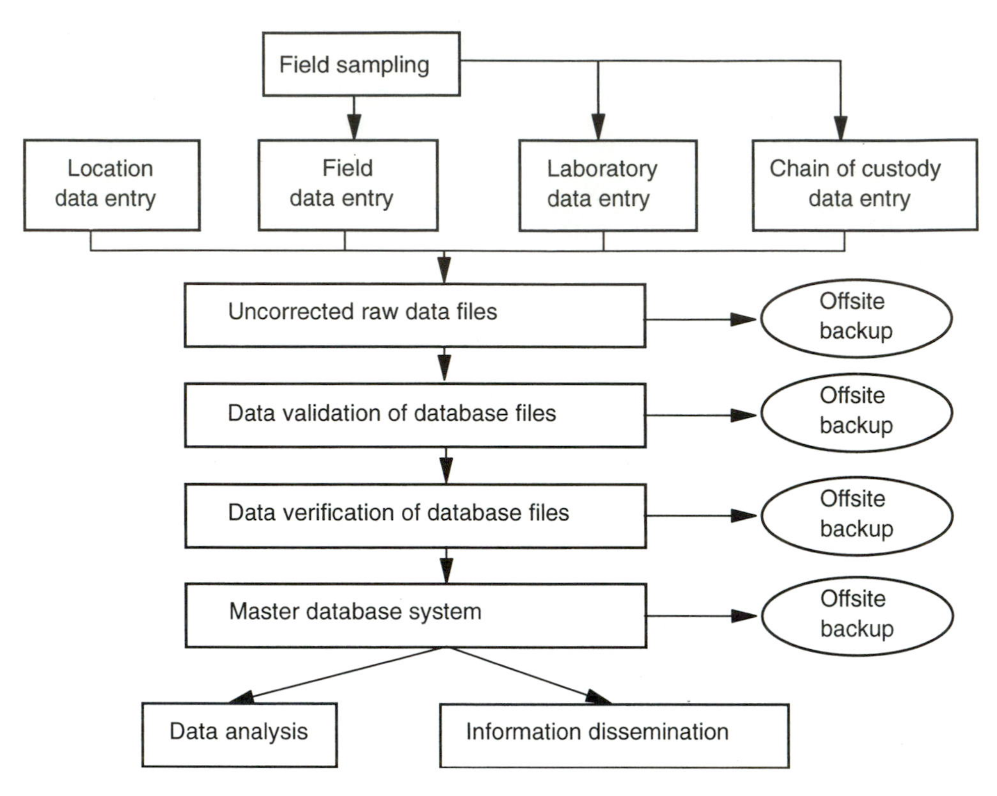

Figure 2.2 Flow diagram of a typical database management scheme, showing how data are handled from collection in the field to storage in the master database. Data to be entered may come from several sources (the laboratory as well as from the field). Data are checked several times and backups are made at several stages along the process, not just when the database is completed (adapted from Baker et al. 1993).

To insure that data are entered correctly, many database systems contain software for checking ranges of numbers and entries of species codes. Range-checking routines generally flag data that fall outside predefined ranges and allow the data entry person to check the original data sheets for correctness. More time consuming, but with a higher level of error detection, are (1) producing printouts of data that are then checked against the data entry sheets, (2) reviewing the data on screen, or (3) entering data twice and then running programs that compare the two sets for anomalies. The cost of such checks must be weighed against the consequences of wrong data for management programs.

2.3 DATA VISUALIZATION

Graphs enhance the comprehension and interpretation of data. Unlike numerical summaries such as sample means, variances, and ranges, which reduce the data to single numbers, graphs can display much of the original data. Such displays provide a great deal of information for interpretation. Manipulating data for alternative graphical displays can be very instructive for the analyst as well.

The goal of graphing is to inform without distorting information (Cleveland 1985). Hundreds of types of graphs have been developed (Tufte 1983, 1990), but most

Box 2.1 Selected Principles of Graph Construction
(adapted from Cleveland 1985)

1. Make the data stand out.
   a. Use visually prominent symbols to represent the data.
   b. Do not clutter the data region (the area inside the axes).
   c. Make overlapped symbols and superposed data sets distinguishable.
   d. Select symbols and type that are large enough to remain legible when graph is reduced for publication.
   e. Avoid heavy axis lines, oversized labels, and disparate lettering sizes that distract attention from the data. The largest lettering need be only twice the size of the smallest.

2. Make graphs understandable.
   a. Write legends that describe the contents of the graph completely and define symbols and abbreviations that are used.
   b. Clearly explain error bars (state precisely what error term has been used).
   c. Always proofread graphs just as you would text and strive for clarity.

3. Use appropriate graph scales.
   a. Choose tick marks to include or nearly include the data range but do not use an excessive number of ticks; 3–10 per axis generally suffice.
   b. Select scales so that data fill as much of the data region as possible. When graphs are to be compared, try to scale them consistently.
   c. Do not begin scales at zero if empty space results near the axes.
   d. Use logarithmic scales when these can improve resolution or facilitate graphing of data over wide ranges.

4. Make efficient use of space.
   a. Place data identification keys within, above, or below a graph—not to the side. Keys placed to the side typically leave a lot of space above or below, and they may force excessive reduction of the graph for publication (pages have less width than height).
   b. Keep axis labels reasonably close to the numbers on both axes.
   c. Constrain the few (1–3) extreme data that would greatly expand a graph by breaking the axis with a space or zigzag and plotting the extremes close to the main data. Break bars and lines on the graph as well.

published graphs are variations of bar charts, histograms, scatter plots, and line graphs (Kennedy and Kennedy 1990). Most graphic software packages include these in many forms. Although creating computer graphics is simplified by user-friendly software, the computer gives little help in the creation of clear, easily interpreted graphs; that job is left to the user. General guidelines for good graph construction are listed in Box 2.1.

2.3.1 Histograms and Bar Charts

Histograms and bar charts are probably the most frequently used graphical techniques for illustrating data. Histograms are used to display essentially continuous data such as measurements (e.g., length and weight). A typical example in fisheries is the length–frequency histogram (Figure 2.3). The continuous data on the horizontal axis are grouped into vertical bars whose heights generally represent the counts or the frequencies of values in the respective groups. The horizontal group intervals are kept constant and are represented by bars of equal width. As shown in Figure 2.3, group intervals can be varied in such a way that alternative, and sometimes erroneous, interpretations of the data are likely.

Bar charts often look like histograms but are used to display categories of data. For example, Figure 2.4 illustrates the response of fish biomass to various angling regulations (no length limit, minimum length limit, and slot length limit) over successive 5-year periods of measurement. Each category, in this case years, is denoted on the horizontal axis and the value measured for that category is shown on the vertical axis.

2.3.2 Pie Charts

Pie charts are another popular way of conveying categorical data but suffer from a serious problem in that most people cannot accurately judge the sizes of the divisions (Cleveland 1985). Most creators of pie charts attempt to alleviate this problem by including the value of the pie "slice" alongside the graph, but this simply reinforces the fact that areas are poor representations of numeric quantities. Pie charts can always be replaced by bar charts, dot charts, or other graphical devices (Cleveland 1985).

2.3.3 Scatter Plots

Scatter plots are used to show the relationship between two variables (Figure 2.5) and are produced by plotting a pair of measurements for each sample unit as a point on a graph. Generally, the vertical (y) axis represents the dependent variable (response variable) and the horizontal (x) axis indicates the independent variable (predictor variable). A typical use of a scatter plot is to display information for two related observations, such as length and weight, effort and yield, or spawners and recruits. In many cases the graphical presentation is accompanied by calculated correlation coefficients. Scatter plots are excellent tools for data exploration. Relationships between variables can be visualized, and patterns may emerge as plots are being constructed or manipulated. Graphical software packages often have numerous options with which to modify scatter plots, such as overlaying the plot with a regression line, varying the size of the plotting symbol to represent the variance of the point estimate, or constructing matrices of multiple scatter plots to show the relationships between several variables in one large table (the SPLOM facility in SYSTAT; Wilkinson 1990). All of these display options help the user visualize patterns and increase understanding of the data.

2.3.4 Line Graphs

Line graphs are most frequently used to display ordered data such as those taken in sequence over a span of time (Figure 2.6). The horizontal axis usually denotes the time units. As in the scatter plot, the vertical axis is scaled in the unit of measurement of the variable being plotted.

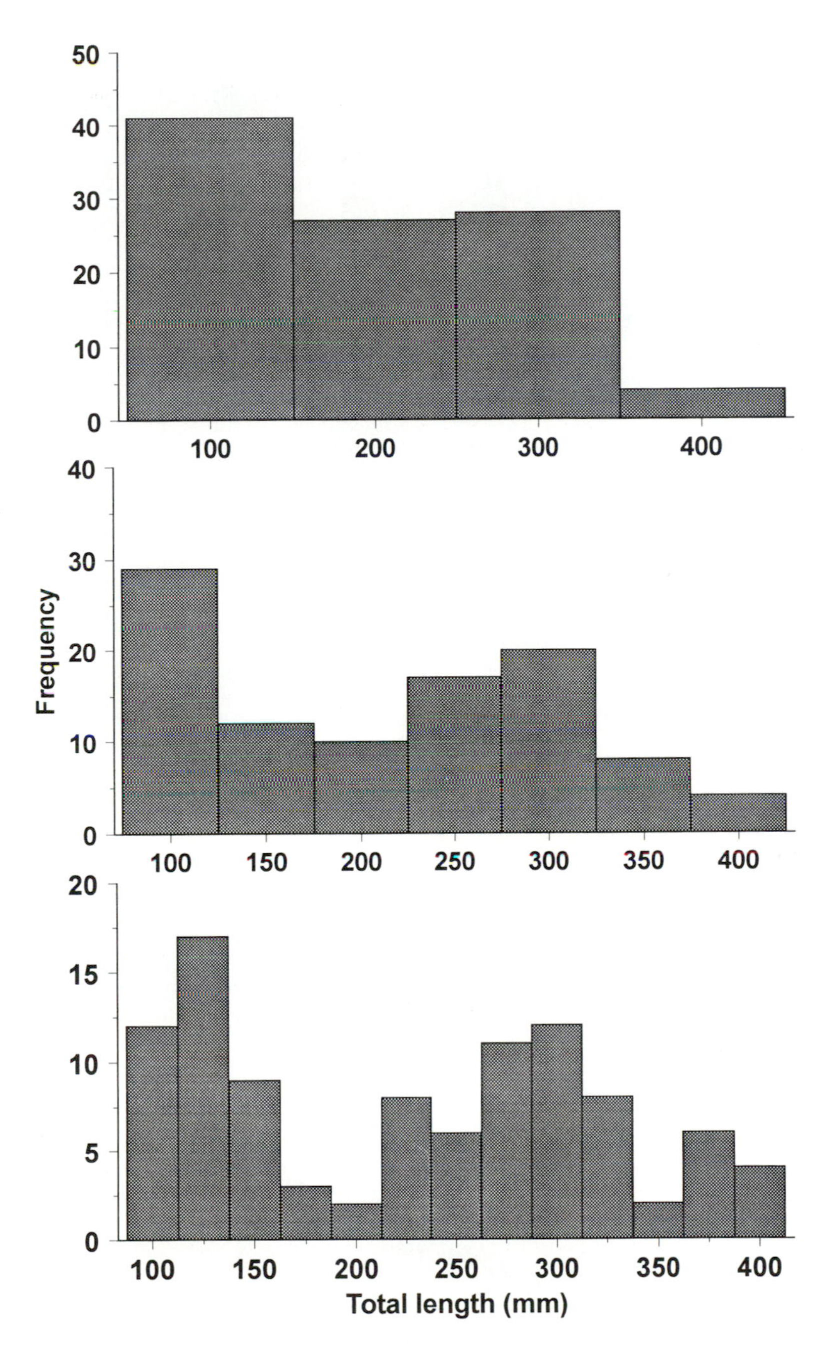

Figure 2.3 Example of length frequency histograms for a largemouth bass sample collected from a reservoir. Note two items in particular. First, the coarseness of the length categories chosen affects the amount of information the graph can reveal about the sample. Second, because their abscissa scales differ, the panels are difficult to compare in detail.

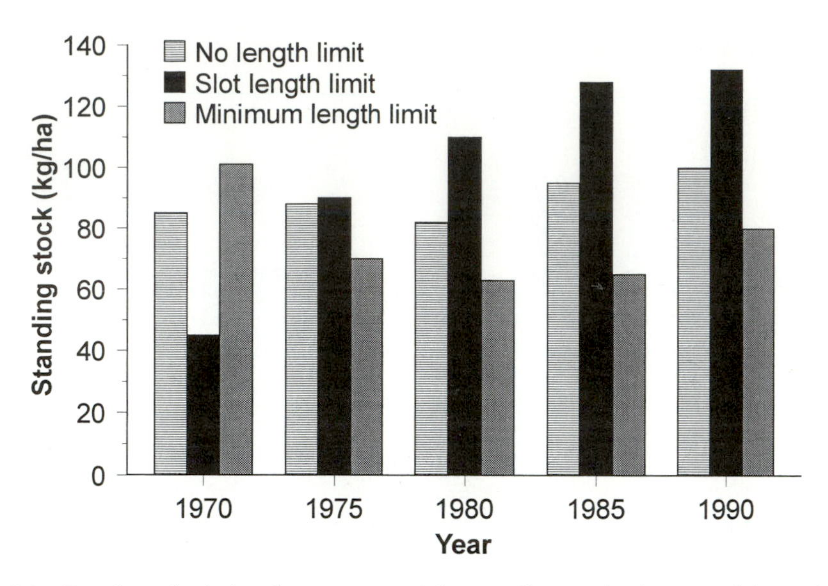

Figure 2.4 Bar chart depicting the response of the standing stock of a sport fish species over successive 5-year periods to three length limit regulations.

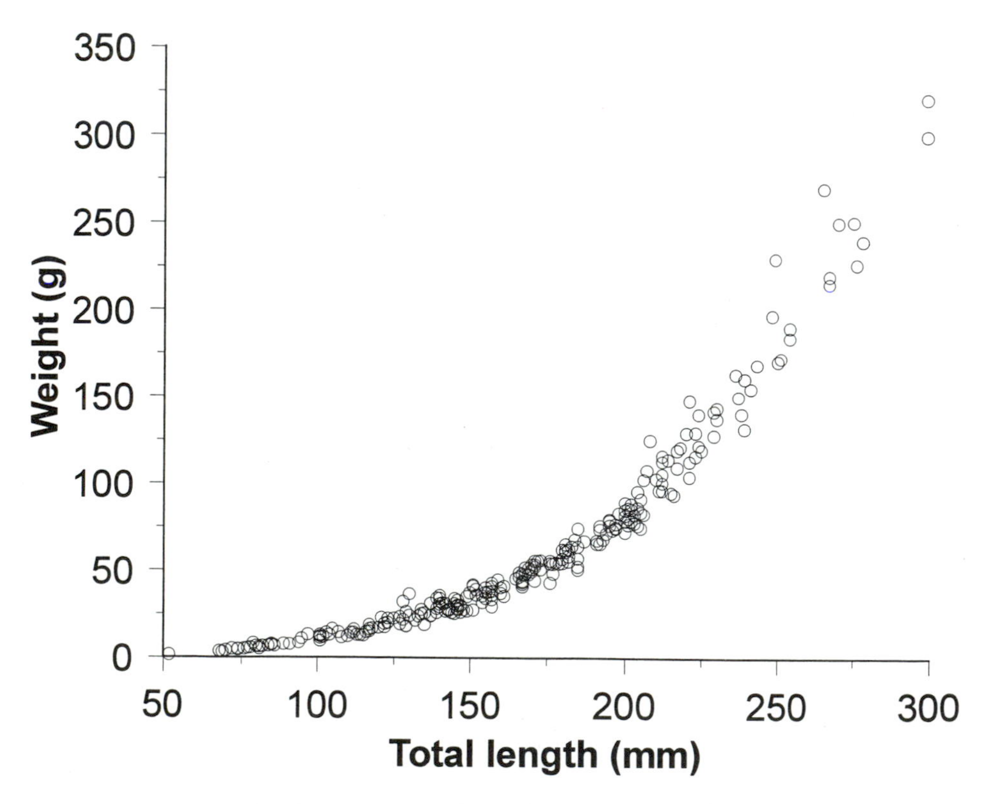

Figure 2.5 A scatter plot (or scatter graph) showing the relationship between weight (y) and length (x). Data points represent the individual weights and lengths for a sample of largemouth bass.

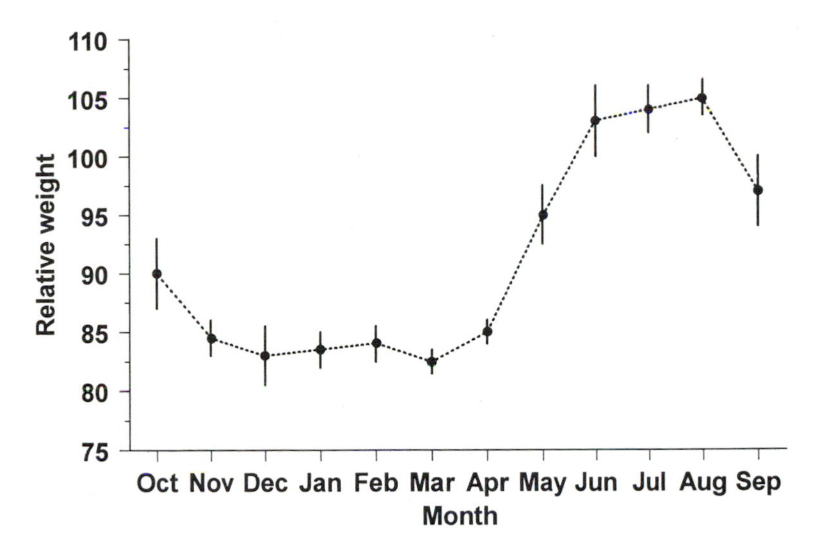

Figure 2.6 Example of a line graph that shows mean relative weight values of stock- to quality-length (200–299-mm) largemouth bass over 1 year. Vertical bars denote standard errors of the mean.

The variety of graphical methods used to display data is limited only by the imagination of the analyst and no doubt will expand as new software and visual display methods are developed. For example, several software packages enable three-dimensional (3-D) plots to rotate on the screen, permitting examination from any angle. Modern computer graphics software tempts people to create artistic 3-D graphs, but these may actually inhibit a biologist's ability to interpret data (Kennedy and Kennedy 1990). Moreover, trends in data are more difficult to see in static 3-D plots unless they are obvious. In general, 3-D graphics should be avoided in reports and other print media. Color imaging of data (expensive in print) and movielike displays of 3-D data arrays will become more common as computer networks improve and complicated data sets can be downloaded and displayed electronically.

2.4 DATA TERMINOLOGY AND CHARACTERISTICS

Basic data terminology includes data set, variables, and cases, which are terms common to both data management and statistics. The data set is the entire collection of information gathered for analysis. These data are frequently arranged in a table format that reads horizontally (rows) and vertically (columns) (Example 2.1). A case is a subset of data consisting of all the assigned, observed, or measured attributes associated with one member of the data set. If the member is a fish, its case might comprise its identification number, species, length, weight, sex, and age. A variable is a descriptive category or set of attributes (e.g., sex or weight) represented in some or all cases.

Example 2.1 Data set consisting of six variables and 67 cases

	Variable					
	Identification number	Species code	Length (mm)	Weight (g)	Sex	Age (year)
Case 1	0001	15	223	131.1	M	1
Case 2	0002	15	298	325.9	F	3
⋮	⋮	⋮	⋮	⋮	⋮	⋮
Case 65	0065	32	137	48.2	F	3
Case 66	0066	32	98	17.3	U	1
Case 67	0067	32	108	20.3	M	2

Data are of two basic kinds: qualitative and quantitative. Qualitative data are categorical: the categories may be nominal or ordinal. Nominal data may be separated according to categories, but do not have any numerical value or relationship; the categories have labels, such as male and female. Ordinal data can be arranged from smallest to largest (forming what is also known as a rank scale); numbers or letters assigned to items within a variable have a categorical significance such as greater than, less than, or equal to. An example is ranking the weights of a group of fish from lightest to heaviest.

Quantitative data provide the amounts of something in a discrete or continuous fashion. Discrete, or discontinuous, data are integers (0, 1, 2, 3, . . .), and no other values can occur between two points in a series. All counts yield discrete data (e.g., number of lateral line scales, number of bluegills in a sample), as do integral measures (e.g., age in years). Continuous data may be measured to any degree of accuracy (e.g., 110 mm, 110.2 mm, 110.23 mm, . . .), limited only by the measurement tool used, and therefore can take any value in a given interval.

Interval and ratio scales are used in the measurement of quantitative data. Both scales have the characteristics of the ordinal scale, plus a unit of measurement. Interval data are discrete (can take on integer values only). The interval scale uses the relative order (ordinal), a zero reference point, and the size of the interval between measures. However, it is not important which measurement is declared to be zero (e.g., Fahrenheit and centigrade scales have different zero temperatures in definitions of 1 degree). Ratio data are measures for which the order, the interval size, and the ratios between measures are meaningful, such as lengths (millimeters, centimeters), or weights (grams, kilograms). Ratio data can be discrete or continuous. The major difference between interval and ratio scales is that the ratio scale uses a true (natural) zero.

2.4.1 Precision, Accuracy, and Bias

The terms precision and accuracy are not synonymous; they define different characteristics of the data. Precision refers to the closeness of repeated measurements of the same item and how well test results can be reproduced (Figure 2.7). Accuracy is the closeness of a measured value to its true value, and it depends on the measurer and the measuring device. Measuring a fish with a measuring board that has been incorrectly scaled may provide precise estimates, but they would be consistently inaccurate in the same manner—a phenomenon called "bias." If an estimate is unbiased and precise, then it is accurate.

Because accuracy may not be determinable, precision is commonly used to provide assurance that no unsuspected bias exists with the measure. Precision cannot be used

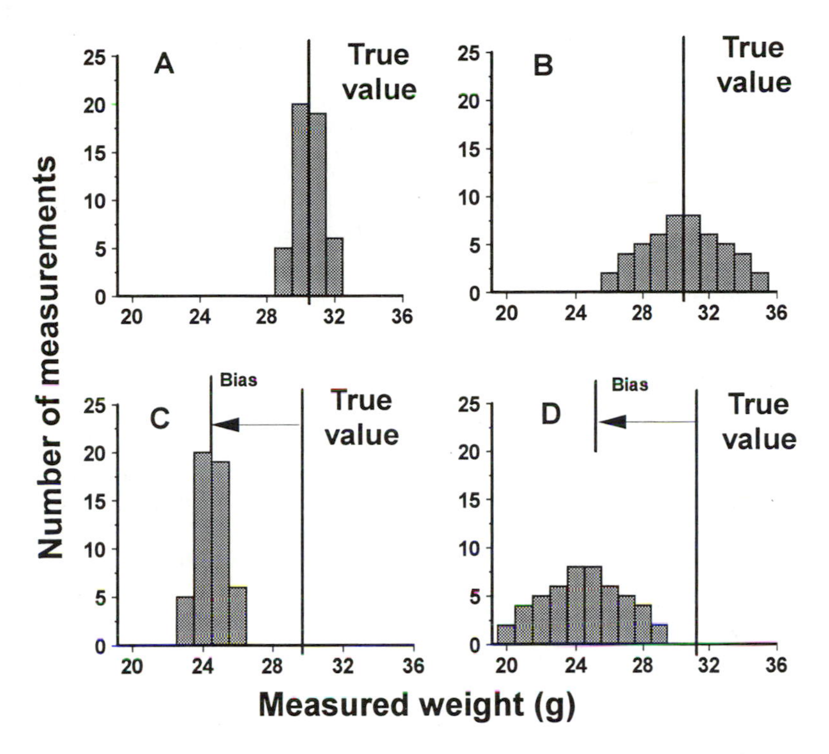

Figure 2.7 Illustration of accuracy, precision, and bias. For each of the four panels, 50 repeated measurements were made of a single variable, in this case the weight of a single fish. Sample **A** is accurate (average weight is close to true weight) and has high precision (little variation in measured weight); **B** is accurate but has low precision; **C** is biased (inaccurate) but has high precision; and **D** is biased and also has low precision.

as assurance if bias is associated with the measuring device itself. For example, if we wished to know the population structure of white crappies in a reservoir, we might collect them with a trap net set near the shoreline (Chapter 6). If we sample during August, when most of the larger white crappies are offshore and not likely to be caught in our nearshore gear, then we will underestimate the population size structure and average size of white crappies in that population. Our sample would be inaccurate (biased). However, our estimate could be precise if repeated samples contained similar size distributions of small white crappies.

Statistics such as variance, standard deviation, and standard error are measures of precision. In general, the lower the variability around an estimate made from a sample, the higher the precision and the more likely a similar result will be obtained from another sample taken in the same manner.

2.4.2 Significant Digits

Continuous data are recorded to some fixed level of accuracy; thus, a measurement's accuracy should be implied by the number of significant digits reported. Greater accuracy warrants more digits. If more than the correct number are used, they are meaningless. The absolute measurement error (not necessarily the percentage error) generally increases as the magnitude of the measurement increases. Sokal and Rohlf (1981) suggested a rule of thumb for determining the number of significant digits: "the number of unit steps from the smallest to the largest measurement in an array should be between 30 and 300." For example, if fish lengths were measured over the range of 21.362 to 51.482 cm, the minimum level of accuracy reported would be approximately 1.0 cm (30.120 ÷ 30) and the maximum level would be about 0.1 cm (30.120 ÷ 300). Based on that determination, it would be necessary to report values to a maximum of one decimal place (e.g., 21.4 and 51.5 cm) or, minimally, as 21 and 51 cm. Even though the time and expense required to obtain higher levels of accuracy may not be warranted in certain circumstances, investigators should strive to achieve maximum levels of accuracy.

2.5 STATISTICS

Statistics is the science of analyzing and interpreting data. Most of a fisheries professional's time is spent in planning a study, analyzing data, and developing recommendations based on the interpretation of those data. A relatively small amount of time is spent actually collecting the data.

Without samples it would be virtually impossible to evaluate fish populations. Thus the biologist seeks to withdraw from the fish population a sample that is representative of the population. Measurements are then made on some variable (attribute) of the fish in the sample, and the nature of this variable in the population is inferred from the sample measurement. The following sections treat the important matters of statistics and sampling.

2.5.1 Descriptive Statistics

In general, fisheries professionals use both descriptive statistics and inferential statistics (section 2.5.3). Descriptive statistics allow reduction of a large set of measurements of one variable to a few summary measures presented in graphic (section 2.3.1) or numeric form. The most common graphical ways to show how a variable is distributed among cases are frequency distributions and histograms. A plot of a frequency distribution often approximates a smooth bell-shape, or normal, curve (Figure 2.8A), provided that the sample is adequate. A frequency histogram resembles stair steps (Figure 2.8B).

The normal curve illustrates two important types of descriptive statistics: (1) statistics of location and (2) statistics of dispersion and variability. Statistics of location include all measurements of central tendency; that is, the tendency of measurements to "clump" about the center of a distribution. The three most common measures of central tendency are the sample mean, median, and mode. The sample mean ($\bar{y}$) is the arithmetic average of all observations, the median (M_d) is the midpoint or middle observation of an ordered distribution, and the mode (M_o) is the value that occurs most frequently in the distribution.

The sample mean is the most commonly used measure of central tendency and is expressed by the equation

$$\bar{y} = \frac{\Sigma y_i}{n}; \qquad (2.1)$$

$\bar{y}$ = the sample mean,
Σy_i = the sum of all measurements (i denotes any one measurement in a series), and
n = the number of measurements made.

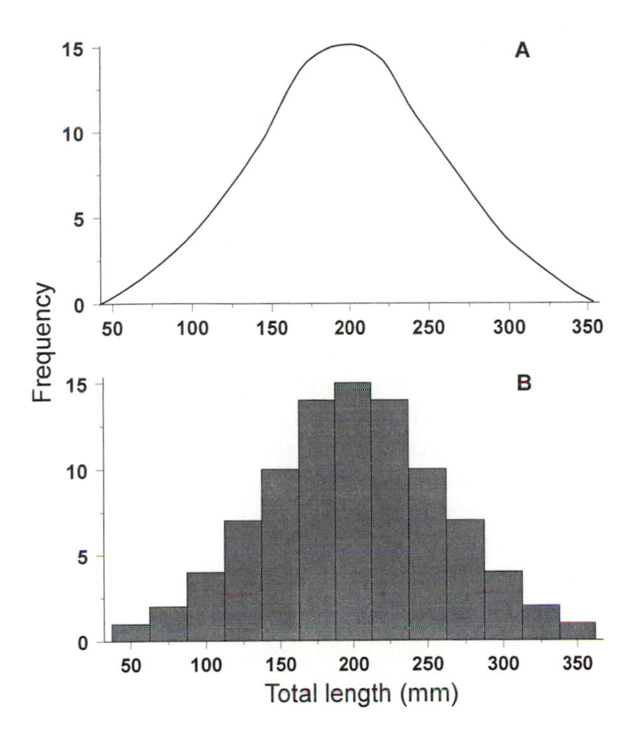

Figure 2.8 A normal distribution of fish lengths illustrated as a curve (**A**) and a frequency histogram (**B**).

Thus, the sample mean is the sum of all measurements divided by the total number of measurements. The sample mean is used to make an inference about the corresponding population mean (μ).

Statistics of variability (dispersion) characterize the spread of sample measurements about a statistic (e.g., $\bar{y}$) used to express central tendency. Common measures of dispersion that describe this sampling variation are the range, sample variance (s^2), and sample standard deviation (s, often denoted SD). The range is the simplest measurement of dispersion and is the difference between the largest and smallest measurements in the set. Variance is a more widely used measure of dispersion than range. The equation for sample variance of continuous data is

$$s^2 = \frac{\Sigma(y_i - \bar{y})^2}{n - 1}; \tag{2.2}$$

for computational simplicity the form commonly used is

$$s^2 = \frac{\Sigma y^2_i - \dfrac{(\Sigma y_i)^2}{n}}{n - 1}. \tag{2.3}$$

Technically, the variance of a set of n measurements, $y_1, y_2, \ldots, y_n$, is the sum of the squared deviation of each measurement from the sample mean $\bar{y}$, divided by n minus

1 (the degrees of freedom). In other words, we calculate the difference between each measurement and the sample mean, square the difference (to treat all deviations from the sample mean as positive values), sum the squared deviations, and divide by $n - 1$ to obtain roughly an average squared deviation. (Dividing by $n - 1$ provides a slightly larger—more "conservative"—estimate of variance than dividing by n.)

In practice the standard deviation is a more commonly used measure of variability. The standard deviation is the positive square root of the variance:

$$s = \sqrt{s^2}. \tag{2.4}$$

This measure is the conservative average difference of all values from the sample mean. A standard deviation of 0 indicates all values are identical; higher standard deviations indicate higher variability. For instance, the distribution of 1, 2, and 3 has a mean of 2 and a standard deviation of 1; the distribution of 10, 20, and 30 has a mean of 20 and a standard deviation of 10. The standard deviation does not depend on the magnitude of the values but upon their dispersion.

Consider the following variance and standard deviation calculations from a set of fish total length measurements. We calculate the sum, squares, and then sum of squares.

Example 2.2 Largemouth bass total length (mm) data

y_i	$y_i - \bar{y}$	$(y_i - \bar{y})^2$	
161	−52	2,704	$\bar{y} = 213$
178	−35	1,225	$M_o = 210$
192	−21	441	$M_d = 210$
210	−3	9	$\Sigma\, y_i = 1{,}917$
210	−3	9	$\Sigma(y_i - \bar{y})^2 = 8{,}650$
221	8	64	
234	21	441	
247	34	1,156	
264	51	2,601	
$\Sigma = 1{,}917$		$\Sigma = 8{,}650$	

Incorporating these values into equation (2.2), we calculate the variance as

$$s^2 = \frac{8{,}650}{9 - 1} = 1{,}081.25.$$

Taking the square root of this value then provides the standard deviation:

$$s = \sqrt{1{,}081.25} = 32.88.$$

We cannot imply greater accuracy than that found in our original measurements; therefore, fractional values such as 32.88 should always be rounded up to provide a more conservative value (i.e., $s = 33$).

2.5.1.1 Degrees of Freedom

We step backwards for a moment to explain the concept of degrees of freedom (here, $n - 1$). A general definition of the degrees of freedom associated with a particular statistic is the number of independent observations in the data set. For example, if we calculate a sample mean and standard deviation for 10 observations, we have 9 degrees of freedom (df) because by knowing any 9 of the data points and the sum of 10 measurements, we can easily calculate the 10th value. Thus, the degrees of freedom depend on sample size. Say, for example, our true population consists of 10,000 elements and we sample only 10 of these elements; the sample is not likely to represent the entire population very accurately, and it would have only 9 df. However, if we obtain a sample of 100 elements, a sample that should better represent the population, the degrees of freedom increase to 99. The increase in degrees of freedom has several important consequences. For one, it is likely to reduce the variance of the sample mean (equation 2.2) and the confidence interval around it (section 2.5.1.2). For another, it increases the sensitivity of statistical tests; if our estimate is compared with another made at a different place or time, more degrees of freedom allow smaller differences between estimates to be judged "significant" (i.e., unlikely to be due to chance).

2.5.1.2 Confidence Intervals

Often we make a point estimate ($\bar{y}$) of the true mean (μ) of some parameter of a population. The sample mean $\bar{y}$ will seldom be exactly equal to the true population parameter. Therefore, it is often wise to try to bracket the true population parameter by expressing the estimated value as a range of values (e.g., $\bar{y} \pm 5.0$) in the form of a confidence interval (CI):

$$\mathrm{CI} = \bar{y} \pm t_{n-1}\left(\frac{s}{\sqrt{n}}\right); \tag{2.5}$$

here, t_{n-1} is a value of Student's t with $n - 1$ degrees of freedom. The t-values (Appendix 2.1) are selected for the desired level of significance (0.05 or 0.01, which translate to 95 or 99% confidence), according to the degrees of freedom in the sample. Calculations of the 95% CI for Example (2.2) follow.

Step 1. Calculate the sample mean: $\bar{y} = 213$.
Step 2. Select the tabular t-value (Appendix 2.1); the t-value for 8 and a 0.05 level of significance (α in two tails) is 2.306.
Step 3. Calculate the standard deviation: $s = 33$.
Step 4. Obtain the sample size: $n = 9$.
Step 5. Calculate the confidence interval:

$$\text{mean} \pm 95\%\ \mathrm{CI} = 213 \pm 2.306\left(\frac{33}{\sqrt{9}}\right) = 213 \pm 25.$$

In this example, the resulting CI means that if we resampled the population 100 times and recalculated the CI, on the average 95 of those CIs will contain the true mean. A 99% CI calculation would be 213 ± 37, or 176 – 250. Note that the more confidence we wish to have in our estimate, the broader our interval becomes.

2.5.1.3 Other Measures of Variation

Take a step back and look at the parenthetical term in the CI equation (2.5). This value is termed the standard error of the mean (SEM, often represented simply as SE):

$$\mathrm{SEM} = s_{\bar{y}} = \frac{s}{\sqrt{n}} = 11. \tag{2.6}$$

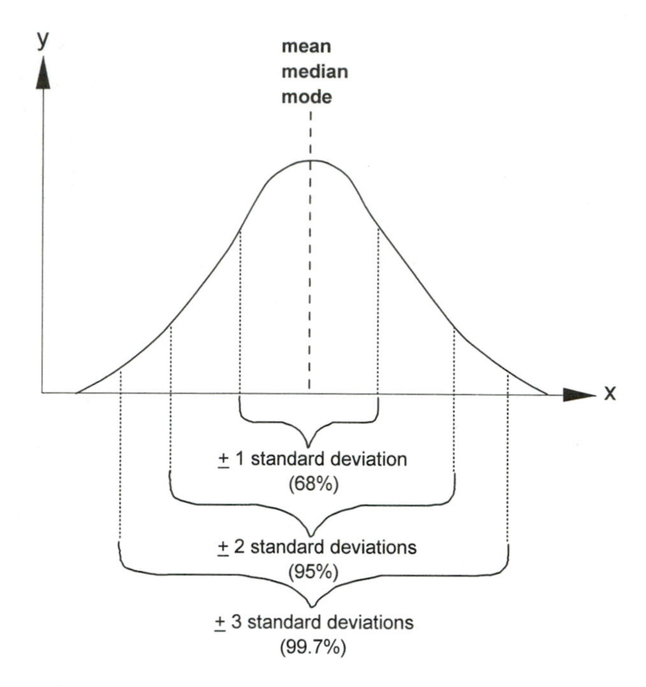

Figure 2.9 Illustration of the empirical rule and statistics of location. If a variable is normally distributed, the mean, median, and mode will be clumped near the center of the distribution. Within one standard deviation of the mean will be approximately 68% of the data, within two standard deviations will be approximately 95%, and within three approximately 99.7%.

This is not a confidence interval but the standard deviation of the sample means. In our example, SEM is 11 (expressed as 213 ± 11 SEM).

A statistic closely related to the standard deviation is the coefficient of variation (CV), or relative standard deviation. This measure provides a measure of precision by expressing the standard deviation as a percentage of the sample mean:

$$\mathrm{CV} = \frac{s}{\bar{y}} \cdot 100. \tag{2.7}$$

Using the sample standard deviation and mean values determined in Example (2.2), we get a CV of (33/213) × 100 = 15%. Because this measure is not influenced by the magnitude of the original measurements, it is a useful tool for comparing variability between two or more samples. This is especially true if the intervals of the values obtained for the samples are quite different.

2.5.1.4 Distributions

In a "normal distribution" (Figure 2.9), the measures of central tendency (mean, median, and mode) are close enough to be considered the same. Data having a normal distribution are as likely to fall above as below the mean. If a normal distribution pertains, approximately 68% of the data will fall within one standard deviation of the mean, 95% will fall within two standard deviations, and 99% will fall within three standard deviations. Although true normal distributions have rigorous mathematical definitions, these empirical rules are commonly applied to any data having a bell-shaped distribution and even to any "mounded" distribution.

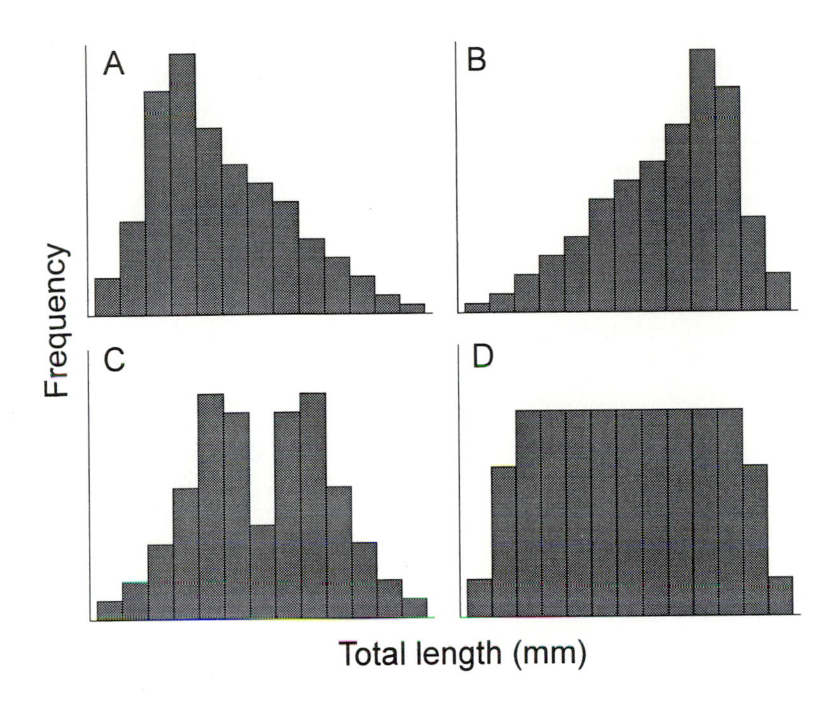

Figure 2.10 Examples of nonnormal distributions. Graphs **A** and **B** depict skewed distributions in which one tail is stretched farther than the other. The direction of the skewness is on the side of the longer tail (e.g., **A** is skewed right and **B** is skewed left). Graph **C** indicates a bimodal distribution in which two different classes have been sampled. Graph **D** represents a uniform distribution in which most values appear in nearly equivalent frequency.

Typically, if the data are continuous and the population sampled is normally distributed, the sample will approximate a normal distribution provided that the data are collected according to a random, nonselective design with an adequate sample size. However, the true distribution of the sampling data is not known until collections are complete. Thus, other types of distributions, similar to those exhibited in Figure 2.10, may be revealed in the early stages of data interpretation.

Count data frequently are not normally distributed. For example, when individuals are sampled over a large area, the distribution of number of individuals found per unit sampling effort or per small unit of area depends largely on the spatial patterning of the individuals. One of the most common spatial patterns observed in nature is a random one. Data collected from random spatial patterns are usually best described by the Poisson statistical distribution (e.g., Elliott 1983). Another fairly common spatial distribution is an aggregate (clumped) pattern. The most common statistical distribution used to describe an aggregate pattern is the negative binomial (e.g., Krebs 1989). The sampling approach and the number of individuals sampled (sample size) will also influence the distribution of these types of data. For example, a characteristic of random spatial data is that as the sample mean value becomes larger the frequency distribution approaches normality.

2.5.2 Sampling Considerations

In addition to describing data sets, as in descriptive statistics, statistical analysis has a second important function: making inferences based on incomplete information.

Statistical inference is a process of concluding something about a population from which only a sample was drawn. For example, we can assume that the growth rates of largemouth bass in two lakes are similar, but it is not practical to measure individual ages and lengths for those entire populations. Therefore, we collect samples from those populations to draw a statistical conclusion. In addition, we might want to know whether apparent differences between two or more sample means and variances are real or due to chance, and testing such differences is an important component of inferential statistics.

Statistical inferences are only as good as the samples on which they are based. Before taking up inferential statistics (in section 2.5.3), we next present some basic elements of sampling design.

2.5.2.1 Populations and Samples

A population comprises all the elements under investigation. A sample comprises some of the elements of a population. Why not measure the whole population? The three primary reasons are the large area or volume over which the population may occur, the time that would be required to measure the entire population, and the cost (time and materials) of comprehensive measurements. Embracing an entire population usually requires an impractical allotment of worker hours and a prohibitive amount of money in relation to project objectives. In lieu of collecting all elements, we draw samples from populations in a way that allows us to describe the biological population statistically.

A crucial point to consider here is the distinction between a biological population and a statistical population. A biological population is a group of organisms of one species living in an area at a particular time. Biological populations are sometimes difficult to delineate—the blue catfish population in the Santee–Cooper Reservoir, say, or the striped bass population in the Arkansas River. The limits of the statistical population may or may not bound the biological population under study; for example, we may be interested in the local striped bass population in the Keystone Reservoir of the Arkansas River. Obviously, migration can occur, and members can enter or leave the local reservoir area. In this scenario, the target population is a subset of the biological population and would be delineated by spatial and temporal scales.

2.5.2.2 Sample Design

The value of sampling data is largely determined by the quality or appropriateness of the sampling plan. Samples are often the sources of mistakes, and so arises the question "When can a sample be trusted?" This depends greatly on the way the sample has been selected. Some of the major considerations that must be addressed in the sampling design are the size of the area to be sampled (boundaries of the population), the number of sampling units in each sample, and the location (distribution) of the sampling units in the sampling area.

Selection of the proper sampling unit is a fundamental aspect of sampling design. A sample is an aggregate of sampling units; therefore, the sampling unit must be defined for the population to be sampled. The sampling unit is an observational unit; it can be a 60-m section of shoreline, a given volume of water or substrate, or a single fish. Certain techniques provide very distinct sampling units, such as collecting benthic invertebrates with a Ponar dredge. Sampling units must not overlap; that is, each member of the population can belong to only one sample unit.

Randomization is another very important aspect of sampling design. In a completely randomized sample each member of the population has an equal chance of being selected. However, samples may be biased due to gear type (gill net, electroshocker, seine), collector, location, time of day, season, and so on. To avert bias we usually increase the number of samples, transform the data (section 2.5.5), or make other statistical adjustments.

Samples may be collected by a variety of designs. The ideal sampling design provides good estimates with small confidence intervals at low cost. Designs should be compared for the time and money each requires before one design is selected. Among the most common sampling designs in fisheries management are random sampling, stratified random sampling, cluster sampling, and systematic sampling.

2.5.2.3 Random Sampling

A sample can be trusted (within limits that can be calculated) if every individual or object in the population has an equal opportunity of being chosen for the sample. A sample that has been chosen in this manner is a random sample, and random sampling prevents subjective selection by an investigator. As with most statistical concepts, random sampling is easier to describe than to apply in field situations. The word random has a specific meaning in statistical work. It does not mean haphazard or aimless. It does not even necessarily imply that such a sample is a typical cross section of the population. It refers only to a particular way of selecting individuals from a population, in which care is taken to see that every individual has the same chance of being included in the sample group. For a sample to be statistically representative of the population, it must be chosen by a random process. This randomness criterion is also built into more complex forms of sampling to guarantee that the accuracy of the data is not diminished by the investigator.

Random sampling can be accomplished with or without replacement of individuals. For example, if trap-netted white crappies were returned to the water in good condition before the next sample were taken, they would have yet another opportunity to be captured. On the other hand, if they were removed there would be no chance that they could be recaptured.

Selection of a random sample requires the use of a random number table or a computer program that will generate random numbers. For example, if a sampling program called for seining logperch in six 10-m sections along a 300-m shoreline, the shoreline would be divided with markers into 30 sites that would be consecutively numbered. Then, by using a random number table or a computer software program, or by simply drawing numbered papers from a hat, six sampling sites would be chosen from among sites 1 to 30. As illustrated in Figure 2.11, sites 2, 3, 7, 15, 19, and 28 were chosen from a random number table.

Cost is the primary drawback associated with random sampling because large samples are needed to account for the high variability that is common in biological populations. There is also an appreciable risk that a small random sample will contain, by chance, a disproportionate number of some minority group whenever there is much variation among individuals. Thus, to minimize that risk the random sample should be relatively large.

2.5.2.4 Stratified Random Sampling

When a simple random sample is acquired from each of several subdivisions of a population the process is termed stratified random sampling. In Figure 2.11, the same 30 sample sites used in the random sampling example were stratified into three

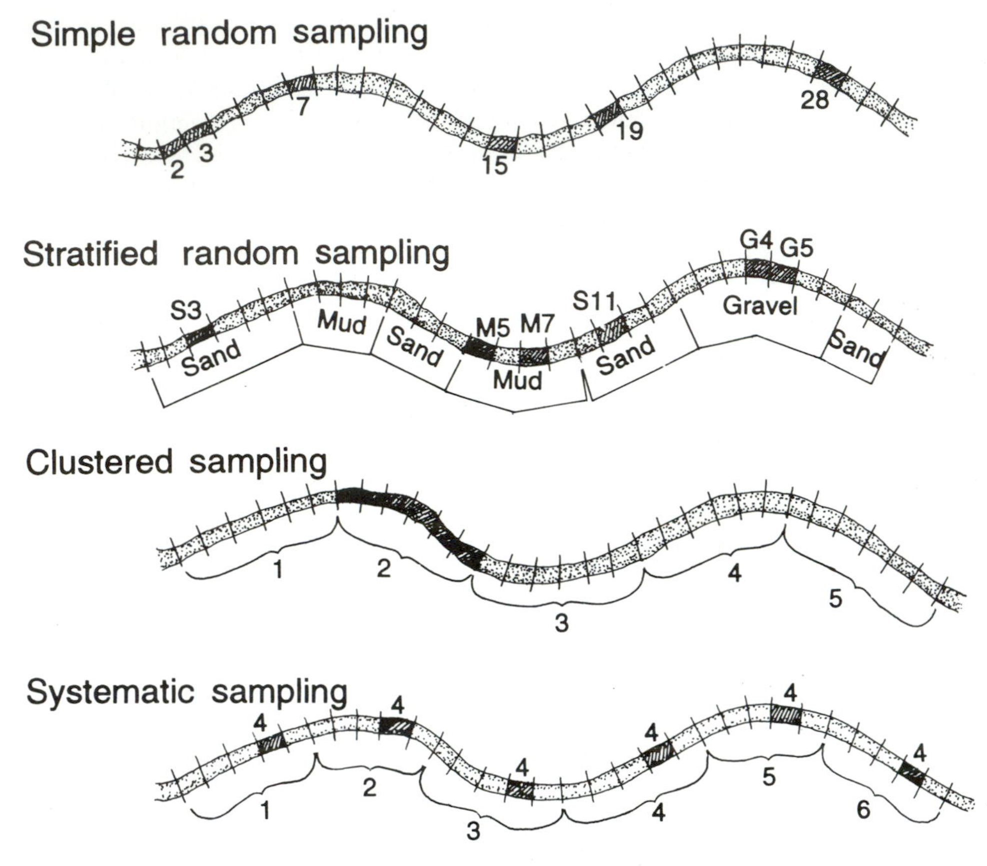

Figure 2.11 Examples of various sampling designs for the selection of six seining locations along a continuous shoreline. All numbers were chosen from a random number table (reprinted from Johnson and Nielsen 1983).

groups based upon substrate type (mud, sand, or gravel). Two sampling sites were then randomly selected within each stratum.

The differences among sampling locations or times are recognized and accounted for in the stratified random sampling procedure. The general idea of this sampling design is to reduce the overall variance of the sample by accounting for the major source(s) of variation among samples. In some cases the need for stratification may not be apparent during the design and data collection stages; sometimes poststratification of the data prior to analysis is a choice.

There are some very good reasons for using stratified random sampling. Sampling problems such as differing efficiencies of gears (see Chapters 6–8) among species or population segments may increase variance considerably. Furthermore, stratification usually increases precision in estimating characteristics of the population. For example, if fish distribution patterns are well known, the precision of population size estimates can be increased by using area strata. Thus, stratified random sampling provides an accurate way of dealing with any group that possesses considerable internal variability, and it is economical because the sample size needed to give a specified degree of precision is much smaller than would be the case with ordinary random sampling.

Strata, or subdivisions, must be distinct so that one cannot draw the same sample from more than one stratum. For maximum efficiency, each stratum should be as homogeneous as possible. Thus, variation within strata should be kept small; variation among strata does not contribute to sampling error for estimates such as the sample mean. The type and number of strata may be apparent from the study, or formal construction of strata (e.g., best characteristic, number, or boundaries) can be determined by specific allocation methods (Cochran 1977).

Stratification of aquatic ecosystems can be based on known physical, chemical, and biological differences. Physical differences in various habitats within a water body affect data and should be stratified. Examples of habitat strata include littoral and limnetic areas; hypo-, meta-, and epilimnion layers; substrate types; and stream characteristics such as pools, riffles, and runs (Chapter 4). In Figure 2.11, we stratified seining locations by type of bottom substrate.

Cyclic seasonal patterns can also impose a seasonal stratification on data. Biological data, such as growth, migration, and spawning, that coincide with seasons can normally be subdivided accordingly. Diurnal variations in photoperiods that affect water quality, fish stomach contents, and species composition can be stratified according to gradations of light, or simply as light and dark.

At times, fishery biologists may sample fishes at locations where and when they expect to catch particular species. This type of sampling has been variously referred to as fixed sampling, subjective sampling, judgement sampling, or intuitive sampling. Krebs (1989) acknowledged that subjective sampling is common in ecological research, but indicated that such sampling is actually stratified sampling. The stratifications in subjective fisheries sampling are commonly based on the habitat in which a target species is expected to occur. To reiterate, however, we caution that samples should be collected randomly within strata to obtain reliable data.

2.5.2.5 Cluster Sampling

Sampling clusters or collections of observations may be done to reduce sampling costs or to obtain data when random sampling is not practical. For example, it may be too costly to reach distant sites on a large reservoir, especially if the data can be collected in a localized portion of the reservoir. In choosing seining locations along a shoreline, it may be too expensive to identify every possible 10-m site simply to obtain six sampling sites. Instead, the area may be divided into groups of (say), six sites each that can be identified more economically. In Figure 2.11, the shoreline was divided into five clusters of six sites. The cluster to be sampled was randomly chosen.

Cluster sampling is accomplished by sampling several close items rather than by acquiring the same number of items that are a greater distance apart. For example, cluster sampling of habitat is done by randomly choosing a location and defining a cluster of sites within the location. All sites within the cluster then are sampled. Cluster size (number of sites) depends upon the inherent pattern of variability in a location. If sites are similar then cluster size can be small; if considerable variability exists then cluster size should be large.

For the same sample size, cluster sampling gives lower precision than simple random sampling. However, the reduced expenses associated with cluster sampling often allow for an increase in the sample size, which can result in better precision than simple random sampling could provide at a similar cost.

2.5.2.6 Systematic Sampling

Systematic sampling is conducted by selecting sampling units at regular intervals. Systematic sampling is also known as uniform sampling. It consists of dividing the sampling locations into equal sections, numbering the sampling units in each section consecutively, and sampling the same numbered unit in each section (Figure 2.11).

An important characteristic of this design is that it is less subject to investigator errors such as preference. Consider the case of a fishery technician gathering weight data from salmon passing through a fish weir. Netting every fish would be impossible and would inhibit their movement through the weir; therefore the technician has the task of selecting and removing fish in a systematic fashion (otherwise the technician might selectively remove only the largest fish). By selecting a random number, say 36, the technician can remove every 36th fish, weigh it, and return it to the water without impeding the passage of other fish.

Even though systematic sampling is simple and can be used in a variety of sampling situations, care must be taken to account for any trends or cycles that occur in the population. If for any reason it is suspected that the population being sampled varies in some cyclical way, it will be necessary to select sampling units at irregular intervals, such as by randomly changing the interval within each group of 20 units. An unbiased estimate of the sampling error is obtained by taking more than one systematic sample. The precision of systematic sampling may be increased by conducting the sampling within strata (Cochran 1977).

The foregoing discussion of sampling designs is only a brief introduction to the subject. More information on alternative designs, along with numerous examples, can be found in texts such as Cochran (1977) and Scheaffer et al. (1990). A variety of additional designs (e.g., nested subsampling, stratified nesting, and double sampling) can be used to obtain relative and absolute estimates. Excellent references for specific techniques used to estimate population parameters such as size, mortality, and growth are Ricker (1975), Everhart and Youngs (1981), Weatherley and Gill (1987), Krebs (1989), and Van Den Avyle (1993). These techniques are built on aspects of the basic designs mentioned above. It is important to remember that different sampling designs have unique assumptions, practical and statistical features, and efficiencies that vary under different situations. As a final caution on sampling designs, a badly chosen sample is worse than useless—it is misleading.

2.5.2.7 Sample Size Determination

The next step of the sampling plan is to determine the sample size. Sample sizes may be limited by a study's design, but an investigator usually can choose sample sizes freely if funding and other study constraints are not severe. For each design that is considered, rough estimates of the size of the sample needed can be made once the required degree of precision is established. The larger the sample, the more accurately it will reflect the characteristics of the population. Thus statistical reliability increases as the number of sampling units increases. The simplest solution is always to take very large samples; however, in practice a compromise must be made between statistical accuracy, time, and labor. The primary factor that determines the size of the sample needed should be the degree of precision required.

Statistically, accuracy increases approximately with the square root of the sample size for a normally distributed, randomly sampled variable, so a sample must be increased 100-fold to achieve a 10-fold increase in accuracy. For example, a sample of 5 largemouth bass may be adequate for determining the mean weight to the nearest 100 g, but 500 fish may be necessary for accuracy to the nearest 10 g.

Appropriate sample size also can be deduced in a stepwise fashion. Take five sampling units at random and calculate the sample mean and confidence intervals. Then take additional five-unit steps (i.e., $n = 5, 10, 15, 20, \ldots$), calculate the sample mean and confidence interval at each step, and continue until the sample mean ceases to fluctuate and the confidence interval is judged to be sufficiently small (Elliott 1983). The major drawback with this method is that calculations can be difficult to do during sample collections. The advantage of this sequential approach is that the sample size is not fixed in advance, and therefore sampling can continue until the desired level of precision is obtained (Mace 1964). Moreover, the method reduces some of the uncertainties associated with field sampling and minimizes the amount of time and money invested in a project. Weithman et al. (1979) reported that this approach was effective in determining the sample size needed to reach a conclusion about the stock structure of largemouth bass populations. The sequential approach minimizes the likelihood of reaching a false conclusion based on insufficient data and it allows attainment of specified confidence in point estimates.

Sample size determination is a fundamental aspect of any sampling plan and will generally be a compromise between precision and the cost of data collection. Beyond the simple examples given above, the definition of sample sizes is a complex task. Readers should consult more comprehensive works for equations to calculate optimum size and number of sampling units for normally distributed data (e.g., Cochran 1977; Krebs 1989; Scheaffer et al. 1990; Box 11.1) and for data that follow binomial, Poisson, and other distributions (Karandinos 1976; Elliott 1983).

2.5.3 Inferential Statistics

2.5.3.1 Hypothesis Formulation and Testing

The traditional decision-making approach of statistical inference is hypothesis testing. The outcomes of experiments generally are not clear-cut and a decision has to be made between competing hypotheses. Hypothesis testing provides an objective framework for making decisions, as opposed to making subjective decisions by simply looking at data. Numerous opinions could be formed by looking at the data, but hypothesis testing provides uniform criteria for decision making that are consistent for all investigators. A statistical hypothesis is simply a statement that something is true. For example, statements such as "the density of the white crappie population is not changing," "there is no effect of selenium on walleye reproduction," and "there is no relationship between largemouth bass growth and threadfin shad stockings" are hypotheses.

The formal hypothesis-testing approach has five components—the null hypothesis (H_0), an alternative or research hypothesis (H_A), a test statistic, a criterion for rejecting hypotheses, and a conclusion (Ott 1984)—and it is based on the concept of proof by contradiction. The null hypothesis is the hypothesis to be tested. It is generally a statement that the population parameter has a specified value or that parameters from two or more populations are similar. For example, we hypothesize that the mean growth rates (μ_1, μ_2) of two fish populations are similar. This may be formally stated as

$$H_0: \mu_1 = \mu_2.$$

Table 2.3 Guidelines for judging the significance of a statistical test probability (*P*-value) when the threshold level of significance has been set at a probability of 0.05 (adapted from Rosner 1982).

If $P > 0.05$	Then the test result is not statistically significant (NS)
If $0.05 < P < 0.10$	Then a tendency toward statistical significance is sometimes noted
If $0.01 \leq P < 0.05$	Then the test result is "significant"
If $0.001 \leq P < 0.01$	Then the test result is "highly significant"
If $0.0001 \leq P < 0.001$	Then the test result is "very highly significant"

The alternative hypothesis is a statement that the parameter differs between populations:

$$H_A: \mu_1 \neq \mu_2.$$

This alternative is called a two-sided hypothesis because it would be true if μ_1 were less than μ_2 or if μ_1 were greater than μ_2. We could have stated a null hypothesis specifying H_0: $\mu_1 \geq \mu_2$ (which, in null terms, is equivalent to stating that μ_1 is not less than μ_2), in which case the alternative hypothesis, H_A: $\mu_1 < \mu_2$, would be one-sided. Whether hypotheses are one- or two-sided affects the way statistical tests are applied to them. Once hypotheses have been formulated, a sample is obtained, a statistical test is selected (Appendix 2.2), and a test statistic is computed from the data. The test statistic follows a statistical distribution characteristic of the test, and a decision to accept the null hypothesis or reject it in favor of the research hypothesis depends on whether or not the calculated test statistic falls within the "rejection region" of its distribution.

If we state that the null hypothesis is correct, what are the odds that we are wrong? Statistical tests allow us to calculate these odds, which are expressed as probability (P) values. The lower the P-value, the greater is the likelihood that the null hypothesis is incorrect.

2.5.3.2 Level of Significance

How low the P-value should be before the null hypothesis is rejected is primarily a matter of judgement. This threshold value is called the level of significance, and it must be established before an inferential test is performed. Most investigators use either the 0.05 or the 0.01 level of significance for hypothesis testing (Table 2.3). The null hypothesis is rejected when the P-value generated from the test statistic is equal to or less than the threshold level. For example, if the threshold level of significance is 0.05 and the P-value of the calculated statistic is 0.05 or less, the null hypothesis can be rejected with only 1 chance in 20 that the decision is wrong.

2.5.3.3 Statistical Errors

Two types of statistical errors, commonly known as type I and type II errors, may be committed during hypothesis testing. The probabilities of making these errors are a measure of the goodness of the statistical test. Suppose we are deciding whether H_0 or H_A is true. If we compute a test statistic and decide that H_0 is true, we will state that we accept H_0, or more properly, that we failed to reject H_0. (Failing to reject H_0 implies only that we did not find sufficient evidence to reject the statement of H_0.) If the test causes us to reject H_0, we will accept H_A as true.

Most statistical analyses of either field or laboratory data result in one of four possible test outcomes: (1) H_0 is accepted and H_0 is really true; (2) H_0 is accepted but

Table 2.4 Four outcomes of statistically testing a null hypothesis (H_0) in relation to the probability of type I and type II errors.

Decision	Actual status of H_0: H_0 true	Actual status of H_0: H_0 false
Accept H_0	Correct decision with probability $(1 - \alpha)$	Type II error with probability β
Reject H_0	Type I error with probability α	Correct decision with probability $(1 - \beta)$

H_A is true; (3) H_0 is rejected but H_0 is true; and (4) H_0 is rejected and H_A is true (Table 2.4). If H_0 is true and we accept H_0 or if H_A is true and we reject H_0, the correct decision has been made. However, if H_0 is true and we reject H_0 or if H_A is true and we accept H_0, then we have committed a statistical error. Decisions cannot be wholly error-free because they are based on sample information and probability. The best we can hope for is to control the probability, or risk, of error.

The probability of a type I error is the probability of rejecting H_0 when H_0 is true. The probability of a type I error is denoted by the Greek character alpha (α) and is commonly referred to as the significance level of a test. A type I error can be costly. For example, if the H_0 being tested is that an experimental fishery regulation has had no positive effect on the abundance of fish in a test population and a type I error is made (H_0 is rejected when it is actually true), management biologists will infer that the regulation is beneficial and their agency may implement it widely at considerable expense for monitoring, enforcement, and administration. If the regulation were applied to a commercial fishery, it could result in lost jobs and business as well. These would be unfortunate consequences of a regulation that actually is ineffective.

The probability of a type II error is denoted by the Greek character beta (β), and it is the probability of accepting H_0 when H_0 is actually false. Again, this type of error could have substantial consequences. For example, a manager might conclude that a regulation has had no effect on a fishery when it actually has brought a positive benefit. If the regulation is abandoned unnecessarily, years of effort and expense may be needed to find an alternative management tool.

Type I and type II errors are controlled by assigning them small probabilities. The most frequently used probabilities are 0.01 and 0.05. The probability assigned to each error depends on the seriousness of the error. Type I and II errors are inversely related; that is, α increases when β decreases, and vice versa. The sample size required can be estimated if α and β are specified before data are collected (Ott 1984). The required sample size increases as the variance increases, as the significance level (α) is reduced, and as the required power $(1 - \beta)$ increases.

2.5.3.4 Power Analysis

Statistical tests that simultaneously minimize α and β would be ideal, but they are prevented by a basic contradiction: making α small means that H_0 will be rejected less often, whereas making β small means that H_0 will be accepted less often. The standard compromise is to fix α at a specified significance level and to use the test that maximizes statistical power (i.e., minimizes β given the designated α).

The power of a test is 1 minus the probability of a type II error, or $1 - \beta$. Power denotes the likelihood of rejecting an H_0 when the H_A is true; that is, of detecting a significant difference between the values compared. If the power is too low, it allows little chance of finding significant differences even when real differences exist. Poor control of power can create problems for scientists and management biologists. For example, of 408 scientific papers examined by Peterman (1990) in two major journals, statistical tests reported in 160 papers failed to reject a null hypothesis. In 83 of these papers, the authors made 142 recommendations or interpretations under the assumption that H_0 was true, yet in only 1 paper was there a calculation of the probability of a type II error. Such shortcomings magnify the chance of making erroneous decisions.

Besides their selections of α and β, biologists can also affect power by properly selecting sample size and the magnitude of the difference to be detected. The choice of sample size and power are closely connected. For this reason, calculations of sample sizes required to detect effects of a particular magnitude are called power analyses. All research studies should include some initial calculations of the power that particular statistical tests can provide given a designated alpha and alternative sample sizes. Such preliminary analyses will not be easy because crucial information such as variances will not be known, and therefore estimates from previous or pilot studies may have to be used.

Several general statements can be made concerning power.

- Power is related to α. Decreasing α (i.e., making the test more stringent) also decreases the power of the test. To regain power, one must increase sample size. For example, Parkinson et al. (1988) showed that decreasing α from 0.20 to 0.05 while β was kept at 0.20 meant that nearly twice as many fish had to be sampled to detect a 10% change in length at age for rainbow trout. If β were reduced to 0.05 (increasing power) with α at 0.05, detecting the 10% biological change would require nearly four times as many fish.
- Larger effects or differences can be detected with higher power than small effects. Again, requiring greater discrimination means requiring more sampling. Parkinson et al. (1988) showed that at the 0.20 level of α and β, detecting a 5% change in rainbow trout length at age took three times as many fish as detecting a 10% change.
- Power is positively related to the reliability of the samples (i.e., inversely related to the variance). Thus, more precise estimates provide greater power than do less precise estimates. Parkinson et al. (1988) illustrated this point as well. The number of days of sampling required to detect a change in angling catch per unit effort was estimated for a rainbow trout fishery. When all sampled days were used in the analysis, the CV was 44%, and approximately 16 d of sampling were required to detect a 40% change. When only days with more than 50 h of angling effort were included, the CV was 34%, and approximately 10 d of sampling were required to detect a 40% change. Thus, all other things being equal, less variable data allow a given power to be achieved with smaller samples, as in this example, or a greater power to be achieved with a given sample size.

Two points need to be made with regard to power analysis. First, power can be calculated only with respect to a specific alternative scenario. The mechanics of calculating power require knowing not only the H_0 but one or more alternative hypotheses as well. For example, if managers were testing an H_0 of no difference in catch per unit effort (CPUE) as a result of a regulation, they would also have to decide upon alternative effects—that CPUE has increased by 20 fish/h, say—before

Table 2.5 Typical nonparametric tests and their parametric counterparts (derived from Conover 1980).

Nonparametric test	Parametric or alternative test
Mann–Whitney test	Two-sample t-test; median test
Kruskal–Wallis test	One-way F-test; median test
Squared ranks test (equal variance)	Two-sample F-test
Spearman's rho	Regression
Kendall's tau	Regression
Wilcoxon test	Paired t-test
Mood test	F-test
Moses test	F-test
Friedman test (two samples)	Two-sample t-test
Friedman test (k samples)	F-test

they could calculate the power of their test. Second, if statistical tests have already rejected the null hypothesis, the calculation of power is irrelevant. Power refers only to the probability of falsely *failing* to reject an H_0.

Comprehensive statistical software packages include procedures for calculating power. Goldstein (1989) reviewed 13 software packages that provide power calculation, and the number of such packages undoubtedly has grown since. As with all software, however, the user must know specifically what to test and consider the ease of use and flexibility of various software packages.

2.5.4 Nonparametric and Parametric Statistics

Once a study has been designed, whether it is a simple comparison between events in two aquaria or a complex study of the effects of environmental change on an aquatic community, a biologist must decide whether to use parametric tests, which are based on normal distributions of data, or nonparametric tests, which are distribution-free (Table 2.5). These decisions are difficult and often confusing, but they can be simplified by studying the reasons why data may not be normally distributed and the assumptions of the various tests that are appropriate for the data.

An assumption underlying most parametric statistical methods, such as a typical t-test, is that the data follow a specific type of distribution, usually a normal distribution. The methods work well in many situations when theoretical analysis of the process under study predicts a normal distribution or previous data suggest that the data follow a normal distribution. Parametric tests often work well even when the assumption of normality has not been tested, but this must be attributed to luck. Nonnormal data often can be transformed to provide a distribution conducive to parametric testing (see section 2.5.7). Frequently, however, normality cannot be assumed for several reasons (Austin 1987; Potvin and Roff 1993). First, the true distribution of the population from which the sample was taken may not fit the normal model. Second, extreme values (outliers), may be present; if they are honest and not "blunders" (Potvin and Roff 1993), they alter the variance and lead to misinterpretations of the data. Third, the variance of the error term may not be the same for all treatments, violating the property of homoscedasticity (equal variance); homoscedasticity is an assumption of parametric statistics often violated in field studies (Potvin and Roff 1993).

The major advantages of nonparametric tests are that they do not depend upon a specific underlying distribution of the data and are not influenced by the effect of outliers in the data. Although nonparametric tests may not be optimal for a given distribution, they are relatively good for most distributions. Parametric tests are more efficient when their assumptions (e.g., normal distribution) are met (Conover 1980). Nonparametric tests are more efficient when the underlying distribution is not normal (Conover 1980; Potvin and Roff 1993). Some parametric tests, such as the analysis of variance (Ott 1984), perform fairly well (they are robust) with nonnormal data whereas others, such as least-squares regression, can be greatly influenced by a single, possibly erroneous, point.

Bart and Notz (1994) listed three additional points to remember in choosing between parametric and nonparametric methods. First, the magnitude of a difference between populations is most easily shown by parametric tests (e.g., *t*-test and associated confidence intervals). For many nonparametric tests, CI determinations are complex and often not used. Second, nonparametric tests use medians rather than sample means. This may lead to confusion when a nonparametric test determines that sample A has a larger value than sample B, but the sample mean is larger for B than for A. Third, when sample size is large, many nonparametric procedures yield test statistics with an approximately normal distribution; as a result, the tables used to determine significance may be the same for parametric as for nonparametric tests. This does not indicate that the population is normally distributed but simply that the test statistic approximates normality and that tables based on normal distributions can be used for assessing the probability level of the statistic.

Nonparametric statistical procedures are now included in most statistical software packages and usually are as simple to run as parametric procedures. However, some complicated designs that are handled with parametric methods may not have well-developed nonparametric counterparts, particularly in commercial software. For further information refer to the detailed discussion of nonparametric methods provided by Hollander and Wolfe (1973), Lehman (1975), Marascuilo and McSweeney (1977), and Conover (1980).

2.5.5 Basic Inferential Tests of Significance

Inferential statistics for interval and ratio-level data are referred to as parametric statistics because they are used to estimate parameters of a population (e.g., mean and standard deviation) from a sample. Inferential statistics for nominal and ordinal data are referred to as nonparametric tests. Because parameters such as the mean and standard deviation are inappropriate for nominal and ordinal data, nonparametric tests are restricted to situations not requiring such estimates. As a general rule, interval and ratio data must conform to more restrictive assumptions than do the ordinal and nominal data used for nonparametric tests (Conover 1980).

Parametric tests include the independent *t*-test, paired (correlated) *t*-test, and the mean *t*-test (Ott 1984). The independent and paired *t*-tests are for two samples, and the mean *t*-test is for one sample. The mean *t*-test, or one-sample *t*-test, is rarely used in fisheries management because it compares point estimates provided by a sample with the true population parameters, which are rarely known. The more commonly used independent and paired *t*-tests are described in the following sections.

2.5.5.1 Independent *t*-Test

A frequent statistical objective is to determine if two unrelated samples are different and, if so, to what degree (inference concerning two independent means). The independent *t*-test is used to compare the estimated means ($\bar{y}$) of two samples to determine whether or not the samples came from the same population—that is, whether or not the two samples differ significantly. This t-test evaluates the difference between two sample means, $\bar{y}_1 - \bar{y}_2$, a difference that is assumed to follow a normal distribution centered on zero when the samples are from the same population. If many sets of two independent samples are randomly drawn from the same population, the average ("expected") value of $\bar{y}_1 - \bar{y}_2$ will be zero, and few differences will be very large either negatively or positively. Thus, as the magnitude of a difference between two means increases, the probability that the samples really did come from the same population decreases. If the probability is less than some preestablished level of significance (say, 0.05), the null hypothesis is rejected and the samples are judged to be different.

The value of a t-statistic is determined by the difference between two sample means ($\bar{y}$), the sample variances (s^2), and the sample sizes (n):

$$t = (\bar{y}_1 - \bar{y}_2)\sqrt{\frac{n_1 n_2}{n_1 + n_2}} \Big/ \sqrt{\frac{(n_1 - 1)s_1^2 + (n_2 - 1)s_2^2}{n_1 + n_2 - 2}}. \quad (2.8)$$

Associated with each t-value are $(n_1 - 1) + (n_2 - 1)$ or $n_1 + n_2 - 2$ degrees of freedom. Whether or not a calculated t-value indicates a significant difference between means at a given level of α depends on the degrees of freedom and hence on the sample sizes.

Consider the following samples of largemouth bass taken from two ponds to determine if the mean growth rates (g/week) differ between the populations. Our null hypothesis is that the mean growth rates are equal (H_0: $\mu_1 = \mu_2$) at $\alpha = 0.05$.

Example 2.3 Largemouth bass growth rates (g/week)

Data	
Pond 1	Pond 2
7.8	7.1
6.5	7.0
7.0	6.2
7.2	6.7
7.5	7.0
	6.4
$\Sigma = 36.0$	6.6
	$\Sigma = 47.0$

	Summary statistics	
	Pond 1	Pond 2
$\bar{y} =$	7.2	6.7
$\Sigma y_i^2 =$	260.2	316.3
$(\Sigma y_i)^2 =$	1,296	2,209
$n =$	5	7
$s^2 =$	0.25	0.12

Using these data, we calculate the t-value by means of equation (2.8) as

$$t = \frac{(7.2 - 6.7)\sqrt{\dfrac{5 \times 7}{5 + 7}}}{\sqrt{\dfrac{(5 - 1)(0.25) + (7 - 1)(0.12)}{5 + 7 - 2}}} = 2.07.$$

From a table of t-values (Appendix 2.1) we find that at the 0.05 level of significance for 10 df, the t-value is 2.228. The calculated t-value must exceed the tabular value if the null hypothesis is to be rejected. Because our calculated statistic (2.07) is less than the tabular value (2.228), we conclude that there is no significant difference between the two samples. But what if n_1 had been 60, and n_2 had been 62, and the two sample means and variances had remained the same? The calculated t-value would become

6.43, and with 120 df the tabular t-value would decrease to 1.98; therefore, we would conclude that the two samples are significantly different. The influence of sample size on our ability to detect real differences between populations is evident.

2.5.5.2 Paired t-Test

The paired t-test is used to determine if statistical differences exist between samples of paired data (inferences concerning two dependent means). Two samples are paired when each data point of the first sample is related to a unique data point of the second sample (Rosner 1982). The paired t-test equation is

$$t = \frac{\bar{d} - \mu_d}{s_d/\sqrt{n}}, \tag{2.9}$$

where $\bar{d}$ is the point estimate of the mean difference and μ_d is the hypothesized mean difference of 0. This t-statistic has $n - 1$ df. The standard deviation of the differences (s_d) is calculated as

$$s_d = \sqrt{\frac{n(\Sigma d^2) - (\Sigma d)^2}{n(n - 1)}}. \tag{2.10}$$

Because of the relationship between the two dependent samples, this equation provides a test of differences between the paired values. An example of paired data is the weight of a fish at two points in time. The weight change in a group of fish can be evaluated from the weight changes of individuals over a period of time. Thus, the two samples are not independent, and each individual serves as its own reference. In the following exercise, we determine whether the mean weight of these fish (as a group) has changed significantly (H_0: $\mu_d = 0$) at $\alpha = 0.05$.

Example 2.4 Weights (g) of five individual largemouth bass at the beginning (y_1) and end (y_2) of an experiment

Fish	y_1	y_2	d	d^2	
1	210	290	80	6,400	$\bar{d} = 210/5 = 42$
2	280	300	20	400	$\Sigma d^2 = 12{,}100$
3	170	210	40	1,600	$(\Sigma d)^2 = 44{,}100$
4	240	250	10	100	
5	270	330	60	3,600	
			$\Sigma = 210$	$\Sigma = 12{,}100$	

The data are incorporated into equations (2.9) and (2.10) to determine the t-value for $n - 1 = 4$ df:

$$s_d = \sqrt{\frac{5\ (12{,}100) - 44{,}100}{5\ (5 - 1)}} = 28.64;$$

$$t = \frac{42 - 0}{28.64/\sqrt{5}} = 3.28.$$

The tabular t-value (Appendix 2.1) for 4 df and an α of 0.05 is 2.776 for a two-tailed test. Because our calculated t-value (3.28) exceeds the tabular value, we reject the H_0 because we have shown that there has been a significant change in the mean weights of these fish. Moreover, because $\bar{d}$ and the calculated t-value are positive, we know the significant change was a weight gain. Indeed, the calculated t-value also exceeds the tabular t-value for a one-tailed test ($t = 2.132$, $\alpha = 0.05$, 4 df).

2.5.5.3 Multivariate Tests

An important group of parametric tests addresses comparisons of several simultaneous measures. For example, we might wish to determine whether several populations differ in multiple ways or if a measured variable is being affected by two or more independent variables. Making such decisions typically requires an analysis of variance, which simultaneously tests for several differences (see Appendix 2.2). Study design has many implications for analysis of variance, and we refer readers to statistics textbooks for complete treatments of the subject (e.g., Ott 1984; Snedecor and Cochran 1989; Montgomery 1991).

2.5.5.4 Chi-Square Test

One of the most commonly used nonparametric tests is the chi-square (χ^2) test. This analysis provides a test of the "goodness of fit" of observed frequencies in a sample to some hypothetical frequency. For example, suppose we stocked 600 fin-clipped bluegills, 300 jaw-tagged bluegills, and 100 disk-tagged bluegills in a pond at the beginning of summer. At the end of the summer we collected a sample of 60 fish, of which 34 were fin-clipped, 20 were jaw-tagged, and 6 were disk-tagged individuals. Our problem is to decide if the distribution of tag types in the sample at the end of the summer differed from the initial distribution. If it did differ, then we may conclude that the type of tag affected the bluegills in the same way (e.g., by altering survival or catchability) or that retention rates may have varied among the tag types.

The general equation for χ^2 is

$$\chi^2 = \Sigma \frac{(O - E)^2}{E}; \tag{2.11}$$

O = the observed frequency in a given class (here, tag type),
E = the expected frequency in a given class,
n = the number of classes (here, 3),
df = $n - 1$ classes (here, 2)

For our example the calculation is as follows:

$$\chi^2 = \frac{[34 - (0.6)(60)]^2}{0.6(60)} + \frac{[20 - (0.3)(60)]^2}{0.3(60)} + \frac{[6 - (0.1)(60)]^2}{0.1(60)}$$

$$= \frac{4}{36} + \frac{4}{18} + \frac{0}{6} = 0.33.$$

From a table of χ^2 values we find the value for an α of 0.05 and 2 df, which is 5.99. Because our calculated χ^2 (0.33) does not exceed the tabular value (5.99), we should not reject our null hypothesis that the ratio at the end of summer did not differ from the ratio at the beginning of summer. We conclude that the type of tag applied did not differentially affect bluegill survival or catchability, and that tag retention rates were the same.

Many other nonparametric statistical tests are available for analysis of data having an unknown distribution. Several of these tests are compared with parametric tests in Table 2.5; a guide to test selection is provided in Appendix 2.2.

Nonparametric tests are assumed not to follow a specific probability distribution. However, this does not mean that no assumptions about distributions are made in nonparametric methods; most tests assume that populations are continuous and have the same shape. Because these procedures are typically based on ranks, nonparametric tests are robust in the face of problems such as outliers. Simple normalizing transformations of nonnormal data (section 2.5.7) often allow the use of a parametric method, which can be important because parametric procedures provide additional information, such as variance, unavailable from nonparametric procedures.

2.5.6 Regression Analysis and Measures of Association

2.5.6.1 Linear Regression

Two variables often are related systematically such that one is a fairly constant multiple of the other (bivariate data). Common examples in fisheries are the relationships of scale length to body length and growth rate to temperature. A linear regression (also called a least-squares best-fit line) can be calculated for bivariate data that have a linear appearance in scatter plots. The result is an equation for a straight line,

$$y = a + bx, \tag{2.12}$$

that quantifies the relationship between an independent variable (x) and a dependent variable (y). This regression model allows for the prediction of the dependent variable given a value of the independent variable. The statistical parameters that compose the simple regression are the intercept (a) and slope (b). To demonstrate the utility of regression, we consider the following data on total body length and scale length collected from white crappies. For this exercise, scales were first magnified by a constant amount and displayed on a microfiche reader, where their images were measured in millimeters. Fish lengths were measured directly.

Example 2.5 Magnified scale lengths (x, mm) and body lengths (y, mm) for 14 white crappies

x	y	xy	x^2	y^2
21	85	1,785	441	7,225
27	104	2,808	729	10,816
34	126	4,284	1,156	15,876
41	146	5,986	1,681	21,316
49	165	8,085	2,401	27,225
56	183	10,248	3,136	33,489
64	205	13,120	4,096	42,025
67	224	15,008	4,489	50,176
74	246	18,204	5,476	60,516
90	264	23,760	8,100	69,696
104	294	30,576	10,816	86,436
110	323	35,530	12,100	104,329
118	346	40,828	13,924	119,716
121	368	44,528	14,641	135,424
$\Sigma x = 976$	$\Sigma y = 3{,}079$	$\Sigma xy = 254{,}750$	$\Sigma x^2 = 83{,}186$	$\Sigma y^2 = 784{,}265$

We estimate b (slope of the line) by the equation

$$b = \frac{n(\Sigma xy) - (\Sigma x)(\Sigma y)}{n(\Sigma x^2) - (\Sigma x)^2}. \tag{2.13}$$

Performing the calculations from the example data provides

$$b = \frac{14\,(254{,}750) - (976)\,(3{,}079)}{14\,(83{,}186) - (952{,}576)} = 2.65.$$

The intercept of the line then is determined by the equation

$$a = \bar{y} - b\bar{x}, \tag{2.14}$$

which, in our example, is 35.19. Inserting the calculated slope and intercept values into the regression model provides an equation for predicting total length from scale length:

$$y = 35.19 + 2.65\,x.$$

To check this model, if we arbitrarily select a scale length (x-value) of 70, the equation predicts a total length of 221 mm. This length is consistent with the original data when allowances are made for natural biological variability. Thus the developed equation appears to be a reasonable predictor of total length for white crappie—provided that scale magnifications and measurement units stay the same. If measurement conventions change, the regression will have to be recalculated.

2.5.6.2 Measures of Association

If the bivariate data appear to be related, the relationship between the two variables has an intensity, or level of association, that is defined as correlation. The correlation calculation for two variables yields a single statistic called Pearson's product-moment correlation coefficient (r). This statistic is applied to interval and ratio data. Another statistic that accomplishes the same task for ordinal data is the Spearman rank correlation coefficient. The value of correlation coefficients range from $+1.0$ to -1.0. A value near $+1.0$ indicates that as one variable increases, the other variable increases in a very similar fashion (Figure 2.12). A correlation coefficient near -1.0 indicates that as one variable increases, the other variable decreases, and the two variables have a strong correlation (Figure 2.12). If the correlation coefficient is low or near zero, the two variables have little correlation. When the correlation coefficient is low, a derived regression equation would have little predictive value.

The correlation of the total length–scale length relationship in Example 2.5 can be determined with the following equation for r:

$$r = \frac{n(\Sigma xy) - (\Sigma x)(\Sigma y)}{\sqrt{n(\Sigma x^2) - (\Sigma x)^2} \cdot \sqrt{n(\Sigma y^2) - (\Sigma y)^2}}. \tag{2.15}$$

Thus, with one more calculation from the raw data ($\Sigma y^2 = 784{,}265$), we can estimate the strength of the relationship as

$$r = \frac{14(254{,}750) - (976)\,(3{,}079)}{\sqrt{14(83{,}186) - (976)^2} \cdot \sqrt{14(784{,}265) - (3{,}079)^2}} = 0.99.$$

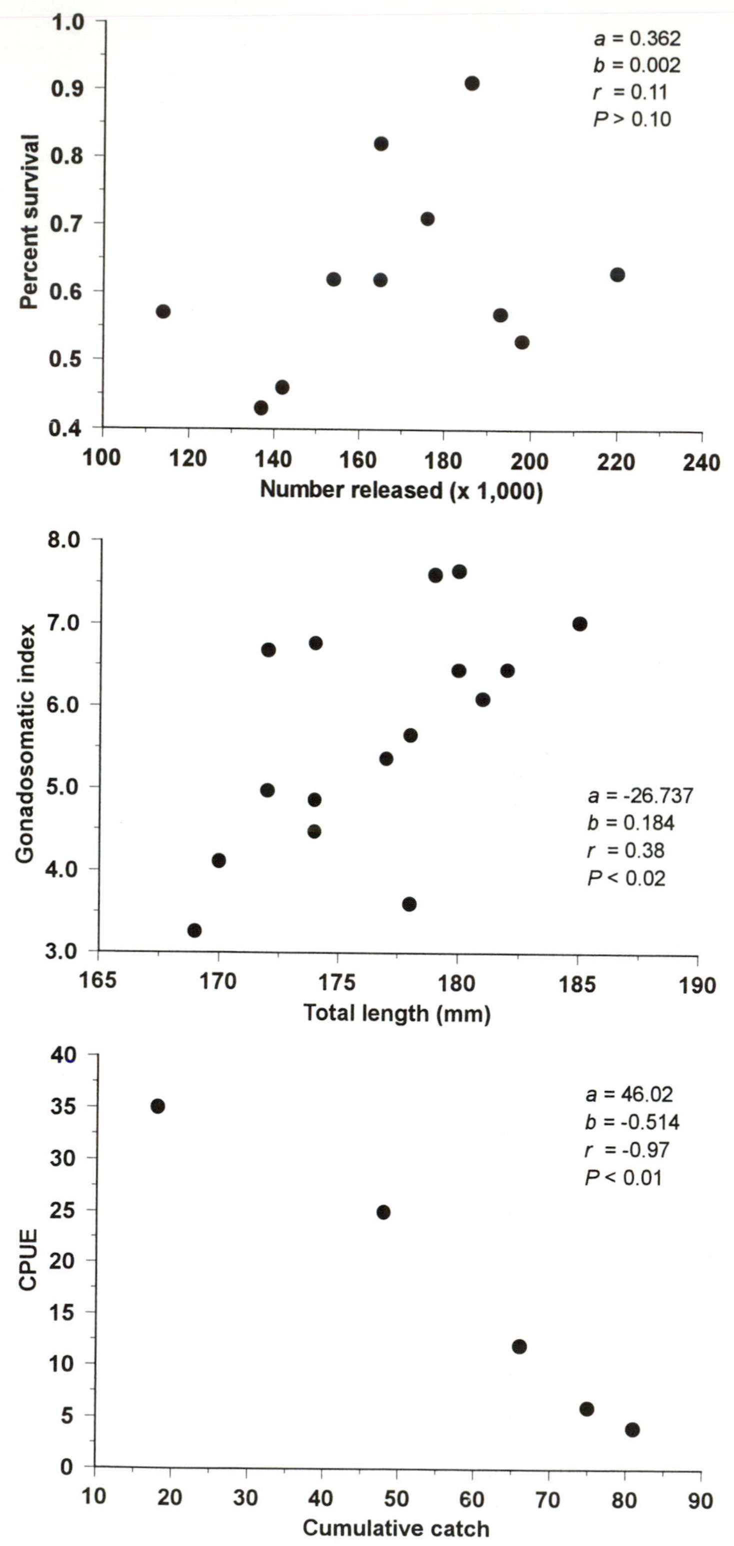

Figure 2.12 Scatter plots showing various strengths of linear relationship (i.e., correlation). The top plot indicates little correlation between the percent survival of released walleye fry and the number of fry released, the middle plot shows a moderate positive correlation between the gonadosomatic index and total length of bluegills, and the bottom plot shows a high negative correlation between catch per unit effort (CPUE) and the cumulative catch. The intercept (a), slope (b), correlation coefficient (r), and probability (P) are often given explicitly on such plots.

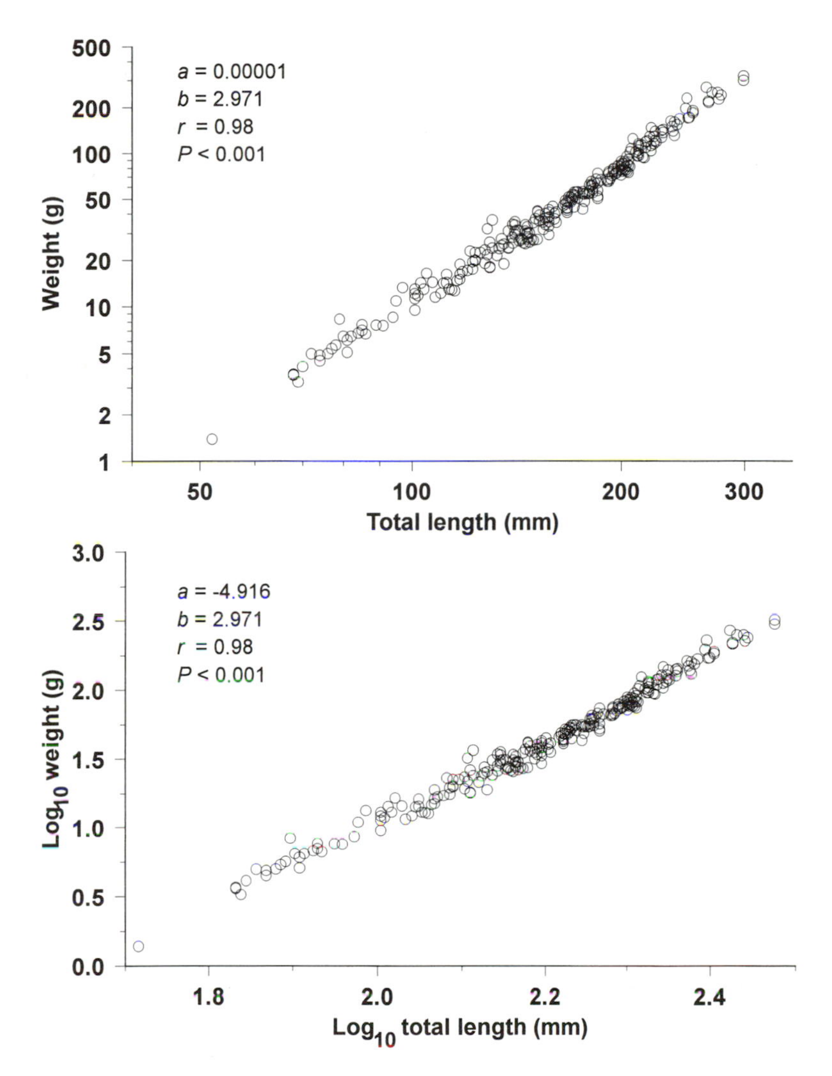

Figure 2.13 Scatter plots showing the linear relation of weight to length with logarithmic axes and untransformed data (top) and after transformation (bottom). The top graph has logarithmic ($\log_{10}$) axes to permit evaluation of the untransformed weight and length values; intercept (a) and slope (b) parameters were derived from the power function $W = aTL^b$, where W = weight (g) and TL = total length (mm). The bottom graph displays the $\log_{10}$-transformed weight–length data. The intercept (a) and slope (b) parameters were derived from the linear equation $\log_{10} W = a + b \cdot \log_{10} TL$.

Relationships such as this one have fairly straightforward interpretations. One must keep in mind, however, that strong correlations between two variables do not establish cause and effect; that is, the increase or decrease in the values of one variable does not necessarily cause a response in the other variable. Investigators often find strong relationships among natural phenomena. However, some relationships arise because both variables are under the strong influence of another variable (e.g., temperature or season) that may not have been considered or measured.

2.5.6.3 Geometric Mean Regression

Central trend lines should be used in place of linear regression when large measurement errors are suspected in both independent and dependent variables, the sample is not random, or the bivariate data are not normally distributed (Ricker 1973). Geometric mean regression (GMR) is often used to describe a central trend line that defines the functional relationship between variables. As with linear regression, the central trend line can be used for both description and prediction.

The GMR uses the calculations from linear regressions to arrive at an equation for a central trend. Therefore, the parameters for linear regression and the correlation coefficient are determined and then applied to the GMR calculation. The GMR slope (v) is the linear regression slope (b) divided by the correlation coefficient (r). The GMR intercept calculation is similar to the intercept calculation for linear regression:

$$a' = \bar{y} - v\bar{x} \tag{2.16}$$

Thus, the GMR equation is given as

$$y = a' + vx, \tag{2.17}$$

which is very similar to equation (2.12).

2.5.7 Data Transformation

Often data require transformation to provide a distribution that is conducive to certain test statistics. The most common include logarithmic (base 10 or base e), square-root, and arcsine transformations. Plotted data often appear to exhibit a slightly curvilinear shape. As fish grow, for example, weight increases at a slightly faster rate than does length, giving the relationship between them a curvilinear or exponential form (Figure 2.5). The best relationship (i.e., highest correlation) between weight and length occurs when the data are transformed by taking the base 10 logarithms ($\log_{10}$) of both variables. This transformation gives the data a linear appearance (Figure 2.13). This form then allows predictability, which is best expressed by the linear equation (equation 2.12). Suppose we obtained the weight measurements for the fish in Example 2.2 and then determined the $\log_{10}$ values and sums required for the regression model.

Example 2.6 Log_{10}-transformed weights (g) and total lengths (mm) of largemouth bass

Total length		Weight			
x	$\log_{10} x$	y	$\log_{10} y$	$\log_{10} (x^2)$	$\log_{10} (xy)$
161	2.207	36.2	1.559	4.871	3.441
178	2.250	52.0	1.716	5.063	3.861
192	2.283	68.4	1.835	5.212	4.189
210	2.322	92.4	1.966	5.392	4.565
210	2.322	93.6	1.971	5.392	4.577
221	2.344	113.8	2.056	5.494	4.819
234	2.369	139.9	2.146	5.612	5.084
247	2.393	170.2	2.231	5.726	5.339
264	2.422	216.5	2.335	5.866	5.655
Σ	20.912		17.815	48.628	41.530

Applying these components to equations (2.13) and (2.14) provides the slope and intercept values of 3.597 and −6.380, respectively. The correlation coefficient was determined to be 0.99.

2.6 CRITICAL STATISTICAL CONSIDERATIONS IN STUDY DESIGN

2.6.1 General Study Designs

Fisheries professionals conduct a variety of studies broadly classified as mensurative (e.g., observational) and manipulative experiments (Hurlbert 1984). In mensurative studies, data are collected by making measurements on populations or communities at intervals in time or space. Therefore, variables associated with time and space are the items of interest and may or may not require tests for statistical significance. The objective of this type of study is often to monitor some process. This is passive observation, and the process is not under the control of an investigator. In some cases the process may not be well understood, so the study may be long term, occurring over a period of several years. Consider, for example, a proposal for a long-term study of a declining stock of anadromous striped bass. The specific cause of the decline is uncertain, and monitoring of various stock components will be necessary to provide knowledge with which to rehabilitate the stock. Data on fishery-dependent and fishery-independent elements (e.g., recruitment, growth, mortality, movement, size structure, disease, and condition) must be collected, analyzed, and interpreted to determine what factors may be affecting the stock. Sampling units for this example might be individual fish, groups of fish (e.g., age- or size-groups), or fish from different tributaries, and measurements will be made at specified points in time. Importantly, no manipulation of the stock will occur unless regulations or habitat are altered or stocking commences.

A simpler mensurative study would be measurement of densities of spawning bluegills within vegetated and nonvegetated habitats. Interest would be in observing the spatial aspect of habitat use by bluegills for a brief period of time. Comparisons also could be made between two or more populations of spawning bluegills and subjected to statistical testing; such an experiment would be replicated, but no manipulation would occur. One of the most critical elements in the design of mensurative experiments is to ensure that replicates are independent of each other by dispersing them appropriately in time or space.

The second type of study involves manipulation of some variable that will supposedly affect a process. Defined alterations of the variable are called treatments; different treatments are assigned to various experimental units. The experimental unit is "the smallest division of experimental material such that any two units may receive different treatments" (Krebs 1989). At least two treatments are necessary for a manipulative experiment, which also requires a control. A control is an experimental unit that is not manipulated and thus serves as a baseline for comparisons. In addition, all other variables associated with the process must be held constant. Consider a hatchery manager who wants to evaluate the response of phytoplankton in earthen nursery ponds to three concentrations of phosphate fertilizer. Except for the addition of urea, all ponds are treated equally. The manager determines the three levels to be used and applies each to a group of ponds; no urea is applied to a fourth group of ponds, which function as the control. The response periodically measured is phytoplankton biomass. In this study each treatment is an experimental unit, individual ponds are sampling units, and each plankton collection is a subsample. The manager has correctly identified three treatments that are each replicated with a group of ponds. These types of controlled experiments typically yield stronger inferences than do mensurative experiments.

Setting up valid controls is often difficult in manipulative field experiments. Sometimes only a before-and-after approach is possible, whereby measurements taken after a manipulation are compared with those obtained prior to the manipulation. In such cases each experimental unit functions as its own control. For example, the effect of a regulation such as a length limit could be determined by evaluating the stock structure of a population before and after the regulation is implemented. However, this kind of control should be used cautiously because changes unrelated to the manipulation may occur in populations over time. Thus it becomes important to collect data on a contemporaneous control as well, such as a population having similar characteristics but which is not manipulated (Krebs 1989). For example, if we wished to determine the effect of a length limit regulation, we might collect data on two populations known to have similar stock structures, impose the regulation on one of those populations, and then follow up by taking the same measurements again on both populations. The difficulty of such a study would be to find two similar populations.

2.6.2 Replication

Providing true replication of treatments and randomly allocating treatments to experimental units are other requirements that must be met in developing a manipulative study having desirable statistical qualities. Replication is repetition of the basic experiment by using multiple experimental units for each treatment. (The sample is the replicate in mensurative studies.) Replication is necessary for several reasons. First, it reduces the influence that a chance event in one experimental unit might have on the inferences drawn from the experiment. In statistical jargon, replication allows estimation of the error that occurs within the experiment (i.e., inherent variability).

Second, increasing the number of replicates allows a statistically more precise measure of the effect to be obtained. The experiment becomes more costly in money and time because of added replication, however. A cost-effective balance between the number of replicates and the number of subsamples allocated per treatment sometimes can be determined, but a statistician should be consulted in such matters to ensure a desired level of precision before any trade-offs are considered. The number of replicates required ultimately depends upon the degree of precision required.

Pseudoreplication is a term used to describe experiments in which treatments are not truly replicated or replicates are not statistically independent (Hurlbert 1984). This problem can occur in either a manipulative or a mensurative experiment. Subsampling in mensurative studies is a common misinterpretation of replication and provides a good example of what constitutes pseudoreplication. A sample is a set of measurements taken at a place or time. If more than one collection is done in the same area at the same time, then subsampling has been performed, not replication of samples.

The experimental units to which the treatments (including controls) are applied should be selected randomly so that independence among treatments is established. Most statistical methods incorporate an assumption that observations are indepen-

Box 2.2 General Rules of Thumb for Data Collection and Analysis (adapted from Krebs 1989)

1. Identify a problem and ask a question.
2. Not everything that can be measured should be.
3. In some situations, the questions may be unanswerable at the present time.
4. Employ nonstatistical knowledge of the problem throughout the planning process.
5. Collect data that will answer the question and meet statistical assumptions.
6. Keep the design and analysis as simple as possible.
7. With continuous data, save time and money by deciding on the number of significant digits needed in the data before the experiment starts.
8. Never report an estimate without some measure of its possible error.
9. Always be skeptical about the results of statistical tests of significance (see Krebs [1989] for further explanation).
10. Recognize the difference between statistical significance and biological significance.
11. Make backup copies of data sheets and database files.
12. Look forward to the final report, in which the quality of planning and data collection will be apparent—garbage in, garbage out.

dent and randomly distributed variables. Randomizing the experiment allows that assumption to be met, and it reduces the effects of extraneous factors present in the experiment. Overall, it is wise to confer with a statistician during experimental design to ensure that these basic principles will be accommodated. In that regard, we leave the reader with some suggested practical rules that should be remembered during the course of data collection and analysis (Box 2.2).

2.7 REFERENCES

Austin, M. P. 1987. Models for the analysis of species response to environmental gradient. Vegetation 69:35–45.

Baker, J. P., H. Olem, C. S. Creager, M. D. Marcus, and B. R. Parkhurst. 1993. Fish and fisheries management in lakes and reservoirs. U.S. Environmental Protection Agency EPA 841-R-93-002, Water Division. Terrene Institute and U.S. Environmental Protection Agency, Washington, DC.

Bart, J., and W. Notz. 1994. Analysis of data. Pages 24–74 *in* T. A. Bookhout, editor. Research and management techniques for wildlife and habitats. The Wildlife Society, Bethesda, Maryland.

Bayley, P. B., and D. J. Austen. 1989. Fisheries analysis system: data management and analysis for fisheries management and research. American Fisheries Society Symposium 6:199–205.

Clark, R. D., Jr., D. L. Garling, Jr., and R. T. Lackey. 1977. Computer use in freshwater fisheries management. Fisheries 2(4):21–23.

Cleveland, W. S. 1985. The elements of graphing data. Wadsworth Advanced Books and Software, Monterey, California.

Cochran, W. G. 1977. Sampling techniques, 3rd edition. Wiley, New York.

Cohen, J. 1988. Statistical power analysis for the behavioral sciences, 2nd edition. L. Erlbaum Associates, Hillsdale, New Jersey.

Conover, W. J. 1980. Practical nonparametric statistics, 2nd edition. Wiley, New York.

Draper, N. R., and H. Smith. 1981. Applied regression analysis, 2nd edition. Wiley, New York.

Elliott, J. M. 1983. Some methods for the statistical analysis of samples of benthic invertebrates. Freshwater Biological Association Scientific Publication 25.

Everhart, W. H., and W. D. Youngs. 1981. Principles of fisheries science, 2nd edition. Cornell University Press, Ithaca, New York.

Goldstein, R. 1989. Power and sample size via MS/PC-DOS computers. American Statistician 43:253–260.

Hollander, M., and D. A. Wolfe. 1973. Nonparametric statistical methods. Wiley, New York.

Hurlbert, S. H. 1984. Pseudoreplication and the design of ecological field experiments. Ecological Monographs 54:187–211.

Johnson, D. L., and L. A. Nielsen. 1983. Sampling considerations. Pages 1–21 *in* L. A. Nielsen and D. L. Johnson, editors. Fisheries techniques. American Fisheries Society, Bethesda, Maryland.

Johnson, R. A., and D. W. Wichern. 1988. Applied multivariate statistical analysis. Prentice-Hall, Englewood Cliffs, New Jersey.

Karandinos, M. G. 1976. Optimum sample size and comments on some published formulae. Bulletin of the Entomological Society of America 22:417–421.

Kennedy, V. S., and D. C. Kennedy. 1990. Graphic and tabular display of fishery data. Pages 33–64 *in* J. Hunter, editor. Writing for fisheries journals. American Fisheries Society, Bethesda, Maryland.

Krebs, C. J. 1989. Ecological methodology. Harper Collins Publishers, New York.

Lehman, E. L. 1975. Nonparametrics: statistical methods based on ranks. Holden-Day Publishing, San Francisco.

Mace, A. E. 1964. Sample-size determination. Reinhold, New York.

Marascuilo, L. A., and M. McSweeney. 1977. Nonparametric and distribution-free methods for the social sciences. Wadsworth, Belmont, California.

Mattison, R. M. 1993. Understanding database management systems. An insider's guide to architectures, products, and designs. McGraw-Hill, New York.

Montgomery, D. C. 1991. Design and analysis of experiments, 3rd edition. Wiley, New York.

Neter, J., W. Wasserman, and M. H. Kutner. 1989. Applied linear regression models. Irwin, Homewood, Illinois.

Ott, L. 1984. An introduction to statistical methods and data analysis, 2nd edition. Prindle, Webber and Schmidt, Boston.

Parkinson, E. A., J. Berkowitz, and C. J. Bull. 1988. Sample size requirements for detecting changes in some fisheries statistics from small trout lakes. North American Journal of Fisheries Management 8:181–190.

Peterman, R. M. 1990. Statistical power analysis can improve fisheries research and management. Canadian Journal of Fisheries and Aquatic Sciences 47:2–15.

Potvin, C., and D. A. Roff. 1993. Distribution-free and robust statistical methods: viable alternatives to parametric statistics. Ecology 74:1617–1628.

Ricker, W. E. 1973. Linear regressions in fishery research. Journal of the Fisheries Research Board of Canada 30:409–434.

Ricker, W. E. 1975. Computation and interpretation of biological statistics of fish populations. Fisheries Research Board of Canada Bulletin 191.

Rosner, B. A. 1982. Fundamentals of biostatistics. Duxbury Press, Boston.

Rothenberg, J. 1995. Ensuring the longevity of digital documents. Scientific American (January 1995):42–47.

Scheaffer, R. L., W. Mendenhall, and L. Ott. 1990. Elementary survey sampling, 4th edition. Prindle, Webber and Schmidt, Boston.

Seiter, C. 1994. Databases that work. Macworld (January 1994):141–146.

Siegel, S. 1956. Nonparametric statistics for the behavioral sciences. McGraw-Hill, New York.

Snedecor, G. W., and W. G. Cochran. 1989. Statistical methods, 8th edition. Iowa State University Press, Ames.

Sokal, R. R., and F. J. Rohlf. 1981. Biometry, 2nd edition. Freeman, San Francisco.

Steele, R. G. D., and J. H. Torrie. 1980. Principles and procedures of statistics, 2nd edition. McGraw-Hill, New York.

Tufte, E. R. 1983. The visual display of quantitative information. Graphics Press, Cheshire, Connecticut.

Tufte, E. R. 1990. Envisioning information. Graphics Press, Cheshire, Connecticut.

Van Den Avyle, M. J. 1993. Dynamics of exploited fish populations. Pages 105–135 *in* C. C. Kohler and W. A. Hubert, editors. Inland fisheries management in North America. American Fisheries Society, Bethesda, Maryland.

Venditto, G. 1992. 9 Multiuser databases: robust and ready to share. PC Magazine 11(6): 289–335.

Weatherley, A. H., and H. S. Gill. 1987. The biology of fish growth. Academic Press, Orlando, Florida.

Weithman, A. S., J. B. Reynolds, and D. E. Simpson. 1980. Assessment of structure of largemouth bass stocks by sequential sampling. Proceedings of the Annual Conference of Southeastern Association of Fish and Wildlife Agencies 33(1979):415–424.

Wilkinson, L. 1990. SYSTAT: the system for statistics. SYSTAT Inc., Evanston, Illinois.

Winer, B. J. 1971. Statistical principles in experimental design, 2nd edition. McGraw-Hill, New York.

Zar, J. H. 1984. Biostatistical analysis, 2nd edition. Prentice-Hall, Englewood Cliffs, New Jersey.

APPENDIX 2.1 CRITICAL VALUES FOR ONE-TAILED AND TWO-TAILED t-TESTS

				α in one tail			
	0.1000	0.0500	0.0250	0.0125	0.0050	0.0025	0.0005
				α in two tails			
df	0.2000	0.1000	0.0500	0.0250	0.0100	0.0050	.0010
1	3.078	6.314	12.706	25.452	63.657		
2	1.886	2.920	4.303	6.205	9.925	14.089	31.598
3	1.638	2.353	3.182	4.176	5.841	7.453	12.941
4	1.533	2.132	2.776	3.495	4.604	5.598	8.610
5	1.476	2.015	2.571	3.163	4.032	4.773	6.859
6	1.440	1.943	2.447	2.969	3.707	4.317	5.959
7	1.415	1.895	2.365	2.841	3.499	4.029	5.405
8	1.397	1.860	2.306	2.752	3.355	3.832	5.041
9	1.383	1.833	2.262	2.685	3.250	3.690	4.781
10	1.372	1.812	2.228	2.634	3.169	3.581	4.587
11	1.363	1.796	2.201	2.593	3.106	3.497	4.437
12	1.356	1.782	2.179	2.560	3.055	3.428	4.318
13	1.350	1.771	2.160	2.533	3.012	3.372	4.221
14	1.345	1.761	2.145	2.510	2.977	3.326	4.140
15	1.341	1.753	2.131	2.490	2.947	3.286	4.073
16	1.337	1.746	2.120	2.473	2.921	3.252	4.015
17	1.333	1.740	2.110	2.458	2.898	3.222	3.965
18	1.330	1.734	2.101	2.445	2.878	3.197	3.922
19	1.328	1.729	2.093	2.433	2.861	3.174	3.883
20	1.325	1.725	2.086	2.423	2.845	3.153	3.850
21	1.323	1.721	2.080	2.414	2.831	3.135	3.819
22	1.321	1.717	2.074	2.406	2.819	3.119	3.792
23	1.319	1.714	2.069	2.398	2.807	3.104	3.767
24	1.318	1.711	2.064	2.391	2.797	3.090	3.745
25	1.316	1.708	2.060	2.385	2.787	3.078	3.725
26	1.315	1.706	2.056	2.379	2.779	3.067	3.707
27	1.314	1.703	2.052	2.373	2.771	3.056	3.690
28	1.313	1.701	2.048	2.368	2.763	3.047	3.674
29	1.311	1.669	2.045	2.364	2.756	3.038	3.659
30	1.310	1.697	2.042	2.360	2.750	3.030	3.646
35	1.306	1.690	2.030	2.342	2.724	2.996	3.591
40	1.303	1.684	2.021	2.329	2.704	2.971	3.551
45	1.301	1.680	2.014	2.319	2.690	2.952	3.520
50	1.299	1.676	2.008	2.310	2.678	2.937	3.496
55	1.297	1.673	2.004	2.304	2.669	2.925	3.476
60	1.296	1.671	2.000	2.299	2.660	2.915	3.460
70	1.294	1.667	1.994	2.290	2.648	2.899	3.435
80	1.293	1.665	1.989	2.284	2.638	2.887	3.416
90	1.291	1.662	1.986	2.279	2.631	2.878	3.402
100	1.290	1.661	1.982	2.276	2.625	2.871	3.390
120	1.289	1.658	1.980	2.270	2.617	2.860	3.373
α	1.2816	1.6448	1.9600	2.2414	2.5758	2.0807	3.2905

APPENDIX 2.2 KEY FOR SELECTING COMMON STATISTICAL TESTS

The following key is intended as a general guide for selecting commonly used statistical tests. Users should consult statisticians or statistical texts to formulate correct statistical hypotheses, identify alternative tests, verify test assumptions, and establish appropriate decision criteria.

The central limit theorem (CLT) referenced in the key states that the sampling distribution of sample means approximates a normal distribution as sample size increases.

1 Two or more variables are under consideration . . . . Go to 5
1′ One variable is under consideration . . . . Go to 2

2 Two or more samples are available . . . . Go to 3
2′ One sample is available
  A Data have an approximately normal distribution and the CLT is not violated
    i Test is of the mean
      a Variance is known . . . . Use the one-sample normal test
      b Variance is unknown . . . . Use the one-sample *t*-test
    ii Test is of the variance . . . Use the one-sample chi-square test for variance
  B Data have a nonnormal distribution and the CLT may be violated
    i Data have a binomial distribution . . . . Use the binomial test
    ii Data do not have a binomial distribution
      a Data are interval . . . . Use the Wilcoxon signed-ranks test
      b Data are ordinal . . . . Use the Quantile test (ordinal data)

3 Three or more samples are available . . . . Go to 4
3′ Two samples are available
  A Data have an approximately normal distribution and the CLT is not violated
    i Test is of means
      a Samples are not independent . . . . Use the paired sample *t*-test
      b Samples are independent
        b′ Sample variances are equal . . . . . . Use the two-sample *t*-test for equal variances
        b″ Sample variances are unequal . . . . . Use the two-sample *t*-test for unequal variances
    ii Test is of variances . . . . Use the two-sample *F*-test
  B Data have a nonnormal distribution and the CLT may be violated
    i Data have a binomial distribution
      a Samples are not independent . . . . Use McNemar's test
      b Samples are independent . . . . . . Use the chi-square test for a 2×2 contingency table

ii Data do not have a binomial distribution

a Samples are not independent

a′ Data are interval .... Use the Wilcoxon signed-ranks test

a″ Data are ordinal .... Use the sign test

b Samples are independent

b′ Data are interval .... Use the randomization test

b″ Data are ordinal .... Use the Mann–Whitney test

4 Data have an approximately normal distribution and the CLT is not violated

A Test is of means .... Use one-way analysis of variance

B Test is of variances ..... Use Bartlett's test for homogeneity of variances

4′ Data have a nonnormal distribution and the CLT may be violated

A Data are nominal .... Use the chi-square test for a row × column contingency table

B Data are ordinal

i Test is of means .... Use the Kruskal–Wallis test

ii Test is of variances .... Use the squared ranks test

5 Three or more variables are to be related ...... Use multiple-regression or multivariate analysis as appropriate

5′ Two variables are to be related

A Both variables are quantitative .... Use linear regression

B One variable is quantitative and the other is qualitative .. Use two-way or n-way analysis of variance

C Both variables are qualitative ... Use n-way contingency table analysis or a log-linear model as appropriate

Chapter 3

Safety in Fisheries Work

CHARLES R. BERRY, JR.

3.1 INTRODUCTION

Fisheries work takes place on land, water, and ice, under water, and in the air. Some activities are hazardous because biologists sometimes use special equipment or chemicals and work in remote locations and inclement weather. An equipment list for any field trip might include items such as boat, motor, gasoline, fixatives, nets, sample jars, and data sheets. However, at the top of the list should be the word SAFE, which is an acronym to remind the biologist to have the Skills, Attitudes, Facts, and Equipment needed to travel and work safely in the field or in the laboratory.

Skills are developed by enrolling in boating, first aid, and other safety courses. Also, learn driving, towing, boating, and other skills by practicing with observers present. Practice driving and towing skills in an empty parking lot. Practice boating skills near shore on a windy day. Experience gained in the practice sessions will remove some of the fears and uncertainties that hamper performance in an emergency situation.

Attitudes about working safely are essential to a safety program. Accidents usually do not just happen but are often caused by deficiencies in one's state of mind (carelessness, fatigue, laziness, disobedience, forgetfulness, or poor judgment) and by faulty conditions (poor lighting, improper equipment, poor maintenance, or inclement weather).

A supervisor should provide safety leadership and foster safety attitudes early in the new employee orientation program (Fisher 1992). Successful supervisors encourage safety attitudes by using words such as "we" and "our safety record," not "You have got to . . . ." A safety orientation might entail (1) a review of the SAFE philosophy, (2) a job hazard analysis of the project, (3) a determination of physical limitations of the employee, (4) a training session, and (5) a signing of a safety awareness contract (Box 3.1). The safety awareness contract forces the supervisor and the employee to talk about specific safety concerns for any project and reinforces in the employee the importance of working safely. Employees and students do have safety responsibilities. They should follow program policies, develop safety awareness, wear proper clothing, and report hazards and accidents.

Facts are the third part of the SAFE philosophy. Training courses, videos, and written materials provide facts. Local sources of safety information include the library, police department, university extension office, Red Cross chapter, or U.S. Coast Guard auxiliary. Safety literature, videos, products, and consultation are

Box 3.1 Safety Awareness Contract

1. I discussed job hazards with my supervisor. Special ones that I might encounter on my project are
   a. chemical (list),
   b. physical (list), and
   c. climatological (list).
2. "SAFE" is an acronym for skills, attitude, facts, and equipment.
   Skills I will need on this project are (list)
   Attitudes I should have are (list)
   Facts I need to know about safety on this project are (list)
   Equipment I will need on this project is (list)
3. I have discussed with my supervisor any physical limitations that should be considered relative to my work.
4. If involved in a vehicle accident,
   a. my medical expenses are covered by ____________________;
   b. my liability to others is covered by ____________________;
   c. my belongings are/are not covered; and state vehicles are/are not exempt from "proof of insurance" laws. An accident reporting card should/should not be kept in the glove compartment.
5. In case of an emergency contact ____________________ .
6. *Pledge*: I have read the following safety materials (a list of reading materials pertinent to the assignment) and will follow safety policies, develop safety awareness, and report hazards and accidents.

____
Employee

Pledge: I have conducted a job hazard analysis, determined the employee's physical fitness for the job, provided training and equipment, and will monitor the employee's safety behavior and abate unsafe practices or conditions.

____
Supervisor

available from commercial suppliers (e.g., Lab Safety Supply, Janesville, Wisconsin), and from the nonprofit National Safety Council (Itasca, Illinois).

Equipment is the last item on the SAFE list. First aid kits, fire extinguishers, and other standard boat, automobile, and field equipment (and knowledge about use) should be available to help prevent or to mitigate the effects of an accident. Scientific supply catalogues have large sections devoted to safety equipment. The very high frequency (VHF) radio and cellular phone are safety equipment. Inexpensive, handheld models can connect a research team with authorities in most places.

There has been no survey of accidents that befall fisheries workers. Other professions have used accident data to help recognize and evaluate the risks taken in fieldwork and to promote training of employees and students (Howell 1990). The information in this chapter briefly covers safety information needed for general fisheries work. The information presented here is not adequate training material. It is presented only as an impetus for the untrained person to seek training and as a refresher for trained individuals. Information on safety can also be found in Chapters 8, 10, and 18.

3.2 FIRST AID

First aid skills are needed because fisheries workers often work far from sources of medical care. It is not true that only medical professionals can help an injured person. In some emergencies time counts more than medical expertise. By acting quickly you can change the outcome of the lives of people. First aid is not complicated; the biggest obstacles are ignorance, fear, and nervousness. It is foolish and negligent to be unprepared for an emergency. Get first aid skills by enrolling in a course conducted by the Red Cross.

In dealing with first aid, be aware of two legal issues. Nearly all states have Good Samaritan laws that protect the rescuer from liability as long as they are not grossly negligent. The other legal issue concerns consent. The victim must give his or her consent; if the victim cannot respond, consent is implied.

The issue of AIDS (acquired immunodeficiency syndrome) in the workplace is important. Get information on AIDS from the U.S. Public Health Service, the Canadian AIDS Society, or any state or provincial communicable disease program. The first aid provider should treat all blood and body fluids as potentially infectious. Universal precautions are (1) wash, (2) wear gloves, (3) seal bloody items in plastic, and (4) clean up with a disinfectant. First aid kits should have plastic gloves and a breathing mask to separate the victim from the first aid provider during cardiopulmonary resuscitation (CPR).

3.2.1 Drowning and Electrical Shock

Drowning or near drowning and electrical shock are accidents to which fisheries biologists are sometimes exposed. Breathing must be restarted within 4–6 min (Eastman 1987). However, recent experiences in treating coldwater (less than 21°C, 70°F) drowning victims have amended this general rule. Now rescuers are advised to aggressively resuscitate anyone who has been submerged for up to an hour. First aid for all drowning victims means a good understanding of CPR procedures. Formal training in CPR techniques is available from the Red Cross, Heart Association, universities, local hospitals, and other organizations. Every field crew should include a person trained in CPR, and every employee should be familiar with the technique.

3.2.2 Temperature Hazards

Working outdoors often exposes biologists to excessive heat or cold. During physical exertion in hot weather, circulation increases in the skin to help reduce heat. High temperatures or long exposure may cause circulatory and nervous system malfunction that results in hyperthermia. Heat exhaustion is a mild disturbance; the extreme condition is heat stroke (Box 3.2).

Hypothermia describes the physical and mental collapse that occurs when the body

Box 3.2 Symptoms and Treatment for Hyper- and Hypothermia

Condition and symptoms or signs	Treatment
Heat exhaustion	
1. Faintness (with a sense of pounding of heart) 2. Nausea, vomiting 3. Headache 4. Restlessness 5. Unconsciousness (the victim who has collapsed and is perspiring surely has heat exhaustion—perspiration rules out the diagnosis of heat stroke)	1. Move victim to a cool place 2. Keep victim lying down, treat for shock 3. If the victim is conscious, water to which 1/2 teaspoon of salt per glass has been added or stimulants such as iced coffee or tea may be given freely
Heat stroke	
1. Headache, dizziness 2. Extreme elevation of body temperature (40–43°C; 105–109°F) 3. Frequent need to urinate 4. Irritability 5. Disturbed vision or unconsciousness 6. Skin hot and dry 7. Pupils constricted 8. Pulse full, strong, pounding	1. Place victim in shade 2. Reduce body temperature as rapidly as possible by (a) pouring cold water over the body, (b) rubbing the body with ice, or (c) covering the body with sheets soaked in ice water 3. Remove clothing 4. Lay victim with head and shoulders slightly elevated 5. Give cool drinks after consciousness returns 6. Do not give stimulants
Hypothermia	
1. Shivering 2. Reduction in or lack of judgement and reasoning power 3. Vague, slow, slurred speech 4. Memory lapse 5. Incoherence 6. Fumbling hands 7. Drowsiness 8. Stumbling, lurching gait 9. Exhaustion (inability to get up after resting)	1. Believe the signs; a victim often denies the problem 2. Shield the victim from wind and rain 3. Remove wet clothing (replace with dry clothes, a blanket, or sleeping bag) 4. Provide warm drinks 5. Keep the semiconscious victim awake 6. Place semiconscious victim between two warm heat donors

is chilled to the core (Hirsh 1975). Wetness, wind, and exhaustion aggravate the chilling, which occurs most often when the air temperature is below 10°C (50°F). Cold kills in two distinct steps: (1) exposure and exhaustion, and (2) hypothermia. The victim may fail to realize what is happening. The potential victim feels fine while moving, but exposure and exhaustion are debilitating. When the victim stops moving, body heat production instantly drops, violent shivering may begin, and the onset of hypothermia occurs in a matter of minutes (Box 3.2). Antiexposure coveralls (U.S. Coast Guard approved or equivalent) are recommended when water temperature is below 10–16°C (50–60°F) and air temperature is below 0°C (32°F).

3.3 BOATING SAFETY

Fishery biologists work on a variety of waters doing a number of tasks that require boats, rafts, and other watercraft. Small, open boats are the most frequently used and are often loaded with equipment and passengers. Several comprehensive references cover small-craft operation (e.g., Andrews and Russell 1964; Griffiths 1966; Chapman 1994; Getchell 1994). Chapman's (1994) *Piloting, Seamanship, and Small Boat Handling*, the "bible of boating," is available as a CD-ROM product (Hearst Marine Books, New York), which also contains interactive practice sessions of boating situations. Two good booklets are available from the National Marine Manufacturers Association (NMMA 1991, 1992).

3.3.1 Power and Capacity Ratings

Most boats have a capacity plate showing the recommended maximum weight in equipment and passengers and the maximum recommended horsepower (Figure 3.1). You should calculate (Box 3.3) the load capacity if the plate is missing or if you attach heavy equipment to the boat. The load weight recommendations on the capacity label or from calculations are for calm water. Rough water decreases the safe carrying capacity. A load should be distributed evenly, kept low in the boat, and be within the capacity rating.

3.3.2 Safe Boating Procedures

Safe boating requires all aspects of the SAFE philosophy. Fisheries biologists might take a course in boating skills that covers seamanship, rules of the "road," aids in navigation, piloting, safe motor boat operation, and boating laws. A self-instruction course is available from the U.S. Coast Guard (Washington, DC). Safe-boating literature is available from the Coast Guard and from the National Marine Manufacturers Association (Chicago, Illinois).

Safe boating is based on good sense, courtesy, and respect for property and life. Operate the boat under control at reasonable speeds. Let other skippers know your intentions and stay away from swimmers, anglers, and dangerous areas. Navigation buoys are the signposts of the waterways. Make sure you are familiar with the buoy marker code.

Passing other boats safely is the skill most often required when under way (Figure 3.2). The maneuver is simple but requires at least visual communication between the two skippers. Rules of the highway do not always apply but, in general, when meeting head on keep to the right and boats on the right have the right-of-way. Sailboats have the right-of-way over powerboats in nearly all cases.

Fire prevention and control are important. Gasoline-powered inboard motorboats of all sizes should have flame arresters and ventilation systems. Outboard motorboats

US COAST GUARD CAPACITY INFORMATION
MAXIMUM HORSEPOWER
MAXIMUM PERSONS CAPACITY (PEOPLE)
MAXIMUM WEIGHT CAPACITY
(PERSONS MOTOR, & GEAR)(POUNDS)
THIS BOAT COMPLIES WITH US COAST GUARD SAFETY STANDARDS IN AFFECT ON THE DATE OF CERTIFICATION
MANUFACTURER:
MODEL:
DESIGN COMPLIANCE WITH THE FOLLOWING BIA CERTIFICATION REQUIREMENTS IS VERIFIED
LOAD AND HP CAPACITY • BASIC FLOTATION
COMPARTMENT VENTILATION
CERTIFIED BIA
BOATING INDUSTRY ASSOCIATIONS

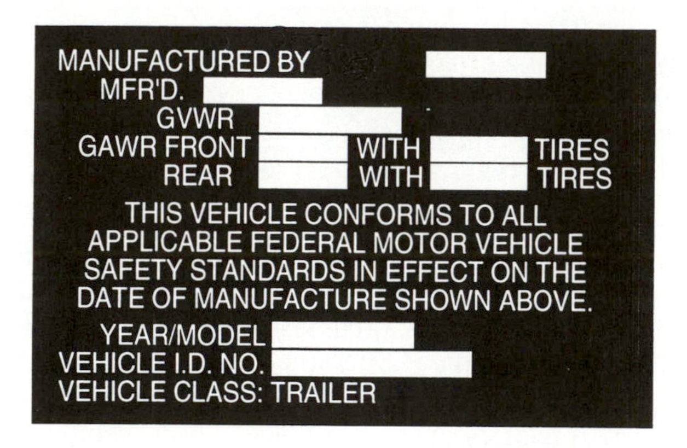

Figure 3.1 Representative capacity plates for boats (top) and trailers (bottom). Specific information will be found stamped in the blank areas of the metal labels. Abbreviations are gross vehicle weight rating (GVWR) and gross axle weight rating (GAWR).

do not require flame arresters, but enclosed fuel compartments should be ventilated. Routinely clean the stern to remove accumulated oils. Use fire extinguishers classified as B-I or B-II (suitable for gasoline, oil, or grease fire). Whistles or horns are suggested for all craft, but they are optional for boats shorter than 6 m (20 ft). Signaling devices, such as red handheld flares, can be used night or day (USCG 1981). Lights are required at night. Motorboats less than 13 m (43 ft) long must display a red port (left) light and a green starboard (right) light that are visible for 1.8 km (1 nautical mile). Remember "port wine is red." A white light at the stern must be visible for 3.6 km (2 nautical miles).

Personal flotation devices (PFDs) should be on board, and many agencies require workers to wear them. The PFDs (or life jackets) must be U.S. Coast Guard approved. Types I and II are recommended because they turn the wearer face up in the water. Type IV devices (throwable cushions) do not meet regulations for any boat as of 1995.

Chest waders are the subject of many drowning myths. If you fall from a boat while wearing waders, you will not be dragged to the bottom (Spurr 1980). Boots and

Box 3.3 Passenger Load and Weight Capacity

Formulas for determining passenger load (A) and weight capacity in pounds (B) of a boat when capacity information is unavailable.

A. Passenger load (PL):

$$\text{Number of passengers} = \frac{\text{length} \times \text{width}}{15}$$

Example: a boat 19 feet × 7 feet can hold 9 passengers:

$$\text{PL} = \frac{19 \times 7}{15} = 9$$

B. Weight capacity (WC): WC = (PL × 141) + 32

Example: the weight capacity of the boat above is

$$\text{WC} = (9 \times 141) + 32 = 1{,}301 \text{ lb}$$

waders weigh less in water than out, and the water that fills the boot adds no extra weight. Trapped air in clothing and waders, and the PFD, will provide enough flotation for you to return to the surface with swimming movements. One disadvantage of waders is that they make swimming kicks almost useless. Use side, breast, or back strokes to swim.

3.3.3 Storms on Water

United States Coast Guard stations, yacht clubs, and many public harbors display flags describing weather conditions. Fisheries biologists should get a weather report, know how to read the flags, and know how to interpret the water and the clouds. A red triangular flag signals small-craft warnings; winds of 7–10 knots (8–11.5 miles per hour) can cause 1-m high waves with scattered whitecaps.

Lightning is a universally feared feature of thunderstorms (Attarian 1992). The best course of action to reduce the danger of thunderstorms is to be ashore. Large boats that have flag masts, radio antennas, or other vertical projections should be outfitted with grounding systems (high-capacity conductors from the highest point to a submerged plate). Engines and fuel tanks also should be grounded (Scott 1993).

Winds and rough water are actually more dangerous than lightning. Reactions to storm emergencies depend on the size of the boat, water, and storm. Start for the shore if in doubt about weather conditions. If reaching the dock is not possible, seek any sheltered location. Do not leave an uncomfortable but safe location until the storm has passed. Making headway in rough water requires experience and skill. When heading into waves, proceed slowly and vary the angle of attack into the waves to attain the best combination of progress and smooth ride. Running in the same direction as large waves requires slight changes in lateral direction, rather than throttle adjustment, to keep the boat ahead of the following wave. Never force the boat up the back of a wave and over the crest or allow the following wave to catch up.

A swamped boat will probably remain floating. Stay with the boat: it can be more readily located than a lone swimmer. Climb onto the boat if possible to avoid losing

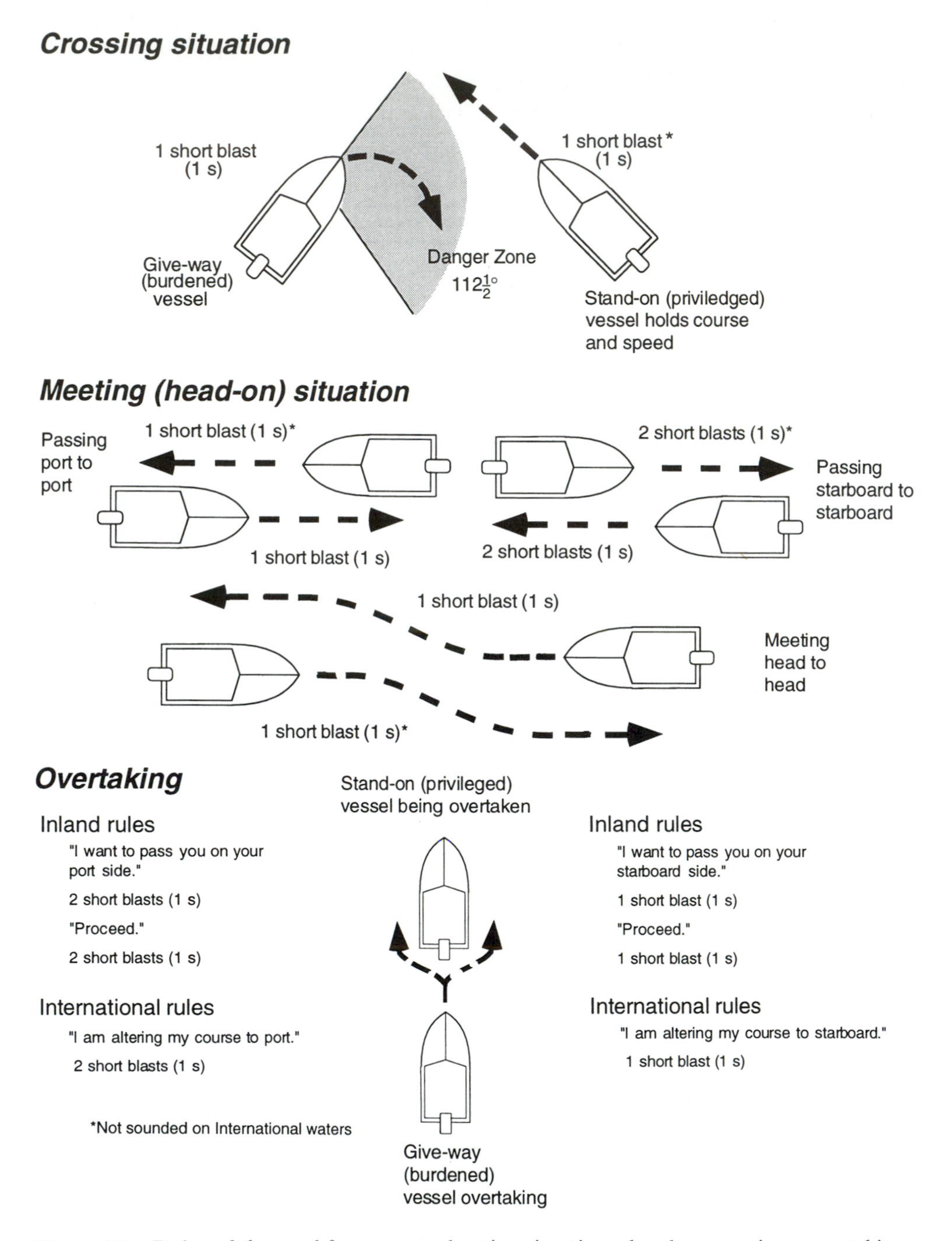

Figure 3.2 Rules of the road for common boating situations: head on meetings, overtaking, and crossing. Protocols for communication with horns for each of the above situations are given.

body heat. Water conducts heat away from the body 25 times faster than does air (USCG 1980). Swimming increases the circulation rate, pumps warm water out of clothing, and further increases heat loss. Remain as still as possible in cold water.

The laws of the sea require that a boater come to the aid of another in distress. There are certain techniques for rescuing someone overboard. Approach from downwind to prevent wind and waves from pushing your boat over people in the water. Shift to neutral while retrieving people from the water. Do not overload the rescue boat.

Occasionally, you must tow another boat. An alongside tow works well in calm waters and when a small boat must tow a larger boat (Chambers 1981). Tightly secure the boats together with bumpers in place, and the two boats will respond as a unit. When using a tow line, tie one end to a solid cleat on the towing boat and attach the other end to the bow (front) eye of the disabled boat. Move passengers to the towing boat and move weight in the towed boat to the stern. In rough waters, however, the disabled boat should be towed behind the tow boat by means of a towing bridle. Make a bridle by attaching a rope to each side of the towing craft as far forward as possible. Run the bridle through or around stern fittings. Attach the tow line so it centers itself during the tow.

3.4 SAFE WADING

Streams have uneven bottoms and currents that make wading hazardous, especially where rocks are slippery. Falls can lead to injury or drowning. Safe wading requires equipment, skills, and knowledge. The secret to safe wading is to think "one foot at a time." Enter a stream where water movement is slow, take short steps to maintain balance, walk into the current, cross fast-water streams at a slight downstream angle, and when turning, turn into the current.

Wear waders for protection from cuts and parasites. Waders should fit and allow mobility. Neoprene waders restrict movement less than do nylon or rubber waders. A belt helps maintain the fit of nylon and rubber waders and will prevent water from entering should a fall occur. Waders with standard soles are best for slow-flowing streams with mud, sand, or pebble substrates. Felt soles or other types of nonslip soles help on slippery rocks.

A wading staff (a "third foot") is a valuable aid for wading in fast water. Hold the staff on the upstream side. Plant it firmly before moving your feet. A personal flotation device should be on the equipment list if wading large streams is part of the job. If you are carrying equipment or an electroshocker on your back, the shoulder and waist straps should have quick-release buckles.

3.5 SAFETY ON ICE

It is difficult to determine when ice is safe for fisheries work. Although 8 cm (3.2 in) of ice on a farm pond is safe, the thickness may vary greatly where there are springs, stumps, plant life, or water currents. The ice may open quickly near shore, on windy days, near pressure ridges, and above springs. Snow cover hides trouble spots making them more dangerous. River ice is especially treacherous because water currents may erode the undersurface, creating dangerously thin ice in proximity to thick, safe ice (Ashton 1979; Sparano 1993).

Ice that makes popping sounds and is shiny and clear is the strongest type of ice. Ice 8 cm thick will support one person on foot; ice 16 cm thick will support a group of people moving single file; and ice 32 cm thick will support a heavy vehicle. Cars traveling across ice produce ice waves that may cause ice to crack. In general, one should drive slower in miles per hour than the water depth in feet (Fales 1983).

Important ice safety tools are the safety line and the ice spud (staff), which should be about 1.5 m (5 ft) long with a chisel on the end. Use the ice spud to test ice thickness by striking ice ahead of you as you walk or use it to dig a test hole. The spud will help hold you partially out of the water should an accident occur. When testing ice thickness, partners should remain 10 m (33 ft) apart and each should have a pole and safety line.

If someone falls through the ice, the victim and the observer(s) should follow certain rescue techniques. The victim should place his or her arms on the ice and roll the upper body onto the ice by pulling and kicking. Biologists who frequently work on ice carry "ice claws" (large nails or awls) that help them gain a handhold on slippery ice in case of an accident. Once on the ice, the victim should roll away from the hole. The rescuer should lie down to distribute weight when approaching the hole. The rescuer should pass the safety line (branch, clothing, or other life line) to the victim and be an anchor while the victim crawls from the hole. Both rescuer and victim should roll away from the hole to avoid breaking through ice again. Leave the area along the same path used to enter.

3.6 SAFETY ON THE ROAD

Safety means adopting an attitude toward operating a motor vehicle that is known as "defensive driving." The defensive driver is not timid or overcautious but uses defensive thinking to avoid traffic accidents. Accident prevention involves three interrelated steps: (1) see the hazard by thinking as far ahead as possible about what might happen; (2) understand the defense for specific hazards and apply the correct defensive maneuver; and (3) act defensively when the hazard is recognized and the defense determined. Employers have to take some responsibility for fostering a safe-driving attitude in employees (Etter 1994). The National Safety Council offers several types of driver safety programs to U.S. employers, including defensive-driving courses. In Canada, contact the Canada Safety Council.

3.6.1 Towing a Trailer

A fisheries biologist frequently has a boat trailer, equipment trailer, or camp trailer in tow. With the trailer attached, the likelihood of a single-car accident increases by a factor of four, especially during long-distance travel (Dark 1971). Safe trailer towing has four requirements: (1) proper trailer and load, (2) proper hitching equipment, (3) a towing vehicle with adequate weight and power, and (4) a driver familiar with trailer towing techniques (Bottomly 1974).

Surveys have indicated that 10% of trailers on highways are overloaded, 30% of towing vehicles are overloaded, and 35% of trailers have excessive weight on the hitch and tongue (Dark 1971). Fisheries workers sometimes overload trailers with electroshocking boats or by carrying nets, net anchors, and other field equipment in boats.

Trailers have their weight capacity information displayed on a capacity label (Figure 3.1). Two weights must be known: the gross trailer weight and the tongue weight. The tongue weight should be 10 to 15% of the gross trailer weight. Determine these weights by weighing the loaded trailer at a public scale (Figure 3.3). The gross

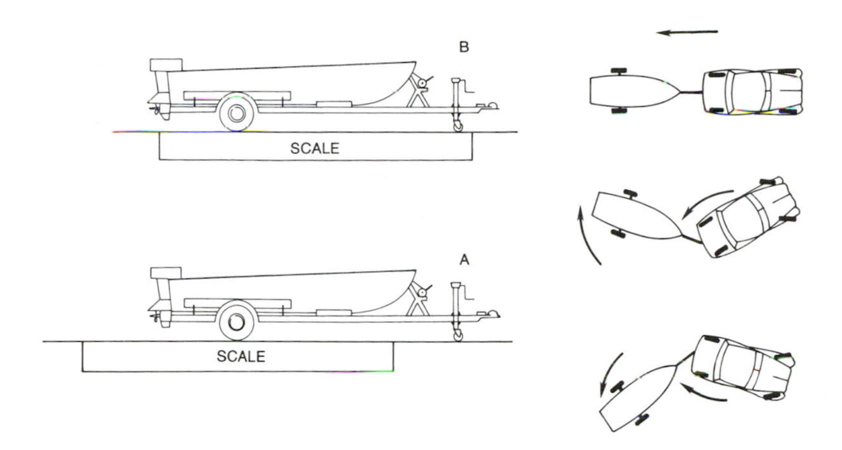

Figure 3.3 Schematic showing how to determine the trailer tongue weight on a public scale (subtract weight A from weight B) and how to steer the towing vehicle when backing up a trailer. The driver should place one hand on the bottom of the steering wheel; the wheel and the trailer move in the same direction.

trailer weight is the weight of the unhitched loaded trailer. Reweigh the loaded trailer with the tongue off the scale to determine the gross axle weight. Determine the tongue weight by subtracting the gross axle weight from the gross trailer weight. Modifying travel trailers and loading equipment trailers should be done to distribute weight evenly (Cavers 1994).

Wheel size and tire condition also determine the capacity of a trailer. Underinflated tires are a problem because they become hot, sway more easily, wear out more quickly, and carry less load safely. The capacity plate on the trailer lists the recommended tire size. Trailers subject to immersion should have bearing protectors (spring-loaded caps that pressurize the bearing grease). Pull the hub off and examine the bearings for corrosion once a year and pump in grease routinely.

Trailer brakes, brake lights, and turn signals extend the braking and signaling ability of the towing vehicle to the trailer. Check the connection and operation of lights often during a trip. Trailer manufacturers recommend unhooking the lights before launching the boat to prevent breaking hot bulbs. Laws differ in requirements for trailer brakes. Most states require brakes on trailers above a certain gross weight, and some states specify the types that are acceptable. Check to ensure that the trailer is in compliance with the laws of your state or province.

The hitching equipment consists of the coupler, ball, safety chains, and hitch. The coupler is the mechanism that attaches the trailer to the ball. The gross trailer weight determines the size of the coupler and ball (Box 3.4). Attach safety chains to the towing vehicle at points separate from the bracket that holds the ball. The hitch type and size depend on the gross trailer weight and tongue weight. A weight-carrying hitch, also called a "bumper" hitch, holds the entire tongue weight of the trailer. A weight-distribution hitch redistributes weight to all four wheels of the towing vehicle as well as to the wheels of the trailer, resulting in better handling, safer operation, and less wear on the tow vehicle than when a bumper hitch is used.

Box 3.4 Matching Hitching Equipment to Trailer Load (from USCG 1979)

Trailer class	*Gross trailer weight: kg (lb)*	*Hitch class*	*Hitch type*[a]	*Ball diameter: cm (in)*	*Ball shank diameter: cm (in)*
I	<910 (<2,000)	I	WC or WD	4.76 (1 7/8)	1.91 (3/4)
II	<1,590 (<3,500)	II	WC or WD	4.76 (1 7/8)	1.91 (3/4)
III	<2,270 (<5,000)	III	WD	5.08 (2)	2.54 (1)
IV	>2,270 (>5,000)	IV	WD (with antisway)	5.87 (2 5/16)	2.54[b] (1)

[a]Weight carrying (WC) and weight distribution (WD) hitches.
[b]Depends on trailer weight: 2.54 cm (1 in) for trailers weighing 2,270 kg (5,000 lb) to 3,405 kg (7,500 lb); 3.18 cm (1 1/4 in) for trailers weighing about 4,540 kg (10,000 lb); and 3.49 cm (1 3/8 in) for trailers weighing about 5,900 kg (13,000 lb).

Safe trailer towing and backing requires knowledge, skill, and caution. Complete the towing checklist one or more days before a trip (Box 3.5). Always remember that a trailer is following. Start slowly, proceed more slowly than usual, and slow down for bumpy areas. Maintain a greater following distance than normal and think twice about passing. Pull over and check occasionally for hot wheel hubs, loose tie-downs, defective lights, and incorrect tire pressure.

Lack of towing experience is the most important factor causing the high accident rate of vehicles towing trailers. Two skills required are backing up and correcting fishtailing. When backing up, place one hand on the bottom of the steering wheel and move the wheel in the direction the trailer must go (Figure 3.3). A trailer that is fishtailing begins to wag the towing vehicle. When the swaying goes beyond a certain point, both usually flip over (Anonymous 1977). Gusty winds, passing trucks, high speeds (especially downhill), and insufficient tongue weight cause fishtailing. If the

Box 3.5 Predeparture Checklist for Safe Trailer Towing

Towing vehicle
- Radiator coolant
- Transmission fluid
- Engine oil
- Shocks–springs
- Mirrors
- Emergency equipment
- Tire pressure

Hitching
- Safety chains
- Locked coupler
- Ball-coupler match

Trailer
- Tire stops
- Tire wear
- Wheels, lugs, and bearing protectors
- Lights
- Load balance
- Tongue weight
- Tie-downs

Driver
- Backing up skills
- Ramp courtesy
- Defensive driving

trailer starts to sway, reduce speed without braking. Keep the wheel straight and hold it tightly; compensatory steering often exaggerates the swaying.

Launching a boat is a critical part of trailer towing (USCG 1979). Prepare the boat away from the ramp for launching. Raise the outboard motor to avoid scraping, install the drain plug, release the tie-downs, and disconnect or remove the lights. Allow the trailer hubs to cool before they are put in the water. Sudden cooling may crack or chip the bearings or allow water entry. On the ramp, never turn off the engine. Set the parking brake and use tire stops. Only the driver should be in the towing vehicle; others should assist the launch by having tire stops ready and by holding the bow line.

3.6.2 Heavy Trucks and Campers

Field fisheries work often requires the use of heavy-duty trucks, vans, and campers. Awareness of the differences between driving a large vehicle and your personal vehicle is important (USDOT 1977). Make sure the total weight of the camper and contents does not exceed the gross vehicle weight and tire rating. When in doubt, weigh the vehicle on a public scale. About 20% of the recreational vehicles on the highway have loads exceeding the tire capacity that is printed on the side of the tire (Dreiske 1971). Overloaded vehicles are unstable, have steering and braking difficulties, and cause overheated tires, which may catch fire or blow out. Oversized mirrors are important auxiliary equipment. Finally, check the tie-down mechanisms on the camper for tightness.

Change your driving habits by making a deliberate and conscious effort to "think big" and "think slow." The size, weight, and center of gravity of the vehicle cause it to be more unstable than a passenger car. Handling, acceleration, turning radius, and stopping distance are affected (NSC 1978). The surface area of a camper or van acts like a sail in the wind. The driver normally compensates, but trouble comes from gusts (natural or from other vehicles) or when the wind is stopped by a passing truck or upon entering a tunnel. Beginning drivers frequently crowd the center lane; instead, watch the shoulder and keep to the right. Remember several points at refueling stops: check for overhanging structures that may be lower than the height of the vehicle, and check the vehicle's engine fluid levels, tire pressure, tie-downs, and load.

The bottled gas system of a camper can be dangerous; read the owner's manual thoroughly. Liquid gas is hazardous because of the possibility of fire, explosion, or asphyxiation caused by leaking gas (NSC 1978). Have gas appliances serviced regularly by trained personnel. Know the location of gas shut-off valves and extinguish pilot lights before traveling. Propane has a chemical additive with an odor that alerts you to close valves, air the vehicle, and correct problems. Find leaks by coating joints with soapy water and watching for bubbles. Tighten loose fittings, but if there is any doubt, have the system serviced. All liquid gas appliances must use outside air for combustion and release exhaust gas outside the vehicle. Never block outside vents of gas furnaces and refrigerators.

Open flames in a closed camper deplete the oxygen and add carbon monoxide (Johnson 1994). Sleeping persons have become victims of carbon monoxide poisoning even with roof and side vents of a camper open. Initial symptoms are headache and sleepiness. A warm sleeping bag is safer than relying on heating devices while sleeping.

Box 3.6 Passenger Safety Checklist for Aerial Surveys

1. Request a safety briefing from the pilot.
2. Brief the pilot on your study.
3. Determine the pilot's experience with the kind of mission.
4. Compare the pilot's sectional charts (showing hazards) with your maps.
5. Know the weight of your equipment and baggage.
6. Approach and depart the plane in view of the pilot.
7. Avoid the propeller area and avoid walking under the wing.
8. Secure loose items.
9. Wear nonflammable clothing and leather boots; avoid synthetics.
10. Avoid interfering with flight controls and switches.

3.7 SAFETY IN AIRPLANES

Biotelemetry studies (Chapter 19) that use aircraft often require low flying (<150 m or 490 ft), low air speeds, and modified aircraft (e.g., attached antennas). Such conditions require extra attention to safety. A pilot is always responsible for aircraft and passenger safety and should follow Federal Aviation Administration or Transport Canada-Aviation rules. However, a passenger can do much to help the pilot accomplish a safe flight (Box 3.6).

The most important safety precaution is preflight communication between the pilot and the biologist. Neither may be experienced in using aircraft for biotelemetry studies. Before departing, both should thoroughly understand the job ahead and safety measures so that good judgements can be made under the subtle pressures (e.g., noise, time constraints, and speed) of airborne work.

Personal protective equipment includes fire-retardant clothing and a flight helmet. If fire-retardant clothing is not available, wear cotton or wool instead of synthetic fibers that melt and stick to the skin if heated. Some agencies require the use of an aviator's protective helmet (Flight Suits Ltd., El Cajon, California). The helmets are compatible with most radiotelemetry receivers used by biologists.

3.8 LABORATORY SAFETY

The SAFE philosophy also applies to the laboratory setting. The laboratory supervisor should promote safety and supply facts, equipment, and training. A laboratory worker may need training in skills such as working with glassware, handling heated materials, carrying chemicals, pipetting liquids, and disposing waste. A safe laboratory is usually reflected in the neatness of the laboratory and conduct of workers. Facts about safe laboratory practices are available in books (e.g., Furr 1989; Walters and Stricoff 1990), pocket guides (NSTA 1990), and journals (e.g., *American Laboratory*). Training videos, pocket cards, wall charts, and other safety information are available from safety supply companies, as is a variety of laboratory safety equipment. There are many ways to make a laboratory safe at no or minimal cost (Box 3.7).

Box 3.7 Thirty-Six Inexpensive Steps to a Safer Laboratory

1. Organize a safety committee and meet regularly.
2. Develop a safety orientation program for workers.
3. Encourage everyone to develop a concern for safety.
4. Give every worker a specific safety responsibility.
5. Provide incentives for safety performance.
6. Require all workers to read certain safety literature and sign a statement that they have done so.
7. Conduct periodic inspections; simulate an Occupational Safety and Health Administration (OSHA) inspection.
8. Make safety part of the science education process.
9. Have discussions of safety and health aspects of laboratory work in laboratory classrooms.
10. Avoid working alone in any lab.
11. Don't allow experiments to run unattended unless fail-safe.
12. When conducting experiments with potential hazards, ask "What are the hazards?" and "How will I deal with them?"
13. Require all accidents and incidents to be reported.
14. Extend the safety program beyond the laboratory.
15. Allow minimum amounts of flammables in each laboratory.
16. Forbid smoking, eating, and drinking in the laboratory.
17. Do not allow food to be stored in chemical refrigerators.
18. Develop plans for emergencies.
19. Display emergency phone numbers.
20. Store acids and bases separately.
21. Store fuels and oxidizers separately.
22. Post warning signs in hazardous areas.
23. Require good housekeeping practices.
24. Maintain a chemical inventory; discard dated materials.
25. Require the use of eye protection.
26. Provide adequate supplies of personal protective equipment.
27. Provide fire extinguishers, showers, and eyewash.
28. Maintain a safety library.
29. Provide guards on vacuum pump belts.
30. Secure compressed gas cylinders.
31. Provide first aid equipment.
32. Require grounded plugs on electrical equipment.
33. Label all chemicals to show nature, degree of hazard, and date purchased.
34. Use safe and ecologically acceptable disposal techniques.
35. Use fireproof cabinets to store flammable chemicals.
36. Use well-ventilated storage for odiferous chemicals.

3.8.1 Polite Laboratory Protocol

Polite laboratory protocol, or PLP, refers to behavior that increases the safety, efficiency, and comfort of colleagues working together in laboratory conditions. This behavior includes (1) clearly labeling rooms, work areas, and containers, (2) cleaning up thoroughly (because it is unsafe for others to clean up after your experiment), (3) communicating the need for space and equipment in shared labs, and (4) transporting samples in double containers. Polite laboratory protocol also includes wearing laboratory apparel only in the laboratory to avoid spreading contaminants, odors, or waste.

3.8.2 Chemical Safety

Much of the current interest in laboratory safety involves the storage, use, and disposal of chemicals (Sanders 1986). University researchers tend to use small quantities of many substances for short periods and at infrequent intervals. Agencies and industry tend to use only a few substances but in large quantities for protracted intervals. In either case a chemical tracking system is recommended (Hans 1993). Because chemicals are used by fisheries biologists in both field and laboratory, some basic safety considerations are mentioned here.

A list of chemicals registered for fisheries work is available (Schnick 1988). All drugs and chemicals used in fisheries operations should be purchased from legitimate sources and used only for those purposes for which the compounds are registered. Covert uses and purchases may cause the sponsor of the label to stop manufacturing the product (Schnick 1988).

Management of chemicals and hazardous waste is an escalating issue for industry, government, and educational and research institutions. A common-sense guideline for fieldwork is to transport liquids in plastic containers. Most organizations have a safety officer or chemical hygiene officer (Mandt 1995) who is knowledgeable about safety and regulations. Contact the officer concerning the ordering, transport, storage, and disposal of chemicals.

An important source of information is the material safety data sheet (MSDS) that accompanies each purchased chemical. An MSDS file should be maintained in a well-known, specific location and be readily accessible. The MSDS describes regulations for the user. It provides information such as toxicity hazards, health hazards, first aid measures, physical data, fire and explosion data, reactivity data, spill cleanup procedures, and handling and storage precautions. As an example, partial information from the MSDS for formaldehyde is given in Box 3.8.

In the past, some academic and industrial laboratories carelessly used drains, garbage bins, or isolated ground for waste disposal. These practices still go on, but awareness of the effects of improper disposal on humans and the environment is improving (Sanders 1986).

Disposal is costly and should be in compliance with regulations. Less is better. There are many ways to reduce the quantity of waste and laboratory trash (ACS 1993). Reduce waste by ordering only needed quantities. Store waste chemicals in their original containers. Recycle where possible. Treat waste for its hazardous characteristics (e.g., neutralized acids and caustics are no longer hazardous waste). In case of an accident, there are a variety of commercial spill kits and absorbants available. A spill report may be required.

Box 3.8 Partial Material Safety Data Sheet (MSDS) for Formaldehyde

Abbreviated information provided on the MSDS for formaldehyde, a commonly used fixative for fish and other biota.

Chemical name: Formaldehyde, 37% solution

Trade name: Formalin

Chemical family: Aldehyde

Formula: HCHO

Appearance: Colorless liquid, pungent odor

Composition: Formaldehyde, ethanol, water

Hazards: Potential cancer hazard—repeated or prolonged exposure increases risk; contact with eye may cause permanent damage; combustible liquid and vapor; if swallowed, may be fatal or cause blindness; harmful if inhaled or absorbed by skin; strong sensitizer—may cause allergic reaction

First aid: Get medical assistance for overexposure; wash skin with soap and water; remove contaminated clothing; flush eyes for 15 min; move to fresh air; give milk, activated charcoal, or water if ingested

Fire fighting: Flash point 140°C (284°F); dry chemical, alcohol foam, carbon dioxide, water spray to cool exposed containers and disperse vapors

Cleanup: Evacuate area; wear protective equipment; eliminate ignition sources; contain release; eliminate source; comply with reporting regulations

Storage: Closed container; protect against physical damage; store at −1.7°C (29°F) minimum; separate from oxidizing and alkaline material

Exposure controls: Ventilation; respiratory protection; protective clothing; eye protection; transfer in approved fume hood

Properties: Boiling point 101°C (214°F); vapor density 1.03; specific gravity 1.08; evaporation rate like water

Stability–reactivity: Stable; polymerization may occur at low temperatures forming paraformaldehyde, a white solid; avoid mixing with acids and oxidizers; avoid heat, spark, flame; avoid contact with HCl (forms potent carcinogen)

Toxicology: Oral lethal dose to 50% test mice, 42 mg/kg; tests on lab animals indicate material may cause tumors and adverse mutagenic and reproductive effects

Disposal: Environmental Protection Agency waste number U122; incineration, fuels blending, or recycling; contact local waste disposal site

Transportation: Department of Transportation identification UN1198

Regulatory: This product is a mixture; all components listed in Toxic Substances Control Act (USA)

Other: National Fire Protection Association hazard ratings; MSDS revision history dates

3.9 CONCLUSION

The SAFE philosophy should be part of all fisheries activities. The responsibility for SAFE is shared between the supervisor (orientation, training, and equipment) and the employee (attitude and compliance). There is no job or service so important or so urgent that we cannot take the time to perform work safely.

3.10 REFERENCES

ACS (American Chemical Society). 1993. Less is better: laboratory chemical management for waste reduction. American Chemical Society, Washington, DC.

Andrews, H., and A. Russell. 1964. Basic boating: piloting and seamanship. Prentice-Hall, Englewood Cliffs, New Jersey.

Anonymous. 1977. Put safety behind you. Family Safety Magazine 36(2):28–30.

Ashton, G. D. 1979. River ice. American Scientist 67:38–45.

Attarian, A. 1992. A lightning safety primer for camps. Camping Magazine 64(6):28–31.

Bottomly, T. 1974. The complete book of boat trailering. Association Press, New York.

Cavers, W. 1994. Calculating changes in tongue weight. Trailer Life 54(4):79–82.

Chambers, V. 1981. Towing—doing it properly and safely. Pennsylvania Angler 50(7):16–17.

Chapman, C. 1994. Piloting, seamanship and small boat handling, 61st edition. Hearst Marine Books, New York.

Dark, H. E. 1971. Know before you tow. Family Safety Magazine 30(2):24–26.

Dreiske, P. 1971. Have home, will travel. Family Safety Magazine 30(3):24–26.

Eastman, P. F. 1987. Advanced first aid afloat. Cornell Maritime Press, Cambridge, Maryland.

Etter, I. 1994. How safe are your drivers? Safety and Health 150(4):3.

Fales, E. 1983. Through the ice. Popular Mechanics Magazine 160(6):88–90, 116–117.

Fisher, G. 1992. Safety begins at the top. Safety and Health 145(6):70–72.

Furr, A. 1989. Handbook of laboratory safety. CRC Press, Boca Raton, Florida.

Getchell, D. R. 1994. The outboard boater's handbook: advanced seamanship and practical skills. McGraw Hill, New York.

Griffiths, G. 1966. Boating in Canada. University of Toronto Press.

Hans, M. 1993. Labs tackle chemical tracking. Safety and Health 147(1):54–58.

Hirsh, T. 1975. The chill that kills. Family Safety Magazine 34(1):5–6.

Howell, N. 1990. Surviving fieldwork: a report of the advisory panel on health and safety in fieldwork. American Anthropological Association Special Publication 26, Washington, DC.

Johnson, R. 1994. Carbon monoxide. Trailer Life 54(9):57–64.

Mandt, D. 1995. The effect of the chemical hygiene law on school biology laboratories. American Biology Teacher 57(2):78–80.

NMMA (National Marine Manufacturers Association). 1991. You and your trailer: a guide to ownership and operation. NMMA, Chicago.

NMMA (National Marine Manufacturers Association). 1992. You and your boat: a guide to powerboat ownership and operation. NMMA, Chicago.

NSC (National Safety Council). 1978. RV camping guide. National Safety Council, Chicago.

NSTA (National Science Teachers Association). 1990. Pocket guides to chemical and environmental safety in schools and colleges. NSTA, Washington, DC.

Sanders, H. J. 1986. Hazardous wastes in academic labs. Chemical and Engineering News 64(5):21–31.

Schnick, R. 1988. The impetus to register new therapeutants for aquaculture. The Progressive Fish-Culturist 50:190–196.

Scott, R. 1993. Shocking. Trailer Boats Magazine 22(7):6.

Sparano, V. T. 1993. Private lessons: ice safety. Outdoor Life 192(6):81.
Spurr, E. 1980. Worthy notes for wise waders. Ducks Unlimited Magazine 40(4):43–45.
USCG (United States Coast Guard). 1979. Trailer-boating: a primer. U.S. Government Printing Office 1979-623-210, Washington, DC.
USCG (United States Coast Guard). 1980. A pocket guide to cold water survival. U.S. Government Printing Office COMDTINST 3131.5, Washington, DC.
USCG (United States Coast Guard). 1981. Visual distress signals for recreational boaters. USCG G-BEL-4, Washington, DC.
USDOT (United States Department of Transportation). 1977. Travel and camper trailer safety. USDOT, National Highway Safety Administration, Washington, DC.
Walters, D. B., and R. S. Stricoff. 1990. Laboratory health and safety handbook. Wiley, New York.

Chapter 4

Aquatic Habitat Measurements

THOMAS E. MCMAHON, ALEXANDER V. ZALE,
AND DONALD J. ORTH

4.1 INTRODUCTION

Imagine yourself as a fishery biologist with the responsibility of evaluating the effectiveness of various habitat improvements to streams in your district. Or imagine you are responsible for assessing the effects of various reservoir hydropower operations on fish populations. How would you design a study to conduct these investigations? Which habitat variables would you measure, and how would you measure them? The purpose of this chapter is to provide initial guidance in designing aquatic habitat studies and in selecting and measuring habitat variables important in fisheries management and research.

Habitat for fish includes all of the physical, chemical, and biological features of the environment needed to sustain life. Examples include suitable water quality, migration routes, spawning grounds, feeding and resting sites, and shelter from predators and adverse environmental conditions (Orth and White 1993). Therefore, habitat quality influences the numbers, sizes, and species of fish that can be sustained in a particular area (Milner et al. 1985), and changes to that habitat (e.g., resulting from perturbations or management actions) can result in altered fishery characteristics. Also, legal mandates (e.g., the U.S. National Environmental Protection Act, 42 U.S.C. §§ 4321 to 4361 and the U.S. Endangered Species Act, 16 U.S.C. §§ 1531 to 1544) require assessment of habitat and prediction of changes to it under different management scenarios. Such legislation spawned the development of habitat models designed to quantify the effect of flow alterations and other habitat changes on fish populations (Terrell 1984; Orth 1987; Fausch et al. 1988; Gore and Nestler 1988). Interest in habitat restoration, manipulation, and assessment will undoubtedly expand in coming years. Accordingly, habitat assessment is an important component of the effective fishery manager's or researcher's repertoire.

This chapter is organized into three major topics. First, considerations in designing habitat assessments and selecting appropriate habitat variables are presented. Second, techniques of measuring habitat information from maps are discussed. Third, measurement techniques for some of the most commonly used habitat variables are described separately for rivers and streams and for lakes and reservoirs. Included are some examples of how particular habitat measurements have been used in fisheries investigations. Although biological features of the environment influence abundance and distribution of fishes through competition, predation, and disease,

Table 4.1 List of aquatic habitat variables described in this chapter.

Stream and river	Lake and reservoir
Watershed variables Area Drainage density Gradient Longitudinal profile Relief ratio River sinuosity Stream order	Morphometric variables Area Shoreline length Shoreline development index Depth Volume Mean depth
Channel variables Cover Channel cross section Channel type Depth Discharge and flow Erosion and sedimentation Gradient Habitat type Habitat complexity Large woody debris Pool:riffle ratio Pool rating Sinuosity Substrate Thalweg profile Velocity Width Width:depth ratio	Physicochemical variables Temperature Dissolved oxygen Transparency Hydrodynamic variables Storage ratio Flushing rate Turnover time Current velocity Wave height Water level Shoreline and substrate variables Substrate Cover Vegetation Trophic state indicators
Bank and riparian area variables Bank angle Bank height Bank stability rating Bank undercut Canopy cover Vegetation overhang	

the emphasis of this chapter will be on measuring abiotic physical and chemical habitat features: those factors that form the structure and media within which a fish exists.

4.2 VARIABLE SELECTION AND STUDY DESIGN

Because aquatic habitats vary greatly in size, configuration, and other characteristics, no standard sampling design or set of all-purpose habitat measurements exists. Investigators must define objectives of their inquiry precisely and measure only those characteristics germane to the problem under consideration. Some of the many potential habitat measurements are listed in Table 4.1. Consult Wetzel (1983), Thornton et al. (1990), Hayes et al. (1993), and Summerfelt (1993) for insights on relevant attributes to measure in lakes and reservoirs. Useful references for selecting variables to measure in rivers and streams include Hynes (1970), Platts et al. (1983, 1987), and Bryan and Rutherford (1993).

Following selection of relevant habitat characteristics to examine, the fishery worker must select the most appropriate methods to measure these characteristics. A major focus in fisheries science has been the development of standardized methods that quantitatively measure specific habitat variables but are flexible and cost-effective enough for wide application (Platts et al. 1983, 1987; APHA 1989; Hawkins et al. 1993; Simonson et al. 1994). Techniques selected must be repeatable and have sufficient accuracy and precision to detect changes over time or differences among sites, yet they must also meet budgetary and personnel constraints. The key is to balance costs with needed levels of accuracy and precision. The extreme accuracy and precision provided by an expensive and time-consuming, state-of-the-art technique may well be superfluous if a more rapid, rudimentary technique can adequately detect differences among samples at a lower cost.

An appropriate sampling design is critical to the success of any habitat measurement program. Chapters 1 and 2, especially Box 2.2, should be reviewed when planning any field habitat measurements. Aquatic habitats are often large and spatially heterogeneous, both vertically and horizontally, and therefore can require substantial sampling effort for adequate characterization. Furthermore, many characteristics can change rapidly over time, thereby necessitating frequent sampling to assess the full range of possible conditions. Care must be taken to define when, where, and how habitat measurements will be made. Exploratory sampling may be needed to estimate potential spatial and temporal heterogeneity in order to devise appropriate, but not excessive, sampling protocols.

In choosing where to sample, it is helpful to think of the total habitat for the species, life stage, or assemblage of interest as a hierarchy of interconnected spatial scales ranging from individual microhabitats to whole lakes and watersheds (Figure 4.1; Frissell et al. 1986). Measurements of aquatic habitat variables at several different scales are often required to characterize fish–habitat relations within a watershed or lake adequately (Bozek and Rahel 1991; Kershner et al. 1991; Sowa and Rabeni 1995). For example, in the question raised at the beginning of this chapter about how to assess habitat improvements to streams, a useful approach would be to compare microhabitat characteristics (e.g., depth and velocity) and habitat types (pools and riffles) in reaches with and without improvements, along with associated measurements of fish abundance (Shuler et al. 1994). Large-scale factors, such as differences in geology among watersheds in the district, should also be considered because differences in habitat characteristics between streams or reaches with and without improvements could be artifacts of different natural sediment loads (Frissell and Nawa 1992).

Dividing sampling into zones, reaches, habitat types, or seasons that have similar characteristics is helpful for defining the spatial and temporal boundaries of sampling strata and for extrapolating measurements made in sampled strata to unsampled strata (Chapter 2). Investigators should be aware of two general types of error commonly associated with habitat sampling. The first is measurement error, or the accuracy of the measured value compared with the true value (Chapter 2). The second is extrapolation error, or the accuracy of extrapolating measurements from sampled areas or time periods to unsampled areas or periods. Habitat studies are typically concerned with minimizing measurement error, but extrapolation error may be the more significant of the two in cases in which the habitat is highly variable or dynamic (Hankin 1984), as it is in many aquatic environments. Minimizing measurement error involves increasing the number of measurements within the area sampled

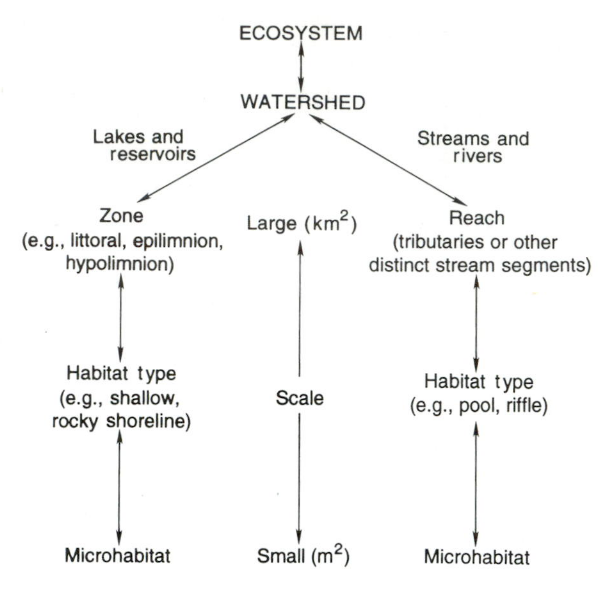

Figure 4.1 Hierarchy of aquatic habitats arranged by spatial scale.

(e.g., increasing the number of depth measurements in a stream reach), whereas minimizing extrapolation error involves increasing the number of areas or strata sampled (increasing the number of reaches sampled). Because it is difficult to know which source of error is more important at the start of an investigation, study designs often must strike a balance between minimizing each source of error.

4.3 HABITAT MAPPING AND MAP INTERPRETATION

4.3.1 Use of Existing Maps

Existing maps and aerial photos provide a wealth of information useful in conducting habitat assessments. Maps and aerial photos give the "bird's eye view" of a study area that is needed to evaluate how factors in the surrounding watershed (also referred to as drainage basin or catchment) affect a stream or lake. Knowledge of land use patterns, ownership, soils, topography, vegetation, and drainage patterns help set study area boundaries at the start of a study and help interpret data at the end. The location of access sites, such as roads, trails, bridges, and boat ramps, and public and private land help identify likely sampling locations. Maps and photos also help locate features such as small springs, irrigation diversions, and low-head dams that might be overlooked during normal ground surveys of the study area.

Topographic maps show landmarks and locations of aquatic habitats and provide contour lines of elevations. "Thematic" maps that display the geology, soils, vegetation, or climate of an area are another source of valuable information. Aerial photos show such factors as snow cover (for estimating water yield), extent of fires, floodplain development, and extent of vegetation along shorelines of large lakes. Aerial photos taken with infrared film can display relative water temperatures around a power plant. A series of past aerial photos provides a means to evaluate habitat changes over time and to place current habitat conditions within a historical

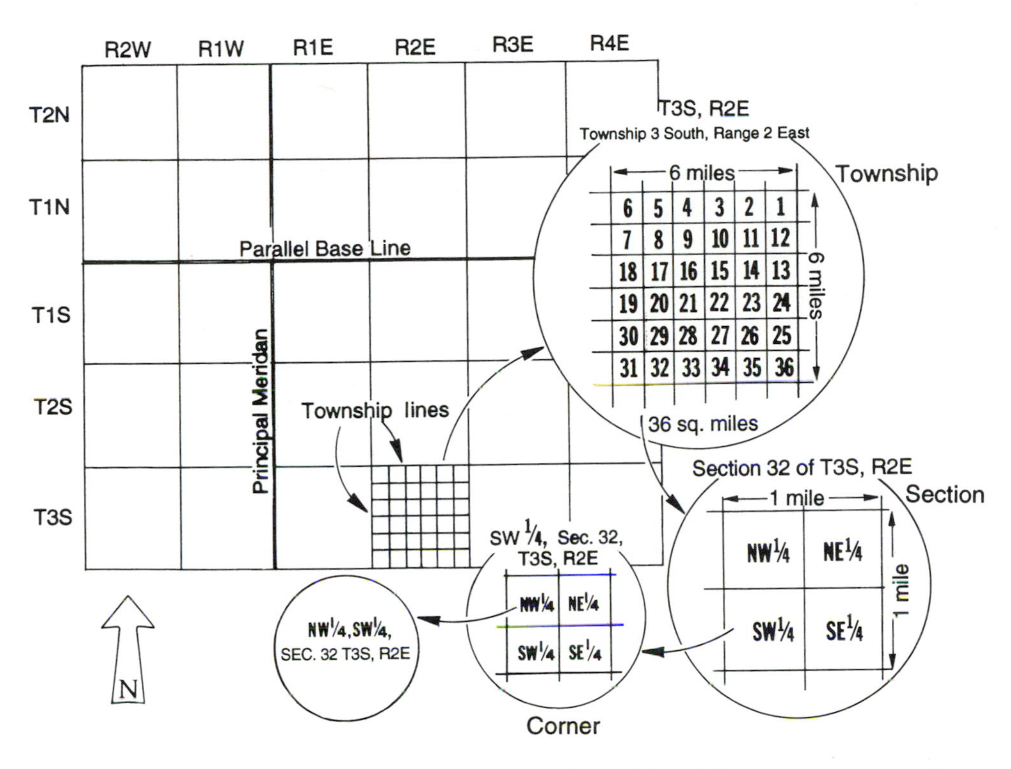

Figure 4.2 Township, range, section, and corner of the U.S. public land map system (from Orth 1983).

context (Platts et al. 1987). Lyons and Beschta (1983) and Grant (1988) provided examples of historical aerial photo analysis for aquatic habitat studies. Aerial photo and satellite images can be obtained through the U.S. Geological Survey.

Sampling locations and other features can be located on a map for future reference by use of latitude and longitude, universal transverse mercator (UTM) coordinates, or township and range coordinates of the public lands system in the United States. The UTM system describes coordinates based on the distance from the equator (in meters) and a reference meridian and is now commonly used in geographic information system (GIS) applications and radiotracking studies (Chapter 19). The U.S. public land system divides land area into 6 mi × 6 mi (9.7 km × 9.7 km) townships (36 mi^2 or 94.1 km^2) identified by a township and range location (Figure 4.2). Each township is divided into 36 sections, each 1 mi^2 (640 acres; or 2.6 km^2 and 260 ha). Square-mile sections are further subdivided into four quarter-mile (0.65 km^2) squares or corners.

4.3.2 Mapping Techniques

Maps of large lakes and reservoirs adequate for most fisheries uses can often be traced from topographic maps available from federal, state, or provincial agencies (e.g., U.S. Geological Survey, Canadian Department of Mines and Technical Surveys, Tennessee Valley Authority, U.S. Bureau of Reclamation, or U.S. Soil Conservation Service). Maps of lakes, reservoirs, ponds, rivers, and large streams can also be traced from aerial photographs; scale is determined by measuring the length

Figure 4.3 Plane table alidade.

of a known object (e.g., a building) in the photograph. However, if aerial photographs are unavailable, or if canopy cover obscures streambanks from above, field surveys are necessary to develop useful shoreline maps. Detailed descriptions of a variety of mapping methods are provided by Welch (1948) and Gordon et al. (1992). In this chapter, we present the plane table alidade, traverse, and deflection angle traverse methods.

4.3.2.1 Plane Table Alidade Method

The plane table alidade method is useful for on-site mapping of small lakes with regular shorelines as well as for mapping segments of medium to large rivers. Points along the shoreline are triangulated from two ends of a baseline of known length. The method requires a plane table mounted on a tripod, an alidade, measuring tape, marking stakes, a brightly marked range pole, and map paper affixed to the plane table. The alidade can be thought of as a straight edge with rifle sights (Figure 4.3). First, stakes are positioned along the water's edge at shoreline inflection points (Figure 4.4A). Two plane table locations (points A and B) are positioned a measured distance apart (e.g., 100 m) to establish the baseline and scale of the map. All of the marking stakes must be visible from both points A and B. After securing the plane table at point A, the baseline is drawn on the map paper along the straight edge by sighting with the alidade to point B. One person then takes the range pole and holds it vertically at each stake along the shoreline while another person sights the pole with the alidade and marks on the paper the line of sight from point A to each stake. After all of the sight lines are drawn from point A, the plane table is moved to point B, the baseline is aligned with point A, and sight lines are drawn on the paper from point B to each marking stake. The intersection of both sight lines for each stake represents the stake's shoreline location; the intersection points can then be connected to form the shoreline map. Because traversing vegetated or boggy shorelines to reach all of the marking stakes can be arduous, surveying during winter may be easier if lakes or rivers freeze.

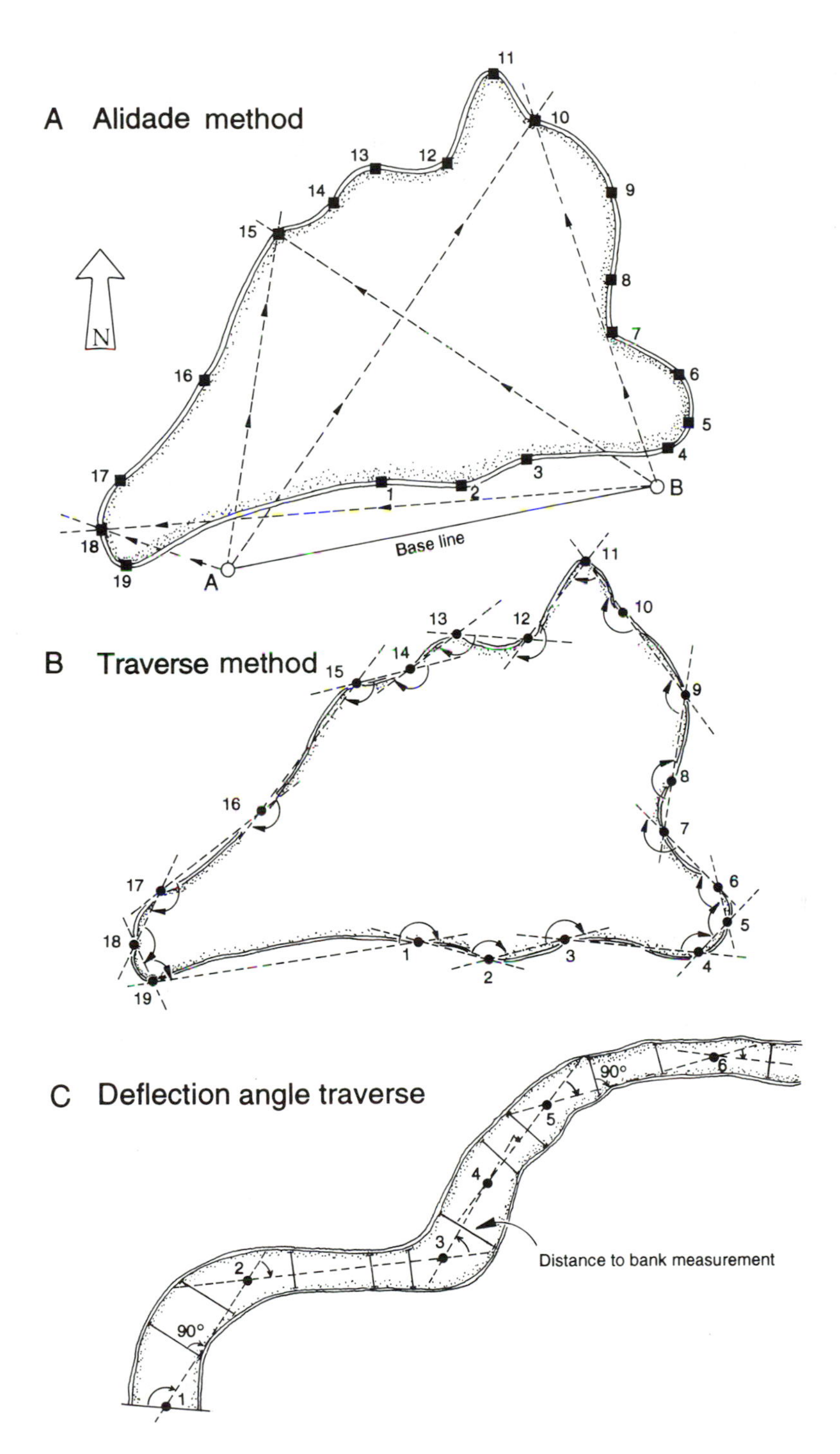

Figure 4.4 Mapping methods: (**A**) plane table alidade, (**B**) traverse, and (**C**) deflection angle traverse (from Orth 1983).

4.3.2.2 Traverse Methods

The traverse method is suitable for large bodies of water with irregular shorelines, but it requires training in the proper operation of an engineer's transit theodolite. This surveying instrument is used to measure vertical and horizontal angles. Starting at point 1 (Figure 4.4B), distance and angle measurements are made serially from point to point along the shoreline. The measurements are then plotted to form the outline of the shoreline. Use of a theodolite also allows mapping of elevations, for instance, along a stream segment.

The deflection angle traverse method is recommended for mapping small streams (Welch 1948). A transit theodolite with crosshairs, stadia rod, and measuring tape are required. Stakes are driven at successive theodolite stations (e.g., points 1, 2, and 3 in Figure 4.4C), marking consecutive longitudinal transects (1–2, 2–3, and so on). The stations are established as far apart as possible while still permitting direct lines of sight within the stream channel from station to station. Distances between stations and deflection angles between successive longitudinal transects are measured using the theodolite and stadia rod. At measured intervals between the stations, distances to each bank are measured at right angles from the longitudinal transect. The measurements are then plotted to form the outlines of the shorelines.

The longitudinal transect method described by Gordon et al. (1992) is a simplification of the deflection angle traverse method. A measuring tape is stretched along each longitudinal transect, and, at recorded distances along the transect, distances to each bank are measured at right angles. Compass readings are used to join successive longitudinal transects.

4.3.2.3 Global Positioning Systems

Global positioning system (GPS; Chapter 19) units can also be used to produce maps of lakes, reservoirs, and rivers, but they lack adequate precision for small streams at present. Most GPS units can describe latitude and longitude of a site in the field with an accuracy of 15–20 m. Readings taken at shoreline inflection points are plotted and connected to form a map of the shoreline.

4.3.3 Measurements from Maps

4.3.3.1 Area and Length

A number of useful watershed habitat variables (Table 4.1) can be derived from topographic maps. Watershed area is one of the more important descriptors of a watershed because it influences water yield, sediment transport, and the number and size of streams (Gordon et al. 1992). The boundaries of a watershed are drawn by joining the highest contour lines on the ridgetops that divide adjacent watersheds. Watershed area can be measured directly from maps by means of an acreage grid, planimeter, or computer digitizing system. An acreage grid consists of a transparent grid overlay. The grid is placed over the map and the number of grids (or dots) that fall within the watershed boundary are counted. For dots or squares bisected by boundaries, count every other one. The number of dots or squares within the watershed boundary or study site multiplied by the area represented by each dot or square equals the watershed area. More accurate area measurements are obtained with planimeters (Figure 4.5) or digitizers. Use of both techniques involves tracing the watershed boundary and converting the measurement to an area. Use of the planimeter is explained in Box 4.1. With planimeters the conversion of measured distance to area is done mechanically, whereas with digitizers this calculation is done

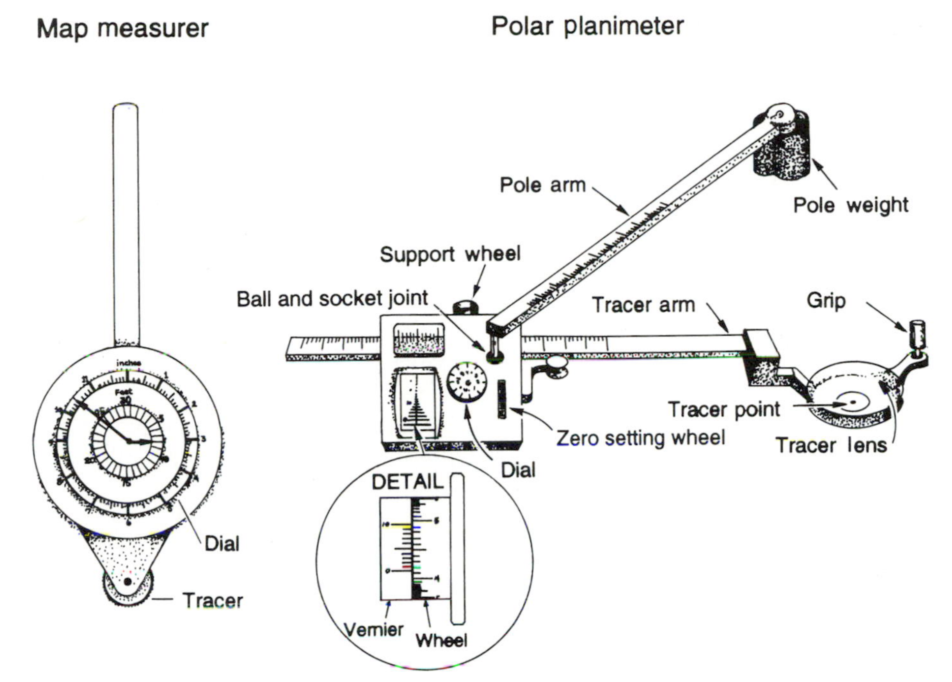

Figure 4.5 Polar planimeter and map measurer (from Orth 1983).

electronically with commonly available computer-assisted drawing software packages. Surface areas of lakes and reservoirs can be determined in a similar fashion. The shoreline and surface area of a proposed reservoir can be determined by drawing a contour line at the projected reservoir elevation(s) to outline the reservoir shoreline. Bathymetric maps of existing reservoirs can be drawn from old topographic maps made before the dam was constructed.

Stream length or shoreline length—factors useful in measuring the amount of habitat available—can be determined with a map measurer (Figure 4.5) or digitizer. To measure a distance, set the digitizer or map measurer's tracing wheel to zero and trace the distance. The true distance can then be determined by using the scale of the map.

4.3.3.2 Geomorphic Features

A number of variables derived from topographic maps help describe the geomorphology (shape) of a watershed. Geomorphic features such as stream gradient, basin size, drainage density, and geologic type influence lake productivity and the composition of stream habitat types within a watershed and, therefore, fish species composition and abundance (Lanka et al. 1987). Geomorphic descriptions of a watershed are useful for predicting the sensitivity of a watershed to disturbance or predicting the success of stream habitat improvement structures. A watershed having steep slopes and unstable soils, for example, may have higher stream flow variability and be more susceptible to sedimentation from logging or road building than might a nearby watershed having gentler slopes and a more stable geology.

Drainage density is calculated by dividing the total stream length of a watershed by the watershed's area. Watersheds with a high drainage density are characteristic of steep, highly dissected watersheds with a diverse network of small streams. Stream

Box 4.1 Use of the Planimeter

The planimeter is used to measure the surface area of an irregularly shaped area on a map. It consists of a measuring wheel that rotates, a tracing arm and point, and a pole arm that rotates around a fixed point or pole weight (Figure 4.5). The planimeter units are read from divisions on the measuring wheel and a dial that counts the number of complete revolutions of the measuring wheel. Conversion of the planimeter units to units of area depends on the length of the tracer arm. If its length is fixed, the planimeter units convert directly to units of area by means of a conversion factor supplied by the manufacturer. If the tracer arm is adjustable, a polygon of known area is supplied by the manufacturer to convert to units of area. To verify your technique, draw a square of known area and carefully trace it three times. The average of the three readings represents the number of planimeter units that equal this known area.

To determine the surface area of a feature on a map, use the following procedures.

1. Spread the map on a smooth, hard surface and fasten it to the surface with adhesive tape.
2. Place the pole weight outside the area to be measured. Make certain that all parts of the map can be reached by the tracer point without moving the fixed point and while maintaining an angle between the tracer arm and pole arm of 15 to 165 degrees. If the area is too large to be traced, divide the area into sections and measure each section separately.
3. Mark a starting point on the map and reset the measuring wheel to zero.
4. Beginning at the starting point, move the tracer point accurately along the outline of the area to be measured, proceeding in a clockwise direction and returning to the starting point.
5. Record the number of planimeter units. The dial shows the number of revolutions of the measuring wheel, and the measuring wheel is divided into intervals. Make a second tracing and average the two readings.
6. Convert the planimeter units to units of area by means of the conversion factor as described above.

The planimeter shown in Figure 4.5 has a mechanical readout. Digital readout and computing planimeters, as well as computer digitizing pads, are available to calculate areas automatically in any units or for any scale map. The principal advantage of a mechanical readout planimeter is low cost. For frequent users, the high-tech planimeters are worth the extra cost.

gradient (slope) of a watershed is the difference in elevation between the stream mouth and source divided by the stream length. Gradient is expressed in meters per kilometer, feet per mile, or percent. It is determined by counting the number of contour intervals crossed by a stream over a given distance. A longitudinal

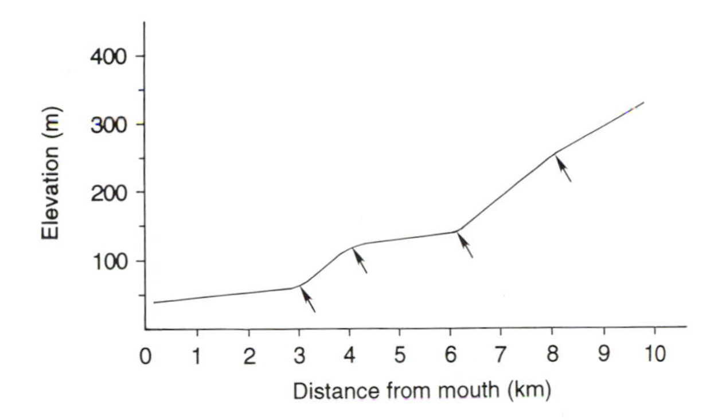

Figure 4.6 Longitudinal profile of stream. Arrows show locations of differences in gradient, which indicate potential study reach boundaries.

profile describes the change in stream gradient graphically as a plot of elevation compared with stream distance (Figure 4.6). By showing abrupt changes in stream gradient, a longitudinal profile is useful for delineating boundaries of study segments and sampling strata. Sinuosity is a descriptor of the general meander pattern of a river. It is calculated by dividing the stream length by the valley length between the same two points. A sinuosity ratio of 1 denotes a straight channel, 1.5 a meandering channel, and 4 a highly meandering channel. Changes in sediment load, from bank instability or upstream erosion for example, may cause changes in channel meander pattern and associated stream habitats (Gordon et al. 1992).

Stream order is a rank of the relative size of streams within a watershed. The most commonly used method (Strahler 1957) of ordering is to designate the smallest unbranched, or headwater, tributaries as first-order streams. A second-order stream is formed when any two first-order streams meet; third order by the junction of any two second-order streams; and so on (Figure 4.7). Note that order is increased only when two tributaries of equal order join. It is important that first-order streams be defined consistently because the smallest tributaries drawn on a map depends on the map scale (Hughes and Omernik 1981). Most fisheries workers define first-order streams as the smallest unbranched tributaries shown as blue lines on 1:31,680- or 1:24,000-scale maps. Stream ordering is useful in describing fish species diversity and abundance patterns over large areas of similar geology and climate (Platts 1979; Fausch et al. 1984). It is a particularly useful method for classifying streams at the start of the study for stratifying potential sampling sites. However, its use is limited where stream order may be independent of stream size, such as large streams that originate from groundwater springs (Hughes and Omernik 1981). See Gordon et al. (1992) for additional descriptions of watershed measurements.

4.4 STREAM AND RIVER MEASUREMENTS

4.4.1 Transect versus Habitat-Based Sampling

Transect and habitat-based sampling are two common approaches used for measuring stream and river habitat features. With the transect method, habitat features are systematically measured or visually estimated at points, sections, or cells along transects placed perpendicular to streamflow (Chapter 2). The number and

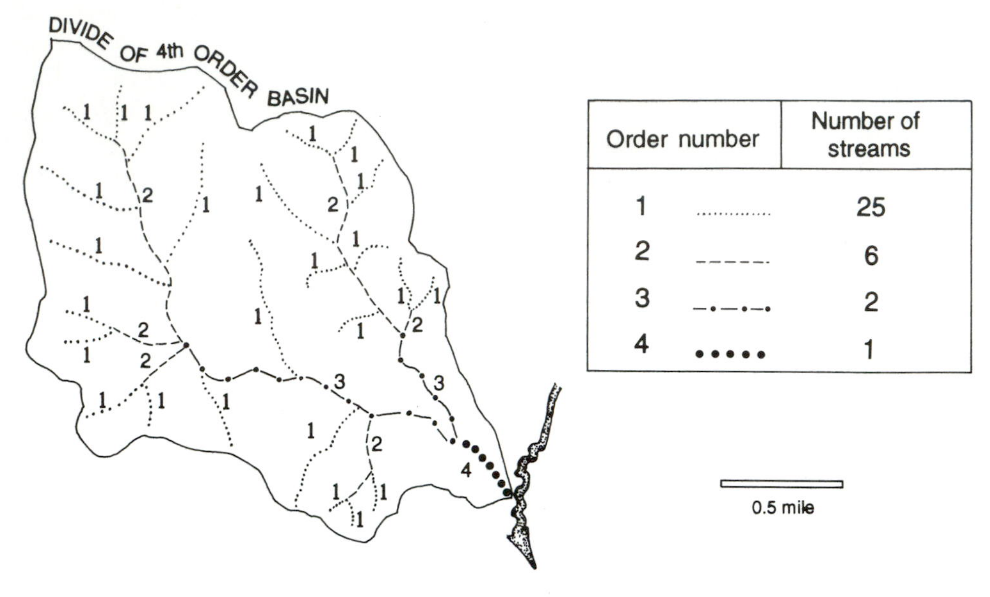

Figure 4.7 Stream ordering system (from Orth 1983).

spacing of transects within a reach varies depending on the resolution needed to characterize the habitat accurately to meet study objectives within time and cost constraints (Platts et al. 1983). Simonson et al. (1994) recommended a minimum of 13–20 transects per reach spaced a distance of 2–3 times the mean stream width apart; the length of reach sampled should be at least 35 times the mean stream width. For example, a reach for a 5-m-wide stream would be 175 m long. Reaches are chosen randomly or selected as representative of the stream of interest.

The habitat-based method divides the entire study area into habitat types (section 4.4.3.8) and visually estimates habitat features (e.g., depth, cover, and substrate) within each habitat unit. At predetermined intervals (e.g., every 5th pool or 10th riffle), actual habitat measurements are also made to correct for bias in visual estimates of habitat features. This method has been deemed the "basinwide survey technique" because it has been developed to characterize habitat over entire watersheds (Hankin and Reeves 1988; Dolloff et al. 1993).

Both approaches have advantages and limitations. The transect method minimizes measurement error (see section 4.2), often resulting in estimates that are precise for the particular reach sampled. However, extrapolation error may be high if the selected reach represents only a small proportion of the total study area. The habitat-based method minimizes extrapolation error and provides a map of the entire study area but may yield a higher measurement error than does the transect method.

4.4.2 Water Quality

Accepted methods for measuring the physical and chemical quality of water are described in *Standard Methods for the Examination of Water and Wastewater* (*Standard Methods*; APHA 1989) and other texts (USEPA 1979; Wetzel and Likens 1991). Various water analysis kits and electronic sensor meters are available also, but they should be calibrated carefully and compared with *Standard Methods* measurements to insure that their accuracy and precision will suffice for the intended application. Kits and meters are typically less precise than are the tests described in *Standard Methods* but may be adequate for many fisheries management applications.

High-gradient rivers are typically well mixed, and stratification or local differences in water quality are not usually evident unless point source pollution discharges are present. Therefore, samples from a few locations may be adequate to describe a long river reach. In larger, lower-gradient rivers, however, differences in water quality often exist between riffles and pools or between surface and bottom waters in pools and backwaters. Also diurnal fluctuation in water quality parameters, such as pH and dissolved oxygen concentration, are likely to occur in sluggish stream segments. Consult Hynes (1970), Walling and Webb (1992), Webb and Walling (1992), and McDaniel (1993) to assist with the interpretation of the physical and chemical quality of running waters.

4.4.3 Stream Channel Measurements

Descriptions of channel morphology are central to any stream habitat survey because they help define the types and frequency of aquatic habitats present. Channel descriptions can vary from simple measurements of width and depth to detailed measurements of channel geometry at different streamflows.

4.4.3.1 Width and Depth

Width and depth measurements are needed to calculate water surface area and volume, which are, in turn, used to determine fish density or standing crop. Stream width is correlated with changes in aquatic community composition and fish distribution and abundance (Vannote et al. 1980; Bozek and Hubert 1992). Stream depth is an important component of fish cover and is used for measuring changes in channel shape. Measurements of both variables represent a potential source of bias because both are dependent on flow. Thus, it is important to measure these variables at specified flows and locations. When measuring width, the distinction should be made between stream, or wetted, width and the bank-full, or channel, width—the bank-to-bank distance (Figure 4.8A). Channel width is often a more standardized descriptor of the stream channel because it is independent of discharge; however, it is more difficult to determine due to the subjectivity involved in estimating where the top of a bank is (Platts et al. 1983; Simonson et al. 1994). Both widths can be measured with a tape or rangefinder.

To measure stream depth in small streams, Platts et al. (1983) recommended that at least three equidistant depth measurements be made across a transect by means of a calibrated wading rod. Mean depth is then calculated by dividing the total by four (three equidistant points equals four sections along the transect). Stream shoreline water depth is used as a measure of available habitat for age-0 fish (Platts et al. 1983). Measurement is made at the shoreline or at the edge of the bank overhanging the shoreline (Figure 4.8A). In larger, unwadeable streams and rivers, depth is measured from a boat or bridge by use of a long rod, sounding weight attached to a wire cable, or depth (echo) sounder. In fast, deep rivers the sounding weight may drift downstream and yield inaccurate depths. True depth can be approximated by measuring the angle of the cable to the vertical and multiplying the cosine of this angle by the length of the cable.

The width:depth ratio provides a useful index of general channel shape and is a sensitive measure of a stream channel's response to changes in bank conditions. Increases in width:depth ratios, for example, indicate increased bank erosion, channel widening, and infilling of pools. Width:depth ratio can be reported either for

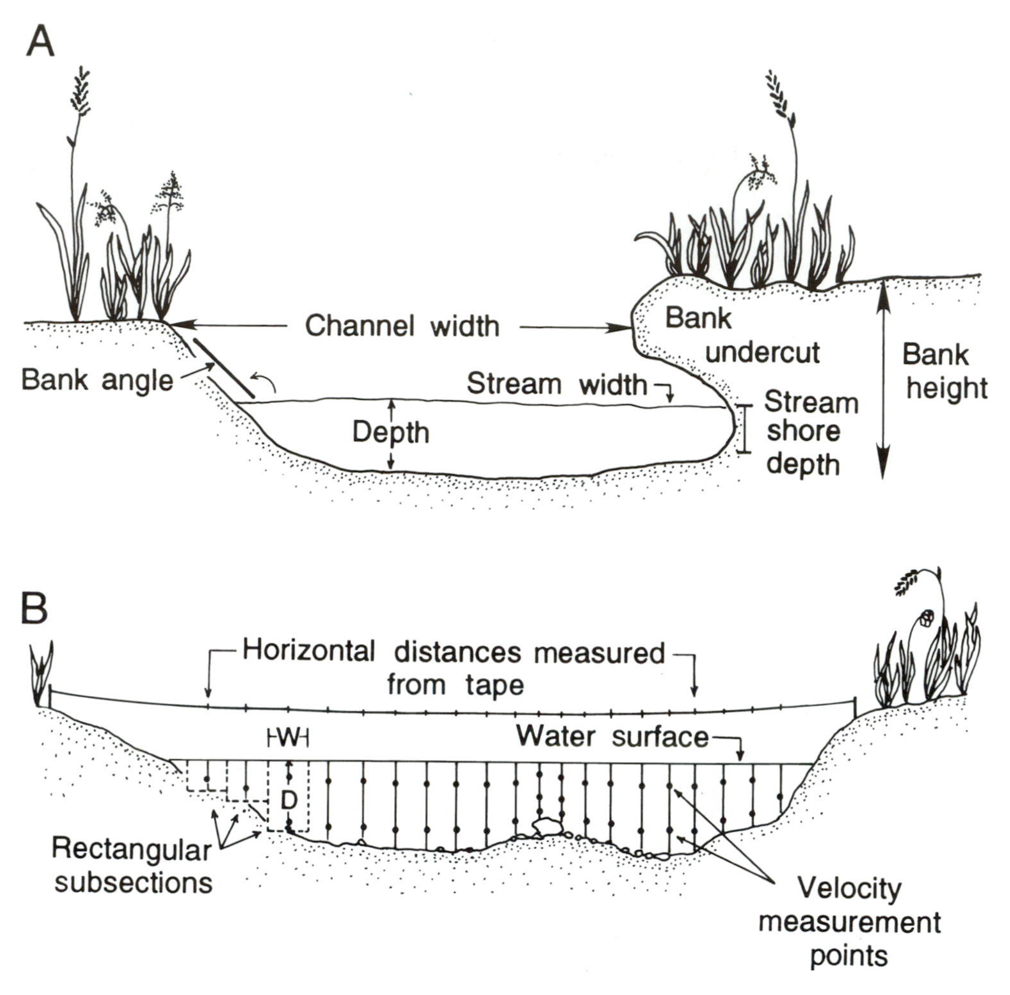

Figure 4.8 Habitat measurements of stream cross sections (from Gordon et al. 1992 with permission of John Wiley & Sons, Inc.). Depth and velocity measurements can be made at specified intervals (**B**).

channel width and depth or for stream, or wetted, width measurements (Lanka et al. 1987; Gordon et al. 1992). The ratio should be determined at several different cross sections over time and the depth calculated as mean rather than maximum depth.

Channel cross sections provide detailed measurements of stream channel width and depth and are a necessary part of calculating streamflow. Cross sections are established by stretching a measuring tape between stakes located on each bank (Figure 4.8B). Depth measurements taken with a stadia rod and level are made across the channel at specified intervals and at breaks in the slope of the channel. Repeated surveys taken at the same locations over time provide a permanent record of changes in streambed elevation and channel width that can be used to quantify changes in channel erosion and sediment deposition (Box 4.2; Olson-Rutz and Marlow 1992). Channel cross sections are also used to measure the wetted perimeter or distance along the streambed in contact with the water. Measuring wetted

Box 4.2 Stream Habitat Evaluation Example[a]

Sediment Monitoring in the South Fork Salmon River, Idaho

The South Fork of the Salmon River in central Idaho is characterized by steep slopes and highly erodible soils. Intensive logging and road construction from the 1940s to 1960s led to high rates of erosion in the watershed. Concern over the effects of fine sediments on the valuable chinook salmon and steelhead populations in the river led to the initiation of a long-term study to monitor trends in sedimentation and its associated effects on spawning and rearing habitats prior to and following a watershed rehabilitation program.

A series of 10 transects was established at each of five major spawning areas. Additional transect sampling sites were established in 47 areas used as rearing habitat. The first transect within a site was selected randomly and the remaining transects located systematically at fixed (91.2 m) intervals. To monitor trends in particle size distribution of the streambed surface, from 1965 to 1985 visual estimates of the dominant particle size were made annually at 0.3-m intervals along each transect. Surface substrate was characterized as the percent of the area sampled occupied by fines, gravel, rubble, or boulder. The composition of the subsurface substrate in spawning areas was monitored by collecting McNeil core samples (Figure 4.12A) annually at 20 randomly selected sites. Changes in streambed elevation (aggradation and degradation) were determined along a series of permanent channel cross sections. Additional monitoring data collected included aerial photo surveys every 5 years; photographs of the river at fixed points to document changes in channel condition; surveys of the depth of fine sediment deposited in pools; snorkel surveys to monitor trends in juvenile fish populations; and monitoring of sediment embeddedness and macroinvertebrate populations in parts of the watershed having different levels of land use activity.

Logging and road construction, in combination with large storm events, resulted in large deposits of fine sediments in spawning and rearing areas. In the early years of the study, 47% of substrate surface area and subsurface volume of cores consisted of fine sediment smaller than 4.75 mm in diameter. Substrate conditions responded dramatically, however, to improvements in watershed condition. With the reduction of sediment delivery to the stream, especially by improved road construction practices, there was a statistically significant decrease in surface and subsurface fines in spawning and rearing areas. The biological response to decreased sedimentation has not been as obvious because of confounding effects of downstream hydroelectric dams on steelhead and chinook salmon populations.

[a]From Platts et al. (1989) and MacDonald et al. (1991).

perimeter at different streamflows is one of several methods used in determining minimum flow requirements for rivers and streams (Gordon et al. 1992).

4.4.3.2 Gradient and Sinuosity

Channel gradient, the drop in elevation per unit length of channel, affects stream velocity and the availability of habitat types and therefore is an important variable for describing species distribution and abundance (Bozek and Hubert 1992). For long channel sections, channel gradient can be determined from topographic maps as described in section 4.3.2.2. Field measurements made using a stadia rod and level provide a more detailed description of channel gradient. The change in angle or slope is read as a percent or in degrees. The recommended procedure is to measure gradient both upstream and downstream of a transect site and average the two values. The observer and rod holder should stand on flat surfaces adjacent to the stream edge and along the same habitat type (e.g., riffle).

Channel sinuosity is similar to river sinuosity (section 4.3.3.2) but is determined in the field for individual reaches by comparing channel length with straight-line distance between two points on a channel (points A and B; Figure 4.9). Low sinuosity is frequently indicative of steep gradients and little pool development, whereas high sinuosity is associated with undercut banks and large, deep pools. Channel length is measured along the thalweg (path of deepest water). Sinuosity should be determined over a distance of at least 20 times the channel width (Platts et al. 1983; Gordon et al. 1992). Repeated measurements of depth along the thalweg can be used to develop a thalweg profile. This measurement is analogous to the longitudinal profile, but it is determined for individual reaches rather than an entire stream section. Whereas cross sections quantify the vertical shape of a channel, thalweg profiles quantify the shape of the channel along its longitudinal axis. Use of a level and stadia rod is recommended for accurate mapping of the thalweg profile, although a measuring tape and calibrated wading rod is sufficient for some applications. Thalweg profiles provide a detailed map of the location and depths of pools and other stream features. They are especially useful for illustrating stream channel changes following habitat alterations.

4.4.3.3 Channel Classification

Channel classification attempts to group stream and river channels (sections of stream typically greater than 1 km in length) having similar morphological features. A classification system in common use among fisheries workers is the channel type classification developed by Rosgen (1994). This system groups channels into broad categories according to gradient and channel shape (Figure 4.10). For example, type A streams are relatively straight with steep gradients and narrow channels whereas type C streams are meandering with low gradients and wide channels. Channel types can be further divided by the dominant substrate particle size (1 = bedrock, 2 = boulder, 3 = cobble, 4 = gravel, 5 = sand, and 6 = silt and clay). Channel type classification is useful for defining sampling strata by reaches when making habitat and fish population measurements, for rating a stream's sensitivity to land disturbance and erosion potential, and for identifying the potential effectiveness of various fish habitat improvement structures for a particular channel type (Myers and Swanson 1992; Rosgen 1994).

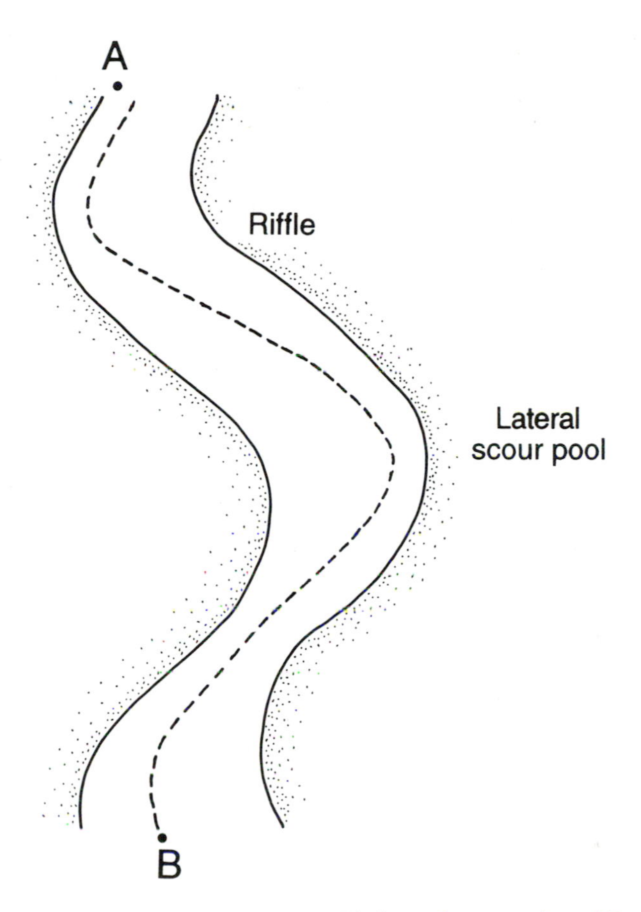

Figure 4.9 Top view of a meandering stream. Thalweg shown as dotted line.

4.4.3.4 Velocity

Different species and life stages often require specific water velocities, and many stream habitat studies have been directed at defining velocity preferences and tolerances in fishes and macroinvertebrates (Gore and Nestler 1988; Lobb and Orth 1991). Current velocity patterns, and hence composition of fish assemblages, can be substantially altered by changes in stream discharge or channel modifications such as channel straightening or debris removal (Hubbard et al. 1993). Fish habitat improvement often involves restructuring the channel to provide current velocity patterns more favorable to desired fish species (Shuler et al. 1994).

Three general methods are available to measure velocity, each differing in accuracy and time involved. The simplest method uses a floating object, such as an orange, pencil, or half-filled bottle, and measures the time it takes the object to travel a known distance downstream. This method is used when no other equipment is available and only rough estimates of velocity are needed. Measurements should be made in straight, uniform reaches with a minimum of surface turbulence. Choose a distance that allows a travel time of at least 20 s. Several measurements should be made to obtain an average. Because surface velocity is greater than is average water column velocity, multiply by 0.8 for a rough bottom and 0.9 for a smooth bottom to

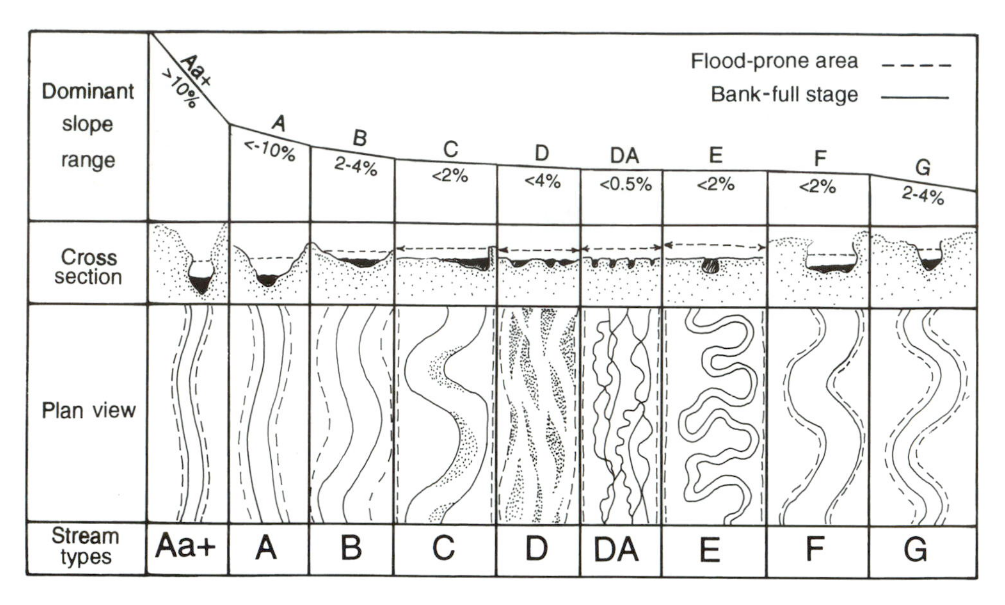

Figure 4.10 Channel type classification (from Rosgen 1994 with permission of Elsevier).

obtain average velocity (Gordon et al. 1992). Another method is to time the movement of dye. This method is typically used for measuring high current speeds when other methods are impractical. One advantage of this method is that an average water column velocity is obtained. The disadvantages are that the dye may color the water some distance downstream and it can be difficult to determine exactly when the dye passes the downstream point if the dye becomes too dispersed.

The most common and accurate method for measuring current velocity is with an electrical current meter (Figure 4.11). Meters are of three general types: propeller, cup, and electromagnetic. Propeller and cup devices come in different sizes; size choice depends on the range of velocities being measured. Generally, the larger the river, the larger the propeller or cup. In cup type models, velocity is determined by counting the number of revolutions of the cups over a fixed time period, usually 40 s. Revolutions can be counted by using a mark on one cup as a reference point or by counting the number of clicks heard through headphones. Use a stopwatch to measure time. The number of revolutions is then converted to current speed by means of a rating curve specific to the particular model of current meter. Current meter digitizers, which convert the number of revolutions to current speed automatically, are also available (Sharp and Rapp 1984). Electromagnetic models offer direct digital or analog readings of velocity. Propeller and cup type meters require care during transport and frequent maintenance to insure accuracy. Electromagnetic current meters measure the change in the magnetic field induced by current around a bulb-shaped detector. These meters are durable and sensitive and provide a direct reading of current velocity.

For most stream survey work, the meter is attached to a graduated metal rod with a sliding support to allow depth adjustment (Figure 4.11). Top-setting wading rods have two support rods, one fixed and one sliding, that allow adjustment of meter depth without removing the meter from the water. A baseplate keeps the rod(s) from sinking into the substrate, and a vane or tail keeps the meter facing into the current.

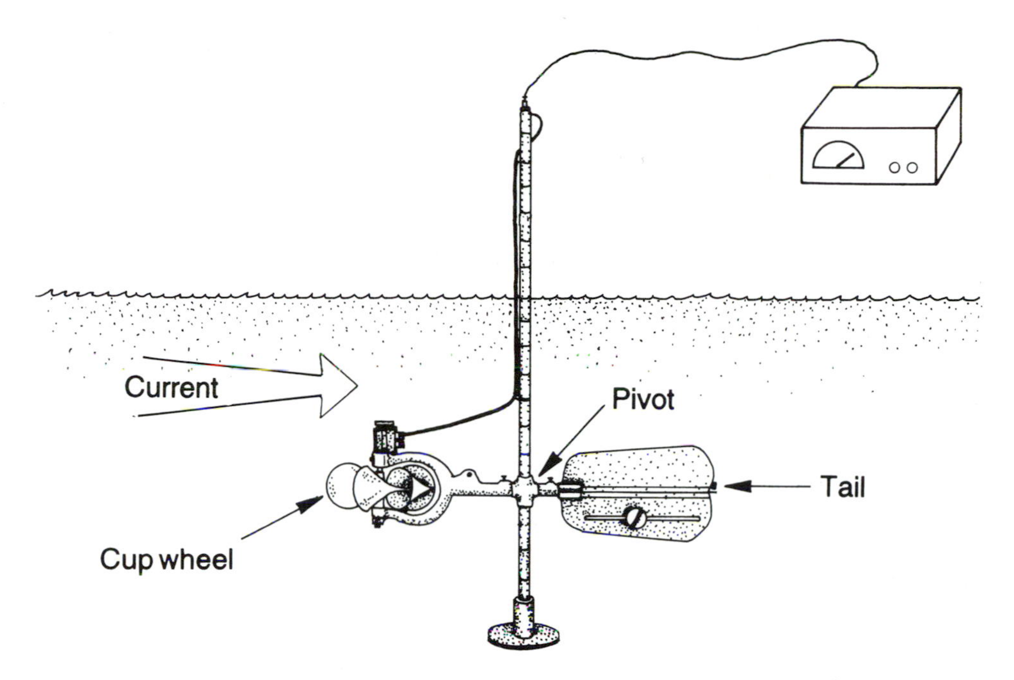

Figure 4.11 Current meter shown mounted on a sliding support that is attached to graduated metal rod (from Orth 1983).

Operators should stand to the side and downstream of the meter to avoid affecting the velocity of water passing the meter. In rivers, the meter is attached to a wire cable and sounding weight with fins and lowered to the desired depth by hand or with a winch and boom mounted to a bridge or boat. The greatest advantage of the electronic current meter is that it allows measurement of velocity at any point in the water column, including a fish's focal position. Such measurements provide descriptions of microhabitat velocity preferences for different species and life stages (Shirvell and Dungey 1983; Shuler and Nehring 1994).

The mean water column velocity is frequently measured in fisheries studies to calculate discharge or to estimate "available" velocities in relation to a fish's "preferred" velocity. In shallow streams, mean velocity is commonly determined by measuring velocity at a point 60% of the depth below the surface (40% of the depth above the streambed; Figure 4.8B). The velocity at this point approximates the mean velocity because water velocity increases exponentially from zero at the stream bottom to a maximum near the stream surface. When water depth exceeds 0.75 m, the mean water column velocity should be calculated from measurements made at 20% and 80% of the total depth (Figure 4.8B). More measurements should be taken when current velocity is nonuniform because of the presence of channel structures such as logs, boulders, and deflector dams.

4.4.3.5 Discharge

Stream discharge or flow is the quantity of water that passes through a cross section of a channel per unit of time. Discharge influences a myriad of factors including the quantity and quality of stream habitats, water quality, and fish passage. At a minimum, discharge measurements are included in stream studies to insure that habitat surveys done at different times are comparable. Repeated discharge measurements are used to prescribe minimum instream flows required to maintain aquatic biota (Gore and Nestler 1988).

Discharge can be determined by a number of different methods. To start, consult state, provincial, or federal water resource agencies for data from stream gauging stations. Gauging stations provide complete and accurate measurements of discharge on a continuous basis. Gauging stations operate by measuring with flumes or weirs the stream height or stage. Stage height is then converted to discharge by a stage–discharge relationship (developed by calculating discharge at a range of different streamflows by means of the velocity–area method described below). Use of gauging station records is limited by their availability. In some areas, hydrographs can be derived indirectly from information on rainfall (Fedora 1987). Other ways to estimate discharge indirectly are described by Gordon et al. (1992).

Discharge at a site is determined by multiplying the cross-sectional area of the water column by the mean current velocity (velocity–area method). Because velocity can vary substantially across a channel, accurate measurements of discharge typically require measurements of velocity and cross-sectional areas within several subsections or cells (Figure 4.8B). The number of subsections chosen depends on the variability of velocities within the channel. As a general rule, subsections should be spaced such that each subsection contains no more than 10% of the total discharge; more measurements should be added if the section is irregular. Total discharge (Q) is then calculated as

$$Q = w_1 d_1 v_1 + w_2 d_2 v_2 + \ldots + w_n d_n v_n,$$

where w is width, d is depth, v is mean velocity for each subsection $(1, 2, \ldots, n)$, and Q is in cubed meters or cubed feet per second. Software programs have been developed for calculating discharge and for predicting width, depth, and velocity as a function of discharge (Milhous et al. 1981; Gordon et al. 1992).

Simple, indirect measurements of relative discharge can be obtained by recording stream height by means of a graduated rod or staff gauge attached to a fence post, bridge abutment, or other fixed object. An estimate of the maximum height of stream level between visits to a site can be obtained by use of a crest gauge. A crest gauge consists of a tube that contains small floatable material, such as ground cork or styrofoam chips. When water level rises, the material floats upward and then adheres to the sides of the tube as the water level drops (Gordon et al. 1992). A simple crest gauge can be constructed of polyvinyl chloride pipe with small holes drilled in the side to allow water to enter. Small-mesh screen is used to keep debris out and buoyant material in the pipe. Readings are taken with a meter stick kept inside the pipe. A reading is taken by measuring the height of the floatable material on the meter stick. The gauge is reset by washing the material back down inside the pipe.

4.4.3.6 Substrate Measurements

Substrate composition determines the quality of spawning habitat and cover for many fish species and influences benthic macroinvertebrate composition and production. Determining the effects of increased sedimentation on stream fishes and macroinvertebrates is a frequent focus of fisheries investigations (Box 4.2).

A common method for estimating surface substrate composition is to classify substrate visually according to particle size and the Wentworth scale (Table 4.2). Reference substrate samples embedded in plastic aid classification of substrate,

Table 4.2 Modified Wentworth classification for substrate particle sizes (Cummins 1962).

Classification	Particle size range (mm)
Boulder	>256
Cobble (rubble)	64–256
Pebble	32–64
	16–32
Gravel	8–16
	4–8
	2–4
Very coarse sand	1–2
Coarse sand	0.5–1
Medium sand	0.25–0.5
Fine sand	0.125–0.25
Very fine sand	0.0625–0.125
Silt	0.0039–0.0625
Clay	<0.0039

especially the smaller particles. Typically, the amount of each particle size is estimated along a transect by using a point, line, or grid system, and data are expressed as a plot of particle size versus frequency as a percent (Platts et al. 1983; Bain et al. 1985; Simonson et al. 1994).

Visual estimates of surface substrates are subject to bias because of variation in results between observers (Platts et al. 1983; Hamilton and Bergersen 1984). A useful alternative to visual estimation in wadeable, gravel bottom streams is the Wolman pebble count method (Wolman 1954; Kondolf and Li 1992). The method involves systematic sampling of the substrate within a grid. Particles are measured under grid points or, more typically, by walking across the stream channel from bank to bank. At each step, the observer reaches down with a finger and touches the particle, or pebble, at the tip of the boot. The size of each particle is recorded according to the Wentworth scale or measured in millimeters across its axis. Traverses are made across the channel until 100 or more points are sampled. To avoid bias towards choosing larger, more visible stones, the eyes should be averted or closed when choosing a particle. A mark on the tip of the boot can also serve as a sampling reference point. Wolman pebble counts have long been used by hydrologists to characterize substrate, but the method has only recently been embraced by fishery biologists as an alternative to visual estimates. Pebble counts sample a larger area, are more reproducible, and take about the same amount of time to perform as visual estimates (Kondolf and Li 1992).

Determination of subsurface substrate composition is often desired to estimate effects of fine sediments on embryo survival in salmon, trout, and other gravel-nesting species (Chapman 1988). Two types of core samplers are employed to obtain a subsurface sample. A McNeil hollow-core sampler (Figure 4.12A) is inserted into the substrate to the desired depth. The enclosed substrate is then excavated and deposited in a basin inside the coring device. A sample of the remaining mixture of water and suspended sediment is removed and allowed to settle in a graduated cone. The volume of the settled fine material is converted to dry weight (Platts et al. 1983). Freeze-core samplers (Figure 4.12B) consist of one or more (three is the standard) hollow probes that are driven into the substrate and injected with liquid nitrogen or carbon dioxide to form a frozen chunk of substrate, which is extracted with a hoist

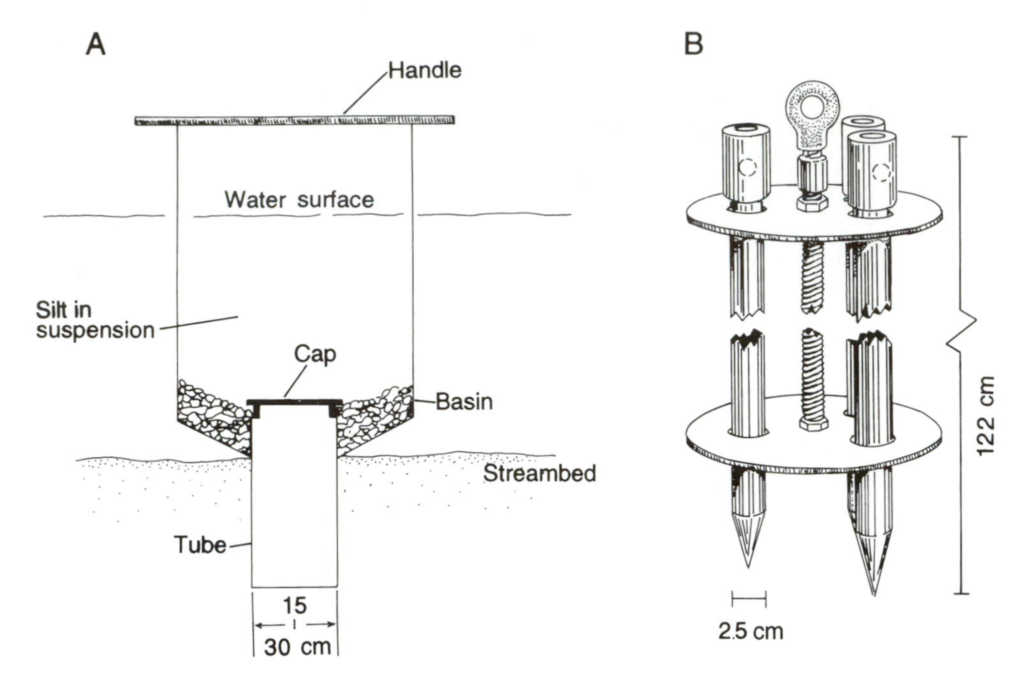

Figure 4.12 McNeil (**A**) and freeze-core (**B**) substrate samplers (from Platts et al. 1987).

(Everest et al. 1980). Both devices have advantages and limitations. The main advantages of freeze-core samplers are an ability to analyze vertical stratification of sediment, a more complete collection of fine sediments, and an ability to sample deeper water than with McNeil samplers. However, McNeil samplers are more portable, less costly, and easier to operate. In some cases, a shovel may be adequate for sampling substrate (Grost et al. 1991).

The percentage composition of the substrate core (obtained with either sampler) is apportioned into size-classes by passing the sample through a series of sieves of various sizes. Wet sieving can be done in the field by flushing the sample with water to sort sediments through the sieves. The trapped sediment in each sieve is either weighed or poured into a water-filled container; the volume of water displaced is used to determine volume of sediment, which is then converted to dry weight (Platts et al. 1983; Hamilton and Bergersen 1984). Dry sieving in the laboratory is recommended to obtain more accurate results. Samples are oven dried, sieved in a mechanical shaker, and contents remaining on each sieve weighed to determine percentage composition by weight. Substrate composition data is displayed as percent size composition or by calculating an index of particle size distribution. Particle size distribution can be characterized by the median diameter (D_{50}), geometric mean diameter (D_g), and the fredle index (f_i) (Platts et al. 1983; Young et al. 1991). D_g is calculated as

$$D_g = D_1^{w_1} \times D_2^{w_2} \times D_3^{w_3} \times \ldots \times D_n^{w_n},$$

where D is the mean diameter in millimeters of particles retained on a given sieve and w is the proportion by weight of particles retained by a given sieve. A short cut formula for D_g is

$$D_g = \sqrt{D_{16} \cdot D_{84}},$$

where D_{16} is the particle size diameter at the 16th weight percentile and D_{84} is the diameter at the 84th weight percentile. The fredle index, f_i is calculated as

$$f_i = D_g/S,$$

where D_g is the geometric mean particle size in millimeters and S the sorting coefficient. A measure of the spread of particle sizes in the substrate, S is calculated as

$$S = d_{75}/d_{25},$$

where d_{75} and d_{25} are the substrate diameters below which 75% and 25% of the sample lie (corresponds to one standard deviation from the mean). The fredle index often is the preferred method because it incorporates measures of both gravel pore size and permeability (Platts et al. 1983; Chapman 1988).

Given the variety of methods available to measure substrate composition, selection of an appropriate method should consider the level of sampling required and type of data needed. In many situations, a combination of methods is desirable to meet study objectives (Box 4.2).

4.4.3.7 Erosion and Sedimentation

Erosion and sedimentation are measured to monitor the effects of road building, logging, grazing, and other land use activities on aquatic habitats. As noted in section 4.4.3.1, repeated measurement of channel cross sections is one method for quantifying channel erosion and sedimentation by showing changes in channel shape over time. Another method employs scour chains to measure the depth of aggradation (filling) and degradation (erosion or scouring) of a stream channel (Platts et al. 1983; Nawa and Frissell 1993). A scour chain consists of a length of chain or beads that is buried into the substrate through the use of a steel pipe and fence post driver. The device is numbered and its position mapped to insure later recovery. The length of chain or number of beads exposed above the substrate after a flood represents depth of scour. Fill is measured as the depth of substrate overlying the chain.

Embeddedness measures sedimentation by determining the degree to which gravel and cobble is embedded or buried by sand and silt. Embeddedness can be visually classified as percent of the surface of particle covered by fine sediment (Platts et al. 1983) or measured by dividing the depth that a rock is buried by silt or sand by its total height (MacDonald et al. 1991).

4.4.3.8 Habitat Type

Streams and rivers encompass a variety of different types of habitats, including riffles, pools, and backwaters (Bisson et al. 1982; Sheehan and Rasmussen 1993). Generally, different habitat types provide different value to different fish species or life history stages (Bisson et al. 1988). By censusing the amount of different habitat types, biologists can quantify habitat availability, identify potentially limiting factors, and estimate fish species abundance. Before-and-after inventories of habitat types can also quantify the effects of habitat alterations and assess effectiveness of stream habitat enhancements. Use of a habitat type classification is also advantageous in that habitat types integrate a number of important habitat features, including depth, velocity, and substrate (Hawkins et al. 1993).

Box 4.3 Stream Habitat Types

Bisson et al. (1982) described a number of types of riffles and pools in streams according to differences in depth, gradient, current velocity, agent of formation, and position in the channel. This classification system is now in wide use for characterizing stream habitat in stream habitat surveys (Hankin and Reeves 1988; Lobb and Orth 1991). Hawkins et al. (1993) modified Bisson's classification system, subdividing habitat types according to the following three-level hierarchy to allow different levels of resolution depending on habitat evaluation objectives.

I. **Fast water:** Riffles; rapid, shallow stream sections with steep water surface gradient.
  A. **Turbulent:** Channel units having swift current, high channel roughness (large substrate), steep gradient, and nonlaminar flow and characterized by surface turbulence.
    1. **Fall:** Steep vertical drop in water surface elevation.
    2. **Cascade:** Series of alternating small falls and shallow pools; substrate usually bedrock and boulders.
    3. **Chute:** Narrow, confined channel with rapid, relatively unobstructed flow and bedrock substrate.
    4. **Rapid:** Deeper stream section with considerable surface agitation and swift current; large boulders and standing waves often present.
    5. **Riffle:** Shallow, lower-gradient channel units with moderate current velocity and some partially exposed substrate (usually cobble).
  B. **Nonturbulent:** Channel units having low channel roughness, moderate gradient, laminar flow, and lack of surface turbulence.
    1. **Sheet:** Shallow water flowing over smooth bedrock.
    2. **Run:** Shallow water flowing over a variety of different substrates; also termed "glide" or "raceway" by some authors.

II. **Slow water:** Pools; slow, deep stream sections with nearly flat water surface gradient.
  A. **Scour pool:** Formed by scouring action of current.
    1. **Eddy:** Formed by circular current pattern created by bank obstruction (e.g., boulder, log jam, or wing deflector dam).
    2. **Trench:** Formed by scouring of bedrock.
    3. **Midchannel:** Formed by channel constriction.
    4. **Convergence:** Formed where two stream channels meet.
    5. **Lateral:** Formed where flow is deflected by a partial channel obstruction (streambank, rootwad, log, or boulder); for example, formed at the outside bends in the channel of meandering streams (Figure 4.8).
    6. **Plunge:** Formed by water dropping vertically over channel obstruction.
  B. **Dammed pool:** Water impounded by channel blockage.
    1. **Debris:** Formed by rootwad and logs.
    2. **Beaver:** Formed by beaver dam.
    3. **Landslide:** Formed by large boulders.
    4. **Backwater:** Formed by obstructions along banks.
    5. **Abandoned channel:** Formed alongside main channel, usually associated with gravel bars.

See Bisson et al. (1982) and Helm (1985) for illustrations of these different habitat types.

At the simplest level, habitat type is classified as pool (slow water) or riffle (fast water). A stream or river can then be described in terms of a pool:riffle ratio or percentages of pools and riffles. The basic categories of fast- and slow-water habitats can be further subdivided into subtypes (Box 4.3). In small- to medium-size streams, the amount of different habitat types can be determined using transects in a fashion similar to estimating cover or substrate (Platts et al. 1983). However, the preferable method is to measure or estimate the length and width of individual habitat units over a large area by means of the habitat-based method described in section 4.4.1. In larger streams, classification becomes more difficult because several habitat types may occur across a stream. As a general rule, measure or include only those habitat units that are at least one channel width in length. Because habitat type classification is based on visual assessment, measurement can be conducted over large areas at a reasonable cost. However, proper training of field crews is necessary to avoid sampling bias and maintain consistency in delineating habitat type boundaries and defining habitat types (Hawkins et al. 1993). See Sheehan and Rasmussen (1993) for an example of a habitat type classification system for large rivers and Rabeni and Jacobson (1993) for a system developed for low-gradient warmwater streams.

4.4.3.9 Cover Measurements

Cover is defined as those instream areas that provide protection from predators and from adverse environmental conditions such as high current velocities. Cover includes instream structures such as aquatic vegetation, boulders, and large woody debris; channel features such as water turbulence and depth; and riparian and bank features such as overhanging vegetation and undercut banks. Numerous studies have shown the critical importance of cover as a fish habitat component (Angermeier and Karr 1984; Wesche et al. 1987; McMahon and Hartman 1989). What constitutes cover is sometimes ambiguous because cover requirements vary by species, life stage, and season. Thus, it is important to define cover type preferences explicitly for the species under study. The most commonly used method to assess cover is to measure dimensions of different cover types and express the total amount of cover as a percentage of the study area. Cover can be measured along a transect or measured at all sites within a study reach (Hamilton and Bergersen 1984; Simonson et al. 1994). Wright et al. (1981) described specific methods for measuring aquatic macrophytes.

A more quantitative and less subjective alternative to the percent cover method is the cover density method (Kinsolving and Bain 1990). This method counts the number of cover surfaces that bisect a plane extending from the substrate to the water surface at 15-cm increments along a transect line. One advantage of the method is that resulting data are more readily analyzed using statistical analysis than are percentage cover data.

Large woody debris (LWD) in the form of logs and rootwads is an important habitat component for a variety of fishes inhabiting coldwater and warmwater streams and rivers. Large woody debris stabilizes channels, forms pools, traps spawning gravel and organic matter, serves as habitat for macroinvertebrates, and provides cover for fish (Angermeier and Karr 1984; Benke et al. 1985; Fausch and Northcote 1992). Measurements of LWD are used to monitor the effectiveness of riparian management practices (Murphy and Koski 1989). The typical definition of LWD is pieces greater than 1–2 m long and 10 cm in diameter. Measurement techniques vary widely in terms of effort required, ranging from a simple count of LWD pieces to detailed descriptions of the characteristics (length, diameter, and volume), location, and function of each piece (Platts et al. 1987). Due to the clumped distribution of debris in larger streams, reliable estimates of debris abundance may require study reaches of 1–2 km (Ralph et al. 1994).

4.4.3.10 Habitat Complexity

Although stream and river habitats are often characterized by individual measurements of depth, water velocity, substrate, and cover, fish species and assemblages respond to a combination of habitat features. Habitat complexity can be defined as the diversity of different habitats available to fishes (Pearsons et al. 1992). A common method of evaluating habitat complexity is the use of classification criteria that rate the value of combinations of several different habitat variables. For example, Platts et al. (1983) presented a pool quality rating for salmonids based on different combinations of pool size, depth, and cover. Large, deep pools with abundant cover are given the highest ranking and small, shallow pools lacking cover the lowest. Another method is to compute a Shannon–Wiener diversity index for different categories of depth, current velocity, and substrate type (Gorman and Karr 1978). Measurement of travel time of released dyes (Pearsons et al. 1992) and retention of plastic strips used to simulate organic matter retention (Speaker et al. 1988) offer promising approaches to quantify habitat complexity over longer stretches of stream.

4.4.3.11 Streambank and Riparian Area Measurements

Streambanks and associated riparian areas play an important role in defining channel morphology, controlling stream temperatures, and creating fish habitat. Measurements of the change in streambank and riparian area variables in response to grazing and logging have been the focus of numerous studies (Hicks et al. 1991; Platts 1991).

Stream shading can be measured by several different methods. A common method is to estimate canopy density with a densimeter, a spherical mirror divided into grids. The number of grid points bisected by vegetation or other structure is counted and converted to a percentage. More detailed techniques include measuring the sun arc and using a solar radiometer (Platts et al. 1987).

Condition, slope, vegetative cover, and amount of overhang of streambanks are important for assessing bank erosion and availability of fish cover. Roots of riparian plants are critical to the development and maintenance of undercut banks, a key element of cover in meandering streams, especially for large fishes (Wesche et al. 1987; Orth and White 1993). Unstable banks support little vegetation and contribute sediment to the stream. Bank stability is estimated by measuring the proportion of the study area with actively eroding banks (Binns and Eiserman 1979) or by using a visual rating system of bank stability at channel cross sections that is based on percentage of vegetative cover (Platts et al. 1983). More direct measures of bank condition include streambank undercut and bank angle and height (Figure 4.8A). Bank angle, defined as the angle of the bank in relation to the water surface, is measured with a clinometer and measuring rod, which are held against the bank or beneath an undercut.

Additional information about stream habitat surveys and habitat measurements can be found in Platts et al. (1983, 1987), Hamilton and Bergersen (1984), Dolloff et al. (1993), and Simonson et al. (1994). Sources of information for predicting future habitat conditions under different management scenarios can be found in Terrell et al. (1982). Hamilton and Bergersen (1984) list equipment and suppliers. Terrell et al.

(1982), Dolloff et al. (1993), and Simonson et al. (1994) provide examples of habitat survey data sheets. Computer programs available for determining channel cross sections, channel type, and discharge are listed by Gordon et al. (1992).

4.5 LAKE AND RESERVOIR MEASUREMENTS

Some of the most important characteristics of reservoirs that affect fisheries may not require measurement by the investigator at all but can be obtained directly from reservoir operators. Investigators should contact reservoir operators during the planning stages of a study to become familiar with reservoir operations and available data. Among the most important factors to be ascertained are purpose of the facility (e.g., flood control, water supply, navigation, or hydropower), water release pattern (run of the river or storage release), and dam structure (type, number, and depth of outlets). These factors dictate water movement patterns and rates in a reservoir, which in turn influence reservoir and tailwater fisheries (O'Brien 1990; Yeager 1993). Reservoir operators often possess large quantities of dam and reservoir operational data (e.g., daily inflows, discharges, reservoir elevations, generation schedules, evaporation rates, and rainfall) and are usually pleased to provide these to interested parties. Accordingly, assess the availability of existing information and data before initiating a sampling program.

4.5.1 Morphometric Measurements

Morphometric measurements commonly used to provide general characterizations of lakes and reservoirs include area, shoreline length, shoreline development index, depth, and volume. Methods for calculating surface area and shoreline length are described in section 4.3.3.1. The shoreline development index (SDI) is a measure of the shape of a body of water and can be indicative of the amount of littoral zone present; littoral zones are important habitats for many species of fish, especially as spawning or nursery habitats. Reservoirs situated in mountain valleys are often dendritic in shape and have high SDI values; however, steep shorelines limit littoral areas (Hayes et al. 1993). The SDI is calculated as

$$\text{SDI} = \frac{L}{2\sqrt{A\pi}},$$

where L is shoreline length and A is area. A perfectly circular lake would have an SDI equal to 1; the greater the irregularity of the shoreline, the greater the SDI.

Depth can be measured using weighted sounding cables as described in section 4.4.3.1. However, electronic echo sounders (depth finders) are far more efficient and can be purchased at relatively low cost. To create a bathymetric map, continuous depth profiles across a lake are made by moving a recording echo sounder along transects at a constant speed. The transects should be marked with flags at each end to keep the boat on course. The number of transects depends on the spatial heterogeneity of depths. In areas where depth changes rapidly, more transects will be needed. Because echo sounders are often inaccurate in shallow waters (<2 m), sounding poles should be used there (Wetzel and Likens 1991). The depth profiles are plotted along the transects on an outline map, and contour lines of equal depth (e.g., at 1-m or 5-m intervals) are drawn by interpolation. If available, contour-mapping computer programs that use depth data digitized by spatial coordinates are more efficient and can also be integrated with other variables that use GIS technologies.

Volume of a body of water may be needed to calculate hydraulic exchange rates, determine the proper amount of fish toxicant to apply, or estimate the total number of larvae present at a depth stratum from density information. To calculate volume, the surface area at each contour must be determined from a bathymetric map. The volume in the stratum between two contour lines (V_s) is calculated as

$$V_s = \frac{h}{3}(A_1 + A_2 + \sqrt{A_1 A_2}),$$

where h is the thickness of the stratum, A_1 is the area at the upper contour line, and A_2 is the area at the lower contour line. The total volume of the body of water is the sum of the volumes of all strata. Mean depth is calculated by dividing the total volume by the surface area. Caution should be exercised when using old bathymetric maps because sedimentation may have reduced depths. Also, the exact elevation of reservoir water level at a specific time should be ascertained when calculating reservoir volume because wide fluctuations in depth, and consequently volume, are possible.

4.5.2 Physicochemical Attributes

Physicochemical attributes of a lake or reservoir, such as temperature, dissolved oxygen (DO) concentration, and transparency, are useful for assessing water quality in the system as well as habitat suitability for sensitive species. For example, fishes have distinct thermal optima (Coutant 1977) and limits (Fry 1971), such that seasonal absence of suitable temperatures can limit success of a species (Box 4.4) or even exclude it entirely. Water temperature can vary greatly in a lake, especially as a function of depth and stratification, but also in relation to thermal features such as tributary inflows, heated effluents, and submerged springs. Therefore, vertical profiles of the entire water column are necessary, and exploratory sampling should be conducted horizontally to detect any thermal anomalies. Sampling throughout a reservoir is especially important because thermal conditions can be markedly different along a reservoir's longitudinal axis from the upper riverine zone, through the middle transition zone, to the lower lacustrine zone (Thornton et al. 1981).

Electronic thermistor temperature probes are accurate (after calibration) and relatively inexpensive, and can be used rapidly and efficiently to collect thermal profiles of precision adequate for fisheries work. Other approaches include mercury thermometers, reversing thermometers, and bathythermographs (Wetzel and Likens 1991). Measurements should be made at 1-m intervals (more frequently at thermal discontinuities) from the surface to the bottom. Some temperature probes now provide convenient readings of depth by use of a pressure sensor, eliminating the need to correct for deviation of the supporting cable from the vertical on windy days.

Concentration and distribution of DO dictate fish distribution in a lake or reservoir and can provide inference on the effects of nutrient enrichment. Oxygen-sensing electrodes are commonly employed in fisheries investigations to collect DO profiles efficiently, usually in tandem with temperature and other measurements, and must be calibrated carefully and frequently. Also, they require continuous flow over the oxygen-permeable membrane (Wetzel and Likens 1991) and should therefore be outfitted with a stirring impeller to insure accuracy and reliability. Chemical

Box 4.4 Reservoir Habitat Evaluation Example[a]

Water Quality Monitoring in Keystone Reservoir, Oklahoma

Limnological conditions and striped bass habitat use were assessed in Keystone Reservoir, Oklahoma, to determine causes of recurring summer die-offs of adult striped bass in the reservoir. Timing of mortality events (late summer and early autumn) suggested that high water temperatures were responsible.

During summer and early autumn in 1986 and 1987 water temperature and dissolved oxygen (DO) profiles were determined at 24 sites throughout the reservoir at 1- to 2-week intervals by means of electronic sensor meters. Extensive preliminary sampling found no submerged springs or cool tributary inflows that could be used as thermal refugia by striped bass. Also, reservoir operators and local biologists (including personnel familiar with the area prior to impoundment) were questioned about the presence of such features, but they knew of none. Water quality characteristics changed gradually along both arms of this V-shaped reservoir that was formed by a dam just below the confluence of the Cimmaron and Arkansas rivers. The sampling sites were therefore spaced at 2–3-km intervals along both arms of the reservoir over the old river channels (located with an echo sounder) at sites relocated by triangulation of shoreline landmarks. Striped bass were captured using gill nets set throughout the reservoir at various depths to encompass the range of water quality conditions present. The depth of capture for each striped bass sampled was verified with an echo sounder, and temperatures and DO concentrations at depth of capture were obtained from profiles made at the capture sites. After investigators realized that the reservoir was chemically stratified because of salinity differences of Cimmaron and Arkansas river inflows, conductivity measurements were added in 1988. Also, striped bass were never found in upstream parts of the reservoir during summer, so sampling was limited to the lower 14 stations in 1988 to save time and money.

Water temperatures and DO concentrations were high in the epilimnion and cooler but anoxic in the hypolimnion below the chemocline (the steep salinity gradient) due to highly mineralized inflow from the Cimmaron River. Intermediate temperatures and DO concentrations were present in only a thin layer at or slightly above the chemocline. Striped bass were found only in this layer, inhabiting the coolest waters available at which DO concentrations exceeded 2 mg/L. When temperatures in this layer exceeded 27°C, the fish stopped feeding, and mortalities ensued in years when exposures to temperatures greater than 27°C were prolonged. Subsequently, dam releases were altered in an effort to produce lower water temperatures in this layer.

[a]From Zale et al. (1990).

Figure 4.13 Niskin water sampler (photo by J. Priscu).

methods, such as the Winkler titration and its modifications (APHA 1989), are more exact but require retrieval of water samples from each depth using Niskin (Figure 4.13), Kemmerer, van Dorn, or Nansen samplers (Lind 1979) and laboratory analysis. Collecting and analyzing sufficient samples to determine DO profiles throughout a large, deep reservoir at frequent intervals would likely be beyond the ability of most fisheries investigators and their budgets.

Light penetration or transparency of water is affected by suspended organic and inorganic particles, as well as by plankton. Low transparency can be indicative of excessive turbidity, which can limit primary production and therefore fish harvests (Jones and Hoyer 1982). High phytoplankton densities resulting from excessive nutrient enrichment can also reduce transparency. Transparency is measured with a Secchi disk, underwater light meter, or submarine photometer. A Secchi disk consists of a circular plate 20 cm in diameter that is painted black in two opposite quarters and white in the other two. The disk is lowered in the water at midday on the shady, calm side of a boat until it disappears from view. The disk is then raised until it reappears. The average of the disappearance and reappearance depths is the Secchi transparency. Transparency can change appreciably over the longitudinal axis of a reservoir as particles suspended in river inflows settle. Also, transparency along wind-swept shorelines can be erroneously high due to local suspension of sediments; readings should be taken well away from shore.

Other water quality attributes, such as pH, specific conductance, biological oxygen demand, alkalinity, color, chlorophyll-*a* concentration, and concentrations of various metals, ions, and nutrients, can be useful in assessing lake or reservoir conditions and problems. Selection of parameters to measure requires careful consideration of the factors affecting the lake or reservoir in question. Consult Wetzel (1983), Thornton et al. (1990), Hayes et al. (1993), and Summerfelt (1993) for insights on relevant attributes to measure. Accepted methods for measuring water quality attributes are described in *Standard Methods* (APHA 1989) and other texts (USEPA 1979; Wetzel and Likens 1991).

4.5.3 Hydrodynamic Variables

Several variables are commonly used to describe the rates at which water moves through a lake or reservoir. High rates of water transport through a reservoir are indicative of systems that resemble rivers, whereas slow passage can cause a reservoir to possess lakelike qualities (Thornton et al. 1990). Recruitment of some species is often reduced in reservoirs with high water transport rates (Walburg 1971; Willis and Stephen 1987), likely because young fish are lost through dam discharges. The storage ratio is the ratio of the average annual volume of the body of water to its annual discharge volume. A reservoir that discharges twice its average volume in a year would have a storage ratio of 0.5. The flushing rate is the proportion of the reservoir volume discharged per unit time and is calculated as discharge divided by volume. It is inversely related to the storage ratio and can be computed on an annual, seasonal, weekly, or daily basis. The annual flushing rate for the reservoir described above would be 2.0. Turnover time is expressed as the number of days required to discharge a volume equivalent to a reservoir's volume; it can be calculated as the storage ratio multiplied by 365. Bear in mind that these terms are often confused. Water discharge rates into or out of reservoirs and lakes are measured as described in section 4.4.3.5.

In some main-stem impoundments, current velocity can be appreciable. Current velocity can be measured with current meters available from limnological and oceanographic supply houses, but care must be taken to acquire equipment with adequate sensitivity; current velocities in reservoirs can be too low to be measured with typical stream-sampling devices. As with most reservoir measurements, current velocity can be expected to vary vertically, horizontally, and longitudinally.

Wave heights may be important because of their relation to shoreline erosion. Wave height is measured as the vertical distance from the crest of a wave to the trough between waves (Wetzel 1983). Small waves can be measured manually with a staff gauge. Permanently installed pressure-sensitive wave gauges are used to record continuous measurements of waves within a wide range of heights. Similarly, water levels can be measured manually with a permanently fixed staff gauge or with continuous recorders. Most reservoirs have continuous water level recorders installed.

4.5.4 Substrate, Vegetation, and Cover Variables

Substrate, vegetation, and other cover types (e.g., flooded timber, crevices, and submerged cliffs) can influence the suitability of a lake or reservoir for habitation by certain species because of specific habitat requirements, especially for spawning (Balon 1975). For example, smallmouth bass reproductive success is enhanced by the presence of gravel interspersed with stumps or boulders in littoral areas whereas spawning northern pike require submerged vegetation. Striped bass have been successfully introduced into reservoirs with extensive open-water areas devoid of cover; an impoundment largely covered by floating and emergent vegetation would likely not be suitable for this species.

In clear, shallow lakes or in littoral areas, substrate can be assessed using techniques described for streams in section 4.4.3.6. At moderate depths (<6 m), substrate can be classified into broad size-classes by probing with a metal rod

(Hamilton and Bergersen 1984). Electrical conduit 2.5 cm in diameter works well. This method requires some practice over known substrates and should be assessed for accuracy occasionally by diving or with grab samplers. In deep water, substrates must be brought to the surface with Ekman, Petersen, or Ponar grab samplers for analysis (Lind 1979; Wetzel and Likens 1991); the Ekman is best suited for soft or fine sediments. Dredge and grab samples also allow collection of aquatic plants (Sliger et al. 1990) and benthic organisms (Chapter 11). Electronic echo sounders can be used to infer the relative hardness (e.g., muck versus bedrock) of substrates in a given body of water because reflected signal intensity is proportional to substrate hardness. Echo sounders are also useful for detecting and mapping cover and vegetation in deep water (Maceina and Shireman 1980; Maceina et al. 1984). In shallow waters, aerial photography can effectively record the distribution of dense aquatic macrophytes (Edwards and Brown 1960), but low densities can be mapped from boat or shore; vegetation mapping procedures are described by Wright et al. (1981). Sampling vegetative biomass requires removal of vegetation from a quadrat surrounded by a box or net (Hiley et al. 1981).

4.5.5 Trophic State

Characterization of a lake or reservoir by its trophic state can be a useful way of providing a general description of its biological qualities and likely fishery characteristics. Nutrient concentrations in oligotrophic lakes are low as are biological production rates. High nutrient concentrations (potentially resulting from human activities in the watershed) and high production rates characterize eutrophic bodies of water. Intermediate lakes are termed mesotrophic. Carlson (1977) described a simple method for evaluating trophic state by using total phosphorus concentration, chlorophyll-*a* concentration, or Secchi disk depth as indicators. High total phosphorus or chlorophyll-*a* concentrations or low Secchi disk depths are indicative of eutrophic conditions. However, turbidity resulting from suspended inorganic particles such as clay will bias results obtained with the Secchi disk method. Shannon and Brezonik (1972) and Porcella et al. (1980) formulated more rigorous indices, but these require collection of additional physicochemical attributes (e.g., primary productivity, nitrogen concentration, specific conductance, morphometric measurements, and macrophyte development).

Trophic state, and the various measurements used as indicators of it, have been used to predict fish production rates, biomasses, and yields. The most widely known estimator of yield is the morphoedaphic index (Ryder 1965; Jenkins 1982), which is calculated as a ratio of total dissolved solids to mean depth. It is based on the observation that among a range of lakes those with higher total concentrations of dissolved solids and shallower mean depths will tend to have higher fish production rates; both of these variables are indicative of eutrophication. Other variables that have been correlated with fish production values include total phosphorus concentration, annual primary production rate, chlorophyll-*a* concentration, summer phytoplankton biomass, macrophyte biomass, Secchi disk depth, and biomass of macrobenthos (Oglesby 1977; Matuszek 1978; Hanson and Leggett 1982; Jones and Hoyer 1982; Durocher et al. 1984; Wiley et al. 1984; Oglesby et al. 1987; Rieman and Myers 1992). Comparison of such values across a range of lakes or reservoirs provides a general idea of how those water bodies may be expected to rank in fish production rates, biomasses, and yields. A useful indicator of fish size distribution in a lake or reservoir is the size structure of sympatric zooplankton (Mills and

Schiavone 1982; Mills et al. 1987). Large zooplankton body sizes are often correlated with large fish body sizes. Zooplankton can be sampled much more easily than can fish (Chapter 11) to provide a quick indication of likely fish assemblage size structure.

4.6 REFERENCES

Angermeier, P. L., and J. R. Karr. 1984. Relationships between woody debris and fish habitat in a small warmwater stream. Transactions of the American Fisheries Society 113:716–726.

APHA (American Public Health Association), American Water Works Association, and Water Pollution Control Federation. 1989. Standard methods for the examination of water and wastewater, 17th edition. APHA, Washington, DC.

Bain, M. B., J. T. Finn, and H. E. Booke. 1985. Quantifying stream substrate for habitat analysis studies. North American Journal of Fisheries Management 5:499–506.

Balon, E. K. 1975. Reproductive guilds of fishes: a proposal and definition. Journal of the Fisheries Research Board of Canada 32:821–864.

Benke, A. C., R. L. Henry, III, D. M. Gillespie, and R. J. Hunter. 1985. Importance of snag habitat for animal production in Southeastern streams. Fisheries 10(5):8–13.

Binns, N. A., and F. M. Eiserman. 1979. Quantification of fluvial trout habitat in Wyoming. Transactions of the American Fisheries Society 108:215–228.

Bisson, P. A., J. L. Nielsen, R. A. Palmason, and L. E. Grove. 1982. A system of naming habitat types in small streams, with examples of habitat utilization by salmonids during low streamflows. Pages 62–73 *in* N. B. Armantrout, editor. Acquisition and utilization of aquatic habitat inventory information. American Fisheries Society, Western Division, Bethesda, Maryland.

Bisson, P. A., K. Sullivan, and J. L. Nielsen. 1988. Channel hydraulics, habitat use, and body form of juvenile coho salmon, steelhead trout, and cutthroat trout in streams. Transactions of the American Fisheries Society 117:262–273.

Bozek, M. A., and W. A. Hubert. 1992. Segregation of resident trout in streams as predicted by three habitat dimensions. Canadian Journal of Zoology 70:886–890.

Bozek, M. A., and F. J. Rahel. 1991. Assessing habitat requirements of young Colorado River cutthroat trout by use of macrohabitat and microhabitat analyses. Transactions of the American Fisheries Society 120:571–581.

Bryan, C. F., and D. A. Rutherford, editors. 1993. Impacts on warmwater streams: guidelines for evaluation. American Fisheries Society, Southern Division, Bethesda, Maryland.

Carlson, R. E. 1977. A trophic state index for lakes. Limnology and Oceanography 22:361–369.

Chapman, D. G. 1988. Critical review of variables used to define effects of fines in redds of large salmonids. Transactions of the American Fisheries Society 117:1–21.

Coutant, C. C. 1977. Compilation of temperature preference data. Journal of the Fisheries Research Board of Canada 34:739–745.

Cummins, K. W. 1962. An evaluation of some techniques for the collection and analysis of benthic samples with special emphasis on lotic waters. American Midland Naturalist 67:477–504.

Dolloff, C. A., D. G. Hankin, and G. H. Reeves. 1993. Basinwide estimation of habitat and fish populations in streams. U.S. Forest Service General Technical Report SE-83.

Durocher, P. P., W. C. Provine, and J. E. Kraai. 1984. Relationship between abundance of largemouth bass and submerged vegetation in Texas reservoirs. North American Journal of Fisheries Management 4:84–88.

Edwards, R. W., and M. W. Brown. 1960. An aerial photographic method for studying the distribution of aquatic macrophytes in shallow waters. Journal of Ecology 48:161–164.

Everest, F. H., C. E. McLemore, and J. F. Ward. 1980. An improved tri-tube cryogenic gravel sampler. U.S. Forest Service Research Note PNW-350.

Fausch, K. D., C. L. Hawkes, and M. G. Parsons. 1988. Models that predict standing crop of stream fish from habitat variables: 1950–85. U.S. Forest Service General Technical Report PNW-213.

Fausch, K. D., J. R. Karr, and P. R. Yant. 1984. Regional application of an index of biotic integrity based on stream fish communities. Transactions of the American Fisheries Society 113:39–55.

Fausch, K. D., and T. G. Northcote. 1992. Large woody debris and salmonid habitat in a small coastal British Columbia stream. Canadian Journal of Fisheries and Aquatic Sciences 49:682–693.

Fedora, M. A. 1987. Simulation of storm runoff in the Oregon Coast Range. U.S. Bureau of Land Management Technical Note 378.

Frissell, C. A., W. J. Liss, C. E. Warren, and M. D. Hurley. 1986. A hierarchical framework for stream habitat classification: viewing streams in a watershed context. Environmental Management 10:199–214.

Frissell, C. A., and R. K. Nawa. 1992. Incidence and causes of physical failure of artificial habitat structures in streams of western Oregon and Washington. North American Journal of Fisheries Management 12:182–197.

Fry, F. E. J. 1971. The effect of environmental factors on the physiology of fish. Pages 1–98 *in* W. S. Hoar and D. J. Randall, editors. Fish physiology, volume 6: environmental relations and behavior. Academic Press, New York.

Gordon, N. D., T. A. McMahon, and B. L. Finlayson. 1992. Stream hydrology: an introduction for ecologists. Wiley, New York.

Gore, J. A., and J. M. Nestler. 1988. Instream flow studies in perspective. Regulated Rivers Research & Management 2:93–101.

Gorman, O. T., and J. R. Karr. 1978. Habitat structure and stream fish communities. Ecology 59:507–515.

Grant, G. E. 1988. The RAPID technique: a new method for evaluating downstream effects of forest practices on riparian zones. U.S. Forest Service General Technical Report PNW-220.

Grost, R. T., W. A. Hubert, and T. A. Wesche. 1991. Field comparison of three devices used to sample substrate in small streams. North American Journal of Fisheries Management 11:347–351.

Hamilton, K., and E. P. Bergersen. 1984. Methods to estimate aquatic habitat variables. U.S. Bureau of Reclamation, Engineering and Research Center, Denver, Colorado.

Hankin, D. G. 1984. Multistage sampling designs in fisheries research: applications in small streams. Canadian Journal of Fisheries and Aquatic Sciences 41:1575–1591.

Hankin, D. G., and G. H. Reeves. 1988. Estimating total fish abundance and total habitat area in small streams based on visual estimation methods. Canadian Journal of Fisheries and Aquatic Sciences 45:834–844.

Hanson, J. M., and W. C. Leggett. 1982. Empirical prediction of fish biomass and yield. Canadian Journal of Fisheries and Aquatic Sciences 39:257–263.

Hawkins, C. P., and 10 coauthors. 1993. A hierarchical approach to classifying stream habitat features. Fisheries 18(6):3–12.

Hayes, D. B., W. W. Taylor, and E. L. Mills. 1993. Natural lakes and large impoundments. Pages 493–515 *in* Kohler and Hubert (1993).

Helm, W. T., editor. 1985. Aquatic habitat inventory: glossary of stream habitat terms. American Fisheries Society, Western Division, Habitat Inventory Committee, Bethesda, Maryland.

Hicks, B. J., J. D. Hall, P. A. Bisson, and J. R. Sedell. 1991. Responses of salmonids to habitat changes. American Fisheries Society Special Publication 19:483–518.

Hiley, P. D., J. F. Wright, and A. D. Berrie. 1981. A new sampler for stream benthos, epiphytic macrofauna and aquatic macrophytes. Freshwater Biology 11:79–85.

Hubbard, W. D., D. C. Jackson, and D. J. Ebert. 1993. Channelization. Pages 135–155 *in* Bryan and Rutherford (1993).

Hughes, R. M., and J. M. Omernik. 1981. Use and misuse of the terms watershed and stream order. Pages 320–326 *in* L. A. Krumholz, editor. The warmwater streams symposium. American Fisheries Society, Southern Division, Bethesda, Maryland.

Hynes, H. B. N. 1970. The ecology of running waters. University of Toronto Press, Toronto.

Jenkins, R. M. 1982. The morphoedaphic index and reservoir fish production. Transactions of the American Fisheries Society 111:133–140.

Jones, J. R., and M. V. Hoyer. 1982. Sportfish harvest predicted by summer chlorophyll-*a* concentration in midwestern lakes and reservoirs. Transactions of the American Fisheries Society 111:176–179.

Kershner, J. L., H. L. Forsgren, and W. R. Meehan. 1991. Managing salmonid habitats. American Fisheries Society Special Publication 19:599–606.

Kinsolving, A. D., and M. B. Bain. 1990. A new method for measuring cover in fish habitat studies. Journal of Freshwater Ecology 5:373–378.

Kohler, C. C., and W. A. Hubert, editors. 1993. Inland fisheries management in North America. American Fisheries Society, Bethesda, Maryland.

Kondolf, G. M., and S. Li. 1992. The pebble count technique for quantifying surface bed material size in instream flow studies. Rivers 3:80–87.

Lanka, R. P., W. A. Hubert, and T. A. Wesche. 1987. Relations of geomorphology to stream habitat and trout standing stock in small Rocky Mountain streams. Transactions of the American Fisheries Society 116:21–28.

Lind, O. T. 1979. Handbook of common methods in limnology, 2nd edition. C. V. Mosby, St. Louis, Missouri.

Lobb, M. D., III, and D. J. Orth. 1991. Habitat use by an assemblage of fish in a large warmwater stream. Transactions of the American Fisheries Society 120:65–78.

Lyons, J. K., and R. L. Beschta. 1983. Land use, floods, and channel changes: Upper Middle Fork Willamette River, Oregon (1936–1980). Water Resources Research 19:463–471.

MacDonald, L. H., A. W. Smart, and R. C. Wissmar. 1991. Monitoring guidelines to evaluate effects of forestry activities on streams in the Pacific Northwest and Alaska. U.S. Environmental Protection Agency EPA-910-9-91-001, Seattle.

Maceina, M. J., and J. V. Shireman. 1980. The use of a recording fathometer for determination of distribution and biomass of *Hydrilla*. Journal of Aquatic Plant Management 18:34–39.

Maceina, M. J., J. V. Shireman, K. A. Langeland, and D. E. Canfield, Jr. 1984. Prediction of submersed plant biomass by use of a recording fathometer. Journal of Aquatic Plant Management 22:35–38.

Matuszek, J. E. 1978. Empirical predictions of fish yields of large North American lakes. Transactions of the American Fisheries Society 107:385–394.

McDaniel, M. D. 1993. Point–source discharges. Pages 1–56 *in* Bryan and Rutherford (1993).

McMahon, T. E., and G. F. Hartman. 1989. Influence of cover complexity and current velocity on winter habitat use by juvenile coho salmon (*Oncorhynchus kisutch*). Canadian Journal of Fisheries and Aquatic Sciences 46:1551–1557.

Milhous, R. T., D. L. Wegner, and T. Waddle. 1981. User's guide to physical habitat simulation system. U.S. Fish and Wildlife Service Research Report FWS/OBS-81/43.

Mills, E. L., D. M. Green, and A. Schiavone, Jr. 1987. Use of zooplankton size to assess the community structure of fish populations in freshwater lakes. North American Journal of Fisheries Management 7:369–378.

Mills, E. L., and A. Schiavone, Jr. 1982. Evaluation of fish communities through assessment of zooplankton populations and measures of lake productivity. North American Journal of Fisheries Management 2:14–27.

Milner, N. J., R. J. Hemsworth, and B. E. Jones. 1985. Habitat evaluation as a fisheries management tool. Journal of Fish Biology 27(A):85–108.

Murphy, M. L., and K. V. Koski. 1989. Input and depletion of woody debris in Alaska streams and implications for streamside management. North American Journal of Fisheries Management 9:427–436.

Myers, T. J., and S. Swanson. 1992. Variation of stream stability with stream type and livestock bank damage in northern Nevada. Water Resources Bulletin 28:743–754.

Nawa, R. K., and C. A. Frissell. 1993. Measuring scour and fill of gravel streambeds with scour chains and sliding-bead monitors. North American Journal of Fisheries Management 13:634–639.

O'Brien, W. J. 1990. Perspectives on fish in reservoir limnology. Pages 209–225 *in* K. W. Thornton, B. L. Kimmel, and F. E. Payne, editors. Reservoir limnology. Wiley, New York.

Oglesby, R. T. 1977. Relationships of fish yield to lake phytoplankton standing crop, production, and morphoedaphic factors. Journal of the Fisheries Research Board of Canada 34:2271–2279.

Oglesby, R. T., J. H. Leach, and J. Forney. 1987. Potential *Stizostedion* yield as a function of chlorophyll concentration with special reference to Lake Erie. Canadian Journal of Fisheries and Aquatic Sciences 44(Supplement 2):166–170.

Olson-Rutz, K. M., and C. B. Marlow. 1992. Analysis and interpretation of stream channel cross-sectional data. North American Journal of Fisheries Management 11:55–61.

Orth, D. J. 1983. Aquatic habitat measurements. Pages 61–84 *in* L. A. Nielsen and D. L. Johnson, editors. Fisheries techniques. American Fisheries Society, Bethesda, Maryland.

Orth, D. J. 1987. Ecological considerations in the development and application of instream flow-habitat models. Regulated Rivers Research & Management 1:171–181.

Orth, D. J., and R. J. White. 1993. Stream habitat management. Pages 205–230 *in* Kohler and Hubert (1993).

Pearsons, T. N., H. W. Li, and G. A. Lamberti. 1992. Influence of habitat complexity on resistance to flooding and resilience of stream fish assemblages. Transactions of the American Fisheries Society 121:427–436.

Platts, W. S. 1979. Relationships among stream order, fish populations, and aquatic geomorphology in an Idaho river drainage. Fisheries 4(2):5–9.

Platts, W. S. 1991. Livestock grazing. American Fisheries Society Special Publication 19:389–423.

Platts, W. S., W. F. Megahan, and G. W. Minshall. 1983. Methods for evaluating stream, riparian, and biotic conditions. U.S. Forest Service General Technical Report INT-138.

Platts, W. S., R. J. Torquemada, M. L. McHenry, and C. K. Graham. 1989. Changes in salmon spawning and rearing habitat from increased delivery of fine sediment to the South Fork Salmon River, Idaho. Transactions of the American Fisheries Society 118:274–283.

Platts, W. S., and 12 coauthors. 1987. Methods for evaluating riparian habitats with applications to management. U.S. Forest Service General Technical Report INT-221.

Porcella, D. B., S. A. Peterson, and D. P. Larsen. 1980. Index to evaluate lake restoration. Proceedings of the American Society of Civil Engineers, Journal of the Environmental Engineering Division 106:1151–1169.

Rabeni, C. F., and R. B. Jacobson. 1993. The importance of fluvial hydraulics to fish–habitat restoration in low-gradient alluvial streams. Freshwater Biology 29:211–220.

Ralph, S. C., G. C. Poole, L. L. Conquest, and R. J. Naiman. 1994. Stream channel morphology and woody debris in logged and unlogged basins of western Washington. Canadian Journal of Fisheries and Aquatic Sciences 51:37–51.

Rieman, B. E., and D. L. Myers. 1992. Influence of fish density and relative productivity on growth of kokanee in ten oligotrophic lakes and reservoirs in Idaho. Transactions of the American Fisheries Society 121:178–191.

Rosgen, D. L. 1994. A classification of natural rivers. Catena 22:169–199.

Ryder, R. A. 1965. A method for estimating the potential fish production of north-temperate lakes. Transactions of the American Fisheries Society 94:214–218.

Shannon, E. E., and P. L. Brezonik. 1972. Eutrophication analysis: a multivariate approach. Proceedings of the American Society of Civil Engineers, Journal of the Sanitary Engineering Division 98:37–57.

Sharp, K. V., and W. L. Rapp. 1984. Operating manual for current meter digitizer (CMD). U.S. Geological Survey, Hydrologic Instrumentation Facility Report 6-84-03, Vicksburg, Mississippi.

Sheehan, R. J., and J. L. Rasmussen. 1993. Large rivers. Pages 445–468 *in* Kohler and Hubert (1993).

Shirvell, C. S., and R. G. Dungey. 1983. Microhabitats chosen by brown trout for feeding and spawning in rivers. Transactions of the American Fisheries Society 112:355–367.

Shuler, S. W., and R. B. Nehring. 1994. Using the physical habitat simulation model to evaluate a stream habitat enhancement project. Rivers 4:175–193.

Shuler, S. W., R. B. Nehring, and K. D. Fausch. 1994. Diel habitat selection by brown trout in the Rio Grande River, Colorado, after placement of boulder structures. North American Journal of Fisheries Management 14:99–111.

Simonson, T. D., J. Lyons, and P. D. Kanehl. 1994. Guidelines for evaluating fish habitat in Wisconsin streams. U.S. Forest Service General Technical Report NC-164.

Sliger, W. A., J. W. Henson, and R. C. Shadden. 1990. A quantitative sampler for biomass estimates of aquatic macrophytes. Journal of Aquatic Plant Management 28:100–102.

Sowa, S. P., and C. F. Rabeni. 1995. Regional evaluation of the relation of habitat to distribution and abundance of smallmouth bass and largemouth bass in Missouri streams. Transactions of the American Fisheries Society 124:240–251.

Speaker, R. W., K. J. Luchessa, J. F. Franklin, and S. V. Gregory. 1988. The use of plastic strips to measure leaf retention by riparian vegetation in a coastal Oregon stream. American Midland Naturalist 120:22–31.

Strahler, A. N. 1957. Quantitative analysis of watershed geomorphology. Transactions of the American Geophysical Union 38:913–920.

Summerfelt, R. C. 1993. Lake and reservoir habitat management. Pages 231–261 *in* Kohler and Hubert (1993).

Terrell, J. W., editor. 1984. Proceedings of a workshop on fish habitat suitability index models. U.S. Fish and Wildlife Service Biological Report 85(6).

Terrell, J. W., T. E. McMahon, P. D. Inskip, R. F. Raleigh, and K. L. Williamson. 1982. Habitat suitability index models: Appendix A. Guidelines for riverine and lacustrine applications of fish HSI models with the habitat evaluation procedures. U.S. Fish and Wildlife Service Research Report FWS/OBS-82/10.A.

Thornton, K. W., R. H. Kennedy, J. H. Carroll, W. W. Walker, R. C. Gunkel, and S. Ashby. 1981. Reservoir sedimentation and water quality—a heuristic model. Pages 654–661 *in* H. G. Stefen, editor. Proceedings of the symposium on surface water impoundments. American Society of Civil Engineers, New York.

Thornton, K. W., B. L. Kimmel, and F. E. Payne. 1990. Reservoir limnology. Wiley, New York.

USEPA (United States Environmental Protection Agency). 1979. Methods for chemical analysis of water and wastes. USEPA Office of Research and Development EPA-600/4-79-020, Washington, DC.

Vannote, R. L., G. W. Minshall, K. W. Cummins, J. R. Sedell, and C. E. Cushing. 1980. The river continuum concept. Canadian Journal of Fisheries and Aquatic Sciences 37:130–137.

Walburg, C. H. 1971. Loss of young fish in reservoir discharge and year-class survival, Lewis and Clark Lake, Missouri River. American Fisheries Society Special Publication 8:441–448.

Walling, D. E., and B. W. Webb. 1992. Water quality I. Physical characteristics. Pages 48–72 *in* P. Calow and G. E. Petts, editors. The rivers handbook, volume 1. Blackwell Scientific Publications, Oxford, UK.

Webb, B. W., and D. E. Walling. 1992. Water quality II. Chemical characteristics. Pages 73–100 *in* P. Calow and G. E. Petts, editors. The rivers handbook, volume 1. Blackwell Scientific Publications, Oxford, UK.

Welch, P. S. 1948. Limnological methods. McGraw-Hill, New York.

Wesche, T. A., C. M. Goertler, and C. B. Frye. 1987. Contribution of riparian vegetation to trout cover in small streams. North American Journal of Fisheries Management 7:151–153.

Wetzel, R. G. 1983. Limnology. Saunders, Philadelphia.

Wetzel, R. G., and G. E. Likens. 1991. Limnological analyses, 2nd edition. Springer-Verlag, New York.

Wiley, M. J., R. W. Gorden, S. W. Waite, and T. Powless. 1984. The relationship between aquatic macrophytes and sport fish production in Illinois ponds: a simple model. North American Journal of Fisheries Management 4:111–119.

Willis, D. W., and J. L. Stephen. 1987. Relationships between storage ratio and population density, natural recruitment, and stocking success of walleye in Kansas reservoirs. North American Journal of Fisheries Management 7:279–282.

Wolman, M. G. 1954. A method of sampling coarse river-bed material. Transactions of the American Geophysical Union 35:951–956.

Wright, J. F., P. D. Hiley, S. F. Ham, and A. D. Berrie. 1981. Comparison of three mapping procedures developed for river macrophytes. Freshwater Biology 11:369–380.

Yeager, B. L. 1993. Dams. Pages 57–113 *in* Bryan and Rutherford (1993).

Young, M. K., W. A. Hubert, and T. A. Wesche. 1991. Selection of measures of substrate composition to estimate survival to emergence of salmonids and to detect changes in stream substrates. North American Journal of Fisheries Management 11:339–346.

Zale, A. V., J. D. Wiechman, R. L. Lochmiller, and J. Burroughs. 1990. Limnological conditions associated with summer mortality of striped bass in Keystone Reservoir, Oklahoma. Transactions of the American Fisheries Society 119:72–76.

Chapter 5

Care and Handling of Sampled Organisms

STEVEN W. KELSCH AND BARBARA SHIELDS

5.1 INTRODUCTION

Techniques for care and handling of sampled organisms are not normally the primary subjects of fisheries research, yet most studies involve working with organisms or tissues and, therefore, either rely on such techniques or address them secondarily. The purpose of this chapter is to assemble an overview of various techniques, to discuss general and specific considerations for care and handling of sampled organisms, and to steer interested readers to more detailed references. Many of these techniques are specific to the study objectives and use of the organism or tissue, so we have organized the chapter into two major sections: "Care and Handling of Live Fish," for those fish that will be kept alive temporarily or until release, and "Care and Handling of Specimens and Tissues," for those fish that will be killed and eventually stored or discarded.

Sampled organisms and tissues are valuable and should be handled properly. At minimum, they are worth the effort and expense of collecting them; they are sources of data and represent a cost to the ecosystem from which they were removed. Poor or careless handling techniques can lead to loss of data, increased cost of sampling, and less definitive research. Care and handling protocols must be clearly defined in advance to minimize handling time, ensure that specimens are in good condition, and avoid the need for resampling.

5.2 CARE AND HANDLING OF LIVE FISH

There are a number of reasons for wanting to keep sampled organisms alive and in good health. Studies concerned with movements, migration, population size, mortality, or behavior require capturing, marking, and releasing live fish that must survive and behave normally after release. Other studies and management practices require the capture of wild broodfish for immediate release after egg removal or retention in hatchery programs. In these situations, it is important to minimize mortality and stress associated with capture, handling, and holding to ensure that the fish survive and tolerate additional stressors that may be uncontrollable or unforeseen. This section is an overview of general strategies and specific techniques for care and handling of live fish.

5.2.1 Experimentation

Fisheries researchers should always be concerned about the treatment of experimental organisms. Several organizations have published guidelines for the care and use of fish in field and laboratory research (NIH 1985; ASIH et al. 1987, 1988; *Animal Behaviour* 1992). In general they suggest that handling and experiments be done according to scientific and ethical principles, including minimization of stress, maintenance of acceptable water quality conditions, and proper use of anesthetics when appropriate. It is recognized that animal care is more difficult in field than in laboratory research and that wild fish have a variety of requirements; therefore, all procedures should be overseen by a trained professional. Many institutions now have committees that oversee the use of animals in laboratory research. Be sure to review guidelines (Appendix 5.1) and obtain required approval from animal care committees prior to conducting research.

5.2.2 Endangered Species

Many aquatic organisms are of special concern because of their limited distributions, declining numbers, or threatened habitats (Meffe 1986; Minckley and Deacon 1991). Carlson and Muth (1993) presented a thorough coverage of endangered species management, including reasons for protection and applicable laws and treaties. It is necessary, however, to continue working with endangered species to provide a sound basis for recovery efforts and to further a general understanding of the systems in which they live (Carlson and Muth 1993); nevertheless, special care must be taken with endangered species to minimize mortality and stress. Field researchers have a responsibility to be aware of the possible presence of protected species in order to avoid incidental mortality or habitat destruction. Also, sensitive species or populations may be present that have not yet been recognized. Accordingly, preferred capture techniques are those that minimize potential mortality and habitat destruction while meeting the objectives of the study. Johnson (1987) reported that 56% of freshwater fishes in the United States and Canada were receiving some sort of protection from responsible agencies. The current statuses of fishes of special concern were reported by Williams et al. (1989) and Warren and Burr (1994).

5.2.3 Methods for Care and Handling of Fish

The most important concern for the care and handling of live fish is to minimize stress, which reduces both survival and capacities to tolerate other stressors. Factors stressful to fish include handling and changes in environmental variables, such as oxygen, temperature, and salinity. Many studies have examined the stress response for fisheries research and aquaculture (Smith 1982; Adams 1990; Wedemeyer et al. 1990). The stress response leads to reduced disease resistance (Wedemeyer et al. 1990), reduced capacity for activity (Schreck 1990; Wedemeyer et al. 1990), increased oxygen consumption (Andrews and Matsuda 1975), increased permeability of membranes that results in osmoregulatory problems (Smith 1982; Wedemeyer et al. 1990), decreased growth (Wedemeyer et al. 1990), decreased reproductive capacity (Wedemeyer et al. 1990), and increased mortality (Smith 1982; Wedemeyer et al. 1990; Fitzsimons 1994). In addition, the effects of stressors are cumulative (Robertson et al. 1987; Wedemeyer et al. 1990). Many activities associated with fisheries research are stressful, including capture (Hatting and van Pletzen 1974; Harrell and Moline 1992), handling (Wedemeyer 1972; Robertson et al. 1987), confinement

(Carmichael et al. 1984; Young and Cech 1993), hauling (Carmichael et al. 1983; Robertson et al. 1987), and physical shock (Fitzsimons 1994).

5.2.3.1 Overview of Strategies

Techniques for the care and handling of live fish are numerous because they have been developed for a variety of study objectives and species requirements. However, several fundamental strategies are common and should apply even to species with unknown environmental requirements: (1) minimize stress; (2) if a change in conditions is to be made, avoid changing any variable in a direction away from the optimum unless suggested by other research; (3) when requirements or optima are unknown, avoid changing conditions from those present in the system from which the organisms were taken; and (4) allow time for acclimation (metabolic adaptation) by making changes gradually when significant changes in environment are required.

Each of these strategies has a metabolic basis: under any set of environmental conditions, an organism pays a certain energetic cost to stay alive (standard metabolism) and it has a maximum amount of energy that it can produce (active metabolism). The difference between the two is the amount of aerobic energy an organism has available for activities such as growth, reproduction, swimming, or surviving stressors. This available energy is called an organism's "scope for activity" (Fry 1947, 1971). It is intuitive that an organism would be most fit if it maximized its scope for activity by selecting appropriate environmental conditions. No studies have tested this hypothesis with regard to integrated environmental variables, but some have shown that fishes select temperatures for maximum scope when other variables are held constant (Fry and Hart 1948; Sullivan 1954; Kelsch and Neill 1990).

The relevance of the four care strategies listed above to scope for activity is apparent from Figure 5.1. This model applies to any environmental variable or combination thereof. Figure 5.1 illustrates the hypothetical effect of a variable (such as temperature or salinity) on scope for activity for a case in which both extremes are deleterious and an optimum exists at some intermediate level—the mode of the scope curve. Hypothetically, a fish cannot survive indefinitely under conditions in which it has no scope. In the (unlikely) absence of any other stressor, a fish typically tolerates a sizable range of an environmental variable. When the fish is otherwise stressed, however, its zone of tolerance for that (or any) variable contracts. Energy a fish uses to cope with stresses is unavailable for other activities (Fry 1947, 1971; Priede 1985; Wedemeyer et al. 1990). Barton and Schreck (1987) reported that even a mild physical stress consumed as much as 25% of available scope in juvenile steelhead. In Figure 5.1, stress is shown to be a function of the environmental variable as it would be with temperature—cold temperatures reduce the susceptibility of poikilothermic fish to a given stressor. The effects of several sublethal stressors can lead to death (Wedemeyer et al. 1990). Thus, strategy 1 above (minimize stress) is paramount.

The effects of moving a fish farther from optimum environmental conditions (counter to strategy 2) are also evident in Figure 5.1. Under any environmental conditions, the energy available for activity and for tolerating additional stressors is proportional to the available scope for activity. The environmental condition that gives the maximum available scope (maximum distance between the scope curve and the stress line) represents an optimum. Moving an organism farther from an optimum in either direction, such as by moving it toward an extreme temperature, reduces available energy and makes the organism more susceptible to additional

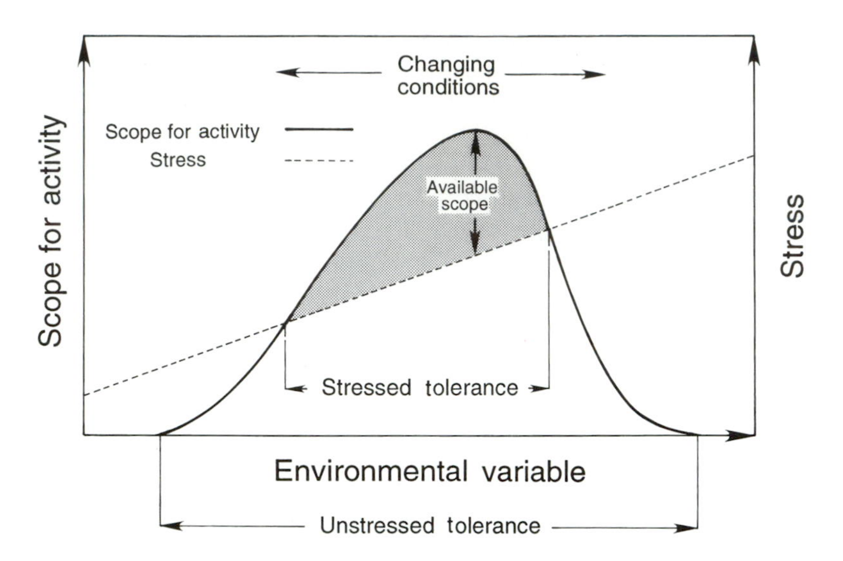

Figure 5.1 Conceptual model showing effects of stress and changing environmental conditions on scope for activity and potential for survival in fish. "Stressed" and "unstressed" tolerance refers to tolerance of the environmental variable with and without an additional stressor.

stressors. If optimum conditions are unknown, the conservative approach is to avoid changing holding conditions from those in the environment from which the organism was collected (because the organism was able to survive those conditions; see strategy 3). However, researchers often need to change the holding conditions from those in which the fish was collected. Substantial changes should be made gradually, allowing time for acclimation to the new conditions (strategy 4). Seasonal changes in environmental variables are slow enough to allow time for acclimation; however, they are too slow to be practical for use in holding fish. The ability to acclimate and rates of acclimation vary among species (Fry 1971). It is generally assumed that fish can acclimate to temperature changes of 1°C per day or faster.

Acclimation requires a series of adaptational changes, some that are rapid, such as hormone secretion and activity and others, such as protein induction or changes in phospholipid concentrations of cell membranes, that may take days or weeks (Hochachka and Somero 1984; Prosser 1986). The stress response—a series of reactions to an environmental change or stressor—begins within seconds, may last for days or weeks (Smith 1982), and is the beginning of adjustments that lead to acclimation. For a thorough discussion of acclimation see Fry (1971) and Hochachka and Somero (1971). In essence, the process of acclimation may lead to a shifting of the scope curve on the axis of the environmental variable (Figure 5.1) resulting in a shift in the optimum and in both high and low levels of tolerance (Kelsch and Neill 1990).

5.2.3.2 Capture

Mortality rates and stress vary substantially with types of gear and techniques used to capture aquatic organisms. Gears that capture fish by entanglement, such as gill or trammel nets (see Chapter 6), cause mortality, physical trauma, and stress during capture and removal. Gears that are fished for longer periods also increase the stress of capture. When live fish are required, high-stress gears should be avoided or fished for shorter duration. For example, gill nets may be effective in the capture of viable fish when nets are checked at hourly intervals or less.

5.2.3.3 Handling

The time spent handling live fish should be kept to an absolute minimum. Studies have shown that all aspects of handling, such as dipnetting, time out of water, and data collection, are highly stressful (Merrick 1990; Thomas and Robertson 1991) and can lead to immediate or delayed mortality (Piper et al. 1982; Stickney and Kohler 1990). Many techniques are stressful by their nature, such as surgical implantation of transmitters (Chapter 18), tissue removal (Harvey et al. 1984; Gold et al. 1991; Wooster et al. 1993), tagging or marking (Parker et al. 1990; LaJeone and Bergerhouse 1991; Chapter 12). In these cases, certain procedures that mitigate the effects of stress should be employed.

Two characteristics of fish, mucous coating on skin and tendency to thrash about when held, make fish difficult to handle and susceptible to injury. When fish thrash about, they are often dropped or squeezed, and both can cause injury. Gloves can greatly improve a handler's grip, and small nets (dip nets or seines) can be used to hold fish safely to reduce the likelihood of injury. These precautions also minimize the possibility of injury to the handler. Sharp spines present on bullhead catfishes, common carp, and spiny-rayed fishes such as centrarchids and percids can easily penetrate wet hands and lead to painful injuries. Siluriform catfishes generally have venom glands in the skin covering dorsal and pectoral spines, and they thrash about with locked spines (Bond 1979). Bullhead catfishes have venom that is painful but not dangerous; of this family, the madtoms have the most painful venom.

Humans have a small risk of contracting disease agents when handling fish. Most so-called zoonotic diseases involve consumption of infected fish, but some infections occur through skin cuts, punctures, or abrasions. In general, the risk of serious infection is low; however, it is important to be aware of the possibility. Bacterial infections pose the greatest risk because they are most readily contracted through damaged skin. Nemetz and Shotts (1993) presented a review of zoonotic diseases, modes of transmission, symptoms, and risks of contraction.

5.2.3.4 Anesthetics

A variety of techniques have been used to anesthetize fish to mitigate the effects of stress. Numerous chemical anesthetics (Piper et al. 1982; Summerfelt and Smith 1990; Brown 1993), electricity (Gunstrom and Bethers 1985; Kolz 1989), carbon dioxide (Post 1979), sodium bicarbonate (Booke et al. 1978; Summerfelt and Smith 1990), and cold (Summerfelt and Smith 1990) have anesthetic effects on fish. Chemical anesthetics generally have an effect that increases with concentration and duration of exposure. Progressive effects of anesthesia include tranquilization, loss of response to external stimuli, loss of equilibrium, cessation of ventilation, and death (Summerfelt and Smith 1990). Properly used, anesthetics can greatly reduce mortality and stress; however, they are inherently stressful and should be used only to mitigate the effects of a greater stressor. Summerfelt and Smith (1990) provided a detailed discussion of anesthesia and anesthetics including the characteristics of those most commonly used. Since 1986 only Finquel and MS-222 (both trademarked formulations of tricaine methanesulfonate) have been registered for use with food fishes by the U.S. Food and Drug Administration, although carbon dioxide and sodium bicarbonate may be used because they are generally recognized as safe (Schnick et al. 1986, 1989; Summerfelt and Smith 1990). Tricaine methanesulfonate is an effective anesthetic; however, a 21-d withdrawal period is required before fish can be used for food or released into a natural system. Several unregistered anesthetics such as quinaldine are effective and apparently may be used, but only when fish will ultimately be buried or incinerated. Studies are being made of the safety and effectiveness of other anesthetics such as benzocaine (Gilderhus et al. 1991) and two related chemicals, metomidate and etomidate (Thomas and Robertson 1991; Ross et al. 1993).

5.2.3.5 Prophylactic Treatments

Prophylactic treatments may be used to minimize the likelihood of infection from bacteria, parasites, fungi, or viruses. Fish have a much greater susceptibility to infection when under stress. Numerous drugs and chemicals have been used to prevent infections, to treat existing infections, and to control unwanted organisms in aquatic systems (herbicides and algacides). These treatments are commonly administered in feeds, by dipping fish in baths containing the treatment, or by adding the treatment to the water. Effectiveness varies according to concentration and duration of exposure. Many treatments are themselves stressful, so one should avoid unnecessary overexposure. Both the U.S. Environmental Protection Agency and the U.S. Food and Drug Administration have responsibility for regulating the use of drugs and chemicals for fishery use. Various authors have presented specific information on drugs or chemicals used for fish with regard to regulations (Schnick et al. 1986, 1989; Mitchell 1991); structure, uses, and dosages (Herwig 1979; Summerfelt and Smith 1990); diagnosis and treatment of diseases, parasites, fungi, and viruses (Herwig 1979; Stickney and Kohler 1990; Stoskopf 1993a); and surgical techniques (Summerfelt and Smith 1990; Stoskopf 1993b).

5.2.3.6 Pollutants

Care should be taken to ensure that fish are not exposed unnecessarily to pollutants that might be present in holding or hauling tanks, in buckets used for transfer, on dip nets, or elsewhere. Possible pollutants include chlorine in city water supplies, detergents used in cleaning, petroleum hydrocarbons from gas or oil, and ammonia remaining from fish previously held in the system. Heath (1987) discussed the effects of these pollutants and others on fish. Pollutants add to the stress response in proportion to their concentration and exposure duration.

5.2.3.7 Tagging and Marking

Special care must be taken when fish are tagged or marked for release. Some population analyses, such as abundance and mortality, require the release and recapture of marked fish for parameter estimation (Ricker 1975; Brunham et al. 1987; Van Den Avyle 1993). Assumptions for these analyses are that marked fish do not have a higher mortality or behave in a fashion that would increase or decrease their likelihood of capture relative to that of the population being sampled. Because the fish are released after marking, delayed mortality can affect the results of the analysis. Lambert (1982) reported that Atlantic mackerel experienced high mortality after surviving 4 d from their handling. An alternative method of handling involving less stressful gear and an anesthetic resulted in high survival. Scale loss during handling apparently was related to the mortality. Similar observations have been made for other species (Smith 1982). Saltwater fishes lost up to 20% of their body weight and freshwater fishes gained weight after scale loss, suggesting that scale loss, either directly or through associated stress, results in severe osmoregulatory problems. This finding has particular implications for handling species with easily lost cycloid scales, such as minnows (Cyprinidae), shads (Clupeidae), and smelts (Osmeridae).

5.2.3.8 Genetic Studies

Care must also be taken when tissues are removed from live fish for genetic analyses. Often fish are valuable, and it is important to keep them alive, such as when the genetic purity of potential broodstock is tested (Harvey et al. 1984; Carmichael et al. 1986; Morizot et al. 1990; Carmichael et al. 1992b). Tissue removal is generally stressful to fish due to time out of water and trauma of surgery. If fish are to be handled for more than a brief period, anesthetics should be used; also be aware that some species such as shads and anchovies (Engraulididae) are much more sensitive to handling than others.

5.2.3.9 Holding and Hauling

Fish are held and transported for a variety of reasons, including aquaculture (Stickney and Kohler 1990), hatchery stocking programs (Heidinger 1993), endangered species management (Carlson and Muth 1993), and research. Numerous studies have shown that holding and hauling cause stress (Robertson et al. 1987; Mazur and Iwama 1993). Accordingly, many techniques have been developed to hold and transfer fish while minimizing mortality and stress. Most of these techniques involve either minimizing the stressor or using a method that mitigates the effects of the stressor. Common stressors associated with holding and hauling fish are low dissolved oxygen concentrations, extreme temperatures, rapid temperature changes, diseases, parasites, viruses, fungi, intense light, and physical shock. Methods of mitigating the effects of stress include use of anesthetics, starvation prior to transport, minimization of crowding, use of baffles in tanks to minimize water sloshing, increased salinity for freshwater fishes or decreased salinity for marine fishes to reduce osmotic costs, and use of cool temperatures.

Winkler (1987) found that devices that decrease water slosh during transport aided in reducing stress. Temperature tolerance of western mosquitofish was not reduced as much when fish were hauled in tanks designed to minimize slosh. Baffles of various types are commonly used in larger transport tanks to reduce slosh. These also aid in the handling and braking of transport vehicles.

Numerous studies have evaluated the effectiveness of various transport and holding systems. Size of the system, density of the fish (numbers per unit volume), and duration of hauling were evaluated for freshwater fishes by Johnson (1979) and Collins (1990). Norris et al. (1960) and Carmichael and Tomasso (1988) surveyed methods used in the transportation of fish. Aeration systems were evaluated by Stickney (1979), Carmichael et al. (1992a), and Fries et al. (1993). Kreiberg (1991) described a collapsible portable field tank for holding fish. Equipment and methods were discussed by Jensen (1990a, 1990b, 1990c).

5.2.3.10 Shipping

Smaller fishes have been transported effectively by shipping them in plastic bags filled with water and pure oxygen (Johnson 1979; Piper et al. 1982). This technique is valuable when a few fish are transported over large distances (Hamman 1981). Fish are normally sealed in plastic bags filled with water and oxygen; bags are placed in

Styrofoam coolers, which are sealed in sturdy cardboard boxes and shipped by commercial air. Regulations for transport of live animals by air can be found in IATA (1992). Upon arrival, bags containing fish should be left intact and floated in the destination tank until the fish are ready for release. This minimizes stress by allowing water temperatures to equilibrate. Bags should not be opened or punctured during thermal equilibration because metabolism during transport increases concentrations of ammonia and carbon dioxide; opening the bag allows carbon dioxide to escape, resulting in higher pH and a corresponding shift of ammonium to highly toxic ammonia (see section 5.2.3.14).

5.2.3.11 Oxygen

It is critical to maintain adequate dissolved oxygen concentrations for fish. The stress response of fish to hypoxia was described by Smith (1982). Progressive signs of hypoxia include increased ventilation rate, gulping air at the water surface, loss of equilibrium, and eventually death. Minimum desirable oxygen concentration for most fishes is approximately 5 mg/L (Boyd 1979). Although species exhibit various sensitivities to hypoxia (Doudoroff and Shumway 1970; Davis 1975), concentrations below 5 mg/L are generally stressful and concentrations below 1 mg/L are lethal with prolonged exposure. Risk of hypoxia is great because fish are often held or transported at high densities for economic reasons, and stress causes increased oxygen consumption. All holding and transport systems require some method of maintaining adequate oxygen concentrations. All are designed to increase diffusion of oxygen into water by increasing surface area for diffusion (agitation and bubbling), decreasing diffusion distance (stirring), or creating a greater concentration gradient from gas to water (using pure oxygen or higher pressure). Most commonly, aeration is accomplished by bubbling air (holding facilities) or pure oxygen (transport tanks) through the water. Diffusers and surface agitators are used to increase surface area for diffusion of oxygen. Demand for oxygen in transport systems can be reduced by use of anesthetics during loading (Johnson 1979), starvation prior to transport (Beamish 1964), or reduced crowding.

5.2.3.12 Temperature

Temperature controls the metabolic rate of fish, and substantial changes in temperature require metabolic adaptation (acclimation) for survival (Fry 1971; Bryan et al. 1990; Kelsch and Neill 1990). Fish have thermal optima and upper and lower lethal extremes that vary with species and thermal history. Adverse effects of temperature can be minimized by avoiding changes in temperature or exposure to extreme temperatures. Fish should not be moved to temperatures farther from their thermal optima, particularly to temperatures warmer than their thermal optima. Fish have limits to their tolerance of both high and low temperatures; temperatures that exceed either limit lead to immediate or eventual death. Lesser temperature changes may lead to reduced tolerance of additional stressors. It has been common practice to "temper" fish when temperature changes must be made (Piper et al. 1982). Tempering involves a gradual change from one temperature to the next. Stickney (1979) recommended tempering for any required temperature change exceeding 2°C, and stated that the rate of tempering should not exceed 5°C/h. Tempering does not allow for complete acclimation to the new temperature in the time usually available, so it probably represents some combination of rapid acclimatory adjustments and reduction of stress. Tempering also enables adjustments to changes in other water quality variables such as salinity and other chemical factors (Piper et al. 1982).

Care should be taken to avoid hauling and holding fish at high temperatures. At high temperatures fish are more active and have greater potential for self-inflicted injury, their oxygen consumption is higher, and the saturation level of oxygen is lower, greatly increasing the possibility of hypoxia. If possible, transport or handling should be restricted to cool times of the day, season, or year. It is also possible to use chillers or ice to reduce water temperatures (Stickney 1979).

5.2.3.13 Salinity

Manipulation of salt concentration in transport tanks can reduce stress. Aquatic animals are subject to osmosis and diffusion of compounds between their body fluids and the environment. Concentrations of dissolved substances in body fluids must remain constant for normal physiological functioning and fish must expend energy to regulate these concentrations. Mechanisms of osmoregulation in fishes were discussed by Smith (1982) and Moyle and Cech (1988). Freshwater fishes exhibit reduced stress when transported at increased salt concentrations, and saltwater species show reduced stress when transported at isotonic concentrations. Because stress is known to lead to greater membrane permeability (Smith 1982; Wedemeyer et al. 1990), controlling salt concentration apparently aids in the reduction of osmoregulatory costs. Johnson (1979) reported that the addition of salt to hauling tanks led to a reduction in scale loss, mucus loss, and skin damage in freshwater fishes. Norris et al. (1960) reported that marine species are often hauled at isotonic concentrations. Robertson et al. (1987) found that salt led to reduced stress in transport of red drum, and Harrell and Moline (1992) found that sodium chloride reduced stress in striped bass broodfish.

5.2.3.14 Ammonia

Care must be taken to ensure that ammonia concentrations do not climb to dangerous levels in closed systems. Fish excrete ammonia as a waste product of protein catabolism. Ammonia is highly toxic to fish (Heath 1987) but only becomes a problem when high fish densities lead to increased ammonia concentrations. Tolerance of fish to ammonia varies with its form. Un-ionized ammonia (NH_3) is toxic to fish, but ammonium (NH_4^+) is not. The proportion of these forms is a function of pH and temperature. High pH and to a lesser degree, high temperature result in higher proportions of NH_3. At 24°C and a pH of 7.0 only 0.5% of ammonia is present in the un-ionized form, compared with 34% at a pH of 9.0 (Boyd 1979; Amend et al. 1982). In addition to ammonia (a base), fish produce carbon dioxide (an acid), which to some degree counteracts the effects of ammonia on pH and mitigates its toxicity. McCraren and Millard (1978) discussed techniques for buffering pH in holding waters. The tolerance of species to un-ionized ammonia varies; concentrations of 0.6 to 2.0 mg NH_3/L were lethal to several species during short-term exposures, and reduced growth and gill damage occurred at lower concentrations (Boyd 1979). Meade (1985) presented a review of allowable ammonia concentrations for fishes, and Amend et al. (1982) presented methods to control ammonia concentrations.

5.3 CARE AND HANDLING OF SPECIMENS AND TISSUES

5.3.1 Identification of Fishes

Unknown specimens must be identified to access available information about the species and to disseminate properly research results. Systematists have classified fishes on the basis of morphological similarity and phylogenetic relationships (Lundberg and McDade 1990). Because of the diversity of fishes, identification of unknown specimens is best achieved with a taxonomic key and preserved specimens in a laboratory rather than with live fish in the field; however, care should be made that threatened or endangered species are not indiscriminately collected for later identification. A taxonomic key is a branching set of dichotomous choices that leads to identification of a specimen. By convention, all counts and measurements (with the exception of circumference counts) should be made on the left side of the body, and destructive sampling (incisions or fin or scale removal) should be done on the right side to preserve key characters. Most taxonomic keys include diagrams and descriptions of characters used in the key; for information describing the methods for making proper counts and measurements see Hubbs and Lagler (1964) and Strauss and Bond (1990). Taxonomic keys are usually limited to species of a given region, so use of inappropriate keys may lead unknowingly to incorrect identification. Identification should be verified by comparison to the species description and known distribution. When in doubt about the identity of a specimen, consult a specialist.

For most studies it is important to collect and preserve voucher specimens. Voucher specimens may consist of whole fish, skeletons, tissue samples, or photographs that are archived in a permanent collection and serve as physical evidence documenting the presence of a species. Voucher specimens also may be used for future studies, including range documentation and genetic analysis.

5.3.2 General Considerations for Preserved Specimens and Tissues

Specimens should be prepared quickly after capture to avoid deterioration of specimen quality. Fish either should be kept alive or maintained on ice until they are prepared to minimize color loss and other postmortem changes.

It is critical that specimens and tissues be clearly marked with unique numbers and accompanied by appropriate documentation. Documentation should include information about the collection (date, precise location, and name[s] of collector[s]), specimen (species, sex, length and weight, relative abundance, specimen type, and preservation or fixation method), and habitat (including substrate and water quality variables) and should include curatorial data (accession number and date and cross-reference to additional specimen preparations) (Cross 1962; Hubbs and Lagler 1964).

To avoid loss of data, only durable writing media should be used for marking specimens and recording information. All inks should be tested for fastness and papers for durability before they are used. The following media are durable in most liquid preservatives: alcohol-resistant marking pens, permanent carbon inks, and pencil on high-quality rag paper (100% rag) or waterproof paper.

5.3.3 Whole-Specimen Preparation

Whole specimens can be prepared for archival storage in several ways. Those most frequently used include fixation, skeleton preparation, freezing, and photography. Due to special equipment requirements or complex techniques, other methods are beyond the scope of this chapter, including clearing and staining (Taylor and Van Dyke 1985), freeze-drying or lyophilization, and radiography (Miller and Tucker 1979).

5.3.3.1 Fixation

Fixation is a process of treating cells and tissues to prevent tissue autolysis and decay that can cause irreversible denaturation of biological macromolecules. It also maintains the structural integrity of the specimen for subsequent voucher or histological purposes and is the standard method for taxonomic voucher specimens. A saturated solution of formaldehyde gas in water, known as formalin (approximately 40% formaldehyde by weight), is the standard base reagent used for fixation of fish (Hubbs and Lagler 1964; Lagler et al. 1977). The standard fixative is a 10% formalin solution that is prepared by diluting one part of concentrated formalin with nine parts water. Because formalin is acidic, it causes decalcification of bony structures unless buffered to a neutral pH. Formalin prepared from seawater is usually sufficiently buffered by sea salts for immediate use (Tucker and Chester 1984); however, formalin prepared from fresh water should be buffered with either 10 g borax/L or 20 g calcium carbonate/L. Calcium carbonate is preferred because borax may clear the specimens (Fink et al. 1979). Like most fixatives, formaldehyde is toxic. It should be used only in well-ventilated areas; eyewear to protect against splashing and a supply of water to rinse in case of contact with skin (even in the field) should be available. Waterproof latex or plastic gloves should be worn when formalin-fixed specimens are handled.

To obtain proper fixation, fish should be fixed by placing them live (preferably after anaesthetization) into buffered 10% formalin. Fish should occupy no more than half the volume of fixative, and their volume should be considered water when concentrations are determined; stronger formalin concentrations may be necessary to yield a 10% solution. Solutions stronger than 10% should be used for large fish and solutions weaker than 10% for small ones (Hubbs and Lagler 1964). After death, an incision should be made on the right ventral side of the abdomen of specimens longer than 15 cm to allow fixative into the body cavity. For specimens larger than 1.5 kg, incisions should be made into the dorsal musculature on both sides of the vertebral column from within the body cavity. This allows fixative to penetrate muscle masses without causing distortion, which can result from external incisions or hypodermic injections. Specimens should remain in formalin for 2–7 d, after which excess fixative is removed by soaking the specimens in water. (Note: formalin and formalin-fixed specimens are considered toxic wastes and must be disposed of properly.) Specimens should be soaked in water for at least 2 d, and water should be changed at least four times during this period.

After fixation and rinsing, specimens are typically preserved for long-term storage by placing them in either 70% ethanol or 40% isopropanol; however, dilute (5–7%) buffered formalin may also be used, especially for the purpose of preserving color. For best long-term storage results, specimens should be transferred to fresh preservative after a few weeks. Further information on the fixation and preservation of fish can be found in Cross (1962), Hubbs and Lagler (1964), Peden (1976), Lagler et al. (1977), and Fink et al. (1979).

Similar methods are used for fixing and preserving ichthyoplankton (including eggs, larval fishes, and zooplankton); however, a 5% solution of buffered formalin is normally used (Ahlstrom 1976; Fink et al. 1979; Markle 1984; see Chapter 9). For further information on invertebrates, see Chapter 11.

5.3.3.2 Skeletonization

Large specimens may require impractical quantities of fixative and space, so partial or complete skeletonization may be beneficial. Before skeleton preparation, all desired information must be recorded. Partial skeletons that retain identifying characters may be prepared from filleted carcasses on which the head, axial skeleton, fins, most of the skin, and viscera have been left intact; filleting reduces mass. Partial skeletons must be frozen, salted, preserved, or fixed (Hubbs and Lagler 1964). If osteological (skeletal) characters are sufficient for identification, complete skeletons may be produced by using the dermestid hide beetle (Hall and Russell 1933; Tiemeier 1940). For this preparation, all skin, viscera, and excess flesh should be removed, and the specimen should be dried in a fly-proof screen box before dermestid cleaning begins. Skeletons are stored dry in boxes.

5.3.3.3 Freezing

Freezing is one of the most convenient methods for storing specimens destined for subsequent genetic analysis, taxidermic mounting, skeleton preparation, or fixation. Freezing is the method of choice for specimens of uncertain use because it is compatible with most subsequent specimen preparations (Scott and Aquino-Shuster 1989; Williams and Rogers 1989). Noteworthy exceptions are tests such as karyotype analysis that require live cells. Fish should be placed on ice as soon after capture as possible to retard color changes and deterioration of the specimen prior to freezing. Each specimen should be tagged with an identification number, wrapped in aluminum foil (preferably heavy-duty thickness) marked with the number, placed in a plastic bag to prevent lyophilization (freezer burn), and frozen on dry ice (in the field) or in a freezer (at the laboratory). Small specimens may be frozen in liquid nitrogen. Specimens destined for genetic analyses should be stored in an ultracold freezer (−60 to −80°C) or liquid nitrogen to prevent deterioration of macromolecules.

5.3.3.4 Photography

Photography is indispensable for creating vouchers of endangered or threatened species because it does not necessitate killing the specimen. Photo vouchers are also useful for documenting life colors for species descriptions and taxidermic mountings because many colors are lost or altered after death and preservation. Photographs may be the only practical method of whole-specimen voucher storage for very large specimens (e.g., large sharks), especially for studies that include many individuals.

Each voucher photograph must include an object to indicate scale, such as a meter stick, so that measurements can be made. The left side of the fish should be photographed and all diagnostic key features should be visible. Also, a unique field collection number must appear in the photograph. An effective device for displaying the field number is a small metal board with magnetic numbers and letters that can be rearranged for each specimen. Several techniques have been developed for photographing fish. For additional information see Randall (1961), Emery and Winterbottom (1980), Baugh (1982), Flescher (1983), and Strauss and Bond (1990).

5.3.4 Tissue Preparation and Storage

Techniques for tissue preparation and storage vary widely. Genetic, histological (Hinton 1990), and health certification studies may require the preparation, preservation, or fixation of tissues from live or fresh specimens. Pathologists interested in histological samples should describe or supply the fixative and specify the type of tissue required for study, because a wide variety of fixatives are routinely used (Greer et al. 1991).

5.3.4.1 Genetic Studies for Fisheries Research

Genetic data have use in many fields of fisheries science, including phylogenetics (Phillips et al. 1989; Billington et al. 1991), population genetics (Allendorf and Ferguson 1990; Arnheim et al. 1990), stock identification (Allendorf and Ryman 1987; Allendorf et al. 1987; Hallerman and Beckman 1988; Wood et al. 1989; Shields et al. 1990, 1992), endangered species management (Ammerman and Morizot 1989), and breeding programs, including hatchery broodstock characterization (Hillel et al. 1990; Allendorf and Phelps 1980). The most common genetic analyses performed on fish include chromosome characterization (Thorgaard and Disney 1990) from live or fixed tissues, allozyme and isozyme analysis (Leary and Booke 1990) from fresh or frozen tissues, and DNA analysis from fresh, frozen, or preserved tissues. The following sections outline procedures for collecting and preserving several types of tissues for a variety of genetic studies. See Dessauer et al. (1990) for a review of methods for collection and storage of tissue for genetic work.

5.3.4.2 General Considerations for Genetic Samples

Because the methods used in genetic analysis are sensitive to tissue contamination, care must be taken to minimize the risk of cross-contamination when one collects and works with samples (Higuchi et al. 1984). Gloves should be worn to prevent contamination with human DNA, and they should be washed with water and alcohol between specimens to remove potential cross-contaminants (blood and slime). All containers for storing tissue (tubes, vials, and envelopes) should be unused and autoclaved prior to use. Clean disposable materials (glass pipettes, cover slips, microscope slides, and razor blades) should be used for dissection when practical. Otherwise, instruments must be cleaned thoroughly between specimens by rinsing with water, wiping with clean paper tissue, rinsing with alcohol, and flaming briefly, if possible, to destroy residual organic material that could cross-contaminate the next sample. A Bunsen burner works well for flaming instruments in the laboratory, and a gas stove or alcohol lamp works well in the field.

Samples should not be fixed with formalin if genetic tests are planned. Formalin damages DNA by cross-linking proteins to the DNA molecules, and unbuffered formalin can cause irreversible damage (Dyall-Smith and Dyall-Smith 1988; Paeaebo 1989; Paeaebo et al. 1990; Greer et al. 1991).

5.3.4.3 Tissues for Genetic Analysis

Required tissues and methods of storage depend on the type of genetic testing to be performed. Karyotype preparations require rapidly dividing tissues to optimize the chances of obtaining mitotic cells with condensed chromosomes. Karyotypes are best prepared in one of two ways: from epithelial tissue (gill lamellae, fins, scales, or gut lining) removed from living fish and placed immediately into a fixative (such as 50% glacial acetic acid or a freshly prepared 1:3 solution of acetic acid to methanol), or from living cell cultures that have been stimulated to undergo mitosis. Blood is a convenient source of live cells for establishing tissue cultures for karyotype analysis and can be collected from live fish with little risk of injury (see section 5.3.4.4). Isozymes are best analyzed from fresh or properly frozen tissue (Murphy et al. 1990; Whitmore 1990). Some enzymes can be assayed from tissues collected by nonlethal methods, such as liver and muscle biopsy (Harvey et al. 1984), blood, fin clips, and external mucus (Robbins et al. 1989; Carmichael et al. 1992b). Nonlethal sampling is important for genetic studies on hatchery broodstock, endangered species, and specimens obtained through catch-and-release fisheries. Analysis of DNA can be performed on any nucleated tissue, including fins, scale epithelia (Whitmore et al. 1992), gametes, organs, muscles, and red blood cells, and it does not require sacrifice of the fish.

5.3.4.4 Blood Drawing

Blood has many uses for fisheries research including histological, pathological (Wooster et al. 1993), and genetic analyses. A simple and safe method for obtaining blood from larger specimens is to draw it from the caudal blood vessel (Stoskopf 1993c). The fish should be placed on its back and restrained with padding (i.e., soft foam pads covered with a thin layer of latex that is then moistened with water) to avoid serious injury to the fish. Insert a hypodermic needle, beveled side towards the vertebral column, between the hemal arches of the ventral side of the caudal region (see Figure 14.4). Use either a Vacutainer tube (Fisher Scientific, Pittsburgh, Pennsylvania) or a standard syringe. Successful positioning of the needle is readily apparent by feel and the fact that blood will rush into the collection tube or syringe.

Cardiac puncture may be used for collecting blood from small specimens but has a higher risk of mortality (Stoskopf 1993c). Blood should be obtained from the ventricle in live specimens and from the sinus venosus in dead ones. The ventricle and sinus venosus lie just anterior and posterior to the cleithrum, respectively. Clotting is prevented by the addition of either heparin (to a final concentration of 10 μg/mL) or EDTA (to a final concentration of 10 mM) directly to the blood. Vacutainer tubes are available with either heparin or EDTA.

Samples should be cooled and stored on ice before they are further processed. Blood to be used for DNA analysis can be stored at 4°C for at least 3 months (Billington and Hebert 1990). Glass vials should never be dropped directly into liquid nitrogen or dry ice because they may shatter. Do not freeze samples destined for karyotype analysis because this will kill the living cells necessary for successful cell culture.

5.3.4.5 Tissue Preservation for Genetic Analyses

Preservation is a method of maintaining the integrity of tissues or specimens and differs from fixation by using less toxic chemicals and by maintaining some macromolecules in their native (not denatured) state. Tissues stored at ambient temperature (dried or preserved) are poor candidates for isozyme electrophoresis, but several methods of tissue preservation have been shown to work well for initial field collection as well as for long-term archival storage of tissues for subsequent DNA analysis.

Freezing. Properly frozen tissues are suitable for a wide array of genetic studies, including isozyme and DNA analysis. Allozymes and isozymes must be studied with frozen or fresh tissue because they must be functional to be detected (Leary and Booke 1990; Murphy et al. 1990; Whitmore 1990). A standard method of collecting samples for genetic studies is to freeze them in a Dewar flask containing liquid nitrogen (−195.8°C) or on dry ice (−57.5°C). Tissue samples or entire small specimens may be wrapped in aluminum foil or placed in shatter-resistant polypropylene vials or cryotubes for freezing. Although aluminum foil is less expensive, cryotubes are easier to organize, are necessary for liquids, and normally have O-rings that prevent cross-contamination of samples.

Several methods of organizing samples in a Dewar flask are available. Dropping samples directly into the flask is fast and saves space, but retrieval is difficult and samples may be lost because the vials froze to the walls of the flask. Samples may be effectively organized by using women's nylon knee-high stockings. Nylon stockings are porous, do not become brittle, occupy less space than conventional organizational methods, and transfer little heat to the system. Samples can be sorted into stockings identified with paper tags and tethered with labeled strings or monofilament lines. To aid in removal, stockings should not be filled to a diameter greater than the throat of the flask and tethers should be kept taut to avoid entanglement.

Dry ice in insulated containers is routinely used for shipping specimens and for freezing and short-term storage of samples during field collections. Dry ice can be purchased, or it can be prepared in the field with compressed carbon dioxide and a dry-ice maker.

Rechargeable "dry shippers" are useful for freezing samples in the field and when Dewar flasks are not practical or are not allowed. The sample chambers of these units are surrounded by a thick layer of an absorbent, high-heat-capacity matrix that is cooled with liquid nitrogen. Even after the liquid nitrogen has been poured off or evaporated, the temperature remains below −100°C for periods of 2 weeks or more.

Charge duration for Dewar flasks and dry shippers can be maximized by storing them in insulated containers in cool locations. The cooling capacity of these systems is lost rapidly during freezing, removing, and handling specimens. Charge duration can be extended by precooling samples by means of ice or dry ice, minimizing handling, and keeping the system closed. Dry shippers should be stored upright to minimize loss of cold.

Long-term storage of frozen specimens and tissues for genetic studies should be in ultracold freezers rather than in standard freezers. Dewar flasks are practical for storing limited numbers of small samples. Ultracold freezers can maintain samples at temperatures from −60°C to −90°C, and have an alarm system to warn of failure. Upright models save floor space and are more readily organized, but chest models are less subject to temperature fluctuations and mechanical failure (Dessauer et al. 1990).

Drying. Dried tissue is an excellent source of DNA for certain genetic studies. Small quantities of fragmented DNA can be extracted and amplified from tissues that were not preserved for genetic studies, such as dried study skins (Higuchi et al. 1984), scales, or fin clips (Shiozawa et al. 1992). Dried samples can be prepared by removing several small, thin (less than 1-mm-thick) slices of tissue from specimens and storing them directly in small, labeled envelopes (such as coin envelopes). The envelopes must be kept dry to prevent the growth of fungi and bacteria; this can be achieved by placing them in airtight containers with a desiccant such as Drierite.

Liquid preservation. Screw-top polypropylene microcentrifuge tubes with silicon O-ring seals are ideal for the storage of tissues in liquid preservatives. Alcohols (Shiozawa et al. 1992) and Queen's buffer (see below) are recommended for preservation and storage of tissue for DNA studies. Proper alcohol preservation maintains the integrity of DNA and disinfects samples by killing most bacteria and many viruses. The quality of DNA may be superior in alcohols that have been buffered by the addition of EDTA to a final concentration of 10 mM (Dessauer et al.

1990). For good DNA recovery, tissues should be fixed and stored in a concentration of 70% ethanol (volume of tissue is calculated as 100% water) or approximately a 1:3.5 volume ratio of tissue to 95% ethanol. Isopropanol can be used at a final concentration of 40% (approximately 1:1 volumes of tissue and isopropanol) with good results, but tissues preserved in isopropanol typically yield less DNA than ethanol-preserved tissues (Shiozawa et al. 1992). Isopropyl alcohol should not contain additives that may cause DNA structural damage (Greer et al. 1991). Alcohol used for initial preservation should be replaced with fresh alcohol for long-term storage of samples (Dessauer et al. 1990).

Queen's buffer (0.25 M EDTA, pH 8; 10 mM tris, pH 8; 20% [volume to volume] dimethyl sulfoxide [DMSO]; and saturated with NaCl; modified from Seutin et al. 1991) is superior to alcohols as a preservative. It inhibits degradative enzymes and bacterial and fungal growth (EDTA), penetrates tissues (DMSO), and stabilizes the DNA structure (tris and NaCl). It is stable at room temperature but sensitive to freezing and degradation in sunlight. The buffer should be stored in an opaque bottle and specimens in boxes or drawers. Initial preservation should be in a minimum 1:4 ratio of tissue to preservative, but after 7 d the tissues can be transferred to a smaller volume of buffer for long-term storage.

5.3.5 Ichthyological Collections

Ichthyological collections are repositories for specimens collected for a variety of purposes. Although each institution has its own procedures, the following is a general overview of the organization of typical ichthyological collections and protocols for visiting, borrowing, and handling specimens.

5.3.5.1 Organization of Ichthyological Collections

Ichthyological collections are of four major types: sorted collections, unsorted collections, type collections, and borrowed collections. Sorted collections include specimens that have been identified and placed into containers by species. Containers in unsorted collections typically hold a variety of species captured at a single sampling site. Type collections include type specimens that have been used for species descriptions. Borrowed collections include specimens temporarily borrowed from other collections.

Sorted and type collections are generally arranged in phylogenetic order; however, they rarely are rearranged to keep pace with revised views of phylogenetic relationships (Lagler et al. 1977; Lundberg and McDade 1990). For recent phylogenetic listings of fishes see Robins et al. (1991) and Nelson (1994). Unsorted collections may be arranged by collection or field number, watershed, or other biogeographical subdivisions.

Ichthyological collections traditionally were catalogued by an index card system, but many institutions now also use various electronic databases. The traditional indexing system for sorted collections is arranged into some 65 phylogenetic "groups," then alphabetically by the species' scientific name. If a species has been collected many times, collections may be subdivided by geographic region. Each index card or computer record contains the group number, scientific name, and unique catalogue number. The records also must contain precise information about the collection locality, collection dates, collectors, method of capture, original preservative or fixative, number of specimens in the container, size range of the specimens, name(s) of the person(s) responsible for specimen identification, and the date of identification. These records may also contain information about water quality, vegetation, substrate, water and air temperature, shore description, current or tide, distance from shore, depth of capture, and condition of specimen (quality of specimen and maturity stage), as well as other remarks taken from the field notes. Electronic systems have the advantage of accessing collections by many criteria, thus facilitating studies of regional communities and species groups.

When individuals are abundant in a collection, they are normally stored and catalogued in lots rather than individually (Cross 1962). If the specimens are all from the same field collection, they may be housed in a single container stating the number of specimens and their size range. If the specimens are from separate collections at the same site or at different sites, each specimen in the lot must be identified. Small specimens may be placed individually in small vials (with a label and alcohol) that are plugged with cotton and sealed in jars of alcohol. Larger specimens may be stored in lots and identified by individual identification tags of tin, plastic, or tough paper sewn to the caudal peduncle or by identification tags inserted into the abdominal cavity or into the opercular chamber of the fish (Cross 1962).

When a specimen or collection has been deposited formally in an ichthyological collection, it is assigned an accession number and becomes the property of the collection housing it. Collections must include field notes, either in original or photocopy form.

5.3.5.2 Use of Ichthyological Collections

Personal safety. Physical contact with fixed specimens or inhalation of fumes can be harmful and should be minimized. Gloves should be worn when specimens are handled to prevent irritation and delayed rashes from sensitization to formaldehyde. Specimens should be handled in areas with adequate ventilation, especially when the work is close. Safety glasses should be worn to prevent irreversible corneal damage that can result from having fixative or preservatives splashed into unprotected eyes. Although fixation destroys most pathogens and venoms, spines on fixed specimens are still sharp and can cause painful wounds. Gloves are easily punctured by spines, allowing fixatives and preservatives to come in contact with skin. Preservatives such as ethanol and isopropanol are flammable and should not be used near open flames.

Care of preserved material. Proper care in handling and storage can extend greatly the useful life of specimens. Proper handling includes minimizing destruction and deterioration of specimens as well as maintaining the organization of the specimens in the collection. Avoid tearing delicate fins or membranes while performing measures or counts. Do not allow specimens or fins to dry out; keep specimens in a pan of water and cover them with a piece of damp cheese cloth or paper towel if examination must be interrupted.

To help maintain specimen organization, do not work from too many open jars or skeleton boxes at one time to avoid inadvertently returning specimens to the wrong container. When examinations are complete, return specimens to a designated area for reshelving by the collection manager or curator.

Fixed, preserved specimens best retain their integrity and appearance when they are housed under conditions that protect them from extremes of light and temperature (Fink et al. 1979). Specimens should be housed in an area with adequate ventilation and good climate control, preferably without windows. If present, windows should be painted or shaded.

Alcohol is more volatile than water and is differentially lost to evaporation during long-term storage. This loss can lead to reduced alcohol concentrations and subsequent softening of the tissues of fixed specimens. Fluid levels and alcohol contents of all specimen containers should be checked at least once a year and replaced as needed with fresh alcohol. An alcohol hydrometer can be used to determine alcohol concentrations. Glass storage containers facilitate visual inspection of fluid levels and are recommended for the storage of most specimens. A stuck lid may be loosened by tapping the inverted jar on a solid surface. Many collections are stored in gasket-sealed Ball canning jars; gaskets should be replaced when worn.

Visits to and loans from collections. Visits and loans should be arranged in advance because permission may be required from the collection manager, curator, and in some cases even the museum director. Every effort should be made to establish dialogue by telephone, but written requests should be submitted for the museum's records. The letter should state professional affiliations and proposed use of the museum collections and give an outline of research plans. Indicate whether the sampling proposed will be nondestructive (e.g., examination of external characters or X rays) or destructive (e.g., examination of gut contents, otolith removal, or tissue sampling). Permission is generally granted for reasonable, nondestructive use of specimens; curators may be hesitant to permit destructive sampling, especially of rare or type specimens.

5.4 CONCLUSIONS

The application of proper techniques for care and handling of sampled organisms requires advance planning. Such techniques may seem an inconsequential part of fisheries research, but the value of specimens and tissues is often related to the quality of their care and handling. Poor techniques or careless handling may lead to loss of data, less definitive research, or the need to resample. It is particularly important for personnel involved in only certain phases of the research, such as sampling, to be aware of the overall objectives of the study and the requirements for proper care and handling of the sampled organisms.

Sampled organisms and tissues often have potential for research that goes beyond the study for which they were collected. Consideration should be given to the feasibility of transferring specimens to other researchers or arranging for specimens to be accessioned into an appropriate museum where they will continue to be of value to research. Such transfers are especially important for rare specimens or species that are difficult or expensive to capture.

5.5 REFERENCES

Adams, S. M. 1990. Status and use of biological indicators for evaluating the effects of stress on fish. American Fisheries Society Symposium 8:1–8.

Ahlstrom, E. H. 1976. Maintenance of quality in fish eggs and larvae collected during plankton hauls. Pages 313–318 *in* H. F. Steedman, editor. Zooplankton fixation and preservation. UNESCO Press, Paris.

Allendorf, F. W., and M. M. Ferguson. 1990. Genetics. Pages 35–63 *in* Schreck and Moyle (1990).

Allendorf, F. W., and S. R. Phelps. 1980. Loss of genetic variation in a hatchery stock of cutthroat trout. Transactions of the American Fisheries Society 109:537–543.

Allendorf, F. W., and N. Ryman. 1987. Genetic management of hatchery stocks. Pages 141–160 *in* N. Ryman and F. Utter, editors. Population genetics and fishery management. University of Washington Press, Seattle.

Allendorf, F. W., N. Ryman, and F. Utter. 1987. Genetics and fishery management: past, present, and future. Pages 1–20 *in* N. Ryman and F. Utter, editors. Population genetics and fishery management. University of Washington Press, Seattle.

Amend, D. F., T. R. Croy, B. A. Goven, K. A. Johnson, and D. H. McCarthy. 1982. Transportation of fish in closed systems: methods to control ammonia, carbon dioxide, pH, and bacterial growth. Transactions of the American Fisheries Society 111:603–611.

Ammerman, L. K., and D. C. Morizot. 1989. Biochemical genetics of endangered Colorado squawfish populations. Transactions of the American Fisheries Society 118:435–440.

Andrews, J. W., and Y. Matsuda. 1975. The influence of various culture conditions on the oxygen consumption of channel catfish. Transactions of the American Fisheries Society 104:322–327.

Animal Behaviour. 1992. Guidelines for the use of animals in research. Animal Behaviour 43:185–188.

Arnheim, N., T. White, and W. E. Rainey. 1990. Application of PCR: organismal and population biology. BioScience 40:174–182.

ASIH (American Society of Ichthyologists and Herpetologists), AFS (American Fisheries Society), and AIFRB (American Institute of Fisheries Research Biologists). 1987. Guidelines for use of fish in field research. ASIH, Carbondale, Illinois.

ASIH (American Society of Ichthyologists and Herpetologists), AFS (American Fisheries Society), and AIFRB (American Institute of Fisheries Research Biologists). 1988. Guidelines for the use of fish in field research. Fisheries 13(2):16–23.

Barton, B. A., and C. B. Schreck. 1987. Metabolic cost of acute physical stress in juvenile steelhead. Transactions of the American Fisheries Society 116:257–263.

Baugh, T. M. 1982. Technique for photographing small fish. Progressive Fish-Culturist 44:99–101.

Beamish, F. W. H. 1964. Influence of starvation on standard and routine oxygen consumption. Transactions of the American Fisheries Society 93:103–107.

Billington, N., R. G. Danzmann, P. D. N. Hebert, and R. D. Ward. 1991. Phylogenetic relationships among four members of *Stizostedion* (Percidae) determined by mitochondrial DNA and allozyme markers. Journal of Fish Biology 39(A):251–258.

Billington, N., and P. D. N. Hebert. 1990. Technique for determining mitochondrial DNA markers in blood samples from walleyes. American Fisheries Society Symposium 7:492–498.

Bond, C. E. 1979. Biology of fishes. Saunders, Philadelphia.

Booke, H. E., B. Hollender, and B. Lutterbie. 1978. Sodium bicarbonate, an inexpensive fish anesthetic for field use. Progressive Fish-Culturist 40:11–13.

Boyd, C. E. 1979. Water quality in warmwater fish ponds. Auburn University, Auburn, Alabama.

Brown, L. A. 1993. Anesthesia and restraint. Pages 79–90 *in* M. K. Stoskopf, editor. Fish medicine. Saunders, Philadelphia.

Brunham, K. P., D. R. Anderson, G. C. White, C. Brownie, and K. H. Pollock. 1987. Design and analysis methods for fish survival experiments based on release–recapture. American Fisheries Society Monograph 5.

Bryan, J. D., S. W. Kelsch, and W. H. Neill. 1990. The maximum power principle in behavioral thermoregulation by fishes. Transactions of the American Fisheries Society 119:611–621.

Carlson, C. A., and R. T. Muth. 1993. Endangered species management. Pages 355–381 *in* C. C. Kohler and W. A. Hubert, editors. Inland fisheries management in North America. American Fisheries Society, Bethesda, Maryland.

Carmichael, G. J., R. M. Jones, and J. C. Morrow. 1992a. Comparative efficacy of oxygen diffusers in a fish-hauling tank. Progressive Fish-Culturist 54:35–40.

Carmichael, G. J., M. E. Schmidt, and D. C. Morizot. 1992b. Electrophoretic identification of genetic markers in channel catfish and blue catfish by use of low-risk tissues. Transactions of the American Fisheries Society 121:26–35.

Carmichael, G. J., and J. R. Tomasso. 1988. Survey of fish transportation equipment and techniques. Progressive Fish-Culturist 50:155–159.

Carmichael, G. J., J. R. Tomasso, B. A. Simco, and K. B. Davis. 1984. Confinement and water quality-induced stress in largemouth bass. Transactions of the American Fisheries Society 113:767–777.

Carmichael, G. J., G. A. Wedemeyer, J. P. McCraren, and J. L. Millard. 1983. Physiological effects of handling and hauling stress on smallmouth bass. Progressive Fish-Culturist 45:110–113.

Carmichael, G. J., J. H. Williamson, M. E. Schmidt, and D. C. Morizot. 1986. Genetic marker identification in largemouth bass with electrophoresis of low-risk tissues. Transactions of the American Fisheries Society 115:455–459.

Collins, C. 1990. Live-hauling warmwater fish. Aquaculture 16:70–76.

Cross, F. B. 1962. Collecting and preserving fishes. University of Kansas Museum of Natural History, Miscellaneous Publication 30.

Davis, J. C. 1975. Minimal dissolved oxygen requirements of aquatic life with emphasis on Canadian species: a review. Journal of the Fisheries Research Board of Canada 32:2295–2332.

Dessauer, H. C., C. J. Cole, and M. S. Hafner. 1990. Collection and storage of tissues. Pages 25–44 *in* D. M. Hillis, C. J. Cole, and M. S. Hafner, editors. Molecular systematics. Sinauer Associates, Sunderland, Massachusetts.

Doudoroff, P., and D. L. Shumway. 1970. Dissolved oxygen requirements of freshwater fishes. FAO (Food and Agriculture Organization of the United Nations) Fisheries Technical Paper 86.

Dyall-Smith, M., and D. Dyall-Smith. 1988. Recovering DNA from pathology specimens—a new life for old tissues. Molecular Biology Reports (Bio-Rad Laboratories)6:1–2.

Emery, A. R., and R. Winterbottom. 1980. A technique for fish specimen photography in the field. Canadian Journal of Zoology 58:2158–2162.

Fink, W. L., K. E. Hartel, W. G. Saul, E. M. Koon, and E. O. Wiley. 1979. A report on current supplies and practices used in curation of ichthyological collections. American Society of Ichthyologists and Herpetologists. Report of the ad hoc Subcommittee on Curatorial Supplies and Practices, Carbondale, Illinois.

Fitzsimons, J. D. 1994. Survival of lake trout embryos after receiving physical shock. Progressive Fish-Culturist 56:149–151.

Flescher, D. D. 1983. Fish photography. Fisheries 8(4):2–6.

Fries, J. N., C. S. Berkhouse, J. C. Morrow, and G. J. Carmichael. 1993. Evaluation of an aeration system in a loaded fish-hauling tank. Progressive Fish-Culturist 55:187–190.

Fry, F. E. J. 1947. Effects of the environment on animal activity. University of Toronto Biological Series 55:1–62.

Fry, F. E. J. 1971. The effect of environmental factors on the physiology of fish. Pages 1–99 *in* W. S. Hoar and D. J. Randall, editors. Fish physiology, volume 6. Academic Press, New York.

Fry, F. E. J., and J. S. Hart. 1948. Cruising speed of goldfish in relation to water temperature. Journal of the Fisheries Research Board of Canada 7:169–175.

Gilderhus, P. A., D. A. Lemm, and L. C. Woods, III. 1991. Benzocaine as an anesthetic for striped bass. Progressive Fish-Culturist 53:105–107.

Gold, J. R., C. J. Ragland, M. C. Birkner, and G. P. Garrett. 1991. A simple procedure for long-term storage and preparation of fish cells for DNA content analysis using flow cytometry. Progressive Fish-Culturist 53:108–110.

Greer, C. E., J. K. Lund, and M. M. Manos. 1991. PCR amplification from paraffin-embedded tissues: recommendations on fixatives for long-term storage and prospective studies. PCR Methods and Applications 1:46–50.

Gunstrom, G. K., and M. Bethers. 1985. Electrical anesthesia for handling large salmonids. Progressive Fish-Culturist 47:67–68.

Hall, E. R., and W. C. Russell. 1933. Dermestid beetles as an aid in cleaning bones. Journal of Mammalogy 14:372–374.

Hallerman, E. M, and J. S. Beckman. 1988. DNA-level polymorphism as a tool in fisheries science. Canadian Journal of Fisheries and Aquatic Sciences 45:1075–1087.

Hamman, R. L. 1981. Transporting endangered fish species in plastic bags. Progressive Fish-Culturist 43:212–213.

Harrell, R. M., and M. A. Moline. 1992. Comparative stress dynamics of brood stock striped bass (*Morone saxatilis*) associated with two capture techniques. Journal of the World Aquaculture Society 23:58–63.

Harvey, W. D., R. L. Noble, and W. H. Neill. 1984. A liver biopsy technique for electrophoretic evaluation of largemouth bass. Progressive Fish-Culturist 46:87–91.

Hatting, J., and A. J. J. van Pletzen. 1974. The influence of capture and transportation on some blood parameters of freshwater fish. Comparative Biochemistry and Physiology 49A:607–609.

Heath, A. G. 1987. Water pollution and fish physiology. CRC Press, Ann Arbor, Michigan.

Heidinger, R. C. 1993. Stocking for sport fisheries enhancement. Pages 309–333 *in* C. C. Kohler and W. A. Hubert, editors. Inland fisheries management in North America. American Fisheries Society, Bethesda, Maryland.

Herwig, N. 1979. Handbook of drugs and chemicals used in the treatment of fish diseases: a manual fish pharmacology and materia medica. C. C. Thomas, Springfield, Illinois.

Higuchi, R., B. Bowman, M. Freiberger, D. A. Ryder, and A. C. Wilson. 1984. DNA sequences from the quagga, an extinct member of the horse family. Nature 312:282–284.

Hillel, J., and six coauthors. 1990. DNA fingerprints applied to gene introgression in breeding programs. Genetics 124:783–789.

Hinton, D. E. 1990. Histological techniques. Pages 191–211 *in* Schreck and Moyle (1990).

Hochachka, P. W., and G. N. Somero. 1971. Biochemical adaptation to the environment. Pages 100–156 *in* W. S. Hoar and D. J. Randall, editors. Fish physiology, volume 6. Academic Press, New York.

Hochachka, P. W., and G. N. Somero. 1984. Biochemical adaptation. Princeton University Press, Princeton, New Jersey.

Hubbs, C. L., and K. F. Lagler. 1964. Fishes of the Great Lakes region. University of Michigan Press, Ann Arbor.

IATA (International Air Transport Association). 1992. Live animal regulations. IATA resolution 620, 19th edition. IATA, Montreal.

Jensen, G. L. 1990a. Transportation of warmwater fish: loading rates and tips by species. Louisiana Cooperative Extension Service, Southern Regional Aquacultural Center, SRAC-393, Baton Rouge.

Jensen, G. L. 1990b. Transportation of warmwater fish: procedures and loading rates. Louisiana Cooperative Extension Service, Southern Regional Aquacultural Center, SRAC-393, Baton Rouge.

Jensen, G. L. 1990c. Transportation of warmwater fish: equipment and guidelines. Louisiana Cooperative Extension Service, Southern Regional Aquacultural Center, SRAC-390, Baton Rouge.

Johnson, J. E. 1987. Protected fishes of the United States and Canada. American Fisheries Society, Bethesda, Maryland.

Johnson, S. K. 1979. Transport of live fish. Texas Agricultural Extension Service, Fish Disease Diagnostic Laboratory Publication FDDL-F14, College Station.

Kelsch, S. W., and W. H. Neill. 1990. Temperature preference versus acclimation in fishes: selection for changing metabolic optima. Transactions of the American Fisheries Society 119:601–610.

Kolz, M. L. 1989. Current and power determinations for electrically anesthetized fish. Progressive Fish-Culturist 51:168–169.

Kreiberg, H. 1991. Collapsible portable tank for fish-handling operations. Progressive Fish-Culturist 53:55–57.

Lagler, K. F., J. E. Bardach, R. R. Miller, and D. R. M. Passino. 1977. Ichthyology, 2nd edition. Wiley, New York.

LaJeone, L. J., and D. L. Bergerhouse. 1991. A liquid nitrogen freeze-branding apparatus for marking fingerling walleyes. Progressive Fish-Culturist 53:130–133.

Lambert, T. C. 1982. Techniques for the capture and handling of the Atlantic mackerel with special reference to the use of quinaldine. Progressive Fish-Culturist 44:145–147.

Leary, R. F., and H. E. Booke. 1990. Starch gel electrophoresis and species distinctions. Pages 141–170 *in* Schreck and Moyle (1990).

Lundberg, J. G., and L. A. McDade. 1990. Systematics. Pages 65–108 *in* Schreck and Moyle (1990).

Markle, D. F. 1984. Phosphate buffered formalin for long term preservation of formalin fixed ichthyoplankton. Copeia 1984:525–528.

Mazur, C. F., and G. K. Iwama. 1993. Handling and crowding stress reduces number of plaque-forming cells in Atlantic salmon. Journal of Aquatic Animal Health 5:98–101.

McCraren, J. P., and J. L. Millard. 1978. Transportation of warmwater fishes. Pages 43–88 *in* Manual of fish culture, section G: fish transportation. U.S. Fish and Wildlife Service, Washington, DC.

Meade, J. W. 1985. Allowable ammonia for fish culture. Progressive Fish-Culturist 47:135–145.

Meffe, G. K. 1986. Conservation genetics and the management of endangered fishes. Fisheries 11(1):14–23.

Merrick, J. R. 1990. Freshwater fishes. Pages 7–15 *in* S. J. Hand, editor. Care and handling of Australian native animals: emergency care and captive management. Macquarie University, New South Wales, Australia.

Miller, J. M., and J. W. Tucker. 1979. X-radiography of larval and juvenile fishes. Copeia 1979:534–538.

Minckley, W. L., and J. E. Deacon, editors. 1991. Battle against extinction—native fish management in the American West. University of Arizona Press, Tucson.

Mitchell, G. A. 1991. Compliance issues and proper animal drug use in aquaculture. Veterinary and Human Toxicology 33(Supplement 1):9–10.

Morizot, D. C., M. E. Schmidt, G. J. Carmichael, D. W. Stock, and J. H. Williamson. 1990. Minimally invasive tissue sampling. Pages 143–156 *in* D. H. Whitmore, editor. Electrophoretic and isoelectric focusing techniques in fisheries management. CRC Press, Boca Raton, Florida.

Moyle, P. B., and J. J. Cech. 1988. Fishes: an introduction to ichthyology. Prentice Hall, Englewood Cliffs, New Jersey.

Murphy, R. W., J. W. Sites, Jr., D. J. Buth, and C. H. Haufler. 1990. Proteins I: isozyme electrophoresis. Pages 45–126 *in* D. M. Hillis, C. J. Cole, and M. S. Hafner, editors. Molecular systematics. Sinauer Associates, Sunderland, Massachusetts.

Nelson, J. S. 1994. Fishes of the world, 3rd edition. Wiley, New York.

Nemetz, T. G., and E. B. Shotts, Jr. 1993. Zoonotic diseases. Pages 214–220 *in* M. K. Stoskopf, editor. Fish medicine. Saunders, Philadelphia.

NIH (National Institutes of Health). 1985. Guide for the care and use of laboratory animals. NIH Publication 85-23, Bethesda, Maryland.

Norris, K. S., F. Borcato, F. Calandrino, and W. M. McFarland. 1960. A survey of fish transportation methods and equipment. California Fish and Game 46:6–33.

Paeaebo, S. 1989. Ancient DNA: extraction, characterization, molecular cloning, and enzymatic amplification. Proceedings of the National Academy of Sciences 86:1939–1943.

Paeaebo, S., D. M. Irwin, and A. C. Wilson. 1990. DNA damage promotes jumping between templates during enzymatic amplifications. Journal of Biological Chemistry 265:4718–4721.

Parker, N. C., A. E. Girgi, R. C. Heidinger, D. B. Jester, Jr., E. D. Prince, and G. A. Winans, editors. 1990. Fish-marking techniques. American Fisheries Society Symposium 7.

Peden, A. E. 1976. Collecting and preserving fishes. Museum Methods Manual 3, British Columbia Provincial Museum, Victoria.

Phillips, R. B, K. A. Pleyte, L. M. VanErt, and S. E. Hartley. 1989. Evolution of nucleolar organizer regions and ribosomal RNA genes in *Salvelinus*. Physiological Ecology Japan 1:429–447.

Piper, R. G., I. B. McElwain, L. E. Orme, J. P. McCraren, L. G. Fowler, and J. R. Leonard. 1982. Fish hatchery management. U.S. Fish and Wildlife Service, Washington, DC.

Post, G. 1979. Carbonic acid anesthesia for aquatic organisms. Progressive Fish-Culturist 41:142–144.

Priede, I. G. 1985. Metabolic scope in fishes. Pages 33–64 *in* P. Tytler and P. Calow, editors. Fish energetics: new perspectives. Johns Hopkins University Press, Baltimore, Maryland.

Prosser, C. L. 1986. Adaptational biology: molecules to organisms. Wiley, New York.

Randall, J. E. 1961. A technique for fish photography. Copeia 1961:241–242.

Ricker, W. E. 1975. Computation and interpretation of biological statistics of fish populations. Fisheries Research Board of Canada Bulletin 191.

Robbins, L. W., D. K. Tolliver, and M. H. Smith. 1989. Nondestructive methods for obtaining genotypic data from fish. Conservation Biology 3:88–91.

Robertson, L., P. Thomas, C. R. Arnold, and J. M. Trant. 1987. Plasma cortisol and secondary stress responses of red drum to handling, transport, rearing density and disease outbreak. Progressive Fish-Culturist 49:1–12.

Robins, C. R., and six coauthors. 1991. Common and scientific names of fishes from the United States and Canada, 5th edition. American Fisheries Society Special Publication 20.

Ross, R. M., T. W. H. Backman, and R. M. Bennett. 1993. Evaluation of the anesthetic metomidate for the handling and transport of juvenile American shad. Progressive Fish-Culturist 55:236–243.

Schnick, R. A., F. P. Meyer, and D. L. Gary. 1989. A guide to approved chemicals in fish production and fishery resource management. U.S. Fish and Wildlife Service and University of Arkansas Cooperative Extension Service MP241, Little Rock.

Schnick, R. A., F. P. Meyer, and D. F. Walsh. 1986. Status of fishery chemicals in 1985. Progressive Fish-Culturist 51:133–139.

Schreck, C. B. 1990. Physiological, behavioral, and performance indicators of stress. American Fisheries Society Symposium 8:29–37.

Schreck, C. B., and P. B. Moyle, editors. 1990. Methods for fish biology. American Fisheries Society, Bethesda, Maryland.

Scott, N. J., Jr., and A. L. Aquino-Shuster. 1989. The effects of freezing on formalin preservation of specimens of frogs and snakes. Collection Forum 5:41–46.

Seutin, G., B. N. White, and P. T. Boag. 1991. Preservation of avian blood and tissue samples for DNA analysis. Canadian Journal of Zoology 69:82–90.

Shields, B. A., K. S. Guise, and J. C. Underhill. 1990. Chromosomal and mitochondrial DNA characterization of a population of dwarf cisco, *Coregonus artedii* LeSueur, in Minnesota. Canadian Journal of Fisheries and Aquatic Sciences 47:1562–1569.

Shields, B. A., A. R. Kapuscinski, and K. S. Guise. 1992. Mitochondrial DNA variation in four Minnesota populations of lake whitefish: utility as species and population markers. Transactions of the American Fisheries Society 121:21–25.

Shiozawa, D. K., J. Kudo, R. P. Evans, S. R. Woodward, and R. N. Williams. 1992. DNA extraction from preserved trout tissues. Great Basin Naturalist 52:29–34.

Smith, L. S. 1982. Introduction to fish physiology. T. F. H. Publications, Neptune, New Jersey.

Stickney, R. R. 1979. Principles of warmwater aquaculture. Wiley, New York.

Stickney, R. R., and C. C. Kohler. 1990. Maintaining fishes for research and teaching. Pages 633–663 *in* Schreck and Moyle (1990).

Stoskopf, M. K., editor. 1993a. Fish medicine. Saunders, Philadelphia.

Stoskopf, M. K. 1993b. Surgery. Pages 91–97 *in* M. K. Stoskopf, editor. Fish medicine. Saunders, Philadelphia.

Stoskopf, M. K. 1993c. Clinical examination and procedures. Pages 62–78 *in* M. K. Stoskopf, editor. Fish medicine. Saunders, Philadelphia.

Strauss, R. E., and C. E. Bond. 1990. Taxonomic methods: morphology. Pages 109–140 *in* Schreck and Moyle (1990).

Sullivan, C. M. 1954. Temperature reception and responses in fish. Journal of the Fisheries Research Board of Canada 11:153–170.

Summerfelt, R. C., and L. S. Smith. 1990. Anesthesia, surgery, and related techniques. Pages 213–272 *in* Schreck and Moyle (1990).

Taylor, W. R., and G. C. Van Dyke. 1985. Revised procedures for staining and clearing small fishes and other vertebrates for bone and cartilage study. Cybium 9:107–119.

Thomas, P., and L. Robertson. 1991. Plasma cortisol and glucose stress responses of red drum (*Sciaenops ocellatus*) to handling and shallow water stressors and anesthesia with MS-222, quinaldine sulfate and metomidate. Aquaculture 96:69–86.

Thorgaard, G. H., and J. E. Disney. 1990. Chromosome preparation and analysis. Pages 171–190 *in* Schreck and Moyle (1990).

Tiemeier, O. W. 1940. The dermestid method of cleaning skeletons. University of Kansas Science Bulletin 26:377–383.

Tucker, J. W., and A. J. Chester. 1984. Effects of salinity, formalin concentration, and buffer on quality of preservation of southern flounder larvae. Copeia 1984:981–988.

Van Den Avyle, M. J. 1993. Dynamics of exploited fish populations. Pages 105–135 *in* C. C. Kohler and W. A. Hubert, editors. Inland fisheries management in North America. American Fisheries Society, Bethesda, Maryland.

Warren M. L., Jr., and B. M. Burr. 1994. Status of freshwater fishes of the United States: an overview of imperiled fauna. Fisheries 19(1):6–18.

Wedemeyer, G. A. 1972. Some physiological consequences of handling stress in the juvenile coho salmon (*Oncorhynchus kisutch*) and steelhead (*Salmo gairdneri*). Journal of the Fisheries Research Board of Canada 29:1780–1783.

Wedemeyer, G. A., B. A. Barton, and D. J. McLeay. 1990. Stress and acclimation. Pages 451–489 *in* Schreck and Moyle (1990).

Whitmore, D. A., editor. 1990. Electrophoretic and isoelectric focusing techniques in fisheries management. CRC Press, Boca Raton, Florida.

Whitmore, D. H., T. H. Thai, and C. M. Craft. 1992. Gene amplification permits minimally invasive analysis of fish in mitochondrial DNA. Transactions of the American Fisheries Society 121:170–177.

Williams, J. E., and seven coauthors. 1989. Fishes of North America endangered, threatened, or of special concern: 1989. Fisheries 14(6):2–20.

Williams, S. L., and S. P. Rogers. 1989. Effects of initial preparation methods on dermestid cleaning of osteological materials. Collection Forum 5:11–16.

Winkler, P. 1987. A method to minimize stress during fish transport. Progressive Fish-Culturist 49:154–155.

Wood, C. C., D. T. Rutherford, and S. McKinnell. 1989. Identification of sockeye salmon (*Oncorhynchus nerka*) stocks in mixed-stock fisheries in British Columbia and southeast Alaska using biological markers. Canadian Journal of Fisheries and Aquatic Sciences 46:2108–2120.

Wooster, G. A., H. M. Hsu, and P. R. Bowser. 1993. Nonlethal surgical procedures for obtaining tissue samples for fish health inspections. Journal of Aquatic Animal Health 5:157–164.

Young, P. S., and J. J. Cech. 1993. Physiological stress responses to serial sampling and confinement in young-of-the-year striped bass, *Morone saxatilis* (Walbaum). Comparative Biochemistry and Physiology 105A:239–244.

APPENDIX 5.1 GUIDELINES FOR USE OF FISH IN FIELD RESEARCH[1]

AMERICAN SOCIETY OF ICHTHYOLOGISTS AND HERPETOLOGISTS (ASIH)

AMERICAN FISHERIES SOCIETY (AFS)

AMERICAN INSTITUTE OF FISHERY RESEARCH BIOLOGISTS (AIFRB)

Introduction

Respect for all forms and systems of life is an inherent characteristic of scientists and managers who conduct field research on fishes. Consistent with our long standing interests in conservation, education, research, and the general well-being of fishes, the ASIH, AFS, and AIFRB support the following guidelines and principles for scientists conducting field research on these animals. As professional scientists specializing in fish biology concerned with the welfare of our study animals, we recognize that guidelines for the laboratory care and use of domesticated stocks of fishes are often not applicable to wild-caught fishes, and in fact may be impossible to apply without endangering the well-being of these fishes. Laboratory guidelines may also preclude techniques or types of investigations known to have minimal adverse effects on individuals or populations (1,2,3), and which are necessary for the acquisition of new knowledge.

The respectful treatment of wild fishes in field research is both an ethical and a scientific necessity. Traumatized animals may exhibit abnormal physiological, behavioral, and ecological responses that defeat the purpose of the investigation. For example, animals that are captured, marked, and released must be able to resume their normal activities in an essentially undisturbed habitat if the purposes of the research are to be fulfilled.

The acquisition of new knowledge and understanding constitutes a major justification for any investigation. All effects of possibly valuable new research procedures (or new applications of established procedures) cannot be anticipated. The description and geographic distribution of newly discovered species justifies studies of organisms that are poorly known. It is impossible to predict all potential observation or collection opportunities at the initiation of most fieldwork, yet the observation or acquisition of unexpected taxa may be of considerable scientific value. Field studies of wild fishes often involve many species, some of which may be unknown to science before the onset of a study. A consequence of these points is that frequently investigators must refer to taxa above the species level, as well as to individual species in their research design.

Because of the very considerable range of adaptive diversity represented by the over 20,000 species of fishes, no concise or specific compendium of approved methods for field research is practical or desirable. Rather, the guidelines presented below build on the most current information to advise the investigator, who will often be an authority on the biology of the species under study, as to techniques that are

[1]Reprinted from Fisheries 13(2):16–23, 1988. The section "Preparation and Revision of These Guidelines" has been omitted.

known to be appropriate and effective in the conduct of field research. Ultimate responsibility for the ethical and scientific validity of an investigation and the methods employed must rest with the investigator. To those who adhere to the principles of careful research these guidelines will simply be a formal statement of precautions already in place.

General Considerations

Research proposals may require approval of an IACUC (see below). In situations requiring such approval, each investigator must provide written assurance in applications and proposals that field research with fishes will meet the following requirements.

a. The living conditions of animals held in captivity at field sites will be appropriate for fishes and contribute to their health and well-being. The housing, feeding, and nonmedical care of the animals will be directed by a scientist (generally the investigator) trained and experienced in the proper care, handling, and use of the fishes being maintained or studied. Some experiments (e.g., competition studies) will require the housing of mixed species in the same enclosure. Mixed housing is also appropriate for holding or displaying certain species.
b. Procedures with animals must avoid or minimize distress to fishes, consistent with sound research design.
c. Procedures that may cause more than momentary or slight distress to the animals should be performed with appropriate sedation, analgesia, or anesthesia, except when justified for scientific reasons in writing by the investigator.
d. Fishes that would otherwise experience severe or chronic distress that cannot be relieved will be euthanized at the end of the procedure, or, if appropriate, during the procedure.
e. Methods of euthanasia will be consistent with the rationale behind the recommendations of the American Veterinary Medical Association (AVMA) Panel on Euthanasia (4), but fishes differ sufficiently that their specific techniques do not apply. The method listed by the Royal Society (5) may be followed.

Additional general considerations that should be incorporated into any research design using wild fishes include the following:

f. The investigator must have knowledge of all regulations pertaining to the animals under study, and must obtain all permits necessary for carrying out proposed studies. Investigators must uphold not only the letter but also the spirit of regulations. (Most applicable regulations are referenced in publications of the Association of Systematics Collections [6,7,8].) Researchers working outside the United States should ensure that they comply with all wildlife regulations of the country in which the research is being performed. Work with many species is regulated by the provisions of the Convention on International Trade in Endangered Species of Wild Fauna and Flora (CITES) (see "CITES" references in 6,7). Regulations affecting a single species may vary with country. Local regulations may also apply.
g. Individuals of endangered or threatened taxa should neither be removed from the wild (except in collaboration with conservation efforts), nor imported or exported, except in compliance with applicable regulations.

h. Investigators must be familiar with the fishes to be studied and their response to disturbance, sensitivity to capture and restraint and, if necessary, requirements for captive maintenance to the extent that these factors are known and applicable to a particular study.
i. Taxa chosen should be well-suited to answer the research question(s) posed.
j. Every effort should be made prior to removal of fishes (if any) to understand the population status (abundant, threatened, rare, etc.) of the taxa to be studied, and the numbers of animals removed from the wild must be kept to the minimum the investigator determines is necessary to accomplish the goals of the study. This statement should not be interpreted as proscribing study and/or collection of uncommon species. Indeed, collection for scientific study is crucial to understanding why a species is uncommonly observed.
k. The number of specimens required for an investigation will vary greatly, depending upon the questions being explored. As discussed later in these guidelines, certain kinds of investigations require collection of relatively large numbers of specimens, although the actual percent of any population taken will generally be very small. Studies should use the fewest animals necessary to reliably answer the questions posed. Use of adequate numbers to assure reliability is essential, as studies based on insufficient numbers of fishes will ultimately require repetition, thus wasting any benefit derived from any animal distress necessarily incurred during the study.

Numerous publications exist that will assist investigators and animal care committees in implementing these general guidelines; a number of such journals, monographs, etc. are listed in Appendix A.

Role of the Institutional Animal Care and Use Committee (IACUC)

Field resources for the care and use of fishes are very different from laboratory resources, and the role of the IACUC necessarily is limited to considerations that are practical for implementation at locations where field research is to be conducted. Prevailing conditions may prevent investigators from following these guidelines to the letter at all times. Investigators must, however, make every effort to follow the spirit of these guidelines to every extent possible. The omission from these guidelines of a specific research or husbandry technique must not be interpreted as proscription of the technique.

The IACUC must be aware that while fishes typically used in laboratory research represent a small number of species with well understood husbandry requirements, the classes Agnatha, Chondrichthyes, and Osteichthyes contain at least 20,000 distinct species with very diverse and often poorly known behavioral, physiological, and ecological characteristics. Therefore, "... in most cases, it is impossible to generate specific guidelines for groups larger than a few closely related species. Indeed, the premature stipulation of specific guidelines would severely inhibit humane care as well as research" (9). The IACUC must note the frequent use of the word "should" throughout these guidelines, and be aware that this is in deliberate recognition of the diversity of animals and situations covered by the guidelines. Investigators, on the other hand, must be aware that use of the word "should" denotes the ethical obligation to follow these guidelines when realistically possible.

Before approving applications and proposals or proposed significant changes in ongoing activities, the IACUC shall conduct a review of those sections related to the care and use of fishes and determine that the proposed activities are in accord with these guidelines, or that justification for a departure from these guidelines for scientific reasons is presented.

When field studies on wild vertebrates are to be reviewed, the IACUC must include personnel who can provide an understanding of the nature and impact of the proposed field investigation, the housing of the species to be studied, and knowledge concerning the risks associated with maintaining certain species of wild vertebrates in captivity. Each IACUC should therefore include at least one institution appointed member who is experienced in zoological field investigations. Such personnel may be appointed to the committee on an ad hoc basis to provide necessary expertise. When sufficient personnel with the necessary expertise in this area are not available within an institution, this ad hoc representative may be a qualified member from another institution.

Field research on native fishes usually requires permits from state and/or federal wildlife agencies. These agencies review applications for their scientific merit and their potential impact on native populations, and issue permits that authorize the taking of specified numbers of individuals, the taxa and methods allowed, the period of study, and often other restrictions that are designed to minimize the likelihood that an investigation will have deleterious effects. Permission to conduct field research rests with these agencies by law, and the IACUC should seek to avoid infringement on their authority to control the use of wildlife species.

If manipulation of parameters of the natural environment (day length, etc.) is not part of the research protocol, field housing for fishes being held for an extended period of time should approximate natural conditions as closely as possible while adhering to appropriate standards of care (10,11). Housing and maintenance should provide for the safety and well-being of the animal, while adequately allowing for the objective of the study.

An increasing body of knowledge (e.g., 12) indicates that pain perception of the many species of vertebrates is not uniform over the various homologous portions of their bodies. Therefore, broad extrapolation of pain perception across taxonomic lines must be avoided. For example, what causes pain and distress to a mammal does not cause an equivalent reaction in a fish (13).

Field Activities With Wild Fishes

Collecting

Field research with fishes frequently involves capture of specimens, whether for preservation, data recording, marking, temporary confinement, or relocation. While certain of these activities are treated separately below, they form a continuum of potential field uses of fishes.

The collection of samples for museum preservation from natural populations is critical to: (1) understanding the biology of animals throughout their ranges and over time; (2) the recording of biotic diversity, over time and/or in different habitats; and (3) the establishment and maintenance of taxonomic reference material essential to understanding the evolution and phylogenetic relationships of fishes and for environmental impact studies. The number of specimens collected should be kept at the minimum the investigator determines necessary to accomplish the goal of a study. Some studies, for example, diversity over geographic range or delineation of variation of new species, require relatively large samples.

Capture techniques. Capture techniques should be as environmentally benevolent as possible within the constraints of the sampling design (14,15). Whenever feasible, the potential for return to the natural environment must be incorporated into the sampling design. Current literature should be reviewed to ascertain when and if capture distress has been properly documented. Those capture techniques (seines, traps, etc.) that have minimal impact on the target fishes are not discussed below. Many capture techniques must mimic those of commercial and recreational [fishers] in order to obtain reliable data on population trends for the regulation of such fisheries.

Gill netting (15,16) and other forms of entangling nets are an accepted practice in fish collecting. Many studies contrast recent and prior sampling and thus repetition of a prior technique is mandated for sampling reliability. Net sets should be examined at a regular and appropriate schedule, particularly in warm water, to avoid excessive net mortality.

Collecting fish using ichthyocides is often the only and by far the most efficient sampling technique (cf. 17). Use of ichthyocides should be accomplished with maximal consideration of physical factors such as water movement and temperature, so as to avoid extensive mortality of natural populations and nontarget species.

Electrofishing is a suitable sampling technique in water of appropriate conductivity inasmuch as fish mortalities will be minimal. Proper adjustment of current will stun fishes and complete recovery is possible. Fish can be returned with minimal adverse impact. Care must be exercised to avoid excessive electric currents that may injure or harm the operators as well as the fish.

Capture of fishes by hooks or spears is an accepted practice of recreational [fishers]. Spearfishing is appropriate to cases in which capture in special environments is necessary, for example, deep reefs, caves, kelp beds, etc., and to provide comparable data for recreational fishing statistics. Similarly, many fishes are most efficiently captured by hooks.

Museum specimens and other killed specimens. The collection of live animals and their preparation as museum specimens is necessary for research and teaching activities in systematic zoology and for many other types of studies. Such collections should further our understanding of these animals in their natural state. Descriptions of ichthyological collecting techniques and accepted practices of collection management have been compiled (18,19), as have references to field techniques. Whenever fishes are collected for museum deposition, specimens should be fixed and preserved so as to assure the maximum utility of each animal and to minimize the need for duplicate collecting. In principle, each animal collected should serve as a source of information on many levels of organization from behavior to DNA-sequencing. Whenever practical, for example, blood and other tissues should be collected for karyotypic and molecular study prior to formalin fixation of the specimen (20).

Formalin fixation of specimens is an acceptable practice; however, fishes that do not die rapidly following immersion in a formalin solution should be killed before preservation by means of a chemical anesthetic such as sodium pentobarbital, hydrous chlorobutanol, MS-222, urethane or similarly acting substances, unless justified in writing by the investigator. When field fixation of formalin resistant fishes without prior introduction of anesthetics is necessary, prior numbing of the specimen in ice water should be considered. Several kinds of anesthetics and their efficacy have been reviewed in the Investigations in Fish Control series (21). Their use requires little additional time and effort and adds little to the bulk or weight of collecting equipment. Urethane has been shown to be carcinogenic; thus, caution should be observed with its use and field disposal.

Live capture. Investigators should be familiar with the variety of ichthyological capture techniques and should choose a method suited to both the species and the study. Capture techniques should prevent or minimize injury to the animal. Care should be exercised to avoid accidental capture or insure field release of nontarget species. The interval between visits to traps and net sets should be as short as possible, although it may vary with species, weather, objectives of the study, and the type of trap or net.

Habitat and population considerations. Whether collecting for future release or for museum preparation, each investigator should observe and pass on to students a strict ethic of habitat conservation. Collecting always should be conducted so as to leave the habitat as undisturbed as possible. The collection of large series of animals from breeding aggregations should be avoided if possible. Systematists should be familiar with extant collections of suitable specimens before conducting field work. If the purpose of an experiment is to alter behavior, reproductive potential, or survivability, the interference should be no more than that determined by the investigator to accurately test the hypothesis.

Restraint and Handling

General principles. Restraint of wild fishes ranges from confinement in an aquarium through various types of physical restrictions to drug-induced immobilization. The decision whether to use physical or chemical restraint should be based upon the design of the experiment, knowledge of behavior of the animals, and the availability of facilities. Investigators must use the least amount of restraint necessary to do the job. When not under study, aggressive species should not be confined with other animals (other than food) which they may injure or may injure them. The well-being of the animal under study is of paramount importance, and we emphasize that improper restraint, especially of traumatized animals, can lead to major physiological disturbances that can result in any of a series of deleterious or even fatal consequences.

Animals should be handled quietly and with the minimum personnel necessary. Darkened conditions tend to alleviate stress and subdue certain species, and are recommended whenever possible and appropriate.

Hazardous species. Sharks and other large or venomous fishes are potentially dangerous to the investigator, and thus require special methods of restraint that must involve a compromise between potential injury to the handlers and injurious restraint of the animal. The particular method chosen will vary with the species and purpose of the project. Adherence to the following general guidelines is recommended when working with hazardous fishes:

a. Procedures chosen should minimize the amount of handling time required and reduce or eliminate contact between handler and animal.
b. One should never work alone. A second person, knowledgeable in capture and handling techniques and emergency measures, should be present at all times.
c. Prior consultation with workers experienced with these species, as well as a review

of the relevant literature, is of particular importance since much of the information on handling dangerous species has not been published, but is simply passed from one investigator to another.

Prolonged distressful restraint should be avoided. In some cases, utilization of general anesthesia for restraint in the field may be advisable. If so, the anesthetic chosen should be a low risk compound that permits rapid return to normal physiological and behavioral status, and the animal must be kept under observation until appropriate recovery occurs. The relatively unpredictable response of some poikilotherms to immobilants or anesthetics under field conditions may contraindicate field use of these chemicals under certain conditions.

Chemical restraint. Many chemicals used for restraint or immobilization of fishes are controlled by the Federal Bureau of Narcotics and Dangerous Drugs/Drug Enforcement Administration (DEA). A DEA permit is required for purchase or use of these chemicals. Extensive information on these substances and their use is available (22,23), and permit application procedures are available from regional DEA offices. Investigators should choose the chemical for immobilization with consideration of the impacts of that chemical on the target organism.

The potent drugs available for wildlife immobilization when properly used are, with the exception of succinyl choline, safe for target animals but can be extremely dangerous if accidentally administered to humans. The degree of danger varies according to the drug, and users must be aware of the appropriate action to take in the event of accident.

Animal Marking

Fish marking, by a variety of techniques, provides one of the most important methods of analyzing fish movements, abundance, and population dynamics (cf. 24). It is basic to all field studies. Important considerations in choosing a marking technique are its effect on behavior, physiology, and survival of the target species or a close relative. Investigators must consider the nature and duration of restraint, the amount of tissue affected, whether distress is momentary or prolonged, whether the animal, after marking, will be at greater than normal risk, whether the animal's desirability as a mate is reduced, and whether the risk of infection or abscess formation is minimal. Careful testing of markers on preserved or captive animals before use on wild animals may reveal potential problems and is recommended. Marking techniques for fishes have been extensively reviewed (25) and are summarized below.

Fin-clipping is relatively easy, may have minimal impact on survival and social structure of the marked fish, and is a recommended procedure for many studies. Fins used for clipping or removal would depend upon the species selected, that is, clipping of the anal fin of poeciliid males would be inappropriate, but removal of the adipose fin of a salmonid would have negligible impact. The importance of fins to the survival and well-being of fishes varies so widely that specific guidelines are not possible.

Marking techniques involving tissue removal or modification (branding, etc.) should be preceded by local anesthetic (aerosols containing benzocaine, such as Cetacaine, may be applied) and followed by the application of topical antiseptic. Chilling of fishes prior to marking may be effective for immobilization.

Electrocauterization of a number, letter, or pattern on the skin, in which deep layers of skin are cauterized to prevent regeneration, provides a marking system that, if performed properly, heals rapidly and seldom becomes infected. Brand marks typically, however, are not visible in captive fishes after a few months. Freeze branding is often the preferred branding technique.

Tattooing and acrylic paint injections have been used with success on fishes. Two potential problems that must be resolved prior to marking are: (1) the selection of a dye which will be visible against the pigmentation of the skin, and (2) the loss of legibility due to diffusion or ultraviolet degradation of the dye.

Tagging is perhaps the most widely used and best investigated means of fish marking. Several logical constraints should be considered in planning any tagging program. Tags that cause projections from the body could produce physical impairment and enhance the risk of entanglement in underwater vegetation. Brightly colored tags may compromise a fish's camouflage. The size, shape, and placement of tags should permit normal behavior of the animal to the greatest extent possible.

Radiotelemetry. Radiotelemetry is a specialized form of animal marking, and the same general procedures apply. Underwater telemetry, however, is primarily limited to acoustic rather than radio frequency transmission. Radio transmission is only practical in fresh water and at relatively shallow depths. Radio transmission is regulated by the Federal Communications Commission, and investigators should inquire about availability of frequencies they plan to use. General telemetry techniques are summarized by Mackay (26), Amlaner and MacDonald (27), and Stasko and Pincock (28).

Many fishes are unsuitable for radiotelemetric studies because of their small size and habit of living in confined spaces. Component miniaturization will undoubtedly facilitate the future use of radiotelemetry in studies of small fish species, particularly with internally implanted transmitters.

Researchers intending to use radiotelemetry on fish species should consider the following guidelines and comments:

a. *Force-Fed and Implanted Transmitters*: Force-fed packages should be small enough to pass through the gut without obstructing the passage of food. Force-fed or implanted packages should be coated with an impervious, biologically inert coating. Residence time of up to several days in the gut is generally long enough to provide useful information on movement and body temperature. Implanted transmitters should not interfere with the function of the organs surrounding them or with the fish's normal behavior. For intracoelomic or subcutaneous implants, the transmitter package may have to be sutured in place to prevent its movement or interference with vital organs.
b. *Externally Attached Transmitters*: Consideration must be given to the effect of an externally attached transmitter package on behavioral interactions between tagged fishes and other individuals. For example, the transmitter should neither conceal nor enhance the appearance of dorsal fins or opercular flaps. Transmitters should be shaped and attached so as to eliminate or minimize the risk of entanglement with underwater vegetation or other obstructions.

Most fishes continue to grow throughout life. External transmitters should be removed or designed to be lost after a time, or they may constrict or irritate the animals. Special consideration must be given to soft-skinned species to prevent abrasion.

Radioisotopes. The use of radioisotopes as markers in natural systems is very valuable, and may be the only means of adequately gathering data on movements of very small species; the technique, however, should be undertaken with caution. Special training and precautions are required of researchers by federal and frequently state law (8). A license, which specifies safety procedures for laboratory use, is required for release of isotopes into natural systems and for disposal of waste material. The pros and cons of using strong emitters must be assessed in terms of possible deleterious effects on the animal, to predators that might ingest isotope-labelled animals, and potential hazard to the public.

When marking with radioisotopes, the animal does not have to be handled for identification, several individuals can be monitored rather quickly, the label is easy to apply, and it can be useful for a limited time if desired. Strong emitters, however, cause extensive tissue necrosis at the implant site, and even weaker ones carry the chance for induction of mutations that may compromise future genetic studies of these populations.

Housing and Maintenance at Field Sites

Because the biological needs of each species and the nature of individual projects vary widely, only the most general recommendations on housing wild vertebrates in the field can be made. When dealing with unfamiliar species, testing and comparing several methods of housing to find the method most appropriate for the needs of the animal and the purposes of the study may be necessary. Restraint and ease of maintenance by animal keepers should not be the prime determinants of housing conditions, though these are certainly important considerations.

Normal field maintenance should incorporate, as far as possible, those aspects of the natural habitat deemed important to the survival and well-being of the animal. Adequacy of maintenance can be judged, relative to the natural environment, by monitoring a combination of factors such as changes in growth and weight, survival rates, breeding success, activity levels, general behavior, and appearance (29). Nutritionally balanced diets should be provided or natural foods should be duplicated as closely as possible. Natural light and temperature conditions should be followed unless alteration of these are factors under investigation.

Frequency of aquarium cleaning should represent a compromise between the level of cleanliness necessary to prevent disease (30,31,32), and the amount of distress imposed by frequent handling and exposure to unfamiliar surroundings. Applied knowledge of animal ethology can assist the investigator in providing optimum care and housing.

Disposition Following Studies

Upon completion of studies, researchers should release wild-caught specimens whenever this is practical and ecologically appropriate. Exceptions are: if national, state, or local laws prohibit release, or if release might be detrimental to the well-being of the existing gene pools of native fishes in a specific geographic area.

As a general rule, field captured fishes should be released only:

a. At the site of the original capture, unless conservation efforts or safety considerations dictate otherwise. Release should never be made beyond the native range of distribution of a fish without prior approval of the appropriate state and/or federal agencies, and approved relocations should be noted in subsequent publication of research results.

b. If their ability to survive in nature has not been irreversibly impaired.
c. Where it can be reasonably expected that the released animal will function normally within the population.
d. When local and seasonal conditions are conducive to survival.
e. When release is not likely to spread pathogens.

Captured animals that cannot be released or are not native to the site of intended release should be properly disposed of, either by distribution to colleagues for further study, or if possible by preservation and deposition as teaching or voucher specimens in research collections.

In both the field and laboratory, the investigator must be careful to ensure that animals subjected to a euthanasia procedure are dead before disposal. In those rare instances where specimens are unacceptable for deposition as vouchers or teaching purposes, disposal of carcasses must be in accordance with acceptable practices as required by applicable regulations. Animals containing toxic substances or drugs (including euthanasia agents like T-61) must not be disposed of in areas where they may become part of the natural food web.

References

1. Young, E. (ed.) 1915. The capture and care of wild animals. Ralph Curtis Books, FL.
2. Guide to the care and use of experimental animals, vols. 1 and 2. Canadian Council on Animal Care, Ottawa, K1P 5H3, ON.
3. Pisani, G. R., S. D. Busack, and H. C. Dessauer. Unpub. Guidelines for use of amphibians and reptiles in field research.
4. Smith, A. W., et al. 1986. Report of the AVMA Panel on Euthanasia. J. Am. Vet. Med. Assoc. 188(13):252–268.
5. The Royal Society and Universities Federation for Animal Welfare. 1987. Guidelines on the care of laboratory animals and their use for scientific purposes. I. Housing and care. Wembley Press. 29 pp.
6. Estes, C. and K. W. Sessions (compilers). 1984. Controlled wildlife, vol. 1: federal permit procedures. ISBN 0-942924-05-3. Assoc. Syst. Coll., Mus. Nat. Hist., Univ. Kansas, Lawrence, KS. 304 pp.
7. ibid. 1983. Controlled wildlife, vol. 2: Federally controlled species. ISBN 0-942924-06-1.
8. King, S. T. and R. S. Schrock. 1985. Controlled wildlife vol. 3: state regulations. ISBN 0-942924-07X. 315 pp.
9. Guidelines for the care and use of lower vertebrates. 17 September 1986. Committee for the Protection of Animal Subjects, University of California, Berkeley, CA 94720. 8 pp.
10. National Institutes of Health guide for grants and contracts, special edition: laboratory animal welfare. 14(8):1–30, June 25, 1985. Superintendent of Documents. U.S. Government Printing Office, Washington, D.C. 20402.
11. ibid., Supplement. 14(8):1–82, June 25, 1985.
12. Green C. J. 1979. Animal handbook #8. Lab. Anim. Ltd., London.
13. Smith, M. M. and P. G. Heemstra. 1986. Smith's sea fishes. J. L. B. Smith Institute of Ichthyology. 1047 pp. (see pp 9–10).
14. Lagler, K. F. 1978. Capture, sampling and examination of fishes. Pages 7–47 *in* Methods for assessment of fish production in fresh waters. T. Bagenal, ed. Blackwell Science Publications, London.
15. Nielsen, L. A. and D. L. Johnson (eds.) 1983. Fisheries techniques. American Fisheries Society, Bethesda, MD. 496 pp.
16. Hanley, J. M. 1975. Review of gill net selectivity. J. Fish. Res. Board Can., 32(11):1943–69.

17. Russell, B. C., F. H. Talbot, G. R. V. Anderson, and B. Goldman. 1978. Collection and sampling of reef fishes. Pages 329–343 *in* D. R. Stoddard and R. E. Johannes, eds. Coral reefs: research methods. UNESCO. Page Bros., Norwich, England.
18. Fink, W. L., K. E. Hartel, W. G. Saul, E. M. Koon, and E. O. Wiley. Unpub. A report on current supplies and practices used in curation of ichthyological collections. Amer. Soc. of Ichthyologists and Herpetologists, Ichthyological Collection Committee. 63 pp.
19. Cailliet, G. M., M. S. Love, and A. W. Ebeling. 1986. Fishes: a field and laboratory manual on their structure, identification, and natural history. Wadsworth, Inc., Belmont. 194 pp.
20. Dessauer, H. C. and M. S. Hafner (eds.) 1984. Collections of frozen tissue. Value, management, field and laboratory procedures, and directory of existing collections. Assoc. Syst. Coll., Mus. Nat. Hist., Univ. Kansas, Lawrence, KS. 74 pp.
21. Investigations in fish control, nos. 22–24, 47–54. U.S. Dept. Interior, Fish and Wildlife Service, Washington, D.C.
22. Code of federal regulations 21: food and drugs, Part 1300 to end. April 1, 1980. Superintendent of Documents, U.S. Government Printing Office, Washington, D.C. 20402.
23. Parker, J. L. and H. R. Adams. 1978. The influence of chemical restraining agents on cardiovascular function: a review. Lab. Anim. Sci. 28:575.
24. Brownie, C., D. R. Anderson, K. P. Burnham, and D. S. Robson. 1978. Statistical inference from band recovery data—a handbook. U.S. Dept. of Interior. Resource Pub. no. 131. 212 pp.
25. Anon. 1974. Marking fishes and invertebrates. Mar. Fish. Rev. 36(7).
26. MacKay, R. S. 1970. Bio-medical telemetry, 2nd ed. John Wiley and Sons, Inc., New York. 533 pp.
27. Amlaner, C. J., Jr. and D. W. MacDonald (eds.) 1980. A handbook on biotelemetry and radio tracking. Pergamon Press, Oxford, England.
28. Stasko, A. B. and D. G. Pincock. 1977. Review of underwater biotelemetry, with emphasis on ultrasonic techniques. J. Fish. Res. Board Can. 34(9):1261–1285.
29. Snieszko, S. F. et al. 1974. Fishes: guidelines for the breeding, care, and management of laboratory animals. ISBN 0-309-02213-4. National Academy of Sciences. 85 pp.
30. Wallach, J. D. and W. J. Boever. 1983. Diseases of exotic animals: medical and surgical management. W. B. Saunders Co., Philadelphia. 1159 pp.
31. Fryer, J. L., et al. 1979. Proceedings from a conference on disease inspection and certification of fish and fish eggs. Oreg. State Sea Grant Coll. Prog. ORESU-W-79-001.
32. Snieszko, S. F. (ed.) 1980. A symposium on diseases of fishes and shellfishes. Am. Fish. Soc. Spec. Publ. 5.

Appendix A Additional References

Canadian Journal of Zoology. National Research Council of Canada, Ottawa, ON K1A 0R6.

Canadian Veterinary Journal, 339 Booth St., Ottawa, ON K1R 7K1.

Journal of the American Veterinary Medical Association, 930 N. Meacham Rd., Schaumburg, IL 60196.

Journal of Wildlife Diseases, Wildlife Diseases Assoc., Box 886, Ames, IA 50010.

Directory, Resources of Biomedical and Zoological Specimens. 1981. Registry of Comparative Pathology, Washington, D.C. 20306.

International Species Inventory, Minneapolis Zoo, Minneapolis, MN. World Geographic and Zoological Institute.

Guidelines and Procedures for Radioisotope Licensing. U.S. Atomic Energy Commission, Isotope Branch—Division of Materials Licensing, Washington, D.C. 20545.

Veterinary Anesthesia, 2nd Edition. 1984. W. V. Lumb and E. W. Jones. Lea & Febiger, Philadelphia, PA. 693 pp.

Copeia. Research Journal of the American Society of Ichthyologists and Herpetologists. Allen Press, Lawrence, KS 66044.

Transactions of the American Fisheries Society. Research Journal of the American Fisheries Society, 5410 Grosvenor Lane, Suite 110, Bethesda, MD 20814.

Chapter 6

Passive Capture Techniques

WAYNE A. HUBERT

6.1 INTRODUCTION

Passive capture techniques involve the capture of fish or other aquatic animals by entanglement, entrapment, or angling devices that are not actively moved by humans or machines while the organisms are being captured (Lagler 1978). The techniques used in passive sampling of fisheries are similar to those used for food gathering over the centuries. Nets and traps have been widely used among primitive peoples, and it is known that many of the currently applied techniques were used by the ancient Egyptians, Greeks, and Romans (Alverson 1963).

Based on their mode of capture, passive sampling devices can be divided into three groups: (1) entanglement, (2) entrapment, and (3) angling gears. Entanglement devices capture fish by holding them ensnared or tangled in a fabric mesh. Gill nets and trammel nets are both entanglement gears. Entrapment devices capture organisms that enter an enclosed area through one or more funnel- or V-shaped openings and once inside cannot find a path to escape. Hoop nets, trap nets, and pot devices are examples of entrapment gears. Angling devices capture fish with a baited hook and line. Trotlines and longlines are examples of passive angling gears.

When describing passive sampling devices, the concepts of gear selectivity and gear efficiency must be addressed. Often these terms are used interchangeably, but they have different, specific definitions. Gear selectivity is the bias of a sample obtained with a given gear. Selectivity or bias for species, sizes, and sexes of fish occur in samples taken with specific types of gear. Species selectivity refers to overrepresentation of particular species within samples from an assemblage of species. Similarly, size or sex selectivity refers to overrepresentation of specific sizes (lengths) or one sex of fish within samples from a stock. By contrast, the efficiency of a gear refers to the amount of effort expended to capture target organisms. It is generally desirable to maximize the efficiency of a sampling gear to save time and money.

6.1.1 Advantages of Passive Gears

Passive gears are relatively simple in their design, construction, and use. They are generally handled without mechanized assistance other than a boat, and they require little specialized training to operate. Passive gears can be used to capture fishes for many purposes and can yield data on relative abundance for many species and types of aquatic habitats. Identical items of passive gear fished in a similar manner and time of year can provide reasonable indices of change in stock abundance. Presampling can be used to estimate sampling variability and sample sizes needed for management and research objectives (see Chapter 2).

Box 6.1 Catch per Unit Effort

Catch per unit effort (CPUE) with passive or active sampling gear can be used as an index of population density. A basic assumption of this method is that CPUE is proportional to stock density. True density of the target species will be unknown (unless estimated by other means), and the value of the proportionality constant, therefore, will also be unknown. If the proportionality can be assumed to be constant (whatever its value), however, changes in CPUE will indicate corresponding changes (increases or decreases) in the species' abundance.

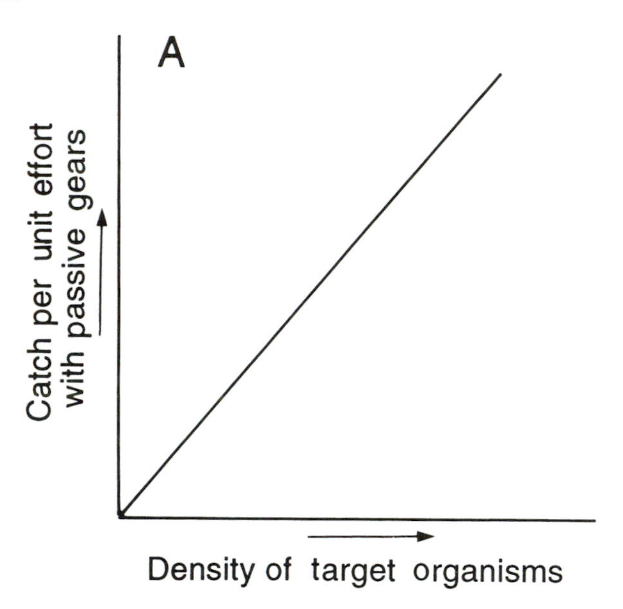

Variability in fish behavior can cause large fluctuations in CPUE and hamper interpretation of CPUE data with respect to relative abundance. To use CPUE, attention must be given to reducing variability by standardizing gear, methods, and sampling design (Welcomme 1975; Fisheries Techniques Standardization Committee 1992). With such controls, passive gears have been fished in a similar fashion and time each year to give reasonable estimates of the abundance of species (Carlander 1953; Walburg 1969; Le Cren et al. 1977).

Passive sampling gears are often used to assess the density of fish or other aquatic animals, but variability in CPUE that occurs due to factors other than abundance of the target organism can confound assessments. One way to reduce variability in CPUE so that changes in CPUE indicate changes in density of target species is to adhere to rigid sampling regimes within specific habitats and time periods. Most agencies and researchers adhere to such protocols (Carlander 1953; Bulkley 1970; Scidmore 1970; Fisheries Techniques Standardization Committee 1992). Sampling is done with the same gear, in the same locations, and at the same times each year. Despite such controls, extensive sampling may be necessary to provide a reasonable estimate of CPUE (Parkinson et al. 1988).

Box 6.1 Continued.

Analysis of CPUE data in relation to densities of aquatic organisms is complicated by a general lack of normal distributions.

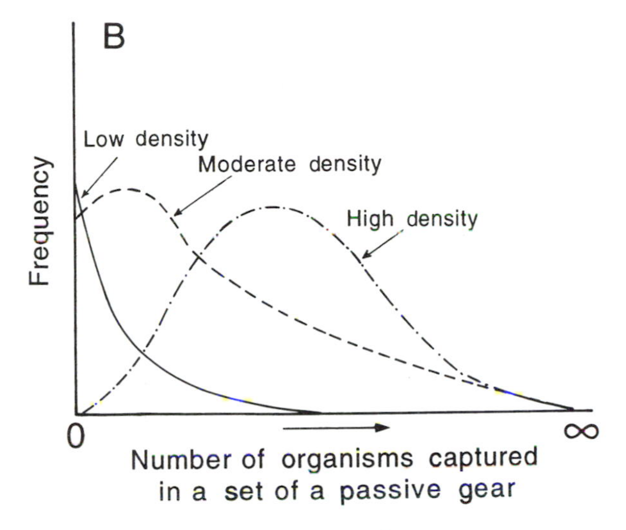

When target species are at very low densities, few of them are caught, and the catch frequency curve approximates a declining logarithmic function. At moderate densities, distributions are highly skewed. Only at very high densities, when the target species are caught on almost every set, do normal distributions tend to occur. As a result of this phenomenon, descriptive statistics derived from normal distributions (i.e., mean, standard deviation, and 95% confidence interval; see Chapter 2) are seldom applicable to CPUE data. Also, because the shape of the distributions change with fish densities, no single data transformation (such as $\log_{10}[x + 1]$) can be applied to generate a more normal distribution (Bagenal 1972; McWilliams and Mayhew 1974).

Management and research biologists have wrestled with the appropriate descriptive statistics and statistical tests to use in assessing differences in CPUE over time. Nonparametric statistics offer equivalents to most procedures (e.g., Student's *t*-test and analysis of variance) that assume normal distributions, and nonparametric statistical procedures are more powerful when the assumption of normality is violated (Sokal and Rohlf 1981; Zar 1984). Because analysis of means depends on normal distributions, a more appropriate descriptive statistic for CPUE is the median (50th percentile). Researchers have suggested use of other descriptive statistics, such as the frequency of zero catches (Bannerot and Austin 1983), but fishery scientists still have no widely accepted protocol for statistically analyzing CPUE data.

6.1.2 Disadvantages of Passive Gears

All passive sampling devices are selective for certain species, sizes, or sexes of animals. Commercial users of passive gear apply their knowledge of gear selectivity to enhance their efficiency in catching targeted species and sizes of aquatic animals (Carter 1954; Starrett and Barnickol 1955). The act of capturing an animal involves several stages: the animal has to encounter the gear, it must be caught by the gear, and finally the animal needs to be retained by the gear until the gear is retrieved. Selectivity occurs at each stage of the capture sequence. A quantitative understanding of gear selectivity is needed to interpret data, but in general, little such information is available for most sampling devices.

Theoretically, the catch per unit effort (CPUE) of a passive sampling gear should be directly proportional to the density of fish in the stock (see Box 6.1), but there have been relatively few studies confirming the assumption (Le Cren et al. 1977; Whitworth 1985; Schorr and Miranda 1991; Borgstrom 1992). Many variables contribute to variability in CPUE in addition to stock density. Some of the more important variables influencing capture efficiency are season, water temperature, time of day (day or night), water level fluctuation, turbidity, and currents.

Changes in animal behavior lead to a great degree of variability in CPUE among species and among age-groups within a species because animal capture with passive gear is a function of animal movement. Many movements are unpredictable because the ways in which environmental factors influence animal behavior are poorly understood. The CPUE of entanglement gear is further influenced by fish morphology. Gill nets and trammel nets tend to be more selective in the capture of species with external protrusions and less selective in the capture of species with compressed bodies and no protrusions.

6.2 ENTANGLEMENT GEARS

6.2.1 Gill Nets

Gill nets are vertical panels of netting normally set in a straight line (Figure 6.1). Fish may be caught by gill nets in three ways: (1) wedged—held by the mesh around the body, (2) gilled—held by mesh slipping behind the opercula, or (3) tangled—held by teeth, spines, maxillaries, or other protrusions without the body penetrating the mesh. Most often fish are gilled. A fish swims into the net and passes only part way through the mesh. When it struggles to free itself, the twine slips behind the gill cover and prevents escape.

6.2.1.1 Construction

A gill net is made of a single wall of webbing held vertically in the water by weights and floats. The hanging ratio—the length of the net mounted on float and lead lines divided by the length of the unmounted, fully stretched net—determines the shape of the mesh (see Chapter 7; Gebhards 1966). The influence of hanging ratio is greatest on species caught by tangling. The lower the ratio, the more the diamond-shaped mesh is elongated vertically and the more efficient the gill net becomes in entangling deep-bodied fish (Welcomme 1975).

The mesh size of gill nets is generally expressed as bar measure or stretch measure. Bar measure (also known as square measure) is the distance between knots (see Figure 7.2). Stretch measure is the length of a single mesh when the net is stretched taut. In general, the units of bar measure are half that of stretch measure. For example, a 5-cm-bar-measure mesh is the same as a 10-cm-stretch-measure mesh.

Gill-net webbing has been made of cotton, linen, nylon, and monofilament twine. Cotton and linen nets have been replaced by multistrand nylon and synthetic monofilament nets because these new materials catch more fish, do not deteriorate rapidly, and require less maintenance (Lagler 1978). Monofilament nets are more difficult to repair and to handle in cold weather than are nets of other materials, but they are less visible to fish, easier to clean, and more durable. The choice of webbing

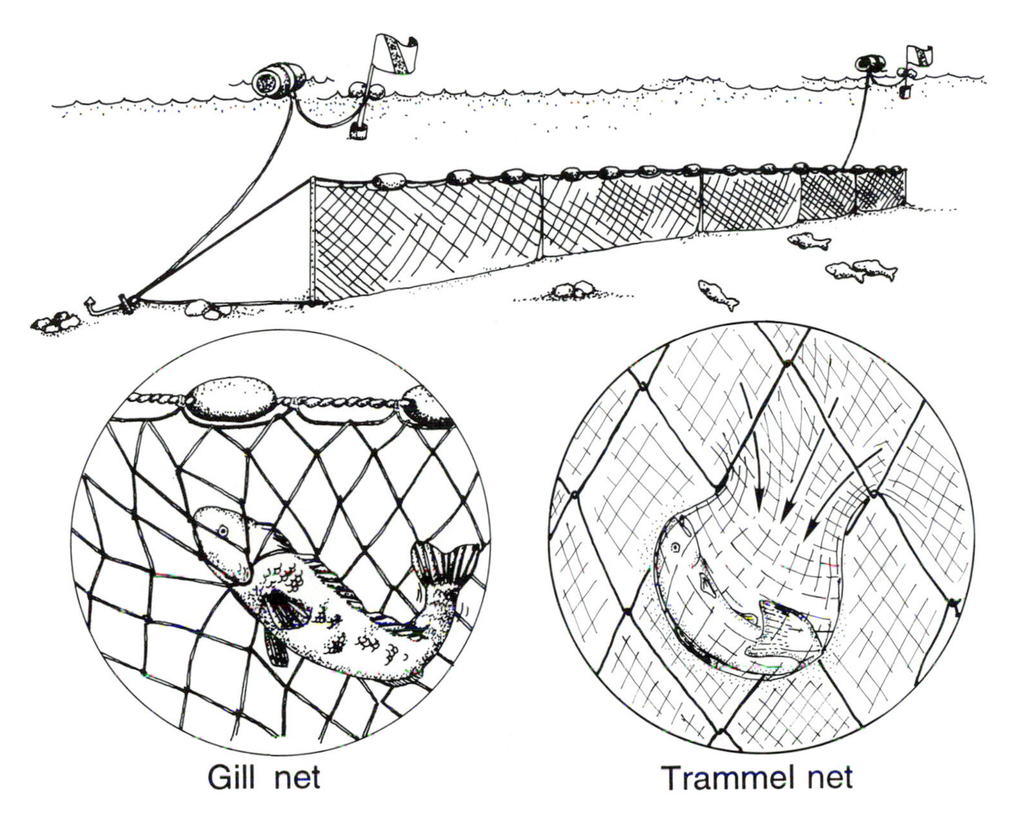

Figure 6.1 Two types of entanglement gear—a gill net and a trammel net—and illustration of a typical bottom set (modified from Dumont and Sundstrom 1961).

material can influence CPUE due to the material's visibility, stiffness, elasticity, smell, and breaking strength (Welcomme 1975; Jester 1977).

Traditionally, wooden, cork, and plastic floats have been strung on the top line to float gill nets. Lead weights have been strung or crimped onto the bottom line to weight the net. Both floats and leads tend to tangle with the meshes and to catch on obstructions. Consequently, foam core float lines and lead core lead lines have become popular.

Experimental gill nets consist of several panels of different mesh sizes to reduce the overall effects of size selectivity (Lagler 1978; Lott and Willis 1991; Hubert and O'Shea 1992). A common design for an experimental gill net used on lakes and reservoirs is a 2-m net depth (height) and 8-m-long (wide) panels of 19-, 25-, 38-, 51-, and 64-mm-bar-mesh webbing. Regier and Robson (1966) suggested that for general sampling purposes it would be more efficient for the mesh sizes to increase in a geometric progression rather than the usual arithmetic progression. Most experimental gill nets are hung with a ratio of 0.5, which results in mesh diamonds that are somewhat elongated in the vertical dimension. However, the effectiveness of the mesh shape varies with the morphology of the fish species being sought (Welcomme 1975).

6.2.1.2 Deployment and Applications

Gill nets can be set in many ways, depending on the species sought and the habitats involved (von Brandt 1964; Nedelec 1975). The most common deployment is the stationary bottom set. The net, anchored at both ends, is set as an upright fence of netting along the bottom (Figure 6.1). Before it is set, the net is rigged with appropriate anchors, lines, and buoys. To set the net, the anchor is dropped over the bow and the boat is backed as the net is set by handling the float line and shaking out tangles in the mesh. Nets set perpendicular to the wind in water less than 5 m deep tend to roll and tangle due to wave action (Scidmore 1970). In winter, rope can be strung for setting gill nets or other sampling gear under the ice by means of a willow stick, Murphy stick, or "jigger" device (Hamley 1980; see Box 6.2).

Gill nets are retrieved by starting at the downwind end and pulling the net over the side of the boat. The net is placed into a wash tub or similar container in coils or figure eights. Fish are generally removed as they come out of the water, but some people prefer to haul in the entire gill net and move to a sheltered area to remove the fish. Lagler (1978) suggested that removal of fish can be made easier if a hook or spatula is used to lift meshes over the opercula and slide them off the body. A simple device can be made by driving a nail lengthwise into the end of a short stick, cutting the head off the nail, and flattening the metal. A small screwdriver also can be used.

Gill nets can be placed at discrete depths to determine depth distributions of fish. A normally bottom-set net can be suspended at midwater depths on droplines from large buoys, or a buoyant net can be held below the surface by lines attached to anchors (Von Brandt 1964). Gill nets can be floated at the surface of lakes.

Gill nets can also be set with the long axis perpendicular to the water surface to determine the vertical distribution of fishes in water up to 50 m deep. A variety of methods have been used to deploy vertical sets, but they generally involve a mechanism similar to a window shade. The nets are wound around a cylindrical column that may also suffice as a float, then unwound to the bottom with periodic placement of lightweight "spreader bars" to hold the net open (Kohler et al. 1979; Negus 1982; Chadwick et al. 1987; Lynch et al. 1989).

Gill nets can be fished by setting them around concentrations of fish or areas suspected of harboring fish. Once the net is set, fish can be driven into it with noise, light, electricity, or chemicals. White (1959) described several encircling net sets made by commercial fishers on the Tennessee River.

Gill nets are used to sample fish in a wide range of habitats. They are generally considered a shallow-water gear, but bottom sets may be made at depths exceeding 50 m. The use of gill nets is generally limited to areas free of obstructions, snags, and floating debris, as well as locations with little or no current. Although gill nets are not considered to be widely applicable in riverine habitats, they have been drifted in the current, set in eddies, used like seines, and anchored at the downstream end of sandbars and allowed to swing in the current. Gill nets are widely used to monitor populations and determine distributions of fish in lakes and reservoirs (Fisheries Techniques Standardization Committee 1992; Hubert and O'Shea 1992). They are commonly used in remote areas where access can be attained only by aircraft, boat, horse, or hiking.

Large gill nets are used in inland and marine commercial fisheries. Anchored or set gill nets, as well as free-floating or drift gill nets, are used to harvest a variety of species (Dumont and Sundstrom 1961). Drift gill nets are used principally for pelagic species such as herrings and salmon (Rounsefell and Everhart 1953).

Box 6.2 Methods for Stringing a Rope Under Ice

Willow stick method. A long, pliable "willow" stick is used to pass a rope down through one hole and up through a second. The process can be repeated several times to extend the rope as far as desired.

Murphy stick method. The Murphy stick is a modern adaptation of the willow stick named for "Murphy's Law," that is, what can go wrong will go wrong (J. Reynolds, Alaska Cooperative Fish and Wildlife Research Unit, personal communication). The method utilizes a 3-m section of 2.5–3.0-cm-diameter aluminum pipe hinged to a second piece of pipe (0.5–2.0 m long) that extends as a probe under the ice. The second piece is fashioned at the far end with a sponge rubber sleeve for flotation and a snap. The probe can vary in length to accommodate different water depths. To operate, the far end of the second pipe is pushed through one hole and maneuvered toward a second hole where a rope attached to the pipe is hooked and pulled up through the second hole. This technique is especially useful in currents or under thick ice.

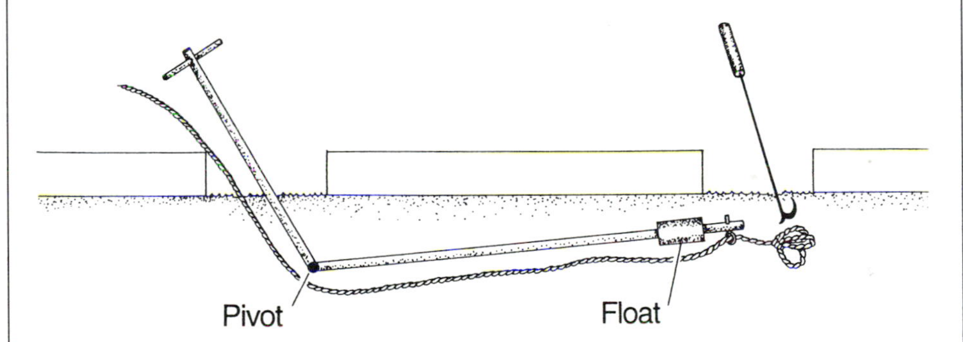

Jigger. The jigger is a floating device that is placed under the ice and maneuvers a rope to a second hole. When the rope is pulled, the metal claw moves the jigger away from the first hole (adapted from Hamley 1980, with permission).

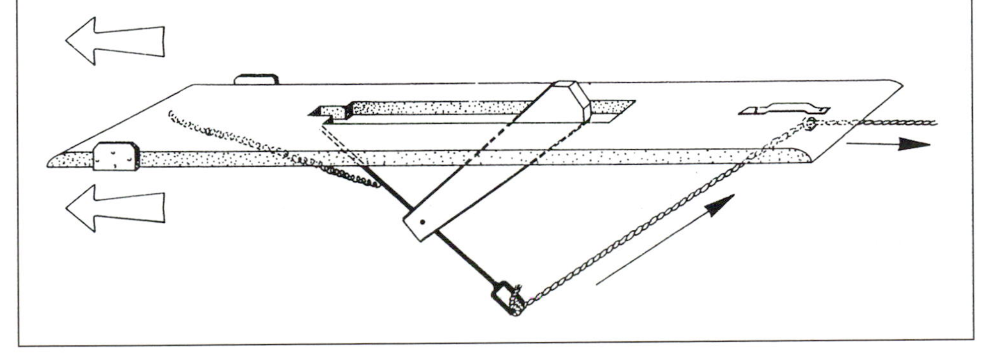

6.2.1.3 Target Organisms

Many fish species are caught in gill nets, but gill nets are especially selective for species that move substantial distances in their daily routines. The species selectivity of experimental gill nets has been described in several studies (Berst 1961; Heard 1962; Trent and Pristas 1977; Yeh 1977; Bronte and Johnson 1983, 1984; Boxrucker and Ploskey 1989).

6.2.1.4 Biases

The size selectivity of various mesh sizes is quite specific with gill nets. For a particular mesh size, fish of a particular size are held most securely; smaller or larger fish are less likely to be caught. Very small fish can swim through the mesh, and very large fish cannot penetrate into the mesh to become entangled. A typical gill-net size selectivity curve is bell shaped; catch frequency declines to zero at both sides of a maximum (Pope et al. 1975). Size selectivity curves for various mesh sizes have been computed for many species (Hamley 1975; Jensen 1986, 1990; Reddin 1986; Winters and Wheeler 1990; Henderson and Wong 1991; Lott and Willis 1991; Wilde 1993). Two generalizations regarding size selectivity are (1) the optimum girth for capture is about 1.25 times the mesh perimeter and (2) fish more than 20% longer or shorter than the optimum length are seldom caught (Hamley 1980). Substantial variation in the shape and magnitude of size selectivity curves has been observed. For example, Hamley and Regier (1973) described a bimodal curve for walleye. The shape of curves can vary within a species by sex and season (Hamley 1975). Methods have been developed to correct for the size selectivity of an array of mesh sizes when the size distribution of fish in a stock is described (Willis et al. 1985; Spangler and Collins 1992).

In general, large fish are more easily captured than small fish in gill nets. Young fish are not the most abundant age-group in most samples collected with experimental gill nets because small fish are not as likely to push themselves into the mesh. Size selectivity can bias estimates of growth rate, mortality, and body condition because larger meshes select larger fish of each age-group (Hamley 1980). The most important factors influencing the size selectivity of a gill net for a particular species are mesh size, elasticity of the twine, hanging ratio, strength and flexibility of the twine, and visibility of the twine (Jester 1977), as well as the time and manner in which the net is fished.

The mesh material can have a substantial influence on gill-net efficiency. In general, nylon mesh yields a higher CPUE than does linen or cotton, whereas monofilament yields the highest CPUE of all materials (McCombie and Fry 1960; Berst 1961; Larkins 1963; Pristas and Trent 1977; Collins 1979; Henderson and Nepsey 1992). Because of the efficiency of monofilament gill nets, most biologists now use this type of mesh material. However, changes in mesh material have confounded long-term monitoring programs by altering the efficiency of the gear.

Species and size selectivity of gill nets, as well as efficiency for particular species, are governed to a great degree by the construction of a net. Monitoring or assessment projects should use gill nets of identical design, material, and construction. Lack of concern for variables as insignificant as the diameter of the monofilament or the color of a net can greatly influence the CPUE and size selectivity of a gill net and the quality of subsequent sampling data. Hansen (1974) compared the size selectivity of monofilament gill nets with different filament diameters and found that nets with smaller filament diameters captured larger fish. The CPUE can also be reduced when fish see a net, so color of the mesh can be important (Jester 1973).

The capture of fish in gill nets is also a function of fish activity. The activity of many fishes is related to the degree of light under water, which leads to diurnal movement patterns. Most species exhibit nocturnal or crepuscular activity peaks that are consistent from day to day within a particular season. Seasonal patterns in movements and distributions occur as a result of spawning activity, habitat requirements, and food availability (Hubert and O'Shea 1992).

Several physical and chemical variables influence the movement and distribution of fish (Berst 1961; Welcomme 1975; May et al. 1976; Pristas and Trent 1977; Craig and Fletcher 1982; Hubert and Sandheinrich 1983; Craig et al. 1986; Mero and Willis 1992; Pope and Willis 1996). These variables include season, weather fronts, currents, water temperatures, water depths, water level fluctuations, turbidity, and thermocline location. Many of these variables can be measured, and predictive relations between factors influencing fish activity and subsequent gill-net catches can be defined.

The duration of a gill-net set also influences sampling results. Catches do not accumulate in a gill net at a uniform rate (Kennedy 1951; Austin 1977; Minns and Hurley 1988). The efficiency of a gill net decreases as fish accumulate in it. Eventually, the number of captured fish can reach a saturation point at which further capture of fish does not occur. Catch is generally not linearly related to the duration of the set ("soak time"), and saturation generally occurs when only a small percentage of the meshes are occupied.

The sampling regime (i.e., season, time of day, location, and duration of sets) influences the CPUE. A standardized sampling scheme can be used from year to year, as well as among water bodies, to minimize the variability that is generated by physical, chemical, and biological variables. A rigidly defined sampling design that identifies the season, time, location, and duration of sets, coupled with precise gear and deployment specifications, can reduce much of the variability among gill-net samples and enable comparisons of CPUE over time or among water bodies (Hubert and O'Shea 1992; Mero and Willis 1992).

Gill nets are used routinely in monitoring programs. The CPUE of fish in gill nets has been related to several factors of interest to managers, such as recruitment (Willis 1987), harvest (Nesler 1986), angler catch rates (Isbell and Rawson 1989), and fish density (Borgstrom 1992).

Because many fish caught by gill nets die in the net or are injured upon removal, gill nets are less appropriate when live fish must be obtained or released. Gill nets stress fish more than do other types of passive gear (Hopkins and Cech 1992).

6.2.2 Trammel Nets

A trammel net generally consists of three parallel panels of netting suspended from a float line and attached to a lead line (Figure 6.1). The two outer panels are of large-mesh netting, whereas the inner panel is of small mesh. The inside small panel has greater depth and hangs loosely between the two outer panels.

Fish generally are captured in a bag or pocket of netting. A fish swimming into the net from either side passes through one of the large-mesh outer panels and hits the small-mesh inner panel; the fish then pushes the small-mesh panel through one of the large openings of the facing large-mesh outer panel. This action forms a bag or pocket in which the fish is entangled. Small fish may be wedged or gilled, whereas large fish may be tangled in the netting.

6.2.2.1 Construction

Trammel nets are generally constructed of cotton or nylon webbing because these materials are highly flexible. Numerous designs of trammel nets have been used by commercial fishers (Dumont and Sundstrom 1961; Nedelec 1975). A typical trammel net used in reservoir and river sampling is a 2-m-deep net with 250-mm-bar-mesh outer panels and a 25-mm-bar-mesh inner panel. The inner panel typically is two-thirds greater in depth (3.3-m-deep netting). All three panels are attached to a single float line with either floats or foam core and a single lead line with either lead weights or a lead core.

6.2.2.2 Deployment and Applications

Trammel nets are set in the same ways as gill nets: as stationary bottom sets or drifting, floating, or encircling sets. Commercial fishers consider trammel nets to be most efficient when the nets are set around an aggregation of fish and the fish are frightened or driven into the nets (White 1959). A typical set of this type involves surrounding an area of aquatic vegetation with trammel nets. Stationary bottom sets can be made in lakes and in backwaters and slow-moving sections of rivers.

Drifted trammel nets have been used to sample channel habitats of large rivers (Hubert and Schmitt 1982a; Hurley et al. 1987). Trammel nets can be drifted in channel habitat free of snags.

Despite their versatility, trammel nets are seldom used in assessments of fisheries. However, the CPUE of commercial species in trammel nets can provide a reliable estimate of total harvest from a water body (Bronte and Johnson 1984).

6.2.2.3 Target Organisms

Trammel nets are selective for certain fish species (Bronte and Johnson 1983, 1984). They are efficient in capturing large, mobile species that occur in shallow water of lakes and reservoirs. Trammel nets are also effective in capturing large fishes in channels of large rivers when the nets are drifted with the current.

6.2.2.4 Biases

The sampling biases that exist with gill nets also occur with trammel nets. However, trammel nets are less size selective than are gill nets. An advantage of trammel nets over gill nets is that observed mortality of captured fish is less with trammel nets.

6.3 ENTRAPMENT GEARS

Aquatic animals enter entrapment gear by their own movement. They are captured when the entrance is in the path of their movement, when they attempt to move around or over a barrier, or when they are attracted to the enclosure by the presence of bait or other animals or by the cover it appears to provide. Once in the entrapment device, an animal may escape through an entryway or be retained until it is removed.

A variety of entrapment devices exist (Dumont and Sundstrom 1961; von Brandt 1964; Nedelec 1975; Everhart and Youngs 1981). Knowledge of migration, cover-seeking habits, escape reactions, and diets of aquatic animals influences the design of entrapment devices used for various organisms. Entrapment devices used in fisheries sampling are often small, portable versions of commercial fishing gear.

6.3.1 Hoop Nets

A hoop net is a cylindrical or conical net distended by a series of hoops or frames covered by web netting. The net has one or more internal funnel-shaped throats that are directed inward from the mouth of the net (Figure 6.2; Nedelec 1975). Local terminology for hoop nets can be confusing because the name often refers to the species selectivity of a particular design. Some names given to hoop nets include buffalo nets, bait nets, fiddler nets, and even fyke nets (Starrett and Barnickol 1955).

6.3.1.1 Construction

Hoop nets are typically made with wooden, plastic, fiberglass, or steel hoops. Hoop diameter can vary from 0.5 m to over 3 m; there can be four to eight hoops in a net. The cotton or nylon webbing tied around the hoops can range from 10-mm to over 100-mm bar mesh. Generally, two funnel-shaped throats are attached, one to the second hoop and a second to the fourth hoop from the mouth of the net. The throats may be of two basic designs, square or finger throats. A square throat is simply a square to circular opening in the constricted end of the funnel. A finger throat is composed of two half-cones of twine on each side of a hoop secured to a back hoop (Hansen 1944). A finger throat is often placed in the second funnel to lessen the chances of fish escaping once they have passed through it. The closed end of the net, where the fish accumulate, is called the cod end or the pot. A drawstring is attached to the cod end for removing captured fish. Hoop nets can be protected from deterioration by periodic dipping in petroleum-based net-coating material.

6.3.1.2 Deployment and Application

In riverine habitats, hoop nets are set with the mouth opening downstream in water that entirely covers the hoops of the net. Hoop nets are held in place by attaching a rope from the hoop net to an anchor or stake driven into the stream bottom. Metal stakes 5–15 mm in diameter and 0.6–1.5 m in length can be driven into the stream bottom in water up to 5 m deep by means of a driver pole. A driver pole is a long pole with a sleeve (2–4 cm diameter) at one end. The pole is used to drive a stake with a rope attached into the stream bottom.

Commercial fishers generally set hoop nets without buoys to protect their gear from vandalism and theft. Landmarks are used to identify the location of nets. Retrieval is achieved by dragging the bottom with a grappling hook until the net or rope is hooked.

Current keeps the hoops separated and the net stretched. During parts of the year hoop nets are often baited with cheese scraps or pressed soybean cake to enhance their efficiency for catfishes and buffaloes (Pierce et al. 1981; Gerhardt and Hubert 1989).

Hoop nets can be modified for use in lakes and reservoirs. For example, Bernard et al. (1991) developed a hoop net to capture burbot in lakes.

Hoop nets most often are used in channel habitats of rivers because they can be set and fished effectively in strong currents without being washed away or becoming clogged with debris. Hoop nets have been used in several reservoir habitats from the headwaters to the lower ends of impoundments. They have been used to assess population structure (Gerhardt and Hubert 1991), evaluate life history (Smith and Hubert 1989; Hubert and O'Shea 1991), and describe habitat associations (Hubert and Schmitt 1982b) of fishes.

Hoop net

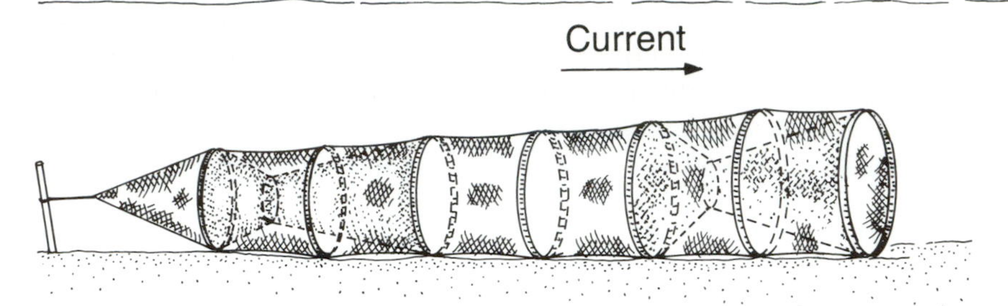

Fyke net

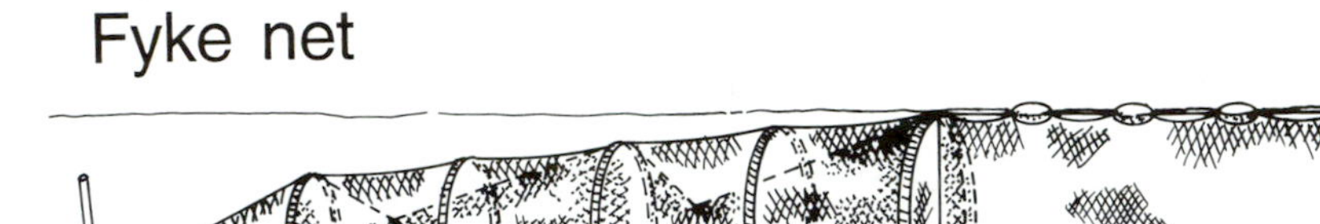

Modified fyke net

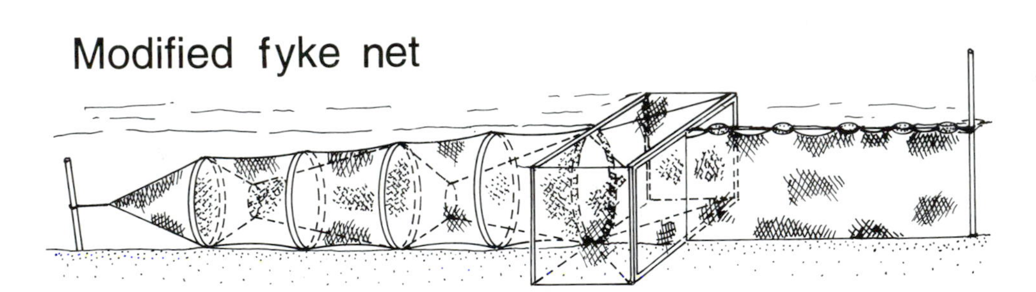

Trap net

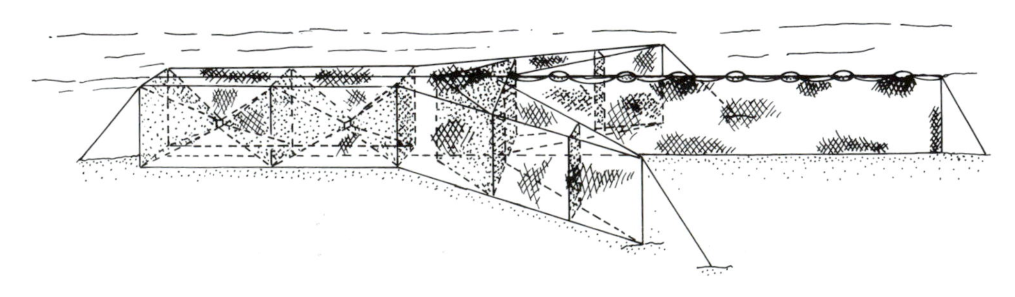

Figure 6.2 Four types of entrapment gears—hoop net, fyke net, modified fyke net, and small trap net—and illustrations of typical sets (modified from Crowe 1950 and Sundstrom 1957).

6.3.1.3 Target Organisms

Hoop nets are selective for fish species attracted by cover, bait, or other fish. In the Mississippi River basin, hoop nets of different designs have been selective for catfishes, buffaloes, and common carp (Starrett and Barnickol 1955). Hoop-netted fish are generally captured unharmed and can be released with little or no injury to the fish.

6.3.1.4 Biases

Net construction (hoop diameter, mesh dimensions, and mouth size) and deployment have a substantial influence on the species and size selectivity of hoop nets (Hubert and Schmitt 1982b; Holland and Peters 1992; Hubert and Patton 1994). For example, nets of three mesh sizes (25-, 32-, and 38-mm bar mesh) yielded significant differences in species composition, length frequencies, and CPUE of fishes in the Platte River, Nebraska (Holland and Peters 1992). In a study on the Mississippi River, when hoop nets of two sizes were deployed, the larger-diameter nets captured twice as many fish and half again as many species as the smaller net with equal sampling effort (Hubert and Schmitt 1982b). The species selectivity and CPUE of hoop nets can also be influenced by bait (Pierce et al. 1981; Gerhardt and Hubert 1989). Escape rates of different species from hoop nets also influence sampling results because some species are more adept at escape than others (Hansen 1944).

Physical, chemical, and biological variables can have significant influences on CPUE. Season, water temperature, current velocity, turbidity, dissolved oxygen, and habitat type all affect CPUE of individual species (Mayhew 1973; Hubert and Schmitt 1982b; Holland and Peters 1992). Of these, the most influential variables seem to be season, water temperature, and turbidity. Catch rates are often high immediately preceding and during the spawning periods of riverine fishes (Smith and Hubert 1989; Hubert and O'Shea 1991). In most cases, as water temperature declines or turbidity increases, the CPUE declines.

As with entanglement gear, the design and construction of hoop nets can be standardized, along with sampling time and location, to reduce sampling variability. However, because of the dynamic nature of rivers, little can be done to control variability of physical factors such as turbidity and current velocity.

6.3.2 Fyke Nets and Trap Nets

Fyke nets are similar to hoop nets, but they have been modified for use in lentic habitats. Fyke nets have one to three leads or wings of webbing attached to the mouth to guide fish into the enclosure (Figure 6.2). The net is set so that the leads or wings intercept moving fish. When fish follow the lead or wing in an attempt to get around the netting, they swim into the enclosure. Fyke nets are also known as wing nets, frame nets, trap nets, and hoop nets. Modified fyke nets have rectangular frames to enhance their stability (Figure 6.2).

A variety of large entrapment devices have been used in coastal-marine and large-lake fisheries (Dumont and Sundstrom 1961; Alverson 1963; Grinstead 1968). These include the pound-net fishery of the Atlantic Coast (Reid 1955) and the deep-trap-net fishery of the Great Lakes (Van Oosten et al. 1946). Modified, scaled-down versions of commercial trap nets have been used by fisheries scientists (Crowe 1950; Beamish 1972).

6.3.2.1 Construction

Fyke nets are hoop nets to which one or more leads are attached (Figure 6.2). A single lead extending from the mouth of the hoop net outward along the axis of the net can be added in lentic habitats to increase catch efficiency (Winkle et al. 1990; Hubert and Guenther 1992; Johnson et al. 1992). Leads are generally of the same height as the hoop net and constructed of similar net material. They are suspended between buoyant and weighted lines much like a gill net.

Modified fyke nets are widely used to sample fish in lakes and reservoirs. Modified fyke nets generally have at least one or two rectangular or square frames at the mouth to prevent the net from rolling on the bottom. The first throat is recessed from the mouth and attached to the second or third frame. Modified fyke nets can have a single lead extending outward along the axis of the net, but two wings are sometimes added at an angle to the lead (often at angles of approximately 45° to the lead). Leads and wings are held in place with poles or anchors.

A portable trap net that weighed 45 kg, including the anchors, was described by Crowe (1950). It was similar to a fyke net except all the frames were rectangular. This net was constructed of 1.5-cm-bar-mesh netting and had a 1.3-m-deep enclosure or "pot" that was 2.6 m long and 2 m wide. The net had two wings leading to an "outside heart" of 50-mm-bar mesh and two wings leading to an "inner heart" of 32-mm-bar mesh. Both hearts were covered with similar netting. A 30-m-long, 1.3-m-deep lead was attached.

6.3.2.2 Deployment and Applications

Generally, fyke nets, modified fyke nets, and trap nets are set in shallow water not much deeper than the height of the lead or wings and the height of the first frame or hoop. However, they can be set in water over 10 m deep. The net and lead are set taut by anchors or poles driven into the bottom. A single-pot set involves one lead and one pot. The end of the lead is set on or near the shore, and the lead is extended perpendicular to shore so that fish cannot swim around or under it. When the net is set from a small boat, it is placed on the bow with the pot on the bottom and the lead on top. The end of the lead is staked or anchored and played out as the boat moves in reverse. When the lead is fully extended, the pot is put overboard, stretched, and staked or anchored in position. Fish are removed by lifting only the pot into the boat.

Fyke nets and modified fyke nets can also be deployed away from shore in pairs with a single lead between them (Nedelec 1975). This type of set is generally made parallel to shore along the outer edge of weed beds or along shallow offshore reefs.

The location of fyke nets, modified fyke nets, or trap nets in lakes or reservoirs, as well as the manner in which they are set, can influence catch (Bernhardt 1960). Seasonal variation in catches is typical (Hansen 1953; Kelley 1953; Guy and Willis 1991). Standardized nets, sampling locations, and times are needed to reduce sampling variability.

Fyke nets, modified fyke nets, and trap nets are generally used in shallow areas of lakes and reservoirs (Crowe 1950). They have been used to sample fish in sections of streams and rivers with slow current velocities, as well as in backwaters and sloughs (Swales 1981). Fyke nets can be used in relatively heavy vegetation or marsh-type habitats. Where dense submerged or emergent vegetation occurs, paths can be cut to place fyke nets, but net damage by aquatic mammals such as muskrats can be substantial in these habitats (Kelley 1953).

Fyke nets, modified fyke nets, and trap nets can be used in water up to 15 m deep over a clean, firm bottom. They are not applicable over soft bottoms because anchors fail to hold.

6.3.2.3 Target Organisms

As with hoop nets, fyke nets, modified fyke nets, and trap nets are selective for certain species and sizes of fish. Cover-seeking, mobile species seem to be the most susceptible to capture (Hoffman et al. 1990). Modified fyke nets are especially efficient in the capture of crappies (Boxrucker and Ploskey 1989; McInerny 1989). Fyke nets, modified fyke nets, and trap nets are very effective in the capture of migratory species that tend to follow shorelines.

6.3.2.4 Biases

As with other types of gear, fyke nets and trap nets have species selectivity (Laarman and Ryckman 1982) and size selectivity (Meyer and Merriner 1976; Milewski and Willis 1991; Kraft and Johnson 1992), but they are less selective than gill nets. They are selective for larger fish of age-groups above the minimum imposed by the physical dimensions of the net (Latta 1959). Some selectivity also occurs because of variable escape rates relative to season, species, and size of fish (Hansen 1944; Patriarche 1968). Trap nets can be modified to allow escapement of small fishes in commercial fisheries (Meyer and Merriner 1976).

Fyke nets and trap nets induce less stress on captured fish than do entanglement gears (Hopkins and Cech 1992), and most captured fish can be released unharmed. However, gilling of small fish in the mesh of fyke nets and trap nets causes some mortality (Schneeberger et al. 1982). Fyke nets and trap nets are widely used in the assessment of fisheries stocks because low mortality of fish is associated with their use, but catch rates with these nets are generally lower than those with gill nets. The CPUE of fish in trap nets has been related to the duration of sets, habitat, and season (Hamley and Howley 1985), as well as the abundance of sport fishes (Ryan 1984).

6.3.4 Pot Gears

Pot gears are portable, rigid traps with small openings through which animals enter. They are used to capture fish and crustaceans. Pot devices vary in designs and dimensions depending on the species sought (Carter 1954; Sundstrom 1957; Beall and Wahl 1959; Dumont and Sundstrom 1961; Alverson 1963; Nedelec 1975; Rounsefell 1975; Schwartz 1986; Perry and Williams 1987). Lobster pots, minnow traps, slat traps for catfishes, eel pots, and crab pots are examples of different types of pot gears (Figure 6.3). Pot gears are generally small enough that many of the devices can be put on a boat or carried by hand.

Pot gears are most efficient in the capture of bottom-dwelling species seeking food or shelter (Everhart and Youngs 1981). To reach a receptacle containing bait, the fish or crustacean must pass through a more or less conical-shaped funnel in most pot designs. In some designs they pass through more than one funnel, making escape more difficult.

An example of a typical pot gear is the half-round lobster pot constructed with a rectangular base and three half-round bows, one at each end and one near the center (Dumont and Sundstrom 1961; Alverson 1963; Everhart and Youngs 1981). The bows are covered with lath, and a door is constructed on one side for removal of American lobsters. The pot contains two inside compartments. The smaller chamber has on each side an opening with a funnel of netting. From the smaller chamber a

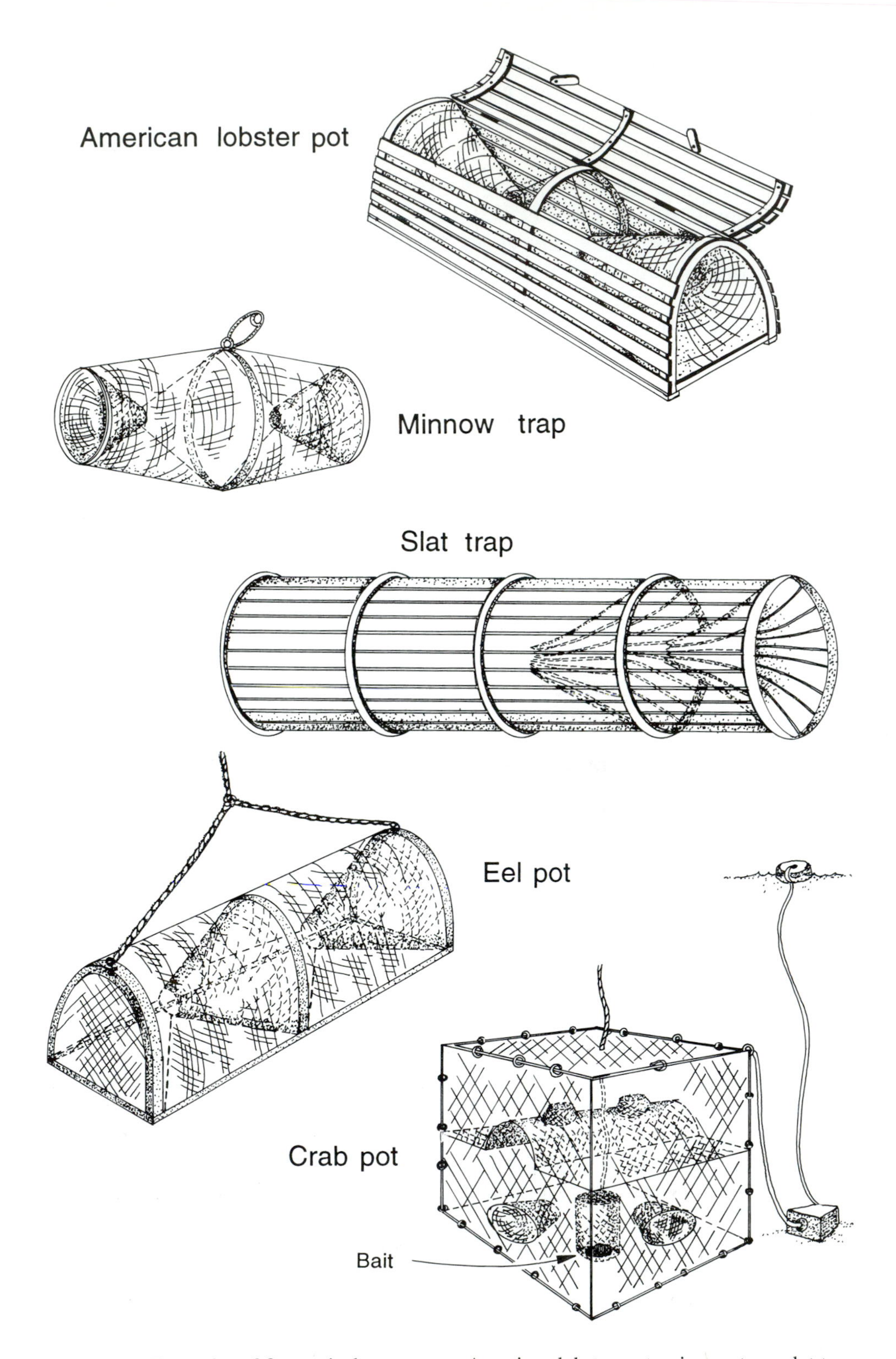

Figure 6.3 Examples of five typical pot gears—American lobster pot, minnow trap, slat trap, eel pot, and crab pot (modified from Sundstrom 1957).

funnel attached to the middle bow leads to the larger parlor. Bait is placed on a hook or in a mesh bag attached to the center bow. The lobster pot is deployed by weighting it with bricks or stones and attaching a buoy line to a lower corner of the chamber end. Commercial fishers deploy the pots in strings of 10–15 spaced 10–20 m apart.

The crayfish industry in North America depends upon trapping crayfish in pot gear, and substantial work has gone into enhancing gear efficiency (Rach and Bills 1987; Romaire and Osorio 1989; Stuecheli 1991; Kutka et al. 1992). As with other passive sampling gears, species and size selectivity with different trap designs and baits have been observed.

Catfishes are commercially harvested with pot gears (Perry 1979; Perry and Williams 1987). A common device is the slat trap (Figure 6.3). Wire cages have also been used to capture catfishes. They have been constructed of 2.5-cm-mesh (bar measure) wire netting rolled into a cylinder (approximately 0.7 m in diameter) about 1.3 m long. Two funnel-shaped throats of wire netting are attached about 20 cm apart at one end of the cylinder. A door is constructed at the opposite end. The bottom of the trap is flattened to prevent rolling. The traps are baited with cheese or soybean cake.

Pot gears have not been used extensively to monitor or assess fisheries. However, one of the most studied pot devices is the Windermere perch trap (Worthington 1950; Bagenal 1972). The trap is constructed with three semicircular wire hoops attached to a 67-cm × 76-cm base and covered with wire netting that has 1.3-cm-diameter hexagonal openings. One end of the trap has a funnel with a 8.5-cm-diameter opening, and the other end has a door for removal of the catch. The traps are set unbaited. The trap is cheap, easy to make, and easy to use and is best suited for estimating age structure and condition of fish. It is not efficient for monitoring relative abundance because of the variability in catch among traps and sets of individual traps.

A problem associated with pot gears, but also characteristic of other passive gears, is continued capture of animals by the gear if it is lost, a process called "ghost fishing" (Guillory 1993). Efforts have been made to modify pot gears by incorporating biodegradable material into their construction (Kumpf 1980).

6.3.5 Weirs

Weirs are barriers built across a stream to divert fish into a trap. They are most suited for capturing migratory fishes as they move upstream or downstream. A variety of weir designs have been used to capture fish; they can be permanent or temporary structures (von Brandt 1964; Welcomme 1975; Craig 1980).

The Wolf-type weir (Wolf 1951) is an efficient design used to capture salmon and trout moving downstream. The basic design has been applied in large salmon streams with extremely variable flow (Hunter 1954). Whelan et al. (1989) provided an improved design for use in rivers with severe and rapid fluctuations in flows.

Salmonid smolt traps that use weirs have been designed to assess emigration in small rivers and through beaver ponds (Tsumura and Hume 1986; Elliott 1992). The traps are relatively inexpensive, easily maintained, and effective in the capture of smolts. In large rivers where weirs cannot be constructed, floating inclined-plane smolt traps have been used (McMenemy and Kynard 1988; DuBois et al. 1991).

Two-way fish traps have been designed for use in small trout streams (Whalls et al. 1955; Twedt and Bernard 1976). Two-way fish traps are effective in relatively constant flows without substantial amounts of debris in the water.

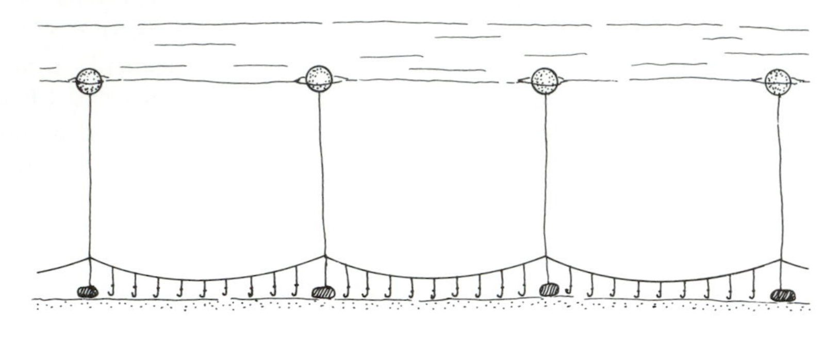

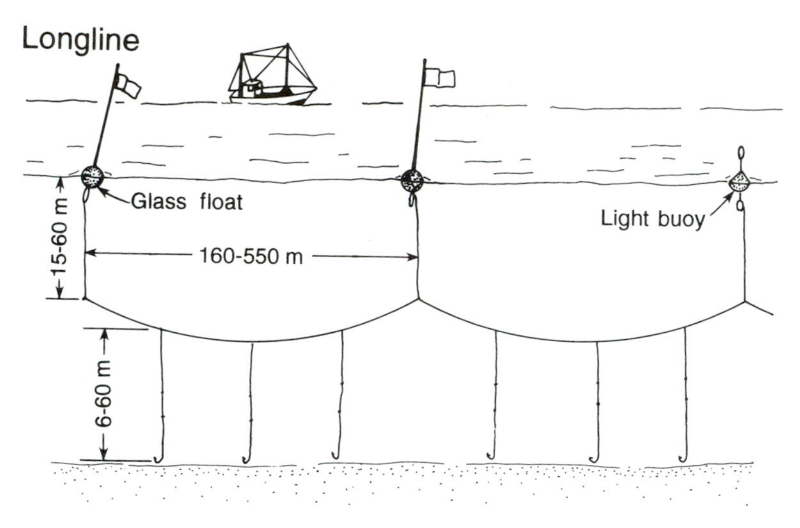

Figure 6.4 Two types of angling gears—trotline and longline (modified Rounsefell and Everhart 1953).

Weirs have been used to gather data on age structure, condition, sex ratio, spawning escapement, smolt production, abundance of sexually mature adults, and migratory patterns of fish. The use of weirs is generally restricted to small rivers and streams because of construction expenses, formation of navigation obstacles, and tendency of weirs to clog with ice and debris, which can cause flooding or collapse of the structure.

6.4 ANGLING GEARS

A gear that has been used worldwide to catch fish is a baited hook and line (von Brandt 1964). Passive angling techniques requiring little human attendance have been developed to capture fishes in both freshwater and marine systems.

Angling gears include devices as simple as a single baited hook attached to a line that is tied to a float or a tree branch. Commercial angling techniques generally involve a main line (ground line), strung horizontally, to which are attached short vertical lines (drop lines), each with a baited hook (Figure 6.4). Terminology for these multiple-hook devices varies with the fisheries in which they are applied, but some of the common terms are trotlines, longlines, drift lines, and trawl lines (Rounsefell and Everhart 1953).

Trotlines are devices used in warmwater inland fisheries (Starrett and Barnickol 1955). There is no standard design of trotlines, but heavy cord usually serves as the main line and lighter cord for the drop lines. The number of hooks on trotlines can be quite variable, and often is regulated by state agencies. Trotlines can be anchored to lie on the bottom of a lake or stream or they can be suspended off the bottom with floats. They are most often used to capture catfishes by use of baits such as minnows, cut fish, crayfish, or clam meat. They can be selective for common carp if dough bait or corn is used. Trotlines have rarely been used to sample or monitor inland fish stocks because they are very selective and catch rates vary widely.

Longlines are used in marine fisheries and are much larger in scale than trotlines. Individual longlines can be several kilometers in length and hold thousands of baited hooks. They are known as longlines or setlines off the Pacific Coast and trawl lines in the North Atlantic (Rounsefell and Everhart 1953; Dumont and Sundstrom 1961). Longlines that are fished without being anchored have been used in commercial fisheries for tunas in the Pacific Ocean. Longlines set on the ocean floor have been used to harvest haddock and halibuts. These devices have not been used as sampling devices by fisheries managers because of their scale, but the catch by commercial anglers can be used to monitor trends in the abundance of marine stocks (Bell 1970; Schaefer 1970; Rounsefell 1975).

6.5 SUMMARY

Passive sampling gears are some of the most useful tools available to fisheries managers and researchers for the appraisal of sport or commercial fisheries or assessment of environmental effects on stocks of aquatic animals (Allen et al. 1960; Hocutt and Stauffer 1980). However, problems with sampling variability and gear selectivity are universal. Standardization of sampling devices and strict sampling protocols are necessary to reduce variation among samples and to detect possible changes in stocks that are the result of management efforts or environmental effects (Fisheries Techniques Standardization Committee 1992).

6.6 REFERENCES

Allen, G. H., A. D. Delacy, and D. W. Gotshall. 1960. Quantitative sampling of marine fishes—a problem in fish behavior and fishing gear. Pages 448–511 *in* E. A. Pearson, editor. Waste disposal in the marine environment. Pergammon Press, New York.

Alverson, D. L. 1963. Fishing gear and methods. Pages 45–64 *in* M. E. Stansby, editor. Industrial fishery technology. Krieger Publishing, New York.

Austin, C. B. 1977. Incorporating soak time into measurement of fishing effort in trap fisheries. U.S. National Marine Fisheries Service Fishery Bulletin 75:213–218.

Bagenal, T. B. 1972. The variability in the number of perch, *Perca fluviatilis* L., caught in traps. Freshwater Biology 2:27–36.

Bannerot, S. P., and C. B. Austin. 1983. Using frequency distributions of catch per unit effort to measure fish-stock abundance. Transactions of the American Fisheries Society 112:608–617.

Beall, H. B., and R. W. Wahl. 1959. Trapping bluegill sunfish in West Virginia ponds. Progressive Fish-Culturist 21:138–141.

Beamish, R. J. 1972. Design of a trap net for sampling shallow water habitats. Fisheries Research Board of Canada Technical Report 305.

Bell, T. H. 1970. Management of Pacific halibut. American Fisheries Society Special Publication 7:209–221.

Bernard, D. R., G. A. Pearse, and R. H. Conrad. 1991. Hoop traps as a means to capture burbot. North American Journal of Fisheries Management 11:91–104.

Bernhardt, R. W. 1960. Effect of fyke-net position on fish catch. New York Fish and Game Journal 7:83–84.

Berst, A. H. 1961. Selectivity and efficiency of experimental gill nets in South Bay and Georgian Bay of Lake Huron. Transactions of the American Fisheries Society 60:413–418.

Borgstrom, K. 1992. Effect of population density on gillnet catchability in four allopatric populations of brown trout (*Salmo trutta*). Canadian Journal of Fisheries and Aquatic Sciences 49:1539–1545.

Boxrucker, J., and G. Ploskey. 1989. Gear and seasonal biases associated with sampling crappie in Oklahoma. Proceedings of the Annual Conference Southeastern Association of Fish and Wildlife Agencies 42(1988):89–97.

Bronte, C. R., and D. W. Johnson. 1983. Occurrence of sport fish in a commercial net fishery in Kentucky. North American Journal of Fisheries Management 3:239–242.

Bronte, C. R., and D. W. Johnson. 1984. Evaluation of the commercial entanglement-gear fishery in Lake Barkley and Kentucky Lake, Kentucky. North American Journal of Fisheries Management 4:75–83.

Bulkley, R. V. 1970. Fluctuation in abundance and distribution of common Clear Lake fishes as suggested by gillnet catch. Iowa State Journal of Science 44:413–422.

Carlander, K. D. 1953. Use of gill nets in studying fish populations, Clear Lake, Iowa. Proceedings of the Iowa Academy of Science 60:623–625.

Carter, E. R. 1954. An evaluation of nine types of commercial fishing gear in Kentucky Lake. Transactions of the Kentucky Academy of Science 15:56–80.

Chadwick, J. W., M. F. Cook, and D. S. Winters. 1987. Vertical gill net for studying depth distribution of fish in small lakes. North American Journal of Fisheries Management 7:593–594.

Collins, J. J. 1979. Relative efficiency of multifilament and monofilament nylon gill net towards lake whitefish (*Coregonus clupeaformis*) in Lake Huron. Journal of the Fisheries Research Board of Canada 36:1180–1185.

Craig, J. F. 1980. Sampling with traps. Pages 55–70 *in* T. Backiel and R. L. Welcomme, editors. Guidelines for sampling fish in inland waters. FAO (Food and Agriculture Organization of the United Nations) EIFAC (European Inland Fisheries Advisory Commission) Technical Paper 33.

Craig, J. F., and J. M. Fletcher. 1982. The variability in the catches of charr, *Salvelinus alpinus* L., and perch, *Perca fluviatilis* L., from multi-mesh gill nets. Journal of Fish Biology 20:517–526.

Craig, J. F., A. Sharma, and K. Smiley. 1986. The variability in catches from multi-mesh gillnets fished in three Canadian lakes. Journal of Fish Biology 28:671–678.

Crowe, W. R. 1950. Construction and use of small trap nets. Progressive Fish-Culturist 12:185–192.

DuBois, R. B., J. E. Miller, and S. D. Plaster. 1991. An inclined-screen smolt trap with adjustable screen for highly variable flows. North American Journal of Fisheries Management 11:155–159.

Dumont, W. H., and G. T. Sundstrom. 1961. Commercial fishing gear of the United States. U.S. Fish and Wildlife Circular 109.

Elliott, S. T. 1992. A trough trap for catching coho salmon smolts emigrating from beaver ponds. North American Journal of Fisheries Management 12:837–840.

Everhart, W. H., and W. D. Youngs. 1981. Principles of fisheries science, 2nd edition. Comstock Publishing Associates, Ithaca, New York.

Fisheries Techniques Standardization Committee. 1992. Fish sampling and data analysis techniques used by conservation agencies in the U.S. and Canada. American Fisheries Society, Fisheries Management Section, Fisheries Techniques Standardization Committee, Bethesda, Maryland.

Gebhards, B. S. 1966. Repairing nets. Pages 110–125 *in* A. Calhoun, editor. Inland fisheries management. California Department of Fish and Game, Sacramento.

Gerhardt, D. R., and W. A. Hubert. 1989. Effect of cheese bait on seasonal catches of channel catfish in hoop nets. North American Journal of Fisheries Management 9:377–379.

Gerhardt, D. R., and W. A. Hubert. 1991. Population dynamics of a lightly exploited channel catfish stock in the Powder River system, Wyoming–Montana. North American Journal of Fisheries Management 11:200–205.

Grinstead, B. S. 1969. Comparison of various designs of Wisconsin-type trap nets in TVA reservoirs. Proceedings of the Annual Conference Southeastern Association of Game and Fish Commissioners 22(1968):444–457.

Guillory, V. 1993. Ghost fishing by blue crab traps. North American Journal of Fisheries Management 13:459–466.

Guy, C. S., and D. W. Willis. 1991. Seasonal variation in catch rate and body condition for four fish species in a South Dakota natural lake. Journal of Freshwater Ecology 6:281–292.

Hamley, J. M. 1975. Review of gill net selectivity. Journal of the Fisheries Research Board of Canada 32:1943–1969.

Hamley, J. M. 1980. Sampling with gill nets. Pages 37–53 *in* T. Backiel and R. L. Welcomme, editors. Guidelines for sampling fish in inland waters. FAO (Food and Agriculture Organization of the United Nations) European Inland Fisheries Advisory Commission Technical Paper 33, Rome, Italy.

Hamley, J. M., and T. P. Howley. 1985. Factors affecting variability of trapnet catches. Canadian Journal of Fisheries and Aquatic Sciences 42:1079–1087.

Hamley, J. M., and H. A. Regier. 1973. Direct estimates of gill net selectivity to walleye (*Stizostedion vitreum vitreum*). Journal of the Fisheries Research Board of Canada 30:817–830.

Hansen, D. F. 1944. Rate of escape of fishes from hoop nets. Transactions of the Illinois State Academy of Science 37:115–122.

Hansen, D. F. 1953. Seasonal variation in hoop net catches at Lake Glendale. Transactions of the Illinois State Academy of Science 46:216–266.

Hansen, R. G. 1974. Effect of different filament diameters on the selective action of monofilament gill nets. Transactions of the American Fisheries Society 103:386–387.

Heard, W. R. 1962. The use and selectivity of small-meshed gill nets at Brooks Lake, Alaska. Transactions of the American Fisheries Society 91:263–268.

Henderson, B. A., and S. J. Nepsey. 1992. Comparison of catches in mono- and multifilament gill nets in Lake Erie. North American Journal of Fisheries Management 12:618–624.

Henderson, B. A., and J. L. Wong. 1991. A method for estimating gillnet selectivity of walleye (*Stizostedion vitreum vitreum*) in multimesh multifilament gill nets in Lake Erie and its application. Canadian Journal of Fisheries and Aquatic Sciences 48:2420–2428.

Hocutt, C. H., and J. R. Stauffer. 1980. Biological monitoring of fish. Heath, Lexington, Massachusetts.

Hoffman, G. C., G. L. Milewski, and D. W. Willis. 1990. Population characteristics of rock bass in three northeastern South Dakota lakes. Prairie Naturalist 22:33–40.

Holland, R. S., and E. J. Peters. 1992. Differential catch by hoop nets of three mesh sizes in the lower Platte River. North American Journal of Fisheries Management 12:237–243.

Hopkins, T. E., and J. J. Cech, Jr. 1992. Physiological effects of capturing striped bass in gill nets and fyke traps. Transactions of the American Fisheries Society 121:819–822.

Hubert, W. A., and P. M. Guenther. 1992. Non-salmonid fishes and morphoedaphic features affect abundance of trouts in Wyoming reservoirs. Northwest Science 66:224–228.

Hubert, W. A., and D. T. O'Shea. 1991. Reproduction by fishes in a headwater stream flowing into Grayrocks Reservoir, Wyoming. Prairie Naturalist 23:61–68.

Hubert, W. A., and D. T. O'Shea. 1992. Use of spatial resources by fishes in Grayrocks Reservoir, Wyoming. Journal of Freshwater Ecology 7:219–225.

Hubert, W. A., and T. M. Patton. 1994. Fish catches with hoop nets of two designs in the Laramie River, Wyoming. Prairie Naturalist 26:1–10.

Hubert, W. A., and M. B. Sandheinrich. 1983. Patterns of variation in gill-net catch and diet of yellow perch in a stratified Iowa lake. North American Journal of Fisheries Management 3:156–162.

Hubert, W. A., and D. N. Schmitt. 1982a. Factors influencing catches of drifted trammel nets in a pool of the upper Mississippi River. Proceedings of the Iowa Academy of Science 88:121–122.

Hubert, W. A., and D. N. Schmitt. 1982b. Factors influencing hoop net catches in channel habitats of Pool 9, upper Mississippi River. Proceedings of the Iowa Academy of Science 88:84–91.

Hunter, J. G. 1954. A weir for adult and fry salmon effective under conditions of extremely variable runoff. Canadian Fish Culturist 16:27–33.

Hurley, S. T., W. A. Hubert, and J. G. Nickum. 1987. Habitats and movements of shovelnose sturgeon in the upper Mississippi River. Transactions of the American Fisheries Society 116:655–662.

Isbell, G. L., and M. R. Rawson. 1989. Relations of gill-net catches of walleyes and angler catch rates in Ohio waters of western Lake Erie. North American Journal of Fisheries Management 9:41–46.

Jensen, J. W. 1986. Gillnet selectivity and the efficiency of alternative combinations of mesh sizes for some freshwater fish. Journal of Fish Biology 28:637–646.

Jensen, J. W. 1990. Comparing fish catches taken with gill nets of different combinations of mesh sizes. Journal of Fish Biology 37:99–104.

Jester, D. B. 1973. Variation in catchability of fishes with color of gillnets. Transactions of the American Fisheries Society 102:109–115.

Jester, D. B. 1977. Effects of color, mesh size, fishing in seasonal concentrations, and baiting on catch rates of fishes in gill nets. Transactions of the American Fisheries Society 106:43–56.

Johnson, S. L., F. J. Rahel, and W. A. Hubert. 1992. Factors influencing the size structure of brook trout populations in beaver ponds in Wyoming. North American Journal of Fisheries Management 12:118–124.

Kelley, D. W. 1953. Fluctuation in trap-net catches in the upper Mississippi River. U.S. Fish and Wildlife Service Special Scientific Report-Fisheries 101.

Kennedy, W. A. 1951. The relationship of fishing effort by gill nets to the interval between lifts. Journal of the Fisheries Research Board of Canada 8:264–274.

Kohler, C. C., J. J. Ney, and A. A. Nigro. 1979. Compact, portable vertical gill net system. Progressive Fish-Culturist 41:34–35.

Kraft, C. D., and B. L. Johnson. 1992. Fyke-net and gill-net size selectivities for yellow perch in Green Bay, Lake Michigan. North American Journal of Fisheries Management 12:230–236.

Kumpf, H. E. 1980. Practical considerations and testing of escape panel material in fish traps. Proceedings of the Gulf and Caribbean Fisheries Institute 32:211–214.

Kutka, F. J., C. Richards, G. W. Merick, and P. W. DeVore. 1992. Bait preference and trapability of two common crayfishes in northern Minnesota. Progressive Fish-Culturist 54:250–254.

Laarman, P. W., and J. R. Ryckman. 1982. Relative size selectivity of trap nets for eight species of fish. North American Journal of Fisheries Management 2:33–37.

Lagler, K. F. 1978. Capture, sampling and examination of fishes. Pages 7–47 *in* T. Bagenal, editor. Methods for assessment of fish production in fresh waters. Blackwell Scientific Publications, Oxford, UK.

Larkins, H. A. 1963. Comparison of salmon catches in monofilament and multifilament gill nets. Commercial Fisheries Review 25:1–11.

Latta, W. C. 1959. Significance of trap-net selectivity in estimating fish population statistics. Papers of the Michigan Academy of Science Arts and Letters 44:123–138.

Le Cren, E. D., C. Kipling, and J. C. McCormack. 1977. A study of the numbers, biomass and year-class strengths of perch (*Perca fluviatilis* L.) in Windermere from 1941 to 1966. Journal of Animal Ecology 46:281–307.

Lott, J. P., and D. W. Willis. 1991. Gill net mesh size efficiency for yellow perch. Prairie Naturalist 23:139–144.

Lynch, W. E., Jr., J. M. Gerber, and D. L. Johnson. 1989. A quickly deployed vertical gill-net system. North American Journal of Fisheries Management 9:119–121.

May, N., L. Trent, and P. J. Pristas. 1976. Relation of fish catches in gill nets to frontal periods. U.S. National Marine Fisheries Service Fishery Bulletin 74:449–453.

Mayhew, J. 1973. Variation in the catch success of channel catfish and carp in baited hoop nets. Proceedings of the Iowa Academy of Science 80:136–139.

McCombie, A. M., and F. E. J. Fry. 1960. Selectivity of gill nets for lake whitefish, *Coregonus clupeaformis*. Transactions of the American Fisheries Society 89:53–58.

McInerny, M. C. 1989. Evaluation of trapnetting for sampling black crappie. Proceedings of the Annual Conference Southeastern Association of Fish and Wildlife Agencies 42(1988): 98–106.

McMenemy, J. R., and B. Kynard. 1988. Use of inclined-plane traps to study movement and survival of Atlantic salmon smolts in the Connecticut River. North American Journal of Fisheries Management 8:481–488.

McWilliams, D., and J. Mayhew. 1974. An evaluation of several types of gear for sampling fish populations. Iowa State Conservation Commission Fisheries Section Technical Series 74-2, Des Moines.

Mero, S. W, and D. W. Willis. 1992. Seasonal variation in sampling data for walleye and sauger collected with gill nets from Lake Sakakawea, North Dakota. Prairie Naturalist 24:232–240.

Meyer, H. L., and J. V. Merriner. 1976. Retention and escapement characteristics of pound nets as a function of pound-head mesh size. Transactions of the American Fisheries Society 105:370–379.

Milewski, C. L., and D. W. Willis. 1991. Smallmouth bass size structure and catch rates in five South Dakota lakes as determined from two sampling gears. Prairie Naturalist 23:53–60.

Minns, C. K., and D. A. Hurley. 1988. Effects of net length and set time on fish catches in gill nets. North American Journal of Fisheries Management 8:216–223.

Nedelec, C., editor. 1975. Catalogue of small-scale fishing gear. Fishing News Ltd., Surrey, UK.

Negus, M. T. 1982. Modified anchoring system for vertical gill nets. North American Journal of Fisheries Management 4:412–414.

Nesler, T. P. 1986. Prediction of ice-fishing harvest of quality-size lake trout from previous gill-net sampling. North American Journal of Fisheries Management 6:277–281.

Parkinson, E. A., J. Berkowitz, and C. J. Bull. 1988. Sample size requirements for detecting changes in some fisheries statistics from small trout lakes. North American Journal of Fisheries Management 8:181–190.

Patriarche, M. H. 1968. Rate of escape of fish from trap nets. Transactions of the American Fisheries Society 97:59–61.

Perry, W. G. 1979. Slat trap efficiency as affected by design. Proceedings of the Annual Conference Southeastern Association of Fish and Wildlife Agencies 32(1978):666–671.

Perry, W. G., and A. Williams. 1987. Comparison of slat traps, wire cages, and various baits for commercial harvest of catfish. North American Journal of Fisheries Management 7:283–287.

Pierce, R. B., D. W. Coble, and S. Corley. 1981. Fish catches in baited and unbaited hoop nets in the upper Mississippi River. North American Journal of Fisheries Management 1:204–206.

Pope, J. A., A. R. Margetts, J. M. Hamley, and E. F. Okyuz. 1975. Manual of methods for fish stock assessment. Part 3: selectivity of fishing gear. FAO (Food and Agriculture Organization of the United Nations) Fisheries Technical Paper 41.

Pope, K. L., and D. W. Willis. 1996. Seasonal influences on freshwater fisheries sampling data. Reviews in Fisheries Science 4:57–73.

Pristas, P. J., and L. Trent. 1977. Comparisons of catches of fishes in gill nets in relation to webbing material, time of day, and water depth in St. Andrew Bay, Florida. U.S. National Marine Fisheries Service Fishery Bulletin 75:102–108.

Rach, J. J., and T. D. Bills. 1987. Comparison of three baits for trapping crayfish. North American Journal of Fisheries Management 7:601–603.

Reddin, D. G. 1986. Effects of different mesh sizes on gill-net catches of Atlantic salmon in Newfoundland. North American Journal of Fisheries Management 6:209–215.

Regier, H. A., and D. S. Robson. 1966. Selectivity of gill nets, especially to lake whitefish. Journal of the Fisheries Research Board of Canada 23:423–454.

Reid, G. K., Jr. 1955. The pound-net fishery in Virginia. Part 1: history, gear description, and catch. Commercial Fisheries Review 17:1–15.

Romaire, R. P., and V. H. Osorio. 1989. Effectiveness of crawfish baits as influenced by habitat type, trap-set time, and bait quantity. Progressive Fish-Culturist 51:232–237.

Rounsefell, G. A. 1975. Ecology, utilization, and management of marine fisheries. C. V. Mosby, St. Louis, Missouri.

Rounsefell, G. A., and W. H. Everhart. 1953. Principles of fishery science. Cornell University Press, Ithaca, New York.

Ryan, P. M. 1984. Fyke net catches as indices of the abundance of brook trout, *Salvelinus fontinalis*, and Atlantic salmon, *Salmo salar*. Canadian Journal of Fisheries and Aquatic Sciences 41:377–380.

Schaefer, M. B. 1970. Management of the American Pacific tuna fishery. American Fisheries Society Special Publication 7:237–248.

Schneeberger, P. J., T. L. Rutecki, and D. J. Jude. 1982. Gilling in trap-net pots and use of catch data to predict lake whitefish gilling rates. North American Journal of Fisheries Management 2:294–300.

Schorr, M. S., and L. E. Miranda. 1991. Catch of white crappie in trap nets in relation to soak time and fish abundance. Proceedings of the Annual Conference Southeastern Association of Fish and Wildlife Agencies 43(1989):198–205.

Schwartz, F. J. 1986. A leadless stackable trap for harvesting common carp. North American Journal of Fisheries Management 6:596–598.

Scidmore, W. J. 1970. Manual of instructions for lake survey. Minnesota Department of Natural Resources Division of Fish and Wildlife Section of Fisheries Special Publication 1.

Smith, J. B., and W. A. Hubert. 1989. Use of a tributary by fishes in a Great Plains river system. Prairie Naturalist 21:27–38.

Sokal, R. R., and F. J. Rohlf. 1981. Biometry, 2nd edition. Freeman, San Francisco.

Spangler, G. R., and J. J. Collins. 1992. Lake Huron fish community structure based on gill-net catches corrected for selectivity and encounter probabilities. North American Journal of Fisheries Management 12:585–597.

Starrett, W. C., and P. G. Barnickol. 1955. Efficiency and selectivity of commercial fishing devices used on the Mississippi River. Illinois Natural History Survey Bulletin 26:325–366.

Stuecheli, K. 1991. Trapping bias in sampling crayfish with baited funnel traps. North American Journal of Fisheries Management 11:236–239.

Sundstrom, G. T. 1957. Commercial fishing vessels and gear. U.S. Fish and Wildlife Service Circular 48:1–48.

Swales, S. 1981. A lightweight, portable fish-trap for use in small lowland rivers. North American Journal of Fisheries Management 12:83–88.

Trent, L., and P. J. Pristas. 1977. Selectivity of gill nets on estuarine and costal fishes from St. Andrew Bay, Florida. U.S. National Marine Fisheries Service Fishery Bulletin 75:185–198.

Tsumura, K., and J. M. B. Hume. 1986. Two variations of a salmon smolt trap for small rivers. North American Journal of Fisheries Management 6:272–276.

Twedt, T. M., and D. R. Bernard. 1976. An all-weather, two-way fish trap for small streams. California Fish and Game 62:21–27.

Van Oosten, J., R. Hile, and F. Jobes. 1946. The whitefish fishery of Lakes Huron and Michigan with special reference to the deep-trap-net fishery. U.S. Fish and Wildlife Service Fishery Bulletin 50:297–394.

von Brandt, A. 1964. Fish catching methods of the world, 1st edition. Fishing News Ltd., London.

Walburg, C. H. 1969. Fish sampling and estimation of relative abundance in Lewis and Clark Lake. U.S. Fish and Wildlife Service Technical Paper 18.

Welcomme, R. L., editor. 1975. Symposium on the methodology for the survey, monitoring, and appraisal of fishery resources in lakes and large rivers. FAO (Food and Agriculture Organization of the United Nations) European Inland Fisheries Advisory Commission Technical Paper 23 (Supplement 1), Rome, Italy.

Whalls, M. J., K. E. Proshiek, and D. S. Shetter. 1955. A new two-way fish trap for streams. Progressive Fish-Culturist 17:103–109.

Whelan, W. G., M. F. O'Connell, and R. N. Hefford. 1989. Improved trap design for counting migrating fish in rivers. North American Journal of Fisheries Management 9:245–248.

White, C. E., Jr. 1959. Selectivity and effectiveness of certain types of commercial nets in the TVA lakes of Alabama. Transactions of the American Fisheries Society 88:81–87.

Whitworth, W. E. 1985. Factors influencing catch per unit effort and abundance of trout in small Wyoming reservoirs. Doctoral dissertation. University of Wyoming, Laramie.

Wilde, G. R. 1993. Gill net selectivity and size structure in white bass. Proceedings of the Annual Conference Southeastern Association of Fish and Wildlife Agencies 45(1991): 470–476.

Willis, D. W. 1987. Use of gill-net data to provide a recruitment index for walleyes. North American Journal of Fisheries Management 7:591–592.

Willis, D. W., K. D. McCloskey, and D. W. Gabelhouse, Jr. 1985. Calculation of stock density indices based on adjustments of efficiency of gill-net mesh size. North American Journal of Fisheries Management 5:126–137.

Winkle, P. L., W. A. Hubert, and F. J. Rahel. 1990. Relations between brook trout standing stocks and habitat features in beaver ponds in southeastern Wyoming. North American Journal of Fisheries Management 10:72–79

Winters, G. H., and J. P. Wheeler. 1990. Direct and indirect estimation of gillnet selection curves of Atlantic herring (*Clupea harengus*). Canadian Journal of Fisheries and Aquatic Sciences 47:460–470.

Wolf, P. 1951. A trap for the capture of fish and other organisms moving downstream. Transactions of the American Fisheries Society 80:41–45.

Worthington, E. B. 1950. An experiment with populations of fish in Windermere, 1938–1948. Proceedings of the Zoological Society of London 120:113–149.

Yeh, C. F. 1977. Relative selectivity of fishing gear used in a large reservoir in Texas. Transactions of the American Fisheries Society 106:309–313.

Zar, J. H. 1984. Biostatistical analysis. Prentice Hall, Englewood Cliffs, New Jersey.

APPENDIX 6.1 REPAIRING NETS[1]

STACY V. GEBHARDS

A properly mended net can mean the difference between catching many fish or none at all. This discussion is intended to acquaint beginners with mending procedures. The techniques described are those used by commercial fisher[s] in the Illinois River Valley.

Trimming

The first step is to trim the hole (Figure 6.5) so it can be rewoven in one continuous operation. Each knot has 4 unbroken strands (quarter meshes) leading from it. Around the edges of a tear you will find knots with 1, 2, or 3 unbroken strands. Trim as follows:

1. Start at the top of the hole and leave one knot with 3 unbroken strands. This will be the starting point for the reweaving.
2. Work down one side of the hole, knot by knot. Hereafter, when finding a knot with 3 unbroken strands, cut out the lower strand. Leave knots with 2 or 4 unbroken strands as they are.
3. Trim down one side to the bottom and then trim the other side. Leave one knot at the bottom with 3 unbroken strands. This will be the last tie in the weaving.
4. The hole is now ready for weaving (Figure 6.5). Each knot around the edge of the hole should have 2 or 4 unbroken strands, except the starting point at the top and the finishing point at the bottom.

Weaving

The twine used for weaving is wound on a shuttle filled by passing the twine beneath the tongue, around the notch at the bottom, up, and beneath the tongue from the opposite side.

The basic knot is the sheet bend. Gill nets which utilize synthetic threads in construction (nylon, orlon, dacron, etc.) require special knots to prevent slippage. Carrothers (1957) describes some of these special knots. Two nonslip knots for nylon are shown in Figure 6.11.

Figure 6.6 illustrates the sequence of knots in weaving. Details are shown in Figure 6.7 through 6.11. Knake (1947) describes variations practiced in New England.

When weaving from left to right, bring the shuttle up through the mesh. Do the same for sider knots. When weaving from right to left, pass the shuttle down through the mesh.

Section Replacement

Sometimes a section of net must be replaced (Figure 6.12). The starting knot and finishing knot begin and end at a 3-strand knot. The remaining knots along the edges are all 2- and 4-strand knots (Figure 6.13). The seaming procedure follows:

---

[1]Adapted from *Inland Fisheries Management*, edited by A. Calhoun and published by the California Department of Fish and Game (Sacramento, 1966), with permission.

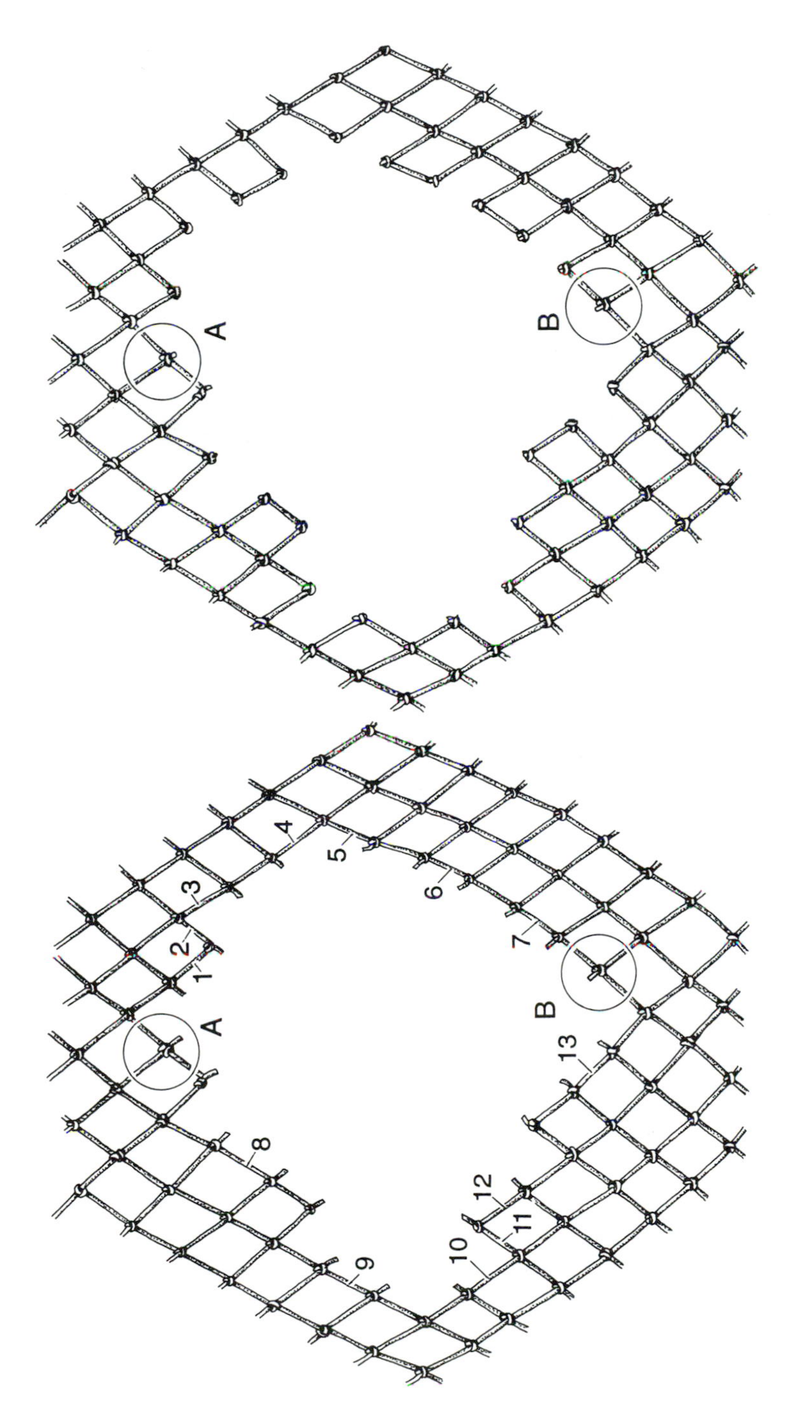

Figure 6.5 Trimming a hole for repair. Left, numbers **1–13** indicate the sequence in cutting. Knots **A** and **B** are the two knots with three unbroken strands which are not cut. Knot **A** is the starting point, and knot **B** is the finishing point. Right, the hole trimmed and ready for weaving. Note that all knots other than **A** and **B** have either two or four unbroken strands.

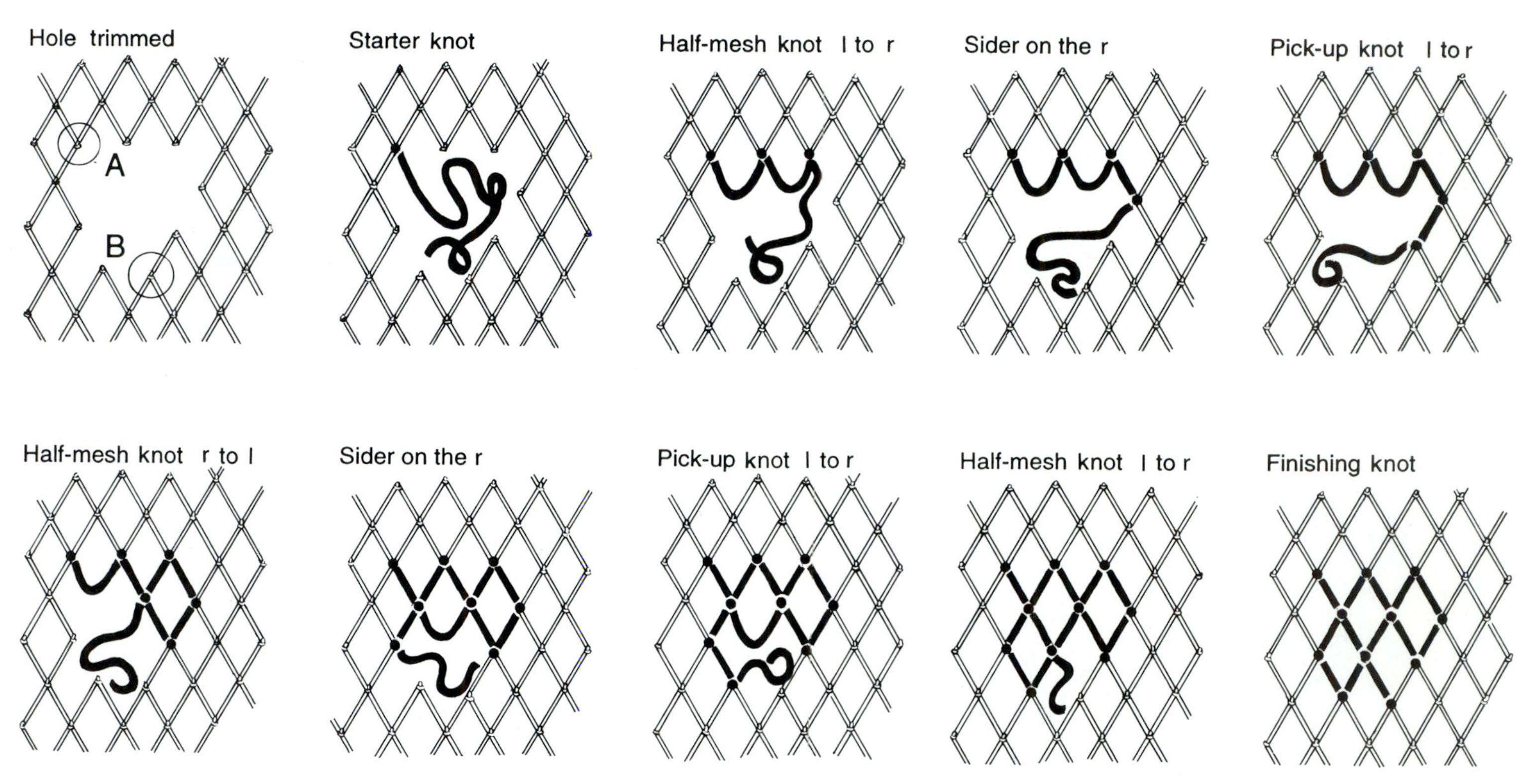

Figure 6.6 Sequence of knots in weaving. Knot-tying steps shown in later figures.

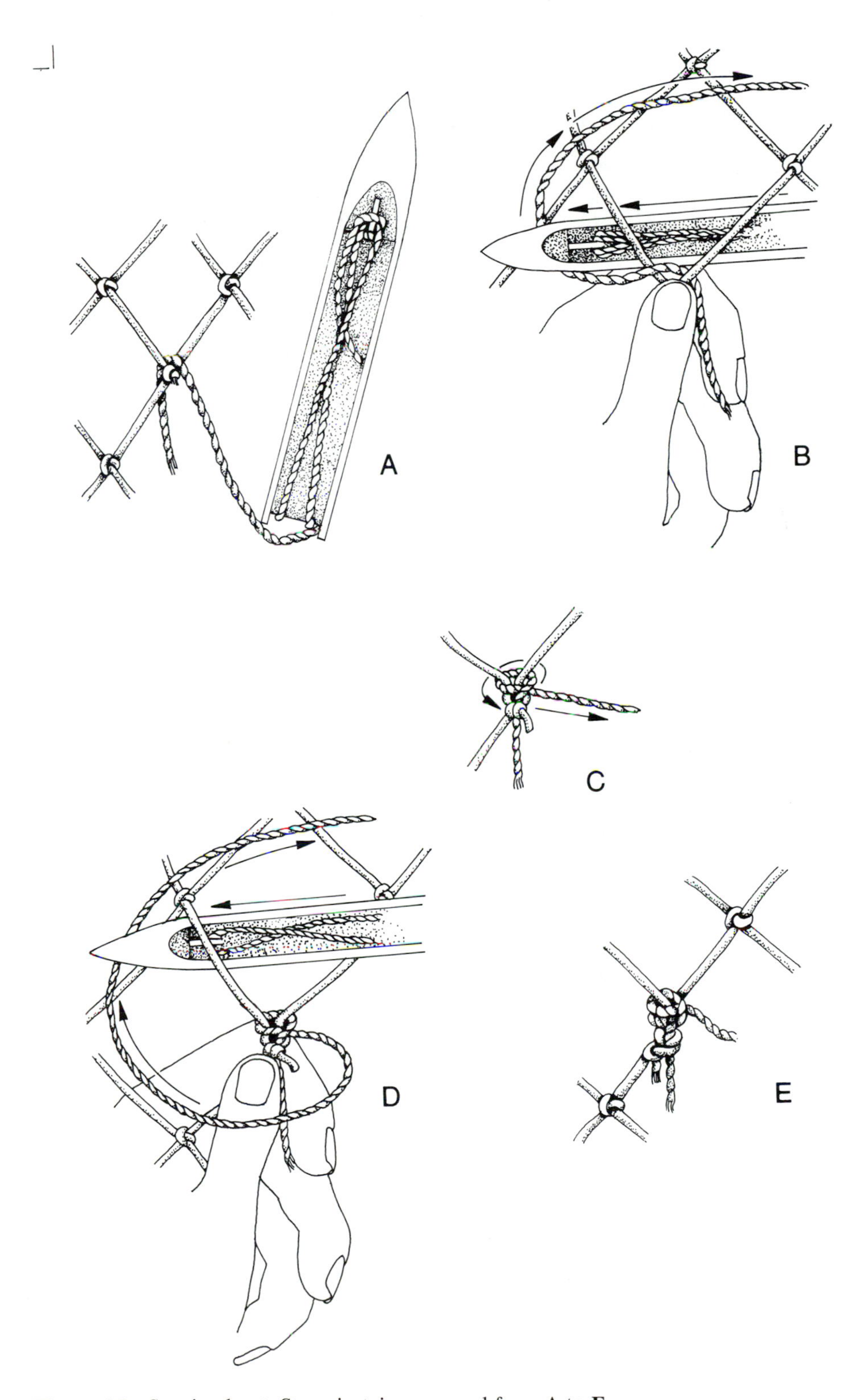

Figure 6.7 Starting knot. Steps in tying proceed from **A** to **E**.

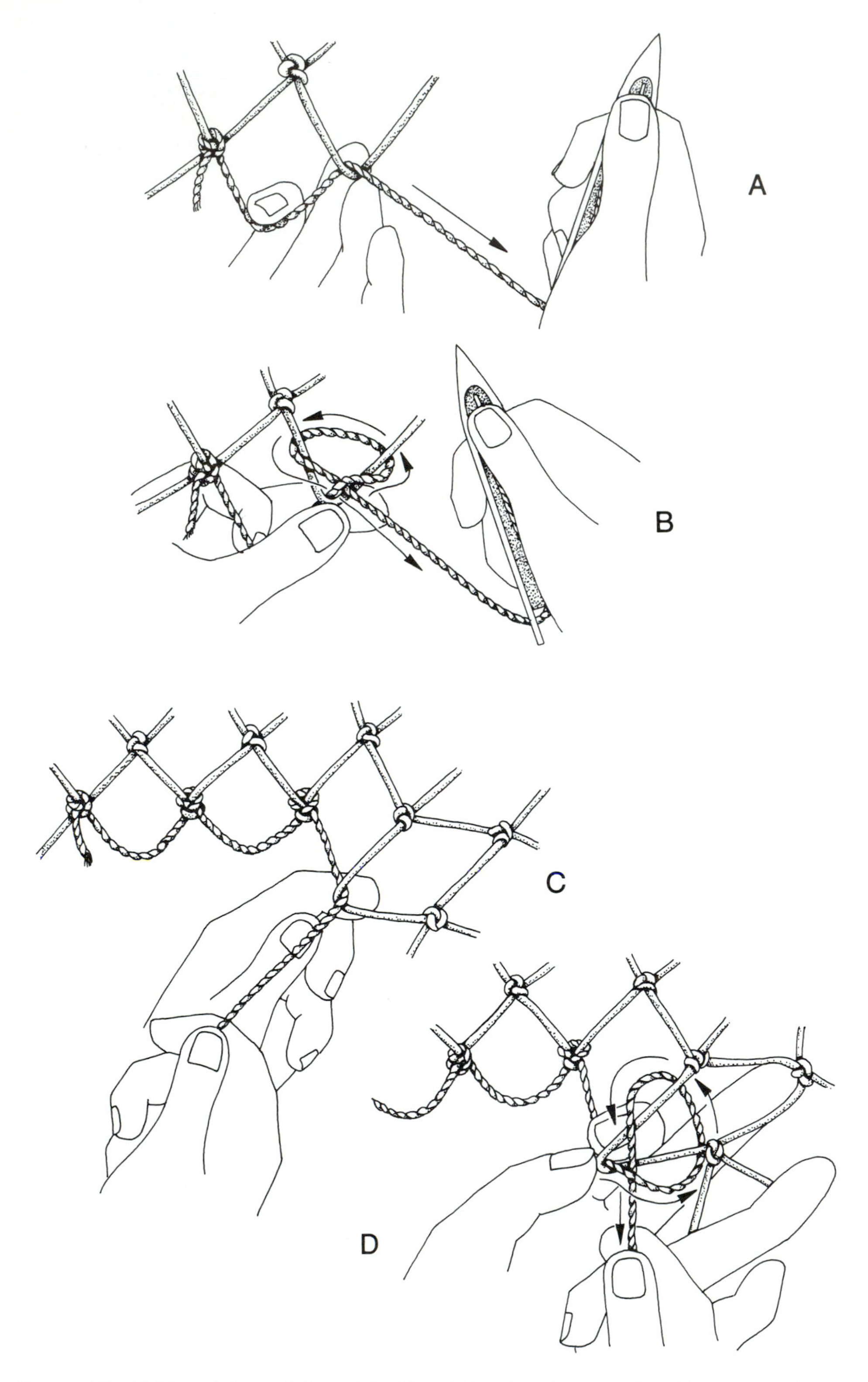

Figure 6.8 Half-mesh knot, left to right (**A** and **B**); sider knot on the right (**C** and **D**).

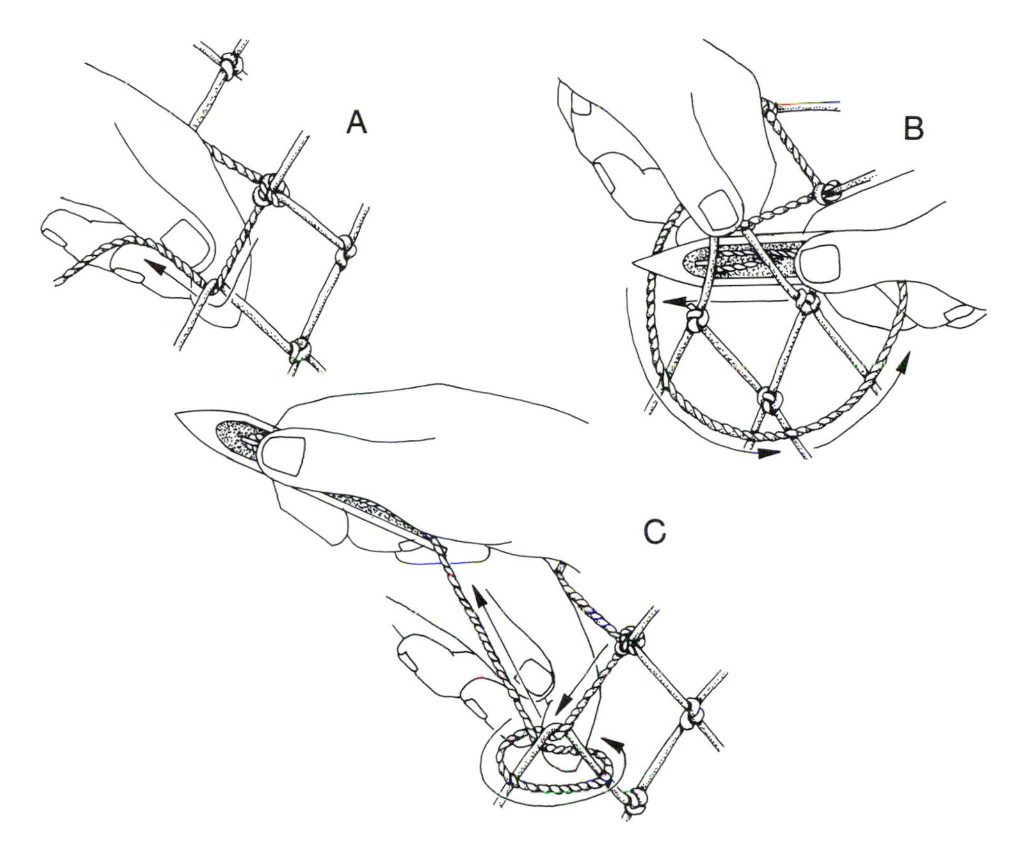

Figure 6.9 Pick-up knot, right to left. Steps in tying proceed from **A** to **C**.

1. Trim each edge of the hole straight, with one continuous row of meshes (Figure 6.12).
2. Cut the new section (Figure 6.12C) to the same depth as the hole and 2 meshes narrower than the original section.
3. Figure 6.13 shows the sequence of knots used in seaming. As in mending, the strands are gaged with the fingers. The beginning and finishing ties are half-meshes; the others are all quarter-meshes.

Maintenance of Nets

Synthetic fibers have greatly reduced the problem of rotting associated with cotton and linen nets. However, some care is still necessary to insure maximum life from synthetic-fiber nets. Copper and creosote-base preservatives for natural fibers are discussed in various sources in the bibliography.

Nylon breaks down in sunlight twice as fast as linen or cotton (Carrothers 1957). Hence, it should be protected from direct sunlight. Dips for synthetics which contain compounds that screen out sunlight can be purchased in heavy and light viscosities. The heavy grades reduce abrasion.

Fish slimes may create acidic conditions which will damage nylon. Carrothers (1957) recommends dipping nylon gill nets in a 2% copper sulphate solution long

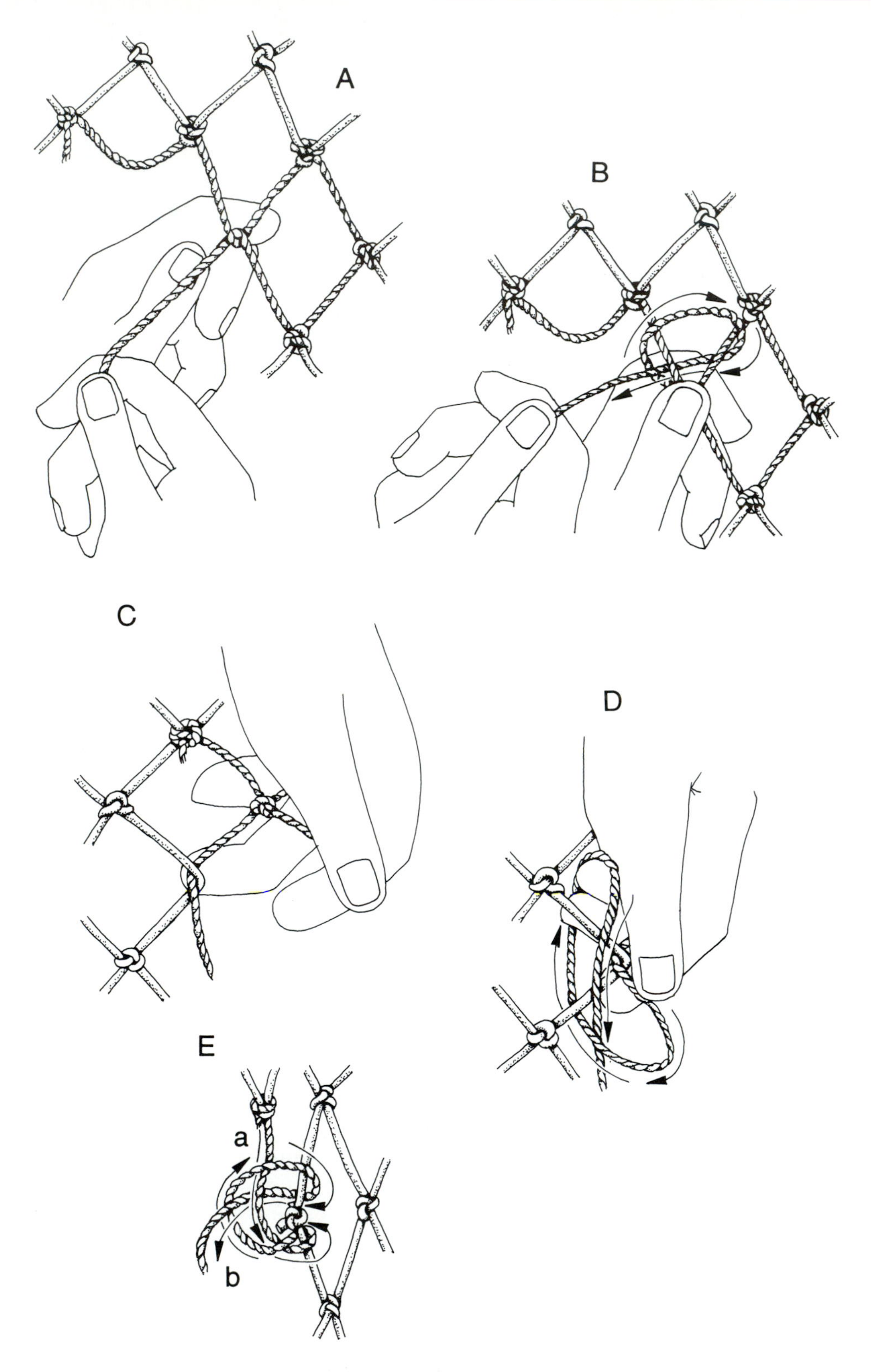

Figure 6.10 Half-mesh knot, right to left (**A** and **B**); sider knot on the left (**C** and **D**). An alternate sider knot is shown at **E**; it is tied the same on right and left sides. Pull the hitch tight below the knot first, then weave and tighten the hitch above the knot.

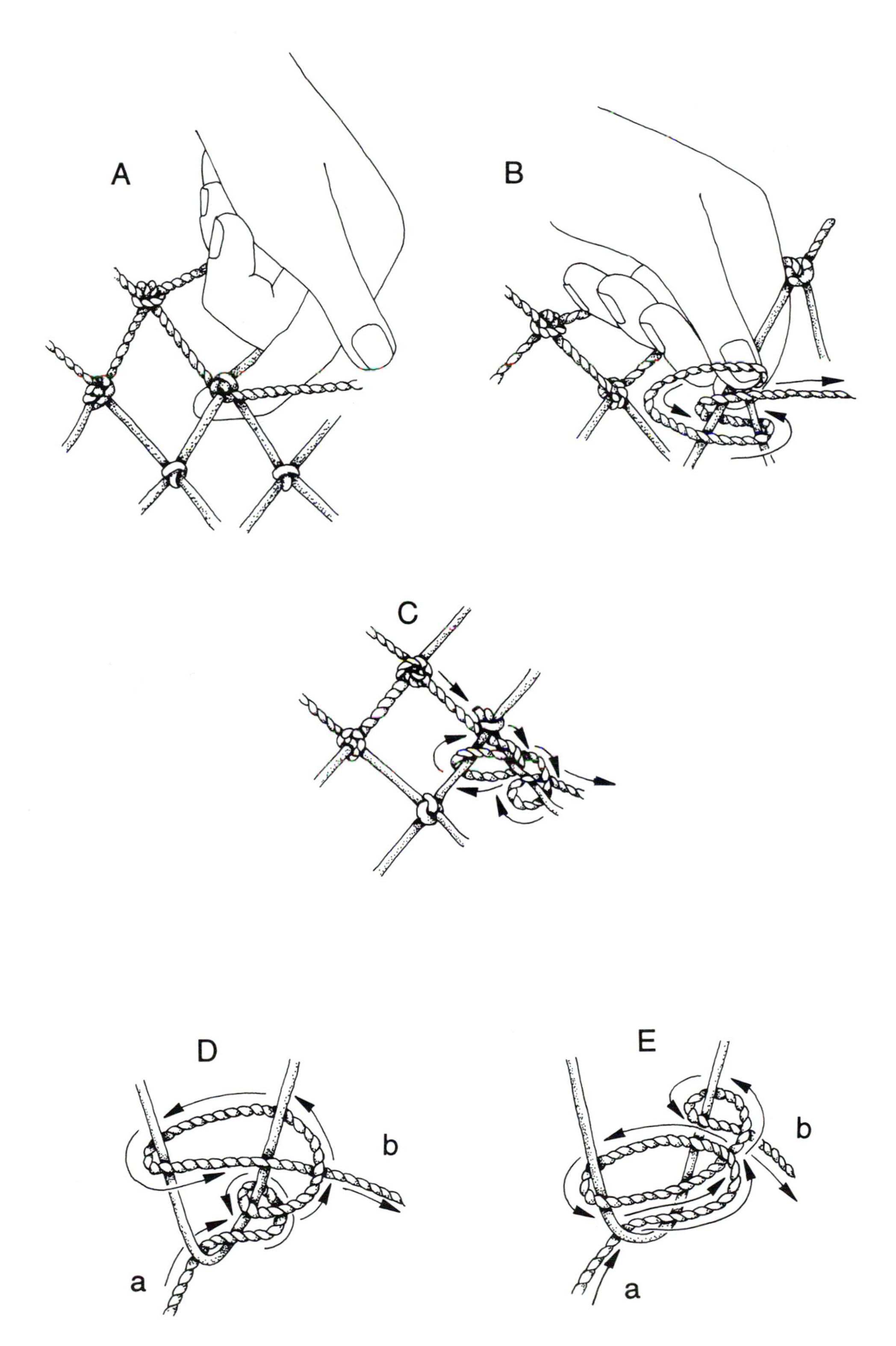

Figure 6.11 Pick-up knot, left to right (**A** and **B**); finishing knot (**C**). Two variations of the "knot-and-a-half" used in hand tying nylon nets are shown at **D** and **E**. The hitch at (**a**) is pulled tight before making the second hitch at (**b**).

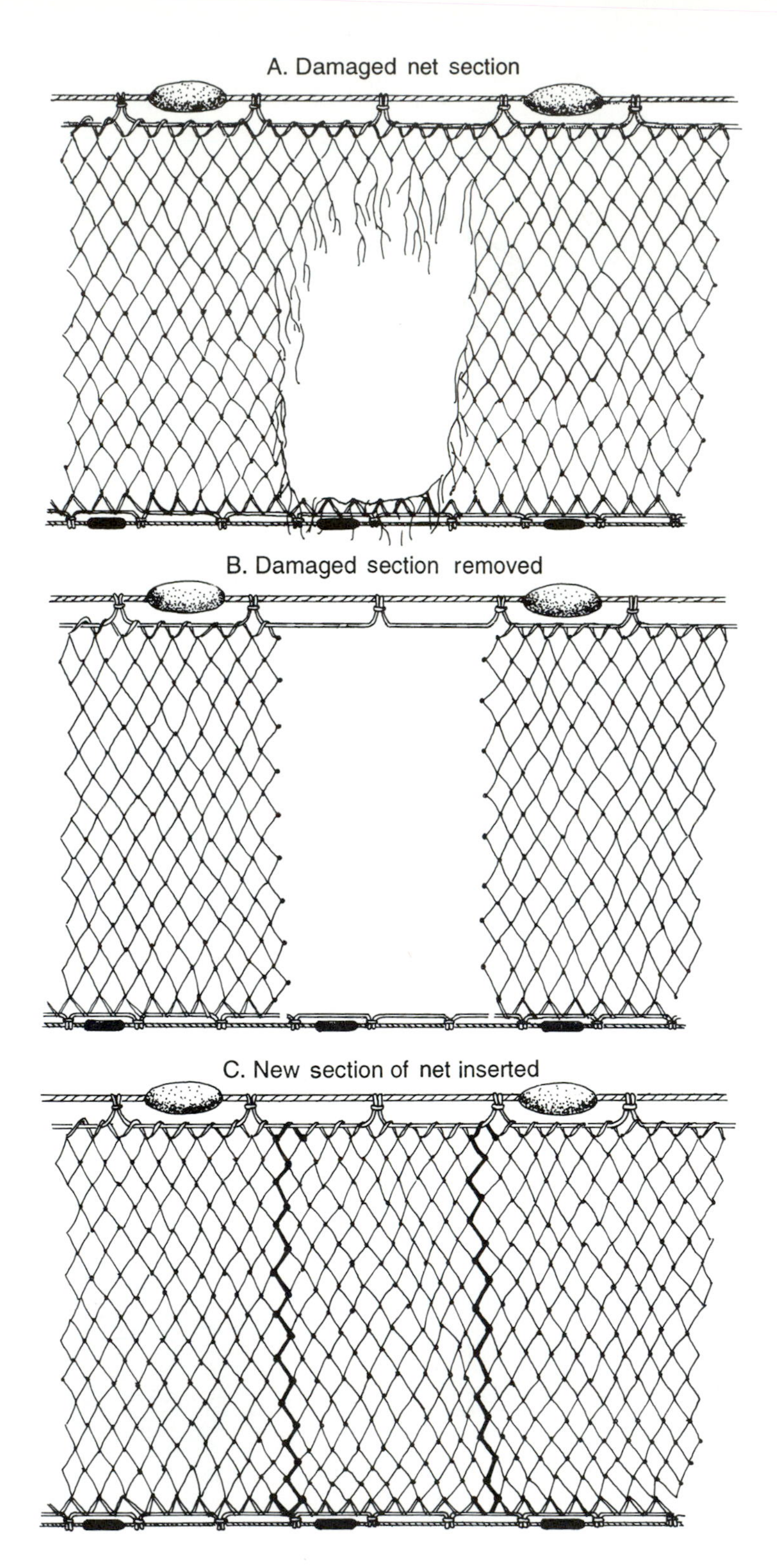

Figure 6.12 Procedures for replacing a damaged section of net. The darkened meshes in **C** indicate meshes woven with twine to join the new and old sections of net.

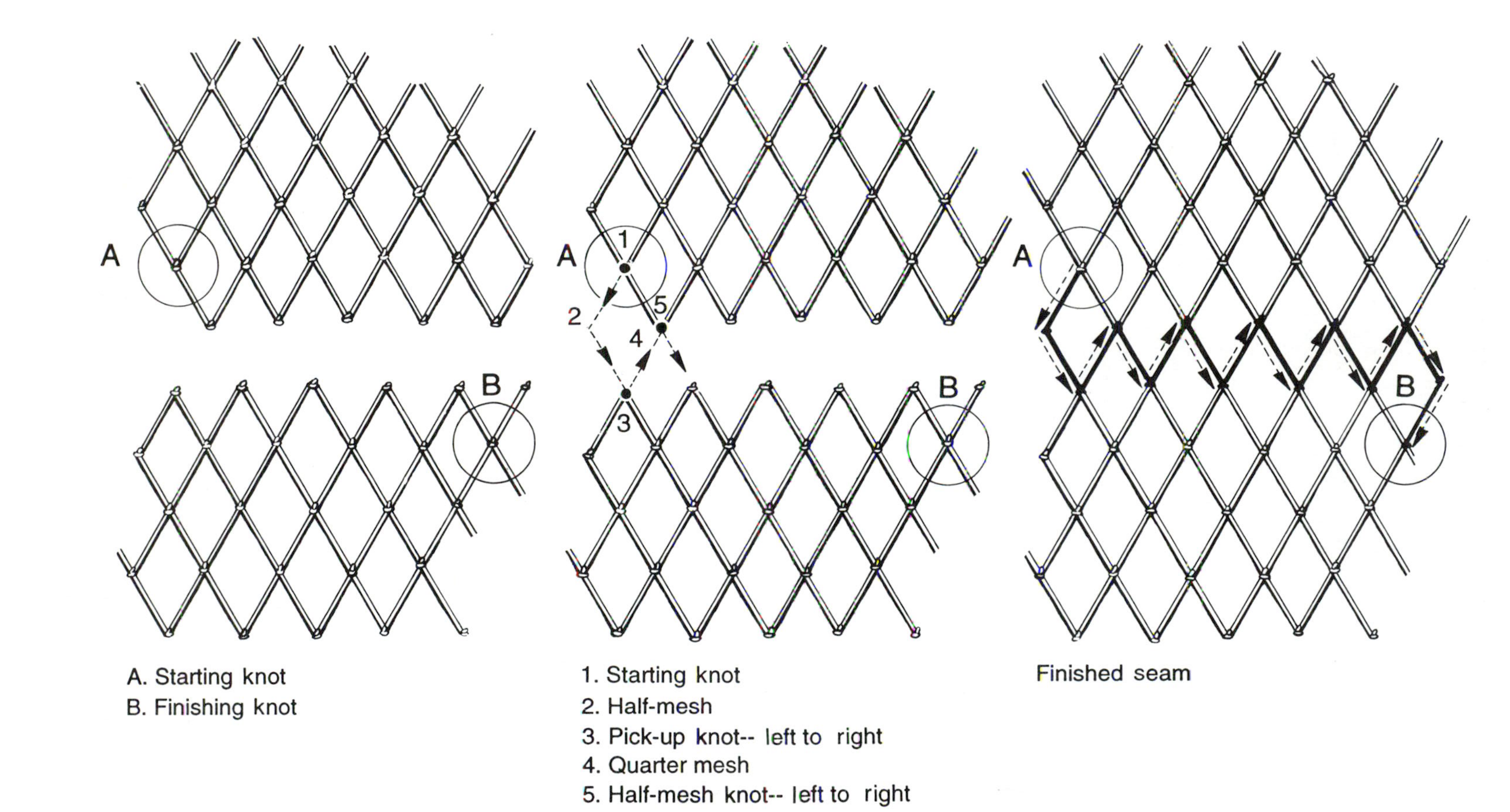

Figure 6.13 Procedure for seaming two net sections together, after trimming as shown in Figure 6.12.

enough to remove the slime and rinsing in clean water before drying. Strong solutions or residues can reduce wet mesh strength.

Chlorine, oxidizing bleaches, and drying oils, such as linseed oil, also damage nylon.

Sunlight damages many other synthetics less seriously than nylon. Orlon is highly resistant. Dacron is resistant to acids but not to alkaline conditions. Excessive heat reduces the tensile strength and elasticity of synthetic fibers.

References

Carrothers, P. J. G. 1957. The selection and care of nylon gill nets for salmon. Fisheries Research Board of Canada, Industrial Memorandum 19, Ottawa.

Knake, B. O. 1947. Methods of net mending—New England. U.S. Fish and Wildlife Service, Fishery Leaflet 241, Washington, DC.

Chapter 7

Active Fish Capture Methods

DANIEL B. HAYES, C. PAOLA FERRERI, AND
WILLIAM W. TAYLOR

7.1 INTRODUCTION

As the name implies, active fish capture methods use moving nets or gears to collect finfish, shellfish, and other macroinvertebrates. These methods are in contrast to passive capture methods that rely on fish movement into a stationary device (Chapter 6). The distinction between the two is not always clear-cut. Because some active fishing methods are described elsewhere (e.g., electrofishing in Chapter 8), and because of the variety of methods used to capture fishes, we suggest you also see other chapters in this book when considering methods of fish capture. In this chapter, we will focus on three major types of active fish capture methods: towed nets, dredges, and surrounding nets. Other active methods, such as angling and cast nets, will also be covered but in less detail.

The major active gears have the advantage of enclosing or "sweeping" a specified geometric space (Figure 7.1) and operate over a specified time, thus allowing an accurately defined unit of effort. Accurately defining sampling effort is particularly important when an index of abundance is being calculated. Some other active fish capture methods, such as angling, and most passive capture techniques do not enclose a specified area and as such do not allow such a precise determination of effective effort. Although passive capture methods are often used to compute indices of abundance, catches are influenced greatly by fish behavior and may not accurately reflect actual abundance.

Another advantage of active gears is that sampling operations are mobile in space and time; samples collected with active gears can typically be obtained in a time span of minutes to hours. This mobility comes at a cost; one of the disadvantages of active sampling gears relative to most passive capture methods is that active sampling often requires a larger vessel and may require two or more people for safe operation. Passive gears, in contrast, can often be set quickly from a small vessel and with a minimum of labor. To capture fish most effectively, however, passive gears often have to remain in place for periods ranging from hours to days. The fact that active gears can be deployed for short periods of time is advantageous for two reasons. First, shorter sampling periods allow for a larger sample size to be taken per time spent sampling. Larger sample sizes increase the statistical precision (Chapter 2) of indices of abundance as well as allow sampling to be done over a wider area or to cover the area being sampled more completely. Thus, a more precise picture of fish spatial distribution and habitat use can often be obtained by using active fishing gears.

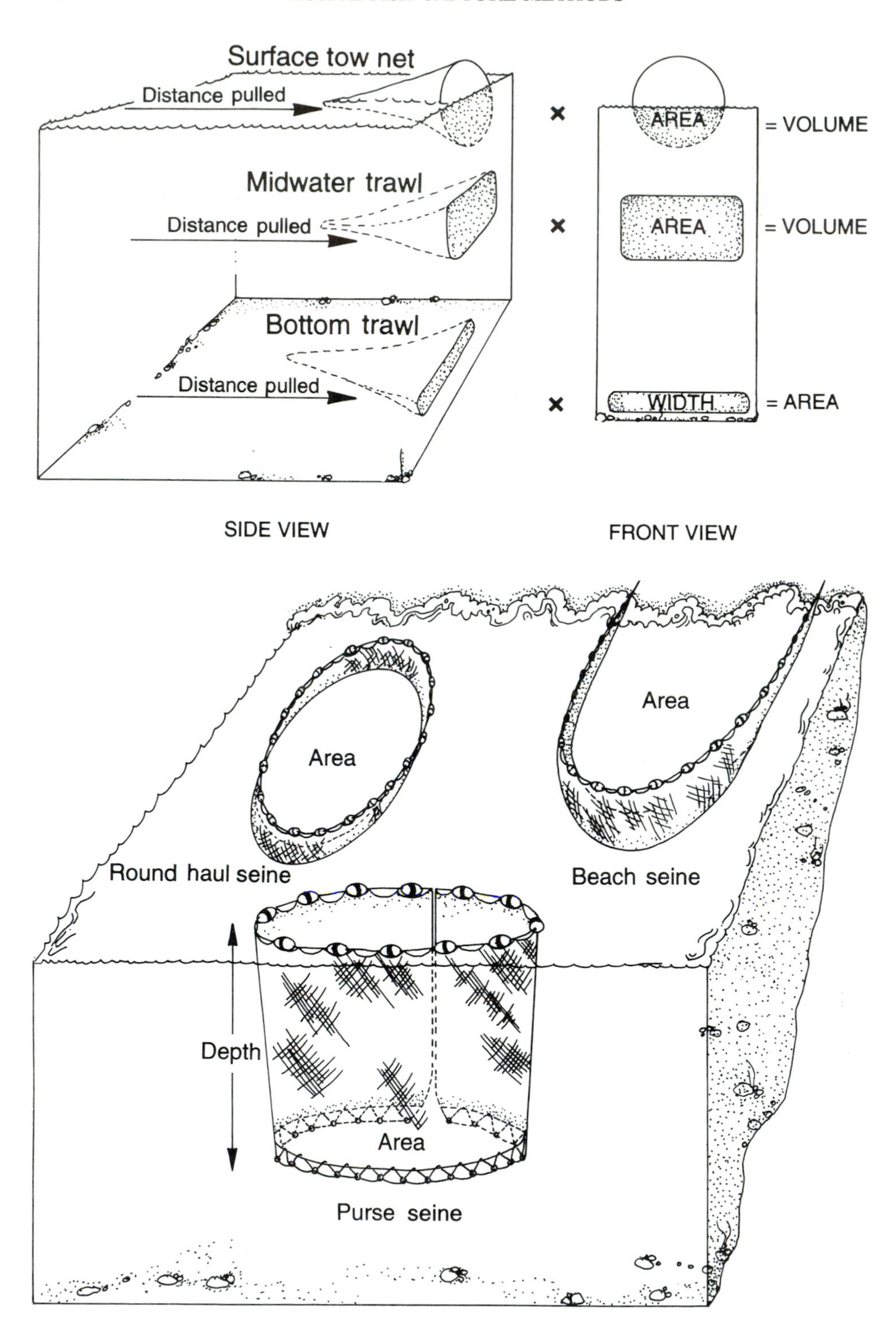

Figure 7.1 Sampling characteristics of active capture methods. Surface tow net, midwater trawl, and purse seine catches are generally reported in terms of volume sampled; bottom trawl and beach or haul seine catches are usually reported in terms of area sampled (Hayes 1983).

However, active gears may catch fewer fish per time spent sampling than do passive gears, which can be a disadvantage if large numbers of fish are needed for a study. The second advantage of having short sampling times with active fishing gears is that the time of capture can be determined more precisely. Knowing the time of capture is particularly important in studies of fish diet, feeding rate, behavior, and movement.

This chapter focuses on scientific or assessment sampling rather than commercial fishery applications (Chapter 21). Fishing gears used in scientific sampling are similar to commercial fishery gears, but are typically scaled down in size. Also, the deployment of a gear for scientific sampling is often standardized. In commercial fisheries, modifications to the gear during or between trips are common. This allows fine-tuning of the gear to maximize catch in the habitat being fished. In scientific applications for which standardized sampling effort is needed, the gear is rarely altered during sampling or between sampling trips, except for needed repairs. Although the catch may not be maximized in such situations, sampling remains representative and allows for comparisons over time and among sites.

7.2 NET MATERIAL AND CONSTRUCTION

Many active fishing gears are assembled using netting. Early nets were constructed of natural fibers such as cotton, hemp, and linen (Hayes 1983). Nets made of these materials needed relatively thick twine and were prone to decay. Synthetic fibers were introduced in the 1950s and are the primary materials used today. The four most common synthetic fibers used in making seines are polyamide, polyester, polyethylene, and polypropylene. New netting materials (e.g., Kevlar and Spectra; Buls 1989) continue to be developed in the trend of stronger and thinner twines. Although these materials are still undergoing refinement for use in netting, their use will allow fishers in the future to tow larger nets with less power. Synthetic fibers are superior to natural fibers in that they are stronger and lighter, do not absorb water, and are not affected by bacteria. Polyamides and polyesters both are more dense than water and naturally sink; polyethylene and polypropylene are both less dense than water and float.

The effective size of the openings in the netting (webbing) is determined by several factors, the mesh size is the most obvious. Mesh size is typically measured in one of two ways. The bar measure of a mesh is the length from the beginning of a knot to the beginning of an adjacent knot (Figure 7.2A). Stretch measure is taken from the beginning of a knot to the beginning of the opposite knot when the net is stretched out (Figure 7.2A). Either measure is adequate, but it is important to have measurements taken under consistent conditions and to be sure to specify which measurement was taken. The mesh size determines the maximum size opening through which a fish can pass. In addition to the mesh size, the hanging ratio (E) also plays an important role in determining the effective size of the openings in a net (Figure 7.2C). The hanging ratio is a measure of the amount of netting attached to the supporting rope and is computed as the length of the supporting rope divided by the stretched length of the netting. The hanging percentage (P) is an alternate way of expressing the hanging ratio and is computed simply as $100(1 - E)$. The hanging ratio determines both the shape of the openings as well as the amount of webbing in the body of the net. Generally, trawls and seines are attached to supporting ropes with a hanging ratio between 0.6 and 0.8. An illustration of the appearance of netting

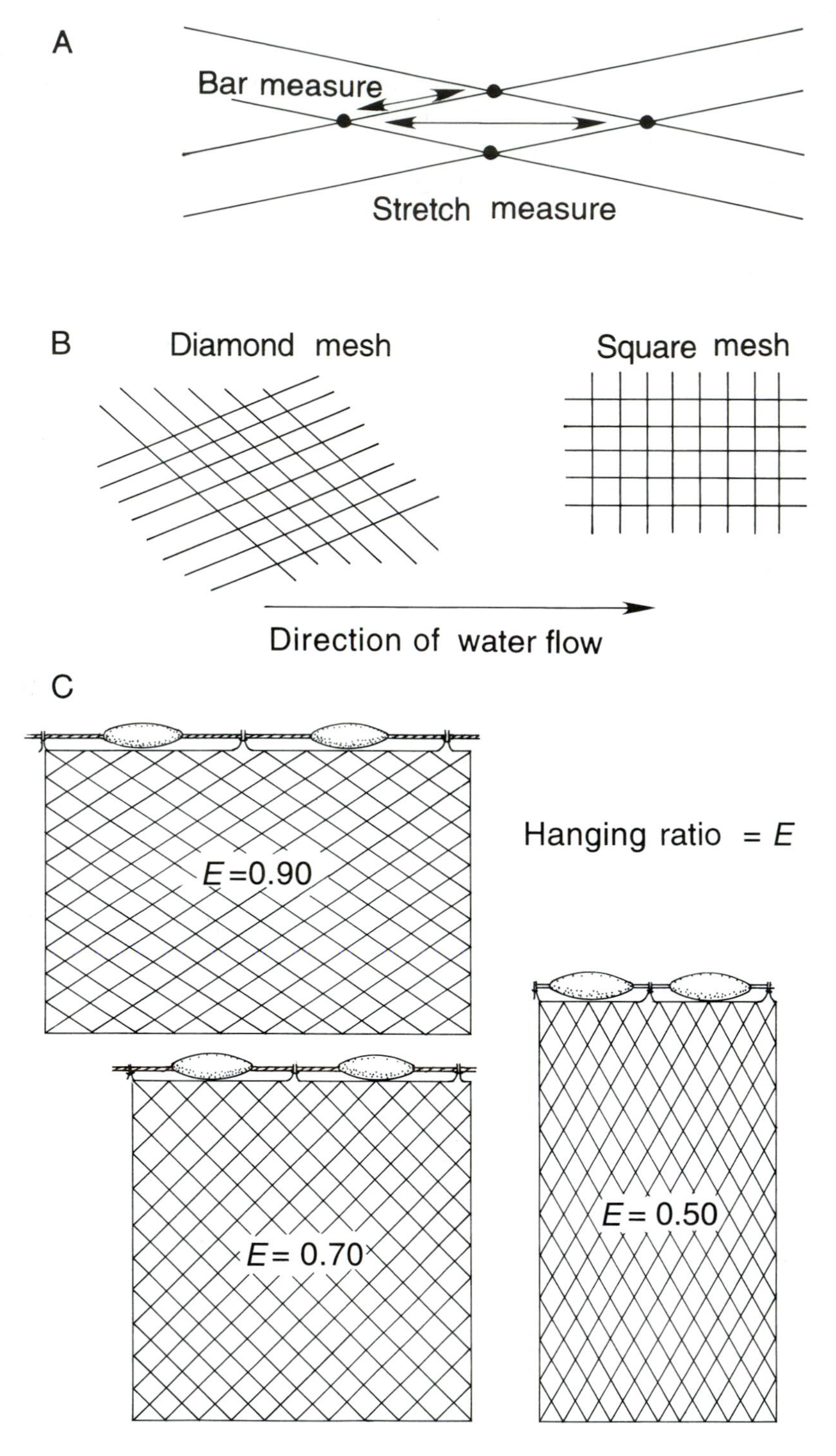

Figure 7.2 Netting measurement and nomenclature: (**A**) measurements of mesh size; (**B**) effect of hanging orientation on net appearance; and (**C**) effect of hanging ratio (*E*) on effective mesh openings.

for various hanging ratios is given in Figure 7.2C. A hanging ratio of 0.67 (P = 33%) results in a mesh that is square in shape.

The orientation of the webbing to the direction of water flow also has an effect on the effective mesh openings and hence the selectivity of trawls for different sizes of fish and species with different shapes. Diamond-mesh trawls have the point of the openings oriented towards the oncoming flow of water, whereas square-mesh trawls have the point of the openings oriented away from the flow of water (Figure 7.2B). Trawls constructed with the netting in a diamond-mesh orientation generally capture smaller, round-bodied species (e.g., haddock and cods) than do trawls with the same size square-mesh (Walsh et al. 1992). Conversely, square-mesh netting selects for smaller, flat-bodied species (e.g., flounders) for a given mesh size. Differences in selectivity between diamond and square mesh occur because flat-bodied species tend to escape more readily from the elongated holes in diamond-mesh nets.

Twine size also plays an important role in determining the effective mesh openings, as well as determining the towing resistance of the net. Finer twine results in less resistance and yields a slightly larger opening for a given mesh size. Modern synthetic fibers, being much stronger than natural fibers used in the past, allow for smaller-diameter twines to be used to give equal strength.

7.3 DRAGGED OR TOWED GEARS: TRAWLS

Trawls are funnel-shaped nets (Figures 7.3 and 7.4) that are towed along the bottom (bottom trawls) or in the water column (midwater trawls). As the net is towed through the water, fish entering the net eventually tire and fall to the end of the net (cod end; Figures 7.3A and 7.4B) where they are held until the net is retrieved. Most trawls are designed so that the cod end is tied shut while fishing and can be opened after retrieval to make removing fish from the net easier. There are numerous variations on trawls, with many adaptations to particular habitats or species's behaviors.

Early trawls used a wooden beam to keep the mouth of the trawl open while the trawl was being dragged along the bottom (Figure 7.4A). The mouth opening of beam trawls are limited by the length of the beam that can be readily handled. Beam trawls are still used commercially and for scientific sampling when a relatively small bottom trawl is needed. An advantage of beam trawls is that the trawl width is fixed by the length of the beam. As such, the area swept by the trawl can be controlled easily by maintaining a constant tow duration and velocity.

During the late 1800s and early 1900s, the otter door, or trawl door (Figure 7.4B, C), was developed as another method for opening the mouth of bottom trawls. After deployment, the wings of the trawl (Figure 7.4B) are spread by otter doors, also known as otter boards. Larger trawls resulting in higher catches can be fished by using otter doors rather than beams to open the net horizontally. Most bottom trawls for scientific programs and commercial fisheries use otter doors. Otter doors for bottom trawls are relatively heavy in order to keep the net on the bottom. Most otter doors are rectangular or oval in shape and keep lateral pressure on the net by having the trawl warp (Figure 7.4B) attached to a bracket (Figure 7.4C) that angles the door outward when being towed. The bottom of the doors typically have metal shoes (Figure 7.4C) to protect the door from abrasion with the substrate.

Trawls in their various forms are probably the most commonly used sampling gear

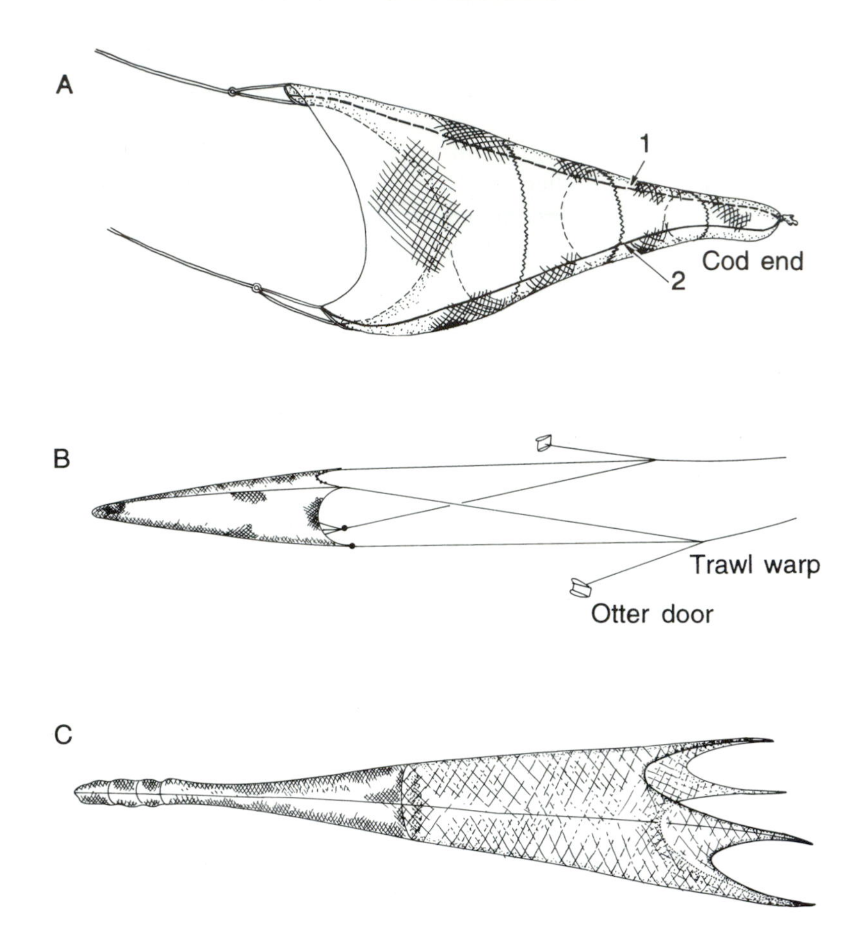

Figure 7.3 Design and nomenclature of midwater trawls: (**A**) two-seam trawl; (**B**) four-seam trawl with pelagic otter doors; and (**C**) four-seam midwater trawl with very large mesh webbing near trawl mouth (von Brandt 1984).

in oceanic and estuarine habitats. They are also used extensively in large lakes and occasionally in large rivers. One reason for the popularity of trawls is that they sample a discrete area or volume over a specified time. As such, they provide quantitative indices of population abundance. At the same time, specimens for age, growth, diet analyses, and tissue samples can also be taken. Live fish suitable for mark–recapture experiments or for use in laboratory studies can also be obtained by trawling, but care must be taken not to injure the fish during the capture operation. To avoid injuring captured fish, tow times should be kept short (e.g., 15 min or less), and trawls should be brought on board as slowly as possible to allow fish time for decompression. Fish survival is enhanced if trawls are taken as shallow as possible; for some species their natural distribution precludes sampling in shallower areas.

Although trawls are quite versatile in the habitats they can sample, they do have limitations. As with many other active sampling gears, trawls cannot be effectively fished where there are objects on which they can get caught. Coral reefs, aquatic macrophyte beds, and rocky outcroppings are examples of habitats that are difficult or impossible to sample with otter trawls. Another limitation of trawls is that they

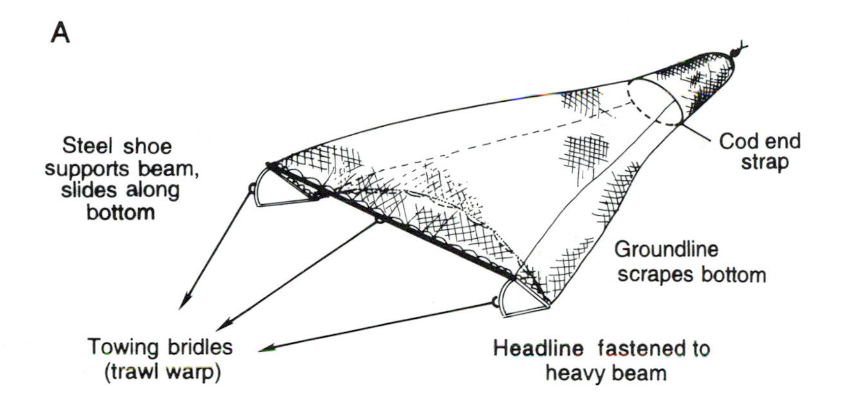

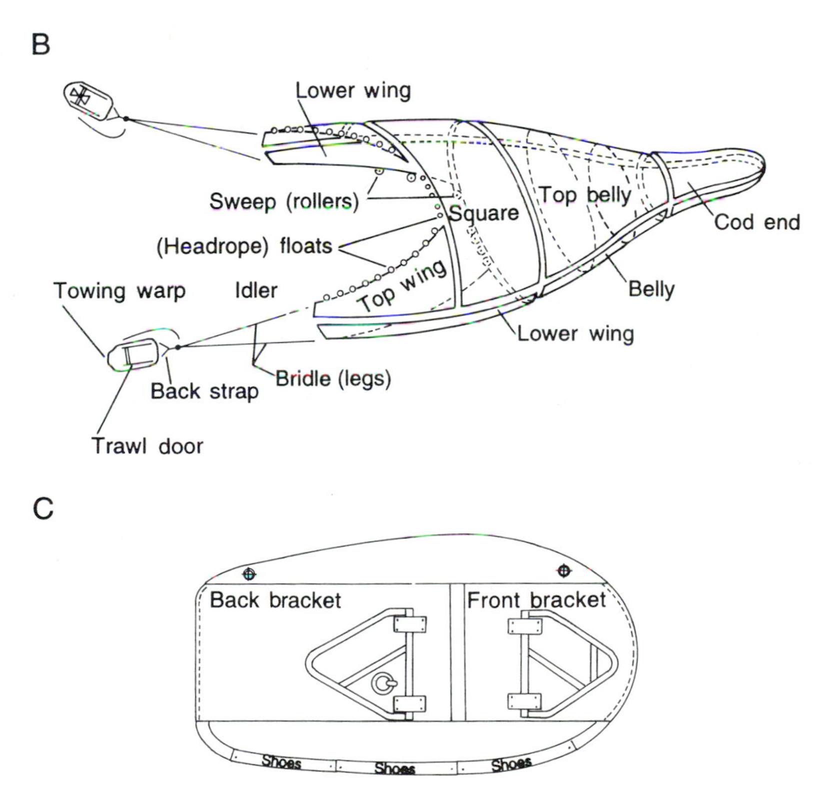

Figure 7.4 Design and nomenclature of bottom trawls: (**A**) beam trawl (Sainsbury 1975); (**B**) otter trawl (Byrne and Nicolas 1989); and (**C**) otter door (Byrne and Nicolas 1989). The towing warp is attached to the front bracket of the door and the bridle is attached to the back bracket.

generally require a relatively powerful vessel to pull them along the bottom or through the water column. Because of this, they are often not suitable for sampling in shallow habitats along shore. Also, trawls generally require two or more people for safe operation.

7.3.1 Variations

Beam trawls. Beam trawls have been used for collecting both fishes and macroinvertebrates. Although beam trawls are suitable for use in developing indices of abundance, otter trawls are more commonly used for this purpose because larger nets can be used to sample larger areas. Beam trawls, however, are sometimes used to collect fish for biological specimens. For example, Nash et al. (1992) used a 2-m beam trawl to collect plaice in Port Erin Bay, Isle of Man. Fish were captured alive, tagged, and released with minimal mortality.

Otter trawls. Otter trawls are commonly used in surveys of demersal species, particularly in marine ecosystems, estuaries, and large lakes. An example of a long-term survey in which an otter trawl is used is that conducted by the National Marine Fisheries Service (NMFS) along the eastern seaboard of North America, from Cape Hatteras, North Carolina, to Nova Scotia, Canada. In addition to providing an index of abundance for numerous species, this survey supplies biological samples for a variety of purposes (e.g., scale, otolith, and stomach content samples). Since its inception, over 400 species of fishes have been collected; more than 60 species are caught every year. Most years a "36 Yankee" trawl, which is a commercial net, has been used for the survey (Azarovitz 1981). The 36 Yankee trawl is approximately 30 m in length and has a headrope 18 m in length and a footrope 24 m in length. Initially, the trawl was fished with wood and steel doors that were oval in shape and weighed about 460 kg. After 1984, these doors were replaced with all-steel doors that were rectangular in shape and weighed about 480 kg. Although these doors are similar in size and weight, catches differed by roughly 50% in paired tow experiments that compared the two doors' fishing performance (Byrne and Forrester 1987). The higher catches are believed to occur because the all-steel doors better maintain the net's mouth opening and contact with the bottom. This difference in catch highlights the need to consider and evaluate changes in fishing power carefully whenever changes are made to the gear configuration.

Otter trawls have also been used extensively in freshwater survey applications. The Great Lakes Laboratory of the National Biological Service initiated a bottom trawl survey in Lake Michigan in the early 1960s (Brown 1972) and has since conducted surveys in all of the Great Lakes (e.g., Argyle 1982; O'Gorman and Schneider 1986; Bronte et al. 1991). The general purpose of these surveys is to provide indices of abundance for forage fishes and provide biological samples for understanding their population dynamics. Sampling designs vary among lakes; however, all are based on a fixed-station sampling design (Chapter 2). Trawls are conducted from as shallow as 5 m up to roughly 150 m in depth. Many of these bottom trawl surveys are combined with midwater trawling, hydroacoustic surveys (Chapter 13), or both to estimate biomass and numbers of fish unavailable to bottom trawls. Comparisons of bottom trawl and midwater trawl catches revealed differences in fish length composition as well as seasonal variation in the proportion of the population vulnerable to bottom trawling (Argyle 1982). This finding highlights the need for considering the spatial distribution of the population as a whole as well as the size-specific distribution because they both affect the selectivity of the gear (see section 7.7.4).

Otter trawls can be altered in numerous ways to perform better under specific conditions. For example, the trawl sweep can be covered with rollers (Figure 7.4B) to prevent the trawl from getting snagged in rocky areas. Adjusting the weight of the doors, the angle of the warp when trawling, and the boat's speed are examples of other factors that affect the performance of the net.

Midwater trawls. These are among the simplest trawls in design but can be difficult to fish effectively in a consistent fashion. Midwater trawls are often constructed along a four-seam design (Figure 7.3B), but numerous adaptations are available. For scientific sampling, midwater trawls often have a small mesh size in the cod end and progressively larger mesh toward the mouth of the net. In commercial applications of midwater trawls, very large mesh webbing may be used at the front of the net to increase the effective mouth size (Figure 7.3C). Pelagic fishes will often avoid crossing through the mesh, even when they are small enough to do so. Consequently, the large mesh effectively forms a behavioral barrier to the fish, and they eventually fall to the cod end of the net where they cannot escape.

The mouth of midwater trawls can be held open in several ways. One method is to tow the net with two boats (pair trawling), thus maintaining pressure on the net laterally. The depth at which the net fishes is controlled by weights attached to the bridles or trawling warps (Figure 7.3B), the amount of cable out, and the boats' speed. For single-boat operation, otter doors are generally used to open the net (Figure 7.3B). Otter doors for midwater trawls are relatively lightweight and are designed to provide lateral pull on the net's mouth. Depressor plates may also be used to help the net fish deeper. The fishing depth depends on the boat's speed and the amount of weight attached to the bridles.

Midwater trawls have been used extensively for sampling pelagic fishes in both marine and freshwater habitats. Frequently, midwater trawls are used in conjunction with hydroacoustic surveys (Chapter 13) to identify hydroacoustic targets and to provide biological samples of fish. Specialized versions of midwater trawls are often used for sampling fish larvae and young juveniles. Plankton nets and bongo nets (Chapter 9) are often used to sample fish larvae. The Isaacs–Kidd trawl is another variation and is frequently used to sample young fish suspended in the water column (Chapter 9).

Midwater trawls were used in a multipurpose study by Moyle et al. (1992) in which they studied the life history and abundance of delta smelt. They fished a 17.6-m-long trawl at a set of standardized stations to determine an index of abundance and the diet of delta smelt. Another example of the use of midwater trawls is the annual juvenile rockfishes survey conducted by the Tiburon Laboratory of the NMFS. The goals of this survey were to provide (1) an index of abundance, (2) biological specimens for understanding the ecology of pelagic rockfish juveniles, and (3) information on environmental conditions (Echeverria et al. 1990). The investigators used a combination of chains and pelagic otter doors to open and weight the net to sample at a specific depth. They observed that changes in construction of the otter doors from wood to steel had a substantial effect on the depth at which the net fished for a given amount of cable out.

7.3.2 Gear Performance: Monitoring and Evaluation

The performance of the gear should be evaluated during any sampling operation, but evaluating the consistency of gear performance is particularly important when catches are used as an index of abundance. The first question to be answered is whether the gear was fishing properly or not. Several things can happen that will prevent a trawl from fishing properly or effectively. A common and easily recognized mishap is that the cod end was not tied shut before deploying the net or the cord untied during fishing operations. Another problem is the net catching on a bottom obstruction. Small snags may not be felt while fishing but may tear the net. Trawl doors may also get crossed while the net is being set or is fishing, causing the net mouth to collapse. For bottom trawls, the metal shoes of the otter door will be worn through contact with the substrate; the wear pattern will indicate whether or not the doors are fishing equally.

In addition to recognizing whether the net was fishing at all, additional equipment may be used to determine if the net was fishing at the desired depth and to monitor net characteristics. A number of acoustic monitoring instruments are available that measure net depth, distance between headrope and footrope, and the spread of the wings. These instruments are used on both midwater and bottom trawls and provide instant feedback on where the net is and how well it is performing. These instruments are often combined with temperature meters or other sensing devices, allowing for the measurement of environmental variables at the depth the net is fishing.

7.4 DRAGGED OR TOWED GEARS: DREDGES

Dredges are heavy-framed samplers designed to collect demersal macroinvertebrates and fishes. Many dredges have teeth, cutting bars, or pressure plates to dig into the substrate or may use hydraulic pumps to remove animals from the substrate (Figure 7.5). There are numerous dredge designs, each intended to collect a different species or operate in a different habitat. As with otter trawls, dredges generally require a vessel and two or more people to operate safely.

7.4.1 Variations

Scallop dredge. Scallop dredges used for scientific sampling are generally scaled-down versions of commercial dredges. Scallop dredges have a rectangular metal opening at the front with a triangular frame attached for towing (Figure 7.5). The frame may have a pressure plate at the top to cause the dredge to dig deeper into the substrate. A bag made of metal rings is attached behind the frame to capture animals that are dislodged. In commercial dredges, the ring size is chosen or regulated to allow smaller scallops to escape. For example, rings used in the sea scallop fishery along the east coast of the United States are commonly between 8.9 cm and 10.2 cm in diameter. With this size rings, the catch of sea scallops with a shell height of less than 7 cm is low. For scientific sampling, smaller rings (e.g., 5.1 cm) are commonly used, and the ring bag may be lined with a small mesh net to capture animals not yet recruited to the fishery (i.e., scallops with a shell height between 3 and 7 cm) . Oyster dredges are similar in design to scallop dredges except that the opening of the metal frame is taller and the chain bag is shorter (Figure 7.5).

Hydraulic dredge (clam dredge). As with the scallop dredge, hydraulic dredges used for sampling are generally based on commercial designs. Hydraulic dredges generally have a heavy "sled" at the front that runs along the bottom. The sled houses the attachment of the hydraulic jet(s) and also has a mechanism that cuts into the substrate and passes animals up into a net or conveyor system (Figure 7.5).

7.4.2 Gear Performance: Monitoring and Evaluation

As with trawls, a number of electronic attachments are available to determine the depth and water temperature where the dredge is operating. Because the mouth opening of a dredge is fixed, determining the area fished by the dredge is easier than

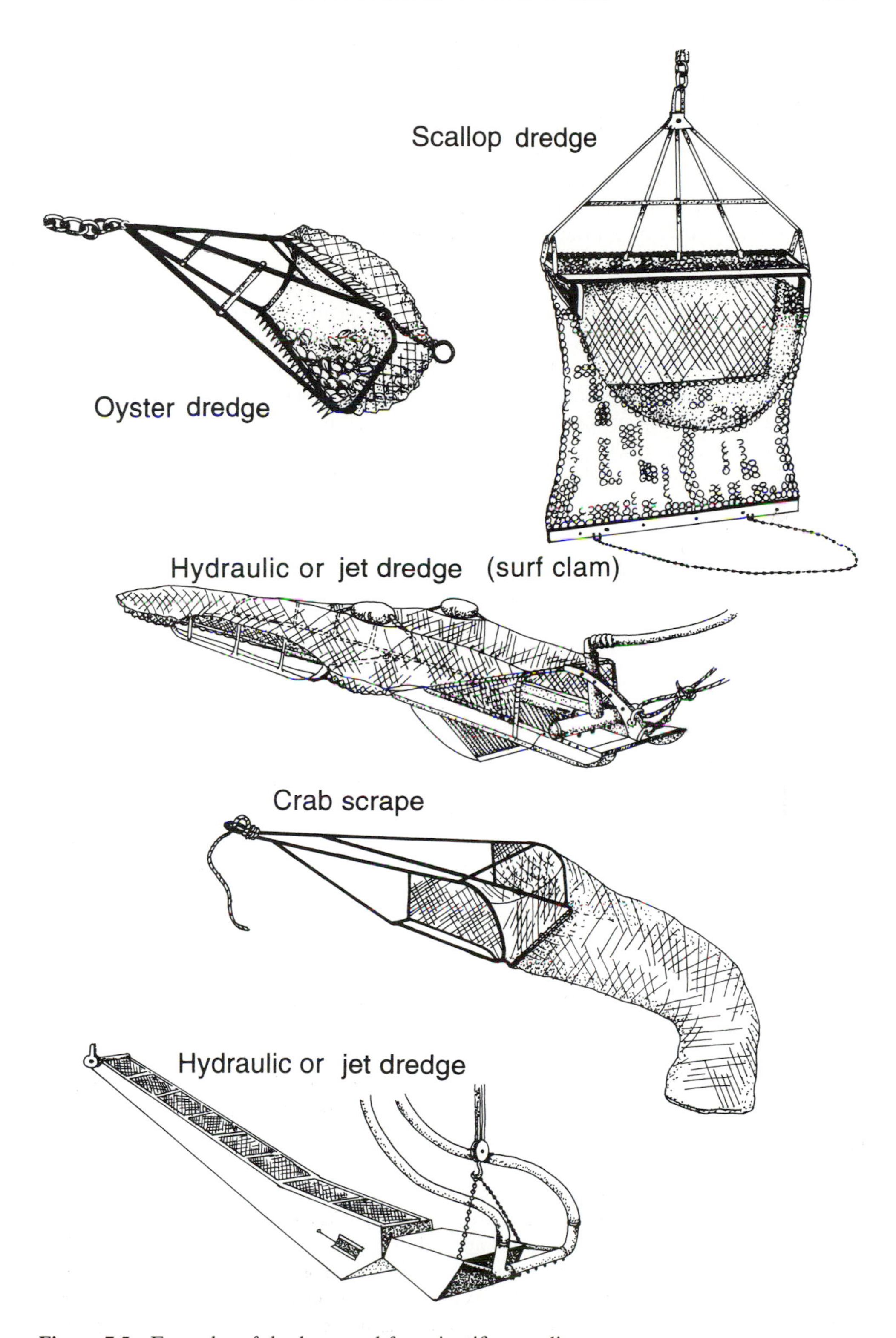

Figure 7.5 Examples of dredges used for scientific sampling.

it is with trawls. The main problems encountered with dredges are hanging up on bottom obstructions and having the dredge fish upside down. To determine if the dredge has been fishing right side up, the shoes on the base of the sled or frame should be examined. Abrasion from the bottom will shine the metal shoes, indicating which side of the dredge was in contact with the bottom. Occasionally, large boulders or timbers will be caught in the dredge—rocks up to 1 m in diameter are possible in offshore sampling with large scallop dredges. Rocks and timbers can cause the dredge to fish poorly in addition to crushing much of the catch.

7.4.3 Dredge Applications

Although biological specimens are often collected with dredges, few examples are available of resource surveys being conducted with dredges. The NMFS uses a scallop dredge to conduct an annual survey of sea scallop abundance along the eastern seaboard of the United States. For these surveys a 2.44-m-wide dredge with 5.1-cm rings and a 3.8-cm polypropylene mesh liner is used (Serchuk and Wigley 1986). This survey is conducted to provide indices of abundance for sea scallops not yet vulnerable to commercial gears (prerecruits) as well as sea scallops large enough to be harvested (recruits). Information on the size composition of the population, shell height–meat weight relationships, and relative fecundity is also obtained from animals collected. Similar surveys are also conducted by the Department of Fisheries and Oceans, Canada, on Georges Bank and in other areas.

7.5 SURROUNDING OR ENCIRCLING NETS

Encircling nets include beach or haul seines, purse seines, and lampara nets (Figure 7.6). Encircling nets are used to trap fishes actively by surrounding them in a fencelike wall of netting. Surrounding nets provide many advantages over other sampling gears because the gear is relatively easy to deploy, sampling is rapid, a large area can be sampled, the limits of the sampling area are precisely defined, and fish are obtained live with minimal trauma. These nets also provide for live release (Pierce et al. 1990). In contrast to trawls and most dredges, beach seines can be fished without a vessel and can be operated by a single person.

All encircling nets have a float line and a lead line, and many have a specially constructed bag (Figure 7.6A) in which the fish are concentrated as the net is hauled. The float line has cork or plastic floats fastened to it, which maintain it on the surface. The lead line has weight attached to it or is made of a lead core line so that it sinks, forming the desired wall of webbing. The bunt (Figure 7.6D) is a section of netting in which fish are concentrated. The bunt section hangs loosely and can be located in any position along the net depending on how the net is hauled. In some seines, the bunt is enlarged to form a bag in the netting. For large nets, hauling can be facilitated by using winches to pull in the net.

7.5.1 Variations

Beach seine. In its simplest form, the beach or haul seine is made of mesh of uniform size and consists of two wings and a bunt section that holds the catch (Figure 7.6A). The wings form a long vertical wall that funnels fish toward the bunt. The wall of netting can be lengthened by using long towing lines or warps. For large catches, a bag between the wings is useful to concentrate the fish into one area of the net. Additional netting may also be included in the bag to form a trap to keep fish

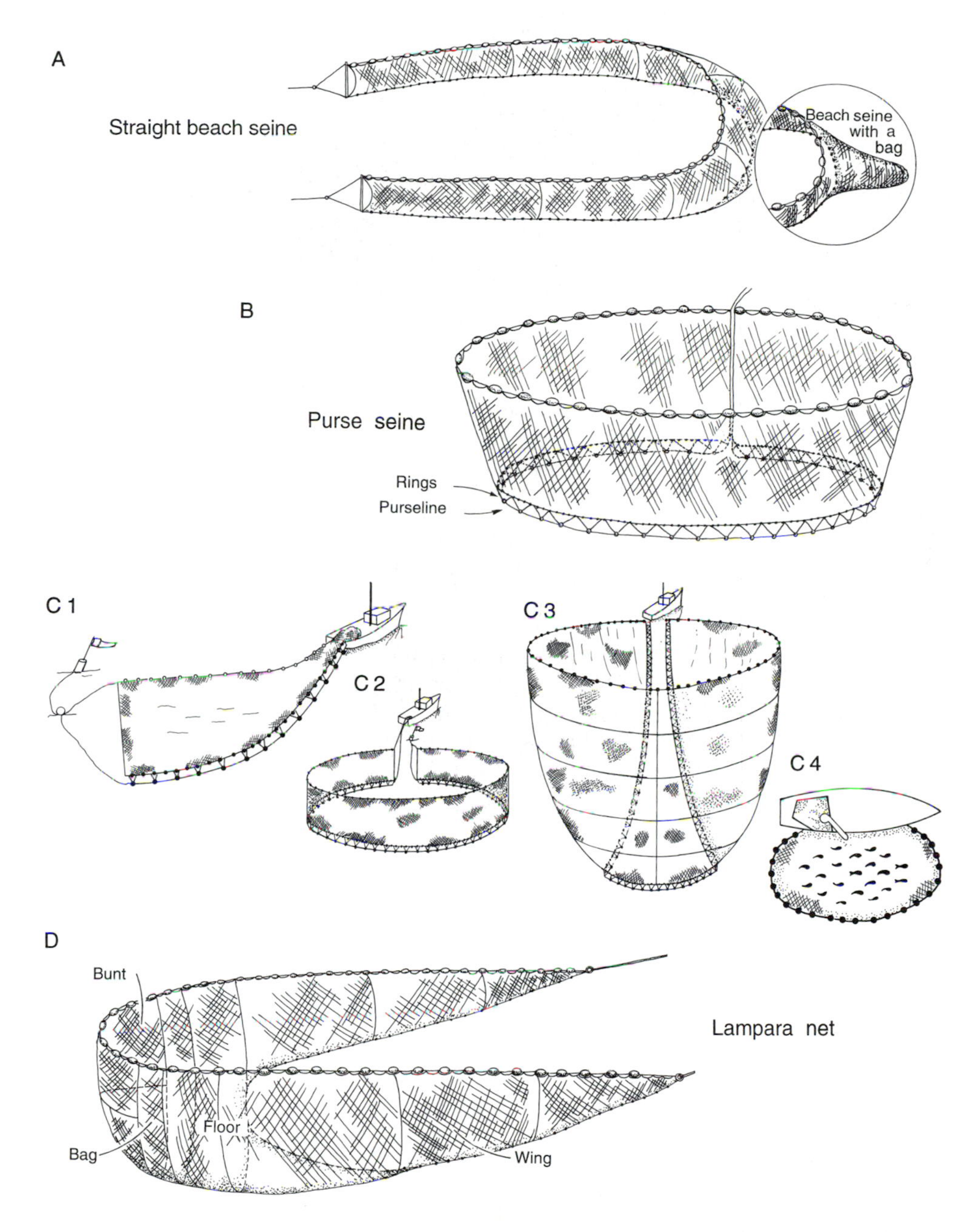

Figure 7.6 Design and nomenclature of surrounding nets: (**A**) straight beach seine; (**B**) purse seine; (**C**) the setting of a purse seine in the one-boat system (Sainsbury 1975); and (**D**) lampara net.

from escaping once they are in the bag (von Brandt 1984). The bag can be located in the middle of the net or closer to one of the sides. If the net is hauled by pulling both towlines simultaneously, the bag should be in the center. If hauling proceeds by pulling primarily one towline, then the bag may be to one side of the net. The wings of the net may be tapered, and larger-size mesh can be used in the end panels to reduce overall drag.

Beach or haul seines are typically used in shallow water where the net wall can extend from the surface to the bottom. They are most effective for nearshore residents or for species that concentrate near shore seasonally or on a daily basis. Seines without bags are frequently used in ponds where catches are small and can be used in rivers to avoid having the bag tangle in the current. Seines with bags are more popular in lakes where currents are not a problem and catches may be large. To prevent fish from escaping under the net, the lead line must be in contact with the bottom and not become entangled or caught on obstructions along the bottom. Thus, the use of beach seines is restricted to stretches of smooth, shallow bottoms with little surf.

Beach seines are generally set in a semicircle around the targeted fish and dragged to shore, herding fish into the net. Setting and hauling beach seines can be done in several ways. One method is for a single vessel or person wading in shallow water to fix a towline on shore and then set a wing and the bunt before turning in to shore and setting the second wing and towline. The net is hauled in by dragging in the towlines until the bunt reaches shore and the catch can be retrieved. A second method requires two vessels or two people wading in shallow water. In this case, the two vessels or people start together at the farthest point from the beach and set the bag. They open the seine parallel to the beach, each setting a wing and towline. Then, the seine is pulled towards the shore, and the wings are brought together. The bag with the catch is held between the vessels or hauled ashore where the catch can be collected. Another common method of seining is for two fishers to set the net perpendicular to shore and then drag the net parallel to shore. At the end of the tow, the person farthest out turns in to shore and both wings are pulled ashore.

Beach seine collection efficiency has been estimated in lakes, reservoirs, and estuaries by using block nets to enclose completely the area being seined. The area is then seined until either no more fish are caught by the seine or a fixed number of hauls is made. The area may be sampled using rotenone to capture the remaining fish or the initial number present estimated by a removal method (e.g., Leslie method; Ricker 1975). Capture efficiency is determined as the ratio of the number of fish caught in a given haul over the total number of fish available to the seine before deployment of the gear (Lyons 1986; Parsley et al. 1989; Pierce et al. 1990; Allen et al. 1992).

Beach seine collection efficiency has been found to vary with the position of each species within the water column. In general, beach seines tend to have higher catch efficiencies for fishes residing in the middle of the water column than for demersal fishes (Lyons 1986). Collection efficiency also varies with fish behavior. For example, Allen et al. (1992) found that capture efficiency was highest for species that stayed well off the bottom or for demersal species that tended to swim ahead of the approaching net (e.g., mullets) as opposed to those demersal species that tried to swim under or around the net (e.g., topminnows and spot). In addition, the capture efficiency of beach seines is related to the bottom structure of the area being sampled. Structures that cause the seine to snag or roll will reduce seine efficiency. However, Pierce et al. (1990) found that collection efficiency was higher in certain types of macrophyte beds. This occurred because fish in extensive macrophyte beds were less agitated by the presence of people and were less likely to search for an escape route. Because of the variability in efficiency, it is recommended that capture efficiency be estimated at least once each season at each study site if the purpose of a study is to estimate density rather than provide an index of abundance.

Many studies in which beach seines are used rely on a single haul to represent the fish community in the area being seined (Allen et al. 1992). In a study to estimate seine capture efficiency of estuarine fishes, comparisons made between the catch in the first haul and the total population in the seined area showed that species richness, species rank, and size distributions of dominant taxa were well represented by the first haul. However, single hauls were not as reliable for rare taxa or for determining the total abundance of all species (Allen et al. 1992). Using capture efficiency corrections should minimize this problem.

Purse seine. Purse seines (Figure 7.6B) are generally used to collect pelagic species swimming near the surface in open water. They can be used to catch demersal species by sinking the wall of netting to the bottom; the substrate, however, must be smooth enough to allow the purse line to close (von Brandt 1984). Purse seining involves setting a long net to enclose the school of fish being targeted. Purse seines are used in open water and have sufficient flotation to suspend the net at the water's surface. Unlike the beach seine, purse seines surround fish both vertically and horizontally (Figure 7.6C). Purse seines are made of long walls of netting with a lead line of equal or greater length than the float line. The purse line, located below the lead line (Figure 7.6B), runs through oval rings that are made of nonrusting metal alloys and connected by short lengths of rope to the lead line. When the net has encircled the fish, the purse line is pulled from one or both ends to close the bottom of the net. This causes part of the net that was initially hanging vertically to be pulled horizontally under the fish, preventing their escape downward. This action traps the fish in an artificial pond of webbing. As the net is retrieved, the holding area is slowly made smaller until the fish are gathered alongside the vessel where they can be taken aboard. The bunt, where the catch is gathered, is often made of smaller mesh and of the strongest material used in the gear. There are two basic types of purse seines: those in which the fish are finally contained in a bunt in the center of the net and those in which the bunt is at one end.

Two different methods of purse seining are the one-boat and two-boat systems. The one-boat system is more popular and economical whereas the two-boat system can deploy larger gear faster. In the one-boat system (Figure 7.6C), the net is fixed at one end to a buoy or to shore, and the vessel sets the rest of the net around the school of fish. The net is then pursed and hauled from one vessel. One way of setting a purse seine by means of a two-boat system is for each boat to carry about half the gear and for both boats to set their part of the seine simultaneously, beginning with the middle of the net. Once the encircling is completed, the net is pursed, and each boat recovers opposite ends of the net. The fish are trapped in the bunt between the two boats. Another way of setting the net in the two-boat system is to have one boat remain stationary while the other boat encircles the area with the net, returning to the stationary boat.

Pearcy and Fisher (1988) used a 495-m purse seine to determine the migratory pattern of juvenile coho salmon off Oregon and Washington. In this study, catch rates and different sizes of fish caught at fixed transects along the coast were used to document the northerly movement of juvenile coho salmon during their first summer in the ocean. Another study used a purse seine to study the food habits of rainbow trout in the offshore areas of Lake Washington (Beauchamp 1990). The vessel deployed a purse seine that was 37-m deep, had a 600-m float line, and was made with 25-mm stretch mesh netting.

Charles-Dominique (1989) compared the catch efficiencies of purse and beach seines used in Ivory Coast lagoons. Both nets in this study were deployed in shallow water so that they enclosed the entire water column. The purse seine was found to be more efficient in limiting escapement than was the beach seine. The pursing action caused the netting to vibrate and may have kept fish away from the net, giving them less opportunity to escape (Charles-Dominique 1989). This study also determined that active avoidance of the purse seine while it is being set is the main factor affecting catch efficiency. Thus, care should be taken in interpreting abundance estimates when collecting species that can easily outswim the net as it is being set.

Lampara net. The lampara net (Figure 7.6D) is used in open water to trap pelagic fishes near the surface of the water. Like a purse seine, a lampara net surrounds fish both vertically and horizontally. Lampara nets are effective when operated over rough ground where haul seine nets and trawls cannot be used. They can also be used in areas where beaches are unsuitable for hauling a gear such as a beach seine (von Brandt 1984). The dustpan shape of the lampara net is attained by having the leadline shorter than the floatline, which causes the bottom part of the net to protrude beyond the top part (Figure 7.6D). To reduce drag, the wings are usually made of larger mesh than is the bag, which is usually small meshed.

Lampara nets are generally fished from a single vessel. The end of one wing is attached to a buoy or to a stationary skiff. The remaining netting is carried around the school of fish by the vessel until it returns to the initial spot. Once the encircling is complete, the two wings are hauled together. This hauling action causes the leadline along the two wing sections to come together, thereby closing the bottom of the net.

There are few examples of scientific collections being made with lampara nets, although they have been used to capture juvenile chinook salmon in the Great Lakes.

7.6 OTHER ACTIVE SAMPLING GEARS

Scientists have used a number of other active sampling gears to collect different life stages of fishes and invertebrates for fishery research and management purposes. Many of these gears are small and easily portable and can be adapted for use in several different habitat types.

Push nets. Push nets are usually made of netting attached to a frame designed to keep the mouth of the net open (Figure 7.7A). Push nets are fished by a person wading in the water and pushing the net forward or by having the net attached to the front of a boat (Kriete and Loesch 1980). Push nets have been used along the bottom to collect shrimps (von Brandt 1984) and near the surface to collect fish fry (Eaton et al. 1992; Chapter 9). Standardization of net design and operating procedure can facilitate the interpretation of results and comparison of push-net estimates through time (Jessop 1985).

Lift nets. Lift nets are generally used to catch small schooling fishes and invertebrates. They are constructed of sheets of netting set horizontally along the bottom or kept suspended in the water column (Figure 7.7B). Fish are caught by lifting the net out of the water after they have concentrated over the net. A variety of frames, from small handheld frames to large round, square, or triangular frames that are lifted mechanically can be used. Bait or lights can be used to concentrate desired species over the lift net. Fish caught in an Indonesian commercial fishery that

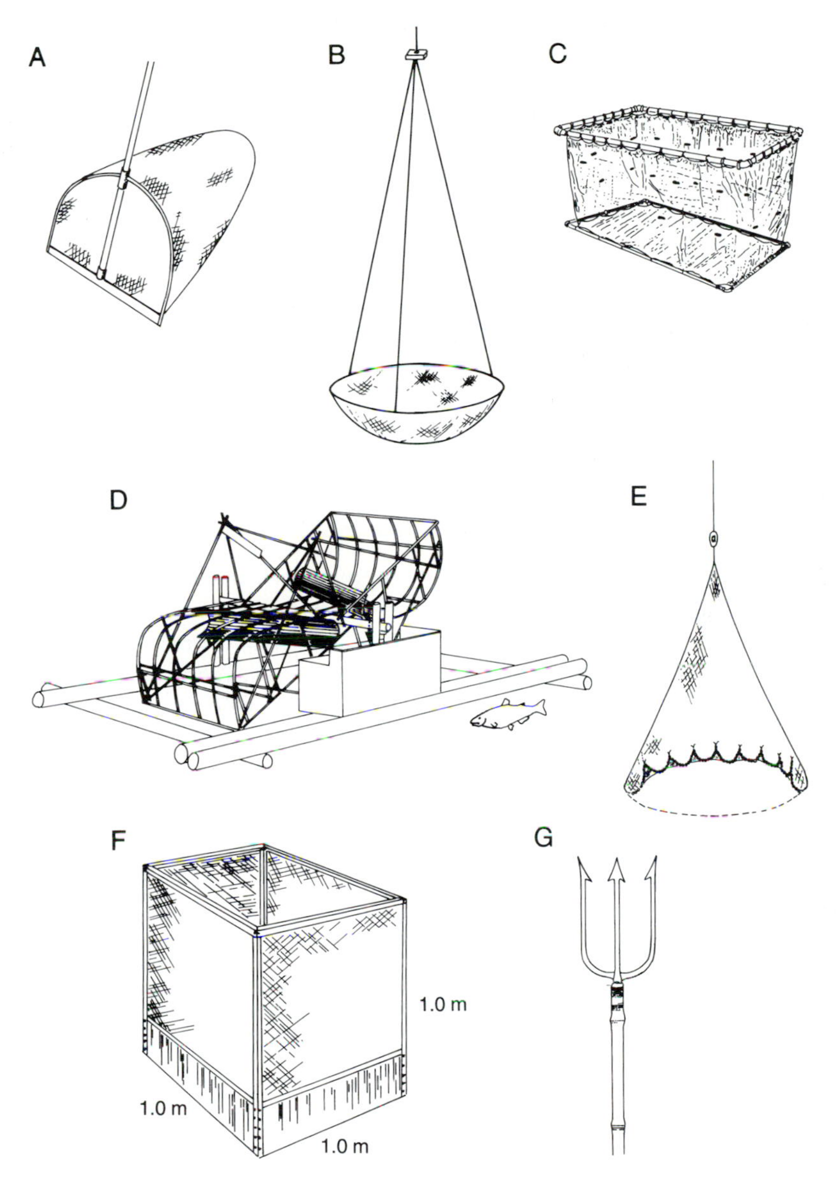

Figure 7.7 Examples of less commonly used active fish capture gears: (**A**) push net (von Brandt 1984); (**B**) lift net (von Brandt 1984); (**C**) pop net (Dewey et al. 1989); (**D**) fish wheel (von Brandt 1984); (**E**) cast net (von Brandt 1984); (**F**) drop net (Chubb 1985); and (**G**) spear (von Brandt 1984).

used lift nets were studied to determine ovarian development (Wright 1992). Conical lift nets operated by a hydraulic winch were used to collect roughtongue kapenta in Lake Kariba, Zimbabwe, to study the effects of harvests on their abundance (Marshall 1988).

Pop nets. Pop nets are a variation of lift nets and are used to sample fishes within a vertical column of water (Figure 7.7C). These nets are set on the bottom and are deployed by a release mechanism. Because pop nets sample a known volume of water, they can be used to provide quantitative estimates of fish abundance (Espegren and Bergersen 1990). A study comparing fish catches with pop nets and seines found that pop nets effectively sampled fish in shallow, nonvegetated areas and were useful in areas of heavy vegetation where seining or electrofishing were difficult. However, the pop-net catch was found to be less diverse than was the seine catch. This could be attributed to the fact that the pop net enclosed a much smaller area than did the seine (Dewey et al. 1989).

Dip nets. Dip nets are bag-shaped nets on a circular frame that is attached to a handle. They are similar to lift nets in the sense that they are used to collect fish by placing the net underneath the fish, which are then caught by lifting the net from the water. However, an active scooping motion, rather than a simple vertical lifting motion, is used to capture fish. Dip nets are often used to recover fish or invertebrates that have been trapped using other methods (e.g., Minello et al. 1989) or stunned by electroshocking or toxicants. Dip nets are also used to catch small fish in lakes and streams (Graham and Vrijenhoek 1988).

Fish wheel. The fish wheel is a large waterwheel with a series of paddles or shovels that rotate in the current of a river (Figure 7.7D). Historically, fish wheels were used by Native Americans and others to harvest anadromous fishes. The fish wheel operates like a lift net: fish swimming over the shovels are lifted out of the water. Generally, a collection box is attached to the fish wheel, and the shovels are slanted to pass fish into the collection box as the shovel comes out of the water. In a scientific application, Eiler et al. (1992) used a fish wheel to collect adult sockeye salmon as they migrated upstream. Fish were then fitted with a radio transmitter to determine the distribution of spawning adults in the river.

Cast nets. Cast nets are circular nets with weighted edges (Figure 7.7E) that are thrown so that they fall over fish which have been sighted. Cast nets are restricted to use in areas that are free of obstacles or plants and have a smooth bottom so that the net does not get caught as it falls, which would allow fish to escape. Bait or light is sometimes used to attract and concentrate fish over suitable bottom types. Much experience is needed to cast the net to fall flat on the water so that it is completely open while sinking. Most hand cast nets have a radius of 2 to 3 m and have a central line for hauling in the net. Hauling in the central line causes the webbing to collapse inward. Some cast nets have pockets along their circumference to trap fish as the net is hauled. Cast nets may also have a purse line so the net can be operated in deeper water to catch schooling fish in midwater. The net is cast in the normal fashion and then pursed and hauled before the net reaches the bottom. Cast nets have been used to collect fish in shallow-water canals surrounding marshes (Meador and Kelso 1990) and to collect nearshore lake species. Taylor and Gerking (1978) used a cast net to collect the Ohrid rifle minnow in depths of 0.5 m to 1.0 m, where the net had a capture efficiency of approximately 10%.

Drop nets. Drop nets are a combination of a net and an enclosure trap. These nets are often mounted on a boxlike frame (Figure 7.7F), and they are set by being lowered or thrown into the area being sampled. Drop nets enclose a known volume of water and therefore allow for estimates of fish density. One study found that a drop net which was slowly lowered into the water was less effective than one that was thrown into the desired area (Kushlan 1981). Drop nets have been used effectively to sample small fish in heavily vegetated marshes (Kushlan 1981) and to provide population estimates of European smelts in reservoirs (Lee and Whitfield 1992).

Angling. Angling with rod and reel is typically thought of as a sportfishing method, but it can be used to acquire samples for scientific research. Angling has been used for a variety of purposes ranging from procuring live specimens for a radiotelemetry study (West et al. 1992) to procuring tissue samples for studies of fish genetics (King and Pate 1992). Angling has also been used to evaluate the mortality caused by catch and release of lake trout (e.g., Loftus et al. 1988), among other species.

Angling can also provide useful information on fish population structure and abundance. For example, the size structure of largemouth bass samples caught by angling has been observed to be well correlated with size structure in electrofishing samples (Santucci and Wahl 1991; Isaak et al. 1992).

Many state natural resource agencies monitor angling tournaments or contests to obtain information on relative abundance and size structure (e.g., Willis and Hartmann 1986; Olson and Cunningham 1989; Quertermus 1991). However, Schramm et al. (1991) cautioned that such data need to be carefully interpreted to provide useful trend information. Gabelhouse and Willis (1986) noted that tournament bass anglers overestimated the proportion of larger fish in a population compared with estimates from electrofishing; Ebbers (1987) and D. M. Green et al. (Cornell University, unpublished data) found that size structure data supplied by carefully screened anglers were comparable to data from electrofishing.

Size selectivity should be considered when angling is used as a sampling gear. A study of the size selection of hook-and-line gear found that small hooks caught considerably more small fish whereas large hooks were more effective in catching larger fish (Ralston 1990). The type of lure or bait used can also influence the size of fish caught (Payer et al. 1989).

Spears. Fishing with spears has been employed for over 10,000 years. Spearing is most effective in calm, shallow waters or through the ice. Fish that are speared can escape from a single point spear by wriggling vigorously. Barbs on the spear (Figure 7.7G) can help prevent escape. Hand spearing has been used to collect the half-banded seaperch in a study of their reproductive pattern (Webb and Kingsford 1992).

A variation on the handheld spear is the speargun (Chapter 18). Spearguns project the spear farther and with more power than does throwing by hand. The shaft of the spear used with a gun is usually shorter than that of a handheld spear and reduces drag underwater. Spearguns are often used for sportfishing and have been used in scientific studies to collect fish specimens in coral reefs (Shpigel and Fishelson 1991).

Harpoons were developed from spears and differ in that the point becomes separated from the shaft when the harpoon penetrates the target. The shaft remains connected to the point by a line and floats to the surface. The floating shaft tires the fish as well as indicates the location of the fish to the fisher, who can follow the shaft to retrieve the catch. Harpoons can be used to kill large fish such as tunas, swordfish, and sharks, or sea mammals such as porpoises and whales (von Brandt 1984). They are sometimes equipped with explosive tips to more quickly disable the animal.

Detonating cord. Concussion sampling with detonating cord has been used as an alternative to rotenone sampling of fish communities. Detonating cord is a flexible, ropelike material containing a core of explosives (usually pentaerythritol tetranitrate), that produces a pressure wave sufficient to kill fish. Detonating cords have been used successfully to sample fish communities in streams (Metzger and Shafland 1986). However, a comparison of the sampling effectiveness of rotenone and detonating cord in a warmwater impoundment found that rotenone was much more effective in this habitat (Bayley and Austen 1988). Using explosives is potentially dangerous and destructive. Further, fish collection can be hampered by habitat structure, and many fishes sink because their air bladders are destroyed. For these reasons, detonating cord should be used with care when other alternatives are not feasible.

Diving. Diving with or without scuba gear can be used to capture or observe fish. Diving is especially useful in studies of habitat availability and behavior; underwater observation is discussed in Chapter 18.

7.7 GEAR SELECTION

A number of considerations must be made before selecting a fish capture method. One of the first is the purpose for the collection and the goals of the study. Another obvious consideration is the habitat being sampled because all gears do not perform equally well in all habitats. The behavior and size of organisms that are targeted are also key concerns when selecting a sampling method, as are the time, budget, personnel, and vessel resources that are available for sampling.

7.7.1 Relationship Between Study Objectives and Gear Selection

There are three broad purposes for fish collections: determining density or relative abundance, collecting live specimens for study or whole-specimen samples, and collecting accessory data from fish (e.g., scale samples, tissue samples, or stomach contents). As indicated earlier, active gears are often chosen because they provide a means of accurately determining relative abundance. Generally, methods providing a measure of fish abundance also allow the investigator to obtain live specimens or ancillary data. Hydroacoustic surveys (Chapter 13), however, are an exception to this because they do not provide specimens.

7.7.2 Influence of Environmental Characteristics on Gear Selection

Because active gears are moving through the water, they are prone to snagging on obstructions present in the water column or on the bottom. Some gears, such as the otter trawl, can be modified to resist catching on obstructions. However, obstructions such as rocks, logs, weeds, or coral beds generally present substantial problems for active fish capture methods.

Water depth will often limit the gears that can be used (Figure 7.8). Beach or haul seines, for example, are limited to relatively shallow depths along the shore and without obstructions. In contrast, otter trawls are often used in deep water. Seines can be modified to fish in deeper water along the bottom (Scottish or Danish seines), but their use is generally limited to commercial fisheries. The slope of the bottom also influences gear choice. Otter trawls, for example, do not perform well where the bottom has a very steep slope or where depth changes irregularly because the gear loses contact with the bottom. However, other methods, such as angling, are well suited for use in these types of areas.

Sampling in streams and rivers presents special problems for sampling with active fish capture methods. Besides obstructions (logs, boulders, and so on), the current in

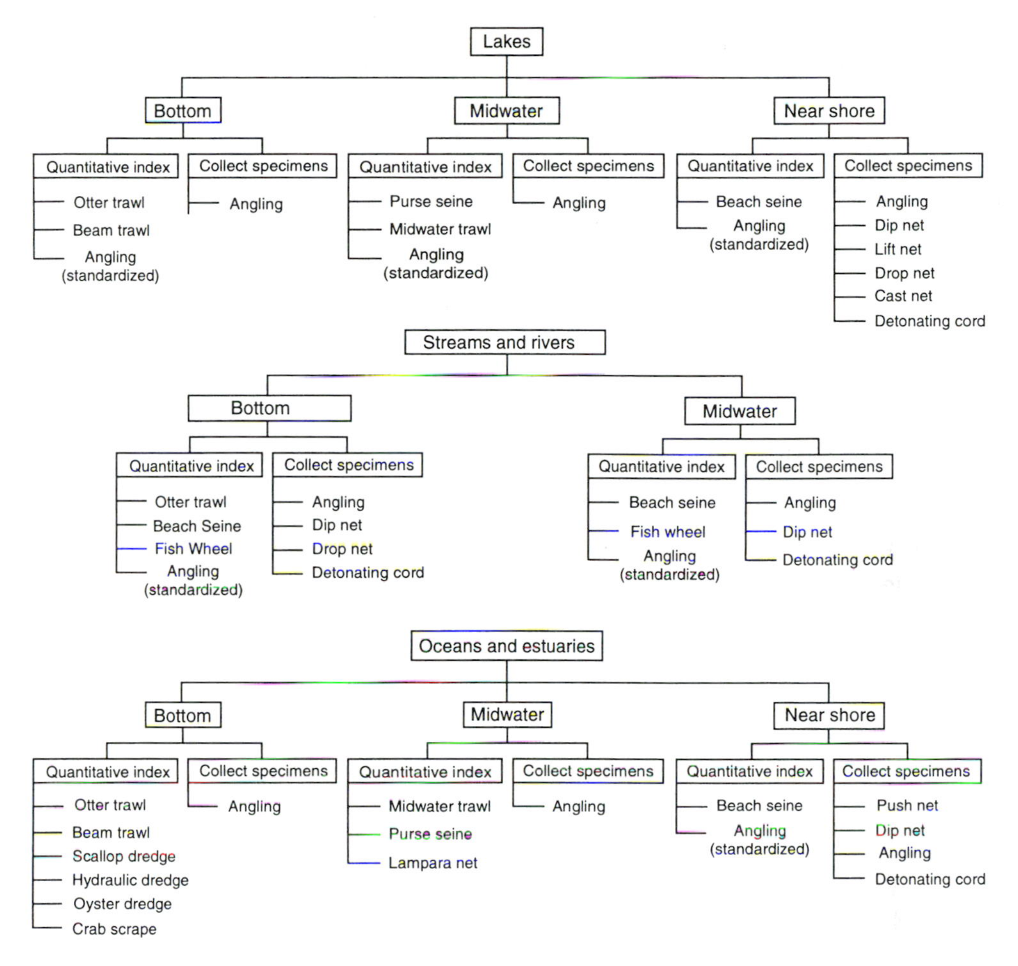

Figure 7.8 Use of active sampling methods in major habitat types. Note that gears used to determine quantitative indices can also be used to collect specimens.

lotic habitats makes it difficult to control fishing gears. Because of these difficulties, electrofishing (Chapter 8) is a common method of actively sampling fish in moving waters.

7.7.3 Influence of Animal Behavior on Gear Selection

An important consideration in choosing a sampling gear is where the fish lives. Some species of fish are associated with the bottom of the lake, river, or ocean. Such species are called demersal or benthic. Other fishes are commonly found suspended in the water column and are known as pelagic. Examples of demersal species include flounders, scallops, and sculpins. Pelagic fishes include groups such as the tunas, herrings, and squids. In addition to their orientation to the bottom, another consideration is whether fish choose habitats near underwater structures or not. Largemouth bass and coral reef fishes are examples of species that often live in or near structures in their habitats. Sampling fishes associated with large structures is a substantial problem for most active fish capture methods. Another element of fish habitat choice that affects gear selection is how close the species lives to shore. For example, beach seines are often used to sample minnows that choose nearshore habitats, but are not used to collect other species, such as the marine goosefishes that are rarely caught near shore. Figure 7.8 outlines the applicability of various gears for various water depths and distances from shore.

An important aspect of fish behavior to consider is how fish distribution and habitat choice varies over the course of a day (diel behavior), between seasons and over the fish's lifetime. For example, in freshwater lakes larval yellow perch are found in offshore habitats from 0 to 5 m below the water surface during the day but are concentrated in the upper 1 to 2 m at night (Clady 1976). As juvenile yellow perch reach 3 to 5 cm, they often move inshore into the nearshore or littoral zone and are associated with bottom habitats or macrophyte beds (Weber and Les 1982). Yellow perch older than age 1 are found in a variety of habitats, depending on the thermal regime of the lake (stratified or unstratified) and food resources (Hayes et al. 1992). In some lakes, adult yellow perch are found in nearshore habitats, whereas in other lakes they are found in the offshore or pelagic zone, suspended off the bottom. Because of the varied set of habitats occupied by yellow perch throughout their life, different gears are necessary to sample their population representatively.

Beyond the "natural" behavior of fishes, biologists also need to consider the response of fishes to the gear being used. This is particularly important if the goal of the sampling program is to determine fish density or provide an index of abundance. When seeing a net, or encountering the pressure head on a moving sampling device, many fishes will react by trying to avoid the gear. Some fishes, particularly large, pelagic predators such as tunas, can outswim most towed gears. Other fishes respond to encounters with gears by rising above the net or diving below the net. Some gears take advantage of these behavioral responses; otter trawls often use groundlines to herd fish into the path of a net, thereby fishing an area larger than the mouth of the net.

7.7.4 Influence of Fish Size on Gear Selection

Nearly all fishing gears are selective in the size of fish and the species they catch. In many cases, size selectivity occurs because the mesh openings in the net allow smaller fish to pass through the net while retaining larger fish (e.g., Rulifson and Cooper 1986). Fish that are most vulnerable to a gear are called fully recruited, and fish less vulnerable are called partially recruited. Besides obvious factors such as the size of mesh openings in a net, fish may be partially recruited to a gear for other reasons. For example, some fishes (e.g., tunas) may reach a size at which they can outswim the net. Changes in spatial distribution over a fish's lifetime also affect how vulnerable different age- and size-classes of fish are to a gear. For example, if the juveniles of a species are distributed near shore and the adults of a species are distributed offshore, the adults would be less vulnerable to beach seines even though the holes in the net are small enough to retain them.

7.8 SAMPLING PROBLEMS

Habitats that have obstructions. Many species of fish are associated with structures, especially in streams, nearshore areas in lakes, and reefs in open waters. Such structures can be as simple as boulders and rocks or as complex as downed trees, woody debris, or rooted aquatic plants. The sampling problems are twofold in this type of habitat. First, underwater structures interfere with sampling efficiency. For instance, it is nearly futile to sample nearshore fishes with a seine in areas with tree stumps because the net will catch often, allowing fish to escape. Second, the sampling gear is unable to sample the habitat where the fish either live or hide. An example of this type of habitat is coral reefs, where many fish species live either within the reef complex (e.g., moray eels) or enter the reef complex for cover.

Sampling freshwater habitat structures is often best done with a fish toxicant or, if the area is shallow enough to retrieve stunned fish, by electrofishing. These methods will be successful only if the fish do not become entangled in the structure before arriving at the surface. In some habitats, bait or light can be used to attract some animals away from structures so that another gear can be used (e.g., nets or traps) to capture them. For small sample sizes, a suction "gun" or spear may be successful in these habitats. Much work is needed in gear development if we are to quantify efficiently and effectively the fish species using these habitats.

Deepwater habitats in small lakes. Sampling deepwater demersal and pelagic fishes in small lakes is an example of the morphometry of a system limiting gear selection. For most demersal fishes, some form of bottom trawling is an effective means of capture. However, in a small body of water, the boat required for successful trawling may be too large for the lake or access to the lake may be difficult. Because of this, stationary sampling devices (e.g., gill nets) are often used in these situations if the behavior of the target species allows. A new technology that may be useful, with appropriate modifications, is the jet ski. These crafts are now powerful enough to pull a small trawl, but the safety and stability of these crafts under such a load is unknown.

Large rivers. Fishes residing in large river ecosystems provide particular sampling challenges. This is especially true for high-current habitats that limit the effectiveness of boat and gear technology. In such areas, active sampling gears are usually ineffective, and stationary sampling devices are difficult to deploy and maintain properly. Fish toxicants are also of limited value due to the volume of toxicants needed and our inability to maintain the toxicant effectively in the sampling area. Radiotelemetry, however, can be effectively used to monitor the behavior of fishes in these systems.

Ice cover. Spearing and angling have been widely used to capture fishes under the ice. Using other types of active sampling gear under ice conditions is difficult because of reduced mobility. The use of nets under the ice is generally restricted to passive gears set through holes in the ice (Chapter 6). However, seines have been used successfully to capture fish under the ice (von Brandt 1984). This was accomplished by augering holes close to one another and passing the ends of the net from one hole to the next until the shore was reached, whereupon the ice was broken and fish were seined onto shore. Likewise, lift nets can be useful, especially with the use of light or bait, providing the hole can be kept open during the capture period. Because fish movements are slow during cold times of the year, another option for collecting fish under the ice is the use of active capture techniques in concert with scuba diving. However, great care must be taken to insure the diver's safety under the ice.

7.9 MAXIMIZATION OF SAMPLING EFFECTIVENESS

One of the key questions when designing or evaluating a sampling program is what constitutes success. The measures of success of a sampling program should be derived from the objectives of the study.

If the objective is to collect live fish suitable as experimental subjects, the fish should be captured with a minimum of stress. The number of fish needed must also be considered. When sampling for live fish, investigators must often make judgements and tradeoffs between the number of fish collected and the stress placed on them; fish are usually more stressed when captured in large numbers at a single time. After comparing the utility of various gears for capturing fish alive, the costs of obtaining enough samples can then be compared.

In studies for which the goal is to determine biological characteristics of a fish population or fish community, the representativeness of the samples is a key question. In this chapter, we have emphasized that sampling gears are selective for the size and species caught. As such, biological samples often do not represent the entire population or the entire fish community. We need to recognize this fact in our sampling designs and analytical methods, as well as in our interpretation of results. For example, we often compute mean length at age to describe the growth of a fish population. If the gear used selects for larger individuals, estimates of mean length will be too high for younger fish. In situations such as these, the most effective sampling program would probably be to use two or more gears to sample different sizes of fish. Likewise, samples taken with a single gear generally do not reflect the true composition of a fish community because some species are underrepresented or absent in the catch.

Two types of criteria are useful for evaluating the effectiveness of a sampling design. First, the sampling variance (Chapter 2) of the design is useful to indicate the amount of variability in the index due to random chance of site selection or other factors affecting catch. A second criterion is how well the index corresponds with information on population abundance from other sources. The sampling variability is a measure of the precision obtained with the given design and sample size, whereas the second criterion reflects the accuracy of the sampling program. In stratified random sampling, for example, the variance due to random site selection can be computed directly (e.g., Cochran 1977). Even if the variance is low, however, an index based on this design may not relate well to population abundance. For example, if fish movement into and out of the area being sampled is substantial, the index may be more reflective of migration rather than actual population abundance.

Three of the most common sampling designs are stratified random, systematic, and fixed-station sampling (see Chapter 2). In stratified random sampling, the area being sampled is divided into smaller areas based on depth, temperature, or species composition. Sampling sites within each smaller area are then selected at random. As indicated above, stratified random sampling allows for computation of sampling variation due to site choice. In systematic sampling, sites are selected in a pattern, such as a grid. Computing variance for systematic sampling is problematic because the sites are not selected independently of one another. In fixed-station sampling, sites are randomly selected or selected because they are believed to be representative sites. These sites are then maintained over time. As with systematic sampling, computing variance (particularly over time) is more difficult in fixed-station sampling, especially if the sites are not chosen randomly. Even though computing sampling variance is more difficult, it has been argued that fixed-station and systematic sampling designs may be more effective (i.e., track actual population abundance better) than is stratified random sampling because of the more uniform or consistent coverage of the sampling area. Questions about the relative performance of these different designs have not been answered yet, and there is still much debate over which design is best.

7.10 ACTIVE FISH CAPTURE METHODS AND FISH STOCK ASSESSMENT

A major function of fish stock assessments is to collect and analyze fishery data to estimate population abundance, size structure, growth, mortality, and recruitment. Information obtained from stock assessments then forms the basis for the scientific management of fisheries. Two basic types of data are used in fish stock assessments. Fishery-dependent data come from commercial or recreational catches, whereas fishery-independent data are obtained by sampling programs that use active or passive fish capture methods. Fishery-dependent data are very useful in providing information on the status of fish populations but are prone to a number of biases and limitations (e.g., Hilborn and Walters 1992). One of the main limitations is that commercial and recreational fishers generally target their effort to areas that have concentrations of fish. Thus, measures of relative abundance such as catch per unit effort from fishery-dependent sources are dependent on fishers' behavior as well as the actual abundance of fish. This being the case, estimates of population abundance based solely on fishery-dependent data need to take account of the behavioral component of the fishing processes.

Fishery-independent data have the advantage that the sampling program can be designed to sample randomly or systematically across the range of the population. Even if sampling is concentrated in areas of higher fish density, the sampling design should take this into account by appropriately weighting the catches in the samples. A critical concern in using data collected for assessments is gear selectivity. As discussed earlier in this chapter, all active fishing gears have some degree of size selectivity. Because of this, total catch and catch by size is partly a function of true fish abundance, but catch also depends on the size composition of the population. Generally, the true size composition of the population is unknown. Furthermore, the size composition of the catch is rarely if ever an accurate reflection of the actual size composition of the population because of the selective nature of the fishing gear. A number of sophisticated models that attempt to resolve the actual population abundance and size (or age) structure have been developed (e.g., Gavaris 1988; Methot 1989). One of the fundamental concepts underlying these models is that it is generally not possible to obtain a true picture of the population from solely a single gear because size selectivity is unknown. Estimates of the size selectivity of a single piece of gear can be obtained only by comparison with catches from one or more other types of gear with different selective patterns. As such, current methods in stock assessment emphasize an integrative approach in which information from a variety of sources (e.g., commercial catch per unit effort, survey abundance indices, and commercial catch at age) are combined. Through appropriate statistical and mathematical models, the size selectivity of all gears (including commercial fleets and recreational anglers) can be evaluated, and a better picture of the actual population abundance and size composition can be achieved.

7.11 REFERENCES

Allen, D. M., S. K. Service, and M. V. Ogburn-Matthews. 1992. Factors influencing the collection efficiency of estuarine fishes. Transactions of the American Fisheries Society 121:234–244.

Argyle, R. L. 1982. Alewives and rainbow smelt in Lake Huron: midwater and bottom aggregations and estimates of standing stock. Transactions of the American Fisheries Society 111:267–285.

Azarovitz, T. R. 1981. A brief historical review of the Woods Hole Laboratory trawl survey time series. Canadian Special Publication Fisheries and Aquatic Sciences 58:62–67.

Bayley, P. B., and D. J. Austen. 1988. Comparison of detonating cord and rotenone for sampling fish in warmwater impoundments. North American Journal of Fisheries Management 8:310–316.

Beauchamp, D. A. 1990. Seasonal and diel food habits of rainbow trout stocked as juveniles in Lake Washington. Transactions of the American Fisheries Society 119:475–482.

Bronte, C. R., J. H. Selgeby, and G. L. Curtis. 1991. Distribution, abundance, and biology of the alewife in U.S. water of Lake Superior. Journal of Great Lakes Research 17:304–313.

Brown, E. H., Jr. 1972. Population biology of alewives, *Alosa pseudoharengus*, in Lake Michigan, 1949–70. Journal of the Fisheries Research Board of Canada 29:477–500.

Buls, B. 1989. 'Super fibers' get mixed reviews from industry. National Fisherman 70(7):73–74.

Byrne, C. J., and J. R. S. Forrester. 1987. Effect of a gear change on a standardized bottom trawl survey time series. Oceans Magazine 2:614–621.

Byrne, C. J., and J. R. Nicolas. 1989. Bottom trawl survey manual. National Marine Fisheries Service, Northeast Fisheries Science Center, Woods Hole, Massachusetts.

Charles-Dominique, E. 1989. Catch efficiencies of purse and beach seines in Ivory Coast Lagoons. U.S. National Marine Fisheries Service Fishery Bulletin 87:911–921.

Chubb, S. L. 1985. Spatial and temporal distribution and abundance of larval fishes in Pentwater Marsh, a coastal wetland on Lake Michigan. Master's thesis. Michigan State University, East Lansing.

Clady, M. D. 1976. Influence of temperature and wind on the survival of early stages of yellow perch, *Perca flavescens*. Journal of the Fisheries Research Board of Canada 33:1887–1893.

Cochran, W. G. 1977. Sampling techniques. Wiley, New York.

Dewey, M. R., L. E. Holland-Bartels, and S. J. Zigler. 1989. Comparison of fish catches with buoyant pop nets and seines in vegetated and nonvegetated habitats. North American Journal of Fisheries Management 9:249–253.

Eaton, J. G., W. A. Swenson, J. H. McCormick, T. D. Simonson, and K. M. Jensen. 1992. A field and laboratory investigation of acid effects on largemouth bass, rock bass, black crappie, and yellow perch. Transactions of the American Fisheries Society 121:644–658.

Ebbers, M. A. 1987. Vital statistics of a largemouth bass population in Minnesota from electrofishing and angler-supplied data. North American Journal of Fisheries Management 7:252–259.

Echeverria, T. W., W. H. Lenarz, and C. A. Reilly. 1990. Survey of the abundance and distribution of pelagic young-of-the-year rockfishes, *Sebastes*, off central California. NOAA (National Oceanic and Atmospheric Administration) Technical Memorandum NMFS (National Marine Fisheries Service), NOAA-TM-NMFS-SWFC-147, Southwest Fisheries Science Center, Tiburon, California.

Eiler, J. H., B. D. Nelson, and R. F. Bradshaw. 1992. Riverine spawning by sockeye salmon in the Taku River, Alaska and British Columbia. Transactions of the American Fisheries Society 121:701–708.

Espegren, G. D., and E. P. Bergersen. 1990. Quantitative sampling of fish populations with a mobile rising net. North American Journal of Fisheries Management 10:469–478.

Gabelhouse, D. W., Jr., and D. W. Willis. 1986. Biases and utility of angler catch data for assessing size structure and density of largemouth bass. North American Journal of Fisheries Management 6:481–489.

Gavaris, S. 1988. An adaptive framework for the estimation of population size. Canadian Atlantic Fisheries Scientific Advisory Committee Research Document 88/29, St. Andrews, New Brunswick.

Graham, J. H., and R. C. Vrijenhoek. 1988. Detrended correspondence analysis of dietary data. Transactions of the American Fisheries Society 117:29–36.

Hayes, D. B., W. W. Taylor, and J. C. Schneider. 1992. Response of yellow perch and the benthic invertebrate community to a reduction in the abundance of white suckers. Transactions of the American Fisheries Society 121:36–53.

Hayes, M. L. 1983. Active fish capture methods. Pages 123–146 *in* L. A. Nielsen and D. L. Johnson, editors. Fisheries techniques. American Fisheries Society, Bethesda, Maryland.

Hilborn, R., and C. J. Walters. 1992. Quantitative fisheries stock assessment: choice, dynamics and uncertainty. Chapman and Hall, New York.

Isaak, D. J., T. D. Hill, and D. W. Willis. 1992. Comparison of size structure and catch rate for largemouth bass samples collected by electrofishing and angling. The Prairie Naturalist 24:89–96.

Jessop, B. M. 1985. Influence of mesh composition, velocity, and run time on the catch and length composition of juvenile alewives (*Alosa pseudoharengus*) and blueback herring (*A. aestivalis*) collected by pushnet. Canadian Journal of Fisheries and Aquatic Sciences 42:1928–1939.

King, T. L., and H. O. Pate. 1992. Population structure of spotted seatrout inhabiting the Texas Gulf Coast: an allozymic perspective. Transactions of the American Fisheries Society 121:746–756.

Kriete, W. H., Jr., and J. G. Loesch. 1980. Design and relative efficiency of a bow-mounted pushnet for sampling juvenile pelagic fishes. Transactions of the American Fisheries Society 109:649–652.

Kushlan, J. A. 1981. Sampling characteristics of enclosure fish traps. Transactions of the American Fisheries Society 110:557–562.

Lee, S., and P. J. Whitfield. 1992. Virus-associated spawning papillmatosis in smelt, *Osmerus eperlanus* L., in the River Thames. Journal of Fish Biology 40:503–510.

Loftus, A. J., W. W. Taylor, and M. Keller. 1988. An evaluation of lake trout (*Salvelinus namaycush*) hooking mortality in the upper Great Lakes. Canadian Journal of Fisheries and Aquatic Sciences 45:1473–1479.

Lyons, J. 1986. Capture efficiency of a beach seine for seven freshwater fishes in a north-temperate lake. North American Journal of Fisheries Management 6:288–289.

Marshall, B. E. 1988. A preliminary assessment of the biomass of the pelagic sardine *Limnothrissa miodon* in Lake Kariba. Journal of Fish Biology 32:515–524.

Meador, M. R., and W. E. Kelso. 1990. Growth of largemouth bass in low salinity environments. Transactions of the American Fisheries Society 119:545–552.

Methot, R. D. 1989. Synthetic estimates of historical abundance and mortality for northern anchovy. American Fisheries Society Symposium 6:66–82.

Metzger, R. J., and P. L. Shafland. 1986. Use of detonating cord for sampling fish. North American Journal of Fisheries Management 6:113–118.

Minello, T. J., R. J. Zimmerman, and E. X. Martinez. 1989. Mortality of young brown shrimp *Penaeus aztecus* in estuarine nurseries. Transactions of the American Fisheries Society 118:693–708.

Moyle, P. B., B. Herbold, D. E. Stevens, and L. W. Miller. 1992. Life history and status of delta smelt in the Sacramento–San Joaquin Estuary, California. Transactions of the American Fisheries Society 121:67–77.

Nash, R. D. M., A. J. Geffen, and G. Hughes. 1992. Winter growth of juvenile plaice on the Port Erin Bay (Isle of Man) nursery ground. Journal of Fish Biology 41:209–215.

O'Gorman, R., and C. P. Schneider. 1986. Dynamics of alewives in Lake Ontario following a mass mortality. Transactions of the American Fisheries Society 115:1–14.

Olson, D. E., and P. K. Cunningham. 1989. Sport-fisheries trends shown by an annual Minnesota fishing contest over a 58-year period. North American Journal of Fisheries Management 9:287–297.

Parsley, M. J., D. E. Palmer, and R. W. Burkhardt. 1989. Variations in capture seine efficiency of a beach seine for small fishes. North American Journal of Fisheries Management 9:239–244.

Payer, R. D., R. B. Pierce, and D. L. Pereira. 1989. Hooking mortality of walleyes caught on live and artificial baits. North American Journal of Fisheries Management 9:188–192.

Pearcy, W. G., and J. P. Fisher. 1988. Migrations of coho salmon, *Oncorhynchus kisutch*, during their first summer in the ocean. U.S. National Marine Fisheries Service Fishery Bulletin 86:173–195.

Pierce, C. L., J. B. Rasmussen, and W. C. Leggett. 1990. Sampling littoral fish with a seine: corrections for variable capture efficiency. Canadian Journal of Fisheries and Aquatic Sciences 47:1004–1010.

Quertermus, C. J. 1991. Use of bass club tournament results to evaluate relative abundance and fishing quality. American Fisheries Society Symposium 12:515–519.

Ralston, S. 1990. Size selection of snappers (*Lutjanidae*) by hook and line gear. Canadian Journal of Fisheries and Aquatic Sciences 47:696–700.

Ricker, W. E. 1975. Computation and interpretation of biological statistics of fish populations. Fisheries Research Board of Canada Bulletin 191.

Rulifson, R. A., and J. E. Cooper. 1986. A method to determine mesh-size selectivity in commercial menhaden purse seines. North American Journal of Fisheries Management 6:359–366.

Sainsbury, J. C. 1975. Commercial fishing methods—an introduction to vessels and gears. Fishing News Books Ltd., Surrey, UK.

Santucci, V. J., Jr., and D. H. Wahl. 1991. Use of a creel census and electrofishing to assess centrarchid populations. American Fisheries Society Symposium 12:481–491.

Schramm, H. L., Jr., and nine coauthors. 1991. Sociological, economic, and biological aspects of competitive fishing. Fisheries 16(3):13–21.

Serchuk, F. M., and S. E. Wigley. 1986. Evaluation of USA and Canadian research vessel surveys for sea scallops (*Placopecten magellanicus*) on Georges Bank. Journal of Northwest Atlantic Fishery Science 7:1–13.

Shpigel, M., and L. Fishelson. 1991. Territoriality and associated behaviour in three species of the genus *Cephalopholis* (Pisces: Serranidae) in the Gulf of Aqaba, Red Sea. Journal of Fish Biology 38:887–896.

Taylor, W. W., and S. D. Gerking. 1978. Potential of the Ohrid rifle minnow, *Alburnoides bipunctatus ohridanus*, as an indicator of pollution. Verhandlungen Internationale Vereinigung fur Theoretische und Angewandte Limnologie 20:2178–2181.

von Brandt, A. 1984. Fish catching methods of the world, 3rd edition. Fishing News Books Ltd., Farnham, UK.

Walsh, S. J., R. B. Millar, C. G. Cooper, and W. M. Hickey. 1992. Codend selection in American plaice: diamond versus square mesh. Fisheries Research 13:235–254.

Webb, R. O., and M. J. Kingsford. 1992. Protogynous hermaphroditism in the half-banded sea perch, *Hypoplectrodes maccullochi* (Serranidae). Journal of Fish Biology 40:951–961.

Weber, J. J., and B. L. Les. 1982. Spawning and early life history of yellow perch in the Lake Winnebago system. Wisconsin Department of Natural Resources Technical Bulletin 130.

West, R. L., M. W. Smith, W. E. Barber, J. B. Reynolds, and H. Hop. 1992. Autumn migration and overwintering of arctic grayling in coastal streams of the Arctic National Wildlife Refuge, Alaska. Transactions of the American Fisheries Society 121:709–715.

Willis, D. W., and R. F. Hartmann. 1986. The Kansas black bass tournament monitoring program. Fisheries 11(3):7–10.

Wright, P. J. 1992. Ovarian development, spawning frequency and batch fecundity in *Encrasicholina heterolba* (Ruppell, 1958). Journal of Fish Biology 40:833–844.

Chapter 8

Electrofishing

JAMES B. REYNOLDS

8.1 INTRODUCTION

Electrofishing, in the strictest sense, is the use of electricity to capture fish. However, electricity has also been used to guide or block the movements of fish and to anesthetize or quickly kill them after capture. In fact, the earliest attempts to use electricity on fish, in the late nineteenth and early twentieth centuries, were often aimed at manipulating fish movements and not at capturing fish. Serious development of electrofishing as a technique for fishery science began after World War II. Research in the 1950s and 1960s was aimed primarily at development of field equipment for capturing fish and at understanding fish reactions in controlled electrical fields (Friedman 1974). In the 1970s and 1980s, much attention was given to the refinement of techniques and the effects of electrofishing on physiology, stress, and injury of fish. The need for a unification of theory and technique should be addressed in the 1990s and beyond (Reynolds 1995).

This chapter offers the reader a primer on electrofishing. It provides some general principles of electricity in circuits and electrical fields in water and guidelines for the understanding, construction, and safe, efficient use of electrical fishing devices. Although it does not give detail on procedures for a given situation, references are provided for additional direction. A symposium-based book, *Fishing with Electricity* (Cowx and Lamarque 1990), is the most detailed, comprehensive text now available on electrofishing. Talking to knowledgeable people before beginning or modifying an electrofishing operation and investing much time and money is very important. Experienced biologists, commercial suppliers, and electricians or electronics technicians are information sources that can steer a project away from costly mistakes.

8.2 PRINCIPLES

8.2.1 Circuit Theory

Terms and relations. All matter consists of particles that attract and repel each other because of the positive or negative charges the particles bear; these particles are called charge carriers. Electricity is the form of energy that results from the dissociation of neutral particles into charged particles (electrons, protons, and ions). An external power source (e.g., mechanical or chemical) is required for this separation of positive and negative electrical charges. Separation results in interactions (attraction or repellence) of charge carriers, producing electricity.

A circuit is a closed path (e.g., insulated wire) along which charge carriers move. In most common circuits, the charge carriers are electrons. The number of charge

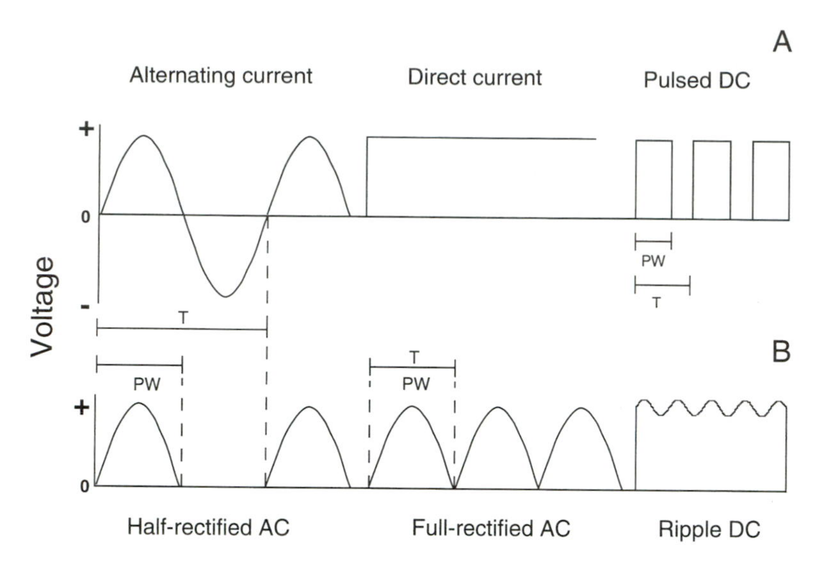

Figure 8.1 Electrical waveforms expressed as voltage over time. One cycle is indicated by T, and PW is pulse width. Duty cycle is PW/T, expressed as a percentage. (**A**) Characteristics of AC, DC, and pulsed DC. (**B**) Waveforms of DC produced by modification of AC.

carriers per unit time is the current, measured in amperes (A). The amount of energy per charge carrier is voltage, measured in volts (V). The ratio of voltage to current in any circuit is a constant called resistance, measured in ohms (Ω). This relationship is given by Ohm's Law:

$$\text{resistance } (\Omega) = \text{voltage (V)/current (A).} \tag{8.1}$$

Conductance, the inverse of resistance, is a measure of the ability of a circuit to carry an electrical current, measured in mhos or siemens (S). (Appropriately, mho is ohm spelled backwards). Electrical power is the product of voltage and current, or energy per unit time, measured in watts (W). One watt of power results when a current of one ampere flows through a resistance of one ohm under the effect of one volt. The flow of current in a circuit is somewhat analogous to the flow of water in a pipe. The pressure (voltage) sustains a flow (current) through the pipe (circuit). The amount of flow the pipe can handle depends on its cross-sectional area, length, and fluid viscosity (resistance). As the flow reaches the end of the pipe, it releases energy to do work at some rate (power).

Electrical currents. There are two basic types of electrical current (Figure 8.1A). Direct current (DC) flows only in one direction because the polarity, or the negative and positive terminals (electrodes) of the circuit, are always the same. Negative charge carriers are repelled from the negative electrode (cathode) and attracted to the positive electrode (anode). The storage battery is a common source of DC. Alternating current (AC) flows alternatingly in both directions because the polarity of the power source reverses continuously on the two electrodes. In AC, voltage is commonly a sine wave, increasing from zero to a maximum and back to zero as current flows in one direction and then repeating the same pattern in the opposite direction. A cycle of AC begins and ends at zero voltage and includes current flow in each direction. Frequency of AC is measured as the number of cycles per second, or hertz (Hz). One hertz equals one cycle per second. Domestic AC is delivered to consumers at 60 Hz. The portable generator is a common source of AC in electrofishing, usually operating at frequencies of 60 or 400 Hz.

Both AC and DC can be modified to produce a waveform called pulsed DC. Pulsed DC gives a unidirectional current with periodic interruptions that result in square waves (pulses) of voltage (Figure 8.1A). Although waveforms are usually depicted by voltage over time, current follows the same pattern in electrofishing applications. Pulse duration, commonly called pulse width, is the "on" time of each pulse, usually measured in milliseconds. The ratio of on time to total time within one cycle (one cycle is measured from the beginning of one pulse to the beginning of the next pulse) is called duty cycle and is expressed as a percentage; a 50% duty cycle means that current flows during half of each cycle. Frequency of pulsed DC is also measured in Hz.

Modification of AC is an inexpensive way to produce a DC waveform (Figure 8.1B). Half-rectified AC results when AC is passed through a rectifier that leaves a sequence of half-sine waves, in the same direction, which are separated by pauses of equal duration; this is pulsed DC at a frequency equal to that of the AC generator. Full-rectified AC is composed of an uninterrupted sequence of half-sine waves in the same direction; this is pulsed DC at a frequency double that of the generator. A pure DC waveform, often called smooth DC, can be generated from AC by electrical filter circuits; when insufficiently filtered, the current has weak sinusoidal variations and is called ripple DC (no pulses). Other waveforms of pulsed DC are too numerous to describe here. Most electrofishing equipment does not produce pulsed DC as the classic square wave. Only an oscilloscope will reveal the actual shape of a waveform; using one to examine output (i.e., voltage waveform) at different settings of voltage, frequency, and pulse width (or duty cycle) can be quite helpful to biologists.

8.2.2 Electrical Fields and Fish

Field theory. The electrodes deliver voltage and current in the water to form a three-dimensional electrical field (Figure 8.2). Current flows between electrodes of opposite polarity (i.e., flux lines), and voltage surrounds each electrode at right angles to the flux lines (i.e., equipotential lines). The field is nonhomogeneous and weakens with distance from the electrode as energy dissipates in the water. The three parameters that applied to the circuit—voltage, current, and resistance—now differ at different points in the water. They must be expressed in spatial terms, using another form of Ohm's Law:

$$\text{resistivity } (\Omega\text{-cm}) = \frac{\text{voltage gradient (V/cm)}}{\text{current density (A/cm}^2\text{)}}. \qquad (8.2)$$

Imagine a 1-cm cube of water in an electrical field (Figure 8.3). Current density is the current that flows through a 1-cm^2 plane. Voltage gradient is the change in voltage over a l-cm distance. Resistivity is a measure of the water's resistance. The inverse of resistivity is conductivity and is measured as mho/cm. The modern unit of conductivity measurement is siemens/cm, or S/cm, which is equivalent to mho/cm. Because freshwater conductivity is measured as μS/cm, its inverse value must be multiplied by 10^6 to calculate resistivity. Conductivity is a measure of the ability of an aqueous solution to carry an electrical current (APHA 1992) and is the most important

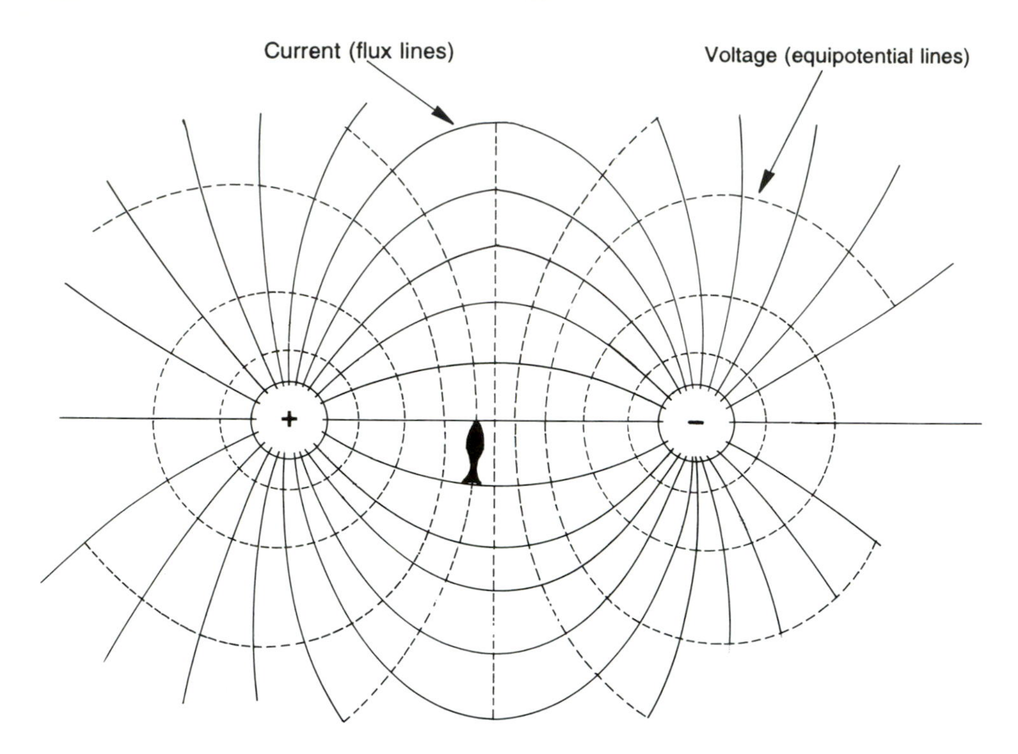

Figure 8.2 An electrical field produced by two electrodes. The solid lines connecting the two electrodes represent constant current. Dashed lines surrounding each electrode represent constant voltage. Voltage gradient and current density are highest near the electrodes and decrease rapidly with linear distance from the electrodes. Voltage gradient is minimized in fish lying along voltage (equipotential) lines and perpendicular to current (flux) lines.

environmental measurement related to electrofishing (Box 8.1). Distilled water has very low conductivity (0.5–3.0 μS/cm). The conductivity of most freshwater bodies is between 50 and 1,500 μS/cm. On average, seawater is 500 times more conductive than is freshwater.

Power density is the product of voltage gradient and current density:

$$\text{power density (W/cm}^3) = \text{voltage gradient (V/cm)} \times \text{current density (A/cm}^2). \quad (8.3)$$

Power density and current density cannot easily be directly measured. However, by substituting conductivity for 1/resistivity in equation 8.2 and then solving and substituting equation 8.2 for current density in equation 8.3, we obtain

$$\text{power density (W/cm}^3) = \text{conductivity (S/cm)} \times [\text{voltage gradient (V/cm)}]^2. \quad (8.4)$$

Water conductivity and voltage gradient can be directly measured in water, making possible estimates of current density and power density. Procedure and equipment for measuring voltage gradient are described in detail by Kolz (1993). Because the

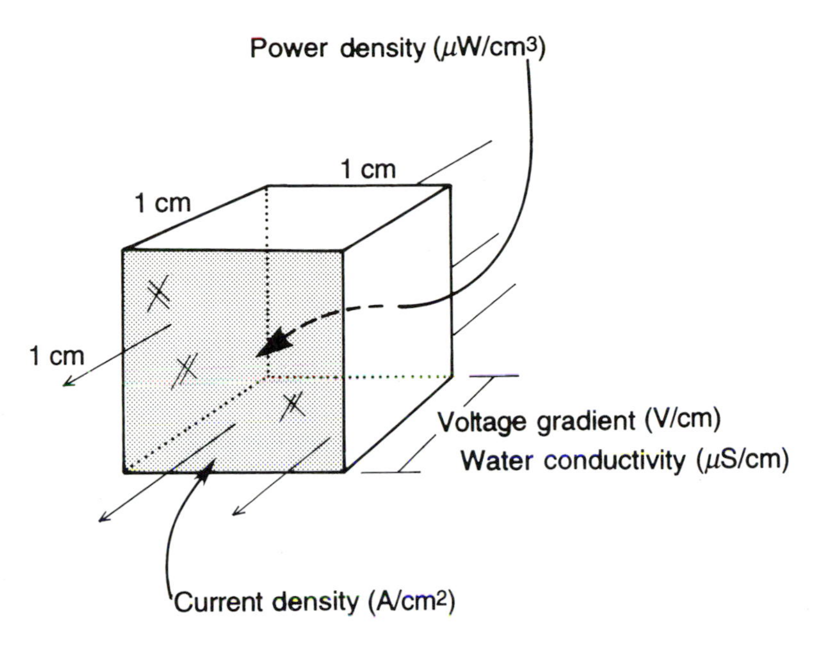

Figure 8.3 A 1-cm cube of water in an electrical field as the basis of four field parameters—voltage gradient, current density, water conductivity, and power density. The cube represents the basic approach to description of electrical fields in water (adapted from Kolz and Reynolds 1989, with permission).

conductivity of freshwater is measured as μS/cm, power density is expressed as μW/cm^3. The logarithmic relationship among water conductivity, voltage gradient, current density, and power density can be expressed in a graphic form (Figure 8.4) that is useful for approximating two of the parameters when the other two are known (Kolz 1989).

Electrofishing effectiveness. Although current and voltage, as indicated by control unit metering, are often used to monitor electrofishing effectiveness, these circuit measurements give little indication of the nature of the electrical field around the electrodes. Ultimately, electrofishing success depends on a transfer of electrical energy from the water into the fish. As long as biologists continue to focus on circuit measurements, an understanding of electrofishing effectiveness and its dependence on energy transfer from water to fish will be difficult. Use of in-water parameters and relationships (equations 8.2, 8.3, and 8.4) are essential to the foundation of electrofishing principles.

Opinions vary on which of the field parameters—voltage gradient, current density, or power density—are most directly related to the effects of electricity on fish. These opinions often differ according to the range of water conductivity encountered while electrofishing. Biologists sampling low-conductivity waters (<100 μS/cm) prefer to use voltage gradient (or circuit voltage) because current is not a responsive indicator of field intensity; those working in waters of high conductivity (500–1,000 μS/cm or more) prefer current density (or circuit current) because voltage is not a responsive indicator. Voltage gradients of 0.1–1.0 V/cm are generally considered effective for stunning fish in waters of intermediate-to-high conductivity (>100 μS/cm), but higher gradients are needed in low-conductivity waters.

Box 8.1 Measuring Water Conductivity

Water conductivity is directly related to total ion concentration and water temperature. Waters containing mostly inorganic compounds conduct electricity better than those containing mostly organic compounds because the latter are less likely to dissociate in solution (APHA 1992). Electronic conductivity (not conductance, a circuit term) meters provide a convenient, dependable method of measurement; they should be a standard component of electrofishing operations. The metal bands on the probe under the removable sleeve should be wiped clean periodically. Standard solutions are available for meter calibration. Most meters are designed to measure specific conductivity at 25°C (consult the owner's manual or manufacturer). The user measures ambient water temperature on site, inputs the temperature into the meter, immerses the meter probe into the water, and reads conductivity (μS/cm) automatically adjusted to 25°C. However, electrofishing success depends on ambient conductivity (conductivity at ambient temperature), not specific conductivity. Because conductivity changes by 2% for every 1°C change in temperature (APHA 1992), ambient conductivity (C_a) can be estimated from data on specific conductivity (C_s), specific temperature (T_s), and ambient temperature (T_a):

$$C_a = C_s/[1.02^{(T_s - T_a)}].$$

Ambient conductivity may be directly measured with a meter by using T_s as input, instead of T_a, thus eliminating adjustment for specific temperature. If the direct method is used, water temperature should be measured nevertheless; it is an important factor affecting electrofishing efficiency (see section 8.4). When reporting results, state clearly the basis of measurement. For example, "ambient conductivity was 100 μS/cm at 17°C" or "specific conductivity was 117 μS/cm at 25°C, measured at 17°C."

Actually, voltage gradient and current density each represent only one-half of our understanding of electrofishing effectiveness. To do work (e.g., stun a fish), power, the product of voltage and current, must be transferred from the water to the fish. Hence, power density may be the field parameter most directly related to the effects of electricity on fish (Kolz 1989). By viewing electrofishing as a power-based technique, many observations regarding electrofishing effectiveness can be explained. For example, the observed effects of water conductivity on fish capture can be explained in terms of power transfer.

Power transfer is dependent on the mismatch between water conductivity and fish conductivity (Figure 8.5). Unlike water conductivity, fish conductivity cannot be directly measured, but it can be estimated (Kolz and Reynolds 1989). When water is less conductive than are fish (i.e., low water conductivity), current tends to flow through the fish, making voltage gradient more indicative of electrical effects than current density. However, when water is more conductive than are fish (i.e., high water conductivity), current tends to flow through water, around the fish, making current density more indicative of electrical effects than voltage gradient. When water and fish conductivity are equal (no mismatch), the electrical field is not

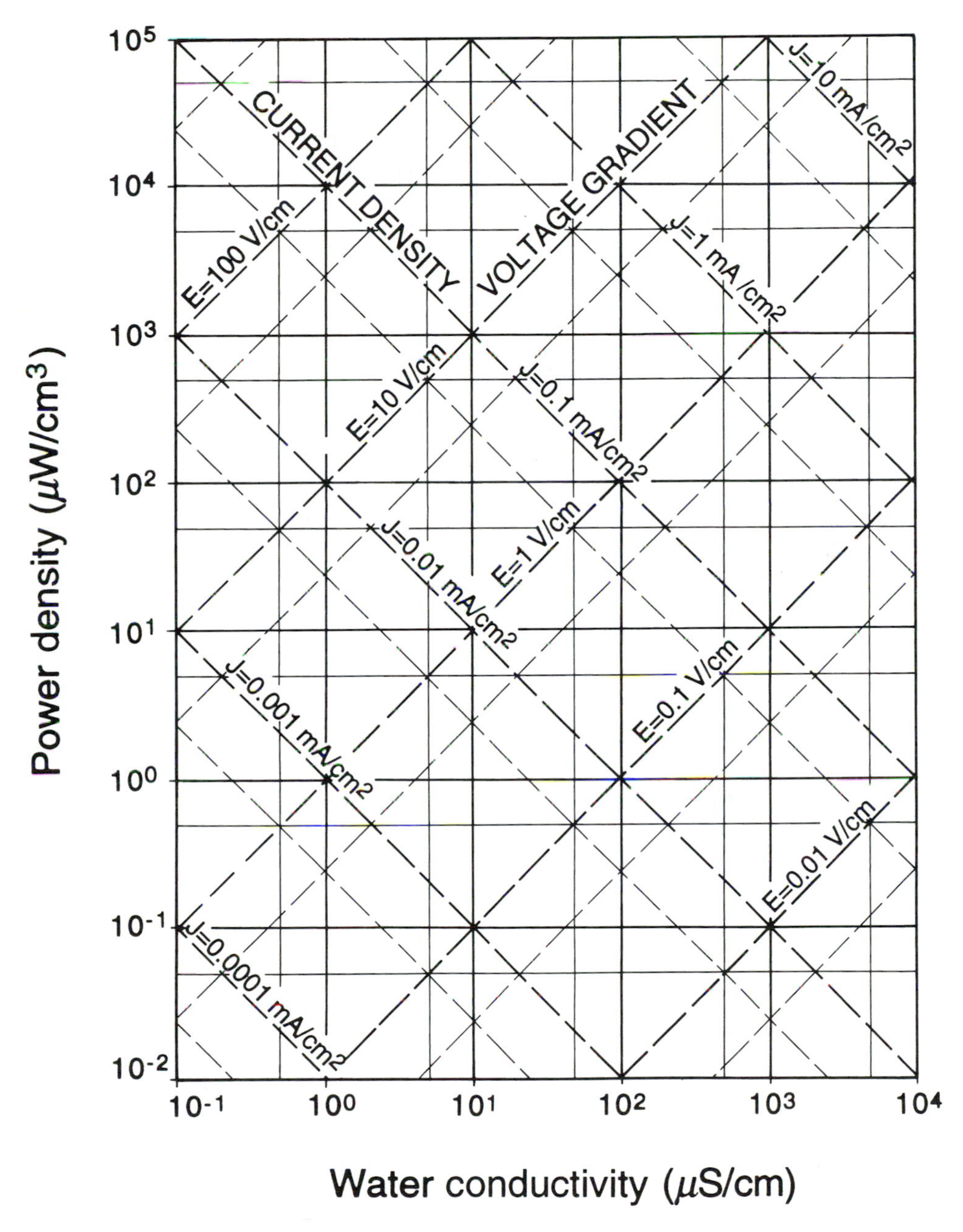

Figure 8.4 Logarithmic relationship among the four parameters of an electric field in water: power density, water conductivity, voltage gradient (E), and current density (J) (adapted from Kolz and Reynolds 1989, with permission). Entering the graph with values of any two parameters will yield approximations of the other two.

distorted, and both voltage gradient and current density are equally indicative of electrical effects. In this matched condition, the maximum amount of power density applied to the water (i.e., energy/volume/time) is transferred to the fish. In either of the mismatch conditions, not all of the power applied to the water is transferred to

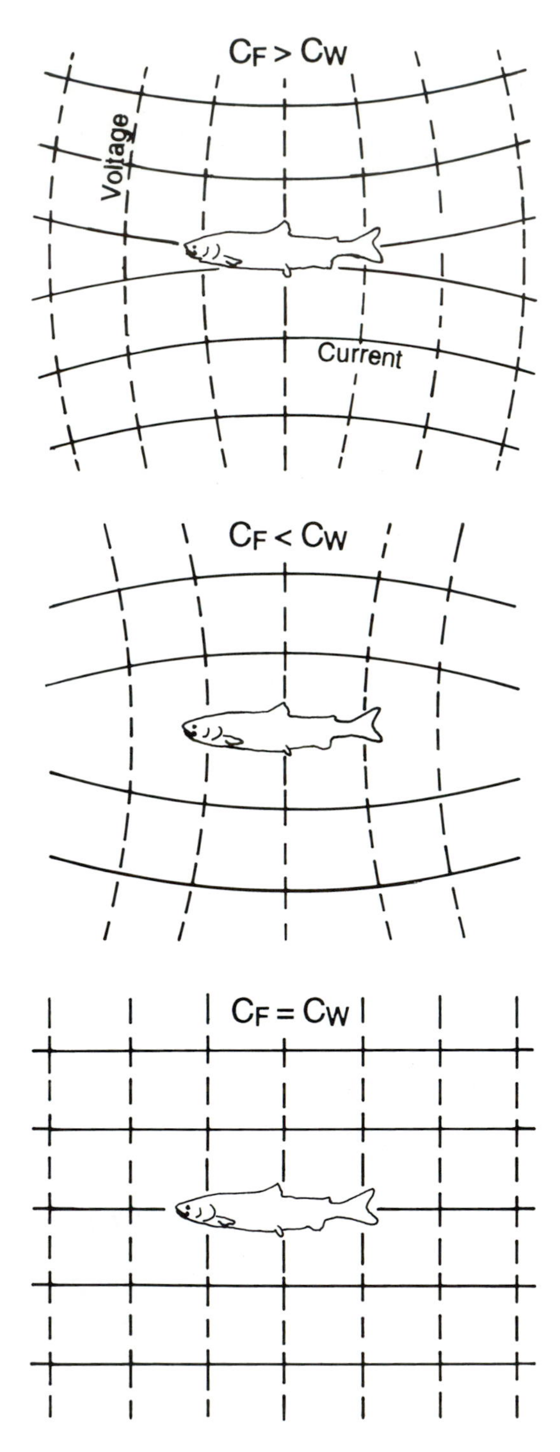

Figure 8.5 Relationship of current (flux lines, solid) and voltage (equipotential lines, dashed) when fish conductivity (C_F) is greater than, less than, and equal to, water conductivity (C_W). Maximum transfer of energy from water to fish occurs when C_F equals C_W.

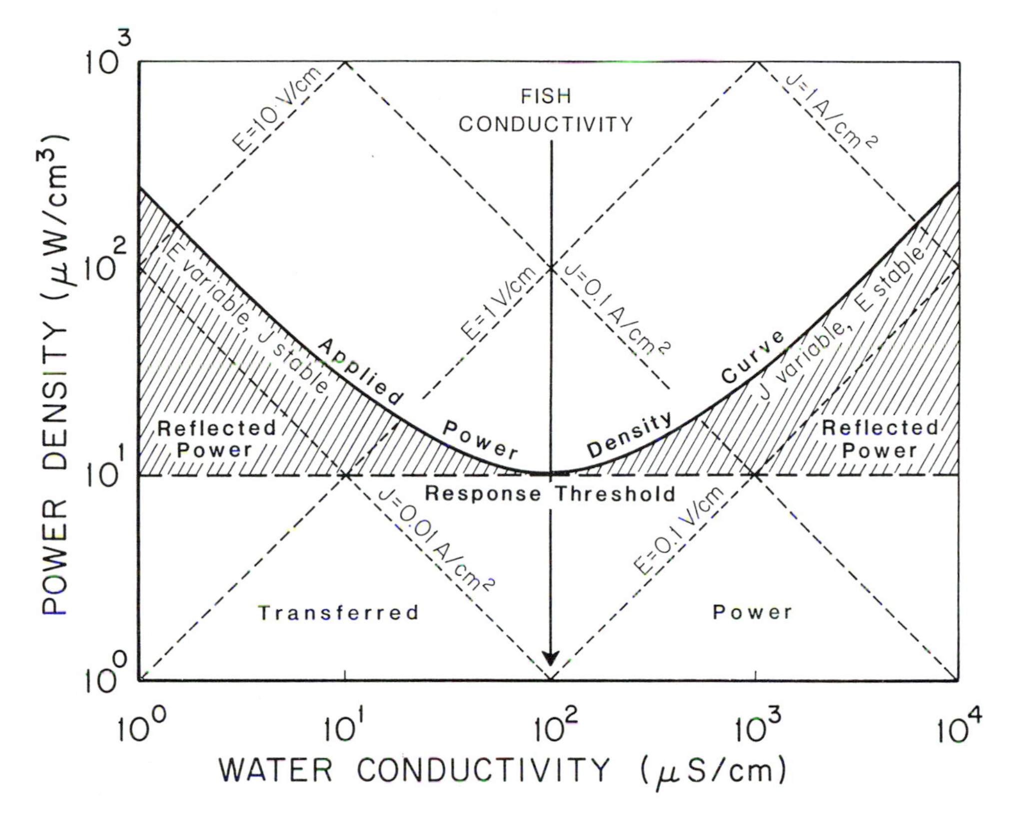

Figure 8.6 Applied threshold of power density, as a function of water conductivity, required to elicit a given response at a response (transfer) threshold of 10 μW/cm^3 in a hypothetical fish with a body conductivity of 100 μS/cm. Voltage gradient is given by E and current density by J (reprinted from Reynolds 1995, with permission).

the fish. This concept is based on the principles of power transfer. Unlike voltage gradient and current density, power density can serve as an indicator of electrofishing effectiveness over the entire range of water conductivity.

Electrofishing has been proposed as a power-related phenomenon based on power transfer theory (Kolz 1989) and laboratory experiments with fish in electrical fields (Kolz and Reynolds 1989). Further testing of this theory is needed, but some fieldwork has provided corroboration (Fisher and Brown 1993; Burkhardt and Gutreuter 1995). Power transfer theory in electrofishing assumes that a fixed threshold of power density must be transferred to a fish in order to elicit a given response and that the power density applied to the water, required to achieve the threshold transfer, is determined by conductivity mismatch of the fish and water. A hypothetical example (Figure 8.6) assumes that a fish with a body conductivity of 100 μS/cm must receive at least 10 μW/cm^3 power density to display a defined response (e.g., loss of equilibrium). If water conductivity is 100 μS/cm, only 10 μW/cm^3 need be applied to the water to achieve fish response because power transfer is 100%. As water becomes more or less conductive than 100 μS/cm, the power density applied to the water in order to achieve a 10 μW/cm^3 transfer to the fish increases logarithmically. Reflected power is that portion of applied power not transferred to the fish; it is needed, even though it is not transferred, to overcome the effect of conductivity mismatch. As water conductivity decreases, the applied power density curve crosses lines of increasing voltage gradient (Figure 8.6; diagonal lines, E, from lower left to upper right) and approaches a stabilized level of current density (diagonal lines, J, lower right to upper left); the reverse occurs as water conductivity increases. That is, at low conductivities voltage gradient (or circuit voltage) is the changing, or responsive, variable; at high conductivities the changing variable is current density (or circuit current). This difference is why biologists electrofishing in low-conductivity waters tend to be "voltage people," and those working in high-conductivity waters tend to be "current people." Power transfer theory explains why small changes in water conductivity can affect electrofishing success, particularly when equipment is operating near the limits of its power-generating capacity.

Power transfer, in measurable terms, is usually limited by insufficient voltage gradient or low water conductivity (equation 8.4); in either case, the power source becomes inadequate. Insufficient voltage gradient can occur at either low or high conductivity. At low conductivity, voltage gradient may be inadequate because power transfer threshold is unattainable. At high conductivity, voltage gradient may be unsustainable because of high current demand (e.g., generator overload). Increasing water conductivity by direct application of salt (ions) to the sampling area has met with little success except in small, restricted waters (Lennon and Parker 1958). Increasing voltage gradient can be accomplished by increasing circuit voltage with a larger power source (e.g., doubling circuit voltage doubles voltage gradient at any point in the field) or by changing the size or shape of an electrode (see section 8.3). Electrode design is one of the most promising, but least used, options exercised by biologists to optimize the use of available power.

Electroshock effects. Fish responses to electroshock can be grouped into two general categories: (1) behavior or reactive movements, and (2) trauma resulting from stress (physiological changes), injury (mechanical damage to tissues), or both. Trauma may lead to death. Although all fish sampling methods cause stress in fish, biologists must use electrofishing with care because fish may be injured or killed, depending on size and species. The goal of electrofishing is to elicit a behavioral response in fish that will lead to their capture while avoiding injury and minimizing stress.

Fish behavior in an electrical field depends on the nature and intensity of the waveform applied. In any electrical field, a threshold of intensity must be reached to achieve a given behavioral response. Higher intensities will elicit a different response if a new threshold is achieved. Lamarque (1990) offers a comprehensive review of fish responses to an electrical field, summarized below.

In responding to AC, a fish tends to assume a position perpendicular to the electrical current (flux lines), thereby minimizing voltage gradient in its body (Figure 8.2). It may undulate in attempted rhythm with the AC cycle, exhibiting oscillotaxis (forced movement without orientation, a "thrashing" motion). At higher field intensity (i.e., closer proximity to an electrode), tetany, or muscle contraction, occurs, and the fish is immobilized. Fish do not orient to a particular electrode in an AC field because of reversing polarities. Therefore, fish responses to AC are less predictable than those to DC.

In a DC field, a fish typically turns toward the anode and exhibits electrotaxis (forced swimming). As the fish nears the anode, a new threshold causes narcosis (muscle relaxation) and loss of equilibrium. While under narcosis, the fish may continue to swim, upside down, toward the anode. Tetany is achieved near the anode. Fish may also exhibit taxis and tetany near the cathode; however, these responses are less predictable. Continuous DC elicits taxis and narcosis in fish if the appropriate threshold is reached. The effects of pulsed DC are less predictable: depending on fish size and species, taxis and narcosis may not occur. Continuous DC requires more power than does pulsed DC, often more than a power source can provide. In a DC field, whether continuous or pulsed, the anode is usually regarded as the capture device; the cathode need not be near the anode (e.g., wading operations with a shore-based power source) and should be no smaller, preferably larger, than the anode to assure that at least half the power is available at the anode for fish capture.

The importance of electrofishing-induced stress is often unrecognized because captured fish may appear normal upon release. However, the stress syndrome results in an abnormal physiological state, including acidosis and reduced respiratory efficiency, that requires hours, even days, for recovery (Schreck et al. 1976). During recovery, fish may be more vulnerable to predation, less competitive, and unable to feed; wild fish require longer recovery periods than do hatchery fish (Horak and Klein 1967; Mesa and Schreck 1989). Stress-related death usually occurs within minutes or a few hours after a fish is shocked; respiratory failure, not cardiac arrest, is usually the cause.

Electrofishing-induced injuries in fish generally take two forms: hemorrhages in soft tissues and fractures in hard tissues (bone). Bruising, often called "branding," occurs when capillaries hemorrhage under the skin producing a dark spot, often chevron-shaped, that corresponds to underlying myomeres. Bruising can be long lasting and may offer sites for bacterial or fungal infections. When a fish is bruised, it has likely sustained internal hemorrhages or fractures, but an unbruised fish caught by electrofishing is not necessarily free of internal injuries. Dark splotches may not be injuries because they result from dilation of chromatophores on the skin; these will disappear in a short time. Gill hemorrhaging is a less frequent, but more severe, external injury caused by electric shock.

Internal injuries can range from minor hemorrhages isolated in dorsal white muscle to complete separation of the spinal column accompanied by rupture of the dorsal aorta. A rating system of apparent severity of internal hemorrhages and spinal damage has been developed for reporting electrofishing-induced internal injuries in fish (Box 8.2). The ratings are used to produce standardized, ranked data; they are not linear measurements, and their actual relation to fish health is not established. Bleeding of internal organs occurs much less frequently, probably because they are less affected by muscle contractions. Fish often recuperate from internal injuries; however, they may not be able to function normally (e.g., grow or reproduce).

Sensitivity of fertilized eggs to electroshock and mechanical shock is similar (Dwyer et al. 1993). Egg mortality is low the first day after fertilization and during water hardening, increases to a maximum at gastrulation, and then decreases to low levels at the eye-up stage just before hatching. The effect of shock on egg viability and spawning behavior of ripe female fish is poorly documented. Use of electrofishing to capture broodfish should be carefully evaluated.

Alternating current has long had a reputation as the waveform most injurious to fish (Hauck 1949); the peak-to-peak (positive-to-negative) range in voltage and reversing polarity are postulated as the causes. Alternating current can be effective for capturing fish when other waveforms are not, but its regular use should be

Box 8.2 Procedures and Rating System for Evaluating Severity of Electrofishing-Induced Injuries in Fish

Internal Hemorrhage

Fish should be killed within 1 h after capture and either frozen or held on ice to allow clotting in blood vessels. Fish should not be filleted immediately after death because fillet-related bleeding will mask injury-related hemorrhages. Fillets should be smoothly cut close to rays and spine and through the ribs and back to the caudal peduncle. Rate the injury from the actual specimen, then photograph the worst side of fish with the fillet inside up (color slides are best for follow-up evaluation). Rate the worst hemorrhage in the muscle mass as follows

0—no hemorrhage apparent
1—mild hemorrhage; one or more wounds in the muscle, separate from the spine
2—moderate hemorrhage; one or more small (≤ width of two vertebrae) wounds on the spine
3—severe hemorrhage; one or more large (> width of two vertebrae) wounds on the spine

Spinal Damage

Fish should be dead or anesthetized to ensure good resolution on X-ray negatives. Photograph the left side of each fish, positioning the fish to include all vertebrae. Photographs from the dorsal aspect may also be necessary to clarify the injury rating. X-rays of two or more fish per plate will save money. Record the position of every affected vertebra, counting the first separate vertebra behind the head as number 1. Rate the worst damage to the spine as follows

0—no spinal damage apparent
1—compression (distortion) of vertebrae only
2—misalignment of vertebrae, including compression
3—fracture of one or more vertebrae or complete separation of two or more vertebrae

discouraged. Direct current or pulsed DC should be used if at all possible. Continuous DC is generally viewed as the least damaging of waveforms (e.g., Spencer 1967); its lack of pulsing current, except when turning the waveform on and off, seems to reduce the intensity and frequency of muscle contractions. The effects of pulsed DC are apparently intermediate to those of AC and DC. However, more recent studies have shown that popular forms of pulsed DC (50–60 Hz, 25–50% duty cycle) cause at least 50% incidence of spinal injury among rainbow trout 30 cm or longer (Sharber and Carothers 1988; Holmes et al. 1990). Concerns about excessive fish trauma have stimulated numerous studies and networking among biologists in the 1990s. This recent activity has confirmed that salmonids are particularly vulnerable to injury from pulsed DC (e.g., McMichael 1993; Hollender and Carline

1994). The incidence of electroshock injuries among other species, especially those in warmwater communities, is not adequately documented; biologists should actively pursue evaluations of injury among non-salmonids. Development of low-energy pulsed DC waveforms offers promise for significant reduction of injury incidence (Sharber et al. 1994).

For now, biologists should continue electrofishing but with caution, particularly when salmonids or endangered species are involved. Present recommendations to reduce fish stress and injury include removing fish from the electrical field quickly, before they get near an electrode; using continuous DC, if possible; using pulsed DC at low frequencies (30 Hz or less) and low duty cycles (near 10%) if larger fish will be affected; adjusting voltage selection to the minimum necessary to achieve adequate sample sizes; holding captured fish under optimal conditions for recovery from stress; and using alternative capture methods when electrofishing unavoidably causes concerns for conservation and animal welfare.

8.3 ELECTROFISHING SYSTEMS

The electrical components of an electrofishing system are classified into six subsystems according to function (Novotny 1990). These are the power source, power conditioner (herein called the control unit), instrumentation (meters), interconnections (wiring), electrodes, and auxiliary equipment (lights, pumps, and so on). For simplicity, in this chapter all components except electrodes are grouped under the category "electrical system." In addition, mechanical systems and safety systems are treated separately.

Electrical systems for electrofishing contain similar components; they differ primarily at the power source (Figure 8.7). Battery-based systems require a DC-to-AC conversion, followed by voltage amplification with a transformer to achieve high-voltage AC. From this point the components of battery-based and generator-based systems perform similar functions, although the power levels may be different, and electrofishing is possible with output from any component (Figure 8.7). Novotny (1990), Goodchild (1991), and USFWS (1992) provide details on the construction, maintenance, and safe use of electrofishing equipment.

8.3.1 Electrofishing Boats

Mechanical system. Flat-bottomed boats at least 4 m long and 1.5 m wide made of heavy-gauge aluminum are commonly used for electrofishing. Other hull types (e.g., pontoons) are used to advantage under various conditions. Wood, rubber, or fiberglass boats may be used but do not provide grounding and cannot be used as cathodes. Outboard motors of moderate horsepower and with a trolling speed are best for electrofishing boats. Seats should not be removed because they contain flotation material and provide rigidity to the boat. The boat should have a minimum load rating of 300 kg and a forward deck large enough to accommodate two standing adults as dipnetters. The work deck must be surrounded by a waist-high bow railing, welded or bolted to the deck. One or two booms, extending forward from the bow for electrode support, should be made of sturdy, nonconductive material—fiberglass is best, but wood is an adequate, cheaper alternative. The booms should be retractable or removable but must not be movable on a metal boat while electrofishing. They should be easy to repair or replace because they occasionally break during use.

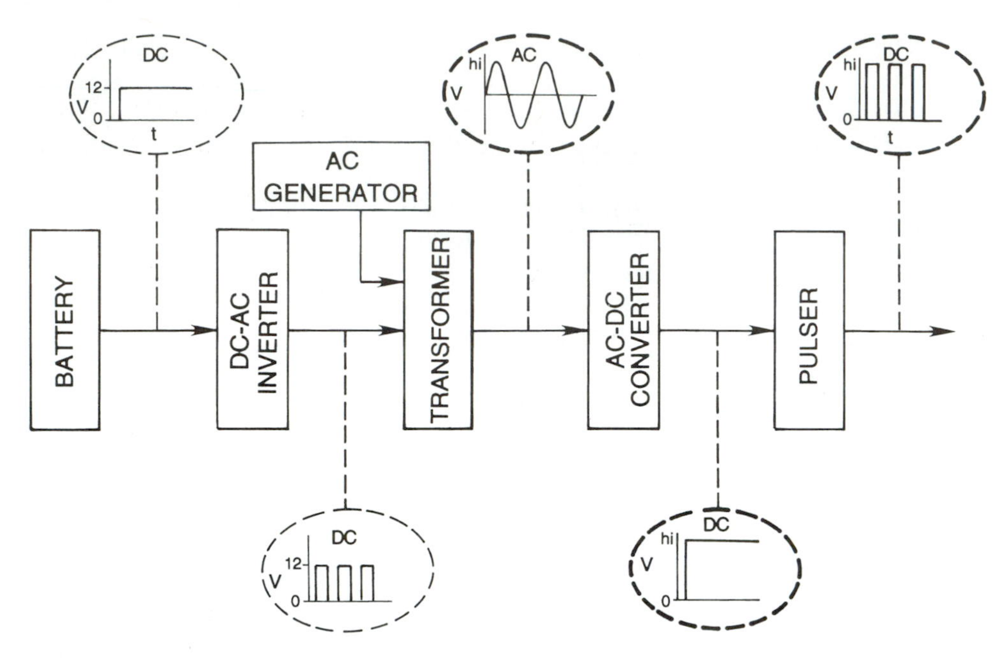

Figure 8.7 Electrical components of battery-based and generator-based electrofishing systems. Arrows indicate direction of current flow. Light ellipses indicate low-voltage output not useful for electrofishing. Heavy ellipses indicate high-voltage output useful for electrofishing.

Electrical system. A biologist standing in an electrofishing boat is like a bird perched on an electrical transmission line; both are safe as long as they remain on an equipotential surface and avoid touching anything at a different potential (voltage). A metal-hull electrofishing boat is an equipotential surface if all large metal surfaces (railings, generator, control unit, outboard engine, and so on) have metal-to-metal electrical contact with the hull. Removable items, such as the generator, can be temporarily wired to the hull. Floating, or isolated, metal can be detected with a multimeter set on the resistance scale; readings above 1 Ω are undesirable. When standing in a boat with an equipotential surface, one should be able to touch any two surfaces in the boat safely, but contact with surfaces outside the boat during electrofishing (e.g., water or overhanging vegetation) may result in a shock. A nonconductive boat hull requires a large piece of metal (e.g., railing in a fiberglass boat or rowing frame in a rubber raft) to which all metal surfaces and electrical components may be wired.

Although many embellishments are possible, the basic layout of an electrofishing boat is quite simple (Figure 8.8). A portable generator is the best power source; it consists of an alternator to produce AC and a gasoline-powered engine to drive the alternator. A single-phase, 230-V generator with 2–5 kW capacity is satisfactory for most electrofishing. Although generators with a 60-Hz frequency are most common, higher frequencies, such as 180 or 400 Hz, are also effective. The higher the frequency, the smaller and lighter the generator. New generators normally have the neutral (also called the internal ground) of the 230-V outlet attached to the generator frame. If an isolation transformer does not exist in the circuit, either as a separate component or in the generator or control unit, the neutral must be

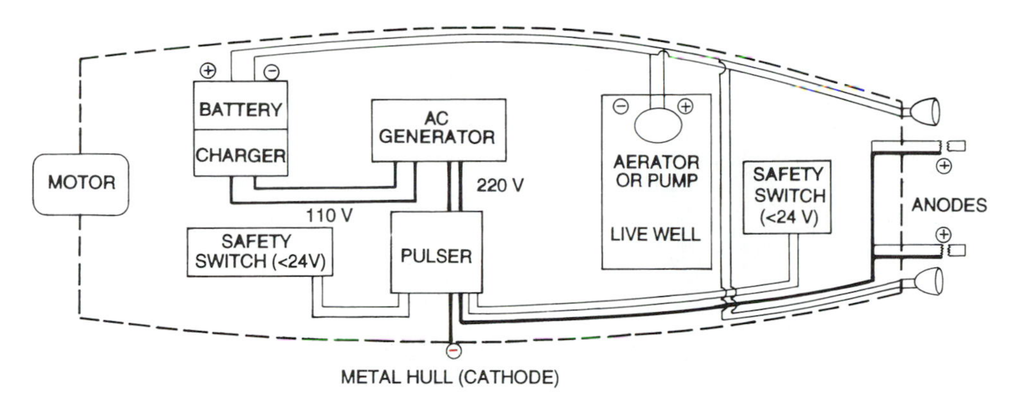

Figure 8.8 Generalized circuit in a generator-powered electrofishing boat with DC output, metal hull serving as the cathode. Heavy circuit lines indicate high voltage for electrofishing. Light circuit lines indicate low voltage for auxiliary equipment. Circled plus and minus signs indicate polarity of electrical connections.

disconnected to avoid the possibility of a potentially dangerous short circuit. The frame of the generator should then be connected to the metal hull to eliminate shock hazard. It is extremely important that you consult your supplier of electrofishing equipment if you are uncertain about the existence of an isolation transformer in your system. A generator should have a circuit breaker to protect it when overloaded.

A good quality of stranded, insulated, gasoline-resistant wire rated for the maximum current must be used; number 10, 12, or 14 wire is desirable for electrofishing boats. The choice of wire depends upon the voltage a circuit will carry (USFWS 1992). All wiring for the electrofishing circuit must be encased in conduit. Wire connections must never be made by splicing and taping. Where connections are necessary, use a plastic wire nut with a rating greater than or equal to the wire rating and protect it in a metal, watertight junction box grounded to the hull. Safety switches should be weatherproof and operate on low-voltage circuits (24 V or less) that are carried in a conduit separate from that carrying high voltage.

A control unit (also called a pulser or electrofisher) between the power source and electrodes is necessary if waveforms other than generator-produced AC are needed (e.g., pulsed DC). These units can be made locally, if experts are available, or purchased from commercial suppliers (Figure 8.9). The unit should be sturdy and provide AC, DC, and pulsed DC for the maximum flexibility in electrofishing. Output voltage selections should be possible, as well as selections for pulsed DC: frequency and duty cycle or pulse width. The unit should be protected by both manual and automatic circuit breakers. Meters to monitor circuit voltages and currents are essential. Control units are susceptible to moisture and should be contained in a hood or protective housing to avoid short circuits and water damage.

Lights, aeration units, and other electrical accessories should be operated with a 12-V DC system, not with the 110–120-V outlet from the generator; using the latter approach robs power for electrofishing and increases the probability of high-voltage shock. A 12-V, "deep-charge" battery (aviation or marine type) connected through a separate circuit to weatherproof outlets will accommodate these accessories. During electrofishing, the battery can be charged continuously through a battery charger driven from a 110–120-V AC outlet on the generator. When the generator

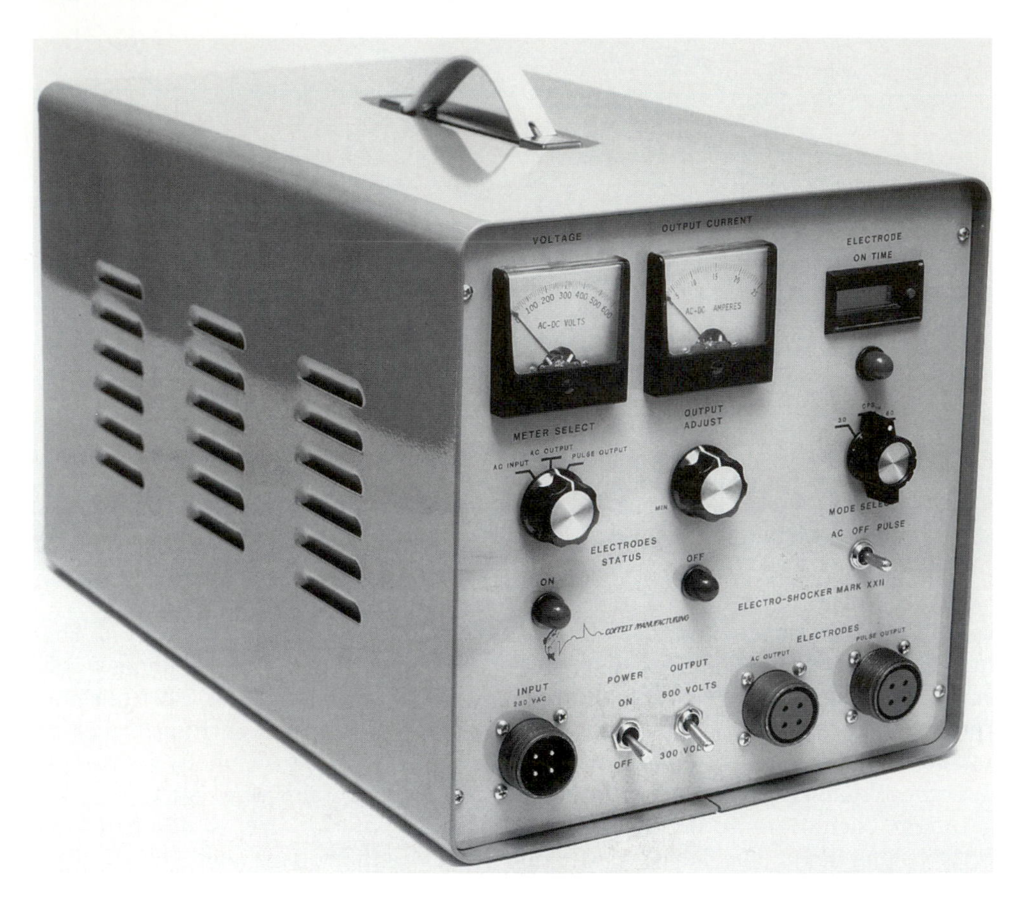

Figure 8.9 Example of a control unit, built by Coffelt Manufacturing, Inc. (Flagstaff, Arizona), commonly used in electrofishing boats. Input is received from a 230-V AC generator at lower left; output is AC or pulsed DC at lower right. Switches and knobs allow one to control and modify output. Meters provide measurements of voltage, current, and on time of the output.

is shut down, the battery can operate essential accessories, such as work lights and aerators, for periods sufficient to meet most fish handling needs. If a battery is inadequate, the 12-V system can be operated through a high-ampere, AC-to-DC converter commonly used in motor homes. Stationary bow lights should be mounted well above the waterline for easy access and replacement. However, waterproof lights mounted on hinged booms permit the use of lights above or below the waterline. Some biologists feel that underwater lights provide better viewing of fish.

Electrode system. Biologists should take advantage of electrode design to optimize available power and achieve sampling objectives. Electrodes provided with commercially made systems are not necessarily the best for a particular purpose. No electrode design is inherently good or bad; any design will support some sampling objectives but not others. Types of metal (copper, aluminum, or stainless steel) differ in their cost and usefulness relative to sampling objectives, but their electrical characteristics are similar for the purposes of electrofishing. In general, diameter, or cross section, of individual electrodes should be as large as the power source will

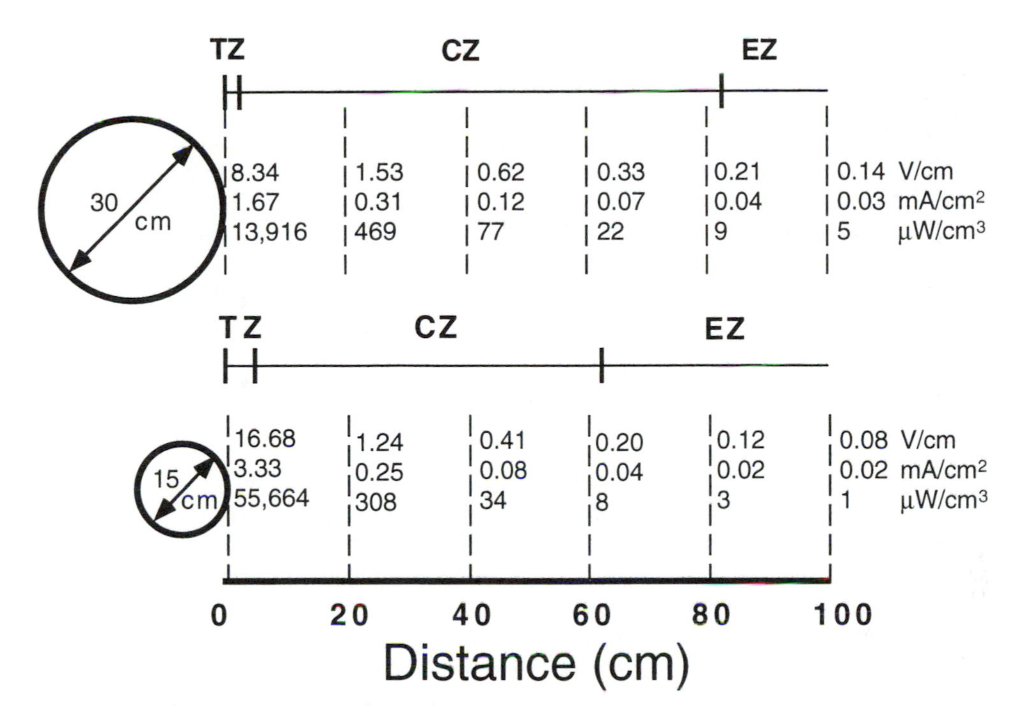

Figure 8.10 Effect of doubling electrode diameter (in this case, a 15-cm sphere versus a 30-cm sphere) on theoretical values of field intensity (voltage gradient, V/cm; current density, mA/cm^2; and power density, μW/cm^3) at 20-cm intervals, 0–100 cm from electrode surface. Applied voltage is 250 V in 200 μS/cm ambient water conductivity. Field intensity is substantially higher at the surface of the 15-cm sphere. Assuming a trauma threshold of 7 V/cm (arbitrary), the width of the trauma zone (TZ) of the 30-cm sphere is about half that of the 15-cm sphere. Assuming an escape threshold of 0.2 V/cm (arbitrary), the boundary between the capture zone (CZ) and escape zone (EZ) extends to 82 cm from the 30-cm sphere but only to 61 cm from the 15-cm sphere.

allow (Figure 8.10). Electrode resistance varies with water conductivity and electrode size and shape. Resistance measurement is useful for matching electrodes to the power source. Although electrode resistance can be estimated, based on electrode theory (Novotny and Priegel 1974), it can be directly measured once at a given water conductivity and then easily recalculated for every new conductivity (refer to Kolz 1993 for details).

Cylinder, sphere, and ring electrodes are most often used on boats. Cables or pipes, usually about 1 m long, serve as cylinder-shaped electrodes. A linear array of three to six cylinders may be suspended from the crossbar on a boom. Spheres, usually about 30 cm in diameter, are more resistant to water flow but have the advantage of directing the field downward as well as to the side. A ring electrode, commonly called a Wisconsin ring, consists of 6–12 "dropper" electrodes suspended at equal intervals from a horizontal aluminum ring 0.5–1 m in diameter. Each dropper electrode is a 15-cm-long metal tube, 0.6–1.2 cm in diameter, attached to a length of insulated wire 30–45 cm long (Novotny and Priegel 1974). The ring electrode simulates a large electrode at a distance (>1 m), especially if each dropper has a diameter 1 cm or more. The large-electrode effect is lost when fish touch, or get close to, an individual dropper. Electrodes of other shapes may be preferred, depending on operational requirements (e.g., blade-shaped electrodes in submerged vegetation).

In an AC operation, each equal-sized electrode produces an equally effective fishing field. A three-phase generator (seldom used in the United States) needs three electrodes to avoid possible damage due to unequal current in its three circuits. In a DC operation, the anodes are in front of the boat, and the cathodes are along either side if the boat hull is made of nonconductive material. Metal boats may be used as cathodes in place of side droppers because less power is wasted at the cathode and metal boats cannot be isolated from the electrical field anyway. Anodes should affect a large area (1-m radius or more) near the water surface to enhance fish capture; individual electrodes with diameters of 1.0–2.5 cm do this best.

Safety system. The boat must have ample room for safe movement from bow to stern. Passageways must be skid proof. Railings must surround the entire bow area to protect dipnetters. Side railings are a very desirable feature. At least two foot-activated switches should be available, one or more in the bow for the dipnetter(s) and one in the stern for the boat operator. The boat operator should also be able to operate the control unit from a seated position while guiding the boat. The face of the control box is normally lit during operation in darkness. All hazardous surfaces, electrical or otherwise, should be clearly marked in words or color codes (e.g., red—danger, yellow—caution). Generator exhaust must be piped to the side of the boat; hot exhaust pipes should be encased in screening or insulation. Hooded generators should be vented—they will overheat without ample air circulation, and explosion of combustible fumes may result. Batteries must be held in an acid-proof, nonmetallic container and placed on the deck in a low-traffic area away from the generator.

8.3.2 Backpack Units

The objective of an equipotential surface in a boat cannot be attained when one wades in water with electrofishing gear. Backpack units and other wading devices require that the user be well insulated to avoid an electrical shock. Dry skin and clothing, covered by rubber insulation, are the best protection against shock during in-water operations.

Mechanical system. Except for leads and electrodes, the entire backpack shocker unit must be housed in a weatherproof container that is fastened securely to a comfortable pack frame (Figure 8.11). The weight of the unit should be on the hips and lower back; a high center of gravity makes wading more difficult and dangerous. Backpack shockers that are safe and effective may be made locally with expert direction. Most units are now purchased from commercial suppliers.

Electrical system. The power source is either a 12-V, deep-charge battery or a 110–120-V AC generator (Figure 8.11). Although both types of units provide similar levels of instantaneous power, generator-powered units provide continuous power for longer periods than do battery-powered units; however, the latter are quieter, and often weigh less, than the former. Gel batteries have less capacity but are safer than acid batteries. The unit should allow selection of output voltage, frequency, and duty cycle if maximum flexibility is required. Meters for circuit voltage and current are essential if output is to be monitored.

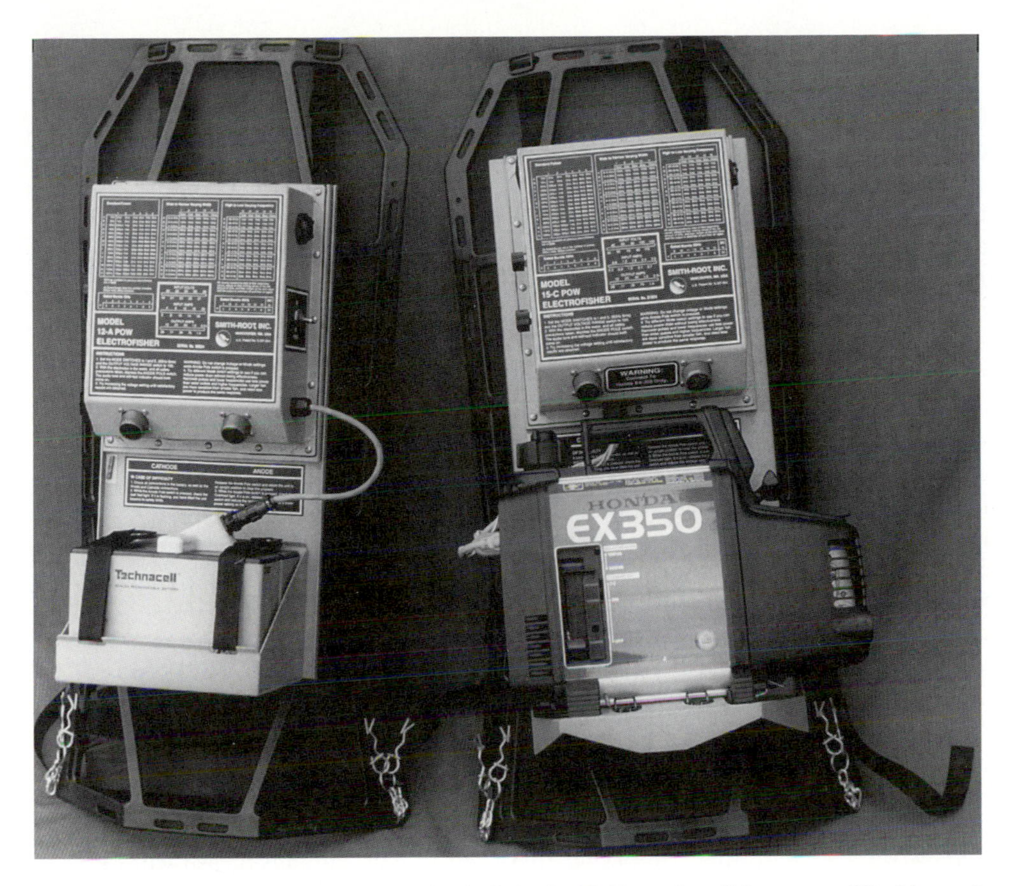

Figure 8.11 Examples of backpack units, built by Smith-Root, Inc. (Vancouver, Washington), commonly used for electrofishing in streams. The battery-powered unit (left) is quieter than the generator-powered unit (right) but is slightly less powerful and delivers 150 min of continuous power before recharging is required. Both units produce DC and pulsed DC at 15–120 Hz and have similar weights (66–69 kg).

Electrode system. The electrodes are handheld and must be insulated from the operator by handles 1.5–2.0 m long, preferably made of fiberglass. It is common to use a horizontal ring electrode attached to the end of the handle. Rake- and diamond-shaped electrodes are also used. Electrode material should be at least 0.6 cm in diameter, preferably 1 cm or greater, to reduce field intensity near the electrode and increase the area of effective field (Figure 8.10). In AC operations, two equal-sized electrodes should be used side by side. In DC operations, the anode is typically used for fishing and the cathode trails behind. The cathode should have as much, or more, surface area than the anode to avoid wasting power at the cathode.

Safety system. Manually activated switches on each electrode handle are essential safety features. The power source should have both automatic and manual circuit breakers. The automatic switch must be tilt activated to prevent shocks if the backpacker falls down; reset should be manual, not automatic. Anodes should not be used as dip nets to capture and move fish; "live" anodes swung inadvertently out of water have caused serious accidents. Anodes used as dip nets are also more likely to harm fish.

8.3.3 Specialized Systems

Shore-based electrofishing is frequently used in large streams and rivers where a backpack shocker produces a field too small and weak to be effective. In a backpack operation, one person carries control equipment in the stream. During shore-based electrofishing, all equipment except long leads and electrodes is on shore. (Equipment in a towed raft is a more mobile version.) Large generators are used for more effective electrofishing, but their use results in additional hazards (section 8.5.1). In AC operations, both electrodes are handheld within 1–2 m of each other. In DC operations, both anode and cathode can be handheld, or the cathode can be placed on the stream bottom or buried in the streambank, usually near the generator. Less power is wasted at the cathode if one achieves a good earth ground. One or more anodes can be handheld for fishing in a wide area; this, too, presents a hazard (see section 8.5.1).

Specialized electrodes have included an interesting array of recent applications. In large rivers, anodes are thrown into shoreline cover and then energized to stun fish with AC (Penczak and Romero 1990) or to retrieve fish by means of the taxis effect of DC, which has been used to draw large trout to the boat. James et al. (1987) reported that small anodes at the tips of long rods are used by divers to immobilize territorial fish selectively in warmwater streams. (The term "diver-operated anode" yields the uncomfortable acronym DOA, well known to paramedics as dead on arrival. Unconventional mobile electrodes, such as those for throwing and diving, must be used with great caution and should never be used on metal boats.) Long, insulated cables with exposed ends are used as boat electrodes to stun fish in deep (8–10 m) water or to "drive" them toward the surface (Grunwald 1983). Electrodes, in configurations forming circles, squares, or rectangles, are placed on the bottom in shallow waters where they are briefly energized to give samples of fish from known areas (Bain et al. 1985; Fisher and Brown 1993). Known as "prepositioned area shockers," these devices are a passive form of quantitative electrofishing that is proving successful for fish habitat preference studies. Refined versions of the electric seine are used to sample fish in warmwater streams (Bayley et al. 1989) or to harvest fish in holding ponds or the wild (Cave 1990). An electric seine consists of a series of cylindrical electrodes suspended from a floating line between two brailles, and energized from a shore-based power source; it is used to drive fish to shore or to a block net for capture. Finally, an electrode net made of metal mesh and energized with only 12-V DC is used to anesthetize large salmonids (out of the water) for mark and release, eliminating the possibilities of fish injury during physical restraint or fish stress during recovery from chemical anesthesia (Gunstrom and Bethers 1985; Orsi and Short 1987).

8.4 EFFICIENCY

Many factors influence electrofishing efficiency. For most factors, efficiency is highest in a given range and drops off at extreme values of the variable. If one understands these effects, data from electrofishing samples can be rendered more useful. For example, sampling variance can be reduced through sample stratification (e.g., keeping day and night samples separate) or standardization of sampling methods (e.g., sampling only when certain conditions exist). Each factor of electrofishing efficiency can be placed in one of three categories: biological, environmental, and technical (Zalewski and Cowx 1990). Many of these factors are interrelated and

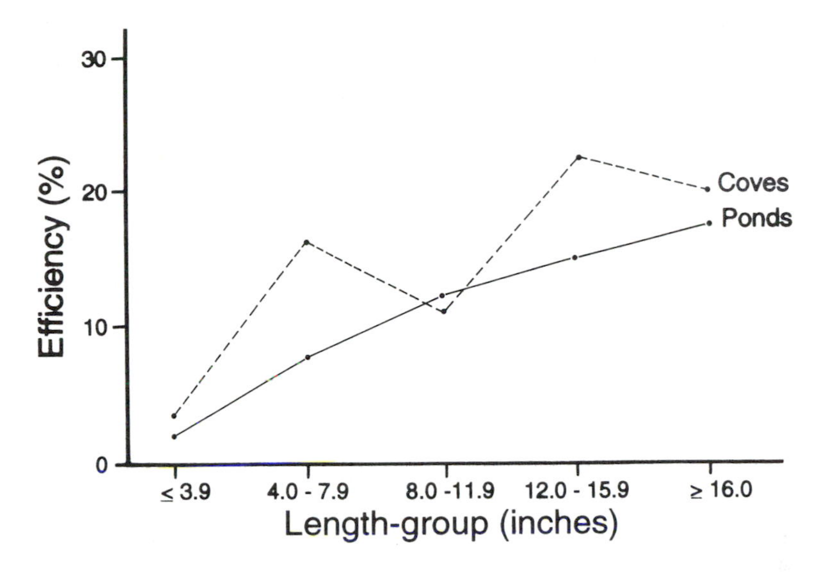

Figure 8.12 Effect of size of largemouth bass on electrofishing efficiency in 35 ponds and 16 reservoir coves in the midwestern United States (adapted from Simpson 1978, with permission).

their combined effects are difficult to isolate. Biologists should conduct an "audit" to determine which factors significantly affect their electrofishing efficiency, then monitor these factors while electrofishing.

8.4.1 Biological Factors

Biological factors affecting efficiency are related to the characteristics of the fish community or population being sampled. Like any other sampling method, electrofishing tends to select for some fish and against others. Thus, electrofishing samples should not be accepted as representative of the community or population until such an assumption is evaluated.

Vulnerability to electrofishing varies among species due to innate differences in morphology, physiology, and behavior. Bony fishes conduct current more readily than do cartilaginous fishes. Among the bony freshwater fishes, some species vary widely in their tissue resistance (Edwards and Higgins 1973); however, those with vestigial scales (e.g., ictalurids) or fine scales (e.g., salmonids) tend to be more vulnerable to electroshock than those with coarse scales (e.g., cyprinids). Because electrofishing is a sampling technique for shallow waters, habitat preference among species will affect their vulnerability (Larimore 1961). Species that seasonally or diurnally inhabit shoreline, such as centrarchids, cyprinids, and percids, are generally more vulnerable than are pelagic or benthic species, such as clupeids, osmerids, and ictalurids.

Fish size is an important determinant of electrofishing efficiency (Figure 8.12). Electrofishing tends to select for larger fish of a species. The most common explanation is that at a given voltage gradient, total body voltage increases with length, resulting in greater electroshock to larger fish. Also, large fish are more visible to dipnetters than are small ones and may be inadvertently selected. Size selectivity of electrofishing has been amply demonstrated in the field (Sullivan 1956; Junge and Libosvarsky 1965; Reynolds and Simpson 1978).

Other fish characteristics are more subtle and have not been adequately documented. Predators and spawners may be more vulnerable than are other fish to electrofishing, not only because of their large size but also because their territorial behavior makes them less likely to avoid an oncoming electrofishing operation. On the other hand, schooling fish may more readily escape capture by group fright response. Cross and Stott (1975) demonstrated that repeated electrofishing at intervals of 2 h reduced the catchability of fish; at intervals of 1 d or more this did not happen. They believed that previously shocked fish had not learned avoidance but, rather, were often too weak to be drawn to the anode another time. However, even samples taken on successive days (or nights) may not be statistically independent if previously shocked fish remain stressed. The abundance of fish may also affect electrofishing efficiency. In a study of factors affecting electrofishing efficiency, Simpson (1978) found that population density of largemouth bass was inversely related to efficiency; he postulated that dense populations may result in offshore distributions, group fright response, or less effective dipnetting and, thus, reduced efficiency.

8.4.2 Environmental Factors

Water conductivity is the most important environmental factor affecting electrofishing efficiency. Extreme conductivity, whether low or high, exceeds the capacity of most power sources and reduces efficiency. Ambient conductivity is a better indicator of efficiency than specific conductivity (see section 8.2.2 and Box 8.1).

Water temperature affects efficiency both directly and indirectly. High temperature increases fish metabolism, resulting in an increased ability to perceive and escape an electrical field; it also increases water conductivity, an advantage in low-conductivity waters. Low temperature decreases flotation rate of stunned fish, making capture less likely. Conversely, stunned fish may be easier to capture at low temperature because they recover less quickly. Temperature indirectly affects electrofishing efficiency by changing fish distribution. Adults typically move shoreward or upstream to spawn in conjunction with a temperature increase; however, very high temperatures force fish into deeper water, away from electrofishing areas.

Water transparency, like temperature, has an optimal, bell-shaped relation to electrofishing efficiency. High transparency may shift fish distribution toward deeper water and may allow fish to see the boat and avoid capture, but low transparency also reduces capture rate by making stunned fish less visible to dipnetters. Kirkland (1965) observed an increase in catch rate of largemouth and spotted bass when normally clear water became turbid; he concluded that the turbidity made it more difficult for bass to detect the sampling operation and thereby avoid capture. In rivers, electrofishing is often most effective at lower flows. Pierce et al. (1985) found that electrofishing catch rates were inversely related to water level in the upper Mississippi River. As water level increased, efficiency decreased due to increased water velocity, decreased water clarity, and greater fish dispersion.

Dissolved oxygen concentration in the hypolimnion of eutrophic waters can be very low in late summer. When this occurred in small midwestern impoundments, Simpson (1978) found that electrofishing efficiency for largemouth bass and bluegill increased. He concluded that the fish had moved shoreward out of deeper, oxygen-poor water.

Morphometry of the water body, including surface area, shoreline slope, and shoreline development (see Chapter 4), has important influences on electrofishing. Fish are less vulnerable to capture in larger water bodies than in smaller ones, perhaps because a greater proportion of the latter is suitable for electrofishing. Fish are also more vulnerable in lakes and impoundments with high shoreline development. Very steep shoreline slopes make electrofishing less effective because of limited habitat for sampling.

Substrate affects electrofishing efficiency. Mud and silt will often reduce the horizontal intensity of an electrical field by drawing electrical current into the substrate; this is not a problem in less-conductive substrate such as gravel. Large-diameter rocks and boulders provide cover for smaller fishes and may prevent their capture if stunned fish remain hidden instead of rising into the water column.

Cover, such as submerged brush, trees, and rooted macrophytes, has a concentrating effect on fish distribution. Its influence on electrofishing is difficult to quantify because cover itself is difficult to describe accurately. Electrofishing in experimental ponds containing known numbers of largemouth bass was always more effective in sections containing artificial structure than in sections without such structure (Simpson 1978).

Rain and wind have the most immediate effects on electrofishing by disrupting surface visibility for dipnetters, but they are less important for small water bodies. Springtime frontal systems may cause adult fish tending nests to abandon their role as parents temporarily (Summerfelt 1975); such cool fronts may cut catch rates by half during electrofishing operations on large reservoirs (Simpson 1978).

The decision to electrofish at night or during the day may be very important. At night, larger, predatory fish are inshore feeding, and all fish seem less apt to avoid capture. As a result, several studies have shown that electrofishing at night catches more species, larger fish, and greater numbers of fish than does electrofishing during the day (Loeb 1958; Witt and Campbell 1959; Sanderson 1960; Kirkland 1965). Sampling dates can also be controlled; they should be selected so as to increase electrofishing efficiency, particularly as efficiency relates to temperature and seasonal events such as spawning.

8.4.3 Technical Factors

Technical factors involve the personnel, equipment, and organization used in an electrofishing operation. Unlike environmental or biological factors, many technical factors can be selected or controlled (e.g., electrical waveforms). Technical considerations are often based on matters related to standardization and efficiency of sampling. Standardization of equipment and its use are essential if sample variance is to be minimized; standardization is relatively easy for one crew operating at different times and places but harder to achieve among two or more crews. Many agencies have developed standard operating procedures, which are written in manuals or taught at workshops. Whether a crew is working independently or coordinating its work with others, equipment must be periodically checked—not only for safety, but for efficiency and constancy as well. For example, electrical output of a control unit should be verified at least annually with an oscilloscope. Also, electrodes should be kept clean during a field season because corrosion will increase electrode resistance, reducing effectiveness of the electrical field.

The organizational aspects of an electrofishing operation will influence its efficiency. The number of crew members, their experience, and the proven reliability of their equipment and procedure can make the difference between success and failure; only trial and error over time can provide solutions to these problems of electrofishing. Well-defined sampling objectives are essential to operational effectiveness. If survey and inventory are needed, all stunned fish should be collected. If, however, single stock assessment has priority, capturing only the appropriate fish will improve efficiency—in time and money, if not in catch rate. The quality and validity of a time series of data (e.g., several years) collected by the same crew can be seriously affected by a large turnover in personnel, unless steps are taken to provide documentation, standardization, and transfer of field wisdom.

8.5 PROCEDURES

8.5.1 Operational Guidelines

Safety aspects. An organization that sponsors electrofishing activity should have a safety program consisting of personnel training, equipment specifications, and operational protocol. Goodchild (1991) developed a comprehensive safety program. A common-sense approach to any aspect of an electrofishing program is an objective application of the self-scrutinizing question: "If an accident occurred, could I defend my practice before an independent investigative body?"

Electrofishing is hazardous work. The batteries and generators used in electrofishing provide more than enough energy to electrocute a person. The current that passes through the body and vital organs is particularly dangerous. Currents as low as 0.0002 A, under certain conditions, can cause serious injury or death. Most people who grasp a wire carrying as little as 0.01–0.02 A cannot let go because of tetany. Death is usually a result of respiratory arrest, asphyxia (caused by contracted chest muscles), or ventricular fibrillation (uncontrolled, irregular heart beats). Cardiopulmonary resuscitation (CPR) can restore breathing, but for ventricular fibrillation, CPR is only a stalling tactic—a defibrillator must be used to restore normal heart rhythm. Location of the nearest defibrillation unit should be known by the crew. Two-way radios or cellular telephones can provide an extra safety margin at many remote locations.

All members of an electrofishing crew should understand the system they are using and the risks involved. Before field operations, new crew members should receive orientation on equipment and procedures; their training can be formally acknowledged in writing (Box 8.3). A practice run with the generator on and electrodes off will improve efficiency and safety. At least two, preferably all, crew members should have CPR and first aid training. An electrofishing system should receive a safety check before a field operation; detailed inspection and maintenance should occur annually (Goodchild 1991; USFWS 1992).

Heavy-duty rubber gloves and knee-high rubber boots or waders should be worn by all personnel working in electrofishing boats. However, requiring lineman's gloves rated at 50,000 V makes little sense, considering that personnel wear unrated boots or waders. Crew leaders should require their personnel to wear Coast Guard-approved life vests if they consider the water to be deep and swift, cold, or turbid (USFWS 1992). All types of electrofishing operations need a large first aid kit and fire extinguisher (ABC type). Mount the fire extinguisher on an open surface away from likely sources of fire (e.g., generator). Gasoline should be stored in spill-proof, vented containers of approved design and not exceeding 19-L capacity. For people in electrofishing boats, continuous generator noise at close range may cause eardrum

Box 8.3 Suggested Text for Acknowledgment of Orientation by Electrofishing Personnel

Acknowledgment of Electrofishing Orientation[a]

I have received instruction and orientation about electrofishing from my employer. As a result, I understand and accept the following conditions

1. Electric fishing (EF) is an inherently hazardous activity in which safety is the primary concern. The electrical energy used in EF is sufficient to cause death by electrocution.
2. During operations, it is critical to avoid contact with the electrodes and surrounding water. The EF field is most intense near the electrodes and can extend 5–10 m outward.
3. The electrodes are energized by the power source, a generator or battery, and controlled by safety switches; these switches must remain off until the signal is given to begin EF.
4. The power source has a main switch that must be turned off immediately if an emergency occurs.
5. The electrodes are usually metal probes suspended in the water. If direct current is used from a boat, the anodes (+) are in front of the boat to catch fish and the cathodes (−) may be suspended from the sides; both can produce electroshock. When a metal boat is the cathode, the boat is safe as long as all metal surfaces inside it are connected to the hull.
6. Moveable anodes on a boat are dangerous, especially on metal boats. All electrodes on a conventional EF boat should be in fixed position during operation.
7. Dry skin and clothing are good protection against electroshock. The body should be fully clothed during EF. Rubber knee boots are minimal foot protection, as are rubber gloves for the hands. A personal flotation device must be worn when the water is considered swift, cold, or deep. Ear protection is necessary for those working near the generator.
8. At least two members of the EF crew must have knowledge of CPR and first aid. A first aid kit and, in an EF boat, a fire extinguisher must be within immediate reach during an operation. Electroshock can cause heart fibrillation or respiratory arrest; CPR can cure only the latter. The EF crew must know the location of the nearest defibrillation unit.
9. A communication system, particularly hand signals, must be available to all members of an EF crew. When multiple anodes are used in a portable EF operation, the buddy system must be used. Above all, NEVER OPERATE ALONE.
10. Stunned fish should be removed from the EF field as soon as possible and not subjected to continuous electroshock by being held in the dip net. Using the anode as a dip net is unhealthy for fish and people and should be avoided.

Box 8.3 Continued.

11. An EF operation should proceed slowly and carefully; avoid chasing fish and other sudden maneuvers. Night activities require bright, bow-mounted headlights. Operations should cease during lightning or thunderstorms; use discretion during rain. Avoid EF too close to bystanders and pets or livestock.
12. All EF crew members must know who their leader is and recognize his or her authority as final in operational decisions. However, every crew member has the right to ask questions or express concern about any safety aspect of an EF operation. A crew member has the right to decline participation in an EF operation, without fear of employer recrimination, if he or she feels unsafe in such participation.

______________________________ __________
Signature of employee Date

I have discussed the above-named conditions with the employee and am satisfied that he or she understands them.

______________________________ __________
Signature of Supervisor Date

[a]Adapted from Reynolds (1995), with permission.

damage. Ear plugs or mufflers are effective protection against noise. Another solution is the battery-powered communication headset; this device is easy and inexpensive to install, cuts out operational noise, and gives clear voice contact between dipnetter and pilot. Avoid cheap communication devices; they operate at a frequency close to that of 60-Hz AC generators and produce too much static to be useful.

As electrofishing is about to begin, several rules should be observed. Never electrofish alone. Conversely, avoid excessively large crews; they cause confusion and lead to mistakes. Every crew should have at least one biologist experienced at electrofishing who is in charge. That person should supervise all aspects of the operation, including initial settings on the equipment, organization of the crew, and decisions requiring judgment.

While under way, some other rules are important. Avoid operating near bystanders, pets, or livestock that are in or on the water or on shore. Although light rain is not a serious problem, a sheen or continuous film of water on surfaces of electrical equipment greatly increases the hazard of shock; operations should cease when this occurs. Electrofishing does not attract lightning strikes. Thunderstorms present the same risk to those in an electrofishing boat as to other boaters on a lake or reservoir: the high point in an open area is a risky position. Electrofishing should proceed slowly and deliberately—avoid chasing fish and other acrobatics. Resist the urge to hand capture a stunned fish; a missed fish is better than electroshock. Always shut down the power source when equipment changes, repairs, or refueling become necessary. Rest often enough to avoid fatigue.

Finally, the members of an electrofishing crew must stay close together during an operation. With boat and backpack shockers this is easy to do because the power source moves with the crew; this is not the case in shore-based operations. Shore-based operations are dangerous because they combine the riskiest elements of the other methods: high voltages of boat shockers and instream work of backpack shockers. As crews get farther from the power source, more spread apart, or both, the probability increases that someone may fall unseen into the water; sustained electroshock is likely if that person has no safety switch (e.g., a dipnetter). The safest procedure is to use the "buddy system" (anode handler and dipnetter) the entire time that a shore-based operation is in effect.

Technical aspects. When using AC, allow electrodes to extend no more than 1–2 m into the water; given a normal generator, they are not effective at greater depths. When using DC, anodes are normally positioned near the surface (usually less than 1 m) to take advantage of electrotaxis. Adjustment for various conductivities may be accomplished by (1) interchanging electrodes—smaller diameter electrodes (≤0.6 cm) in waters of very high or very low conductivity, or by (2) exposing more or less of each electrode, either by varying submergence or using a movable rubber sleeve. However, one should avoid changing electrodes if sampling standardization is important.

Certain accessories enhance sampling efficiency. Dip nets with long, fiberglass handles are essential to any electrofishing operation; carry extras. Do not use excessively deep net bags because they splash water onto electrical and mechanical surfaces and increase the hazard. A large holding tank should be placed within reach behind the dipnetters with no equipment between the dipnetters and the tank. The tank should be equipped with an aerator (closed system) or circulating pump (open system) to oxygenate the water. There should be enough space on either side of the tank for safe passage. Small stock-watering tanks with corrugated walls are readily available, inexpensive, sturdy, and of sufficient capacity. In backpack or shore-based operations, a holding tank must be kept near the dipnetters. Cages, made of framed metal mesh or hardware cloth, make excellent instream live cars; they permit water flow while shielding the fish they contain from electroshock (Sharber and Carothers 1987). At night, a 12-V DC searchlight permits the boat operator to see fish missed by the dipnetters and either make note of the fish or retrieve them with a short-handled net. If the latter is done, an additional holding tank or small tub must be within reach. The search light also has obvious safety advantages not inherent in fixed working lights. A conductivity meter is essential, if one expects to monitor electrofishing effectiveness. The meter should be checked periodically for accuracy.

Before an operation begins, the amount of effort either by distance, time, or desired sample size should be agreed upon at least tentatively. Duration of electrofishing time is generally a reliable index of effort; it can be measured either as the difference between time of start and stop or the amount of time that the electrodes are energized as indicated by a circuit clock. Hours or minutes are the best units for effort by time. Distance of shoreline sampled is also a good effort index, but it requires shoreline measurements for each sample unless permanent stations are repeatedly used. The best unit for effort by distance is a minimum that is always sampled, such as 100 m or 1 km. A useful variation of the distance index is one complete lap of the shoreline of a small lake or blocked cove of a lake (Simpson 1978); effort should be roughly proportional to surface area of the water body sampled. In the early development of a sampling program, both distance and time for each sample should be measured. The variable most predictive of catch, both in terms of linearity and confidence limits, is the one to use consistently as an effort index. With either measure of effort, it is preferable to report catch per unit effort by interpolation, not extrapolation (e.g., if 15 fish are captured in 0.25 h or 500 m, express adjusted catch as 1 fish/min or 3 fish/100 m, not as 60 fish/h or 30 fish/km).

Proper care of fish after capture will reduce handling mortality (see Chapter 5). Fish held in a dip net while others are gathered will be subjected to prolonged electroshock. Likewise, the longer that fish are held in the tank with other fish, the more likely that stress-induced mortality will occur. Therefore, fish should be processed at frequent intervals; one way is to stop often enough to handle and release fish before they weaken; a collapsible pen attached to the side of the boat allows a final check on recovery of fish before their release. Another way is to process fish while under way. The processor sits amidship within reach of the tank; a hooded light and foot- or voice-activated tape recorder with microphone permits the measuring, weighing, and marking of fish as they are caught. If the catch rate is more than the processor can handle, stops become necessary. Fish are released to the side and rear; if electroshock is not severe, they recover quickly and seldom swim into the electrical field again. By processing fish while under way, sample sizes can be reduced through sequential sampling (see section 8.5.3).

8.5.2 Sampling Guidelines

Like any other sampling method, electrofishing is best deployed with consistency and objectivity. Unless the sampling objective is species- or stock-specific, dipnetters should collect all individual fish possible; otherwise, selectivity may occur while dipnetting. Even when selective dipnetting and all-inclusive dipnetting do not produce different estimates of population characteristics, improvements in efficiency can result from careful decisions about dipnetting tactics (Twedt et al. 1992). These guidelines, therefore, are starting points only; they are based on the assumption that all stunned fish, irrespective of species and size, are to be collected.

During general surveys, the control unit should be set for just enough voltage to obtain the desired response (e.g., narcosis) among all target species and sizes of fish. Lower voltage may still electroshock fish but is selective for larger ones. Higher voltage will often injure fish, particularly if AC is used. If higher voltage is necessary, and fish injury is to be minimized, use pulsed DC at the lowest frequencies and duty cycles possible.

If sample data from different times and places are to be compared, standardization of electrical output is desirable. Many biologists use constant voltage or constant current, as indicated by control unit meters, to standardize output; their choice often depends on the water conductivity they typically encounter (see section 8.2.2). However, constant power may be the best indicator of output for standardizing catches across a broad range of water conductivities, assuming that electrofishing efficiency is based on the transfer of power from water to fish (Kolz 1989). Standardized power output has been tested and adopted by cooperating fisheries agencies on the upper Mississippi River because it reduced catch variance by 12 to 15% with no additional sampling cost (Burkhardt and Gutreuter 1995). Standardized power output requires meters for monitoring output of voltage and current, an assumed value for fish conductivity, and measurement of water temperature and conductivity.

Low pulse rate (5–40 Hz) seems effective for spiny-rayed fishes such as walleye, white bass, yellow perch, and bluegill (Novotny and Priegel 1974); very low pulse rate (3–5 Hz) is effective for large catfishes (Gilliland 1988). Higher pulse rates (40–120 Hz) have a greater effect on young fish and soft-rayed fishes (e.g., salmonids, cyprinids, and ictalurids) but are injurious to large salmonids (see section 8.2.2). Higher pulse rates may also be necessary for strong swimmers such as large, predatory fish. Duty cycle does not seem as important as pulse rate: 25% may be as effective as 50% and conserves power; 10% may not be effective (Novotny and Priegel 1974). Some waveforms are more effective than others when electrofishing in high-conductivity waters (e.g., Hill and Willis 1994); those with narrow pulse widths (e.g., 1–3 ms) permit higher peaks of pulsed DC.

Standing waters. Electrofishing is best done from boats in lakes and impoundments. Spring, when adults and young are mixed in the warmer inshore waters, is the best season for sampling; autumn may be the next best time because adult fish move inshore as lake waters cool. Electrofishing is most productive at night or twilight, particularly just after sunset as predators move inshore to feed. One lap of the shoreline of a small lake or blocked cove is the most complete unit; record both distance and time of the lap. In large lakes, or when a full lap is not possible, randomly sample fixed lengths of shoreline and record the time or sample several lengths of shoreline for fixed periods. To minimize variance and reduce bias, more small samples are better than a few large ones. To ensure consistency, avoid changing the waveform after it is selected. A slow, deliberate capture style is safer than fish chasing and also yields more representative samples. Catches can be increased in areas of submerged cover by moving in with the power to electrodes turned off and then energizing electrodes for an element of surprise. Alternatively, the cover can be surrounded with a block net and then sampled inside the net. In areas with extensive cover, use DC if possible. If representative samples are sought, try to avoid frightening fish out of, or into, the sampling area.

Flowing waters. Power limitations will usually determine the type of electrofishing to use in streams of various sizes. Boat shockers can be used in large streams, backpack shockers in small ones, and shore-based units in either. During night operations, use a boat or restrict wading to shallow, low-flow areas. Avoid electrofishing during high water. For safety, effectiveness, and consistency, restrict sampling to normal or low water stages. Adult fish can often be sampled effectively in spring, soon after cessation of high waters, while they are spawning in shallow mainstream areas or in tributaries. In boat shockers, many biologists float downstream with the current, using the motor mainly for maneuvering near shore and in rapid water; this technique keeps the boat next to the stunned fish. When wading, upstream movement is often preferable for sampling, with dipnetters beside electrode handlers to minimize fish avoidance. Direction of sampling relative to streamflow is not a fixed rule, however, and can vary according to local conditions and sampling objectives. Streams, like lakes, should be covered methodically based on a sampling design incorporating randomness. One side of a stream should be electrofished for a distance sufficient to ensure that all habitats associated with the inside and outside of bends are sampled. Record effort as time fished over a fixed distance.

8.5.3 Data Analysis

Most fish sampling is aimed at one or more of a few, basic objectives; electrofishing is no different. These objectives deal with characterizing a fish community by its species composition or assessing a fish population through estimates of abundance, size and age structure, and dynamic rates (reproduction, growth, and mortality). Electrofishing data, alone, are useful for achieving some objectives. For other objectives, electrofishing data should not be relied on as the sole basis for decisions.

Species composition. For a given type of electrofishing operation (and other gear types, for that matter), species detection is a useful statistic for deciding whether or not to use electrofishing in surveys and inventories. Species detection is the percentage of sampled water bodies known to contain a given species in which at least one specimen was captured. Generally, if detection success is high for a given species (75–100%), the sampling method is useful for population assessments wherever the species occurs. Species detection of centrarchids by boat electrofishing was high in small midwestern impoundments (Reynolds and Simpson 1978). However, because of the aforementioned species selectivity, electrofishing samples alone should not be used to assess fish community structure (i.e., species composition).

Population abundance. Catch per unit effort by electrofishing is a useful, easily obtained index to the abundance of the populations of many fish species (see section 8.5.1 and Chapter 6). This index is more useful when applied to adult fish because of the selectivity of electrofishing for larger fish. Trends in the abundance of young fish can be monitored if electrofishing equipment and procedures are modified (e.g., higher pulse rates) with this objective in mind.

Estimates of absolute abundance based on electrofishing samples may be seriously biased, but it is often difficult to ascertain in which direction the bias occurs. Bohlin et al. (1990) gave an excellent review of electrofishing techniques for abundance estimates. Mark–recapture estimates may be too high because marked (i.e., previously electroshocked) fish are more vulnerable than are unmarked ones (Grinstead and Wright 1973), or the reverse may be true (Cross and Stott 1975). The practice of "importing" marked fish into a blocked cove from other areas of a reservoir is also questionable. For example, Simpson (1978) found that electrofishing efficiency, on the average, was about 18% for these fish, compared with 12% for marked residents of a cove. It is probably better to capture fish by another method for marking and release and then use electrofishing for a single-census estimate; multiple-census estimates by electrofishing at frequent intervals (e.g., every few hours) are probably quite biased (Cross and Stott 1975). Catch depletion estimates may be less seriously affected by electrofishing samples because recaptured fish are not used in the estimate.

Population structure. Because of size selectivity, length–frequency data from electrofishing samples should be regarded with caution. This is especially true of the data regarding the relative abundance of small fish; they are probably biased too low. Length frequencies of stock-size fish (harvestable fish) based on electrofishing data, however, are reasonably accurate and can provide the basis for indices of stock structure (see Chapter 15).

The fish-by-fish nature of capture in electrofishing makes the method suitable for sequential sampling (Weithman et al. 1980). Measurement of stock-size fish continues while electrofishing proceeds. When a category or index of stock structure can be assigned at an acceptable level of confidence, sampling stops. In this way, sample size can often be reduced by almost 50%. If electrofishing continues until 100 stock-size fish have been measured, and no decision about stock structure can be reached, sampling stops and an endpoint estimate is made using the binomial distribution.

Population dynamics. The size selectivity of electrofishing is the greatest hindrance to its use for estimating the dynamic rates of a population—reproduction, growth, and mortality. Rates of reproduction or recruitment are severely compromised because sampling efficiency is low for juvenile, particularly age-0, fish. Estimates of growth rate are less suspect, but they may be biased too high if electrofishing tends to take larger individuals of a year-class, resulting in samples of faster-growing fish. Mortality rates based either on catch curves or mark–recapture data may be biased; in the former, smaller, younger fish will be disproportionately low in electrofishing samples, causing an underestimate of mortality. Recapture data can give biases in either direction, depending upon whether marked (previously shocked) fish are more or less vulnerable to electrofishing than are unmarked ones. Studies of population dynamics should not depend on electrofishing as the sole source of data.

8.6 REFERENCES

APHA (American Public Health Association), American Water Works Association, and Water Environment Federation. 1992. Standard methods for the examination of water and wastewater, 18th edition. APHA, Washington, DC.

Bain, M. B., J. T. Finn, and H. E. Booke. 1985. A quantitative method for sampling riverine microhabitats by electrofishing. North American Journal of Fisheries Management 5:489–493.

Bayley, P. B., R. W. Larimore, and D. C. Dowling. 1989. Electric seine as a fish-sampling gear in streams. Transactions of the American Fisheries Society 118:447–453.

Bohlin, T., T. G. Heggberget, and C. Strange. 1990. Electric fishing for sampling and stock assessment. Pages 112–139 *in* Cowx and Lamarque (1990).

Burkhardt, R. W., and S. Gutreuter. 1995. Improving electrofishing catch consistency by standardizing power. North American Journal of Fisheries Management 15:375–381.

Cave, J. 1990. Trapping salmon with the electronet. Pages 65–69 *in* I. G. Cowx, editor. Developments in electric fishing. Fishing News Books, Oxford, UK.

Cowx, I. G., and P. Lamarque, editors. 1990. Fishing with electricity, applications in freshwater fisheries management. Fishing News Books, Oxford, UK.

Cross, D. G., and B. Stott. 1975. The effect of electrical fishing on the subsequent capture of fish. Journal of Fish Biology 7:349–357.

Dwyer, W. P., W. Fredenberg, and D. A. Erdahl. 1993. Influence of electroshock and mechanical shock on survival of trout eggs. North American Journal of Fisheries Management 13:839–843.

Edwards, J. L., and J. D. Higgins. 1973. The effects of electric currents on fish. Georgia Institute of Technology Engineering Experiment Station Final Technical Report, Projects B-397, B-400 and E-200-301, Atlanta.

Fisher, W. L., and M. E. Brown. 1993. A prepositioned areal electrofishing apparatus for sampling stream habitats. North American Journal of Fisheries Management 13:807–816.

Friedman, R. 1974. Electrofishing for population sampling: a selected bibliography. U.S. Department of the Interior, Office of Library Services, Bibliography Series 31.

Gilliland, E. 1988. Telephone, micro-electronic, and generator-powered electrofishing gear for collecting flathead catfish. Proceedings of the Annual Conference Southeastern Association of Fish and Wildlife Agencies 41(1987):221–229.

Goodchild, G. A. 1991. Code of practice and guidelines for safety with electric fishing. EIFAC (European Inland Fisheries Advisory Commission) Occasional Paper 24.

Grinstead, B. G., and G. L. Wright. 1973. Estimation of black bass, *Micropterus* spp., population in Eufaula Reservoir, Oklahoma, with discussion of techniques. Proceedings of the Oklahoma Academy of Science 53:48–52.

Grunwald, G. L. 1983. Modification of alternating current electrofishing gear for deep water sampling. Pages 177–187 *in* Proceedings of the Annual Meeting, Upper Mississippi River Conservation Committee 39. U.S. Fish and Wildlife Service, Rock Island, Illinois.

Gunstrom, G. K., and M. Bethers. 1985. Electrical anesthesia for handling large salmonids. Progressive Fish-Culturist 47:67–69.

Hauck, F. R. 1949. Some harmful effects of the electric shocker on large rainbow trout. Transactions of the American Fisheries Society 77:61–64.

Hill, T. D., and D. W. Willis. 1994. Influence of water conductivity on pulsed AC and pulsed DC electrofishing catch rates for largemouth bass. North American Journal of Fisheries Management 14:202–207.

Hollender, B. A., and R. F. Carline. 1994. Injury to wild brook trout by backpack electrofishing. North American Journal of Fisheries Management 14:643–649.

Holmes, R., D. N. McBride, T. Viavant, and J. B. Reynolds. 1990. Electrofishing-induced mortality and injury to rainbow trout, Arctic grayling, humpback whitefish, least cisco, and northern pike. Alaska Department of Fish and Game Fishery Manuscript 9, Anchorage.

Horak, D. L., and W. D. Klein. 1967. Influence of capture methods on fishing success, stamina, and mortality of rainbow trout (*Salmo gairdneri*) in Colorado. Transactions of the American Fisheries Society 96:220–222.

James, P. W., S. C. Leon, A. V. Zale, and O. E. Maughan. 1987. Diver-operated electrofishing device. North American Journal of Fisheries Management 7:597–598.

Junge, C. O., and J. Libosvarsky. 1965. Effects of size selectivity on population estimates based on successive removals with electrical fishing gear. Zoologicke Listy 14:171–178.

Kirkland, L. 1965. A tagging experiment on spotted and largemouth bass using an electric shocker and the Petersen disc tag. Proceedings of the Annual Conference Southeastern Association of Game and Fish Commissioners 16(1962):424–432.

Kolz, A. L. 1989. A power transfer theory for electrofishing. U.S. Fish and Wildlife Service Fish and Wildlife Technical Report 22:1–11.

Kolz, A. L. 1993. In-water electrical measurements for evaluating electrofishing systems. U.S. Fish and Wildlife Service Biological Report 11.

Kolz, A. L., and J. B. Reynolds. 1989. Determination of power threshold response curves. U.S. Fish and Wildlife Service Fish and Wildlife Technical Report 22:15–24.

Lamarque, P. 1990. Electrofishing of fish in electric fields. Pages 4–33 *in* Cowx and Lamarque (1990).

Larimore, R. W. 1961. Fish population and electrofishing success in a warm-water stream. Journal of Wildlife Management 25:1–12.

Lennon, R. E., and P. S. Parker. 1958. Applications of salt in electrofishing. U.S. Fish and Wildlife Service Special Scientific Report—Fisheries 280.

Loeb, H. A. 1958. A comparison of estimates of fish populations in lakes. New York Fish and Game Journal 5:66–76.

McMichael, G. A. 1993. Examination of electrofishing injury and short-term mortality in hatchery rainbow trout. North American Journal of Fisheries Management 13:229–233.

Mesa, M. G., and C. B. Schreck. 1989. Electrofishing mark–recapture and depletion methodologies evoke behavioral and physiological changes in cutthroat trout. Transactions of the American Fisheries Society 118:644–658.

Novotny, D. W. 1990. Electric fishing apparatus and electric fields. Pages 34–88 *in* Cowx and Lamarque (1990).

Novotny, D. W., and G. R. Priegel. 1974. Electrofishing boats: improved designs and operational guidelines to increase the effectiveness of boom shockers. Wisconsin Department of Natural Resources Technical Bulletin 73.

Orsi, J. A., and J. W. Short. 1987. Modifications in electrical anesthesia for salmonids. Progressive Fish-Culturist 49:144–146.

Penczak, T., and T. E. Romero. 1990. Accuracy of a modified catch-effort method for estimating fish density in a large river (Warta River, Poland). Pages 191–196 *in* I. G. Cowx, editor. Developments in electric fishing. Fishing News Books, Oxford, UK.

Pierce, R. B., D. W. Coble, and S. D. Corley. 1985. Influence of river stage on shoreline electrofishing catches in the upper Mississippi River. Transactions of the American Fisheries Society 114:857–860.

Reynolds, J. B. 1995. Development and status of electric fishing as a scientific sampling technique. Pages 49–61 *in* G. T. Sakagawa, editor. Assessment methodologies and management. Proceedings of the World Fisheries Congress, theme 5. Oxford & IBH Publishing Co., New Delhi.

Reynolds, J. B., and D. E. Simpson. 1978. Evaluation of fish sampling methods and rotenone census. Pages 11–24 *in* G. D. Novinger and J. G. Dillard, editors. New approaches to the management of small impoundments. American Fisheries Society, North Central Division, Special Publication 5, Bethesda, Maryland.

Sanderson, A. E. 1960. Results of sampling the fish population of an 88-acre pond by electrical, chemical and mechanical methods. Proceedings of the Annual Conference Southeastern Association of Game and Fish Commissioners 14(1960):185–198.

Schreck, C. B., R. A. Whaley, M. L. Bass, O. E. Maughan, and M. Solazzi. 1976. Physiological responses of rainbow trout (*Salmo gairdneri*) to electroshock. Journal of the Fisheries Research Board of Canada 33:76–84.

Sharber, N. G., and S. W. Carothers. 1987. Submerged, electrically shielded live tank for electrofishing boats. North American Journal of Fisheries Management 7:450–453.

Sharber, N. G., and S. W. Carothers. 1988. Influence of electrofishing pulse shape on spinal injuries in adult rainbow trout. North American Journal of Fisheries Management 8:117–122.

Sharber, N. G., S. W. Carothers, J. P. Sharber, J. C. de Vos, Jr., and D. A. House. 1994. Reducing electrofishing-induced injury of rainbow trout. North American Journal of Fisheries Management 14:340–346.

Simpson, D. E. 1978. Evaluation of electrofishing efficiency for largemouth bass and bluegill populations. Master's thesis. University of Missouri, Columbia.

Spencer, S. L. 1967. Internal injuries of largemouth bass and bluegills caused by electricity. Progressive Fish-Culturist 29:168–169.

Sullivan, C. 1956. The importance of size grouping in population estimates employing electric shockers. Progressive Fish-Culturist 18:188–190.

Summerfelt, R. C. 1975. Relationship between weather and year-class strength of largemouth bass. Pages 166–174 *in* H. Clepper, editor. Black bass biology and management. Sport Fishing Institute, Washington, DC.

Twedt, D. J., W. C. Guest, and B. W. Farquhar. 1992. Selective dipnetting of largemouth bass during electrofishing. North American Journal of Fisheries Management 12:609–611.

USFWS (U.S. Fish and Wildlife Service). 1992. Electrofishing. Chapter 6, part 241. Safety operations, occupational safety and health. USFWS, Washington, DC.

Weithman, A. S., J. B. Reynolds, and D. E. Simpson. 1980. Assessment of structure of largemouth bass stocks by sequential sampling. Proceedings of the Annual Conference Southeastern Association of Fish and Wildlife Agencies 33(1979):415–424.

Witt, A., Jr., and R. S. Campbell. 1959. Refinements of equipment and procedures in electrofishing. Transactions of the American Fisheries Society 88:33–35.

Zalewski, M., and I. G. Cowx. 1990. Factors affecting the efficiency of electric fishing. Pages 89–111 *in* Cowx and Lamarque (1990).

Chapter 9

Collection, Preservation, and Identification of Fish Eggs and Larvae

WILLIAM E. KELSO AND D. ALLEN RUTHERFORD

9.1 INTRODUCTION

This chapter is an introduction to methods for collecting, processing, and identifying early life stages of fishes and summarizes the diversity of both marine and freshwater studies on fish eggs and larvae. We review various gears used to collect eggs and larvae, their relative effectiveness in different sampling situations, and the effects of physicochemical characteristics and larval behavior (e.g., vertical migration and phototaxis) on sampling design. We also include a discussion of sample preservation and processing, as well as discussions of early life stage terminology and techniques and taxonomic guides used for egg and larval identification.

Early investigators studying the growth, reproduction, and mortality of fish populations documented the critical importance of early life stages (eggs and larvae; see section 9.6.3) to overall abundance (Hjort 1914; May 1974; Hempel 1979). Fishes typically have high fecundity, and fish populations generally have high egg and larval mortality (>90%) and substantial year-to-year variation in survival of early life stages. Ichthyoplankton mortality is usually attributed to inherited defects, egg quality, starvation, disease, and predation. In addition, periods of high mortality may be associated with critical events in early ontogeny (e.g., hatching, first feeding, and initiation of swimbladder function; Blaxter 1988). Because the timing and duration of these critical periods are closely tied to physicochemical conditions, environmental variability can have substantial effects on egg and larval growth and mortality and ultimately on recruitment to adult stocks (Thorisson 1994).

The importance of early life stage survival to population abundance and fisheries harvests (Smith and Morse 1993) has led to numerous studies on distributions of fish eggs and distributions and behavior of larvae. Egg and larval collections have been used to identify spawning and nursery areas (Heath and Walker 1987) as well as temporal and spatial differences in spawning characteristics of exploited populations (Graham et al. 1984). Larval fish studies have also yielded information on ontogenetic changes in movement patterns (Lough and Potter 1993) and foraging behavior (Brown and Colgan 1985). Because larval survival may be closely linked to rapid growth, several recent investigations have focused on the use of RNA:DNA ratios (Bulow 1987) and larval otoliths to assess growth rates (e.g., Karakiri and von Westernhagen 1988; Zhang et al. 1991; see Chapter 16). Improvements in egg and larval rearing techniques (Hunter 1984) have increased the number of fish species that can be cultured and have provided material for a diversity of physiological experiments (see Hoar and Randall 1988) as well as taxonomic and systematic studies (Moser et al. 1984). In addition, because abundance and survival of fish eggs and larvae may be closely tied to environmental changes, studies of fish early life stages have been important in the assessment and reduction of anthropogenic effects (e.g., entrainment; Dempsey 1988) on aquatic systems.

Because eggs and larvae of marine and freshwater fishes differ in size, vertical and horizontal distribution, temporal availability, and susceptibility to various gears, effective collection techniques are critical to the design of a sampling program. In addition, a well-designed study requires proper handling, preservation, and identification of collected organisms (see Chapter 5). Because of characteristics unique to early life stages (e.g., small size and patchy distribution), techniques used for collection and identification of eggs and larvae are considerably different from those devised for adult fishes.

9.2 COLLECTION OF FISH EGGS AND LARVAE

The diversity of fish reproductive modes combined with species-specific differences in spawning habitats, larval growth, and behavior have resulted in the development of a diverse array of egg and larval collecting gears. Most gears designed for collecting pelagic eggs and larvae involve filtration of water through fine-mesh material, whereas harvest of demersal or attached eggs and larvae usually involves the use of artificial substrates and traps. Many modifications of traditional collecting gears can be found in the literature, and it is frequently necessary to alter the design of a particular gear to fit specific sampling conditions. Regardless of the sampling problem involved, choice of a particular gear must include consideration of the advantages and disadvantages of each gear type (Table 9.1). In addition to the sampling characteristics of each gear, characteristics of the organisms being sought and the habitat being sampled must also be considered. Marcy and Dahlberg (1980) and Bowles et al. (1978) summarized the major mechanical problems associated with sampling fish eggs and larvae, some of which can be minimized by choice of sampling gear.

9.2.1 Active Collecting—Low-Speed Gears

Use of plankton nets (Figure 9.1A) to collect ichthyoplankton can be traced to 1828 (Fraser 1968). Since that time, many modifications of gear design and sampling methodology have been developed to increase accuracy and precision of ichthyoplankton abundance estimates. Choice of sampling gear should be based on consideration of expense, ease of use, relative effectiveness, and sampling bias.

9.2.1.1 Plankton Nets

Conical plankton nets (Figure 9.1) with mouth diameters ranging from about 0.1 m to over 1 m in diameter have been used extensively to sample fish eggs and larvae. Larger nets (>0.5 m in diameter) are usually towed at speeds under 2 m/s for periods ranging from 30 s to an hour, depending primarily on ichthyoplankton density and the abundance of debris (clogging). Nets typically consist of a nylon mesh cone or a cylinder–cone combination attached at the proximal end to a steel or brass ring, which is in turn connected to the towing cable with a three-stranded bridle (Figure 9.1A). The distal end of the net usually ends in a collection bucket (Duncan 1978;

Table 9.1 Advantages and disadvantages of various types of collecting gears for fish eggs and larvae.

Gear type	Examples	Advantages	Disadvantages
Low-speed nets			
Vertical tows	Buoyant net	Reduced net avoidance Useful in shallow, vegetated habitats	Small volume filtered
Horizontal tows	Meter nets	Large water volume in short time period	Clogging of mesh reduces efficiency
	Bongo nets	Inexpensive	Efficiency variable in turbulence
	Benthic sleds	Can be towed or anchored	Net avoidance increases with size of larvae
	Tucker trawl	Only small vessels required	Without modifications, limited use in littoral areas
	Neuston nets	Small conical nets can be towed by hand	
	Henson nets		
High-speed nets	Miller high-speed sampler	Reduced net avoidance	Extrusion, damage to collected organisms
	Jet net	Large water volume can be sampled over extensive areas	Larger vessels required
	Gulf III		Winches required for deployment and retrieval
	LOCHNESS		
Plankton recorders	Hardy plankton recorder	High-speed sampling possible	Damage to sampled organisms
	Longhurst–Hardy plankton recorder	Discrete samples at depth possible	Extrusion of organisms through mesh
			Sampling bias due to variations in passage through net
Midwater nets	Isaacs–Kidd midwater trawl	Pelagic sampling	Large nets, increased personnel needs
	Tucker trawl	Large volume of water	Constant boat speed important
	MOCHNESS	Multiple samples	Increased net handling times
	RMT 1–8		
Pumps	Centrifugal pumps	High-volume samples from turbulent areas	Damage to organisms
	Trash pumps	Low personnel needs	Handling problems with large pumps
		Easy replication	Reduced efficiency for some species
			Avoidance by larger larvae

Graser 1978; Figure 9.1A) into which organisms are washed after net retrieval (although Miller 1973 reported that the use of a 333-μm-mesh cod end bag reduced damage to collected larvae). Paired nets mounted on a rigid frame attached to the towing cable are called bongo nets (Figure 9.1B). These nets have the advantage of not obstructing water flow with the towing bridle and also provide replicate samples to determine sample variability (Smith and Richardson 1977; Colton et al. 1980; Choat et al. 1993).

Flow meters should be mounted in the net mouth (Figure 9.2A) to determine sample volumes (Gehringer and Aron 1968) and should be positioned to measure

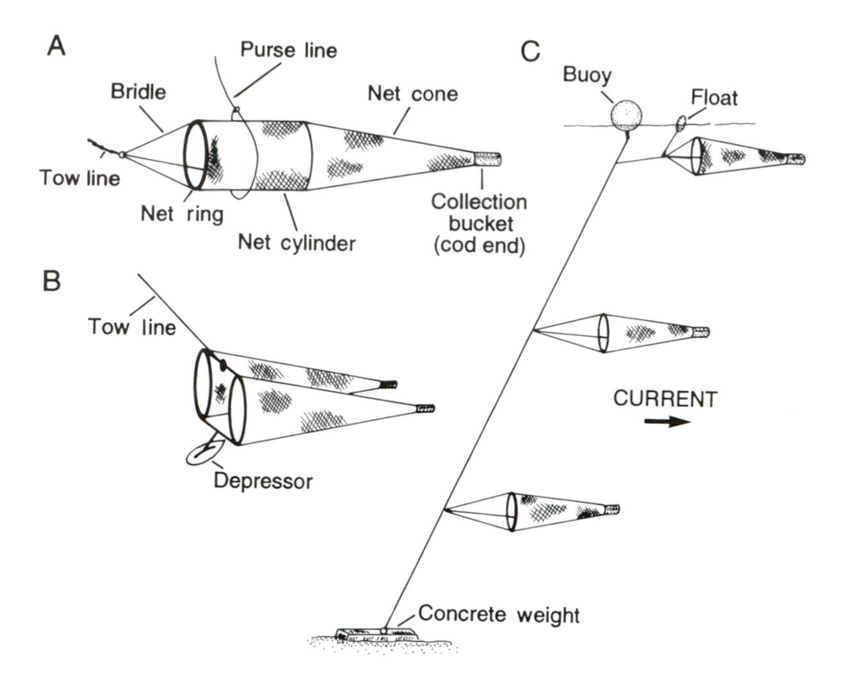

Figure 9.1 Several methods of plankton net sampling: **(A)** simple cylinder–cone plankton net with a purse line for sampling at discrete depths; **(B)** paired bongo nets fitted with a depressor to maintain the nets at a prescribed depth; **(C)** three plankton nets rigged vertically to sample drifting eggs and larvae in lotic habitats.

average velocity through the net. For unbridled nets, the center of the mouth provides a reasonable estimate of average velocity (Mahnken and Jossi 1967); for bridled nets, the flow meter should be located in the net mouth about 20% of the net diameter from the net edge (Figure 9.2A; Trantor and Smith 1968).

The ratio of the open area of a net (exclusive of net material) to its mouth area has been termed the open-area ratio. Studies indicate that filtering efficiency of a net with an open-area ratio exceeding 3 would approach 85%, whereas a ratio of 5 or more would result in up to 95% efficiency (Trantor and Smith 1968). Net efficiency can be increased with a mesh cylinder (40% of the total gauze area) ahead of the conical net (60% of the gauze area); the cylinder acts as additional filtering surface and is much less susceptible to clogging (Figure 9.1A; Smith et al. 1968; Schnack 1974). Additionally, use of a mouth-reducing cone (see Figure 9.4) increases filtering efficiency by creating a low-pressure area that draws a column of water larger than the mouth diameter into the net (Trantor and Smith 1968).

Numerous modifications of standard plankton nets have been reported for specific sampling conditions. Brown and Langford (1975) developed a frame-mounted plankton sled fitted with buoyant floats for sampling ichthyoplankton near the surface. Netsch et al. (1971) incorporated a depressor weight (Figure 9.1B) mounted just ahead of a meter net (a net with a 1-m-diameter mouth) for horizontal towing at depth, and Bath et al. (1979) used a similar design to collect simultaneous bottom, middepth, and surface samples with 0.5-m nets (dimension is mouth diameter). Faber (1968) eliminated the towing bridle and incorporated a purse line to close the net and allow sampling at discrete depths (Figure 9.1A). Nester (1987) developed a

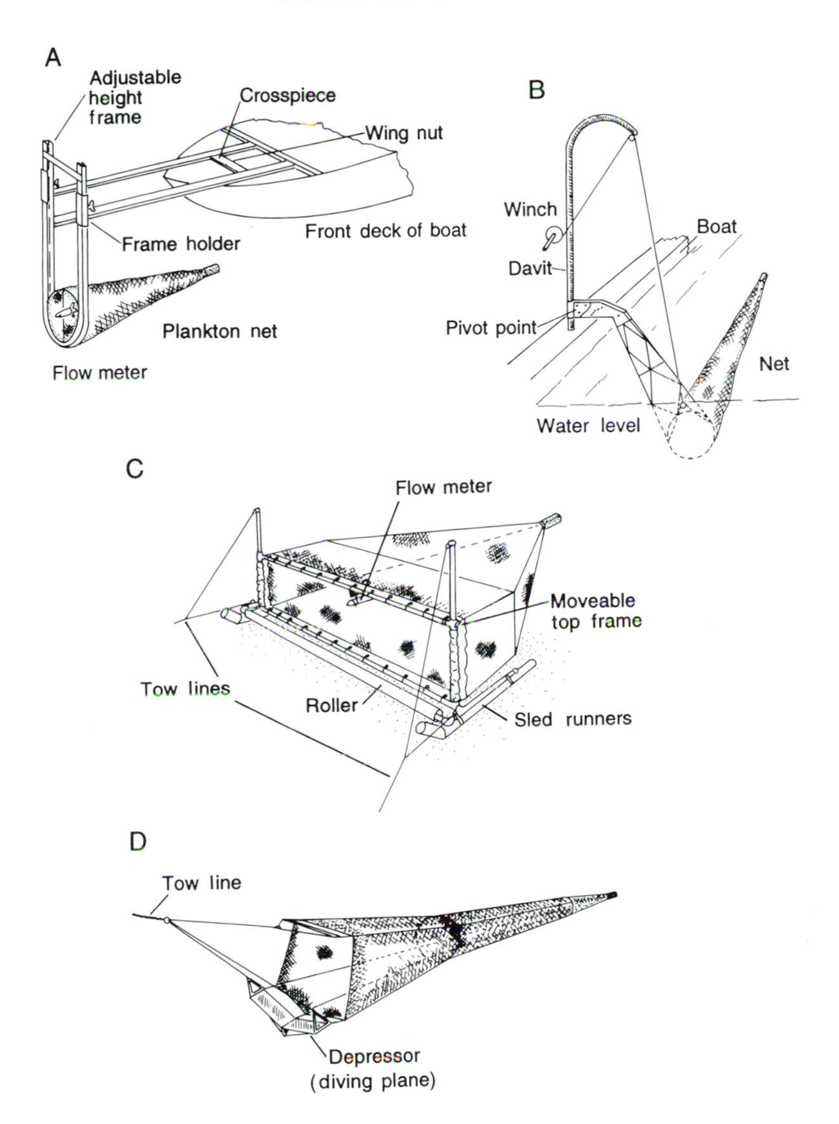

Figure 9.2 Designs for towed or pushed nets: **(A)** a vertically adjustable push net; **(B)** a side-mounted ichthyoplankton net; **(C)** a benthic sled for shallow-water sampling; and **(D)** an Isaacs–Kidd midwater trawl fitted with a depressor for improved performance at depth (adapted from Meador and Bulak 1987 and La Bolle et al. 1985, with permission).

0.5-m cylinder–cone net and depressor weight mounted on a fixed frame to study vertical distribution of Great Lakes ichthyoplankton. Contamination of samples during vertical retrieval was prevented by collapse of the net over the frame. Dovel (1964), Cooper (1977), Tarplee et al. (1979), and Hermes et al. (1984) described side-mounted frames for sampling surface ichthyoplankton with 1.0-m and 0.5-m nets (Figure 9.2B). Miller (1973) used paired nets suspended from the bow to sample ichthyoplankton in shallow bays in Hawaii.

9.2.1.2 Benthic Plankton Samplers

The need to sample eggs and larvae on or just above the bottom has led to the development of benthic plankton sleds (Figure 9.2C). Frolander and Pratt (1962) mounted a cylindrical net (Clarke and Bumpus 1950) on a benthic skimmer for sampling demersal organisms in a lake. Dovel (1964) used a much larger net mounted on an aluminum benthic sled for sampling estuarine ichthyoplankton associated with bottom currents of more saline, higher-density water; however, the net was positioned 0.28 m from the bottom of the sled. Yocum and Tesar (1980) mounted a 0.5-m plankton net in a rectangular sled frame and sampled within 5 cm of the bottom in littoral areas of Lake Michigan. This study and that of Madenjian and Jude (1985) indicated that, relative to a standard plankton net, the plankton sled provided better abundance estimates for fish eggs and demersal fish larvae. The rectangular sled developed by La Bolle et al. (1985) included an adjustable net that could effectively fish the entire water column in depths ranging from 0.15 to 0.70 m. Phillips and Mason (1986) developed a sled fitted with a self-adjusting grate to sample demersal adhesive and nonadhesive fish eggs on irregular coastal substrates.

9.2.1.3 Pelagic Trawls

Several low- to moderate-speed (0.5–3 m/s) midwater trawls have been developed to sample zooplankton and early life stages of pelagic fishes. The Isaacs–Kidd midwater trawl (Isaacs and Kidd 1953; Figure 9.2D) is of simple design and has been used extensively to sample large larvae and small juveniles in pelagic areas (Pearcy 1980). A steel-framed trawl (1.8 m × 1.8 m) developed by Tucker (1951; Figure 9.3A) was used by Haldorson et al. (1993) to monitor ichthyoplankton abundance in an Alaskan bay and was modified by Houser (1983) to include a diving plane for maintenance of position in the water column without the use of ballast. Siler (1986) reported Tucker trawl estimates of juvenile and adult threadfin shad abundance in a North Carolina lake were superior to those obtained with rotenone samples. However, Choat et al. (1993) found the Tucker trawl was ineffective for estimating the density or size composition of pelagic reef fish larvae, particularly small individuals. Clarke (1969) described a rectangular (2.8 m × 4 m) trawl that could be opened and closed acoustically, and Baker et al. (1973) used a similar design for the RMT 1–8 (rectangular midwater trawl) which incorporated two opening–closing nets for sampling at depths up to 2,000 m. The MOCHNESS trawl (multiple opening–closing net and environmental sensing system; Wiebe et al. 1976) was similar to that reported by Frost and McCrone (1974) and incorporated nine sequentially opening and closing nets (1 m × 1.4 m × 6 m long, 333-μm mesh) as well as sensors to monitor depth, temperature, specific conductance, flow, net angle, and net deployment. Sameoto et al. (1977) incorporated a depressor and rigid net frames in a similar 10-net (1-m^2 mouth area, 243-μm mesh) sampler (BIONESS; "—NESS" denotes a net with environmental sensing systems) that could be towed at speeds up to 3 m/s. The BIONESS net proved to be superior to a Tucker trawl for sampling small (<10 mm) Atlantic cod larvae, although the trawl was most effective for larger larvae and juveniles (Suthers and Frank 1989). Because of the small mouth size of the BIONESS nets and the dependence of effective mouth areas on towing speed in the RMT 1–8 and MOCHNESS nets (higher speeds caused the bottom of the net mouth to move back and up, reducing the effective mouth area), Dunn et al. (1993) developed the LOCHNESS sampler. This sampler was designed for larger organisms (2-mm mesh) and incorporated five 2.3-m square nets and environmental sensors for simultaneous collection of organisms and environmental data.

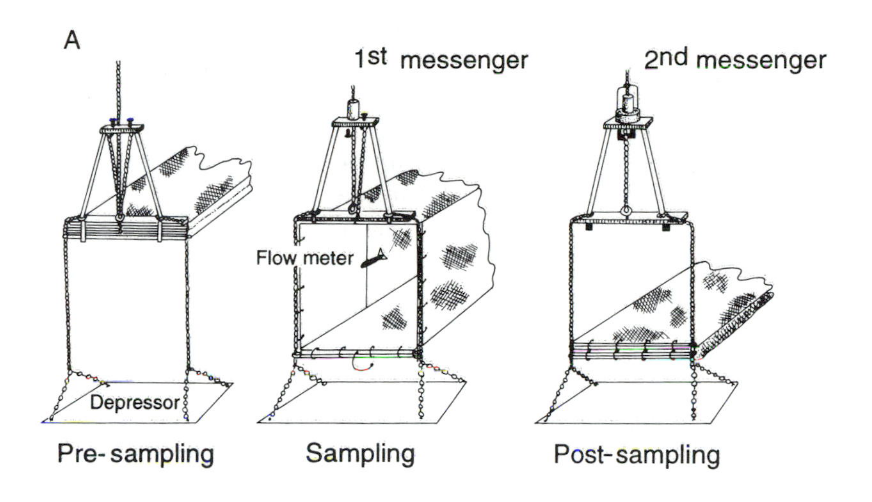

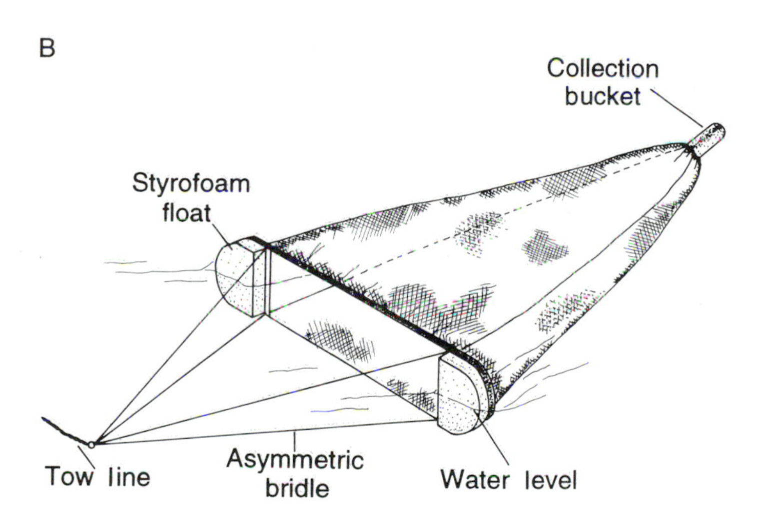

Figure 9.3 Two gears for sampling specific depth strata: **(A)** a modified Tucker trawl for sampling at depth, and **(B)** a neuston net for sampling surface eggs and larvae (adapted from Sameoto and Jaroszynski 1976 and Brown and Cheng 1981, with permission).

9.2.1.4 Neuston Nets

Several nets that are towed with the top edge of the net above the water surface have been developed to sample neustonic organisms (Figure 9.3B; Hempel and Weikert 1972; Lippincott and Thomas 1983). Eldridge et al. (1978) tested 4.9-m- and 8.5-m-long Boothbay neuston nets (pipe frame 2 m wide × 1 m high, 947-μm mesh) at speeds from 1 to 3 m/s and found that the 4.9-m net was easier to handle and caused less damage to collected specimens. Hettler (1979) modified the Boothbay net for stationary sampling in a tidal current and incorporated a wooden collection box for retrieval of live larvae. Brown and Cheng (1981) developed the Manta net, which was designed with fixed wings and asymmetrical towing cables to maintain the net at the surface and away from the boat. The authors reported that the net was superior to other neuston net designs for sampling in choppy waters (waves higher

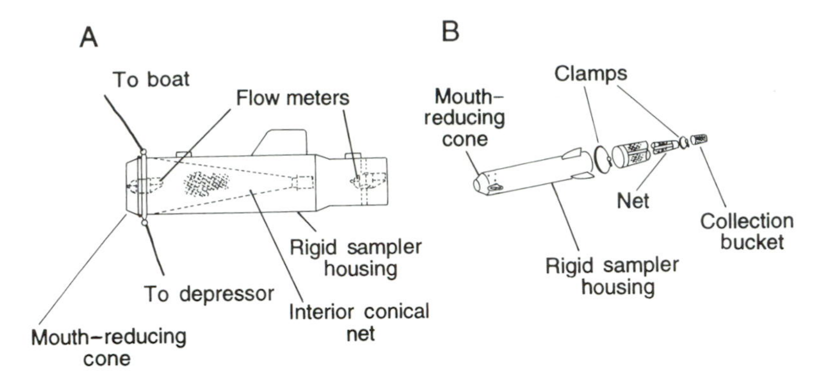

Figure 9.4 High-speed samplers: **(A)** a cutaway of the Gulf III sampler (note the fore and aft flow meters) and **(B)** exploded view of a Miller high-speed sampler (adapted from Gehringer 1952 and Miller 1961, with permission).

than 10 cm). Shenker (1988) also found the Manta net to be an effective sampler for larval fishes and crabs off the Oregon coast, although a larger neuston trawl (3.5 m wide × 1 m deep) was more effective for larger juveniles.

9.2.2 Active Collecting—High-Speed Gears

Conical plankton nets mounted inside hollow cylinders fitted with mouth-reducing nose cones (Figure 9.4) have been used extensively as high-speed samplers in studies of both marine and freshwater ichthyoplankton (Gehringer and Aron 1968). These samplers are typically large (>50 kg; the Miller high-speed net is an exception) and require power winches for deployment and retrieval. Although high-speed samplers reduce net avoidance by mobile larvae (larval avoidance can still occur; Bjørke et al. 1974) and can sample large volumes of water over extended distances in short periods of time, the net-within-cylinder design appears to be more subject to clogging (but see the unencased "Nackthai" high-speed net in Schnack 1974). The Gulf 1-A high-speed sampler (12-cm-diameter tube, 4-cm opening) was described by Arnold (1952) and was modified by Smith et al. (1964) to sample at speeds up to 9 m/s. The Gulf III net (Gehringer 1952) incorporated a 0.5-m net in a rigid housing to sample a greater volume of water than the Gulf 1-A (Figure 9.4A). Bridger (1958) reported that a reduction in nose cone diameter from 40 cm to 20 cm substantially improved net efficiency and resulted in increased diurnal catches of larval Atlantic herring. Beverton and Tungate (1967) modified the Gulf III net to sample larval pleuronectids, phytoplankton, and zooplankton simultaneously.

In freshwater systems, Miller high-speed nets (Miller 1961) have been widely used to sample fish eggs and larvae (Figure 9.4B). These samplers are lightweight and can be operated by a single person from a small boat. Noble (1970) evaluated Miller high-speed samplers attached to side-mounted 3-m poles and found that sampling performance was improved by increasing speed, sampling nocturnally, incorporating an electroshocking grid in front of the samplers, and using clear rather than opaque materials to construct the sampler. Coles et al. (1977) incorporated a small pump to continuously empty the contents of a high-speed Miller-type net that was used to study spatial heterogeneity of Eurasian perch larvae. Such a design seems particularly well suited for studying vertical and horizontal patchiness in egg and larval distributions. An alternative design for a high-speed sampler was the jet net, which was developed to reduce damage to collected organisms by slowing the velocity of water as it moved through the sampler into the collecting net (Clarke 1964).

Opening–closing high-speed samplers have also been developed for sampling at discrete depths (Bé 1962). The Clarke–Bumpus sampler (Clarke and Bumpus 1950; Trantor and Heron 1965) has been used extensively in zooplankton studies. It uses a messenger-operated closing gate (Paquette and Frolander 1957) to eliminate sample contamination. Kinzer (1966) modified a Gulf III net with a messenger-activated, spring-loaded closing mechanism, and Bary and Frazer (1970) incorporated a similar electrically activated closing mechanism and an improved flow meter design on the Catcher II, a modification of the Catcher sampler developed by Bary et al. (1958).

An alternative method for obtaining discrete plankton samples was developed by Longhurst et al. (1966) from a Hardy plankton recorder (Hardy 1936). The Longhurst–Hardy plankton recorder incorporated, at the end of a plankton net, a unique collection box that continuously filtered collected organisms through a gauze strip, which was overlaid by a second strip, both of which were wound up in the box. The plankton recorder has been used extensively and has been reported to be more effective than meter nets (Colton et al. 1961), MOCHNESS, and pump samplers (Brander and Thompson 1989) for assessing the distribution of larval Atlantic herring. However, the recorder may not be effective at low concentrations of larvae ($<0.1/m^3$; Colton et al. 1961), and there may be problems with extended residence of organisms in the netting before capture in the recorder (Haury et al. 1976).

9.2.3 Other Active Gears

9.2.3.1 Shallow-Water Nets

Gear modifications have also been developed for collecting larvae in shallow areas not easily sampled with towed nets. For qualitative surveys, fine-mesh (505 μm or less) dip nets can be used to obtain presence–absence data and to collect eggs and larvae from structurally complex areas. Seines can also be used in areas with smooth bottoms but are of limited use in vegetated habitats (Dewey et al. 1989). Although these gears are easy to use, removal of larvae from seines and dip nets may be time consuming and may result in considerable damage to collected specimens. In addition, because of difficulties in quantifying seine haul and dip-net effort (depth, speed, habitat differences, or amount of water filtered), data obtained with these gears probably should not be analyzed quantitatively without careful assessment and standardization of techniques.

Other shallow-water gears have been designed that incorporate nets in fixed or adjustable boat-mounted frames. Hodson et al. (1981) used side-mounted meter nets to obtain replicate samples of ichthyoplankton from surface waters. Bryan et al. (1989) mounted paired 0.5-m nets on vertically adjustable side frames braced with support wires; the design permitted discrete sampling at two depths up to 4 m at speeds up to 1.3 m/s. Holland and Libey (1981) and Meador and Bulak (1987) used 0.5-m nets in adjustable bow-mounted frames to sample shallow littoral areas (Figure 9.2A). Burch (1983) mounted two 0.5-m nets on a bicycle-type push net to be manually operated in depths up to 1.5 m. Ennis (1972) used a diver-operated device consisting of a 0.5-m net attached to two underwater towing vehicles to sample larvae in shallow coastal areas.

An alternative design developed by Bagenal (1974) for sampling shallow-water areas incorporated a buoyant net ring to allow vertical sampling of larvae in shallow, vegetated areas. These nets, sometimes called pop nets, are deployed with an anchor or weighted frame to take the net to the bottom. After a period of time (10–30 min), a release mechanism is triggered, and the cylindrical net rises to the surface. Although the volume of water sampled is small, Bagenal (1974) indicated no avoidance of the net by cyprinid larvae, and Dewey et al. (1989) and Dewey (1992) found that pop nets provided quantitative estimates of juvenile fish abundance in vegetated habitats of a Minnesota lake. Portable drop nets have also been used to collect larval and postlarval fishes (Kushlan 1981). One design incorporates a square or rectangular frame with a surrounding net suspended along the top. In contrast to the buoyant net, the frame is set in place for a period of time, and then the net is released to fall quickly to the bottom (Dewey 1992). Alternatively, lightweight frames with mesh on all four sides can be thrown to the sample location and collected fishes can be removed by dip nets (Kushlan 1981). La Bolle et al. (1985) used a rectangular drop sampler constructed of clear plexiglass (to reduce visual avoidance) to sample larval fishes in littoral areas of the Columbia River. Similarly, Baltz et al. (1993) used a circular (1.2-m diameter) plexiglass drop sampler deployed from a boat-mounted boom to study microhabitat use by marsh fishes. After deployment, organisms were retrieved from the sampler by means of rotenone and dip nets.

9.2.3.2 Pumps

Centrifugal pumps have been used since about 1887 to collect demersal eggs and larvae and also to study the spatial distribution of pelagic ichthyoplankton (Aron 1958). Most systems involve pumping a target volume of water from an intake hose into a net (Figure 9.5) or a filtering drum (to reduce damage to collected larvae). Such a system has several advantages: depth of sampling and volume of water through the system (duration of pumping) can be easily controlled, discrete quantitative samples can be obtained by intermittent collection of organisms from the filtering surface, and the system can be operated from a stationary or moving platform. Conversely, pumping volumes can be small, pump intakes and filtering screens can be subject to clogging, the effective pumping area of most systems is limited to several centimeters from the pump intake—avoidance by mobile larvae can be significant (e.g., threadfin and gizzard shad; Petering and Van Den Avyle 1988), and most larvae are killed or damaged during sampling (Gale and Mohr 1978).

Aron (1958) reported that pump collections of pelagic fish eggs in Puget Sound were quite similar to collections taken concurrently with a 0.5-m net, although abundance estimates for several copepod taxa differed between the two gears. Harris et al. (1986) found a high-volume pump was particularly useful for sampling larvae and associated food organisms from discrete depths. A pump system was used successfully by Manz (1964) to collect viable walleye eggs from Lake Erie, although pump performance was reduced over mud, silt, and sand substrates because of clogging. Leithiser et al. (1979) found the abundance of fish larvae (particularly those longer than 5 mm) taken from a coastal power plant intake was significantly higher in pump samples than in concurrent samples taken with 0.5-m and 1.0-m plankton nets. Stauffer (1981) tested three pump designs for collecting lake trout eggs and early life stages and reported adequate collections only with a system incorporating a diver-directed intake. Novak and Sheets (1969) also used scuba divers to direct the pump intake for collecting smallmouth bass larvae.

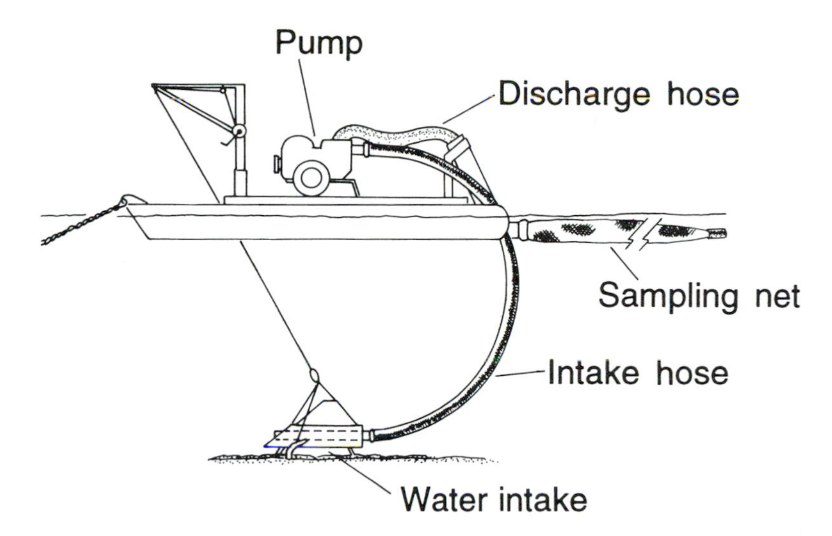

Figure 9.5 A towable pump sampler with adjustable intake for sampling at discrete depths (adapted from Gale and Mohr 1978, with permission).

Other types of pumping devices have also been used to collect eggs and larvae. Dorr et al. (1981) used diver-operated underwater diaphragm pumps (Flath and Dorr 1984) to sample alewife and lake trout eggs as well as age-0 sculpins. Vogele et al. (1971) used compressed air from a scuba cylinder in a portable, diver-operated suction device for collection of centrarchid (sunfish) eggs and larvae in Arkansas reservoirs.

9.2.3.3 Electrofishing Gear

Electrofishing equipment has not been widely employed to sample fish larvae. However, battery- or generator-powered electrofishing gear is particularly well suited for sampling fishes in shallow, structurally complex areas by wading or from small boats and may be much less species selective than are nets or pumps. Electrofishing was successfully used by Braem and Ebel (1961) to sample sea lamprey ammocoetes in Great Lakes tributaries. The electrofishing unit was battery powered and 20-cm-square, wire-mesh dip nets mounted on 1.2-m handles were used as electrodes; the authors reported that interruption of the current was most effective in attracting larvae from their burrows. For sampling lamprey larvae in deeper waters, McLain and Dahl (1968) developed an electrified plankton sled (pulsed DC) that successfully collected larval (>30 mm) sea, American brook, northern brook, and silver lampreys. Copp and Peňáz (1988) used electrofishing for sampling larval and juvenile fishes in floodplain habitats of the upper Rhône River. The electrofishing unit was modified to include a 10-cm diameter anode to create a steep voltage gradient; at 200 V and 400 Hz, the battery-charged unit created a voltage gradient ranging from 3.6 to 0.13 V/cm (minimum effective gradient for galvanotaxis of 20-cm trout; Cuinat 1967) at 10 and 30 cm from the anode, respectively. The unit collected 1,048 larvae representing 12 species, and larval lengths ranged from 5 to 22 mm. Modified electrofishing gear probably deserves increased use for collection of larval and juvenile fishes. However, further studies are needed on the effects of fish size, water chemistry, electrode design, voltage gradient, current level, and pulse width and shape on sampling effectiveness.

9.2.3.4 Other Active Sampling Methods

Many situations unique to various fish species require other methods to sample eggs and larvae. Many species spawn on vegetation (e.g., northern pike), and collection of epiphytic eggs or larvae may require clipping and examination of submerged macrophytes. Collection of rocks or debris from the bottom may be best for species that shed adhesive demersal eggs over such substrates (e.g., darters and sculpins). Demersal eggs and larvae in or on the substrate can be collected with dredges or corers (although damage to larvae from these gears can be substantial), and eggs and early larvae of benthic-nesting species can sometimes be retrieved with small suction devices such as pipettes or slurp guns. Snorkeling, scuba diving, and underwater video photography (Chapter 18) can be used to gather important data on spawning locations, egg deposition, and larval behavior (Aggus et al. 1980).

9.2.4 Passive Collecting Gears

9.2.4.1 Egg Traps

Although pumps (Nigro and Ney 1982), net tows (Haug et al. 1984), drift nets (Johnston and Cheverie 1988; Pitlo 1989), and scuba surveys (Newsome and Aalto 1987) have been used in studies of fish egg distribution and abundance, egg traps have also been used extensively to capture and protect demersal eggs as they are spawned in the water column. Gammon (1965) used a simple wooden frame fitted with a fiberglass screen bottom and a 6.4-mm screen top (protecting eggs from minnow predation) to collect esocid eggs. Eggs of cavity spawners such as the channel catfish can be collected easily with spawning containers (Moy and Stickney 1987); other types of samplers have been designed for various other substrate-spawning fishes. Downhower and Brown (1977) used slate tiles to collect egg masses of the mottled sculpin in an Ohio creek, and Fridirici and Beck (1986) devised a series of stacked plastic plates for laboratory collections of eggs of crevice-spawning spotfin shiners. Stauffer (1981) used bucket samplers filled with substrate material to collect lake trout eggs in Lakes Michigan and Superior. Buckets were placed in holes in the substrate that had been excavated by scuba divers and were retrieved after lake trout had spawned. Additional gears used in studies of lake trout egg deposition have included nets (Horns et al. 1989) and egg traps (Marsden et al. 1991) strung together on collection lines. Egg traps proved to be a more effective gear than nets in terms of both the number of eggs retrieved and the percentage of undamaged eggs in the samplers (Marsden et al. 1991).

9.2.4.2 Drift Samplers

Drifting eggs and larvae are usually collected with stationary sets of standard plankton nets, although nets with mouth-reducing cones have also been used (Franzin and Harbicht 1992). Mesh size depends on the size of the target organisms and mesh clogging tendencies but typically ranges from 116 μm (Lindsay and Radle 1978) to over 1 mm (Graham and Venno 1968). Horizontal (Carter et al. 1986; Winnell and Jude 1991) and vertical location of drift nets in the water column depends on the drift characteristics of the species under study. Drift nets have been set to sample ichthyoplankton throughout the water column (Johnston and Cheverie 1988), as well as at the surface (Gale and Mohr 1978; Lindsay and Radle 1978; Carter et al. 1986), middepth (Franzin and Harbicht 1992), surface and bottom (Clifford 1972), and at several depths simultaneously (Dovel 1964). Graham and Venno (1968) attached drift nets to swivel-mounted vanes for sampling larval Atlantic herring in tidal areas of the Gulf of Maine. The weight of the cod end of the net collapsed the net at slack tides to prevent escape of larvae, and the vane ensured that the net was aligned with the current when tides were flowing (Figure 9.6A). Winnell and Jude (1991) obtained replicate samples of drifting fishes and macroinvertebrates from paired nets attached at several depths to fixed poles. In addition to summer sampling, the authors were able to deploy the gear through the ice during winter. Lewis et al. (1970) developed a versatile net to capture Atlantic menhaden in estuarine channels. The net resembled a bag seine (3-mm mesh in the wings, 500-μm mesh in the bag) and was held stationary by a 1-m × 3-m frame that could be moved up and down in the water column on fixed poles (Figure 9.6B). Pitlo (1989) used rigid drift nets (15-cm × 46-cm frame; 61-cm bag) constructed of window screening (6.1 meshes/cm) to collect walleye and sauger eggs and define habitat boundaries in the Mississippi River.

9.2.4.3 Emergence Traps

Fishes such as salmonids (trouts) that deposit their eggs in gravel nests below the surface of the substrate provide the opportunity to capture offspring from individual spawnings as larvae emerge. Several investigators have developed traps to sample larvae as they leave the nest. Phillips and Koski (1969) used a covering net with an attached collecting bag to sample emerging larvae from individual coho salmon redds and reported near 100% trap efficiency. Porter (1973) designed an oval-shaped mesh and canvas trap with a downstream collecting box; the box was designed to reduce water velocity and resulted in 100% survival of rainbow trout larvae (Figure 9.6C). Gustafson-Marjanen and Dowse (1983) used the Porter trap to study emergence of Atlantic salmon, and Field-Dodgson (1983) developed a larger Porter trap and successfully captured emerging chinook salmon larvae. A slightly different trap was used by Bardonnet and Gaudin (1990) to study emergence of grayling larvae from artificially spawned eggs placed in the trap. The trap incorporated a downstream compartment that was filled with gravel and connected by pipe to an upstream above-gravel compartment; the trap was oriented at 45° into the current. Collins (1975) and Stauffer (1981) used pyramidal emergence traps (Figure 9.6D) that relied on vertical migration by larvae into the trap to study emergence of lacustrine salmonids. Although the trap was designed to sample redds, it also collected larvae of broadcast-spawning lake whitefish after emergence from demersal eggs.

Trap design (e.g., trap size) for emergence studies depends on the species being studied, as well as on the characteristics of the water body (e.g., water velocity and substrate composition). Temporal emergence patterns of the target fishes may also be important in the design of trapping studies. Most salmonid larvae emerge at night (but see Bardonnet and Gaudin 1990 for grayling), and the bulk of emergence occurs over restricted (approximately 10-d) periods (Gustafson-Marjanen and Dowse 1983; Brännäs 1987). More importantly, de Leaniz et al. (1993) and others have reported within-gravel movements of salmonid larvae, which could substantially bias results of trapping studies unless traps are large relative to the magnitude of lateral movements of larvae from the nest or trap aprons are buried deep enough in the substrate to minimize within-gravel dispersal.

9.2.4.4 Activity Traps

Several investigators have developed traps for free-swimming larvae and juveniles in littoral habitats. A simple trap consisting of two mesh cones mounted inside a

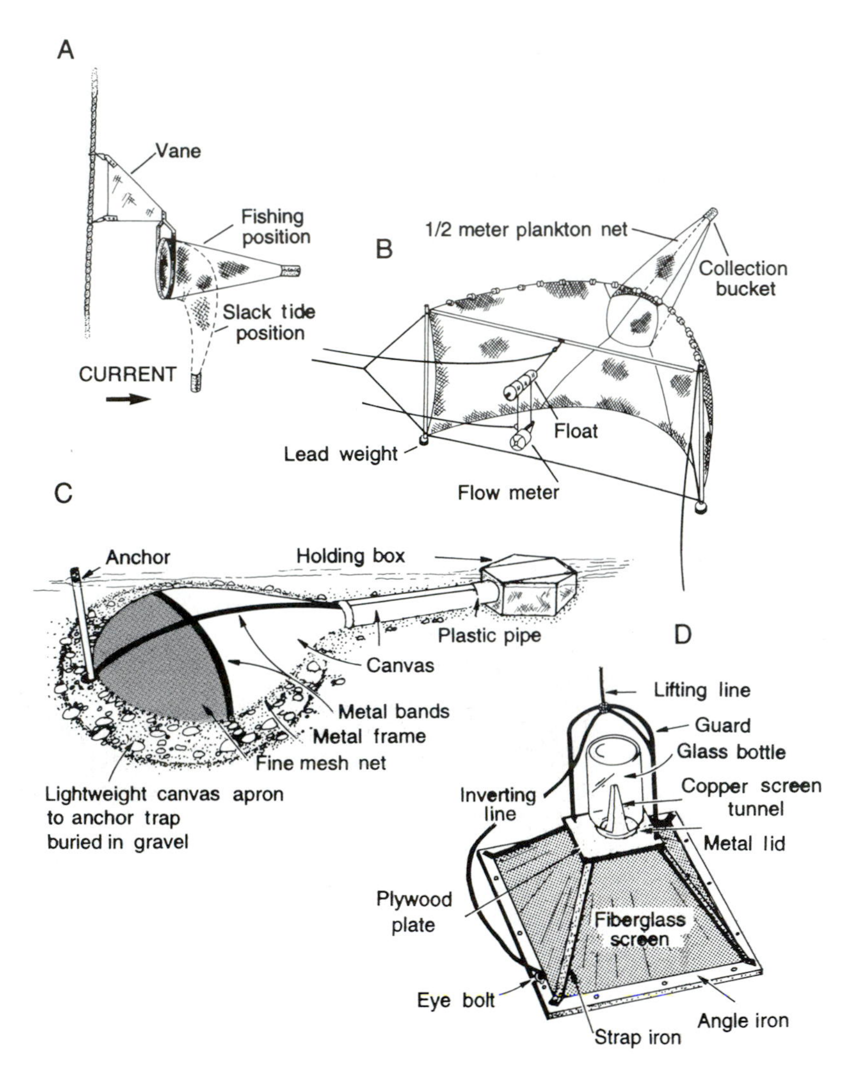

Figure 9.6 Various types of passive collecting gears: **(A)** a plankton net attached to a vane for sampling in tidal currents; **(B)** a channel net developed to collect larvae in flowing currents; and **(C)** and **(D)** emergence traps for demersal larvae (adapted from Graham and Venno 1968; Lewis et al. 1970; Porter 1973; and Collins 1975, with permission).

mesh cylinder was used by Baugh and Pedretti (1986) to catch 8- to 60-mm fishes in a shallow desert spring. Breder (1960) constructed a box trap of clear plexiglass that had removable wings which directed fish to a slot in the interior of the trap (Figure 9.7A). This design served as the basis for traps developed by Casselman and Harvey (1973) and a collapsible plexiglass and net trap designed by Trippel and Crossman (1984). These traps are low cost, easy to build, and highly adaptable to various sampling conditions. In particular, the size of the entrance slots into the trap can be adjusted to capture small larvae or larger juveniles, and the traps can be fished at various depths and positions in the water column, depending on the behavior of the target fishes. Small versions of fyke nets (Beard and Priegel 1975) and trap nets (Beamish 1973) have also been used to sample larval and juvenile fishes in lentic

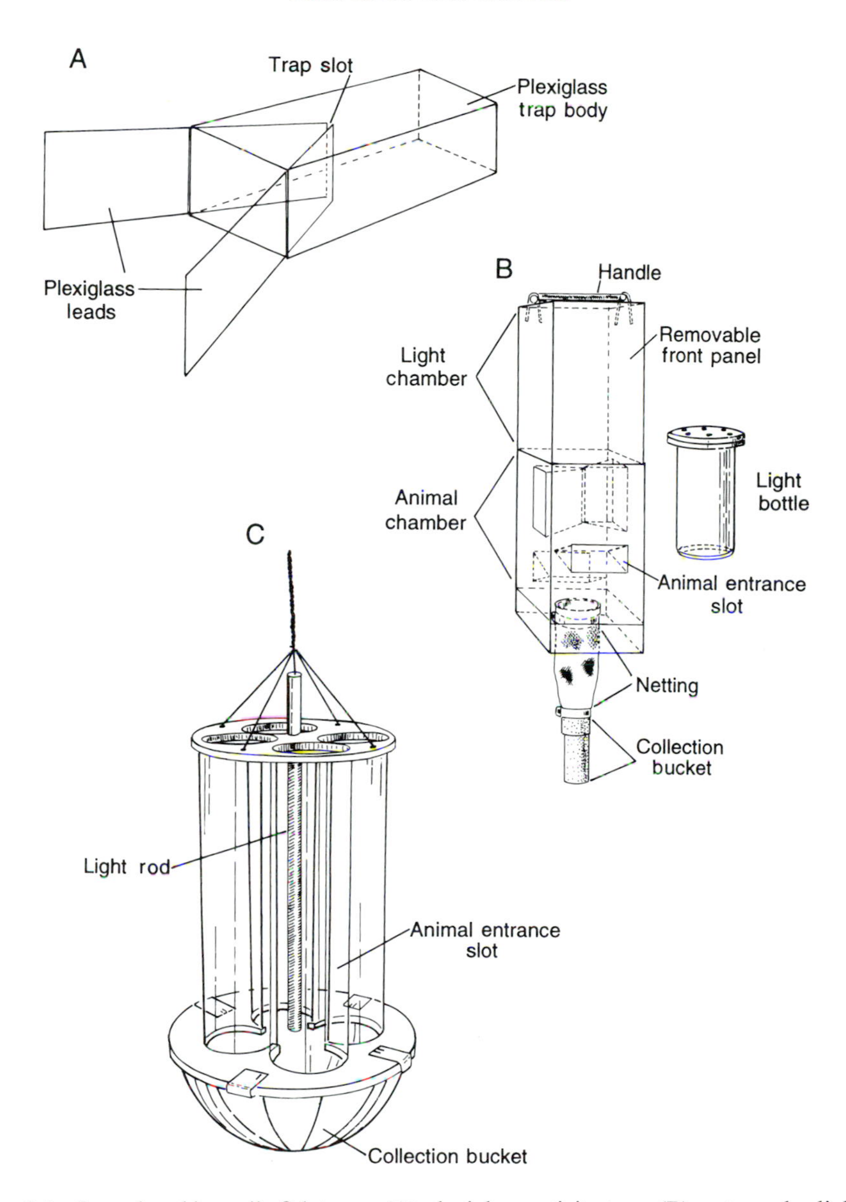

Figure 9.7 Larval and juvenile fish traps: **(A)** plexiglass activity trap; **(B)** rectangular light trap with horizontal and vertical entrance slots; and **(C)** the Quatrefoil trap (adapted from Breder 1960; Faber 1982; and Floyd et al. 1984a, with permission).

systems. Plexiglass or net traps are easy to set and are particularly appropriate for vegetated habitats, as long as the taxa and life stages under study are mobile and tend to move laterally along a visible or invisible barrier. However, because of interspecific differences in larval susceptibility to traps (due to differences in mobility, behavior, and microhabitat preferences), comparison of catch-per-unit-effort data among species may not accurately reflect relative species composition.

9.2.4.5 Light Traps

Larvae and juveniles of many fishes are positively phototactic, and the use of artificial light sources in nocturnally fished traps has been an effective method for collecting larvae of some species. Dennis et al. (1991) used a light mounted above a lift net to assess abundance of 12 ichthyoplankton taxa in reef, seagrass, and mangrove habitats in Puerto Rico. Paulson and Espinosa (1975) used cylindrical, wire mesh (6.4-mm mesh) light traps to collect juvenile (approximately 40-mm) threadfin shad from midwater depths in Lake Mead, Nevada; Kindschi et al. (1979) used a similar design to assess spatial and temporal trends in ichthyoplankton abundance in Rough River Lake, Kentucky. Faber (1981) designed a box-shaped plexiglass light trap (Figure 9.7B) that was used by Faber (1982) and Gregory and Powles (1988) to determine seasonal abundance patterns of larval fishes in vegetated habitats of Lac Heney, Québec, and Chemung Lake, Ontario, respectively. In both studies, collected taxa represented about 50% of the species in the lakes; collections indicated differences in trap susceptibility among species and larval developmental stages. Muth and Haynes (1984) developed a smaller, floating plexiglass trap that incorporated plexiglass leads to guide larvae to the trap entrance slots. Field trials indicated the trap effectively sampled larvae and juveniles from 11 to 60 mm and captured three taxa not found in concurrent seine samples.

Floyd et al. (1984a) developed a trap with a central light-distributing rod surrounded by four plexiglass cylinders milled to three-fourths of a full circle (the Quatrefoil trap, Figure 9.7C). Advantages of this trap design include large trapping slots relative to the size of the trap as well as easy adjustment of trap size (length of the plexiglass cylinders) and sampling depth. Floyd et al. (1984b) used Quatrefoil traps, seines, and drift nets to collect larvae and juveniles in a small Kentucky stream. Of the 28 taxa collected during the study, the Quatrefoil trap collected 25, compared with 21 in seine hauls and 11 in drift nets; the light trap was particularly effective for cyprinid larvae. The Quatrefoil trap was modified by Secor et al. (1992) to include a chemical light source, a flotation device, and a collection bucket and proved to be an effective trap for pond-reared larval and juvenile striped bass (7–35 mm).

Because of differences in movement patterns, microhabitat preferences, and phototactic behavior among species, light traps are probably best suited for determining species presence or absence, as opposed to providing estimates of species relative species composition. Light traps are also useful for investigating intraspecific patterns of temporal or spatial abundance through time (Doherty 1987); however, changes in phototactic behavior with increasing larval size (e.g., Bulkowski and Meade 1983) must be considered in interpretation of temporal data. Light traps may be particularly effective for early larval stages (but see Doherty 1987 and Choat et al. 1993) and typically provide larvae in excellent condition if traps are checked at frequent (e.g., 1-h) intervals (Faber 1981). Gregory and Powles (1988) found light traps to be much more effective than Miller high-speed nets for 2.5–7.5-mm Iowa darter larvae, whereas length distributions of yellow perch and pumpkinseed larvae captured by the two gears were similar.

9.3 SAMPLING CONSIDERATIONS

Formulation of specific research objectives is the first step in selection of egg and larval sampling methods. Budget, personnel, equipment, and time limitations will affect the study design (see Chapter 1), as will numerous physicochemical, ecological, biological, and statistical factors. Larval fish are morphologically and behaviorally distinct from juveniles and adults (Snyder 1990), and knowledge of fish reproductive life history and larval behavior and ecology are important factors in the choice of collecting methods, gear types, sampling periodicity, and sampling habitat. There is a large volume of information on fish reproduction (Balon 1975). Data on sampling-related topics such as reproductive habitat preferences, seasonality, and diel periodicity can be found in Wootton (1990); temperature preferences and substrate and flow requirements are discussed in Breder and Rosen (1966), Carlander (1969, 1977), and Potts and Wootton (1984). Additional information is available in numerous regional summaries (Bigelow and Schroeder 1953; Scott and Crossman 1973; Lee et al. 1980; Becker 1983; Jenkins and Burkhead 1993), and larval fish bibliographies (Smith and Richardson 1979; Lathrop and Snyder 1986; Simon 1986; Hoyt 1988 [available as an ASCII file from American Fisheries Society Computer User Section]; Richards 1990).

9.3.1 Spatial and Temporal Effects on Sampling Design

Distributions of fish eggs and larvae vary both temporally and spatially, and this variability must be accounted for in the study design. Duration of the spawning season for various fishes ranges from a few days to several months (Potts and Wootton 1984), and although spawning times of some species are remarkably constant (Cushing 1970), spawning activity may vary temporally both within and between years based on seasonal and annual variability in rainfall, temperature, or other physicochemical variables (Bye 1984). Several studies have focused on temporal succession of larval fishes over various time periods (Amundrud et al. 1974; Gallagher and Conner 1980; Floyd et al. 1984b), and although initiation, cessation, and frequency of egg and larval sampling depend on study objectives, sampling typically commences just prior to spawning of the target species and continues at hourly to biweekly intervals until catches cease or decline to low levels.

9.3.1.1 Marine Systems

Numerous reports attest to the horizontal and vertical patchiness of egg and larval distributions due to passive and active aggregations in marine systems (Haug et al. 1984), which can have substantial effects on abundance estimates (Wiebe and Holland 1968). Vertical patterns of distribution depend on egg and larval buoyancy and larval behavior and can be affected by factors such as temperature and current patterns (Bailey 1980), salinity, light (Cada et al. 1980), and the distribution and movement of predators and prey. Eggs of Atlantic halibut are most abundant at intermediate depths, where temperature and salinity render the eggs neutrally buoyant (Haug et al. 1984). In contrast, in Mobile Bay, Alabama, fish eggs were most abundant in bottom strata during periods of high river discharge and were uniformly distributed in the water column during periods of low river discharge (Marley 1983). Demersal eggs typically show disjunct distributions related to depth and substrate (Newsome and Aalto 1987; Marsden et al. 1991), and spawning habitat specificity may play a large role in determining effective egg sampling methods.

The importance of tidal transport of marine fish eggs and larvae to estuarine nursery grounds has led to several studies on spatial and temporal variability in the distribution of estuarine ichthyoplankton (Miller et al. 1984; Pietrafesa and Janowitz 1988). In Gulf of Maine estuaries, Graham and Venno (1968) found differences in the vertical distribution of Atlantic herring larvae between flood (greatest numbers at 10–15 m) and ebb (greatest numbers at the surface) tides. Similarly, Atlantic croaker larvae were found to occupy lower, inward-flowing stratified waters of the Chesapeake Bay (Norcross 1991). Rijnsdorp et al. (1985) found demersal plaice larvae moved into the water column during nocturnal flood tides. Diel variability in catch was evident for 41 of 47 taxa of marine ichthyoneuston (0–50 cm in depth) studied by Eldridge et al. (1978); 29 were more abundant nocturnally, whereas 12 were more abundant diurnally. Larval clupeids (herrings; Harris et al. 1986), gadids (cods; Lough and Potter 1993), and cupelin (Fortier and Leggett 1983) were patchily distributed vertically, and abundance patterns were related to diel period, thermocline position, and larval length. Larval fishes collected in Narragansett Bay were found to exhibit an overall downward migration at night from surface waters, whereas eggs were found predominantly (72%) at 0–3-m depths (Bourne and Govoni 1988). The importance of identifying larval microhabitat preferences was demonstrated by Lindsay and Radle (1978) in their study of inland, Atlantic, and rough silverside larvae. Silversides were abundant in only the 0–5-cm surface layer in mesohaline areas of the Delaware River and were virtually absent from deeper strata.

9.3.1.2 Freshwater Rivers and Streams

In freshwater lotic systems, spatial and temporal variability in ichthyoplankton distributions have been reported for many taxa. Sager (1987) reported increased density of gizzard shad larvae in backwater areas relative to main-stem habitats of the Cape Fear River, North Carolina. In the Susquehanna River, Gale and Mohr (1978) found limited numbers of larval fishes near the bottom during the day, peak drift of cyprinid and catostomid larvae between 2400 and 0300 h, and a night:day drift abundance ratio of 3.8:1. Carter et al. (1986) found similar increases in nocturnal abundance of drifting larval fishes in the upper Colorado River but reported significantly higher abundance of larvae along the shoreline compared with the midchannel surface zone. Gallagher and Conner (1980) also found the highest larval densities at turbulent, nearshore stations in the lower Mississippi River, whereas larval fishes in the St. Mary's River, Michigan–Ontario, were more abundant in the upper half of the water column in midchannel (Winnell and Jude 1991). Larval distribution may also be related to temporal changes in stream physicochemistry. The diel pattern of larval drift of anadromous blueback herring in an Atlantic coastal stream was temperature dependent; higher abundances of larvae occurred nocturnally at temperatures less that 13°C, and abundance of larvae increased diurnally at temperatures over 15°C (Johnston and Cheverie 1988).

9.3.1.3 Freshwater Lakes

Similar spatial and temporal heterogeneity in larval fish distributions has been reported in freshwater lentic systems. Conrow et al. (1990) found substantial differences in distributions of larval fishes in areas of open water, panic grass *Panicum* spp., *Hydrilla* spp., and floating and emergent vegetation in Orange Lake, Florida. Larval clupeids typically exhibit diel vertical movements, with highest densities at dusk in open water near the surface, whereas freshwater drum larvae have been reported to be most abundant at depths of 3–6 m during the day and to move to deeper waters at night (Tuberville 1979). Ontogenetic changes in larval distribution (e.g., from limnetic to littoral habitats [Werner 1969] or from high-density to low-density macrophyte beds [Gregory and Powles 1985]) must also be considered in the sampling design. Golden shiner larvae (5–10 mm) were reported by Faber (1980) to be scattered in shallow (10 cm) shoreline areas, whereas larger individuals (10–30 mm) schooled among floating and emergent macrophytes at depths of 0.25–1 m. These ontogenetic changes in habitat preferences, combined with increasing size and decreasing susceptibility to ichthyoplankton samplers, often result in biased abundance estimates for late larvae and early juveniles.

9.3.2 Fish Density and Sample Volume Effects on Sampling Design

Studies cited above attest to the importance of considering vertical, horizontal, and temporal discontinuities in ichthyoplankton distributions when a sampling program is designed. Sampling a large volume of water increases the chances of encountering patches of eggs and larvae; 100 m^3 is generally accepted as a target volume in fresh water, whereas sample volumes of 250 to 1,500 m^3 have been filtered in studies of marine eggs and larvae (Marcy and Dahlberg 1980). If objectives include a description of ichthyoplankton vertical and horizontal patchiness (Leslie 1986), the study design should include filtration of target volumes at discrete depths with opening–closing gears or pumps (Harris et al. 1986). In contrast, vertical tows (e.g., Smith et al. 1989) or oblique tows (deployment to specified depth and continuous sampling as the gear is towed to the surface at a constant tow angle; Smith and Richardson 1977) are commonly used if the study is designed primarily for collection of presence–absence data or assessment of temporal trends in egg and larval abundance (Rijnsdorp and Jaworski 1990). In addition to the towing path, towing speed can have a direct effect on the species and size composition of the catch, although the effects may vary among species (Aron and Collard 1969).

9.3.3 Statistical Considerations

Because most studies of ichthyoplankton involve estimates of egg and larval abundance and analyses of larval length distributions, care should be taken to ensure that data are both accurate and precise (see Chapter 2). Succeeding sections on sampling eggs and larvae discuss potential sources of bias associated with various collection methods. In particular, length distributions of sampled taxa may be biased due to extrusion of small larvae through net meshes and net avoidance by larger larvae; several formulas have been proposed to correct egg abundance estimates (D'Amours and Grégoire 1992) and larval length–frequency distributions due to avoidance (Barkley 1972; Murphy and Clutter 1972), extrusion (Lenarz 1972), or both (Somerton and Kobayashi 1989).

Because eggs and larvae of many fishes typically exhibit patchy distributions, data obtained from replicate samples may be subject to high variability and low precision. A set of observations (e.g., counts) at a single time and place (site) constitutes a sample; true replicates are similar independent and randomly collected samples (see Chapter 2). Replication allows for estimation of between-sample variance at a particular site, which is the basis for parametric tests of significant differences (e.g., mean length or abundance) between sites (see Waters and Erman 1990 for a discussion of statistical design). Accuracy of ichthyoplankton data depends on the ability of the sampling design to describe egg and larval characteristics effectively (e.g., abundance, distribution, or size composition). Precision is strongly affected by ichthyoplankton patchiness, and a given level of precision is dependent on the number of replicates taken. Cyr et al. (1992) reported that most larval fish surveys were based on low numbers (approximately four) of high-volume (approximately 300-m^3) replicates. The authors found that half of the published studies on larval fish abundance could detect only order-of-magnitude differences among sites or time periods; 33 replicates were needed to detect a 50% change in population density ($\alpha = 0.05$) at larval abundances of 10 per replicate. Many ichthyoplankton studies have been based on two or three replicates at each site (e.g., paired net tows). Although increasing the number of replicates may be problematic, larger numbers of lower-volume samples may be needed to increase the probability of detecting significant differences in egg and larval abundances between sites.

9.3.4 Effects of Gear Characteristics on Sampling Design

Passive avoidance of gears by eggs and larvae can be due to clogging of nets (Smith et al. 1968; Williams and Deubler 1968) or pump intakes. For towed nets (Aron et al. 1965) and filter nets used with pumps, clogging is primarily a function of gauze material, mesh size, density of organisms and debris in the water column, and duration of the sampling. Clogging can be a particular problem during oblique or vertical tows. Progressive clogging as the net is hauled up through the water column can lead to unequal sampling at different depths and inaccurate abundance estimates if eggs or larvae are not uniformly distributed (Schnack 1974). The magnitude of clogging can be assessed with flow meters mounted inside and outside the net, and clogging can be reduced by increasing the net area:mouth area ratio to at least 3:1 (preferably 5:1), incorporating mouth-reducing cones or prenet cylinders, and reducing the duration of sampling (Trantor and Smith 1968).

Extrusion of collected organisms through the mesh (Vannucci 1968) and damage to organisms in the net are primarily related to the size and shape of the collected taxa, mesh size (Lenarz 1972; Houde and Lovdal 1984), towing speed (Colton et al. 1980), tow duration, and water temperature. Thayer et al. (1983) reported catch of larvae in a modified Miller high-speed sampler (Figure 9.4B) increased with towing speed up to 7–8 m/s but declined at higher speeds due to either extrusion of larvae through the net or deflection of larvae by pressure waves ahead of the net mouth. Gregory and Powles (1988) also reported extrusion of small (<6.0 mm) Iowa darters through Miller high-speed nets. Extrusion of larvae and eggs can be reduced with smaller-mesh nets, but smaller mesh is more susceptible to clogging, reduced filtration, and increased net avoidance as larvae are deflected by pressure waves in front of the sampler.

Damage to collected organisms is more evident with high-speed samplers (towing speeds greater than 2 m/s) and is particularly important if it prevents identification. Several papers have documented changes in larval fish morphology due only to the effects of netting. Theilacker (1980) observed a decrease of up to 19% in the standard length of northern anchovy captured in plankton nets, and Hay (1981) and McGurk (1985) found up to 18% shrinkage in net-captured Pacific herring larvae. McGurk (1985) also found that netting effects on larval morphology were not consistent; that is, body depth and head width increased as standard length decreased, resulting in inaccurate assessment of larval condition

Mesh sizes of nets used in ichthyoplankton studies typically range from 333 to 505 μm, and several studies have focused on gauze construction (Heron 1968) and selectivity of plankton net mesh (Saville 1958; Barkley 1972). Southward and Bary (1980) used high-speed samplers to study Atlantic mackerel egg abundance and concluded that previous studies employing larger-mesh nets had underestimated egg abundance due to egg loss through the net mesh. Loss of threadfin shad and gizzard shad larvae through 500-μm mesh during sampling and net washdown was documented by Tomljanovich and Heuer (1986), and studies by Leslie and Timmins (1989) in the Great Lakes indicated losses of 26 and 13% of larval fishes from nets with 1,000- and 480-μm mesh, respectively; all larvae were retained with a mesh size of 250 μm. O'Gorman (1984) compared catches of larval alewife and rainbow smelt among 0.5-m nets with 355-, 450-, 560-, and 750-μm mesh and reported significantly fewer larvae with the latter three mesh sizes. A change in material and mesh size from 550-μm silk to 505-μm nylon was found to increase retention of larval northern anchovy in towed meter nets from 60% to near 100%, certainly an important factor affecting interpretation of data (Lenarz 1972).

Choice of mesh size in ichthyoplankton studies depends on a number of factors, including gear type, water velocity through the gear, and the size of the target organisms. Smith et al. (1968) concluded that capture of an organism depended on whether its width exceeded the mesh diagonal, and although this may be a conservative criterion (Lenarz 1972), it is likely a good guideline. Choosing the largest mesh that will collect the desired sizes of target organisms should maximize sampling effectiveness and minimize clogging problems and reductions in net performance. For small organisms in systems with large amounts of organic or inorganic debris, this may require several tows of short duration.

Gear failure can occur due to mechanical problems, operator inexperience, and collisions with debris or the substrate. Mechanical problems were particularly common with early opening–closing nets and have resulted in numerous gear modifications (Paquette and Frolander 1957; Bary and Frazer 1970). Several types of flow meters have been employed to measure sampled water volumes; meter accuracy may be dependent on towing speed and should be verified with frequent calibration. Contamination of samples can result from incomplete washing of nets and collection buckets (Figure 9.1A) between tows or sampling outside the target strata. Opening–closing nets can eliminate the latter problem and can be of a simple purse line design (Currie and Foxton 1956; Figure 9.1A).

9.3.5 Effects of Fish Behavior on Sampling Design

Fish behavior can have important effects on where, when, and how early life stages are collected (Bowles et al. 1978; Marcy and Dahlberg 1980). Active avoidance of towed nets and pumps is related to larval size (Lenarz 1973) and position relative to the net (Barkley 1964), light levels, physical characteristics of the sampling gear, and the velocity of the gear or water flow entering the gear. Visual signals (Clutter and Anraku 1968) and hydrostatic pressure waves may trigger avoidance responses in larval fishes, which can cause significant underestimates of the abundance of larger larvae. Conversely, slow growth of larval fishes may prolong their vulnerability to plankton nets, which can result in overestimates of larval fish abundance through time if the size distribution of the catch is not analyzed (Hamley et al. 1983). Noble (1971) used high-speed samplers to evaluate the effectiveness of a meter net for sampling yellow perch and walleye larvae, and reported avoidance of the meter net by larvae 10 mm and larger. Thayer et al. (1983) found the abundance of 10–16-mm spot and 19–26-mm Atlantic menhaden to be significantly underestimated by a 20-cm bongo net towed at 2 m/s.

Active avoidance of sampling gear by larvae has been examined with collection of diurnal and nocturnal samples, and many studies have documented increased catches of larvae at night (Bridger 1957). Cada and Loar (1982) found diurnal avoidance of a low-volume pump (compared with a 0.5-m Hensen net) by 5–10 mm clupeid larvae, and Graham and Venno (1968) reported that larval Atlantic herring were able to visually avoid diurnally fished stationary nets. Comparisons of pump and meter net samples by Leithiser et al. (1979) also indicated visual avoidance of a slowly towed (up to 0.4 m/s) meter net by larvae larger than 5 mm; however, higher towing speeds may have made catches of the two gears more comparable. Murphy and Clutter (1972) compared efficiencies of a meter net and a miniature purse seine (Hunter et al. 1966; Kingsford and Choat 1985) for sampling larval engraulids. Meter net samples contained substantially fewer larvae greater than 5 mm, and improvements in nocturnal catch rates indicated that avoidance was primarily visual. In general, nocturnal sampling with low-velocity gears results in substantially higher catch rates compared with diurnal sampling (Cole and MacMillan 1984), although this phenomenon could be due to larval movements and temperature effects as well as reduced net avoidance (Marcy and Dahlberg 1980). Use of high-speed samplers can decrease active avoidance by larvae, although extrusion and damage of larvae may increase.

If study objectives include assessment of larval length–frequency distributions or growth, two or more types of gears can be used to improve accuracy and reduce mechanical or biotically related bias (Suthers and Frank 1989). Gallagher and Conner (1983) used a meter net and paired 0.5-m push nets to collect larvae in the Mississippi River and found that the relative effectiveness of the two gears varied by habitat (main stem versus slackwater) and time of day. If a study is designed to assess larval mortality (e.g., entrainment; Dempsey 1988) it is important that mortality due to sampling be quantified; larval mortality in nets has been found to be a direct function of water velocity (O'Conner and Schaffer 1977; Cada and Hergenrader 1978; see McGroddy and Wyman 1977 for a low-mortality collection device developed for entrainment sampling). Regardless of the type or number of gears used, sampling duration, gear characteristics (e.g., mesh size), sampling speed, and diel sampling periodicity should be quantified for each gear and be as consistent as possible among samples. In addition, interspecific variability in spatial distribution and susceptibility to various gear types must be considered when data are used to assess differences in relative species composition. Investigators must consider whether differences in the numbers of various taxa collected reflect true relative abundance or are a result of interspecific differences in swimming ability, behavior, and microhabitat preferences.

9.4 SAMPLE PRESERVATION

Maintenance of morphological integrity of eggs and larvae during initial fixation and long-term preservation is important for both taxonomic and ecological studies (e.g., length frequency and condition factors). Chemicals used for fixation and preservation should prevent microbial degradation and minimize autolysis and cellular damage due to osmotic changes (Jones 1976). The degree of specimen degradation (e.g., shrinkage or structural or pigmentation deterioration) from fixation and preservation depends primarily on developmental stage (Hay 1982), chemical concentration (Hay 1982), and osmotic strength (see studies in Tucker and Chester 1984). Fixation and preservation of eggs and larvae thus involve trade-offs between prevention of microbial-induced degradation and fixation-induced alterations.

9.4.1 Fixation and Preservation

Most methods for both fixation and preservation of ichthyoplankton (Lavenberg et al. 1984) involve the use of aldehyde-based solutions (e.g., formaldehyde and glutaraldehyde), which are excellent fixatives because they combine with tissue proteins and prevent proteins from reacting with other reagents (Pearse 1968). These effects can be reversed by washing, and washing after fixation is not recommended (Taylor 1977). Although alcohol is used as a long-term preservative for some larval fish collections (DeLeon et al. 1991), it is not recommended because alcohol solutions cause significant specimen shrinkage and deformation due to dehydration. Formaldehyde is typically preferred to glutaraldehyde because it is less noxious and less expensive and is regarded to have superior long-term stability (Steedman 1976). In attempts to improve preservation qualities, formalin-based solutions have been mixed with buffers, acids and alcohols (e.g., Bouin's and Davidson's fluids; DeLeon et al. 1991).

Historically, fish eggs and larvae have been fixed in 5–10% formalin and preserved in 3–5% buffered formalin (Ahlstrom 1976; Smith and Richardson 1977). Formalin-based solutions are acidic (pH 2.5–5.0), and their acidity may increase over time through oxidation of formaldehyde to produce formic acid (Steedman 1976). Long-term storage of larval specimens in acidic formalin solutions can result in decalcification and demineralization of bone (Taylor 1977), including otoliths. (For age and growth studies, it is recommended that larvae be frozen after capture to eliminate preservative-related changes in otolith structure.) To prevent acid-induced degradation of specimens, a long-term preservative should have a neutral pH (7.0–7.5; Tucker and Chester 1984) and should be buffered with sodium borate (borax; Ahlstrom 1976), calcium carbonate (marble chips or limestone powder; Steedman 1976), sodium phosphate (Markle 1984), or sodium acetate (Tucker and Chester 1984). However, addition of buffers can raise the pH to levels (above ~8.0) that can increase larval transparency (clearing) and loss of pigmentation (sodium borate buffer; Taylor 1977), formation of calcium carbonate crystals on specimens (Tucker and Chester 1984), and precipitation of sodium phosphate on specimens (Markle 1984).

Larval fishes. Choice of a fixative and preservative may depend on the goals of the study. For long-term storage, Tucker and Chester (1984) found little shrinkage (97.4–100.7% of live standard length) and good pigment retention in larval southern flounder preserved in 4% unbuffered formalin in freshwater, although this preservative caused decalcification (pH < 6.0). They concluded that the best long-term preservative was 4% formalin in distilled water buffered with 1% sodium acetate. Alternatively, good preservation qualities have been reported for 5% formalin buffered with sodium phosphate (1.8 g sodium phosphate monobasic, 1.8 g anhydrous sodium phosphate dibasic [0.013 M, pH 6.8] in 1L of 5% formalin; Markle 1984).

Eggs. Oocytes have traditionally been fixed and preserved with formalin (4–10%) or modified Gilson's fluid (100 mL 60% methanol or ethanol, 880 mL of water, 15 mL of 80% nitric acid, 18 mL glacial acetic acid, and 20 g of mercuric chloride; Bagenal and Braum 1978). However, formalin fixation typically hardens ovarian tissue and makes oocyte separation difficult, and use of Gilson's fluid tends to result in degeneration of hydrated oocytes (Brown-Peterson et al. 1988) and oocyte shrinkage (15–24%; DeMartini and Fountain 1981). To avoid these problems and the toxicity of mercuric chloride (West 1990), Lowerre-Barbieri and Barbieri (1993) recommended physical separation of oocytes before fixation and preservation in 2% buffered formalin.

Fish eggs are not subject to the same preservation problems associated with larval fishes (i.e., decalcification and clearing), but eggs are often fixed and preserved similarly (Ahlstrom 1976; Smith and Richardson 1977). Inadequate preservation of fish eggs has been reported for buffered (sodium phosphate and sodium borate) formalin solutions typically used in larval fish preservation (Ahlstrom 1976; Markle 1984; Gates et al. 1987). Markle (1984) recommended the use of 5% unbuffered formalin and Klinger and Van Den Avyle (1993) recommended the use of 4–7% unbuffered formalin for preservation of fish eggs.

Alternative preservation methods. Cooling or freezing of ichthyoplankton samples may be practical alternatives to formalin-based fixation if specimens are to be either processed quickly or used in biochemical studies. Genetic studies generally require that specimens be frozen rapidly in liquid nitrogen and stored at −76°C to ensure retention of the biochemical properties of proteins and DNA. If the effects of formalin-based fixation (i.e., shrinkage and clearing) are problematic for goals of a study, specimens should be placed on ice immediately after collection and processed quickly. Long-term freezing of specimens will result in some shrinkage (generally less than formalin-based solutions) and may cause cellular damage (Halliday and Roscoe 1969; Jones and Geen 1977).

Color preservation. Little research has been conducted on the addition of antioxidants to ichthyoplankton samples to maintain natural coloration. Brown and black melanins are generally well preserved in acidic to neutral formalin-based solutions without the use of antioxidants. Because color is not a commonly used character in identification of larval fishes, addition of antioxidants is not necessary. In collections in which natural color preservation is required (e.g., reds in larvae of tunas), addition of 0.2–0.4% solutions of IONOL CP-40 (40% butylated hydroxytoluene) in formalin-based preservatives has been recommended (Berry and Richards 1973; Scotton et al. 1973). Use of other antioxidants has been reported for adult (Gerrick 1968) and larval fishes (Ahlstrom 1976). If color attributes are important for larval fish identification, specimens should be examined immediately after collection, before fixation (Ahlstrom 1976).

9.5 SAMPLE PROCESSING

Processing egg and larval samples typically begins on site immediately upon collection. Depending on study objectives, fish eggs and larvae are either fixed in a formalin-based solution or frozen. To ensure specimen integrity, immediate processing is important (Ahlstrom 1976; Hay 1981). Samples are typically returned to the laboratory for sorting, enumeration, identification, measurement, and other study-specific analyses (e.g., age determination, gut analysis, or electrophoresis). Illustration of larval and egg characteristics (Faber and Gadd 1983; Lindsey 1984; Sumida et al. 1984) can be an important part of taxonomic studies, and use of photomicrography and photoimaging technology can greatly improve illustrative efforts. All data associated with egg and larval collections (e.g., date, location, collection personnel, and physicochemical data) should be stored with the specimens, which should be protected from excessive heat and sunlight and placed in either a teaching or museum collection for long-term curation (Lavenberg et al. 1984). These collections and related data provide important systematic references for future researchers.

9.5.1 Subsampling

Because fish eggs and larvae usually compose less than 1% of a plankton sample (Scotton et al. 1973), subsampling is typically not recommended. Time saved by subsampling may be outweighed by potential specimen damage and sampling bias (Griffiths et al. 1976), and subsamples may not adequately represent the abundance of rare taxa and developmental stages in a sample. However, if densities of organisms in samples are high subsampling (Lewis and Garriott 1971; Smith and Richardson 1977; Herke 1978) may be necessary.

9.5.2 Sorting

Most sampling methods collect fish eggs and larvae as well as other organisms and organic and inorganic debris. Thus, sorting should efficiently separate all eggs and larvae from the unwanted portion of a sample. The time required to sort a sample is dependent on the quantity and size of eggs, larvae, and debris.

For worker safety, sample fixatives or preservatives are typically removed, and samples are washed in water (hold in water for a minimal period of time to avoid fixation problems; Taylor 1977) and sorted in a well-ventilated area (this is particularly important when working with formaldehyde, which is carcinogenic). Samples may be placed in a petri dish marked on the bottom with a grid (or a side-lit sorting chamber; Dorr 1974), magnified, and sorted with reflected or transmitted light against a high-contrast background. Small specimens may require low-power microscopy, whereas larger specimens can be sorted with a dissecting microscope or lighted magnifying lens. For identification, dissecting microscopes with polarizing filters are often used to aid in counting larval characteristics; transmitted polarized light is often used to discern muscle structure (see section 9.6.4). Because initial sorting may miss up to 20% of the larvae (Scotton et al. 1973), samples are often re-sorted to ensure accuracy. Eggs and larval tissues are fragile and easily damaged, and rigid handling tools should be avoided; specimens should be handled with pipettes, wire loop probes, and flexible forceps.

Use of biological stains has been shown to reduce sorting time and increase sorting efficiency for ichthyoplankton samples (Mason and Yevich 1967; Mitterer and Pearson 1977). However, staining of fish larvae tends to obscure muscle myomeres and other morphometric characters used in the identification of larval fishes and is generally not recommended. Stains that have been used for sorting fish eggs and larvae include rose bengal (1% solution; Mitterer and Pearson 1977), a combination of eosin and biebrich scarlet (1:1; Klinger and Van Den Avyle 1993), phloxine b (Mason and Yevich 1967), and Lugol's iodine counterstained with chlorazol (Williams and Williams 1974).

9.6 TERMINOLOGY AND IDENTIFICATION

To date, eggs and larvae of less than 20% of North America's freshwater and anadromous fishes (Snyder and Muth 1988, 1990) and 10% of the world's marine fishes (Smith and Richardson 1977; Kendall and Matarese 1994) have been adequately described. Identification of fish eggs and larvae and subsequent analyses of meristic and morphometric characters are always subject to error, and all efforts should be made to limit or quantify these errors. Identification of eggs and larvae to species, or even higher taxonomic levels, is often impossible; thus, taxonomic assignments (e.g., species, genera, and family) should be made only when considerable evidence is available. In addition to morphological and ecological characteristics, valuable taxonomic information can often be gained through the use of individual and comparative descriptions, regional keys or manuals (Table 9.2), reference collections, and taxonomic experts.

Table 9.2 Regional guides for the identification of fish eggs and larvae. Most manuals are illustrated and include regional notes on the distribution and ecology of fish eggs and larvae and on adult spawning.

Author(s) and publication date	Region	Comments
Fish 1932	Lake Erie	62 species accounts; several misidentified
Mansueti and Hardy 1967	Chesapeake Bay	45 species accounts; Acipenseridae through Ictaluridae
May and Gasaway 1967	Oklahoma	18 species accounts; photographs and key
Colton and Marak 1969	Northeast coast North America	27 species accounts
Scotton et al. 1973	Delaware Bay	56 species accounts
Lippson and Moran 1974	Potomac River	88 species accounts; keys
Hogue et al. 1976	Tennessee River	32 species descriptions; photographs and keys
Russell 1976	British Isles marine fishes	40 families; taxonomic characters and methods
Hardy et al. 1978	Mid-Atlantic Bight	278 species accounts; includes tidal freshwater zones
Drewry 1979	Great Lakes region	Key to yolk-sac larvae; illustrations
Wang and Kernehan 1979	Delaware estuaries	113 species accounts; keys
Miller et al. 1979	Hawaiian Islands	46 taxonomic accounts; illustrations
Elliott and Jimenez 1981	Beverly–Salem Harbor, Massachusetts	47 species accounts
Snyder 1981	Upper Colorado River, Colorado	19 species accounts; Cyprinidae and Catostomidae; keys
Wang 1981	Sacramento–San Joaquin estuary, California	74 species accounts; keys and comparison tables
Auer 1982	Great Lakes basin	148 species accounts; keys
Garrison and Miller 1982	Puget Sound, Washington	124 species accounts
Fahay 1983	Western North Atlantic	290 species accounts; comparison table
Leis and Rennis 1983	Indo-Pacific	49 family accounts
McGowan 1984	South Carolina	11 families, 18 species; illustrations
Conrow and Zale 1985	Florida	18 species accounts
Wang 1986	Sacramento–San Joaquin estuary, California	125 species accounts; keys and comparison tables
McGowan 1988	North Carolina reservoirs	10 families, 22 species
Leis and Trnski 1989	Indo-Pacific	54 family accounts
Matarese et al. 1989	Northeast Pacific	232 species accounts; keys and illustrations
Holland-Bartels et al. 1990	Upper Mississippi River	19 illustrated families, 63 unillustrated species
Wallus et al. 1990	Ohio River basin	24 species accounts; Acipenseridae through Esocidae
Olivar and Fortuno 1991	Southeast Atlantic	127 taxonomic accounts; illustrations
Kay et al. 1994	Ohio River basin	21 species accounts; Catostomidae
Ditty and Shaw 1994	Western central Atlantic	21 genera, 55 species; Sciaenidae
Farooqi et al. 1995	Western central Atlantic	7 genera, 28 species; Engraulidae

9.6.1 Egg Developmental Stages

Egg development is a dynamic process and is usually assumed to encompass the time period from ovulation until hatching. Egg structure consists of an outer membrane or chorion, perivitelline space, an inner egg membrane (present in some fishes), and yolk (Ahlstrom and Moser 1980; Kendall et al. 1984; Figure 9.8). Most fishes are oviparous: ovulation is followed by release of eggs to the external environment to be fertilized by sperm from associated males. Upon fertilization, eggs

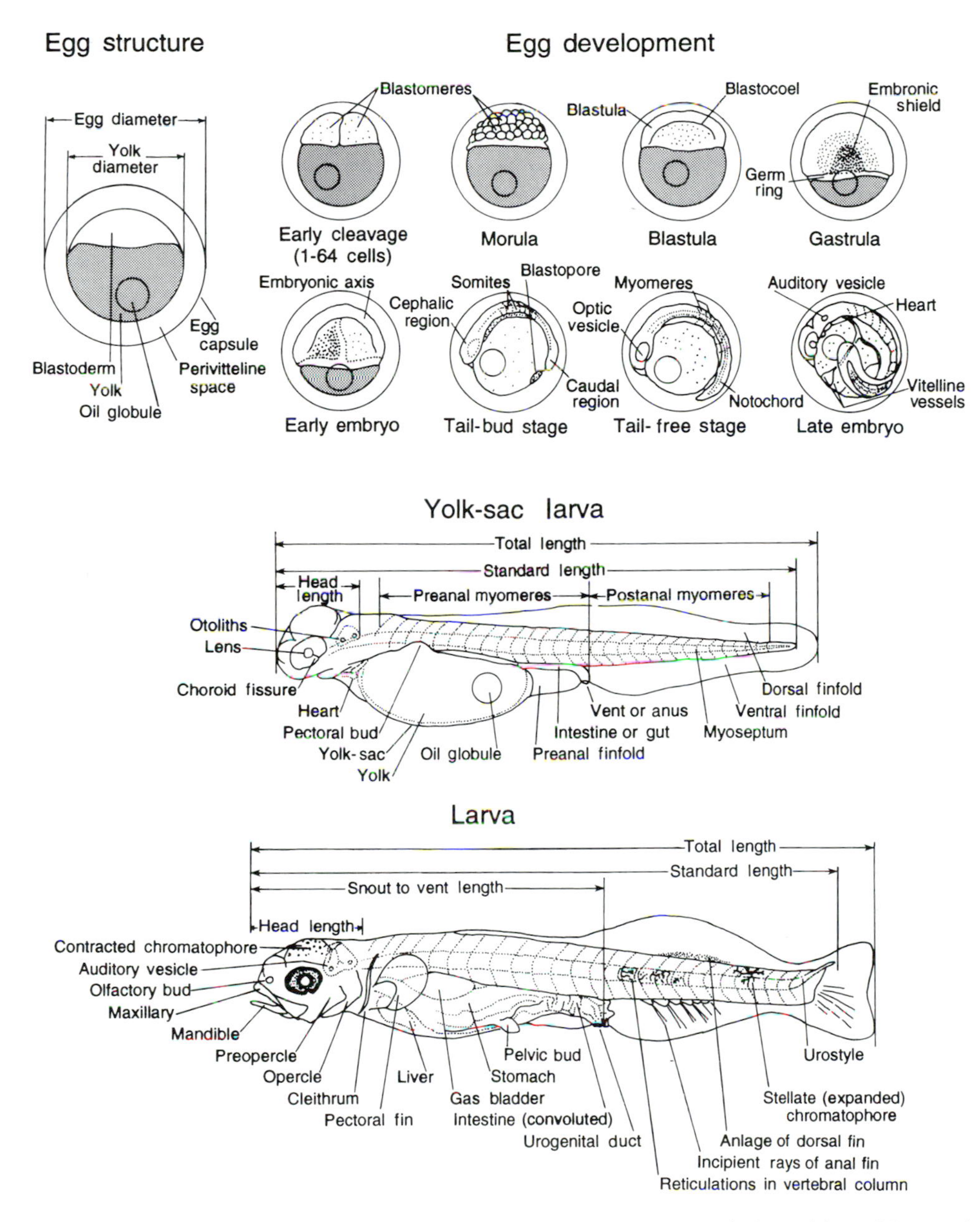

Figure 9.8 Developmental stages of a typical teleost egg and larva (adapted from Mansueti and Hardy 1967, with permission).

undergo changes in structure and function (egg activation) that prevent multiple fertilization (polyspermy), harden the chorion (water hardening; Redding and Patino 1993), and begin embryonic development. Cell division (cleavage) of the egg in fishes is most commonly meroblastic but is occasionally holoblastic (e.g., lampreys) or intermediate (e.g., South American lungfish, sturgeons, gars, and bowfin; Blaxter 1969; Lagler et al. 1977). For a general review of fish reproduction (e.g., gonad maturation, gonosomatic indices, and fecundity estimates) see Crim and Glebe (1990).

Several workers have described various stages of egg and embryo development useful in egg identification. Matarese and Sandknop (1984) adopted developmental stages proposed by Ahlstrom and Ball (1954), which included the following categories: early development, from fertilization to closure of the blastopore; middle development, from closure of the blastopore to tail bud lifting off the yolk; and late development, from tail bud lifting off the yolk to hatching. Mansueti and Hardy (1967) presented a more detailed description of egg development (see Figure 9.8): (1) early cleavage, 1–64 cells; (2) morula, blastomeres that form a cluster of cells; (3) blastula, formation of the blastocoel; (4) gastrula, differentiation of cells into ectoderm, mesoderm, and endoderm; (5) early embryo, formation of the embryonic axis; (6) tail-bud stage, prominent caudal bulge and cephalic development; (7) tail-free stage, separation of the tail from yolk; and (8) late embryo, embryo has developing characteristics of its hatching stage.

9.6.2 Egg Identification

Various egg types are found in both marine and freshwater fishes. Overall, fish eggs average 1 mm in diameter (although coelocanths have 90-mm eggs; Potts and Wootton 1984). Eggs are typically translucent but may be dark (e.g., paddlefishes, sturgeons, and gars), they may be buoyant (pelagic) or nonbuoyant (demersal), adhesive or nonadhesive, and they may have modifications to aid in attachment or flotation. Eggs are typically spherical but may be ovoid or irregularly shaped. Oil globules may be present or absent, and may vary in number, size, color, position, and pigmentation. Yolk may have a characteristic segmentation, color, pigmentation, and circulation. The chorion may vary in surface topography, ornamentation, thickness, color, coatings, and micropyle size. The width of the perivitelline space and the location, presence, or absence of the inner egg membrane may also be variable among taxa. These characters, along with morphological information on the developing embryo, collection information (e.g., location, water temperature, season, and collection gear), and mode of reproduction are often used to aid in egg identification (Newell and Newell 1963; Hempel 1979; Ahlstrom and Moser 1980; Matarese and Sandknop 1984; Balon 1985; Blaxter 1988; Table 9.2). Difficulties in identifying fish eggs by means of morphological characters have led researchers to explore biochemical techniques to aid in identification, such as immunodiffusion and immunofluorescence (Johnson et al. 1975), molecular degradation (Valcarce et al. 1991), gas chromatography (Knutsen et al. 1985), isoelectric focusing (Mork et al. 1983), protein electrophoresis (Scobbie and Mackie 1990), and mitochondrial DNA analyses (Graves et al. 1989).

9.6.3 Larval Developmental Stages

Numerous terminologies have been proposed to describe early life history stages of developing larvae. None of the proposed terminologies is without problem because any attempt to categorize a dynamic and often species-specific process into a static framework is difficult (see Snyder 1976; Kendall et al. 1984; Snyder and Holt 1984; Balon 1985; and Blaxter 1988 for summaries of early terminologies). Most researchers agree that the term "embryo" encompasses development from fertilization to hatching (for exception, see Balon 1984), the "juvenile period" begins with acquisition of an adult body form and ends at sexual maturation, and the "larval period" is in between. Although the term "fry" has been used frequently in the larval fish literature for larvae from hatching to flexion of the notochord, there is little consensus on a precise definition for this term, and we have restricted our use of early life stage terminology to larvae and juveniles. There are currently three commonly accepted terminologies used to categorize the phases of larval fish development.

1. Mansueti and Hardy (1967) and Hardy et al. (1978) described three phases of larval fish development based on the presence or absence of yolk material and fin ray development:
   Yolk-sac larvae: Phase between hatching and yolk absorption.
   Larvae: Phase between yolk absorption and the acquisition of adult fin ray complement.
   Pre-juvenile or transitional: Intermediate phase between larval and juvenile forms that begins with acquisition of the minimum adult fin ray complement and terminates in a more adultlike juvenile form.
2. Ahlstrom et al. (1976) described three phases of larval fish development based primarily on changes in the homocercal caudal fin:
   Preflexion larvae: Phase between hatching and upward flexing of the tip of the notochord or appearance of the first caudal rays.
   Flexion larvae: Phase characterized by upward flexion of the notochord (this phase terminates with formation of all principal caudal rays and the first appearance of secondary caudal rays).
   Postflexion larvae: Phase beginning after upward flexion of the tip of the notochord and terminating with a complete complement of fin rays. For some species prejuvenile or transitional phases are applied.
3. Snyder (1976, 1981) described three developmental phases based on morphogenesis of the median finfold and fins:
   Protolarvae: Phase between hatching and appearance of the first median fin ray or spine (dorsal, anal, or caudal fins).
   Mesolarvae: Phase beginning with the appearance of the first median fin ray or spine and terminating with acquisition of the pelvic fins or fin buds and a full complement of principal soft rays in the median fins.
   Metalarvae: Phase beginning with acquisition of pelvic fins or fin buds and a full complement of principal soft rays in the median fins and terminating with the loss of all finfolds and acquisition of the adult complement of spines and rays (including some ray segmentation) in all fins.

Each of these terminologies has been used successfully, and although there have been attempts to standardize larval fish terminology, no one method currently predominates in the early life history literature (Snyder 1976). Lack of standardization may be due to historical inertia as well as the broad array of topics (e.g., ontogeny, taxonomy, physiology, and ecology) addressed by early life history studies. Any terminology adopted to describe larval fish development should be inclusive of the diversity of forms, have some morphological and functional significance in the life history of the fish, and have observable and well-defined endpoints for each phase (Kendall et al. 1984). Although Snyder's (1976, 1981) terminology is typically used in studies of freshwater larval fishes in North America, several marine researchers recommend the terminology of Ahlstrom et al. (1976; see above) because of the functional importance of caudal fin development and associated changes in body shape and fin ray development and the terminology's simplicity and generality (Kendall et al. 1984; Blaxter 1988). Flexion of the notochord is a major developmental landmark that leads to increased larval mobility for pursuit of prey and avoidance of predators. Ultimately, a combination of the terminologies (e.g., postflexion mesolarvae with yolk or yolk-sac mesolarvae; Snyder and Holt 1984) may prove most useful for standardization of terminology and definitions.

9.6.4 Larval Fish Identification

Many meristic, morphometric, and composite characters (e.g., pigmentation patterns, shape, size, and osteological development) have been used to aid in identification of larval fishes (Kendall et al. 1984; Figure 9.8). Morphological characters used for adult fish taxonomy are often not useful for larvae because of developmentally related structural differences. Some larvae possess specialized structures (e.g., eye stalks, elongated dorsal fins, sucker discs, unique spines, trailing gut, or photophores) that are unique to the larval stage and may be useful at some taxonomic level. More commonly used characters include myomere counts (total, preanal, and postanal) and the size, shape, and position of the gut, air bladder, yolk sac, oil globules, mouth, finfolds, and fins. Patterns of melanophore pigmentation have proven to be of particular value for species identification (Berry and Richards 1973; Snyder 1981; Kendall et al. 1984). In general, taxonomic characters tend to vary throughout the larval period; thus all meristic, morphometric, and composite characters must be related to size or developmental stage.

Myomere counts. Structurally, body muscles of postembryonic fishes are divided into myomeres, which are chevron-shaped serial segments separated by connective tissue (myosepta). Myomeres are conspicuous morphological features that approximate the number and position of vertebrae (typically number of myomeres minus one), although vertebral numbers are less variable (Snyder 1979; Fuiman 1982). Myomere counts (e.g., total, preanal, and postanal) are useful taxonomic characters because they are relatively consistent throughout the larval period. Total myomere counts include all myomeres from the first myomere, posterior to the occiput, to the urostylar myomere, posterior to the last myosepta (Fuiman 1982). Preanal myomere counts include all myomeres anterior to the posterior margin of the anus and include the myomere bisected by an imaginary vertical line drawn from the posterior margin of the anus; postanal myomeres are counted from this line posteriorly (Siefert 1969). Other partial myomere counts may be useful to reference the location of important structural features.

Morphometric analyses. Morphometric characters used for taxonomic purposes generally describe body form (e.g., body depth or eye width; Figure 9.8). Many morphometric characters are allometric (i.e., as larval fish grow there is a systematic change in shape). Differential growth rates are characteristic of the ontogeny of most organisms, but allometry is magnified during the larval period because of rapid growth. For comparative purposes, morphometric characters are often reported as ratios (proportions or percentage) in an attempt to remove the effect of body size (e.g., standard length, notochord length, or total length) from variation in body shape. Researchers have reported statistical problems with the use of ratios (e.g., inflated sampling errors, nonnormal frequency distributions, and erroneous character correlations; Atchley et al. 1976), and several regression methods have been applied to avoid statistical problems associated with ratios (see Strauss and Bond 1990).

Recent advances in morphological analyses may be particularly applicable to studies of larval fishes. Truss networks have been used to characterize differences in shape for adult fishes (Strauss and Bookstein 1982; Strauss and Bond 1990;

Bookstein 1991), but few studies have dealt with developmental changes in larval morphology (Strauss and Fuiman 1985). Truss analysis quantifies the shape (oblique, longitudinal, and vertical) of an organism with distance measurements among anatomical landmarks (see Douglas 1993 for application of videoimaging technology to truss analysis). Even though the number of landmarks in larval fishes is limited, several prominent morphological features (e.g., snout tip, bone articulations, tip of urostyle, and so on) can be identified to divide larvae into functional units. Landmarks are chosen to produce a series of contiguous quadrilaterals (anterior to posterior), with the landmarks forming the boundary of each quadrilateral (truss cell). Within a truss cell six pairwise measurements are made. The strength of truss analysis is based on the assumption that the anatomical landmarks are homologous among species. Multivariate methods (e.g., principal components analysis, sheared components principal analysis, and discriminant function analysis) are often used with morphometric character sets developed from the truss protocol to describe size and shape differences (Humphries et al. 1981; Bookstein et al. 1985; Strauss and Fuiman 1985; Strauss and Bond 1990).

Taxonomic guides. Because of the dynamic nature of anatomical characters during the larval period, comprehensive larval fish keys are difficult to prepare and thus are not common. Generally, keys or identification guides include only limited developmental ranges, distributional areas, and taxonomic groups (e.g., Fuiman 1979; Fuiman et al. 1983; Nishikawa and Rimmer 1987; Ditty 1989; Ditty et al. 1994; Richards et al. 1994). Because larval identification is typically based on a collection of anatomical (e.g., meristic and morphometric), ecological (e.g., adult spawning season), and zoogeographic (adult distribution patterns) characteristics that vary regionally, numerous regional identification guides have been developed (Table 9.2).

Supplemental identification techniques. Alternative techniques have been devised to aid in resolving larval fish taxonomic problems. Osteological features (Dunn 1984) often yield valuable taxonomic information, and methods for skeletal disarticulation (Mayden and Wiley 1984), whole organism clearing and staining (Taylor 1967; Galat 1972; Brubaker and Angus 1984; Potthoff 1984; Snyder and Muth 1990), X-ray radiography (Miller and Tucker 1979; Tucker and Laroche 1984), and histology (Govoni 1984) may enhance examination of internal structures (e.g., vertebrae). Scanning electron microscopy can be used to discern external characteristics (Boehlert 1984). As with fish eggs, biochemical techniques have been used to resolve taxonomic problems in some larval fishes (for a general review see Leary and Booke 1990; Beckenbach 1991; Park and Moran 1994). Most biochemical studies are designed to identify genetic differences between either closely related species (Morgan 1975; Sidell and Otto 1978; Comparini and Rodino 1980) or within stocks (Heath and Walker 1987; Graves et al. 1989; Grewe et al. 1994).

9.7 REFERENCES

Aggus, L. R., J. P. Clugston, A. Houser, R. M. Jenkins, L. E. Vogele, and C. H. Walburg. 1980. Monitoring of fish in reservoirs. Pages 149–175 *in* C. H. Hocutt and J. R. Stauffer, Jr., editors. Biological monitoring of fish. Lexington Books, Lexington, Massachusetts.

Ahlstrom, E. H. 1976. Maintenance of quality in fish eggs and larvae collected during plankton hauls. Pages 313–321 *in* H. F Steedman, editor. Zooplankton fixation and preservation. Monographs on Oceanographic Methodology 4.

Ahlstrom, E. H., and O. P. Ball. 1954. Description of eggs and larvae of jack mackerel

(*Trachurus symmetricus*) and distribution and abundance of larvae in 1950 and 1951. U.S. National Marine Fisheries Service Fishery Bulletin 56:209–245.

Ahlstrom, E. H., and H. G. Moser. 1980. Characters useful in identification of pelagic marine fish eggs. California Cooperative Oceanic Fisheries Investigations Reports 21:121–131.

Ahlstrom, E. H., J. L. Butler, and B. Y. Sumida. 1976. Pelagic stromateoid fishes (Pisces, Perciformes) of the eastern Pacific: kinds, distributions, and early life histories and observations on five of these from the northwest Atlantic. Bulletin of Marine Science 26:285–402.

Amundrud, J. R., D. J. Faber, and A. Keast. 1974. Seasonal succession of free-swimming larvae in Lake Opinicon, Ontario. Journal of the Fisheries Research Board of Canada 31:1661–1665.

Arnold, E. L., Jr. 1952. High speed plankton samplers. U.S. Fish and Wildlife Service Special Scientific Report—Fisheries 88:2–6.

Aron, W. 1958. The use of a large capacity portable pump for plankton patchiness. Journal of Marine Research 16:158–173.

Aron, W., and S. Collard. 1969. A study of the influence of net speed on catch. Limnology and Oceanography 14:242–249.

Aron, W., E. H. Ahlstrom, B. M. Bary, A. W. H. Bé, and W. D. Clarke. 1965. Towing characteristics of plankton sampling gear. Limnology and Oceanography 10:333–340.

Atchley, W. R., C. T. Gaskins, and D. Anderson. 1976. Statistical properties of ratios. I. Empirical results. Systematic Zoology 25:137–148.

Auer, N. A., editor. 1982. Identification of larval fishes of the Great Lakes Basin with emphasis on the Lake Michigan drainage. Great Lakes Fishery Commission Special Publication 82-3, Ann Arbor, Michigan.

Bagenal, T. B. 1974. A buoyant net designed to catch freshwater fish larvae quantitatively. Freshwater Biology 4:107–109.

Bagenal, T. B., and E. Braum. 1978. Eggs and early life history. IBP (International Biological Programme) Handbook 3:165–201.

Bailey, K. M. 1980. Recent changes in the distribution of hake larvae: causes and consequences. Reports of the California Cooperative Oceanic Fisheries Investigations 21:167–171.

Baker, A. C., M. R. Clarke, and M. J. Harris. 1973. The N.I.O. combination net (RMT 1+8) and further developments of rectangular midwater trawls. Journal of the Marine Biological Association of the United Kingdom 53:167–184.

Balon, E. K. 1975. Reproductive guilds of fishes: a proposal and definition. Journal of the Fisheries Research Board of Canada 32:821–864.

Balon, E. K. 1984. Reflections on some decisive events in the early life of fishes. Transactions of the American Fisheries Society 113:178–185.

Balon, E. K. 1985. The theory of saltatory ontogeny and life history models revisited. Pages 13–30 *in* E. K. Balon, editor. Early life histories of fishes, new developmental, ecological and evolutionary perspectives. Dr. W. Junk Publishers, Dordrecht, Netherlands.

Baltz, D. M., C. Rakocinski, and J. W. Fleeger. 1993. Microhabitat use by marsh-edge fishes in a Louisiana estuary. Environmental Biology of Fishes 36:109–126.

Bardonnet, A., and P. Gaudin. 1990. Diel pattern of emergence in grayling (*Thymallus thymallus* Linnaeus, 1758). Canadian Journal of Zoology 68:465–469.

Barkley, R. A. 1964. The theoretical effectiveness of towed-net samplers as related to sampler size and to swimming speed of organisms. Journal du Conseil International pour l'Exploration de la Mer 29:146–157.

Barkley, R. A. 1972. Selectivity of towed-net samplers. U.S. National Marine Fisheries Service Fishery Bulletin 70:799–820.

Bary, B. M., and E. J. Frazer. 1970. A high-speed, opening–closing plankton sampler (Catcher II) and its electrical accessories. Deep-Sea Research 17:825–835.

Bary, B. M., J. G. DeStefano, M. Forsyth, and J. van den Kerkhof. 1958. A closing, high-speed plankton catcher for use in vertical and horizontal towing. Pacific Science 12:46–59.

Bath, W. B., J. A. Hernandez, T. Rippolon, and G. McCarey. 1979. Technique for simultaneous sampling of planktonic fish eggs and larvae at three depths. Progressive Fish-Culturist 41:158–160.

Baugh, T. M., and J. W. Pedretti. 1986. The penny fry trap. Progressive Fish-Culturist 48:74–75.

Bé, A. W. H. 1962. Quantitative multiple opening and closing plankton samplers. Deep-Sea Research 9:144–151.

Beamish, R. J. 1973. Design of a trapnet with interchangeable parts for the capture of large and small fishes from varying depths. Journal of the Fisheries Research Board of Canada 30:587–590.

Beard, T. D., and G. R. Priegel. 1975. Construction and use of a 1-foot fyke net. Progressive Fish-Culturist 37:43–46.

Beckenbach, A. T. 1991. Rapid mtDNA sequence analysis of fish populations using the polymerase chain reaction (PCR). Canadian Journal of Fisheries and Aquatic Sciences 48:95–98.

Becker, G. C. 1983. Fishes of Wisconsin. The University of Wisconsin Press, Madison.

Berry, F. H., and J. W. Richards. 1973. Characters useful to the study of larval fishes. Pages 48–65 *in* A. L. Pacheco, editor. Proceedings of a workshop on egg, larvae, and juvenile stages of fish in Atlantic coast estuaries. U.S. National Marine Fisheries Service, Middle Atlantic Coastal Fisheries Center Technical Publication 1, Highland, New Jersey.

Beverton, R. J. H., and D. S. Tungate. 1967. A multi-purpose plankton sampler. Journal du Conseil International pour l'Exploration de la Mer 31:145–157.

Bigelow, H. B., and W. C. Schroeder. 1953. Fishes of the Gulf of Maine. Fishery Bulletin of the Fish and Wildlife Service 74 volume 53. U.S. Government Printing Office, Washington, DC.

Bjørke, H., O. Dragesund, and Ø. Ulltang. 1974. Efficiency test on four high-speed plankton samplers. Pages 183–200 *in* J. H. S. Blaxter, editor. The early life history of fish. Springer-Verlag, New York.

Blaxter, J. H. S. 1969. Development: eggs and larvae. Pages 177–252 *in* W. S. Hoar and D. J. Randall, editors. Fish physiology, volume 3: reproduction and growth, bioluminescence, pigments, and poisons. Academic Press, New York.

Blaxter, J. H. S. 1988. Pattern and variety in development. Pages 1–58 *in* W. S. Hoar and D. J. Randall, editors. Fish physiology, volume 11(A): the physiology of developing fish, eggs and larvae. Academic Press, New York.

Boehlert, G. W. 1984. Scanning electron microscopy. Pages 43–48 *in* Moser et al. (1984).

Bookstein, F. L. 1991. Morphometric tools for landmark data: geometry and biology. Cambridge University Press, New York.

Bookstein, F. L., B. Chernoff, R. L. Elder, J. M. Humphries, G. R. Smith, and R. E. Strauss. 1985. Morphometrics in evolutionary biology: the geometry of size and shape change, with examples from fishes. Special Publication Academy of Natural Sciences Philadelphia 15.

Bourne, D. W., and J. J. Govoni. 1988. Distribution of fish eggs and larvae and patterns of water circulation in Narragansett Bay, 1972–1973. American Fisheries Society Symposium 3:132–148.

Bowles, R. R., J. V. Merriner, and G. C. Grant. 1978. Factors associated with accuracy in sampling fish eggs and larvae. U.S. Fish and Wildlife Service FWS/OBS-78/83, Ann Arbor, Michigan.

Braem, R. A., and W. J. Ebel. 1961. A back-pack shocker for collecting lamprey ammocoetes. Progressive Fish-Culturist 23:87–91.

Brander, K., and A. B. Thompson. 1989. Diel differences in avoidance of three vertical profile sampling gears by herring larvae. Journal of Plankton Research 11:775–784.

Brännäs, E. 1987. Influence of photoperiod and temperature on hatching and emergence of Baltic salmon (*Salmo salar*). Canadian Journal of Zoology 65:1503–1508.

Breder, C. M., Jr. 1960. Design for a fry trap. Zoologica 45:155–159.

Breder, C. M., Jr., and D. E. Rosen. 1966. Modes of reproduction in fishes. The Natural History Press, Garden City, New York.

Bridger, J. P. 1957. On day and night variation in catches of fish larvae. Journal du Conseil International pour l'Exploration de la Mer 22:42–57.

Bridger, J. P. 1958. On efficiency tests made with a modified Gulf III high-speed tow-net. Journal du Conseil International pour l'Exploration de la Mer 23:357–365.

Brown, D. J. A., and T. E. Langford. 1975. An assessment of a tow net used to sample coarse fish fry in rivers. Journal of Fish Biology 7:533–538.

Brown, D. M., and L. Cheng. 1981. New net for sampling the ocean surface. Marine Ecology Progress Series 5:225–227.

Brown, J. A., and P. W. Colgan. 1985. Interspecific differences in the ontogeny of feeding behaviour in two species of centrarchid fish. Zeitschrift für Tierpsychologie 70:70–80.

Brown-Peterson, N., P. Thomas, and C. R. Arnold. 1988. Reproductive biology of the spotted seatrout, *Cynoscion nebulosus*, in south Texas. U.S. National Marine Fisheries Service Fishery Bulletin 86:373–388.

Brubaker, J. M., and R. A. Angus. 1984. A procedure for staining fishes with alizarin without causing exfoliation of scales. Copeia 1984:989–990.

Bryan, C. F., R. D. Hartman, and J. W. Korth. 1989. An adjustable macroplankton gear for shallow water sampling. Northeast Gulf Science 10:159–161.

Bulkowski, L., and J. W. Meade. 1983. Changes in phototaxis during early development of walleye. Transactions of the American Fisheries Society 112:445–447.

Bulow, F. J. 1987. RNA–DNA ratios as indicators of growth in fish: a review. Pages 45–64 *in* R. C. Summerfelt and G. E. Hall, editors. Age and growth of fish. Iowa State University Press, Ames.

Burch, O. 1983. New device for sampling larval fish in shallow water. Progressive Fish-Culturist 45:33–35.

Bye, V. J. 1984. The role of environmental factors in timing of reproductive cycles. Pages 187–205 *in* G. W. Potts and R. J. Wootton, editors. Fish reproduction: strategies and tactics. Academic Press, New York.

Cada, G. F., and G. L. Hergenrader. 1978. An assessment of sampling mortality of larval fishes. Transactions of the American Fisheries Society 107:269–274.

Cada, G. F., and J. M. Loar. 1982. Relative effectiveness of two ichthyoplankton sampling techniques. Canadian Journal of Fisheries and Aquatic Sciences 39:811–814.

Cada, G. F., J. M. Loar, and K. Deva Kumar. 1980. Diel patterns of ichthyoplankton length–density relationships in upper Watts Bar Reservoir, Tennessee. Pages 79–90 *in* L. A. Fuiman, editor. Proceedings of the fourth annual larval fish conference. U.S. Fish and Wildlife Service Biological Services Program FWS/OBS-80/43.

Carlander, K. D. 1969. Handbook of freshwater fishery biology, 3rd edition, volume 1. Iowa State University Press, Ames.

Carlander, K. D. 1977. Handbook of freshwater fishery biology, volume 2. Iowa State University Press, Ames.

Carter, J. G., V. A. Lamarra, and R. J. Ryel. 1986. Drift of larval fishes in the upper Colorado River. Journal of Freshwater Ecology 3:567–577.

Casselman, J. M., and H. H. Harvey. 1973. Fish traps of clear plastic. Progressive Fish-Culturist 35:218–220.

Choat, J. H., P. J. Doherty, B. A. Kerrigan, and J. M. Leis. 1993. A comparison of towed nets, purse seine, and light-aggregation devices for sampling larvae and pelagic juveniles of coral reef fishes. U.S. National Marine Fisheries Service Fishery Bulletin 91:195–209.

Clarke, G. L., and D. F. Bumpus. 1950. The plankton sampler—an instrument for quantitative plankton investigations. Limnological Society of America Special Publication 5.

Clarke, M. R. 1969. A new midwater trawl for sampling discrete depth horizons. Journal of the Marine Biological Association of the United Kingdom 49:945–960.

Clarke, W. D. 1964. The jet net, a new high-speed plankton sampler. Journal of Marine Research 22:284–287.

Clifford, H. F. 1972. Downstream movements of white sucker, *Catostomus commersoni*, fry in a brown-water stream of Alberta. Journal of the Fisheries Research Board of Canada 29:1091–1093.

Clutter, R. I., and M. Anraku. 1968. Avoidance of samplers. Pages 57–76 *in* D. J. Trantor and J. H. Fraser, editors. Zooplankton sampling. United Nations Educational, Scientific, and Cultural Organization Monographs on Oceanographic Methodology 2.

Cole, R. A., and J. R. MacMillan. 1984. Sampling larval fish in the littoral zone of western Lake Erie. Journal of Great Lakes Research 10:15–27.

Coles, T. F., G. N. Swinney, and J. W. Jones. 1977. A technique for determining the distribution of pelagic fish larvae. Journal of Fish Biology 11:151–159.

Collins, J. J. 1975. An emergent fry trap for lake spawning salmonines and coregonines. Progressive Fish-Culturist 37:140–142.

Colton, J. B., Jr., J. R. Green, R. R. Byron, and J. L. Frisella. 1980. Bongo net retention rates as effected by towing speed and mesh size. Canadian Journal of Fisheries and Aquatic Sciences 37:606–623.

Colton, J. B., Jr., K. A. Honey, and R. F. Temple. 1961. The effectiveness of sampling methods used to study the distribution of larval herring in the Gulf of Maine. Journal du Conseil International pour l'Exploration de la Mer 26:180–190.

Colton, J. B., Jr., and R. R. Marak. 1969. Guide for identifying the common plankton fish eggs and larvae of continental shelf waters, Cape Sable to Block Island. U.S. Bureau of Commercial Fisheries Biological Laboratory, Laboratory Reference 69-9, Woods Hole, Massachusetts.

Comparini, A., and E. Rodino. 1980. Electrophoretic evidence from two species of *Anguilla* leptocephali in the Sargasso Sea. Nature (London) 287:435–437.

Conrow, R., and A. V. Zale. 1985. Early life history stages of fishes of Orange Lake, Florida, an illustrated identification manual. Florida Cooperative Fish and Wildlife Research Unit Technical Report 15, Gainsville.

Conrow, R., A. V. Zale, and R. W. Gregory. 1990. Distributions and abundances of early life stages of fishes in a Florida lake dominated by aquatic macrophytes. Transactions of the American Fisheries Society 119:521–528.

Cooper, J. E. 1977. Durable ichthyoplankton sampler for small boats. Progressive Fish-Culturist 39:170–171.

Copp, G. H., and M. Peňáz. 1988. Ecology of fish spawning and nursery zones in the flood plain, using a new sampling approach. Hydrobiologia 169:209–224.

Crim, L. W., and B. D. Glebe. 1990. Reproduction. Pages 529–553 *in* C. B. Schreck and P. B. Moyle, editors. Methods for fish biology. American Fisheries Society, Bethesda, Maryland.

Cuinat, R. 1967. Contribution to the study of physical parameters in electrical fishing in rivers with direct current. Pages 131–173 *in* R. Vibert, editor. Fishing with electricity, its application to biology and management. Fishing News Books, Surrey, UK.

Currie, R. I., and P. Foxton. 1956. The Nansen closing method with vertical plankton nets. Journal of the Marine Biological Association of the United Kingdom 35:483–492.

Cushing, D. H. 1970. The regularity of the spawning season of some fishes. Journal du Conseil International pour l'Exploration de la Mer 33:81–97.

Cyr, H., J. A. Downing, S. Lalonde, S. B. Baines, and L. M. Pace. 1992. Sampling larval fish populations: choice of sample number and size. Transactions of the American Fisheries Society 121:356–368.

D'Amours, D., and F. Grégoire. 1992. Analytical correction for oversampled Atlantic mackerel *Scomber scombrus* eggs collected with oblique plankton tows. U.S. National Marine Fisheries Service Fishery Bulletin 90:190–196.

de Leaniz, C. G., N. Fraser, and F. Huntingford. 1993. Dispersal of Atlantic salmon fry from a natural redd: evidence for undergravel movements? Canadian Journal of Zoology 71:1454–1457.

DeLeon, M. F., R. O. Reese, and W. J. Conley. 1991. Effects of fixation and dehydration on shrinkage and morphology in common snook yolk-sac larvae. Pages 121–128 *in* R. D.

Hoyt, editor. Larval fish recruitment and research in the Americas. NOAA (National Oceanic and Atmospheric Administration) Technical Report NMFS (National Marine Fisheries Service) 95.

DeMartini, E. E., and R. K. Fountain. 1981. Ovarian cycling frequency and batch fecundity in the queenfish, *Seriphus politus*; attributes representative of serial spawning fishes. U.S. National Marine Fisheries Service Fishery Bulletin 79:547–560.

Dempsey, C. H. 1988. Ichthyoplankton entrainment. Journal of Fish Biology 33(A):93–102.

Dennis, G. D., D. Goulet, and J. R. Rooker. 1991. Ichthyoplankton assemblages sampled by night lighting in nearshore habitats of southwestern Puerto Rico. Pages 89–97 *in* R. D. Hoyt editor. Larval fish recruitment and research in the Americas. NOAA (National Oceanic and Atmospheric Administration) Technical Report NMFS (National Marine Fisheries Service) 95.

Dewey, M. R. 1992. Effectiveness of a drop net, a pop net, and an electrofishing frame for collecting quantitative samples of juvenile fishes in vegetation. North American Journal of Fisheries Management 12:808–813.

Dewey, M. R., L. E. Holland-Bartels, and S. J. Zigler. 1989. Comparison of fish catches with buoyant pop nets and seines in vegetated and nonvegetated habitats. North American Journal of Fisheries Management 9:249–253.

Ditty, J. G. 1989. Separating early larvae of sciaenids from the western North Atlantic: a review and comparison of larvae off Louisiana and Atlantic coast of the U.S. Bulletin of Marine Science 44:1083–1105.

Ditty, J. G., E. D. Houde, and R. F. Shaw. 1994. Egg and larval development of spanish sardine, *Dardinella aurita* (Family Clupeidae), with a synopsis of characters to identify clupeid larvae from the northern Gulf of Mexico. Bulletin of Marine Science 54:367–380.

Ditty, J. G., and R. F. Shaw. 1994. Preliminary guide to the identification of the early life stages of sciaenid fishes from the western central Atlantic. NOAA (National Oceanic and Atmospheric Administration) Technical Memorandum NMFS-SEFSC-349.

Doherty, P. J. 1987. Light-traps: selective but useful devices for quantifying the distributions and abundances of larval fishes. Bulletin of Marine Science 41:423–431.

Dorr, J. A., III. 1974. Construction of an inexpensive lighted sorting chamber. Progressive Fish-Culturist 36:63–64.

Dorr, J. A., III, D. V. O'Conner, N. R. Foster, and D. J. Jude. 1981. Substrate conditions and abundance of lake trout eggs in a traditional spawning area in Lake Michigan. North American Journal of Fisheries Management 1:165–172.

Douglas, M. E. 1993. Analysis of sexual dimorphism in an endangered cyprinid fish (*Gila cypha* Miller) using video image technology. Copeia 1993:334–343.

Dovel, W. L. 1964. An approach to sampling estuarine macroplankton. Chesapeake Science 5:77–90.

Downhower, J. F., and L. Brown. 1977. A sampling technique for benthic fish populations. Copeia 1977:403–406.

Drewry, G. E. 1979. A punch card key to the families of yolk sac larval fishes of the Great Lakes region. Drewry Publishing Company, Waldorf, Maryland.

Duncan, T. O. 1978. Collection bucket for use with tow nets for larval fish. Progressive Fish-Culturist 40:118–119.

Dunn, J., R. B. Mitchell, G. G. Urquhart, and B. J. Ritchie. 1993. LOCHNESS—a new multi-net midwater sampler. Journal du Conseil International pour l'Exploration de la Mer 50:203–212.

Dunn, J. R. 1984. Developmental osteology. Pages 48–50 *in* Moser et al. (1984).

Eldridge, P. J., F. H. Berry, and M. C. Miller, III. 1978. Diurnal variations in catches of selected species of ichthyoneuston by the Boothbay neuston net off Charleston, South Carolina. U.S. National Marine Fisheries Service Fishery Bulletin 76:295–297.

Elliott, E. M., and D. Jimenez. 1981. Laboratory manual for the identification of ichthyoplankton from the Beverly–Salem Harbor area. Massachusetts Division of Marine Fisheries, Boston.

Ennis, G. P. 1972. A diver-operated plankton collector. Journal of the Fisheries Research Board of Canada 29:341–343.

Faber, D. J. 1968. A net for catching limnetic fry. Transactions of the American Fisheries Society 97:61–63.

Faber, D. J. 1980. Observations on the early life of the golden shiner, *Notemigonus crysoleucas* (Mitchill), in Lac Heney, Québec. Pages 69–78 *in* L. A. Fuiman, editor. Proceedings of the fourth annual larval fish conference. U.S. Fish and Wildlife Service Biological Services Program FWS/OBS-80/43.

Faber, D. J. 1981. A light trap to sample littoral and limnetic regions of lakes. Internationale Vereinigung für Theoretische und Angewandte Limnologie Verhandlungen 21:776–781.

Faber, D. J. 1982. Fish larvae caught by a light-trap at littoral sites in Lac Heney, Québec, 1979 and 1980. Pages 42–46 *in* C. F. Bryan, J. V. Conner, and F. M. Truesdale, editors. Fifth annual larval fish conference. Louisiana Cooperative Fish and Wildlife Research Unit, Louisiana State University, Baton Rouge.

Faber, D. J., and S. Gadd. 1983. Several drawing techniques to illustrate larval fishes. Transactions of the American Fisheries Society 112:349–353.

Fahay, M. P. 1983. Guide to the early stages of marine fishes occurring in the Western North Atlantic Ocean, Cape Hatteras to the Southern Scotian Shelf. Journal of Northwest Atlantic Fishery Science 4:1–423.

Farooqi, T., R. F. Shaw, and J. G. Ditty. 1995. Preliminary guide to the identification of the early life history stages of anchovies (family Engraulidae) of the western central Atlantic. NOAA (National Oceanic and Atmospheric Administration) Technical Memorandum NMFS-SEFSC-358.

Field-Dodgson, M. S. 1983. Emergent fry trap for salmon. Progressive Fish-Culturist 45:175–176.

Fish, M. P. 1932. Contributions to the early life histories of sixty-two species of fishes from Lake Erie and its tributary waters. U.S. Bureau of Fisheries Bulletin 47:293–398.

Flath, L. E., and J. A. Dorr, III. 1984. A portable, diver-operated, underwater pumping device. Progressive Fish-Culturist 46:219–220.

Floyd, K. B., W. H. Courtenay, and R. D. Hoyt. 1984a. A new larval fish light trap: the Quatrefoil trap. Progressive Fish-Culturist 46:216–219.

Floyd, K. B., R. D. Hoyt, and S. Timbrook. 1984b. Chronology of appearance and habitat partitioning by stream larval fishes. Transactions of the American Fisheries Society 113:217–223.

Fortier, L. and W. C. Leggett. 1983. Vertical migrations and transport of larval fish in a partially mixed estuary. Canadian Journal of Fisheries and Aquatic Sciences 40:1543–1555.

Franzin, W. G., and S. M. Harbicht. 1992. Test of drift samplers for estimating abundance of recently hatched walleye larvae in small rivers. North American Journal of Fisheries Management 12:396–405.

Fraser, J. H. 1968. The history of plankton sampling. Pages 11–18 *in* D. J. Trantor and J. H. Fraser, editors. Zooplankton sampling. United Nations Educational, Scientific, and Cultural Organization, Monographs on Oceanographic Methodology 2.

Fridirici, C. T., and L. T. Beck. 1986. A technique for hatching eggs of crevice-spawning minnows. Progressive Fish-Culturist 48:228–229.

Frolander, H. F., and I. Pratt. 1962. A bottom skimmer. Limnology and Oceanography 7:104–106.

Frost, B. W., and L. E. McCrone. 1974. Vertical distribution of zooplankton and myctophid fish at Canadian weather station P, with description of a new multiple net trawl. Pages 159–165 *in* Proceedings of the International Conference on Engineering in the Ocean Environment. Institute of Electrical and Electronics Engineers Inc., New York.

Fuiman, L. A. 1979. Descriptions and comparisons of catostomid fish larvae: north Atlantic drainage species. Transactions of the American Fisheries Society 108:560–603.

Fuiman, L. A. 1982. Correspondence of myomeres and vertebrae and their natural variability during the first year of life in yellow perch. Pages 56–59 *in* C. F. Bryan, J. V. Conner, and

F. M. Truesdale, editors. Proceedings of the fifth annual larval fish conference. Louisiana Cooperative Fish and Wildlife Research Unit, Louisiana State University, Baton Rouge.

Fuiman, L. A., J. V. Conner, B. F. Lathrop, G. L. Buynak, D. E. Snyder, and J. J. Loos. 1983. State of the art identification for cyprinid fish larvae from eastern North America. Transactions of the American Fisheries Society 112:319–322.

Galat, D. L. 1972. Preparing teleost embryos for study. Progressive Fish-Culturist 34:43–48.

Gale, W. F., and H. W. Mohr, Jr. 1978. Larval fish drift in a large river with a comparison of sampling methods. Transactions of the American Fisheries Society 107:46–55.

Gallagher, R. P., and J. V. Conner. 1980. Spatio-temporal distribution of ichthyoplankton in the lower Mississippi River, Louisiana. Pages 101–115 *in* L. A. Fuiman, editor. Proceedings of the fourth annual larval fish conference. U.S. Fish and Wildlife Service Biological Services Program FWS/OBS-80/43.

Gallagher, R. P., and J. V. Conner. 1983. Comparison of two ichthyoplankton sampling gears with notes on microdistribution of fish larvae in a large river. Transactions of the American Fisheries Society 112:280–285.

Gammon, J. R. 1965. Device for collecting eggs of muskellunge, northern pike, and other scatter-spawning species. Progressive Fish-Culturist 27:78.

Garrison, K. J., and B. S. Miller. 1982. Review of the early life history of Puget Sound fishes. University of Washington Fisheries Research Institute, Seattle.

Gates, D. W., J. S. Bulak, and J. S. Crane. 1987. Preservation of striped bass eggs collected from a low-hardness freshwater system in South Carolina. Progressive Fish-Culturist 49:230–232.

Gehringer, J. W. 1952. High speed plankton samplers. U.S. Fish and Wildlife Service Special Scientific Report - Fisheries 88:7–12.

Gehringer, J. W., and W. Aron. 1968. Field techniques. Pages 87–104 *in* D. J. Trantor and J. H. Fraser, editors. Zooplankton sampling. United Nations Educational, Scientific, and Cultural Organization, Monographs on Oceanographic Methodology 2.

Gerrick, D. J. 1968. A comparative study of antioxidants in color preservation of fish. The Ohio Journal of Science 68:239–240.

Govoni, J. J. 1984. Histology. Pages 40–42 *in* Moser et al. (1984).

Graham, J. J., B. J. Joule, C. L. Crosby, and D. W. Townsend. 1984. Characteristics of the Atlantic herring (*Clupea harengus* L.) spawning population along the Maine coast, inferred from larval studies. Journal of Northwest Atlantic Fishery Science 5:131–142.

Graham, J. J., and P. M. W. Venno. 1968. Sampling herring from tidewaters with buoyed and anchored nets. Journal of the Fisheries Research Board of Canada 25:1169–1179.

Graser, L. F. 1978. Flow-through collection bucket for larval fish. Progressive Fish-Culturist 40:78–79.

Graves, J. E., M. J. Curtis, P. A. Oeth, and R. S. Waples. 1989. Biochemical genetics of southern California basses of the genus *Paralabrax*: specific identification of fresh and ethanol-preserved individual eggs and early larvae. U.S. National Marine Fisheries Service Fishery Bulletin 88:59–66.

Gregory, R. S., and P. M. Powles. 1985. Chronology, distribution, and sizes of larval fish sampled by light traps in macrophytic Chemung Lake. Canadian Journal of Zoology 63:2569–2577.

Gregory, R. S., and P. M. Powles. 1988. Relative selectivities of Miller high-speed samplers and light traps for collecting ichthyoplankton. Canadian Journal of Fisheries and Aquatic Sciences 45:993–998.

Grewe, P, M. C. C. Krueger, J. E. Marsden, C. F. Aquadro, and B. May. 1994. Hatchery origins of naturally produced lake trout fry captured in Lake Ontario: temporal and spatial variability based on allozyme and mitochondrial DNA data. Transactions of the American Fisheries Society 123:309–320.

Griffiths, F. B., A. Fleminger, B. Limor, and M. Vannucci. 1976. Shipboard and curating techniques. Pages 17–33 *in* H. F. Steedman, editor. Zooplankton fixation and preservation. Monographs on Oceanographic Methodology 4.

Gustafson-Marjanen, K. I., and H. B. Dowse. 1983. Seasonal and diel patterns of emergence from the redd of Atlantic salmon (*Salmo salar*) fry. Canadian Journal of Fisheries and Aquatic Sciences 40:813–817.

Haldorson, L., M. Pritchett, D. Sterritt, and J. Watts. 1993. Abundance patterns of marine fish larvae during spring in a southeastern Alaskan Bay. U.S. National Marine Fisheries Service Fishery Bulletin 91:36–44.

Halliday, R. G., and B. Roscoe. 1969. The effects of icing and freezing on the length and weight of groundfish species. International Commission for the Northwest Atlantic Fisheries Redbook part 3.

Hamley, J. M., T. P. Howley, and A. L. Punhani. 1983. Estimating larval fish abundances from plankton net catches in Long Point Bay, Lake Erie, in 1971–1978. Journal of Great Lakes Research 9:452–467.

Hardy, A. C. 1936. The continuous plankton recorder. Discovery Reports 11:457–510.

Hardy, J. D., G. E. Drewry, R. A. Fritzche, G. D. Johnson, P. W. Jones, and F. D. Martin. 1978. Development of fishes of the mid-Atlantic bight, an atlas of eggs, larvae, and juvenile stages, volumes 1–6. U.S. Fish and Wildlife Service FWS/OBS-78/12.

Harris, R. P., L. Fortier, and R. K. Young. 1986. A large-volume pump system for studies of the vertical distribution of fish larvae under open sea conditions. Journal of the Marine Biological Association of the United Kingdom 66:845–854.

Haug, T., E. Kjørsvik, and P. Solemdal. 1984. Vertical distribution of Atlantic halibut (*Hippoglossus hippoglossus*) eggs. Canadian Journal of Fisheries and Aquatic Sciences 41:798–804.

Haury, L. R., P. H. Wiebe, and S. H. Boyd. 1976. Longhurst–Hardy plankton recorders: their design and use to minimize bias. Deep-Sea Research 23:1217–1229.

Hay, D. E. 1981. Effects of capture and fixation on gut contents and body size of Pacific herring larvae. Conseil International pour l'Exploration de la Mer Rapports et Proces—Verbaux des Reunions 178:395–400.

Hay, D. E. 1982. Fixation shrinkage of herring larvae: effects of salinity, formalin concentration, and other factors. Canadian Journal of Fisheries and Aquatic Sciences 39:1138–1143.

Heath, M. R., and J. Walker. 1987. A preliminary study of the drift of larval herring (*Clupea harengus* L.) using gene-frequency data. Journal du Conseil International pour l'Exploration de la Mer 43:139–145.

Hempel, G. 1979. Early life history of marine fish: the egg stage. Washington Sea Grant Program, Seattle, Washington.

Hempel, G., and H. Weikert. 1972. The neuston of the subtropical and boreal Northeastern Atlantic Ocean. A review. Marine Biology 13:70–88.

Herke, W. H. 1978. Subsampler for estimating the number and length frequency of small preserved nektonic organisms. U.S. National Marine Fisheries Service Fishery Bulletin 76:490–494.

Hermes, R., N. N. Navaluna, and A. C. del Norte. 1984. A push-net ichthyoplankton sampler attachment to an outrigger boat. Progressive Fish-Culturist 46:67–70.

Heron, A. C. 1968. Plankton gauze. Pages 19–25 *in* D. J. Trantor and J. H. Fraser, editors. Zooplankton sampling. United Nations Educational, Scientific, and Cultural Organization, Monographs on Oceanographic Methodology 2.

Hettler, W. F. 1979. Modified neuston net for collecting live larval and juvenile fish. Progressive Fish-Culturist 41:32–33.

Hjort, J. 1914. Fluctuations in the great fisheries of northern Europe viewed in the light of biological research. Conseil International pour l'Exploration de la Mer Rapports et Proces—Verbaux des Reunions 20:1–228.

Hoar, W. S., and D. J. Randall, editors. 1988. Fish physiology, volume 11: the physiology of developing fish. Part A: eggs and larvae. Academic Press, New York.

Hodson, R. G., C. R. Bennett, and R. J. Monroe. 1981. Ichthyoplankton samplers for simultaneous replicate samples at surface and bottom. Estuaries 4:176–184.

Hogue, J. J., Jr., R. Wallus, and L. K. Kay. 1976. Preliminary guide to the identification of larval fishes in the Tennessee River. Tennessee Valley Authority Technical Note B19, Norris, Tennessee.

Holland, L. E., and G. S. Libey. 1981. Boat attachments for ichthyoplankton studies in small impoundments. Progressive Fish-Culturist 43:50–51.

Holland-Bartels, L. E., S. K. Littlejohn, and M. L. Huston. 1990. A guide to larval fishes of the upper Mississippi River. U.S. Fish and Wildlife Service National Fisheries Research Center, La Cross, Wisconsin, and Minnesota Extension Service, University of Minnesota, St. Paul.

Horns, W. H., J. E. Marsden, and C. C. Krueger. 1989. Inexpensive method for quantitative assessment of lake trout egg deposition. North American Journal of Fisheries Management 9:280–286.

Houde, E. D., and J. A. Lovdal. 1984. Seasonality of occurrence, foods and food preferences of ichthyoplankton in Biscayne Bay, Florida. Estuarine Coastal and Shelf Science 18:403–419.

Houser, A. 1983. Diving plane for a Tucker midwater trawl. Progressive Fish-Culturist 45:48–50.

Hoyt, R. D. 1988. A bibliography of the early life history of fishes, volumes 1 and 2. Western Kentucky University, Department of Biology, Bowling Green.

Humphries, J. M., F. L. Bookstein, B. Chernoff, G. R. Smith, R. L. Elder, and S. G. Poss. 1981. Multivariate discriminations by shape in relation to size. Systematic Zoology 30:291–308.

Hunter, J. R. 1984. Synopsis of culture methods for marine fish larvae. Pages 24–27 *in* Moser et al. (1984).

Hunter, J. R., D. C. Aasted, and C. T. Mitchell. 1966. Design and use of a miniature purse seine. Progressive Fish-Culturist 28:175–179.

Isaacs, J. D., and L. W. Kidd. 1953. Isaacs–Kidd mid-water trawl. Scripps Institute of Oceanography Equipment Report 1:1–18.

Jenkins, R. E., and N. M. Burkhead. 1993. Freshwater fishes of Virginia. American Fisheries Society, Bethesda, Maryland.

Johnson, A. G., F. M. Utter, and H. O. Hodgins. 1975. Study of the feasibility of immunochemical methods for identification of pleuronectid eggs. Journal du Conseil International pour l'Exploration de la Mer 36:158–161.

Johnston, C. E., and J. C. Cheverie. 1988. Observations on the diel and seasonal drift of eggs and larvae of anadromous rainbow smelt, *Osmerus mordax*, and blueback herring, *Alosa aestivalis*, in a coastal stream. Canadian Field Naturalist 102:508–514.

Jones, B. C., and G. H. Geen. 1977. Morphometric changes in elasmobranch (*Squalus acanthias*) after preservation. Canadian Journal of Zoology 55:1060–1062.

Jones, D. 1976. Chemistry of fixation and preservation with aldehydes. Pages 155–171 *in* H. F. Steedman, editor. Zooplankton fixation and preservation. Monographs on Oceanographic Methodology 4.

Karakiri, M., and H. von Westernhagen. 1988. Apparatus for grinding otoliths of larval and juvenile fish for microstructure analysis. Marine Ecology Progress Series 49:195–198.

Kay, L. K., R. Wallus, and B. L. Yeager. 1994. Reproductive biology and early life history of fishes in the Ohio River drainage. Volume 2: Catostomidae. Tennessee Valley Authority, Chattanooga, Tennessee.

Kendall, A. W., Jr., and A. C. Matarese. 1994. Status of early life history descriptions of marine teleosts. U.S. National Marine Fisheries Service Fishery Bulletin 92:725–736.

Kendall, A. W., Jr., E. H. Ahlstrom, and H. G. Moser. 1984. Early life history stages of fishes and their characters. Pages 11–22 *in* Moser et al. (1984).

Kindschi, G. A., R. D. Hoyt, and G. J. Overmann. 1979. Some aspects of the ecology of larval fishes in Rough River Lake, Kentucky. Pages 139–166 *in* R. D. Hoyt, editor. Proceedings of the third symposium on larval fish. Western Kentucky University, Department of Biology, Bowling Green.

Kingsford, M. J., and J. H. Choat. 1985. The fauna associated with drift algae captured with a plankton-mesh purse seine net. Limnology and Oceanography 30:618–630.

Kinzer, J. 1966. An opening and closing mechanism for the high-speed plankton sampler HAI. Deep-Sea Research 13:473–474.

Klinger, R. C., and M. J. Van Den Avyle. 1993. Preservation of striped bass eggs: effects of formalin concentration, buffering, stain and initial stage of development. Copeia 1993: 1114–1119.

Knutsen, H., E. Moksnes, and N. B. Vogt. 1985. Distinguishing between one-day-old cod (*Gadus morhua*) and haddock (*Melanogrammus aeglefinus*) eggs by gas chromatography and SIMCA pattern recognition. Canadian Journal of Fisheries and Aquatic Sciences 42:1823–1826.

Kushlan, J. A. 1981. Sampling characteristics of enclosure fish traps. Transactions of the American Fisheries Society 110:557–562.

La Bolle, L. D., Jr., H. W. Li, and B. C. Mundy. 1985. Comparison of two samplers for quantitatively collecting larval fishes in upper littoral habitats. Journal of Fish Biology 26:139–146.

Lagler, K. F., J. E. Bardach, R. R. Miller, and D. R. M. Passino. 1977. Ichthyology. Wiley, New York.

Lathrop, B. F., and D. E. Snyder. 1986. Bibliographies on fish reproduction and early life history. ELHS Newsletter 7(2):29–36. American Fisheries Society, Early Life History Section, Bethesda, Maryland.

Lavenberg, R. J., G. E. McGowen, and R. E. Woodsum. 1984. Preservation and curation. Pages 57–59 *in* Moser et al. (1984).

Leary, R. F., and H. E. Booke. 1990. Starch gel electrophoresis and species distinctions. Pages 141–170 *in* C. B. Schreck and P. B. Moyle, editors. Methods for fish biology. American Fisheries Society, Bethesda, Maryland.

Lee, D. S., C. R. Gilbert, C. H. Hocutt, R. E. Jenkins, D. E. McAllister, and J. R. Stauffer, Jr. 1980. Atlas of North American freshwater fishes. North Carolina State Museum of Natural History, Raleigh.

Leis, J. M., and D. S. Rennis. 1983. The larvae of Indo-Pacific coral reef fishes. New South Wales University Press, Sydney, Australia, and University of Hawaii Press, Honolulu.

Leis, J. M., and T. Trnski. 1989. The larvae of Indo-Pacific shorefishes. University of Hawaii Press, Honolulu.

Leithiser, R. L., K. F. Ehrlich, and A. B. Thum. 1979. Comparison of a high volume pump and conventional plankton nets for collecting fish larvae entrained in power plant cooling systems. Journal of the Fisheries Research Board of Canada 36:81–84.

Lenarz, W. H. 1972. Mesh retention of larvae of *Sardinops caerulea* and *Engraulis mordax* by plankton nets. U.S. National Marine Fisheries Service Fishery Bulletin 70:839–848.

Lenarz, W. H. 1973. Dependence of catch rates on size of fish larvae. Conseil International pour l'exploration de la Mer, Rapports et Proces-Verbaux des Reunions 164:270–275.

Leslie, J. K. 1986. Nearshore contagion and sampling of freshwater larval fish. Journal of Plankton Research 8:1137–1147.

Leslie, J. K., and C. A. Timmins. 1989. Double nets for mesh aperture selection and sampling in ichthyoplankton studies. Fisheries Research 7:225–232.

Lewis, R. M., W. F. Hettler, Jr., E. P. H. Wilkins, and G. N. Johnson. 1970. A channel net for catching larval fishes. Chesapeake Science 11:196–197.

Lewis, S. A., and D. D. Garriott. 1971. A modified Folsom plankton splitter for analysis of meter net samples. Proceedings of the Annual Conference Southeastern Association of Game and Fish Commissioners 24(1970):332–337.

Lindsey, C. C. 1984. Fish illustrations: how and why. Environmental Biology of Fishes 11:3–14.

Lindsay, J. A., and E. R. Radle. 1978. A supplemental sampling method for estuarine ichthyoplankton with emphasis on the Atherinidae. Estuaries 1:61–64.

Lippincott, B. L., and R. F. Thomas. 1983. Neuston net for sampling surface ichthyoplankton. Progressive Fish-Culturist 45:188–190.

Lippson, A. J., and R. L. Moran. 1974. Manual for identification of early developmental stages of fishes of the Potomac River estuary. Martin Marietta Corporation, Environmental Technology Center, PPSP-MP-13, Baltimore, Maryland.

Longhurst, A. R., A. D. Reith, R. E. Bower, and D. L. R. Seibert. 1966. A new system for the collection of multiple plankton samples. Deep-Sea Research 13:213–222.

Lough, R. G., and D. C. Potter. 1993. Vertical distribution patterns and diel migrations of larval and juvenile haddock *Melanogrammus aeglefinus* and Atlantic cod *Gadus morhua* on Georges Bank. U.S. National Marine Fisheries Service Fishery Bulletin 91:281–303.

Lowerre-Barbieri, S. K., and L. R. Barbieri. 1993. A new method of oocyte separation and preservation for fish reproduction studies. U.S. National Marine Fisheries Service Fishery Bulletin 91:165–170.

Madenjian, C. P., and D. J. Jude. 1985. Comparison of sleds versus plankton nets for sampling fish larvae and eggs. Hydrobiologia 124:275–281.

Mahnken, C. V. W., and J. W. Jossi. 1967. Flume experiments on the hydrodynamics of plankton nets. Journal du Conseil International pour l'Exploration de la Mer 31:38–45.

Mansueti, A. J., and D. J. Hardy, Jr. 1967. Development of fishes of the Chesapeake Bay region: an atlas of egg, larval, and juvenile stages, part 1. University of Maryland, Natural Resources Institute, Baltimore.

Manz, J. V. 1964. A pumping device used to collect walleye eggs from offshore spawning areas in western Lake Erie. Transactions of the American Fisheries Society 93:204–206.

Marcy, B. C., Jr., and M. D. Dahlberg. 1980. Sampling problems associated with ichthyoplankton field-monitoring studies with emphasis on entrainment. Pages 233–252 *in* C. H. Hocutt and J. R. Stauffer, Jr., editors. Biological monitoring of fish. Lexington Books, Lexington, Massachusetts.

Markle, D. F. 1984. Phosphate buffered formalin for long term preservation for formalin fixed ichthyoplankton. Copeia 1984:525–528.

Marley, R. D. 1983. Spatial distribution patterns of planktonic fish eggs in lower Mobile Bay, Alabama. Transactions of the American Fisheries Society 112:257–266.

Marsden, J. E., C. C. Krueger, and H. M. Hawkins. 1991. An improved trap for passive capture of demersal eggs during spawning: an efficiency comparison with egg nets. North American Journal of Fisheries Management 11:364–368.

Mason, W. T., and P. P. Yevich. 1967. The use of phloxine b and rose bengal stains to facilitate sorting of benthic samples. Transactions of the American Microscopical Society 86:221–223.

Matarese, A. C., and E. M. Sandknop. 1984. Identification of fish eggs. Pages 27–31 *in* Moser et al. (1984).

Matarese, A. C., A. W. Kendall, Jr., D. M. Blood, and B. M. Vinter. 1989. Laboratory guide to early life history states of northeast Pacific fishes. NOAA (National Oceanic and Atmospheric Administration) Technical Report NMFS (National Marine Fisheries Service) 80.

May, E. B., and C. R. Gasaway. 1967. A preliminary key to the identification of larval fishes of Oklahoma, with particular reference to Canton Reservoir, including a selected bibliography. Oklahoma Department of Wildlife Conservation, Oklahoma Fishery Research Laboratory Bulletin 5, Norman.

May, R. C. 1974. Larval mortality in marine fishes and the critical period concept. Pages 3–19 *in* J. H. S. Blaxter, editor. The early life history of fish. Springer-Verlag, Berlin.

Mayden, R. L., and E. O. Wiley. 1984. A method of preparing disarticulated skeletons of small fishes. Copeia 1984:230–232.

McGowan, E. G. 1984. An identification guide for selected larval fishes from Robinson Impoundment, South Carolina. Carolina Power and Light Company, Ecological Services Section, Biology Unit, New Hill, North Carolina.

McGowan, E. G. 1988. An illustrated guide to larval fishes from three North Carolina piedmont impoundments. Carolina Power and Light Company, Ecological Services Section, Biology Unit, New Hill, North Carolina.

McGroddy, P. M., and R. L. Wyman. 1977. Efficiency of nets and a new device for sampling living fish larvae. Journal of the Fisheries Research Board of Canada 34:571–574.

McGurk, M. D. 1985. Effects of net capture on the postpreservation morphometry, dry weight, and condition factor of Pacific herring larvae. Transactions of the American Fisheries Society 114:348–355.

McLain, A. L., and F. H. Dahl. 1968. An electric beam trawl for the capture of larval lamprey. Transactions of the American Fisheries Society 97:289–292.

Meador, M. R., and J. S. Bulak. 1987. Quantifiable ichthyoplankton sampling in congested shallow-water areas. Journal of Freshwater Ecology 4:65–69.

Miller, D. 1961. A modification of the small Hardy plankton sampler for simultaneous high-speed plankton hauls. Bulletin of Marine Ecology 5:165–172.

Miller, J. M. 1973. A quantitative push-net system for transect studies of larval fish and macrozooplankton. Limnology and Oceanography 18:175–178.

Miller, J. M., and J. W. Tucker. 1979. X-radiography of larval and juvenile fishes. Copeia 1979:535–538.

Miller, J. M., J. P. Reed, and L. J. Pietrafesa. 1984. Patterns, mechanisms, and approaches to the study of estuarine dependent fish larvae and juveniles. Pages 209–225 *in* J. D. McCleave, G. P. Arnold, J. J. Dodson, and W. H. Neill, editors. Mechanisms of migration in fishes. Plenum, New York.

Miller, J. M., W. Watson, and J. M. Leis. 1979. An atlas of common nearshore marine fish larvae of the Hawaiian Islands. University of Hawaii Sea Grant College Program, Sea Grant Miscellaneous Report UNIHI-SEAGRANT-MR-80-02, Honolulu.

Mitterer, L. G., and W. D. Pearson. 1977. Rose bengal stain as an aid in sorting larval fish samples. Progressive Fish-Culturist 39:119–120.

Morgan, R. P., II. 1975. Distinguishing larval white perch and striped bass by electrophoresis. Chesapeake Science 16:68–70.

Mork, J., P. Solemdal, and G. Sundnes. 1983. Identification of marine fish eggs: a biochemical genetics approach. Canadian Journal of Fisheries and Aquatic Sciences 40:361–369.

Moser, H. G., W. J. Richards, D. M. Cohen, M. P. Fahay, A. W. Kendall, and S. L. Richardson, editors. 1984. Ontogeny and systematics of fishes. American Society of Ichthyologists and Herpetologists Special Publication 1. Allen Press, Lawrence, Kansas.

Moy, P. B., and R. R. Stickney. 1987. Suspended spawning cans for channel catfish in a surface-mine lake. Progressive Fish-Culturist 49:76–77.

Murphy, G. I., and R. I. Clutter. 1972. Sampling anchovy larvae with a plankton purse seine. U.S. National Marine Fisheries Service Fishery Bulletin 70:789–798.

Muth, R. T., and C. M. Haynes. 1984. Plexiglass light trap for collecting small fishes in low-velocity riverine habitats. Progressive Fish-Culturist 46:59–62.

Nester, R. T. 1987. Horizontal ichthyoplankton tow-net system with unobstructed net opening. North American Journal of Fisheries Management 7:148–150.

Netsch, N. F., A. Houser, and L. E. Vogele. 1971. Sampling gear for larval reservoir fishes. Progressive Fish-Culturist 33:175–179.

Newell, G. E., and R. C. Newell. 1963. Marine plankton, a practical guide. Hutchinson Educational Limited, London.

Newsome, G. E., and S. K. Aalto. 1987. An egg-mass census method for tracking fluctuations in yellow perch (*Perca flavescens*) populations. Canadian Journal of Fisheries and Aquatic Sciences 44:1221–1232.

Nigro, A. A., and J. J. Ney. 1982. Reproduction and early-life accommodations of landlocked alewives to a southern range extension. Transactions of the American Fisheries Society 111:559–569.

Nishikawa, Y., and D. W. Rimmer. 1987. Identification of larval tunas, billfishes and other scombroid fishes (Suborder Scombroidie): an illustrated guide. Commonwealth Scientific and Industrial Research Organization, Marine Research Laboratories Report 186, Hobart, Australia.

Noble, R. L. 1970. Evaluation of the Miller high-speed sampler for sampling yellow perch and walleye fry. Journal of the Fisheries Research Board of Canada 27:1033–1044.

Noble, R. L. 1971. An evaluation of the meter net for sampling fry of the yellow perch, *Perca flavescens*, and walleye, *Stizostedion v. vitreum*. Chesapeake Science 12:47–48.

Norcross, B. L. 1991. Estuarine recruitment mechanisms of larval Atlantic croakers. Transactions of the American Fisheries Society 120:673–683.

Novak, P. F., and W. F. Sheets. 1969. Pumping device used to collect smallmouth bass fry. Progressive Fish-Culturist 3:240.

O'Conner, J. M., and S. A. Schaffer. 1977. The effects of sampling gear on the survival of striped bass ichthyoplankton. Chesapeake Science 18:312–315.

O'Gorman, R. 1984. Catches of larval rainbow smelt (*Osmerus mordax*) and alewife (*Alosa pseudoharengus*) in plankton nets of different mesh sizes. Journal of Great Lakes Research 10:73–77.

Olivar, M. P., and J. M. Fortuno. 1991. A guide to ichthyoplankton of the southeast Atlantic (Benguela Current region). Journal of Marine Science 55:1–383.

Paquette, R. G., and H. F. Frolander. 1957. Improvements in the Clarke–Bumpus plankton sampler. Journal du Conseil International pour l'Exploration de la Mer 22:284–288.

Park, L. K., and P. Moran. 1994. Developments in molecular genetic techniques in fisheries. Reviews in Fish Biology and Fisheries 4:272–299.

Paulson, L. J., and F. A. Espinosa, Jr. 1975. Fish trapping: a new method of evaluating fish species composition in limnetic areas of reservoirs. California Fish and Game 61: 209–214.

Pearcy, W. G. 1980. A large, opening–closing midwater trawl for sampling oceanic nekton, and comparison of catches with an Isaacs–Kidd midwater trawl. U.S. National Marine Fisheries Service Fishery Bulletin 78:529–534.

Pearse, A. G. E. 1968. Theoretical and applied histochemistry, volume 1. Little Brown, Boston.

Petering, R. W., and M. J. Van Den Avyle. 1988. Relative efficiency of a pump for sampling larval gizzard and threadfin shad. Transactions of the American Fisheries Society 117:78–83.

Phillips, A. C., and J. C. Mason. 1986. A towed, self-adjusting sled sampler for demersal fish eggs and larvae. Fisheries Research 4:235–242.

Phillips, R. W., and K. V. Koski. 1969. A fry trap method for estimating salmonid survival from egg deposition to fry emergence. Journal of the Fisheries Research Board of Canada 26:133–141.

Pietrafesa, L. J., and G. S. Janowitz. 1988. Physical oceanographic processes affecting larval transport around and through North Carolina inlets. American Fisheries Society Symposium 3:34–50.

Pitlo, J., Jr. 1989. Walleye spawning habitat in Pool 13 of the upper Mississippi River. North American Journal of Fisheries Management 9:303–308.

Porter, T. R. 1973. Fry emergence trap and holding box. Progressive Fish-Culturist 35:104–106.

Potthoff, T. 1984. Clearing and staining techniques. Pages 35–37 *in* Moser et al. (1984).

Potts, G. W., and R. J. Wootton. 1984. Fish reproduction: strategies and tactics. Academic Press, New York.

Redding, M. J., and R. Patino. 1993. Reproductive physiology. Pages 503–534 *in* D. H. Evans, editor. The physiology of fishes. CRC press, Boca Raton, Florida.

Richards, W. J. 1990. List of the fishes of the western central Atlantic and the status of early life history information. NOAA (National Oceanic and Atmospheric Administration) Technical Memorandum NMFS-SEFC-267.

Richards, W. J., and six coauthors. 1994. Preliminary guide to the identification of the early life history stages of lutjanid fishes of the western central Atlantic. NOAA (National Oceanic and Atmospheric Administration) Technical Memorandum NMFS-SEFSC-345.

Rijnsdorp, A. D., and A. Jaworski. 1990. Size-selective mortality in plaice and cod eggs: a new method in the study of egg mortality. Journal du Conseil International pour l'Exploration de la Mer 47:256–263.

Rijnsdorp, A. D., M. van Stralen, and H. H. van der Veer. 1985. Selective tidal transport of North Sea plaice larvae *Pleuronectes platessa* in coastal nursery areas. Transactions of the American Fisheries Society 114:461–470.

Russell, F. S. 1976. The eggs and planktonic stages of British marine fishes. Academic Press, New York.

Sager, D. R. 1987. Distribution of larval gizzard shad in the upper Cape Fear River, North Carolina. American Fisheries Society Symposium 2:174–178.

Sameoto, D. D., and L. O. Jaroszynski. 1976. Some zooplankton net modifications and developments. Canadian Fisheries and Marine Service Technical Report 679.

Sameoto, D. D., L. O. Jaroszynski, and W. B. Fraser. 1977. A multiple opening and closing plankton sampler based on the MOCNESS and N.I.O. Nets. Journal of the Fisheries Research Board of Canada 34:1230–1235.

Saville, A. 1958. Mesh selection in plankton nets. Journal du Conseil International pour l'Exploration de la Mer 23:192–201.

Schnack, D. 1974. On the reliability of methods for quantitative surveys of fish larvae. Pages 201–212 *in* J. H. S. Blaxter, editor. The early life history of fish. Springer-Verlag, New York.

Scobbie, A. E., and I. M. Mackie. 1990. The use of dodecyl sulphate-polyacrylamide gel electrophoresis in species identification of fish eggs. Comparative Biochemistry and Physiology 96B:743–746.

Scott, W. B., and E. J. Crossman. 1973. Freshwater fishes of Canada. Fisheries Research Board of Canada Bulletin 184, Ottawa.

Scotton, L. N., R. E. Smith, K. S. Price, and D. P. de Sylva. 1973. Pictorial guide to fish larvae of the Delaware Bay, with information useful for the study of fish larvae. University of Delaware, College of Marine Studies, Delaware Bay Report Series 7.

Secor, D. H., J. M. Dean, and J. Hansbarger. 1992. Modification of the Quatrefoil light trap for use in hatchery ponds. Progressive Fish-Culturist 54:202–205.

Shenker, J. M. 1988. Oceanographic associations of neustonic larval and juvenile fishes and dungeness crab megalopae off Oregon. U.S. National Marine Fisheries Service Fishery Bulletin 86:299–317.

Sidell, B. D., and R. G. Otto. 1978. A biochemical method for distinction of striped bass and white bass perch larvae. Copeia 1978:340–343.

Siefert, R. E. 1969. Characteristics for separation of white and black crappie larvae. Transactions of the American Fisheries Society 98:326–328.

Siler, J. R. 1986. Comparison of rotenone and trawling methodology for estimating threadfin shad populations. Pages 73–78 *in* G. E. Hall and M. J. Van Den Avyle, editors. Reservoir fisheries management: strategies for the 80's. American Fisheries Society, Southern Division, Reservoir Committee, Bethesda, Maryland.

Simon, T. P. 1986. A listing of regional guides, keys, and selected comparative descriptions of freshwater and marine larval fishes. American Fisheries Society, Early Life History Section Newsletter 7(1):10–15.

Smith, P. E., R. C. Counts, and R. I. Clutter. 1968. Changes in filtering efficiency of plankton nets due to clogging under tow. Journal du Conseil International pour l'Exploration de la Mer 32:232–248.

Smith, P. E., and S. L. Richardson. 1977. Standard techniques for pelagic fish egg and larva surveys. FAO (Food and Agriculture Organization of the United Nations) Fisheries Technical Paper 175.

Smith, P. E., and S. L. Richardson. 1979. Selected bibliography on pelagic fish egg and larva surveys. FAO (Food and Agriculture Organization of the United Nations) Fisheries Circular 706.

Smith, P. E., H. Santander, and J. Alheit. 1989. Comparison of the mortality rates of Pacific sardine, *Sardinops sagax*, and Peruvian anchovy, *Engraulis ringens*, eggs off Peru. U.S. National Marine Fisheries Service Fishery Bulletin 87:497–508.

Smith, R. E., D. P. de Sylva, and R. A. Livellara. 1964. Modification and operation of the Gulf I-A high-speed plankton sampler. Chesapeake Science 5:72–76.

Smith, W. G., and W. W. Morse. 1993. Larval distribution patterns: early signals for the collapse/recovery of Atlantic herring *Clupea harengus* in the Georges Bank area. U.S. National Marine Fisheries Service Fishery Bulletin 91:338–347.

Snyder, D. E. 1976. Terminologies for intervals of larval fish development. Pages 41–60 *in* J. Boreman, editor. Great Lakes fish egg and larvae identification; proceedings of a workshop. U.S. Fish and Wildlife Service FWS/OBS-76/23.

Snyder, D. E. 1979. Myomere and vertebra counts of the North American cyprinids and catostomids. Pages 53–69 *in* R. D. Hoyt, editor. Proceedings of the third symposium on larval fish. Western Kentucky University, Bowling Green.

Snyder, D. E. 1981. Contributions to a guide to the cypriniform fish larvae of the upper Colorado River system. United States Bureau of Land Management, Biological Sciences Series 3, Denver, Colorado.

Snyder, D. E. 1990. Fish larvae—ecologically distinct organisms. Pages 20–23 *in* M. B. Bain, editor. Ecology and assessment of warmwater streams: workshop synopsis. U.S. Fish and Wildlife Service Biological Report 90(5).

Snyder, D. E., and J. G. Holt. 1984. Terminology workshop. American Fisheries Society Early Life History Newsletter 5(2):14–15.

Snyder, D. E., and R. T. Muth. 1988. Descriptions and identification of June, Utah, and mountain sucker larvae and early juveniles. Utah State Division of Wildlife Resources Publication 88-8, Salt Lake City.

Snyder, D. E., and R. T. Muth. 1990. Descriptions and identification of razorback, flannelmouth, white, bluehead, mountain, and Utah sucker larvae and early juveniles. Colorado Division of Wildlife Technical Publication 38.

Somerton, D. A., and D. R. Kobayashi. 1989. A method for correcting catches of fish larvae for the size selection of plankton nets. U.S. National Marine Fisheries Service Fishery Bulletin 87:447–455.

Southward, A. J., and B. M. Bary. 1980. Observations on the vertical discrimination of eggs and larvae of mackerel and other teleosts in the Celtic Sea and on the sampling performance of different nets in relation to stock evaluation. Journal of the Marine Biological Association of the United Kingdom 60:295–311.

Stauffer, T. M. 1981. Collecting gear for lake trout eggs and fry. Progressive Fish-Culturist 43:186–193.

Steedman, H. F. 1976. General and applied data on formaldehyde fixation and preservation of marine zooplankton. Pages 103–154 *in* H. F. Steedman, editor. Zooplankton fixation and preservation. Monographs on Oceanographic Methodology 4.

Strauss, R. E., and C. E. Bond. 1990. Taxonomic methods: morphology. Pages 109–140 *in* C. B. Schreck and P. B. Moyle, editors. Methods for fish biology. American Fisheries Society, Bethesda, Maryland.

Strauss, R. E., and F. L. Bookstein. 1982. The truss: body form reconstructions in morphometrics. Systematic Zoology 31:113–135.

Strauss, R. E., and L. A. Fuiman. 1985. Quantitative comparisons of body form and allometry in larval and adult Pacific sculpins (Teleostei: Cottidae). Canadian Journal of Zoology 63:1582–1589.

Sumida, B. Y., B. B. Washington, and W. A. Laroche. 1984. Illustrating fish eggs and larvae. Pages 33–35 *in* Moser et al. (1984).

Suthers, I. M., and K. T. Frank. 1989. Inter-annual distributions of larval and pelagic juvenile cod (*Gadus morhua*) in southwestern Nova Scotia determined with two different gear types. Canadian Journal of Fisheries and Aquatic Sciences 46:591–602.

Tarplee, W. H., Jr., W. T. Bryson, and R. G. Sherfinski. 1979. Portable push-net apparatus for sampling ichthyoplankton. Progressive Fish-Culturist 41:213–215.

Taylor, W. R. 1967. An enzyme method of clearing and staining small vertebrates. Proceedings of the United States National Museum 122:1–17.

Taylor, W. R. 1977. Observations on specimen fixation. Proceedings of the Biological Society of Washington 90:753–763.

Thayer, G. W., D. R. Colby, M. A. Kjelson, and M. P. Weinstein. 1983. Estimates of larval-fish abundance: diurnal variation and influences of sampling gear and towing speed. Transactions of the American Fisheries Society 112:272–279.

Theilacker, G. H. 1980. Changes in body measurements of larval northern anchovy, *Engraulis mordax*, and other fishes due to handling and preservation. U.S. National Marine Fisheries Service Fishery Bulletin 78:685–692.

Thorisson, K. 1994. Is metamorphosis a critical interval in the early life of marine fishes? Environmental Biology of Fishes 40:23–36.

Tomljanovich, D. A., and J. H. Heuer. 1986. Passage of gizzard and threadfin shad larvae through a larval fish net with 500-μm openings. North American Journal of Fisheries Management 6:256–259.

Trantor, D. J., and A. C. Heron. 1965. Filtration characteristics of Clarke–Bumpus samplers. Australian Journal of Marine and Freshwater Research 16:281–291.

Trantor, D. J., and P. E. Smith. 1968. Filtration performance. Monographs on Oceanographic Methodology 2:27–56.

Trippel, E. A., and E. J. Crossman. 1984. Collapsible fishtrap of plexiglass, bristles, and netting. Progressive Fish-Culturist 46:159.

Tuberville, J. D. 1979. Vertical distribution of ichthyoplankton in upper Nickajack Reservoir, Tennessee, with comparison of three sampling methodologies. Pages 185–203 *in* R. D. Hoyt, editor. Proceedings of the third symposium on larval fish. Western Kentucky University, Department of Biology, Bowling Green.

Tucker, G. H. 1951. Relation of fishes and other organisms to the scattering of underwater sound. Journal of Marine Research 10:215–238.

Tucker, J. W., Jr., and A. J. Chester. 1984. Effects of salinity, formalin concentration and buffer on quality of preservation of Southern Flounder (*Paralichthys lethostegma*) larvae. Copeia 1984:981–988.

Tucker, J. W., Jr., and J. L. Laroche. 1984. Radiographic techniques in studies of young fishes. Pages 37–39 *in* Moser et al. (1984).

Valcarce, R., D. Stevenson, G. G. Smith, and J. W. Sigler. 1991. Discriminant separation of fishes: a preliminary study of fish eggs by Curie-point Pyrolysis-mass spectrometry-chemometric analysis. Transactions of the American Fisheries Society 120:796–802.

Vannucci, M. 1968. Loss of organisms through the meshes. Monographs on Oceanographic Methodology 2:77–86.

Vogele, L. E., R. L. Boyer, and W. R. Heard. 1971. A portable underwater suction device. Progressive Fish-Culturist 33:62–63.

Wallus, R., T. P. Simon, and B. L. Yeager. 1990. Reproductive biology and early life history of fishes in the Ohio River drainage, volume 1: Acipenseridae through Esocidae. Tennessee Valley Authority, Chattanooga, Tennessee.

Wang, J. C. S. 1981. Taxonomy of the early life stages of fishes—fishes of the Sacramento–San Joaquin estuary and Moss Landing–Elkhorn Slough, California. Ecological Analysts Incorporated, Concord, California.

Wang, J. C. S. 1986. Fishes of the Sacramento–San Joaquin estuary and adjacent waters, California. Interagency Ecological Study Program for the Sacramento–San Joaquin Estuary Technical Report 9, Sacramento, California.

Wang, J. C. S., and R. J. Kernehan. 1979. Fishes of the Delaware estuaries—a guide to the early life histories. Ecological Analysts Incorporated, Towson, Maryland.

Waters, W. E., and D. C. Erman. 1990. Research methods: concept and design. Pages 1–34 *in* C. B. Schreck and P. B. Moyle, editors. Methods for fish biology. American Fisheries Society, Bethesda, Maryland.

Werner, R. G. 1969. Ecology of limnetic bluegill (*Lepomis macrochirus*) fry in Crane Lake, Indiana. American Midland Naturalist 81:164–181.

West, G. 1990. Methods of assessing ovarian development in fishes: a review. Australian Journal of Marine and Freshwater Research 41:199–222.

Wiebe, P. H., and W. R. Holland. 1968. Plankton patchiness: effects on repeated net tows. Limnology and Oceanography 13:315–321.

Wiebe, P. H., K. H. Burt, S. H. Boyd, and A. W. Morton. 1976. A multiple opening/closing net and environmental sensing system for sampling zooplankton. Journal of Marine Research 34:313–326.

Williams, A. B., and E. E. Deubler. 1968. A ten-year study of meroplankton in North Carolina estuaries: assessment of environmental factors and sampling success among bothid flounders and penaeid shrimps. Chesapeake Science 9:27–41.

Williams, D. D., and N. E. Williams. 1974. A counter-staining technique for use in sorting benthic samples. Limnology and Oceanography 19:152–14.

Winnell, M. H., and D. J. Jude. 1991. Northern large-river benthic and larval fish drift: St. Marys River, USA–Canada. Journal of Great Lakes Research 17:168–182.

Wootton, R. J. 1990. Ecology of teleost fishes. Chapman and Hall, New York.

Yocum, W. L., and F. J. Tesar. 1980. Sled for sampling benthic fish larvae. Progressive Fish-Culturist 42:118–119.

Zhang, Z., N. W. Runham, and T. J. Pitcher. 1991. A new technique for preparing fish otoliths for examination of daily growth increments. Journal of Fish Biology 38:313–315.

Chapter 10

Sampling with Toxicants

PHILLIP W. BETTOLI AND MICHAEL J. MACEINA

10.1 HISTORICAL PERSPECTIVES ON THE USE OF TOXICANTS IN FISHERIES

Toxicants have been used by fisheries biologists to sample fish communities and remove undesirable fish species since the 1930s. The use of toxicants to meet a variety of objectives enjoyed widespread acceptance throughout North America and other parts of the world, but recent trends indicate those attitudes have changed. Public opinion, scientific advances, increased costs, and regulatory pressures have combined to limit when and where toxicants are used today. Although toxicants are used less frequently now than only 10 years ago, their use remains an important option still available to research and management biologists in many North American locales.

Rotenone was the first of several piscicidal compounds used by fisheries biologists in North America. Rotenone is a naturally occurring compound (see section 10.2) that can be synthetically produced. It enjoyed widespread use over many decades. The goal of many of the earliest applications of rotenone was to control nuisance species so sport fishes could be restocked. The terms "renovation" and "reclamation" have been used interchangeably to describe the use of toxicants to eliminate all fishes in an aquatic ecosystem. Fisheries managers often responded to the public's desire to increase sport fish production by removing potential competitors and predators. For example, fishing pressure on marginal trout streams in Michigan increased dramatically after the streams were treated with rotenone to remove white suckers and cyprinids and then stocked with rainbow trout or brown trout fingerlings (Trimberger 1975). When Adirondack ponds were treated with rotenone to eliminate suckers and yellow perch, stocked brook trout thrived and viable sport fisheries developed (Flick and Webster 1992). Stocking nonnative sport fish species often followed reclamation, and stream and river reclamation projects based on rotenone applications were commonplace in the 1950s and 1960s (Lennon et al. 1971). Reclamations routinely occurred on a small scale, as when 23 km of Abrams Creek in the Great Smoky Mountains National Park were treated in 1957 to benefit sport fishes. Rotenone was also applied to rivers on a grand scale. In the mid-1950s, over 400 km in the Russian River watershed in California were treated with rotenone (Pintler and Johnson 1958). In 1962, over 700 km of the Green River and its tributaries were treated with rotenone prior to the closing of Flaming Gorge Dam in Utah (Holden 1991). In these and many other examples, the rotenone application was followed by stocking nonnative rainbow trout. Partial reclamations of lake and reservoir systems were also undertaken to regulate the dynamics of overabundant, undesirable species, such as gizzard shad, or to reduce overcrowding of largemouth bass and bluegill (Swingle et al. 1953). In the north-central United States, toxicants were often applied to ponds that had suffered winterkill in order to remove undesirable species (e.g., bullheads) that are tolerant of low dissolved oxygen conditions. Lopinot (1975) estimated that piscicides were applied to over 47,000 ha of water in 12 states in the midwestern United States between 1963 and 1972; nearly 6,800 km of streams and rivers were also treated with piscicides during the same 10-year interval. The tremendous increase in toxicant use in North America between the 1930s and 1970s coincided with a boom in construction of large impoundments, private ponds, and flood control reservoirs built by the U.S. Soil Conservation Service (Cumming 1975).

Although it was once a common practice, most state fisheries agencies no longer devote substantial resources to control common carp. This exotic, much-maligned species often composed a large portion of the fish community biomass in many locales and was the object of many toxicant applications in previous decades. During spring, when common carp congregated to spawn, shallow, littoral areas were treated with either rotenone or antimycin (see section 10.2). Wetlands managers, wanting to protect valuable wetland plants, also sought to control common carp by use of toxicants (Moyle and Kuehn 1964). Although large numbers of common carp were often killed, the benefits of such programs rarely persisted, and the practice is no longer common.

The discovery of the piscicidal properties of the antibiotic antimycin in the early 1960s (Derse and Strong 1963) provided biologists with another chemical that could be used in reclamation and sampling activities. Antimycin had several advantages over rotenone (section 10.2) and was quickly preferred by many biologists working in stream systems (Rinne and Turner 1991).

In the 1950s, a synthetic compound, 3-trifluoromethyl-4-nitrophenol (abbreviated TFM), that could selectively kill sea lamprey larvae was discovered (Applegate et al. 1961). Between 1958 and 1978, TFM and another lampricide, Bayluscide, (a nitrosalicylanilide salt also known as Bayer 73) were applied over 1,200 times to more than 300 tributaries of the three upper Great Lakes (Smith and Tibbles 1980). The success of the control program was soon obvious, which fueled the desire by management agencies to expand programs for chemically controlling other undesirable fish species elsewhere.

Treating streams with toxicants to remove nonnative species so that native species can be restored is an increasingly common technique, particularly in the western United States. The object of many of these stream reclamations is the rainbow trout, which was widely stocked in earlier decades, although other introduced species, such as red shiners and mosquitofish, have also been targeted for eradication (Minckley et al. 1991).

Using toxicants to sample (as opposed to reclaim) fish communities in flowing and standing waters gained momentum in the 1950s. Rotenone and antimycin could be applied at concentrations lethal to all fish species and sizes of fish; thus, unbiased estimates of the abundance and composition of fish communities could potentially be generated by applying toxicants in areas of known size. By the late 1950s, biologists were discussing experimental designs for rotenone surveys (Lambou and Stern 1958), and the use of rotenone to estimate standing crops of fishes was widespread by the mid-1960s. Annual rotenone surveys were included in the routine monitoring programs of many agencies, and in 1965 over 150 biologists from 24 agencies and universities cooperated in the first large-scale evaluation of the technique. In that study, a 46-ha embayment of Douglas Reservoir, Tennessee, was subdivided into cove and open-water areas and treated with rotenone (Hayne et al. 1967). Use of rotenone to sample reservoir fish communities reached a pinnacle in 1978 when a multiagency task force sampled an 85-ha embayment of Barkley Lake, Tennessee (Chapter 1). Four hundred people from 34 agencies and universities participated in the Barkley Lake study (Summers and Axon 1980); over 3 million fish, totaling over 83,000 kg, were processed, and the results were presented at a symposium during the 33rd annual conference of the Southeastern Association of Fish and Wildlife Agencies in 1979.

With advances in technologies such as hydroacoustics (Chapter 13) and trawls, the use of rotenone by research biologists to estimate standing crops of some species has waned since the 1980s. For instance, Siler (1986) noted that trawling provided much more precise estimates of threadfin shad standing crops, at less cost, than did traditional cove rotenone sampling (described in section 10.4.1). Nevertheless, sampling fishes in shallow habitats with toxicants is still common because there are few alternatives. In recent years, for example, rotenone has been used to sample large coves (Bettoli et al. 1993) and small, enclosed areas in vegetated habitats (Miller et al. 1991) to meet specific research objectives. Researchers have also relied upon toxicant treatments for many years to validate estimates of abundance, size structure, and species diversity derived from use of other gears (e.g., Reynolds and Simpson 1978; Bayley et al. 1989; Allen et al. 1992). Finally, sampling with toxicants allows researchers to estimate how biomass is apportioned among different species and sizes of fish; these data can then be used to examine predator–prey relationships based on models such as the "available prey:predator model" developed by Jenkins and Morais (1978).

Rotenone was also applied as recently as 1972 to collect fish in coral reefs in the Gulf of Mexico (Bright et al. 1974). Small patches of reef were first covered with a clear plastic sheet to form a tent; divers within the tent then released small amounts of liquid rotenone and collected fishes and invertebrates that succumbed. This technique was very successful in providing information on the fauna inhabiting living coral reefs at a time when little information on these unique communities was available. Given the fragile nature and precarious condition of present-day living coral reefs, regulatory and ecological considerations now prevent the use of this technique throughout much of the world.

In this chapter, the different chemical techniques used to sample fish communities are reviewed. The application of toxicants in management-related activities (i.e., reclaiming entire systems) is briefly discussed because many of the problems associated with reclaiming waters are also present when toxicants are used to sample fish communities. Also, techniques recently developed for controlling fish populations by means of toxicants might be applicable to sampling those same populations. The reader is referred to Wiley and Wydoski (1993) for a more thorough discussion of using toxicants to manage fish communities.

In preparing this review of toxicant use, a short questionnaire was mailed to every state and provincial fishery management agency in the United States and Canada, as well as several U.S. territories. Trends in the use of toxicants to sample and reclaim fish communities over the past 10 years were identified, as well as restrictions on using toxicants in various locales. The results are presented in several sections of this chapter.

10.2 TOXICANTS: PAST AND PRESENT

It is illegal to apply any piscicidal compound not registered for that use with the appropriate government regulatory agency. In the United States, only four compounds are registered, or are being reregistered, for use in fisheries: rotenone, antimycin, TFM, and Bayluscide (Marking 1992). In Canada, TFM, Bayluscide, and rotenone are currently used, and rotenone applications are limited almost exclusively to reclaiming fish communities in a few provinces. In the United States, costs associated with registering these compounds with the Environmental Protection Agency and the Food and Drug Administration are high (Schnick 1991), and cooperative efforts have been undertaken to reregister some of these compounds (Sousa et al. 1987). The regulatory atmosphere surrounding the deliberate introduction of toxicants into aquatic systems is dynamic, and the registration status of these compounds, or other potential compounds, might change in the future. Although other toxicants have been investigated, private industry will not sponsor the cost associated with registering minor-use chemicals because of low market potential (Marking 1992).

In the 1950s and 1960s other compounds were investigated for their general piscicidal properties (reviewed by Lennon et al. 1971; Cumming 1975; Marking 1992), including copper sulfate ($CuSO_4$), sodium cyanide (NaCN), and toxaphene (a chlorinated hydrocarbon). Sodium cyanide was particularly well suited to sampling and renovating streams because the effects on fish were reversible, the compound was easy and inexpensive to use, and a detoxicant was not required downriver of the treatment site (Wiley 1984). Although these and other compounds are extremely toxic to fish, they are no longer approved for use in North America due to their persistence in the environment, negative effects on nontarget species, or threats to human health. In the early 1960s, toxaphene was widely used by fisheries agencies throughout the United States and Canada, but its high toxicity to birds and mammals and its environmental persistence led to its ban by the U.S. Department of the Interior in 1963 (Lennon et al. 1971).

Other compounds were actively sought and investigated for their selective toxicity against other nuisance species. MacPhee and Ruelle (1969) reported on the piscicidal properties of Squoxin, a compound developed to control squawfishes in the western United States. Although Squoxin was reported to be 3 to 100 times more toxic to squawfishes than to salmonids, it has not been registered for use in the United States or Canada.

Several toxicants not registered for use in North America are being used or investigated in other parts of the world to control aquatic species. Treating waters with toxicants such as pyrethroids or copper sulfate to eliminate aquatic snails is still one of the most effective methods available for controlling schistosomiasis (Ebele et al. 1990; Singh and Agarwal 1990). Schistosomiasis is caused by a blood fluke that uses snails as an intermediate host. Schistosomiasis is a considerable human health issue in many parts of the world because the disease is easily transmitted, debilitating, and often fatal. Synthetic pyrethroids are also suitable for controlling crayfishes and other pests in aquaculture operations (Bills and Marking 1988).

10.2.1 Lampricides

Of the four toxicants registered or being reregistered for use as piscicides, TFM and Bayluscide are intended only for sea lamprey sampling or control (Marking 1992). Bayluscide, originally developed as a molluscicide, is often used in small amounts to reduce the amount of TFM needed to kill sea lamprey larvae. When coated on sand particles, Bayluscide has also been used alone in deep water where TFM is difficult to deliver. The compound is an irritant to the larval sea lampreys (ammocetes), and when it is applied, ammocetes leave the substrate and are easily collected. Concentrations of TFM and Bayluscide necessary to kill sea lamprey are below the threshold concentrations toxic to most other fish species and many invertebrates. Deleterious effects on nontarget organisms have been described as minor relative to benefits derived from lampricide applications (Gilderhus and Johnson 1980; DuBois and Blust 1994). The sea lamprey is a nonnative species in the Great Lakes that has had a devastating effect on native fish stocks; the sea lamprey's unpopularity provides some latitude to biologists charged with controlling its numbers. However, alternative methods of control are constantly being investigated to reduce the reliance on toxicants (Great Lakes Fishery Commission 1994, unpublished data). The goal of the Great Lakes Fishery Commission, which coordinates sea lamprey control efforts in the Great Lakes, is to reduce reliance on lampricides to 50% of current levels (Great Lakes Fishery Commission 1992).

The minimum concentrations of TFM, Bayluscide, and both compounds combined needed to kill 100% of sea lamprey ammocetes and no more than 25% of several nontarget organisms were presented by Seelye et al. (1988). A maximum of 2% of Bayluscide can be added to TFM treatments in Canada and the United States in order to reduce the amount of TFM required. Sea lamprey ammocetes are more sensitive to TFM than are two other genera of native lampreys (King and Gabel 1985). Like rotenone and antimycin, TFM is most toxic in soft, acidic waters. "Safe and effective" concentrations of TFM necessary to kill sea lampreys range from 0.8 mg/L at low (40 mg/L) alkalinities to 7.0 mg/L at high (200 mg/L) alkalinities. Both lampricides are readily adsorbed by organic sediments, which might explain why treatments sometimes fail (Dawson et al. 1986).

Treatments of TFM can be applied with metering pumps or delivered in blocks of a surfactant matrix impregnated with TFM (Gilderhus 1985). Bayluscide is applied as a wettable powder or as granules. The toxicities of these lampricides vary markedly with the physical and chemical characteristics of the receiving waters (Seelye et al. 1988); therefore, on-site bioassays with caged sea lamprey ammocetes are recommended to determine the concentration that insures a complete kill but minimizes the effects on nontarget species. Lampricide treatments are often conducted below barriers on tributary streams, and the treatments usually extend all the way to the mouth of each stream; therefore, no steps are taken to detoxify the lampricides. The application goal is to deliver appropriate concentrations of TFM and Bayluscide for contact times of 8–10 h. The use of lampricides is tightly regulated, and they can be applied only by trained personnel certified by the U.S. Fish and Wildlife Service, the Canadian Department of Fisheries and Oceans, or several state conservation agencies.

10.2.2 Rotenone

The piscicidal properties of rotenone have long been known to native peoples in many parts of the world. Rotenone is found in the roots of several species of trees in the genera *Derris* and *Lonchocarpus*. Its properties were discussed thoroughly in reviews by Lennon et al. (1971), Schnick (1974), and Haley (1978).

Rotenone is extremely toxic to fish; its mode of action is through disruption of cellular respiration. In field situations, a 1.0-mg/L concentration of a 5% rotenone formulation (trade name, Noxfish; AGREVO Environmental Health, Inc., Montvale, New Jersey) usually kills all but the most resistant species of fish (e.g., gars, bullheads, and bowfin); concentrations as low as 0.05–0.10 mg/L kill sensitive species such as shads and grass carp. Liquid rotenone formulations include various carriers and solvents to increase dispersal and penetrate thermoclines, but fish readily detect liquid rotenone and display strong avoidance reactions. Although the effects of rotenone sometimes can be reversed by treating fish with an oxidant such as methylene blue (Bouck and Ball 1965), experience has shown that success in reviving fish is highly variable and unpredictable.

Although rotenone is toxic to a wide array of aquatic organisms at high doses, most nontarget organisms are not expected to be affected when recommended doses are used. Many aquatic invertebrates (e.g., insects, molluscs, and crayfishes) are less sensitive to rotenone than are fish (Chandler and Marking 1982). Crustacean zooplankton are often eliminated immediately after a rotenone treatment, but populations usually rebound quickly (see review by Schnick 1974). Some turtle species may succumb during routine rotenone applications (Fontenot et al. 1994). Most adult amphibians and reptiles should not be killed when rotenone is applied at normal concentrations (Farringer 1972), but tadpoles and metamorphosing amphibians are vulnerable to concentrations (5% formulation) as low as 0.1 mg/L (Hamilton 1941).

Rotenone toxicity is affected by many environmental factors, especially turbidity, temperature, and pH, which explains why the amount of rotenone applied varies among field situations. Rotenone is most toxic in warm (>20°C), acidic, clear waters that have little aquatic vegetation. A 0.5-mg/L concentration of bentonite clay decreases the toxicity of rotenone sevenfold (Gilderhus 1982), and abundant vegetation can adsorb much of the rotenone applied.

Rotenone does not persist in the environment. Above 23°C, the half-life of rotenone in water is less than 1 d, and concentrations in sediments normally fall below detectable limits within 24 h (Gilderhus et al. 1988). Rotenone remains toxic in sediments and water for many days or weeks at temperatures below 10°C. Rotenone should be applied so as to limit the amount of time that toxicity exists; therefore, the use of rotenone at temperatures below 15–20°C is not recommended if rapid recolonization of the treated system or area is desired. Rotenone is easily rendered nontoxic by the addition of a strong oxidizing agent such as potassium permanganate.

A rotenone formulation with 2.5% active ingredient is also commercially available under the trade name Nusyn-Noxfish. This formulation is about 25–30% cheaper per unit volume than the 5% active ingredient formulation. The 2.5% formulation is nearly as toxic as the 5% formulation because synergists are added to it; however, its use is recommended only in waters where the pH is circumneutral or acidic (i.e., pH <7.5).

Most early uses of rotenone in lakes and reservoirs relied on a powdered form of rotenone, which was dispersed in sacks or as a slurry pumped from boats. However, handling large amounts of finely powdered rotenone was difficult and potentially hazardous to those applying it. Liquid formulations of rotenone became readily available by the 1960s and were then used more often because of ease of application. In recent years, however, the powdered form has been used successfully in large-scale treatments, and its use is expected to increase for several reasons. Powdered rotenone is much cheaper; at a cost of US $4.40 or less per kilogram, a 3-mg/L concentration of the powder (5% active ingredient) can be established in 1,000 m^3 of water for about $13, compared with about $35 if the 5% liquid formulation is used. Also, the carriers incorporated into the liquid formulation to improve its efficacy are absent from the powder, which reduces environmental concerns. Finally, new systems have been developed to mix and deliver the large amounts of the powder as a slurry (see section 10.4.1; Thompson et al., undated).

10.2.3 Antimycin

Antimycin also interferes with cellular respiration, and it is even more toxic to fish than is rotenone. This antibiotic is produced by a mold (much like penicillin) and can cause 100% mortality of some species at concentrations as low as 1 μg/L (Derse and Strong 1963), although field applications normally require 5–10 μg/L (Gresswell 1991; Rinne and Turner 1991). The toxicity of antimycin varies among fish species. Although most freshwater species succumb at concentrations of 5 μg/L active ingredient (when pH is 8.5 and temperature 15°C), 100–120 μg/L may be necessary under similar conditions to insure a complete kill of resistant species such as bullheads (Antonioni and Baumann 1975). Antimycin toxicity is not greatly affected by temperature, but it is much less toxic and persistent in alkaline (pH >8) and turbid waters (Lee et al. 1971; Marking and Dawson 1972; Gilderhus 1982). In many situations, particularly stream treatments, antimycin is preferred over rotenone because less contact time is required to cause death and fish do not sense (i.e., avoid) antimycin. Antimycin oxidizes even faster than does rotenone and degrades rapidly when exposed to air (or in turbulent water or waterfalls), warm temperatures, and high pH. Potassium permanganate readily oxidizes (detoxifies) antimycin when applied at 1 mg/L.

As with rotenone, antimycin exhibits low toxicity to mammals, and its degradation products are devoid of toxicity (Herr et al. 1967). Protective eyewear and gloves must be used when antimycin is mixed and applied. Currently, antimycin can be purchased only as a liquid solution (trade name, Fintrol; Aquabiotics Corp., Bainbridge Island, Washington). The antimycin solution is shipped as a concentrate (20% active ingredient) and is mixed on site with an acetone-based adjuvant (trade name, Diluent) to achieve a solution with 10% active ingredient; this 10% solution is then mixed with at least 20 L of water before it is applied to lentic waters. The concentrated solution of antimycin (Fintrol Concentrate) can be used in preparations of pelleted feed (see section 10.6.1) and in drip stations (see section 10.4.3). Fintrol Concentrate can also be bound to sand particles by means of carbowax (Coating Place, Inc., Verona, Wisconsin). As the sand particles sink, the antimycin leaches into the water column, and the rate at which the leaching occurs can be adjusted to suit specific situations (e.g., shallow versus deepwater applications).

The cost to treat a water body with antimycin is comparable to the cost associated with using 5% liquid rotenone. Antimycin is sold in individual units ($330), each of which is capable of achieving a 1-μg/L concentration of active ingredient in 46,854 m^3 (38 acre-feet) of water. To achieve a 5-μg/L concentration of active ingredient in 1,000 m^3 of water, the cost is about $35. The quantity of antimycin needed to achieve a given concentration can be calculated by knowing that 10.8 mL of Fintrol Concentrate in 1,000 m^3 yields a 1-μg/L concentration of active ingredient.

10.3 PUBLIC RELATIONS AND REGULATORY CONCERNS

The use of any toxicant to control or sample fishes can lead to bad public relations. In recent decades, citizens have been educated to the adverse effects of pollution on our waterways and the flora and fauna that inhabit them. It is only natural that they might question the intentional release of known toxicants into aquatic ecosystems. Providing advance notice of a treatment program is the very least that biologists can do to minimize misunderstandings or unfounded concerns. Clearly specifying the objectives and expected benefits of a toxicant treatment and providing the public with an opportunity to comment are also prerequisites to conducting a thorough, professional treatment program. In some circumstances, a formal environmental assessment of the proposed treatment may be required for review by various state or federal agencies (Wiley and Wydoski 1993). Other considerations, both ethical and practical, germane to any proposal to collect or remove fish by means of toxicants were presented by ASIH et al. (1988).

Filipek (1982) noted that negative opinions toward large sampling or reclamation projects changed when the public realized that these efforts usually resulted in improved fishing. Conversely, irreparable damage to agency credibility can result if a toxicant application is poorly designed, ill conceived, and carelessly performed. Intentionally killing fish in large-scale reclamation or sampling projects will garner the attention of the public; such situations provide biologists with an excellent opportunity to exhibit their professionalism. Pretreatment publicity will be required to justify a large-scale toxicant application, but how fish are treated during and after sampling with toxicants can affect the public's perceptions of the treatment. Every effort should be made to process fish in an orderly manner and dispose of them properly. Concerns over using toxicants can be alleviated if the public perceives that important information is being collected and will be used to protect or enhance a fishery.

The application of piscicides to aquatic ecosystems is more tightly regulated now than ever before, and this trend will probably continue. Manufacturers of rotenone in the United States will no longer supply their product directly to agency or university personnel without proof that the personnel are certified to apply pesticides or have been trained in the use of the manufacturer's product. Such training is usually provided by state agriculture departments or extension services. Most (77%) of the states surveyed responded that certified personnel must apply the piscicides; however, the number of states requiring a permit to apply piscicides equaled the number that do not (or require only a blanket approval). Although fish collected with rotenone have traditionally been given to indigent persons or nonprofit organizations for consumption, this is not an approved procedure. Efforts are currently under way to seek approval from the U.S. Food and Drug Administration for this use.

10.4 USE OF TOXICANTS IN RESEARCH AND MANAGEMENT SURVEYS

10.4.1 Lacustrine Habitats

10.4.1.1 Cove Sampling

Much information on lake and reservoir fish communities can be obtained by treating small embayments or coves with toxicants. Rotenone has been used for most cove sampling because of its greater availability. Sampling coves or entire systems with rotenone provides data on standing crops (the total weights of fish populations at a particular time) that can be used to build empirical models of the relationship between fish biomass and abiotic factors. In the 1980s, standing crop data obtained from sampling coves with rotenone were modeled by Jenkins (1982) and others as a function of the morphoedaphic index (MEI = total dissolved solids/mean depth). Although the validity of the MEI has subsequently been questioned (e.g., Jackson et al. 1990), the availability of large databases on fish standing crops derived from cove sampling allowed investigators to identify important relationships between fish biomass and other factors, such as phosphorous concentration (e.g., Yurk and Ney 1989). Much of what is known about lake and reservoir fish communities has resulted from extensive sampling with toxicants over many decades.

Collecting cove samples in lakes and reservoirs is still commonplace in the southeastern United States, perhaps because some systems have been sampled with rotenone for many years, and biologists are loathe to abandon these long-term databases. But even in states in which surveys with toxicants have been occurring routinely for decades, this technique is used less frequently now than it was 10 years ago; 66% of the agencies surveyed indicated that their use of toxicants to sample fishes has decreased. Several factors can account for the declining use of toxicants in routine management surveys: public opinion, regulatory pressures, and the high costs associated with collecting enough samples to produce statistically reliable estimates of abundance (section 10.5 and Chapter 2). The challenge facing management biologists today is determining when the benefits derived from sampling with toxicants outweigh the costs and negative public reaction that can be associated with their use.

Cove sampling is traditionally conducted in mid or late summer; when water temperatures are high, the toxic effect of rotenone is heightened and its degradation is hastened. Also, biologists get an immediate indication of reproductive success that year for important species. At lower latitudes, waters are warm enough to apply rotenone in late spring, before aquatic weed growth might become problematic. Although reproductive success for that year cannot be learned from spring samples, the strength of the previous year-class can be assessed by examining the number of age-1 fish in the samples.

Prospective cove sites in a reservoir must be chosen and surveyed in advance of any rotenone application. The coves should be of sufficient size and depth to allow the best representation of the fish community. Block nets, which are stretched across the mouths of coves before treatment, can be ordered in almost any size or configuration. A typical block net is about 100 m long and 6–9 m deep; if the cove mouth is wide, two nets can be tied together. The mesh size can also vary, but it is usually about 6-mm bar measure, which will contain most small species and juvenile sport fishes. It is best to use a net with floats that are enclosed in a sleeve of netting material so that dead, floating fish will not drift over the net and out of the cove. The points of land at which the block net will start and end should be marked with flagging tape, and the maximum depth to which the net will extend must be determined. An electronic depth meter or fish locator is better than a plumb line for sounding because it allows the bottom where the net will be deployed to be examined more closely for irregularities or obstructions. The inner (back) end of a cove may be so shallow that it is inaccessible to boats, in which case another net should be used to block it off.

With aerial photography, an alidade table, or other surveying techniques, map the cove and determine the surface area. Map the depth contours in the cove by running six or eight transects across it and recording depths by means of a plumb line or electronic fathometer (see Chapter 4). If the cove is envisioned in cross section as a cone, the volume in each 1-m-deep layer (known as a "frustum") can be calculated with the formula given in Box 10.1. The volumes for each frustum are summed to generate an accurate estimate of the total volume of water in the cove. An easy way to remember how much rotenone to add is to consider that treating 1,000 m^3 with 3 L (or 1 acre-foot with 1 gallon) of a 5% or 2.5% rotenone formulation yields about a 3-mg/L concentration of the formulation. A concentration of 3 mg/L usually is lethal to all individuals of all species, although concentrations up to 6 mg/L might be required in marsh or swamp habitats. If lower concentrations are desired, the amount of rotenone needed for 3 mg/L is divided by (for example) 2 to yield 1.5 mg/L or by 3 to yield 1.0 mg/L.

To provide better estimates of pelagic species abundance, many investigators set the net across the cove at dusk, or they tie the lead line up to the float line after the net is deployed and then drop the lead line at night. If the net is to be set over irregular bottom or across stumps or other obstructions, scuba divers can secure the net to the bottom. Rotenone should be applied early in the day to allow the maximum time for fish to be picked up.

A gas-powered water pump is used to apply the rotenone to the surface and below the surface of the water (Figure 10.1). Ideally, the pump has one hose with a nozzle for spraying rotenone at the surface and a second, weighted hose that dispenses rotenone well below the surface. Although the liquid rotenone formulation disperses horizontally and vertically through the water column, it is customary to pump some rotenone below the thermocline if one is present. Rotenone can also be applied along the shoreline with hand sprayers. Fish readily detect and try to avoid rotenone, so the chemical should be applied first at the net and then back into the cove. The treatment should not be rushed; the more time taken to apply the rotenone, the longer fish will be surfacing and susceptible to capture.

Dilute 1 part rotenone with about 10 parts water to facilitate even dispersal. Save some rotenone near the end of the application and spray shallow littoral areas. In shallow systems, a simple venturi pump may adequately deliver and mix liquid rotenone (Figure 10.2). The flared nozzle of the venturi pump (which has no moving parts) is attached to the lower unit of an outboard motor. When the engine is running in forward gear, a vacuum is created that pulls rotenone out of a tub on the boat and mixes it in the prop wash.

Thompson et al. (Utah State Department of Natural Resources, unpublished data) described a very efficient means of delivering powdered rotenone by means of a high-pressure water pump and aspirator. The system is housed on a barge that holds four 454-kg bulk bags of powdered rotenone; the powder is suctioned out of each bag at a rate of 45 kg/min, mixed in the aspirator with water, and pumped overboard as

Figure 10.1 Applying rotenone by means of a gas-powered pump.

a slurry at a concentration of about 0.12 kg/L of water. The bulk bags are specially constructed so that a forklift can be used to load the bags onto the barge. This barge system was designed for large-scale treatments and requires three people to operate, but it might be modified for use in smaller cove sampling efforts.

Figure 10.2 A venturi pump mounted on the lower unit of an outboard motor.

Box 10.1 Calculating the Volume of Rotenone to be Added to a Cove

When fish population sizes in a reservoir cove are to be estimated, a rotenone formulation is mixed into the cove to achieve a typical average concentration (of the formulation, not of the active ingredient) of 3 mg/L throughout. (The 5% and 2.5% rotenone formulations have about the same toxicity because of synergists added to the latter.) Calculating the volume of a commercial rotenone solution that must be put in the cove to achieve this concentration requires knowledge of the cove's water volume. Volume is determined from a bathymetric map, which is constructed from transect surveys of cove depth superimposed on a good planimetric map (from which areas of the cove's surface and various depth strata can be estimated). The following example is based on the bathymetric map shown below (depths are in meters).

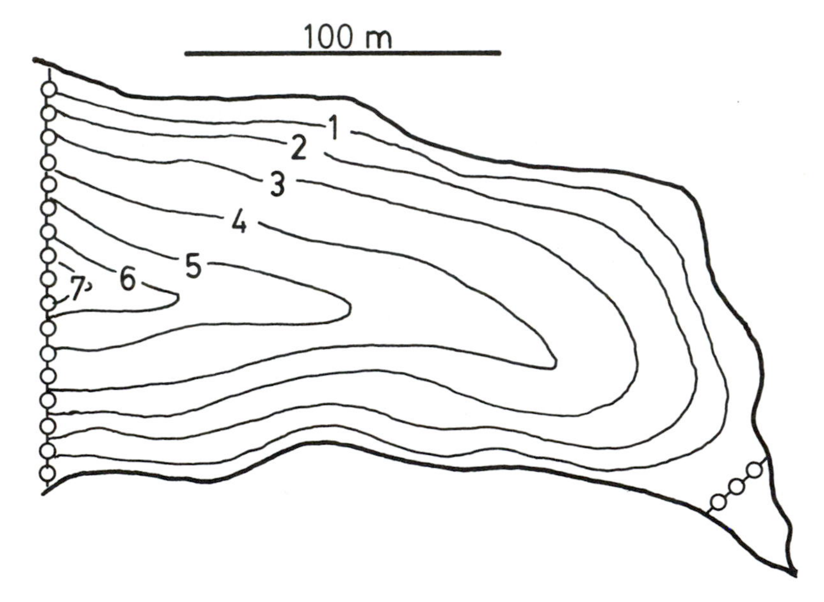

Areas associated with successive depths are:

Depth (m)	Area at depth (m^2)
0 (surface)	21,664
1	18,512
2	15,488
3	11,104
4	5,936
5	2,032
6	400
7	64
7+ (deepest)	0

Box 10.1 Continued.

Volumes of 1-m depth strata can be calculated from the geometry of cones, as shown below.

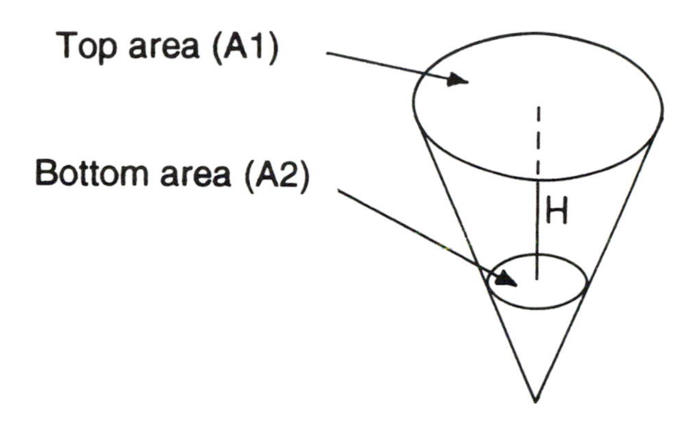

The volume of water (V) between areas A1 and A2 (e.g., between depths 0 and 1 m) is

$$V = (H/3)\,[\mathrm{A1} + \mathrm{A2} + (\mathrm{A1} \cdot \mathrm{A2})^{1/2}].$$

For example, the volume of water in the top 1 m of the sampled portion of the cove (H = 1 m) is

$$V = (1/3)\,[21{,}664 + 18{,}512 + (21{,}664 \cdot 18{,}512)^{1/2}] = 20{,}067 \text{ m}^3.$$

The volumes of successive 1-m depths therefore are

Depth interval (m)	Volume (m^3)
0–1	20,067
1–2	16,977
2–3	13,235
3–4	8,386
4–5	3,814
5–6	1,111
6–7	208
7–7+	21
Total	63,819

A 3-mg/L concentration of rotenone formulation is (closely enough) 3 L/1,000 m^3 (1 L = 10^6 mg; 1,000 m^3 = 10^6 L). The volume of rotenone formulation (X) required to achieve 3 mg/L is

$$\frac{X}{63{,}819} = \frac{3}{1{,}000};$$

$$X = 191.5 \text{ L}.$$

In English units (1 gal = 3.8 L), a bit more than 50 gal of rotenone formulation would be required.

Table 10.1 Recovery of tagged fish of selected species in nine coves sampled during the Lake Barkley rotenone experiment, 1978. Fish were tagged with Floy tags (type FD-68B). Data are from Table 6 of Axon et al. (1980).

Species	Number tagged	Number recovered	Percent
Spotted sucker	52	47	90
Smallmouth buffalo	14	14	100
Channel catfish	7	7	100
White bass	54	49	91
Bluegill	238	217	91
Longear sunfish	317	281	89
Largemouth bass	128	112	88
White crappie	13	11	85
Freshwater drum	72	60	83
Common carp	54	50	93

Water currents created by springs, runoff, or wind action disperse rotenone outside the block net. Potassium permanganate can be applied to detoxify rotenone outside the net; for each liter of 5% rotenone, place about 1 kg of potassium permanganate in fine-meshed bags (or plastic bags with holes punched in them) and drag them outside the block net. Because many fish species can detect and avoid rotenone before they succumb, fish kills outside the block net are often restricted to very susceptible species, such as shads. If shad are abundant in the system, consider distributing potassium permanganate in the cove to a concentration of about 1 mg/L after the block net is retrieved; this will reduce the likelihood that a fish kill will occur after the sampling crew has left the site.

Some fish in the sampling area do not surface after rotenone is applied. The standard way to adjust for nonrecovery is to release tagged fish into the cove before the rotenone is applied and to adjust biomass and density estimates upwards depending on how many tagged fish are recovered (Table 10.1; Barr and McDonough 1978). Bayley and Austen (1990) tested sampling efficiency (proportion of marked fish recovered) in ponds and coves by marking different sizes and species of fish before rotenone was applied. Efficiency was high (80–100%) for large fish (>150 mm total length) in warm water (>25°C) and low for small fish in cool water; efficiency declined somewhat as sampling area increased, but it did not differ among species. If rigorously conducted, a mark–recapture experiment yields good estimates of sampling efficiency; however, many caveats are associated with this technique. Low recovery of tagged fish could be due to several factors. Fish collected elsewhere and transported to the sampled cove are more likely to escape or to be preyed upon when released (Axon et al. 1980); therefore, fish for tagging need to be collected from the cove to be treated. Handling stress should be minimized, particularly for small fish, so that postrelease mortality due to predation is negligible. The tag or mark used must be readily visible or detectable (such as a wire microtag) even on fish that have partially decomposed. Enough people must be available to examine each recovered fish of all tagged species for marked fish. If all of these requirements cannot be met, do not attempt to adjust for nonrecovery; instead, concentrate on sampling and processing fish in a consistent manner from site to site and year to year.

Fish usually begin to surface before a rotenone application is finished; they appear sooner in warm than in cool water. Personnel in pickup boats work particular areas of the cove, usually in circular patterns, netting fish and placing them in buckets.

Some people should walk along the shoreline to retrieve dead fish from brushy or shallow areas not accessible by boat. Retrieved fish are processed on shore.

The size of coves treated generally ranges from 0.8 to 2.0 ha. Within this range, Hayne et al. (1967) found that larger coves provided results with less bias and smaller variances than did smaller coves. Maceina et al. (1994) reported that densities of harvestable largemouth bass were higher in coves larger than 1.0 ha. In moderately productive waters in which fish biomass may exceed 200 kg/ha, between 8 and 15 people are necessary to sample each cove properly. Dead fish may surface for several days after treatment, buoyed by gases of decomposition. Decomposition rate increases with water temperature, allowing faster collection of fish. The total time necessary to collect all fish that will surface is about 4 d at 15–19°C, 3 d at 20–28°C, and 2 d when temperatures exceed 28°C.

The primary purpose of cove sampling is to estimate the total number and total weight of each species collected; a secondary goal is to describe the size structure of each population. All fishes are sorted into species on large tables set near the shoreline, typically after fish retrieval has ended for the day. The sorting tables usually have inlaid meter sticks for individually measuring fish, which are assigned to 25-mm (or other) size-classes. Abundant species such as shads and bluegill may fill several large tubs or garbage cans, and other species need their own repositories, so plenty of containers are needed. If a species is not abundant, each individual can be assigned to a length-class, and the total number and weight of fish in each length-class can be determined. When a species is abundant, it must be subsampled after the total weight of that species is determined. Many agencies have established protocols for subsampling, such as stipulating that at least 5 or 10% of the total weight of a species should be sorted to size, counted, and weighed in order to estimate (by extrapolation) the total number and biomass of that species. Biologists take advantage of a rotenone sample to collect scales or otoliths for aging fish. Food habits of fishes collected with toxicants should never be examined because predators often gorge on stressed prey fishes before they, too, succumb. On the second and third day after rotenone treatment, fish are counted and measured but usually not weighed. Instead, average weights of fish in each length-class are inferred from first-day weights. Projecting weights and summarizing cove rotenone results are facilitated by computer software such as that described by Bivin et al. (1991).

We have observed that many data measured during cove sampling are never used. Length distribution data in particular are often relegated to an appendix in an annual report (if that), yet they were the most time-consuming to collect. The emphasis on length distributions of all species in a cove sample grew from the data needs of predator–prey biomass models used in the 1950s, 1960s, and 1970s (e.g., Swingle 1950; Jenkins and Morais 1978), but those models are used less frequently now. We suggest that biologists consider bulk weighing and counting incidental species and those species whose abundance may be substantial (e.g., gars, carp, bullheads, and suckers) but whose size structures are of little management interest. Protocols for subsampling should be reconsidered if they require processing of unnecessarily large subsamples. For instance, rotenone treatment of a 1-ha cove might yield several thousand threadfin shad, belonging to only two or three length-groups, on the first day. It is unnecessary to count and measure hundreds of fish in a large subsample to estimate the total number of threadfin shad collected and their distribution among only a few length-classes. If slight effort is taken to insure that a small subsample is randomly selected for processing, the total number and size structure of that species in the sample can be established accurately (Box 10.2). We encourage biologists to investigate techniques that reduce the personnel and time needed to collect a cove sample. Such reductions will enable them to collect more samples or to examine target species (usually sport fish species) more thoroughly. For instance, cataloging largemouth bass and their gizzard shad prey to 10-mm or 5-mm length intervals can provide more insight into population structures than traditional coarser intervals offer (Figure 10.3).

10.4.1.2 Block Net Sampling

Natural lakes often do not have coves that can be sampled with rotenone. Instead, biologists can enclose an area of water (usually about 0.4 ha) with a block net that is about 64 m on a side. Each corner of the block net must be anchored with a heavy weight to ensure a square configuration when the whole net is deployed. The ends of the net should overlap 5 or 10 m to trap fish within the enclosure. For this small an area, a single depth measurement in the middle of the enclosure is usually sufficient to determine the volume to be treated and the amount of rotenone needed.

Block nets can be set where aquatic vegetation is emergent or submersed or has floating leaves. Densities of small fish can be extremely high in these habitats, particularly in submersed vegetation. Shireman et al. (1981) found that sampling 0.08-ha enclosures provided estimates of fish density and biomass in aquatic vegetation similar to those from 0.4-ha enclosures. Use of a smaller net reduces unit sampling costs and (for the same budget) allows more samples to be taken, which improves statistical precision.

After rotenone is applied to block-net enclosures, pickup and processing procedures are similar to those used for cove rotenone sampling (section 10.4.1.1). On large water bodies, wind-driven currents can move a great deal of rotenone outside the enclosure, which can cause high fish mortality even when potassium permanganate is used. Therefore, sampling block-net enclosures with rotenone on windy days is not recommended.

A creative use of block nets to evaluate the effect of selective grass carp removal on other fish was described by Colle et al. (1978). They set block nets in an 80-ha Florida lake before treating the whole lake with a low concentration (0.1 mg/L) of 5% rotenone, shown in preliminary tests to kill grass carp but few other species. Following the low-concentration rotenone treatment, each block net enclosure was treated with a 2-mg/L concentration of rotenone to achieve total kills. The 0.1-mg/L treatment did kill fish besides grass carp throughout the lake. The thorough sampling of enclosures, however, showed that the low concentration had had little effect on the populations of most species.

10.4.1.3 Wegener Ring

The Wegener ring was developed to sample fish in both vegetated and unvegetated shallow-water habitats (usually less than 1.0 m deep; Wegener et al. 1974). This circular net (diameter 2.26 m) encompasses 0.0004 ha and weighs about 14 kg (Figure 10.4). The original design had a net depth of 0.5 m, but deeper nets (1.8 m) have also proven effective. The Wegener ring can be thrown by two people 5–6 m from the shoreline or from a boat. After it is thrown, about 10–15 mL of rotenone (5% active ingredient) is sprayed into the enclosed area to kill all fish within the net. Fish are dipnetted with a fine-mesh net while they surface; then, the net is gathered up like a seine to collect the rest of the fish within the enclosure. Up to 15 samples can be collected by two people in an 8-h workday. Wegener rings are best suited to

Box 10.2 Subsampling

Sometimes a species or age-group is so abundant in a sample that measuring every individual is impractical. Subsampling the group—if it is done correctly—yields valid information.

In a 1994 cove-rotenone sample from Normandy Lake, Tennessee (P. W. Bettoli, Tennessee Tech University, unpublished data), threadfin shad were exceptionally abundant, and the following subsampling protocol was applied. All the threadfin shad collected during the first day of sampling were placed into a tub. A handheld scoop was used to grab a subsample from the tub. Length measurements of the subsample gave the following distribution:

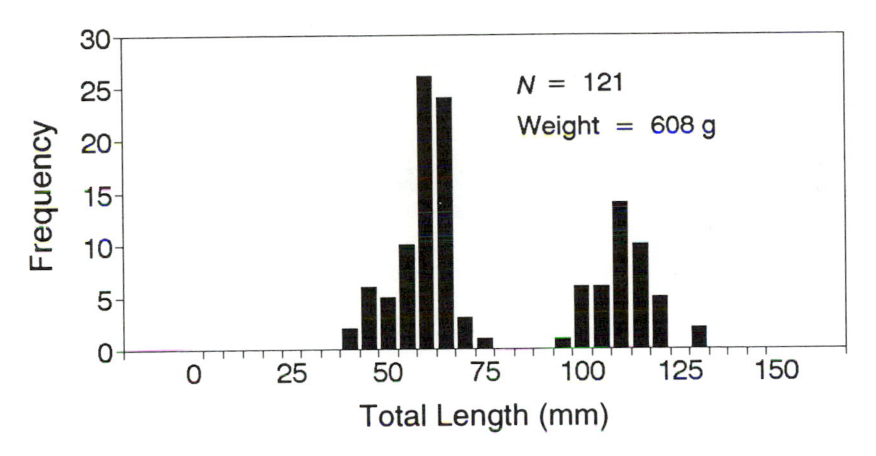

The subsample also yielded gross statistics about the sample:

total subsample weight	608 g;
number of fish in subsample (N)	121;
average weight of fish in subsample	608/121 = 5.02 g.

The rest of the fish in the tub were weighed in bulk, 6,500 g. The average weight of a fish is known (5.02 g), so the total number of fish in the bulk sample is

$$\text{bulk weight/average weight} = 6{,}500/5.02 = 1{,}295 \text{ fish.}$$

The total number and weight of threadfin shad collected in the first day were:

$$\text{total number} = \text{bulk sample} + \text{subsample} = 1{,}295 + 121 = 1{,}416 \text{ fish;}$$
$$\text{total weight} = \text{bulk sample} + \text{subsample} = 6{,}500 + 608 = 7{,}108 \text{ g.}$$

It is important that the subsampling procedure produce a random representation of the fish captured in the sample, because statistical analyses typically require that individuals have equal probabilities of inclusion in a sample or subsample. In this example, preliminary tests showed that threadfin shad did not settle in the sampling tub according to size (e.g., that large fish did not settle preferentially to the bottom of the tub), and that sweeping a handheld scoop through the tub did not result in a biased subsample. The result was not only the gross statistics but a useful length–frequency histogram as well.

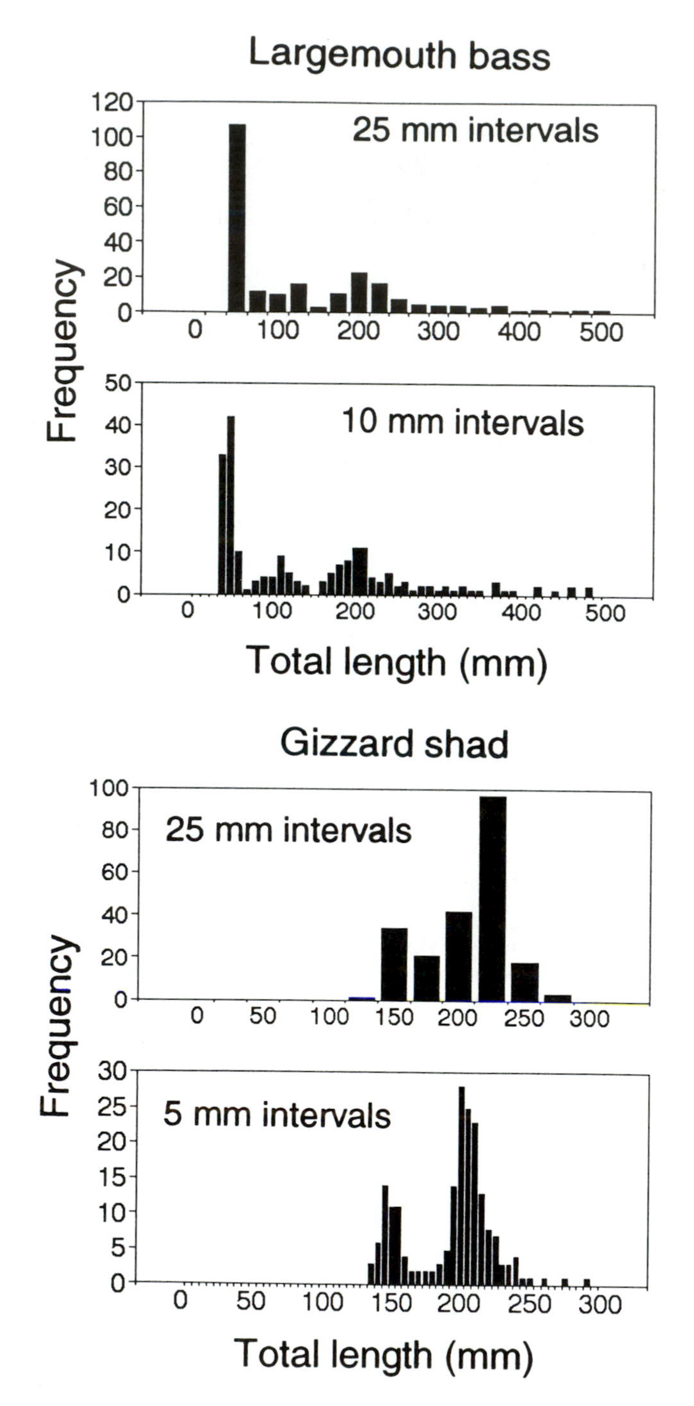

Figure 10.3 Length–frequency histograms for subsamples of largemouth bass and gizzard shad showing the improved resolution of population structure allowed by finer measurement intervals. The 10-mm length-class intervals for largemouth bass better depict the breaks between age-0 fish (<60 mm), age-1 fish (60-150 mm), and age-2 and older fish than do 25-mm intervals. For gizzard shad, 5-mm length-classes reveal two modes in the distribution (age 1, 130–175 mm; age 2 and older, >175 mm) much more clearly than 25-mm classes. Data are for cove rotenone samples taken in Normandy Lake, Tennessee, August 1994. Fish ages were confirmed by otolith analyses.

Figure 10.4 A Wegener ring, used to sample fishes in shallow, usually vegetated habitats.

collecting fish less than 8 cm in total length, and they allow more precise estimates of fish abundance and size distribution in heavily vegetated habitats than do 0.08-ha block nets (Miller et al. 1991).

10.4.1.4 Shoreline Sampling

Young fish of several species occupy shoreline habitats, where they can be sampled with rotenone. Shoreline sampling with rotenone is used primarily to sample juvenile largemouth bass in southeastern U.S. reservoirs. Beach seines cannot be used in many reservoirs because of stumps and other obstacles that were drowned by impounded waters.

Shoreline rotenone sampling is intermediate between beach seining and large-cove rotenone sampling. An area of measured size is enclosed with a small-mesh (≤5 mm bar measure) block net. The net is loaded onto the bow of a boat, one end is anchored to shore, and the net is paid out as the boat moves in reverse and arcs back to the shoreline a short distance away (Figure 10.5). Block nets can be any length but usually are 30–40 m long and 2–3 m deep and enclose 0.01–0.02 ha. Once the net is deployed, rotenone is sprayed within the enclosure to achieve a concentration of 1 mg/L. After the spraying, fish surface quickly, and every effort is made to collect them before they sink. After a few minutes have passed since the last fish has surfaced, the net is retrieved as if it were a seine; someone lifts the lead line over any obstructions while it is hauled. A three-person crew can collect many samples in one day, which facilitates stratification of sampling effort over time and space. This technique has usually been used to assess abundance, survival, and growth of young-of-the-year largemouth bass (Miranda et al. 1984), and it was described in detail by Timmons et al. (1979). Concern has been raised that ontogenetic shifts in habitat use by age-0 fish can bias results obtained from shoreline rotenone samples (Hightower et al. 1982). The concern is valid, because young fish change habitat requirements rapidly. To make good use of shoreline sampling, one must understand the biology of the target

Figure 10.5 Spraying rotenone in a 0.02-ha shoreline sample with dipnetters standing by.

species, understanding that can be obtained (in part) from concurrent sampling with other gear in the same places and in other places in the system under study.

If the desired sampling area is larger, about 0.2 ha of littoral zone can be sampled with a block net that is 100–200 m long. The shoreline serves as one edge of the square or rectangle that the net will form. Such areas often are too deep to wade safely (or at all). Then, fish are collected from a boat along the offshore edge of the net while some dipnetters wade the shoreline. Fish surface for several days; therefore, larger nets are left in place until all fish have surfaced. If the boat carrying the net is large enough to hold three larger nets, a four-person crew can sample three sites in a single day, then revisit each site for the next day or two.

Johnson et al. (1988) described a shoreline rotenone sampling technique in which a large (0.15-ha) quadrat is isolated along shore by means of a plastic barrier. Small-mesh nets with a known area are distributed evenly on the bottom of the enclosure before it is treated with rotenone (1 mg/L) in order to collect age-0 shads, which sink to the bottom and decompose quickly once they succumb. The nets are retrieved after fish stop surfacing, and the catch is extrapolated to estimate the total number of fish in the enclosure.

10.4.2 Navigation Locks

Navigation locks at mainstream dams and along large rivers provide an enclosed area that can be sampled with toxicants, usually rotenone. This technique has been employed on the Ohio River and several of its tributaries (Ohio River Valley Water Sanitation Commission 1988, unpublished data). The primary constraint to sampling these areas is coordinating closure of the lock with the responsible management agency for 2 d. Most navigation locks range from 0.7 to 1.5 ha in area. The lock is treated when water is at the lower elevation to minimize the amount of rotenone that is required.

Procedures for rotenone sampling in locks are similar to those for treating large coves. Crews in their boats enter the lock chamber and begin applying rotenone when the lower door to the lock has closed. Fish pickup continues until all fish that have surfaced are netted. When the collection of fish ends, the lower lock door is opened, sampling personnel and boats proceed to the shoreline to process the catch, and the lower lock door is then closed. The doors of the lock remained closed to all traffic until the following morning when the second-day pickup of fish begins. This sampling is best conducted during periods of low flow because there will be less dispersal of the rotenone out of the lock when the doors are reopened.

One key limitation to this technique is that samples cannot be easily replicated. Also, most dams have only one lock and scheduling its closure for at least 24 h can be problematic. One approach has been to sample a lock repeatedly over many years in an attempt to detect long-term trends. In some instances, biomass can exceed 1,000 kg/ha; therefore, many more people may be required to collect and process lock samples than are needed for sampling a cove of similar size in a lake or reservoir.

10.4.3 Rivers and Streams

Sampling rivers and streams with toxicants was common in past decades. Descriptions of the gear and techniques used to block off sections of large streams, apply rotenone, detoxify it, and collect fishes were provided by Johnson and Pasch (1976) and Ober (1981). However, routine sampling of rivers and streams with toxicants is no longer a common technique in North America. Five Canadian provinces and two states do not allow streams to be sampled or renovated with toxicants. Of 46 states and provinces that allow use of toxicants in flowing waters, according to our survey, only 1 indicated that the technique was used more than five times annually in its jurisdiction.

The potential for disastrous fish kills downstream of treated sections is always present, and the information that might be collected is often not worth the risks—with a few exceptions. In recent years, rotenone has been applied in streams and reservoir tailwaters to verify the efficiency of other gears, such as electric seines (Bayley et al. 1989). Bayluscide is still routinely used to survey sea lamprey populations and, in combination with TFM, is still used extensively for sea lamprey control (section 10.2).

Although use of toxicants in flowing waters has declined overall, the technique is still a valuable option in certain situations. Also, reclamation of streams with toxicants is becoming more common in certain locales (see section 10.1), and the techniques for renovating and sampling are essentially the same. Because reclamations are more common, we discuss the techniques used for that purpose.

Because of the risks, treating streams with toxicants requires close supervision by personnel experienced with the technique. Training leading to certification for pesticide application does not prepare a biologist for conducting a thorough, controlled stream treatment. It is imperative for those planning to use toxicants in flowing water to participate first in several stream treatments to become familiar with the technique. The following overview of the technique, and a description of problems inherent in using toxicants in flowing water, is based on methods described in Rinne and Turner (1991) and Gresswell (1991).

Techniques for applying and detoxifying toxicants in streams are generally the same whether rotenone or antimycin are used. Antimycin is used in most current stream renovations, and we focus on its use. Streams are often renovated above natural or artificial barriers, which prevents rapid recolonization of treated reaches. The goal is to expose all resident fishes to a sufficient antimycin concentration (8–10 μg/L) for a sufficient amount of time (1–10 h); concentration and time depend upon water chemistry and stream morphology. If the target species incubates its eggs in the substrate, the stream should be treated before or after the reproductive season. Antimycin is dripped into the stream at a constant rate to achieve the target concentration immediately downstream; various dispensers can be used for this purpose. Although stream discharge can be measured accurately with standard stream surveying techniques and the appropriate delivery rate of antimycin can be calculated, the downstream dispersal and toxicity of the antimycin cannot be accurately described. Therefore, complete reclamation (i.e., all fishes killed) in a single treatment is rarely realized, and repeat treatments are usually required. Antimycin toxicity varies with the chemistry of the receiving waters; therefore, controlled in-stream assays on target fish species must be performed before a full-scale treatment is attempted. Even when antimycin toxicity in situ has been carefully defined, the amount of antimycin remaining some distance downstream cannot be easily quantified because antimycin oxidizes so readily. In steep-gradient streams (>3% slope) with many falls or cascades, booster drip stations supplying about 4-μg/L concentrations of antimycin must be installed every 50–200 m to maintain effective toxicant concentrations. Booster stations also are required on low-gradient streams but at greater intervals. Caged specimens of target species are often placed above booster stations to indicate whether or not more toxicant should be delivered at that station.

A detoxifying agent must be administered at the downstream end of the section being treated. Caged specimens should always be present below the detoxifying station and continuously monitored to insure that the toxicant is rendered harmless. Potassium permanganate at concentrations of about 1-2 mg/L is universally used to oxidize and detoxify both rotenone and antimycin if the oxidation demand of the treated waters is low. If potassium permanganate concentrations much greater than 4 mg/L are necessary to oxidize the toxicant, permanganate toxicity will have to be minimized by applying sodium thiosulfate, and the potential for unexpected fish kills will increase. Monitoring the dispersal of antimycin and its arrival at the detoxifying station is aided by using dyes such as rhodamine or fluorescein. Both of these dyes are compatible with antimycin and are not toxic at the low concentrations (<0.1 mg/L) used to detect movements of the toxicant (Marking 1969). Equipment such as a fluorometer can be used to measure dye concentrations. In the planning stages of the treatment, the dyes can also be used to identify spring seeps, backwaters, or other areas that might not receive the full concentration of toxicant. The rates at which a toxicant moves down a stream can also be assessed by a salt resistivity technique described by Lennon et al. (1971); a known amount of salt is added upstream and its concentration is monitored downriver with a conductivity meter.

10.4.4 Estuarine Habitats

Sporadic use of rotenone to collect fishes in estuarine habitats has been reported in the literature (Hegen 1985). In general, the procedures employed are the same as those used in standing bodies of fresh water, and estuarine species succumb to rotenone as readily as do freshwater species. Matlock et al. (1982) collected cove samples and quadrat samples along shorelines in Texas estuaries in the late 1970s. They established 2-mg/L concentrations of 5% rotenone in enclosed areas and 4-mg/L levels of potassium permanganate in adjacent areas. Tidal flows can make sampling estuarine habitats as problematic as sampling flowing waters in that extensive kills outside the target area are possible. Another problem in estuaries is extensive bird predation on dead and dying fish. Matlock et al. (1982) noted that public opposition to poisoning estuarine habitats restricted toxicant use to remote areas. Future rotenone use in estuaries may be limited to assessing the efficiency of other sampling gear (Allen et al. 1992).

10.5 DATA ANALYSIS AND BIASES

Toxicant sampling is most often used to generate absolute estimates of fish biomass (standing crop) or density. This is one of the few sampling techniques available that can estimate the weight or numbers of a species per unit area. Data obtained with toxicant use are statistically treated in the same way as data collected by any other technique (Chapter 2). The expense to collect samples and the large volume of data generated in routine surveys sometimes convince biologists (regrettably) that toxicant-derived data can be treated differently. As with any estimate, some measure of precision should accompany reported mean values for standing crop and density. Otherwise, there is no way to tell whether differences shown by future samples represent real changes in a fish community or population. Fish are frequently distributed in a clumped manner, and rotenone sampling readily detects this pattern; therefore, data transformations (Chapter 2) should be considered before parametric analyses are used or, alternatively, nonparametric tests should be used.

In many instances of long-term cove sampling, the same coves are sampled every year. If fixed sites are repeatedly sampled over time and statistical comparisons are made, then repeated-measures analysis of variance should be employed to test for differences in fish abundance over time (Maceina et al. 1994). Individual coves often display unique fish population characteristics that persist from year to year; a consideration of these characteristics by use of a repeated-measures design is, therefore, recommended.

In the 1960s and 1970s, there was great interest in using cove rotenone samples to estimate species composition and abundance throughout a lake or reservoir. To this end, entire embayments were treated with rotenone (see Chapter 1) in elaborate experiments conducted by multiagency task forces on two Tennessee reservoirs (Douglas Lake, Hayne et al. 1967; Barkley Lake, Summers and Axon 1980). Valuable information was obtained, including confirmation that the whole-lake abundances of many species were over- or underestimated in cove samples. Based on the biomass observed in cove samples, adjustment factors were calculated and used to estimate biomass of species in offshore, open-water habitats (Table 10.2). The adjustment factors for each species varied markedly between the two reservoirs because of the different morphometries of each system. Douglas Lake is a steep-sided tributary impoundment that stratifies thermally; Barkley Lake is a relatively shallow mainstream reservoir in which thermal stratification is weak or nonexistent. Further experiences with cove rotenone sampling have convinced many biologists that the relationship between cove and open-water abundances is site specific and that there is no single appropriate adjustment factor for all lakes and reservoirs (Davies and Shelton 1983). Therefore, expansions of cove data to represent systemwide abundances is no longer recommended. Rather, the trend has been to view cove samples as representative of the habitats sampled (i.e., the littoral zone) and only at the time of year the samples are collected. Many lacustrine species display distinct, seasonal inshore–offshore movements (Bettoli et al. 1993), and such movements must be considered when rotenone samples are evaluated.

Table 10.2 Adjustment factors for estimating biomass of selected species in offshore habitats of Douglas and Barkley lakes based on inshore cove rotenone samples. To estimate biomass in offshore habitats, values observed in cove samples were multiplied by the appropriate adjustment factors, which vary according to species, size of fish, and type of system. Data are modified from Table 10.1 of Davies and Shelton (1983).

Species	Length interval (cm)	Douglas Lake	Barkley Lake
Gizzard shad	2–41	1.00	4.20
Threadfin shad	2–30	1.00	21.09
Catfishes and bullheads	2–14	0.64	2.74
	15–23	1.62	3.29
	24–102	1.04	1.87
Bluegill and sunfishes	2–8	0.50	0.21
	9–14	1.00	0.59
	15–33	1.22	0.60
Black basses	2–10	0.50	0.08
	11–24	0.90	0.16
	25–71	1.34	0.24
Crappies	2–8	0.97	5.84
	9–18	1.38	26.02
	19–51	2.42	3.28

Probably the biggest hurdle to overcome in designing any lake or reservoir sampling program that uses toxicants is the cost to collect a sufficient number of samples. Sample sizes necessary to achieve certain levels of precision are often quite high (Chapter 2). Using Wegener rings or shoreline rotenone techniques allows one to collect an adequate number of samples; however, the cost to collect a sufficient number of large, "traditional" cove samples can be prohibitive. Simply treating more, but smaller, coves with toxicants may not meet objectives because of the biases inherent in sampling smaller areas (discussed above). We caution anyone contemplating large cove sampling to consider alternative techniques to collect the desired data. Perhaps more than any other sampling program, cove rotenone sampling demands a thorough review of objectives and rationales before it is undertaken or continued.

10.6 USE OF TOXICANTS IN RECLAMATION AND FISH CONTROL ACTIVITIES

Except in small, private impoundments, the use of toxicants to eliminate or reduce the abundance of nongame species so that sport fish can be restocked is now uncommon in North America, especially in lotic systems. Nongame fish populations often recover quickly after streams are treated, and managers have been urged to consider habitat enhancements in order to increase the abundance of desirable species (Moyle et al. 1983). Threatened and endangered species inhabit several systems that were once the object of reclamation efforts (Holden 1991). Reclamation activities on large, public bodies of water are likely to require extensive documentation of environmental impacts, which would increase the cost and logistics associated with any large reclamation project. Additional considerations were discussed by Wiley and Wydoski (1993).

Ironically, toxicants have been used often in recent years to eliminate resident stream fishes, but the rationale is now different. The toxicants are being applied to eliminate nonnative species, often formerly stocked sport fish species such as rainbow trout, before native, endangered species are reintroduced. The transition from stream reclamations to benefit sport fishes to the current emphasis on removing nonnative, introduced species began in 1970 with the enactment of various U.S. federal regulations such as the National Environmental Policy Act (42 U.S.C §§ 4321 to 4361) and the Endangered Species Act (16 U.S.C. §§ 1531 to 1544) (Rinne and Turner 1991). In many watersheds, biologists are mandated to conserve indigenous species (which by themselves often support viable sport fisheries), and more stream reclamation projects and species reintroductions are likely in the future.

10.6.1 Selective Removal of Target Species

Sea lamprey control with the lampricides TFM and Bayluscide in tributaries of the Great Lakes was discussed in section 10.2. Efforts to control the abundance of nuisance species such as shads have usually relied on applications of rotenone. Compared with many other species, shads are acutely sensitive to rotenone, and concentrations less than 0.1 mg/L cause mortality. However, whole-lake, low-level treatments with rotenone to reduce shad (particularly gizzard shad) biomass were much more common in the past than they are today.

Recent interest has focused on controlling common carp by means of poisoned feed pellets. Fajt and Grizzle (1993) mixed rotenone in a pelleted feed to induce mortality in common carp; a lethal dose was estimated at 11.6 mg rotenone/kg of fish. Biologists in Florida are experimenting with rotenone-laced feed (trade name, Fish Management Bait) to remove grass carp after these fish have achieved vegetation control (Mallison et al. 1994). Grass carp are first lured to feeders that dispense untreated pelleted food; once they are observed to feed routinely there, treated pellets are dispensed. Antimycin-laced bait has also been investigated for its efficacy in controlling common carp. Rach et al. (1994) reported that doses as low as 0.35 mg antimycin/kg of fish were sufficient to kill common carp.

In the southern United States, rotenone is sometimes applied to the shorelines of ponds to reduce the density of overabundant juvenile largemouth bass or bluegills (Swingle et al. 1953). The objectives can be to improve growth or recruitment by bluegills and crappies (McHugh 1990). To reduce the abundance of young largemouth bass, 500 mL of 5% rotenone is applied along every 100 m of shoreline. An initial treatment is made immediately after hatching is complete, followed 2–4 weeks later by a second treatment if juvenile largemouth bass density is very high.

10.6.2 Whole-Lake Reclamation

When large lentic systems are now considered candidates for reclamation, it is usually because they have been severely degraded and the fish community has changed concurrently. One recent example is the Lake Chicot renovation, discussed by Filipek et al. (1993). Lake Chicot is a Mississippi River oxbow lake (1,417 ha) in Arkansas that suffered from excessive sedimentation and pesticide pollution over many decades. After corrective measures were undertaken in the watershed, the lake was partially reclaimed with rotenone in 1985 and then restocked (Filipek et al. 1993). This renovation demonstrated the importance of undertaking remedial measures before a lake is treated with a toxicant; without such measures, the fish community quickly reverts to its former, undesirable condition. For Lake Chicot, the remedial measures included a partial drawdown to allow compaction of bottom sediments and seeding of shorelines and construction of a pump station to divert turbid, pesticide-laden inflows around the lake during periods of high runoff. An even larger whole-lake renovation was undertaken in 1990. The 4,800-ha Strawberry Reservoir, Utah, was treated with nearly 400,000 kg of powdered rotenone to eliminate fishes such as suckers and common carp that dominated that system and competed with more desirable sport fish species.

Despite such recent large-scale efforts, the frequency with which toxicants are used to renovate entire fish communities has decreased since the 1980s. Of 55 fisheries departments in North America responding to our survey of toxicant use, 64% reported that their use of rotenone or antimycin to reclaim fish communities has decreased or ceased in the past 10 years. Only one state indicated that reclamation efforts have increased in the past 10 years. Several agencies indicated that toxicant use has remained the same over the past 10 years because the technique had been used only sparingly beforehand.

10.7 COMMENTS ON FUTURE USE

Despite the general decline in use of toxicants to sample fish communities, toxicant applications will continue (in jurisdictions that allow them) when alternative techniques are unavailable or inappropriate. No other sampling technique produces estimates of fish abundance that are less biased in terms of size distributions and species composition. For instance, seine catches are usually biased towards smaller individuals that cannot escape the sweep of the net, and the susceptibility of fish to electrofishing differs markedly among sizes and species of fish. A key constraint of sampling with toxicants is that fish communities in the typically shallow or slow-flowing waters amenable to this approach may poorly represent fish abundance in an entire river, lake, or reservoir—or in the same habitat at a different time. To understand fish ecology, biologists who sample with toxicants must integrate their findings with data obtained from other habitats with other gear at the same and other times.

Registration of TFM, Bayluscide, and rotenone is currently secure (i.e., they are federally approved for use in the United States), and these compounds will be available for use in the foreseeable future. Future reregistration of antimycin is uncertain due to lack of funding to get this compound through regulatory hurdles in the United States and Canada; at present, antimycin can be used in the United States only because the Environmental Protection Agency granted a temporary extension of the original registration. As discussed in the opening section of this chapter, the regulatory atmosphere surrounding the intentional introduction of toxic compounds into waterways is dynamic and subject to change at any time.

10.8 REFERENCES

Allen, D. M., S. K. Service, and M. V. Ogburn-Matthews. 1992. Factors influencing the collection efficiency of estuarine fishes. Transactions of the American Fisheries Society 121:234–244.

Antonioni, M. E., and P. C. Baumann. 1975. Antimycin as a management and sampling tool. EIFAC (European Inland Fisheries Advisory Council) Technical Paper Number 23 (Supplement 1)1:266–286.

Applegate, V. C., J. H. Howell, J. W. Moffett, B. G. H. Johnson, and M. A. Smith. 1961. Use of 3-trifluoromethyl-4-nitrophenol as a selective sea lamprey larvicide. Great Lakes Fishery Commission Technical Report 1.

ASIH (American Society of Ichthyologists and Herpetologists), AFS (American Fisheries Society), and AIFRB (American Institute of Fisheries Research Biologists). 1988. Guidelines for use of fishes in field research. Fisheries 13(2):16–23.

Axon, J. R., L. Hart, and V. Nash. 1980. Recovery of tagged fish during the Crooked Creek Bay Rotenone Study at Barkley Lake, Kentucky. Proceedings of the Annual Conference Southeastern Association of Fish and Wildlife Agencies 33(1979):680–687.

Barr, W. C., and T. A. McDonough. 1978. Recovery of marked fish in cove rotenone samples. Proceedings of the Annual Conference Southeastern Association of Fish and Wildlife Agencies 30(1976):230–233.

Bayley, P. B., and D. J. Austen. 1990. Modeling the sampling efficiency of rotenone in impoundments and ponds. North American Journal of Fisheries Management 10:202–208.

Bayley, P. B., R. W. Larimore, and D. C. Dowling. 1989. Electric seine as a fish sampling gear in streams. Transactions of the American Fisheries Society 118:447–453.

Bettoli, P. W., M. J. Maceina, R. L. Noble, and R. K. Betsill. 1993. Response of a reservoir fish community to aquatic vegetation removal. North American Journal of Fisheries Management 13:110–124.

Bills, T. D., and L. L. Marking. 1988. Control of nuisance populations of crayfish with traps and toxicants. Progressive Fish-Culturist 50:103–106.

Bivin, W. M., M. L. Armstrong, and S. P. Filipek. 1991. Computer assisted techniques for standardized fisheries data collection and analysis. Proceedings of the Annual Conference Southeastern Association of Fish and Wildlife Agencies 43(1989):206–215.

Bouck, G. R., and R. C. Ball. 1965. The use of methylene blue to revive warmwater fish poisoned by rotenone. Progressive Fish-Culturist 17:134–135.

Bright, T. J., and nine coauthors. 1974. Biotic zonation on the West Flower Garden Bank. Pages 4–53 *in* Biota of the West Flower Garden Bank. The Gulf Publishing Company, Houston, Texas.

Chandler, J. H., Jr., and L. L. Marking. 1982. Toxicity of rotenone to selected aquatic invertebrates and frog larvae. Progressive Fish-Culturist 44:78–80.

Colle, D. E., J. V. Shireman, R. D. Gasaway, R. L. Stetler, and W. T. Haller. 1978. Utilization of selective removal of grass carp (*Ctenopharyngodon idella*) from an 80-hectare Florida lake to obtain a population estimate. Transactions of the American Fisheries Society 107:724–729.

Cumming, K. B. 1975. History of fish toxicants in the United States. Pages 5–21 *in* P. H. Eschmeyer, editor. Rehabilitation of fish populations with toxicants: a symposium. American Fisheries Society, North Central Division, Special Publication Number 4, Bethesda, Maryland.

Davies, W. D., and W. L. Shelton. 1983. Sampling with toxicants. Pages 199–213 *in* L. A. Nielsen and D. L. Johnson, editors. Fisheries techniques. American Fisheries Society, Bethesda, Maryland.

Dawson, V. K., D. A. Johnson, and J. L. Allen. 1986. Loss of lampricides by adsorption on bottom sediments. Canadian Journal of Fisheries and Aquatic Sciences 43:1515–1520.

Derse, P. H., and F. M. Strong. 1963. Toxicity of antimycin to fish. Nature 200:600–601.

DuBois, R. B., and W. H. Blust. 1994. Effects of lampricide treatments, relative to environmental conditions, on abundance and sizes of salmonids in a small stream. North American Journal of Fisheries Management 14:162–169.

Ebele, S., A. A. Oladimeji, and J. A. Daramola. 1990. Molluscidal and piscicidal properties of copper(II)tetraoxosulfate(VI) on *Bulinus globosus* (Morelet) and *Clarias anguillaris* (L.). Aquatic Toxicology 17:231–238.

Fajt, J. R., and J. M. Grizzle. 1993. Oral toxicity of rotenone for common carp. Transactions of the American Fisheries Society 122:302–304.

Farringer, J. E. 1972. The determination of the aquatic toxicity of rotenone and Bayer 73 to selected aquatic organisms. Master's thesis. University of Wisconsin-Lacrosse, LaCrosse.

Filipek, S. 1982. Survey and evaluation of Arkansas' chemical rehabilitation of lakes. Proceedings of the Annual Conference Southeastern Association of Fish and Wildlife Agencies 34(1980):181–192.

Filipek, S. P., J. D. Ellis, W. J. Smith, D. R. Johnson, W. M. Bivin, and L. L. Rider. 1993. The effect of the Lake Chicot renovation project on the fishery of a Mississippi River oxbow lake. Proceedings of the Annual Conference Southeastern Association of Fish and Wildlife Agencies 45(1991):206–215.

Flick, W. A., and D. A. Webster. 1992. Standing crops of brook trout in Adirondack waters before and after removal of non- trout species. North American Journal of Fisheries Management 12:783–796.

Fontenot, L. W., D. G. Noblet, and S. G. Platt. 1994. Rotenone hazards to amphibians and reptiles. Herpetological Review 25:150–156.

Gilderhus, P. A. 1982. Effects of an aquatic plant and suspended clay on the activity of fish toxicants. North American Journal of Fisheries Management 2:301–306.

Gilderhus, P. A. 1985. Solid bars of 3-trifluoromethyl-4-nitrophenol: a simplified method of applying lampricide to small streams. Great Lakes Fishery Commission Technical Report 47:6–12.

Gilderhus, P. A., V. K. Dawson, and J. L. Allen. 1988. Deposition and persistence of rotenone in shallow ponds during cold and warm seasons. U.S. Fish and Wildlife Service Investigations in Fish Control 95.

Gilderhus, P. A., and B. G. H. Johnson. 1980. Effects of sea lamprey (*Petromyzon marinus*) control in the Great Lakes on aquatic plants, invertebrates, and amphibians. Canadian Journal of Fisheries and Aquatic Sciences 37:1895–1905.

Great Lakes Fishery Commission. 1992. Strategic vision of the Great Lakes Fishery Commission for the decade of the 1990s. Report by the Great Lakes Fishery Commission, Ann Arbor, Michigan.

Gresswell, R. E. 1991. Use of antimycin for removal of brook trout from a tributary of Yellowstone Lake. North American Journal of Fisheries Management 11:83–90.

Haley, T. J. 1978. A review of the literature on rotenone 1,2,12,12a-tetrahydro-8,9-dimethoxy-2-(1-methylethenyl)-1-benzopyrano[3,5-b]furo[2,3-h][1]benzopyran-6(6h)-one. Journal of Environmental Pathology and Toxicology 1:315–337.

Hamilton, H. L. 1941. The biological action of rotenone on freshwater animals. Proceedings of the Iowa Academy of Science 48:467–479.

Hayne, D. W., G. E. Hall, and H. M. Nichols. 1967. An evaluation of cove sampling of fish populations in Douglas Reservoir, Tennessee. Pages 244–297 *in* Reservoir fishery resources symposium. American Fisheries Society, Southern Division, Reservoir Committee, Bethesda, Maryland.

Hegen, H. E. 1985. Use of rotenone and potassium permanganate in estuarine sampling. North American Journal of Fisheries Management 5:500–502.

Herr, F., E. Greselin, and C. Chappel. 1967. Toxicological studies of antimycin, a fish eradicant. Transactions of the American Fisheries Society 96:320–326.

Hightower, J. E., R. J. Gilbert, and T. B. Hess. 1982. Shoreline and cove sampling to estimate survival of young largemouth bass in Lake Oconee, Georgia. North American Journal of Fisheries Management 2:257–261.

Holden, P. B. 1991. Ghosts of the Green River: impacts of Green River poisoning on management of native fishes. Pages 43–54 *in* W. L. Minckley and J. E. Deacon, editors. Battle against extinction: native fish management in the American West. University of Arizona Press, Tucson.

Jackson, D. A., H. H. Harvey, and K. M. Somers. 1990. Ratios in aquatic sciences: statistical shortcomings with mean depth and the morphoedaphic index. Canadian Journal of Fisheries and Aquatic Sciences 47:1788–1795.

Jenkins, R. M. 1982. The morphoedaphic index and reservoir fish production. Transactions of the American Fisheries Society 11:133–140.

Jenkins, R. M., and D. I. Morais. 1978. Prey–predator relations in the predator-stocking-evaluation reservoirs. Proceedings of the Annual Conference of the Southeastern Association of Fish and Wildlife Agencies 30(1976)141–157.

Johnson, B. M., R. A. Stein, and R. F. Carline. 1988. Use of a quadrat rotenone technique and bioenergetics modeling to evaluate prey availability to stocked piscivores. Transactions of the American Fisheries Society 117:127–141.

Johnson, T. L., and R. W. Pasch. 1976. Improved rotenone sampling equipment for streams. Proceedings of the Annual Conference Southeastern Association of Game and Fish Commissioners 29(1975):46–56.

King, E. L., Jr., and J. A. Gabel. 1985. Comparative toxicity of the lampricide 3-trifluoromethyl-4-nitrophenol to ammocetes of three species of lampreys. Great Lakes Fishery Commission Technical Report 47:1–5.

Lambou, V. W., and H. Stern, Jr. 1958. An evaluation of some of the factors affecting the validity of rotenone sampling data. Proceedings of the Annual Conference Southeastern Association of Game and Fish Commissioners 11(1957):91–98.

Lee, T. H., P. H. Derse, and S. D. Morton. 1971. Effects of physical and chemical conditions on the detoxification of antimycin. Transactions of the American Fisheries Society 100:13–17.

Lennon, R. E., J. B. Hunn, R. A. Schnick, and R. M. Burress. 1971. Reclamation of ponds, lakes, and streams with fish toxicants: a review. FAO (Food and Agriculture Organization of the United Nations) Fisheries Technical Paper 100.

Lopinot, A. C. 1975. Summary of the use of toxicants to rehabilitate fish populations in the Midwest. Pages 1–4 *in* P. H. Eschmeyer, editor. Rehabilitation of fish populations with toxicants: a symposium. American Fisheries Society, North Central Division, Special Publication 4, Bethesda, Maryland.

Maceina, M. J., P. W. Bettoli, and D. R. DeVries. 1994. Use of a split-plot analysis of variance design for repeated-measures fishery data. Fisheries 19(3):14–20.

MacPhee, C., and R. Ruelle. 1969. A chemical selectively lethal to squawfish (*Ptychocheilus oregonensis* and *P. umpquae*). Transactions of the American Fisheries Society 98:676–684.

Mallison, C. T., R. S. Hestand, III, and B. Z. Thompson. 1994. Removal of triploid grass carp using Fish Management Bait (FMB). Pages 65–71 *in* Proceedings of the 2nd Grass Carp Symposium. U.S. Army Corps of Engineers, Waterways Experiment Station, Vicksburg, Mississippi.

Marking, L. L. 1969. Toxicity of rhodamine B and fluorescein sodium to fish and their compatibility with antimycin A. Progressive Fish-Culturist 31:139–142.

Marking, L. L. 1992. Evaluation of toxicants for the control of carp and other nuisance fishes. Fisheries 17(6):6–12.

Marking, L. L., and V. K. Dawson. 1972. The half-life of biological activity of antimycin determined by fish bioassay. Transactions of the American Fisheries Society 99:100–105.

Matlock, G. C., J. E. Weaver, and A. W. Green. 1982. Sampling nearshore estuarine fishes with rotenone. Transactions of the American Fisheries Society 111:326–331.

McHugh, J. J. 1990. Response of bluegills and crappies to reduced abundance of largemouth bass in two Alabama impoundments. North American Journal of Fisheries Management 10:344–351.

Miller, S. J., and six coauthors. 1991. Comparisons of Wegener Ring and 0.08-hectare block net samples of fishes in vegetated habitats. Proceedings of the Annual Conference Southeastern Association of Fish and Wildlife Agencies 44(1990):67–75.

Minckley, W. L., G. K. Meffe, and D. L. Soltz. 1991. Conservation and management of short-lived fishes: the cyprinodontoids. Pages 247–282 *in* W. L. Minckley and J. E. Deacon, editors. Battle against extinction: native fish management in the American West. University of Arizona Press, Tucson.

Miranda, L. E., W. L. Shelton, and T. D. Bryce. 1984. Effects of water level manipulations on abundance, mortality, and growth of young-of-year largemouth bass in West Point Reservoir, Alabama. North American Journal of Fisheries Management 4:314–320.

Moyle, J. B., and J. H. Kuehn. 1964. Carp, a sometimes villain. Pages 635–642 *in* J. P. Linduska, editor. Waterfowl tomorrow. U.S. Fish and Wildlife Service, Washington, DC.

Moyle, P. B., B. Vondracek, and G. D. Grossman. 1983. Responses of fish populations in the north fork of the Feather River, California, to treatments with fish toxicants. North American Journal of Fisheries Management 3:48–60.

Ober, R. D. 1981. Operational improvements for sampling large streams with rotenone. Pages 364–369 *in* L. A. Krumholz, editor. The warmwaters streams symposium. American Fisheries Society, Southern Division, Bethesda, Maryland.

Pintler, H. E., and W. C. Johnson. 1958. Chemical control of rough fish in the Russian River drainage, California. California Fish and Game 44:91–124.

Rach, J. J., J. A. Louma, and L. L. Marking. 1994. Development of an antimycin-impregnated bait for controlling common carp. North American Journal of Fisheries Management 14:442–446.

Reynolds, J. B., and D. E. Simpson. 1978. Evaluation of fish sampling methods and rotenone census. Pages 11–24 *in* G. D. Novinger and J. G. Dillard, editors. New approaches to the management of small impoundments. American Fisheries Society, North Central Division, Special Publication 5, Bethesda, Maryland.

Rinne, J. N., and P. R. Turner. 1991. Reclamation and alteration as management techniques, and a review of methodology in stream renovation. Pages 219–243 *in* W. L. Minckley and J. E. Deacon, editors. Battle against extinction: native fish management in the American West. University of Arizona Press, Tucson.

Schnick, R. A. 1974. A review of the literature on the use of rotenone in fisheries. U.S. Fish and Wildlife Service, Fish Control Laboratory, LaCrosse, Wisconsin.

Schnick, R. A. 1991. Registration status for fishery compounds. Fisheries 17(6):12–13.

Seelye, J. G., D. A. Johnson, J. G. Weise, and E. L. King, Jr. 1988. Guide for determining application rates of lampricides for control of sea lamprey ammocetes. Great Lakes Fishery Commission Technical Report 52:1–23.

Shireman, J. V., D. E. Colle, and D. F. DuRant. 1981. Efficiency of rotenone sampling with large and small block nets in vegetated and open-water habitats. Transactions of the American Fisheries Society 110:77–80.

Siler, J. R. 1986. Comparison of rotenone and trawling methodology for estimating threadfin shad populations. Pages 73–78 *in* G. E. Hall and M. J. Van Den Avyle, editors. Reservoir fisheries management: strategies for the 80's. American Fisheries Society, Southern Division, Reservoir Committee, Bethesda, Maryland.

Singh, A., and R. A. Agarwal. 1990. Molluscidal properties of synthetic pyrethroids. Journal of Medical and Applied Malacology 2:141–144.

Smith, B. R., and J. J. Tibbles. 1980. Sea lamprey (*Petromyzon marinus*) in Lakes Huron, Michigan, and Superior: history of invasion and control, 1936–1978. Canadian Journal of Fisheries and Aquatic Sciences 37:1780–1801.

Sousa, R. J., F. P. Meyer, and R. A. Schnick. 1987. Re-registration of rotenone: a state/federal cooperative effort. Fisheries 12(4):9–13.

Summers, G. L., and J. R. Axon. 1980. History and organization of the Barkley Lake rotenone study. Proceedings of the Annual Conference Southeastern Association of Fish and Wildlife Agencies 33(1979):673–679.

Swingle, H. S. 1950. Relationships and dynamics of balanced and unbalanced fish populations. Alabama Agricultural Experiment Station Bulletin 274, Auburn.

Swingle, H. S., E. E. Prather, and J. M. Lawrence. 1953. Partial poisoning of overcrowded fish populations. Alabama Agricultural Experiment Station Circular 113:1–15, Auburn.

Timmons, T. J., W. L. Shelton, and W. D. Davies. 1979. Sampling of reservoir fish populations with rotenone in littoral areas. Proceedings of the Annual Conference Southeastern Association of Fish and Wildlife Agencies 32(1978):474–484.

Trimberger, E. J. 1975. Evaluation of angler-use benefits from chemical reclamation of lakes and streams in Michigan. Pages 60–65 *in* P. H. Eschmeyer, editor. Rehabilitation of fish populations with toxicants: a symposium. American Fisheries Society, North Central Division, Special Publication 4, Bethesda, Maryland.

Wegener, W. D., D. Holcomb, and V. Williams. 1974. Sampling shallow water fish populations using the Wegener ring. Proceedings of the Annual Conference Southeastern Association of Fish and Wildlife Agencies 27(1973):663–673.

Wiley, R. W. 1984. A review of sodium cyanide for use in sampling stream fishes. North American Journal of Fisheries Management 4:249–256.

Wiley, R. W., and R. S. Wydoski. 1993. Management of undesirable fish species. Pages 335–354 *in* C. C. Kohler and W. A. Hubert, editors. Inland fisheries management in North America. American Fisheries Society, Bethesda, Maryland.

Yurk, J. J., and J. J. Ney. 1989. Phosphorous-fish community biomass relationships in southern Appalachian reservoirs: can lakes be too clean for fish? Lake and Reservoir Management 5:83–90.

Chapter 11

Invertebrates

CHARLES F. RABENI

11.1 INTRODUCTION

This chapter introduces techniques for collecting and analyzing data on invertebrates—both macroinvertebrates and zooplankton—that are relevant to fisheries concerns. Data on invertebrates are often necessary in fisheries research and management projects for which objectives include determining the prey base for sport fishes, evaluating habitat improvement efforts, determining the biological integrity of a water body, or documenting pollution or other degradations.

Several stages are involved with any project on invertebrates. As with any research, the data will be useful only if all steps are taken intelligently. All steps should be considered in the initial planning phase, once the objectives have been clearly stated (Krueger and Decker 1993); they include determining what sampling gear to use, where to sample, when to sample, how many samples to take, how to preserve and catalog invertebrates, how best to sort invertebrates from debris, how best to enumerate and identify invertebrates, and how best to analyze the data.

The admonition by the author's fourth-grade teacher that "anything worth doing is worth doing well" is appropriate for readers of this chapter because the invertebrate aspect of a fisheries project is often considered important but ancillary to a larger objective. Thus, the effort and resources required for a complete job are often lacking. Incomplete data on invertebrates, or data from samples that have not been properly collected and handled, are usually worse than no data because erroneous conclusions may be drawn from them. Time spent at the inception of a project to determine time, effort, and level of knowledge required to complete the project will be repaid many times over. A preliminary, or pilot, study is often required to evaluate characteristics of both the biota and the samplers and will aid in assessing variability and estimating sampling effort. Many common mistakes can be avoided by investigating the literature, consulting with experienced collectors about the life histories and habitat requirements of the biota, and learning about the problems associated with sampler bias and operator error.

11.2 DEVICES FOR COLLECTING INVERTEBRATES

The investigator must first decide the smallest organism that needs to be collected. The net mesh size used during collection, preservation, or sorting will determine what is retained for further study. The smaller the mesh, the more organisms are collected—either smaller taxa or smaller life stages—but small meshes also collect more fine debris and substantially increase sample-processing time. In addition, smaller meshes increase the likelihood of clogging, which reduces gear efficiency (Chapter 9). Commonly used mesh sizes (openings) are 80–253 μm for zooplankton and 0.595 mm (U.S. Standard Number 30) for macroinvertebrates.

11.2.1 Collecting Macroinvertebrates

11.2.1.1 Sampling Ponds, Wetlands, and Littoral Zones of Lakes

The D-frame aquatic net (Figure 11.1) is the basic tool for macroinvertebrate collections. Its versatility, portability, and low cost allow it to be used in most habitats, and a comprehensive qualitative sample can be quickly obtained with it. The net can be pushed through vegetation or the upper layers of the substrate to collect macroinvertebrates. A quantitative sample—one that can yield the number of taxa or individuals present in some designated area—is obtained by operating the net for a given period of time, within a specified area, or a combination of both. This simple method provides statistical precision of estimates comparable to those of some more sophisticated methods, and it is recommended if attention is paid to variation in mesh sizes among nets and to differences in operator methodology.

Various corers have been used for sampling macroinvertebrates; they can be as simple as pieces of stovepipe or coffee cans (Figure 11.1). They all work on the principle of isolating a given volume of water and area of substrate. The substrate and any invertebrates can be scooped out and sieved, or the substrate and any vegetation can be agitated, suspending the invertebrates in water where they can be caught with a hand net. Corers are particularly suited for standing water with or without vegetation. If invertebrates specifically associated with vegetation are to be collected, the vegetation is clipped near the substrate and carefully removed. A more sophisticated version of the basic corer is the Brown benthos sampler (Figure 11.1; Brown et al. 1987), which incorporates a 12-V pump to agitate the bottom and circulate suspended materials through an in-line filter.

11.2.1.2 Shallow Streams

The D-frame net is commonly used in streams, where it is held on the bottom while the substrate directly upstream is disturbed by foot. (This is sometimes referred to as the crawdad shuffle; music is optional.) Used this way, the sampler is commonly called a "kick net." If used systematically (i.e., in each discerned habitat), kick nets generally collect a better representation of the invertebrate community than do smaller quantitative samplers because they can sample a wider variety of small habitats.

Quantitative samples from shallow streams have traditionally been taken by using a Surber sampler (Figure 11.1). A Surber sampler is placed with its opening upstream; its horizontal rim delineates 929 cm^2 (a square foot) in area. All substrate within the rim is removed or agitated to dislodge organisms, which drift into the collection bag. The sampler was important in advancing quantitative studies in streams, but, because it was found to have serious biases—caused by loss of organisms around the collection bag, backwash in high velocities, and inefficiency in all but very shallow water—the only reason for using one now would be to make comparisons with an earlier study.

The Hess (Figure 11.1) or other similar sampler combines the attributes of a corer and the Surber sampler and is essentially a mesh corer of a particular diameter attached to a collection bag. Its operation is the same as that of a Surber sampler, but it can be used in deeper water and does not have the backwash problem. The Brown

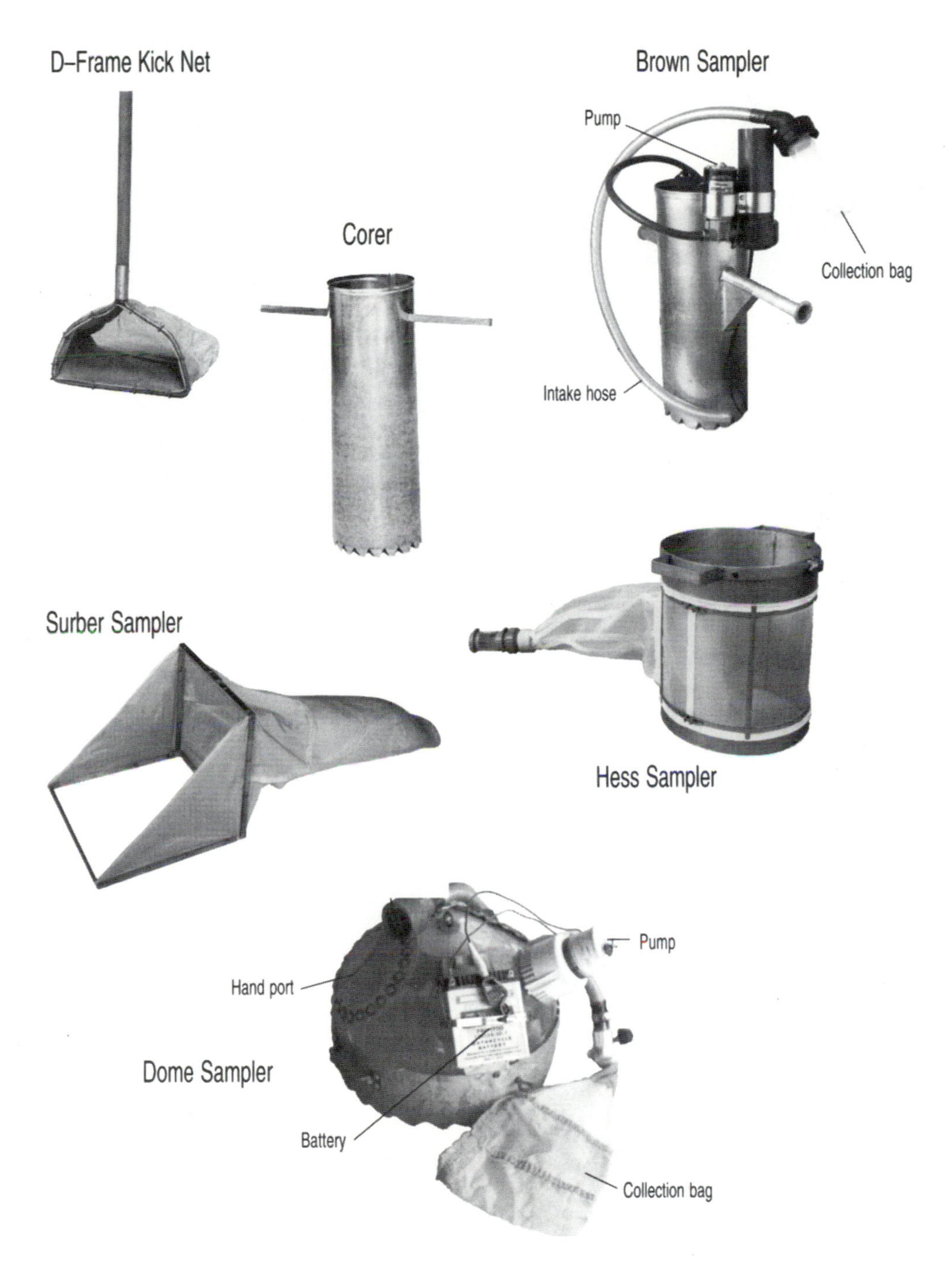

Figure 11.1 Devices for collecting and concentrating invertebrates. Photographs of the Surber sampler, Hess sampler, Ekman dredge, Petersen dredge, multiplate sampler, Kemmerer water sampler, Schindler–Patalas trap, zooplankton nets, and wash bucket were provided by the Wildlife Supply Company (WILDCO); the Brown sampler by Arthur Brown; and the dome sampler by Brian Mangan. All other photographs were provided by Missouri University's Academic Support Center.

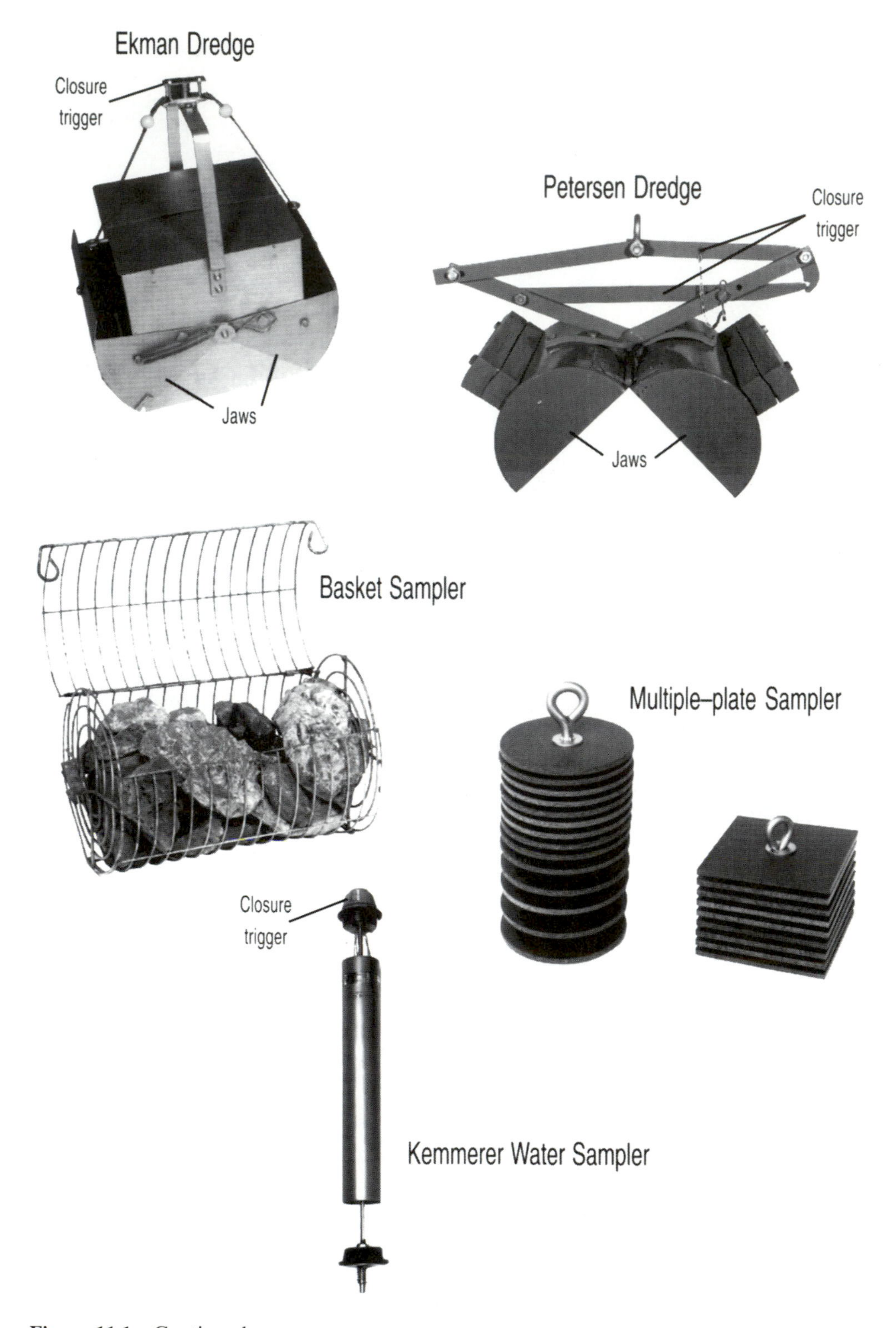

Figure 11.1 Continued.

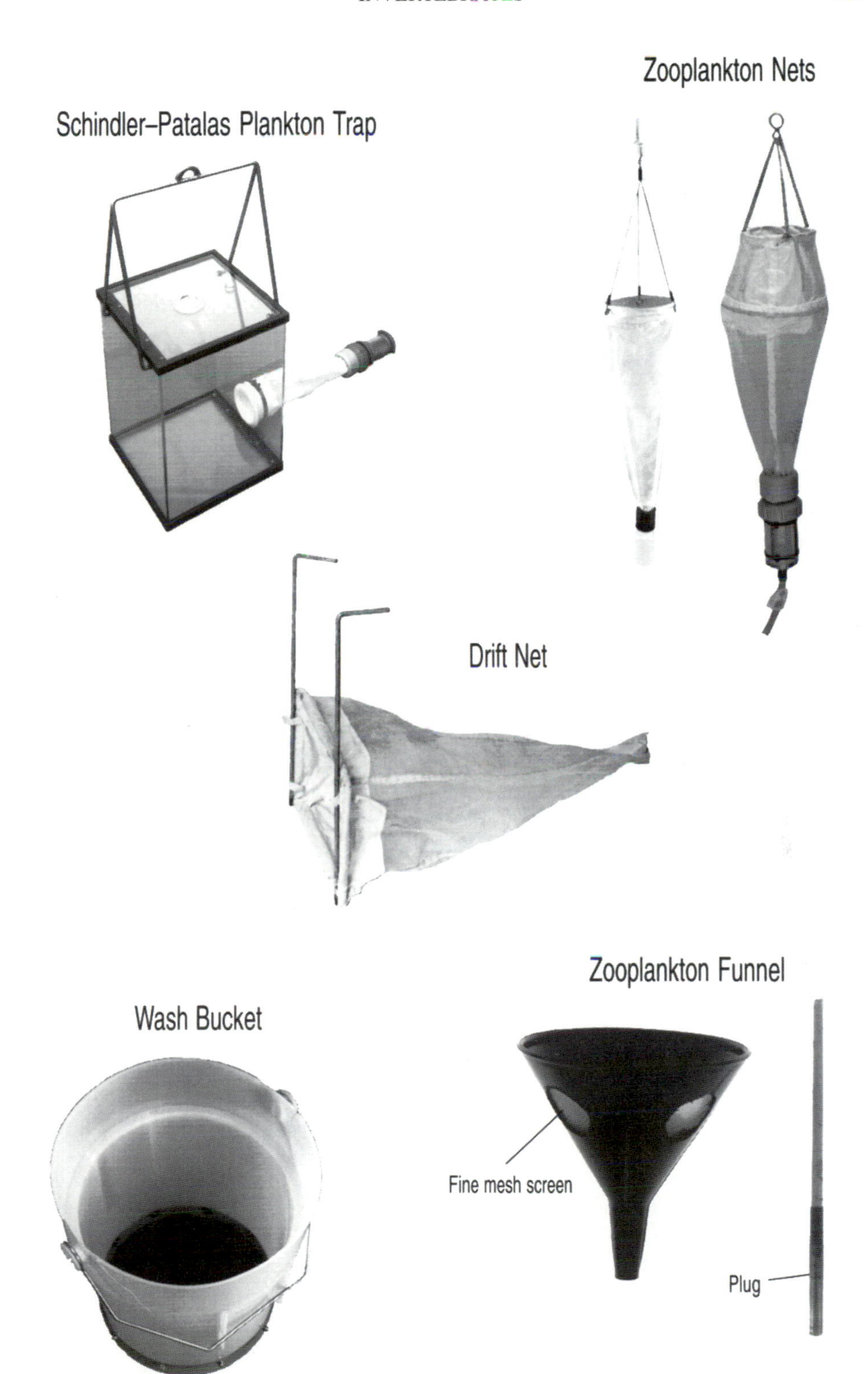

Figure 11.1 Continued.

sampler is also recommended for quantitative stream samples. In one test, it was superior to the Hess sampler in collection efficiency (Brown et al. 1987). Further information on sampling small streams was provided by Peckarsky (1984) and Klemm et al. (1990).

11.2.1.3 Deep Streams and Rivers

Major complications arise when sampling is conducted in water deeper than one's hip waders. The primary concern then becomes safety, and no sampling should be done alone and without formal training. Deepwater (>1 m) sampling can be done with either a modification of a shallow-water sampler or a specially designed sampler. Scuba or other underwater aids are generally required (Chapter 18). Deepwater samplers are totally enclosed except for the bottom. The sampler is set on the substrate, which is then agitated either by hand or by a pump-driven water jet; specimens are collected in a collection bag mounted on the downstream side of the sampler. The Hess and Brown samplers can be modified for use by divers. A specially designed, diver-operated deepwater sampler is the dome sampler (Gale and Thompson 1975; Figure 11.1). Where the bottom material is sand, mud, or clay and the current velocity is low, the grabs and dredges described in the next section may be useful.

11.2.1.4 Lakes

When sediments are soft and water is deep, grabs such as the Ekman or the more rugged Petersen (Figure 11.1) models (often called "dredges") are frequently used for quantitative sampling of benthic macroinvertebrates. The Ekman grab is the most popular type. It is essentially a metal box with jaws on the bottom. The device is lowered by rope to the bottom with its jaws open. At the bottom it sinks into the substrate. A metal weight, termed a messenger, is sent down the line and trips the jaws shut, then the sampler and its contents are retrieved. The sampler is appropriate only for soft sediments without debris. The Petersen grab is heavier than the Ekman design; its two large jaws are closed not by a messenger (which is not always reliable) but by an upward pull on the retrieval rope. It is heavy enough that a boat winch is recommended. All grabs should be used with care, because unexpected closure of the jaws can seriously injure hands and fingers. Further information on these devices was provided by Downing (1984).

11.2.1.5 Introduced Substrates

Introduced (artificial) substrates are materials placed into an aquatic environment, whereupon they are colonized by invertebrates (see review by Rosenberg and Resh 1982). Introduced substrates sometimes yield better samples than many "active" samplers of natural substrates. Study objectives will dictate whether introduced substrates are appropriate. Their primary advantage is a reduction in variation among sampling units placed in the same area. This makes it easier to detect significant differences in communities in different habitats, such as upstream and downstream of an effluent pipe (see Chapter 2 for the influence of variance on statistical decisions). Their primary disadvantage is that they often do not well represent the taxonomic richness, relative species abundances, or species densities of the habitat's community. If study objectives call for an evaluation of water quality effects on invertebrates, introduced substrates should be considered. If objectives call for an estimate of invertebrate biomass available to a sport fish, introduced substrates may provide misleading or erroneous information.

Two commonly used introduced samplers are the rock-filled basket and the multiplate sampler (Figure 11.1). The basket can be filled with any material, but if it contains material similar to the natural substrate, a better representation of the true community will be obtained than with a multiplate sampler, which is often suspended in the water column. The multiplate sampler requires less work to process, which allows more sample replicates (sampling units) to be taken and processed.

11.2.2 Collecting Zooplankton

Numerous samplers have been designed for zooplankton, and they fall into four basic types. They all work by either collecting a volume of water containing organisms or by filtering the water and entrapping organisms (Lind 1979; Wetzel and Likens 1979).

Closing samplers such as the Van Dorn or Kemmerer bottles (Figure 11.1) are cylinders with caps. Held open at each end as they are lowered to a particular depth, they are closed by a messenger and thus obtain a quantitative sample of water from a chosen depth. For an integrated sample at all depths, a long tube, open at both ends, is pushed vertically into the water from a boat or other platform. When the desired depth has been reached, the tube is closed at the top (creating a vacuum at the plug) and the tube is retrieved until the lower orifice can be stoppered (beneath the water surface). The greater the depth plumbed, the smaller the tube diameter must be; otherwise, the top plug cannot sustain the vacuum against the weight (mass) of the water in the tube, and the sample drains away. If the diameter of a sampling tube becomes too small (for physical reasons) to obtain representative samples, multiple samples will have to be gathered. If the depth to be sampled from a floating platform is considerable, flexible tubing (heavily weighted to retain verticality) or joinable tube sections (with pressure-resistant joints) must be used. Closing samplers may produce a biased collection. Some agile zooplankters can actively avoid the relatively small opening.

Traps were developed to minimize the zooplankter avoidance problem. Schindler–Patalas (Figure 11.1) and Juday traps are commonly used, and their operation is essentially the same as the closing sampler. However, the volume collected is greater and the openings are larger than those found in a closing sampler, which reduces avoidance by the larger or more agile zooplankters. Additionally, one haul of a trap both collects and concentrates the zooplankton. Traps are generally thought to be superior to closing samplers.

Pumps are often used in conjunction with a hollow flexible tube or hose. Pumps allow a large volume of water to be filtered, and, in general, the more water that can be filtered the better the estimate of abundance or community composition. Tubes with pumps can obtain a sample from a particular depth or be rigged to provide a depth-integrated sample. A flowmeter or calibrated container is required for a quantitative estimate. Agile taxa may avoid the relatively small openings of tubes and hoses, but comparisons have shown this method to be at least as efficient as traps and nets.

Zooplankton nets are widely used, and the most popular is the Wisconsin model (Figure 11.1). Their ease of use, relatively low cost, and ability to collect both qualitative and quantitative samples make them a favorite. Nets are towed or pulled horizontally or vertically for a particular distance or time. The volume of water filtered is easily estimated by multiplying distance towed by the area of the net opening. This estimate is subject to some error, which has been addressed by development of the Clark–Bumpus sampler, which consists of a net with an attached flowmeter; alternatively, one may fit a flowmeter to the Wisconsin-type net. If the sampling device has not already done so, zooplankton can be concentrated in a zooplankton net, bucket, or funnel (Figure 11.1).

Operator error can be substantial when nets are used. The retrieval must be steady but not so fast as to cause a back pressure by which water is diverted around instead of through the net. Efficiency is also affected by mesh size and clogging (also see Chapter 9).

11.2.3 Sampling in Specialized Habitats

Vegetation. Sampling invertebrates associated with vegetation is often done with a D-frame net swept through the vegetation for qualitative samples. For quantitative samples, a corer (e.g., stovepipe sampler) or bag is placed over the vegetation, which is then clipped near the substrate and removed. Invertebrates associated with vegetation may cling to the plant or burrow within the plant. Reports should clearly state whether densities are for the volume of water, area of substrate, or area of plant surface and whether just the vegetation or vegetation and substrate were sampled.

Woody debris. Invertebrates on submerged wood are often collected by sawing off pieces of the wood and bringing them to the surface, where the invertebrates are scraped or picked off. Mobile or loosely attached forms are saved by placing the wood into a collection bag or a specially designed sampler (Thorp et al. 1992) before it is brought to the surface. Quantitative samples are obtained by assuming the wood piece is a cylinder, or other appropriate shape, and calculating its surface area. Wood too large to be removed may be scraped over a designated area into a collection bag.

Stream drift. Analysis of benthic invertebrates that drift in the water column is popular with fisheries biologists because drifting animals constitute the primary food for many fish species. Drift nets (Figure 11.1) are usually anchored to the stream bottom and left to collect passing insects for a certain period of time. Information on the current velocity at the mouth of the sampler and the area of the sampler opening allows quantification of the collection as the number of drift organisms per some volume of water. Drift is a diel phenomenon, and the common pattern of maximum drift at sunset and sunrise should be considered in any collection scheme.

Large substrates. Substrates composed of large cobble, boulders, or bedrock are difficult to sample. Some samplers, such as the Hess sampler or Brown benthos sampler, can be modified with a spongelike material placed on the bottom edge. The material molds itself to an irregular surface, which can be useful under certain circumstances. Large substrates contain numerous microhabitats for invertebrates—on the boulder, under the boulder, and between boulders—which makes obtaining a representative sample very difficult.

Hyporheos (within-substrate habitat). The discovery that large numbers of invertebrates sometimes exist long distances—vertically and sometimes horizontally—from substrate surfaces was important for stream ecological theory but certainly complicated sampling protocols. Most success in sampling this habitat has been by using "freeze-coring" devices. A single or connected group of metal tubes is driven into the substrate; after a reacclimation period, liquid nitrogen is poured down the tubes, which solidifies the nearby substrate and everything in it. The tubes and the frozen substrate are pulled from the stream bottom, usually with the aid of a winch, and the vertical distribution of invertebrates may then be examined (Pugsley and Hynes 1983).

11.3 COLLECTION STRATEGIES

Knowing when is the best time to sample, where in a particular habitat to sample, and how many samples to take requires some knowledge of invertebrate ecology and life history. A good sampling strategy results from reviewing the extensive literature on invertebrate ecology and conducting a pilot study (see review by Resh 1979). Two probable outcomes of a poorly designed sampling program are insufficient samples for the problem at hand and inappropriate sampling. Rarely are too many samples taken. Whereas it is simple to discard an unneeded sample before it is processed in the laboratory, it is impossible to process a needed sample that was not taken. Because sampling rarely accounts for more than 5% of the time for total processing of a sample, collecting sufficient samples ought not to be a problem.

11.3.1 When to Sample

Deciding when to sample requires knowledge of the temporal variation of the biota. A typical life cycle involves an egg stage; a large number of small, immature individuals; and (through time) a progressively smaller number of larger immature individuals and adults. Each stage likely occupies a somewhat different microhabitat and thus is differentially susceptible to capture by particular gears. In addition, some species produce many generations a year, whereas others may take a year or more to complete one life cycle. Community composition changes by season and even time of day. Conclusions from the data depend very much upon when, and the frequency at which, data were collected.

Aspects of invertebrate biology that influence a sampling scheme are whether the release of eggs occurs all at once or over a period of time, the size range of the organism as growth progresses, changes in habitat preferences as the organism grows, changes in mobility during the life cycle, differences in behavior or size between sexes, and the possession of an inactive (resting) stage.

11.3.2 Where to Sample

Deciding where to sample requires knowledge of variation in the spatial distribution of the organism. Invertebrates are not equally distributed among all habitat types and not equally distributed within a habitat. They tend to be clumped rather than randomly or uniformly distributed. One may safely assume that different habitat types in the same aquatic system—say pool versus riffle, littoral versus profundal, or vegetated versus nonvegetated—possess greatly different types and densities of invertebrates. Focusing on a particular habitat or dividing the sampling effort among two or more well-defined habitat types (e.g., stratified sampling design, see Chapter 2) will usually result in more precise estimates for the measurement of interest.

11.3.3 Appropriate Sample Size

The term sample, in its most useful context, refers to a number of collections—often referred to as sampling units, replicate sampling units, or replicate samples—each taken one after another, in the same manner, and in the same habitat. For example, four zooplankton net tows taken in sequence off the back of a boat in a lake would constitute a sample, as would three concurrent Surber sampler collections across a riffle in a stream. The number of sampling units taken or needed is often referred to as the sample size. (See Chapter 2 for a discussion of treatment replication.) Replicate sampling units are used to determine variation—for example, standard deviation or variance. A value without a measure of variance is not very useful because the precision of the estimate cannot be determined.

Sampling invertebrates is usually conducted to determine either how many taxa or taxon are present, or if there is a difference in the number of taxa or individuals between times or between places. In all cases, the required number of sampling units depends upon several factors, the most important being the ecology of the organism (how dense and how aggregated, or clumped, its distribution) and the acceptable error. Each situation differs from all others, and will be additionally influenced by the habitat sampled, operator efficiency, mesh size used, and sorting efficiency. Whereas density and degree of aggregation are uncontrollable, the necessary sample size can be influenced somewhat by the precision the investigator is willing to accept—the less precision, the fewer sampling units that will be necessary.

The objectives of the study will guide both the selection of the appropriate sampling device and the sampling design. Let us assume that the objective is to determine the number of taxa or individuals of a taxon at a particular location. First decide how much of an error is acceptable. Most experienced biologists are satisfied to be 95% confident that the mean number they estimate is within about 20% of the true mean. The less error one can accept, or the more precise one wants to be, the more samples are necessary. However, the "law of diminishing returns" quickly comes into play, and small gains in precision may require substantial increases in effort (see Box 11.1).

If one is estimating the density of a particular taxon or the entire community (numbers per some area), one should collect a few sampling units, calculate the mean density and the sample variance (var), and then decide the percent error to be tolerated. The sample size (N) needed to be within a particular percentage of the true mean may be estimated with the general formula given by Elliott (1971):

$$N = \frac{\text{var}}{(\text{allowable error})^2\,(\text{mean density})^2}. \tag{11.1}$$

The allowable error is how precise, or how close, one wants to be to the true mean. For a 20% allowable error, $N = \text{var}/(0.20)^2\,(\text{mean})^2 = 25\ \text{var}/\text{mean}^2$.

A biologist may be interested not just in how many individuals or taxa there are at a site but whether there are differences between sites or habitats or whether a particular habitat improvement or perturbation will change invertebrate abundance or diversity. Comparisons of this type often require that N be composed of treatment replicates to avoid pseudoreplication (see Chapter 2). Again, the biologist must decide the probability at which changes should be detected. Stated another way, given a level of sampling, how much of a difference or a change must exist before it can be detected. The general formula of Parkinson et al. (1988), adapted from Snedecor and Cochran (1981), gives a rough estimate of the number of sample units needed to detect a particular change.

$$N = \frac{100^2\,(K)(\text{SD}/\text{mean})^2}{p^2} \tag{11.2}$$

N is the sample size for each place or time period to be compared (equal samples between habitats are assumed; $2N$ is the total number of sample units for two sites

Box 11.1 How Many Samples?

In this simulated, but realistic, example, a series of eight Hess sampler collections from the stream bottom resulted in a sample with a mean density of 55 mayflies and a standard deviation (SD) of 30. Curve A was developed from equation (11.1) and gives the percent error (precision) for any sampling effort. With the eight individual collections of this example, the biologist is 95% confident of being within about 20% of the true mean—that is, that the actual mean number is between 44 and 66 mayflies. Curve A indicates the effort required for different precisions. For the first few collections, an additional collection results in substantial decrease of the error, but as the sample size increases, each additional sample results in a smaller decrease in percent error. For example, going from 3 to 4 collections reduces the error an additional 4% but going from 10 to 11 collections reduces the error only an additional 1%. For this, and any other data set, halving the error quadruples the necessary sample size.

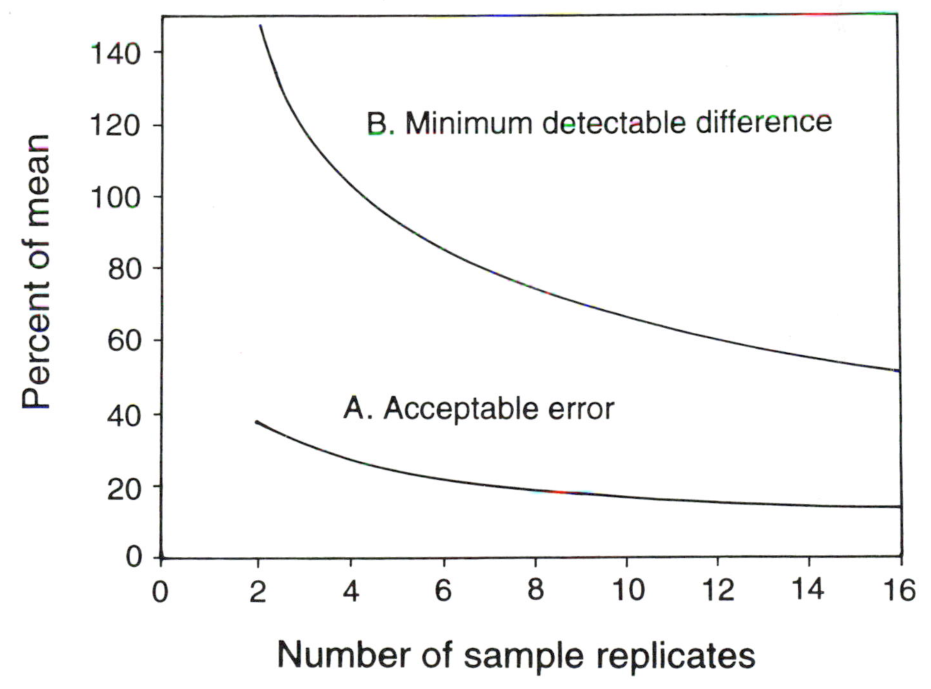

If the biologist were interested in whether the mayflies in our example respond to habitat improvement, the question then becomes, given a particular sampling effort, how much of a change can be statistically detected? Curve B was developed from equation (11.2) and indicates how the number of samples relates to the minimum detectable change as a percent of the mean.

Curve B indicates that a change of less than about 40% of the mean will never be detected, regardless of the sample size. With the eight samples taken, no change less than 75% of the mean will be detected. So, even if the habitat improvement actually caused the invertebrate population to increase by, say, 70%, the biologist could not conclude that habitat improvement increased invertebrate abundance.

or times); SD is the sample standard deviation; p is the minimum detectable change (as a percentage of the mean); and K relates sample size to sample variability given α (probability of type I error) and β (probability of type II error; see Chapter 2). The complement of α $(1 - \alpha)$ represents how confident one wants to be that the difference detected is real and not a random event (e.g., an alpha of 0.05 means one is 95% confident). The complement of β $(1 - \beta)$ represents the ability to detect a difference (e.g., a beta of 0.2 means one has an 80% chance of detecting a difference). A typical value for K in the above equation is 12.3 when alpha is 0.05 and beta is 0.20 (see Parkinson et al. [1988] for a table of K-values). Equation (11.2) can also be solved for N to determine the sample sizes necessary for a given level of precision.

These formulas and others like them (Chutter 1972; Resh 1979; Sheldon 1984; Norris and Georges 1986; and those in Chapter 2) should serve only as guides to give a general estimate of sampling requirements. The necessary number of sampling units will likely change over time due to changes in invertebrate life history characteristics and population dynamics that influence their densities and distribution patterns.

An interesting and often useful approach is sequential sampling, which is an analysis that cumulatively compares values of interest, such as density or species diversity, from individual samples and bases sampling thresholds upon specific levels of precision (Resh and Price 1984). Although many samples are taken, concurrent processing and analysis ensures that only the number of samples needed to reach a particular level of precision are sorted and identified.

11.4 PRESERVATION, CONCENTRATION, AND STORAGE OF SAMPLES

One of the best general preservatives for invertebrates is a formalin solution, sold as a saturated solution of formaldehyde gas in water. Formalin is a suspected carcinogen and has been designated as a hazardous material. Therefore, common sense and an understanding of federal guidelines for handling, storage, and transportation are required. Recommendations for preserving invertebrates usually call for a 3–5% formalin solution. This material is generally available as a concentrated 37% formaldehyde solution (100% formalin). Thus, 5 mL of the concentrate in 95 mL of water produces a 5% formalin solution. Formalin is an excellent fixative and does its job in a matter of days. Once processing of the sample in the laboratory begins, preferably under a fume hood, the sample should be washed completely free of formalin and the organisms should be transferred to 80% ethanol.

11.4.1 Macroinvertebrates

Collections obtained by any means are best prepared for analysis by first concentrating them in the field in a mesh-bottomed bucket (Figure 11.1). Animals and debris are washed and concentrated into one area of the bucket and then removed into a collection jar by hand or by careful washing through the bottom of the inverted bucket. Once in the storage container, 37% formaldehyde should be added to produce about a 3–5% formalin solution. A label, as discussed in section 11.7, becomes part of the sample.

Sorting organisms from debris is time-consuming, and it is very important to avoid thoughtless errors at this stage. Studies have shown definite, often statistically significant, differences among the sorting efficiencies of different personnel and large differences related to the level of microscope magnification used.

Attempts to shorten the time involved in sorting include adding stains, such as rose bengal, that differentially color organisms—but hopefully not much else. "Flotation," which is a technique where sugar or other chemicals are added to change the specific gravity of the water and cause light organic materials such as invertebrates to rise, also works sometimes (Anderson 1959). Centrifugal flotation of samples high in detritus is also an option. Elutriation devices (e.g., Whitman et al. 1983) employ agitation by introducing air into the bottom of a cone-shaped device and "bubbling" out the organisms.

11.4.2 Zooplankton

Zooplankters are easily concentrated by washing the sample through a Wisconsin zooplankton net bucket or a locally made apparatus (zooplankton funnel) as pictured in Figure 11.1. Zooplankton are best preserved by adding 37% formaldehyde to a sample in a known volume of water to produce a 3–5% formalin solution. Formalin may distort soft-bodied animals (e.g., rotifers) or cause Cladocera to balloon and lose eggs and embryos, so special techniques may be necessary in some situations (see De Bernardi 1984).

11.5 SUBSAMPLES

Subsampling large samples may significantly reduce sorting and identification times. Several devices are available to subsample large samples of benthic invertebrates and zooplankton, and all have the same caution. The investigator must be certain that the subsample is representative of the entire sample. This is accomplished if organisms are randomly distributed throughout the sample and if a sufficient subsample is taken. One can judge subsampling adequacy by comparing a series of subsamples against each other and against an entirely processed sample or by using a simple statistical procedure recommended by Elliott (1971).

Subsampling benthic invertebrates can be done with a gridded pan or gridded sieve in which the sample is uniformly distributed but only a portion is processed (Mason 1991; Cuffney et al. 1993a). More elaborate machines involving motors and rotating subsamples have also been developed (e.g., Waters 1971).

Subsampling zooplankton is best done by first washing the sample in a funnel such as that pictured in Figure 11.1 to remove formalin and to concentrate the animals. Then the sample is diluted with water to a known volume, often 100 mL. The sample is gently stirred with a figure-eight motion, a 1-mL aliquot is removed with a large-bore pipette, and the subsample is placed in a Sedgewick–Rafter or other counting cell for analysis under appropriate magnification.

11.6 IDENTIFICATION AND REFERENCES

Identifications of invertebrates for any project should be done by using appropriate taxonomic keys and by establishing a reference collection. A basic taxonomic library is listed in Box 11.2. It is unrealistic, and probably unnecessary, to have every taxon verified by a taxonomic specialist, but the collection should be examined by someone with some taxonomic expertise. Several excellent regional taxonomic keys are available for the genus and species levels of identification.

Box 11.2 Some Basic Taxonomic References

Merritt, R. W., and K. W. Cummins. 1996. An introduction to the aquatic insects of North America, 3rd edition. Kendall/Hunt, Dubuque, Iowa.
Pennak, R. W. 1989. Fresh-water invertebrates of the United States: protozoa to mollusca, 3rd edition. Wiley, New York.
Thorp, J. H., and A. P. Covich, editors. 1991. Ecology and classification of North American freshwater invertebrates. Academic Press, San Diego, California.
Ward, H. B., and G. C. Whipple, editors. 1959. Freshwater biology (revised by W. T. Edmondson). Wiley, New York.

11.7 RECORD KEEPING

Procedures to keep track of samples at every step of the process are simple but essential in establishing the credibility of resulting data. A master logbook (with waterproof paper pages) should be maintained and include data on each sample and a sample identification code (SIC) that specifies the project, study area, habitat, replicate number, date, sampler type used, sample number, and collection jar sequence (see Figure 11.2). The SIC is assigned when the sample is taken in the field and is used to track that sample through the entire project. A label with the SIC should be placed inside the collection jar and another attached to the outside—but not to the cap. India ink or hard lead pencil used on high-rag-content paper make good permanent labels. Entries in the master log must be photocopied and filed when sampling personnel return from the field.

It is important that the person sorting and identifying the sample not know its origin. Therefore, when the sample is being prepared for sorting and counting, only the sample number part of the sample identification code should be attached.

11.8 EVALUATION OF DATA QUALITY

Thus far this chapter has emphasized the collection of invertebrates. Before examining some ways in which the collected data can be used, we should review our collection procedures because valid conclusions can only be made from a properly collected data set. Box 11.3 lists some queries concerning aspects of assembling a data set on invertebrates. If all the questions can be answered appropriately, the biologist can be confident that the basis exists for further analysis.

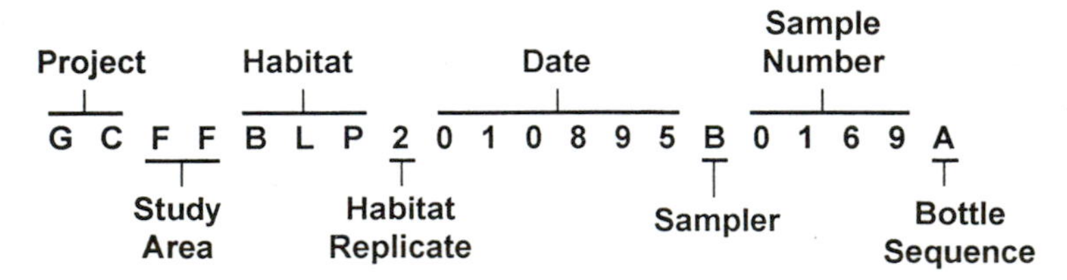

Figure 11.2 Example of a generalized sample identification code (SIC) used to track samples (after Cuffney et al. 1993b).

Box 11.3 Are Your Data Credible?

Are you sampling in the right place? You are if you initially collected and processed samples from different parts of the same habitat or different habitats and compared results.

Are you sampling during the correct time? You probably are if you researched the literature about the life history of your organism(s).

Are you taking enough samples? You are if you determined your sampling variance, established your desired level of precision, and calculated the number of samples required.

Do you understand the error involved in the sampler you are using? You do if you checked it against another sampler by taking and processing samples from the same habitat with both and if you investigated the literature.

Are the mesh sizes on your sampling, sorting, and washing equipment appropriate to the project? They are if you compared two or more mesh sizes on the same sample.

When the sample is sorted, are most invertebrates being found? They are if you have a second person re-sort a small percentage, say 10%, of the samples.

Is your sorting procedure biased toward large organisms? Not if you sorted a few samples with the unaided eye and then again under a binocular scope.

Are you confident of your identification? You are if you established a reference collection, had it verified by someone with formal training in identification, and have identified only to the level of your competence on complete specimens.

11.9 ANALYSIS

Analyses are guided by study objectives, which should be determined before sampling begins. Some objectives may require knowing the number of taxa or population densities in a particular habitat. These are straightforward determinations already discussed. Often the basic data must be further analyzed to extract additional information.

Information on biomass (the weight of material in a group of organisms) of invertebrates is often required in projects dealing with secondary production or fish feeding. The most common method of determining biomass is gravimetric—direct weighing. Both "wet" and "dry" weights are reported. Wet weights are presumed to be the live weight of the organism(s). Excess water is removed by blotting or centrifuging before animals are weighed. Dry weights are determined after the animals are heated in a drying oven at 60°C for 24 h. Dry weights are recommended because they are usually less variable and more ecologically meaningful. Gravimetric methods depend upon good equipment and consistent laboratory procedures. Preservatives such as formalin and alcohol cause a substantial change in organism weight over time. Using freshly collected or freshly frozen individuals to determine wet-weight biomass are better alternatives.

An indirect method for estimating biomass is to use regression equations relating measured length to a predetermined weight for a particular taxon (McCauley 1984). Published data such as Smock (1980) for benthic invertebrates and Dumont et al. (1975) for zooplankton are available. Alternatively, an investigator may develop regression equations from a subset of the collected animals. Having to measure only length, or some other body part, saves a substantial amount of time—especially when a computer and caliper system rapidly converts a linear measurement to a biomass estimate and processes, stores, and retrieves whatever data are desired (see Sprules et al. 1981).

Invertebrates are often used as an index (i.e., a number used as an indicator of a condition) to such things as fishing quality, ecological integrity, and degree of pollution. The abundance of large cladocerans *Daphnia* spp. was used to predict the survival of rainbow trout and fishing quality in small Michigan lakes (Galbraith 1975). Zooplankton size structure was used as an indicator of predator-to-panfish ratio in New York lakes and was related to other aspects of fish community structure (Mills et al. 1987).

Benthic invertebrates are more widely used than are zooplankton as indices of environmental quality. Whereas their initial use was limited to pollution monitoring, more recent studies have used invertebrates to evaluate aspects of ecosystem management (Reice and Wohlenberg 1993)—ecological integrity, ecosystem stability, food chain dynamics, and aquatic system function.

Indices may focus on a single, or sentinel, taxon often called an indicator organism, or they may be based on all collected taxa—the community. Community analyses traditionally focused on numerous diversity indices, which examined both species richness and the distribution of the number of animals among the species. Greater species diversity was equated with greater aquatic health. However, a lack of theoretical soundness and biological basis has diminished the use of diversity indices (Cairns and Pratt 1993).

Biotic indices have been successfully used because they incorporate a biological response to environmental conditions. Taxa are assigned scores according to their tolerance of a particular (or group of) environmental factor(s). The sum of the abundances of the taxa multiplied by their individual scores provides a number by which to judge the condition of the water body (Rosenberg and Resh 1993). This approach has been successfully used to classify lakes in studies of eutrophication and acid rain and to monitor stream water quality. Another approach based on all collected taxa are comparison indices—similarity or dissimilarity indices. Here, comparisons of community structure between a reference (i.e., best case) site and a site to be evaluated are made. Both presence and absence data and relative abundances may be used.

The recent increased use of invertebrates in biological monitoring related to state, provincial, and federal regulations has forced development of indices that are relatively easily and quickly obtained; such indices have been given the name rapid bioassessment (Plafkin et al. 1989). This approach correlates the presence, absence, or relative abundance of certain macroinvertebrate taxa or functional groups with standards from an unaffected site. Numerous metrics are in use from the aforementioned measures of taxon richness, diversity, and similarity to many others based on ratios of taxa, guilds, or functional attributes. The usefulness of many of the protocols has yet to be thoroughly evaluated (Kerans et al. 1992).

The use of indices may relieve the investigator of some of the problems inherent in sampling invertebrates. For example, sampling for a particular taxon allows the investigator to focus equipment, times, and places of sampling, and relative abundances may be more easily obtained. Yet no investigator should ignore the basic obligation to understand some invertebrate ecology and the necessity to replicate and determine the precision of the sampling scheme.

11.10 REFERENCES

Anderson, R. O. 1959. A modified flotation technique for sorting bottom fauna samples. Limnology and Oceanography 4:223–225.

Brown, A. V., M. D. Schram, and P. P. Brussock. 1987. A vacuum benthos sampler suitable for diverse habitats. Hydrobiologia 153:241–247.

Cairns, J. C., Jr., and J. R. Pratt. 1993. The history of biological monitoring using benthic invertebrates. Pages 10–27 *in* D. M. Rosenberg and V. H. Resh, editors. Freshwater biomonitoring and benthic macroinvertebrates. Chapman and Hall, New York.

Chutter, F. M. 1972. A reappraisal of Needham and Usinger's data on the variability of a stream fauna sampled with a Surber sampler. Limnology and Oceanography 17:139–141.

Cuffney, T. F., M. E. Gurtz, and M. R. Meador. 1993a. Methods for collecting benthic invertebrate samples as part of the national water quality assessment program. U.S. Geological Survey Open-File Report 93-406, Washington, DC.

Cuffney, T. F., M. E. Gurtz, and M. R. Meador. 1993b. Guidelines for the processing and quality assurance of benthic invertebrate samples collected as part of the National Water Quality Assessment Program. U.S. Geological Survey Open-File Report 93-407, Washington, DC.

De Bernardi, R. 1984. Methods for the estimation of zooplankton abundance. Pages 59–86 *in* Downing and Rigler (1984).

Downing, J. A. 1984. Sampling the benthos of standing waters. Pages 87–130 *in* Downing and Rigler (1984).

Downing, J. A., and F. H. Rigler, editors. 1984. A manual on methods for the assessment of secondary productivity in fresh waters. Blackwell Scientific Publications, London.

Dumont, J. F., I. Vande Velde, and S. Dumont. 1975. The dry weight estimate of biomass in a selection of Cladocera, Copepoda, and Rotifera from the plankton, periphyton and benthos of continental waters. Oecologia 19:75–97.

Elliott, J. M. 1971. Some methods for the statistical analysis of samples of benthic invertebrates. Freshwater Biological Association Scientific Publication 25, Ambleside, UK.

Galbraith, M. G. 1975. The use of large *Daphnia* as indices of fishing quality for rainbow trout in small lakes. Internationale Vereinigung für Theoretische und Angewandte Limnologie Verhandlungen 19:2485–2492.

Gale, W. F., and J. D. Thompson. 1975. A suction sampler for quantitatively sampling benthos on rocky substrates in rivers. Transactions of the American Fisheries Society 104:398–405.

Kerans, B. L., J. R. Karr, and S. A. Ahlstedt. 1992. Assessing invertebrate assemblages: spatial and temporal differences among sampling protocols. Journal of the North American Benthological Society 11:377–390.

Klemm, D. J., P. A. Lewis, F. Fulk, and J. M. Lazorchak. 1990. Macroinvertebrate field and laboratory methods for evaluating the biological integrity of surface waters. U.S. Environmental Protection Agency EPA/600/4-90/030, Environmental Monitoring Systems Laboratory, Cincinnati, Ohio.

Krueger, C. C., and D. J. Decker. 1993. The process of fisheries management. Pages 33–54 *in* C. C. Kohler and W. A. Hubert, editors. Inland fisheries management in North America. American Fisheries Society, Bethesda, Maryland.

Lind, O. T. 1979. Handbook of common methods in limnology, 2nd edition. C. V. Mosby, St. Louis, Missouri.

Mason, W. T. 1991. Sieve sample splitter for benthic invertebrates. Journal of Freshwater Ecology 6:445–449.

McCauley, E. 1984. The estimation of the abundance of biomass of zooplankton in samples. Pages 228–265 *in* Downing and Rigler (1984).

Merritt, R. W., and K. W. Cummins. 1996. An introduction to the aquatic insects of North America, 3rd edition. Kendall/Hunt, Dubuque, Iowa.

Mills, E. L., D. M. Green, and A. Schiavone, Jr. 1987. Use of zooplankton size to assess the community structure of fish populations in freshwater lakes. North American Journal of Fisheries Management 7:369–378.

Norris, R. H., and A. Georges. 1986. Design and analysis for assessment of water quality. Pages 555–572 *in* P. De Deckker and W. D. Williams, editors. Limnology in Australia. Dr. W. Junk, Dordrecht, The Netherlands.

Parkinson, E. A., J. Berkowitz, and C. J. Bull. 1988. Sample size requirements for detecting changes in some fisheries statistics from small trout lakes. North American Journal of Fisheries Management 8:181–190.

Peckarsky, B. L. 1984. Sampling the stream benthos. Pages 131–160 *in* Downing and Rigler (1984).

Pennak, R. W. 1989. Fresh-water invertebrates of the United States: protozoa to mollusca, 3rd edition. Wiley, New York.

Plafkin, J. L., M. T. Barbour, K. D. Porter, S. K. Gross, and R. M. Hughes. 1989. Rapid bioassessment protocols for use in streams and rivers: benthic macroinvertebrates and fish. U.S. Environmental Protection Agency EPA/440/4-89/001, Assessment and Watershed Protection Division, Washington, DC.

Pugsley, C. W., and H. B. N. Hynes. 1983. A modified freeze-core technique to quantify the depth distribution of fauna in stony streambeds. Canadian Journal of Fisheries and Aquatic Sciences 40:637–643.

Reice, S. R., and M. Wohlenberg. 1993. Monitoring freshwater benthic macroinvertebrates and benthic processes: measures for the assessment of ecosystem health. Pages 287–305 *in* D. M. Rosenberg and V. H. Resh, editors. Freshwater biomonitoring and benthic macroinvertebrates. Chapman and Hall, New York.

Resh, V. H. 1979. Sampling variability and life history features: basic considerations in the design of aquatic insect studies. Journal of the Fisheries Research Board of Canada 36:290–311.

Resh, V. H., and D. G. Price. 1984. Sequential sampling: a cost effective approach for monitoring benthic invertebrates in environmental impact statements. Environmental Management 8:75–80.

Rosenberg, D. M., and V. H. Resh. 1982. The use of artificial substrates in the study of benthic macroinvertebrates. Pages 175–266 *in* J. Cairns, Jr., editor. Artificial substrates. Ann Arbor Science, Stoneham, Michigan.

Rosenberg, D. M., and V. H. Resh. 1993. Freshwater biomonitoring and benthic macroinvertebrates. Chapman and Hall, New York.

Sheldon, A. L. 1984. Cost and precision in a stream sampling program. Hydrobiologia 111:147–152.

Smock, L. A. 1980. Relationship between body size and biomass of aquatic insects. Freshwater Biology 10:375–383.

Snedecor, G. W., and W. G. Cochran. 1981. Statistical methods, 7th edition. Iowa State University Press, Ames.

Sprules, W. G., L. B. Holtby, and G. Griggs. 1981. A microcomputer based measuring device for biological research. Canadian Journal of Zoology 59:1611–1614.

Thorp, J. H., and A. P. Covich, editors. 1991. Ecology and classification of North American freshwater invertebrates. Academic Press, San Diego, California.

Thorp, J. H., M. D. Delong, and A. E. Black. 1992. Perspectives on biological investigations of large rivers: results and techniques from Ohio River studies. Pages 1–8 *in* Biological assessment in large rivers. North American Benthological Society, 5th Annual Technical Information Workshop, Louisville, Kentucky.

Ward, H. B., and G. C. Whipple, editors. 1959. Freshwater biology (revised by W. T. Edmondson). Wiley, New York.

Waters, T. F. 1971. Subsampler for dividing large samples of stream invertebrate drift. Limnology and Oceanography 14:813–815.

Wetzel, R. G., and G. E. Likens. 1979. Limnological analyses. Saunders, Philadelphia, Pennsylvania.

Whitman, R. L., J. M. Inglis, W. J. Clark, and R. W. Clary. 1983. An inexpensive and simple elutriation device for separation of invertebrates from sand and gravel. Freshwater Invertebrate Biology 2:159–163.

Chapter 12

Tagging and Marking

CHRISTOPHER S. GUY, H. LEE BLANKENSHIP,
AND LARRY A. NIELSEN

12.1 INTRODUCTION

The difference between a tag and mark is subtle. A mark is anything external, internal, or incorporated into the integument and used for recognition purposes (Jones 1979). A tag is usually attached externally or internally and contains specific identification information. Despite these subtle differences the terms tag and mark are often used interchangeably. Tagging and marking fish have been conducted for hundreds of years. For example, Izaak Walton documented that in 1653 stream watchers tied ribbons to the tails of Atlantic salmon (McFarlane et al. 1990). Numerous tags and marks have evolved since the 1600s. Within the last 20 years, tagging and marking technologies have made considerable advances; in general, reducing the adverse effects of the tag or mark on fish has driven technological developments.

Tagging and marking fish are essential techniques for any fisheries biologist. In general, three broad categories of information are obtained from tagging and marking studies. First, marking labels animals for specific handling. For example, in aquacultural operations broodstock may be marked to determine breeding histories. Second, marking allows animals to be identified as they move and mingle with other animals. Numerous studies have described the movement and migration patterns of fish through mark–recapture studies. Third, marking provides an avenue for collecting population statistics. Population abundance estimates, direct assessment of growth, and estimates of fishing and natural mortality are a few of the population statistics that can be obtained from tagging studies. Biotelemetry (i.e., the use of radio and sonic tags) is a highly specialized tagging technique and is discussed in Chapter 19.

12.2 ASSUMPTIONS ASSOCIATED WITH TAGGING

The primary assumption associated with fish tagging programs is that tagged fish can be recognized as such. This assumption has two aspects: all tagged fish retain their tags and each tagged fish is recognized and reported.

Tag retention can be influenced by the type of tag (i.e., size and shape), attachment location, and species being tagged. Several researchers have found tag retention to differ among tag types (Ebener and Copes 1982; Dunning et al. 1987; Franzin and McFarlane 1987). Muoneke (1992) found that double-tagging white bass resulted in higher return rates than did single tagging. Retention of tags can be increased by choosing the proper tag for the fish.

Tag recognition often depends on the visibility of the tag. Brightly colored tags (e.g., orange, red, and yellow) are more likely to be recognized than are colors that blend with the fish. Tag color is particularly important for tagging studies that rely on commercial fishers or recreational anglers to report captured tagged fish. Underreporting usually occurs in tagging studies that rely on public tag recognition (Matlock 1981; Green et al. 1983). In addition to highly visible tags, publicizing the study and offering rewards for returned tags are critical components for increasing reporting rates.

Other assumptions are important for particular techniques. In mark–recapture studies, for example, it is important that mortality rates be equal between tagged and untagged fish because the technique requires that the proportion of marked to unmarked animals be constant (Gutherz et al. 1990; Van Den Avyle 1993; Box 12.1). The assumption is less critical for movement studies because only comparative data are used. However, the higher the mortality rate of tagged individuals, the lower the probability of recapturing them.

High mortality of tagged fish can occur if the tagging process is stressful. Mortality related directly to handling usually occurs soon after tagging; delayed mortality is typically caused in indirect ways. Haegele (1990) estimated that 60% of tagged Pacific herring in one evaluation died from the tagging process. Dunning et al. (1987) suggested that handling before tagging was the primary source of striped bass mortality. Tags themselves may prevent tagging wounds from healing properly, and they may make it difficult for fish to swim well and avoid predators (Matthews and Reavis 1990; Mattson et al. 1990b; Moring 1990). Mortality can be reduced by handling animals gently and choosing the appropriate size of tag. Tagging fish underwater likely increases survival (Matthews and Reavis 1990).

It is important that tagging not influence the growth of tagged fish, particularly in studies of growth. Reduced growth is usually an indirect measure of physiological and behavioral problems. For example, a tag may alter feeding habits because it is too large or visible to prey. Early fish tags were attached to the jaw and often interfered with feeding; today these tags are used only in special circumstances. Studies documenting the effects of tags on fish growth have been contradictory. For example, several authors have documented slower growth rates for tagged than for untagged fish (Carline and Brynildson 1972; Gunn et al. 1979; McFarlane and Beamish 1990; Scheirer and Coble 1991); other studies, however, have shown no effect of tags on growth rates (Jensen 1967; Tranquilli and Childers 1982; Eames and Hino 1983).

An important assumption for all fisheries studies is that a particular technique used by the fisheries biologist does not alter fish behavior. This assumption is indirectly related to previous assumptions regarding growth and mortality and is fundamental because behavior affects all life history characteristics. The size and color of the tag can stimulate aggressive or passive behavior by or towards the tagged fish. Tagged fish may also alter their movement and habitat use patterns. For example, channel catfish may be prevented from using preferred woody cover if the tag impairs their ability to maneuver in that habitat.

12.3 EXTERNAL TAGS AND MARKS

External tags are the oldest and historically the most popular technique for marking fish. McFarlane et al. (1990) found that 65% of the 900 papers they reviewed

Box 12.1 Effects of Tag Loss on Population Estimates

Consider two small (0.01-ha) ponds, A and B, that each contain 400 fish. At each pond, a team of investigators attempts to estimate the size of the fish population (unknown to either team) by the same mark–recapture method. Each team starts by catching and tagging 100 fish with T-bar anchor tags. Tagged fish are released back to their ponds and allowed to mix with untagged fish for 1 week. Each team then returns to its pond, collects 200 fish, and examines each fish for the presence of a tag.

The team working at pond A is experienced in anchor tagging fish, and all its tags are retained by the fish in pond A. The pond B team is inexperienced, and 20% of its tags are lost from the fish. Both teams estimate population size in their respective ponds by the modified Petersen formula (Van Den Avyle 1993)

$$N = \frac{M(C + 1)}{R + 1},$$

for which N is the population estimate, M is the number of fish initially tagged, C is the number of fish captured in the second sample, and R is the number of tagged (recaptured) fish in the second sample. Sampling statistics and estimates are:

Pond A	Pond B
$M = 100$	$M = 100$
$C = 200$	$C = 200$
$R = 50$	$R = 40$
$N = 394$	$N = 490$

Team A, the experienced group, came very close to the true number of fish in pond A. Team B, however, greatly overestimated the population size in pond B. Because 20% of team B's tags had been lost, it recaptured 20% fewer tagged fish than did team A, and the effect was to inflate the apparent proportion of never-marked fish and thus the overall population.

on marking involved external tags. External tags have been placed on all sorts of aquatic organisms and in numerous locations. Figure 12.1 shows the commonly used tags and their respective tagging locations.

12.3.1 Fin Marks

Fisheries biologists have been marking fish by fin clipping since the 1800s (McFarlane et al. 1990). This technique has been widely accepted and is the most basic marking procedure. Fin clipping is simple and quick; however, the identification of individual fish is limited because only a few unique clipping combinations can be used (McFarlane et al. 1990).

The first step is to identify the fin to be marked. The scissors or wire cutters used should be large enough to clip the fin in one motion. For partial pelvic or pectoral fin

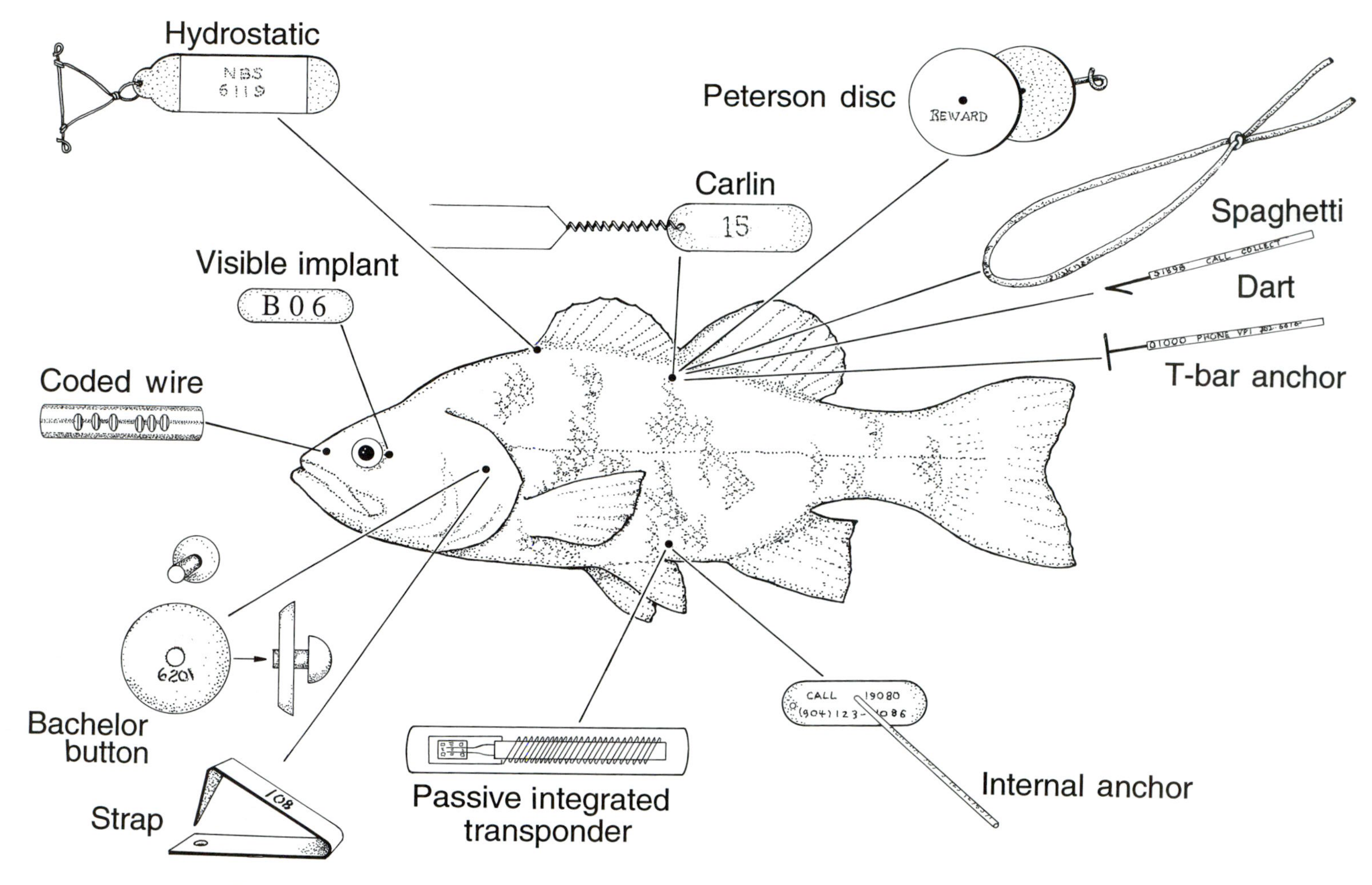

Figure 12.1 Commonly used tags and their attachment sites (adapted from Wydoski and Emery 1983, with permission).

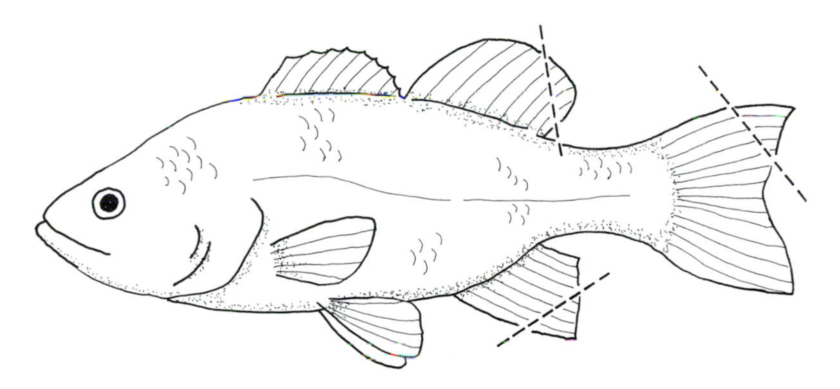

Figure 12.2 Angled clips of dorsal, anal, or caudal fins are most recognizable before and after regeneration (reprinted from Nielsen 1992).

clips, the cut should be perpendicular to the principal fin ray not more than halfway from the base of the fin (Eipper and Forney 1965). For partial clips of the dorsal, caudal, and anal fins, the clip should be made at an angle to increase later recognition (Figure 12.2). An alternative to clipping fins edge to edge is to hole-punch or notch them, in which case several fin rays should be cut (Figure 12.3).

Typically only a portion of a fin is removed by clipping several fin rays. Clipped fins usually regenerate, but regenerated fins typically show recognizable distortions that can be visible for years (Figures 12.3, 12.4). Minor cuts such as small notches or punched holes, however, may leave no discernible trace on a regenerated fin and are best used for short-term experiments (Figure 12.3). Regeneration is usually prevented if the entire fin is removed; thus, the mark lasts the life of the fish. However, it is generally not desirable to remove any fins completely, except an adipose fin,

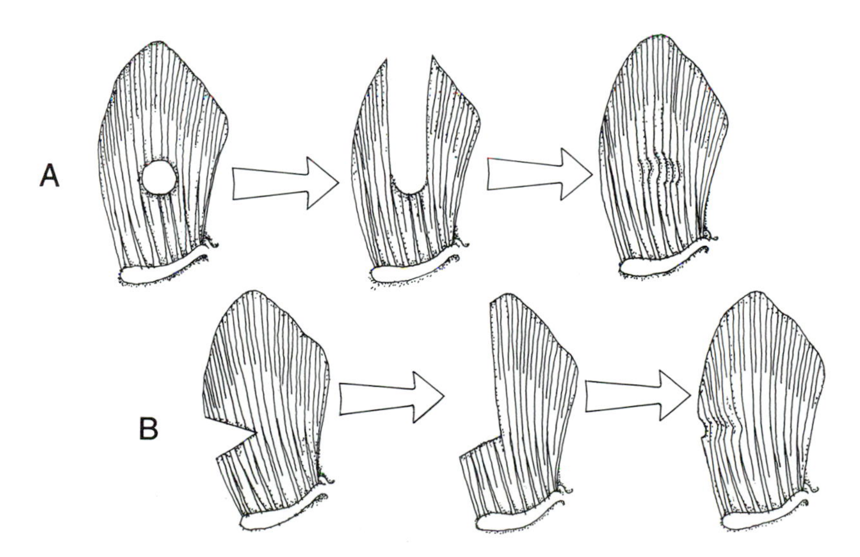

Figure 12.3 The appearance of punched **(A)** and notched fins **(B)**. Fin rays distal to a hole or notch will drop away, and regenerated rays will show irregularities (reprinted from Wydoski and Emery 1983).

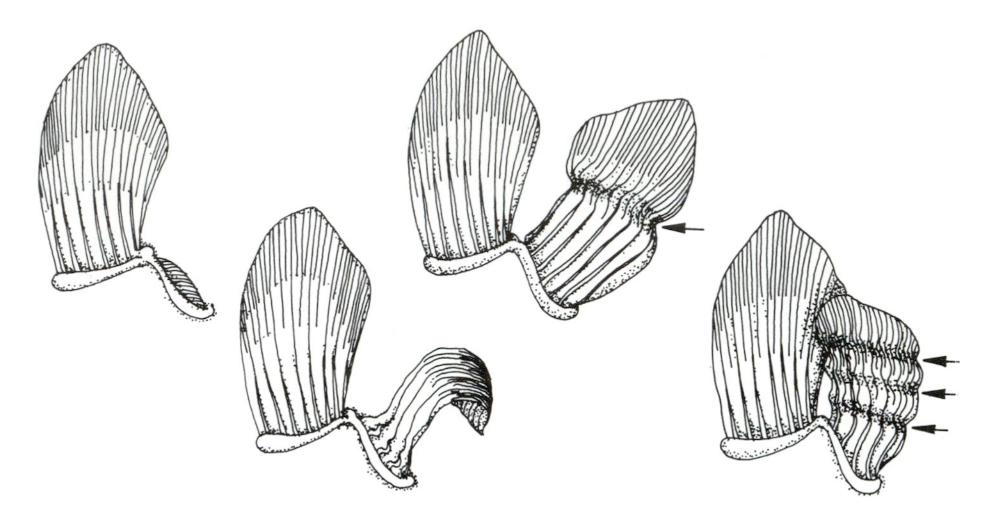

Figure 12.4 Appearance of regenerated fins after fin clipping. Note that fins are distorted after regeneration, allowing for recognition after initial clip. Arrows indicate multiple fin clips at different distances from the base, producing many lines of distortion (reprinted from Wydoski and Emery 1983).

because fins are critical for maneuvering. Radcliffe (1950) found that partial fin clipping did not affect the swimming ability of goldfish or coho salmon.

The advantage of fin clipping is that it is easy and quick. Managers often fin-clip large batches of hatchery fish so they can monitor the success of stocking programs. The equipment used for fin clipping, such as sharp scissors, wire cutters, and punches, is relatively inexpensive. The wounds caused by fin clipping usually heal quickly.

Identifying individual fish by fin clips is almost impossible for large batches of fish. Thus, the utility of fin clips is basically limited to identifying groups of fish. Another disadvantage associated with fin clips is the inability of most people to identify fin clips. Fins can be naturally missing or deformed, which can falsely be attributed to marking. Conversely, underreporting of fin clips can be a problem because of fin regeneration. A third disadvantage of fin clipping relates to public relations. Anglers do not want to see fish with mutilated fins, and they can be particularly upset if a trophy fish is altered. Informing the public of tagging studies and conducting these studies skillfully is extremely important.

12.3.2 Dart and T-Bar Anchor Tags

Anchor tags are the most popular external tags used today (Nielsen 1992); the two principal types are the dart tag and the T-bar tag. These tags are commonly made of plastic or thin wire and penetrate only one side of the fish. The head of the tag, or anchor (the end imbedded in the fish body), is arrowlike (dart tag) or T shaped (T-bar or Floy® tag). The protruding portion is called the shaft and is usually a vinyl tube that displays information. The T-bar tag was derived from tags used to attach prices to clothing and has practically replaced the dart tag (Dell 1968).

In general, dart and T-bar tags are inserted just below the dorsal fin (Figure 12.1). Tags should be inserted in white muscle (Gutherz et al. 1990). White muscle has

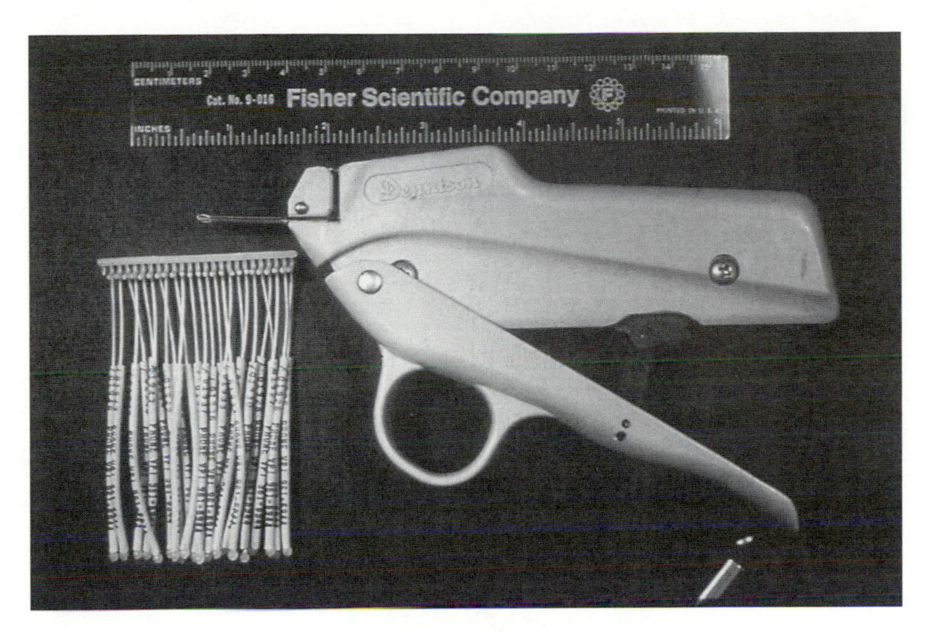

Figure 12.5 Semiautomated T-bar anchor-tagging gun and tags (reprinted from Nielsen 1992).

fewer blood vessels than does red muscle, so tagging in white muscle causes less bleeding. Tags should be inserted at an acute angle to the body axis so that they lie next to the body when the fish swims.

Dart and T-bar tags require applicators for insertion. Dart tags are inserted with an open steel shaft (see Gutherz et al. 1990), whereas T-bar tags are applied from a continuously feeding tagging gun (Figure 12.5) that allows quick and efficient tagging operations. Several tag applicators with a variety of needle sizes should be available to deal with various sizes and species of fish. Tagging equipment does not need to be sterilized, but it should be cleaned before and after use.

The insertion point must be free of scales; if the species to be tagged has scales, one or more of them must be removed. Then the steel shaft (dart tag) or tagging gun needle (T-bar tags) is inserted so that the tag head passes the midline of the fish but does not penetrate the opposite side. It is important to place the anchor (tag head) properly behind the pterygiophores or neural spines to prevent excessive tag loss (Figure 12.6; Waldman et al. 1990). The applicator is twisted 90° to dislodge the tag, and the shaft or needle is withdrawn. The emplaced tag is gently tugged to determine if it feels loose or pulls free. If a tag feels loose, that fish (identified by its tag number) should not be counted as tagged. Poorly tagged fish should not be retagged, the additional trauma will only increase mortality. Tagged fish can be treated with antibiotics to reduce infection. Several tagged animals should be held for 24 h to assess the short-term effects of tagging. These fish should be enclosed in the receiving water body to simulate postrelease conditions.

Dart and T-bar tags have several advantages. First, they can be applied quickly and easily and thus used in a wide variety of circumstances. Second, they can carry a considerable amount of information on the surface of the tag shaft, such as

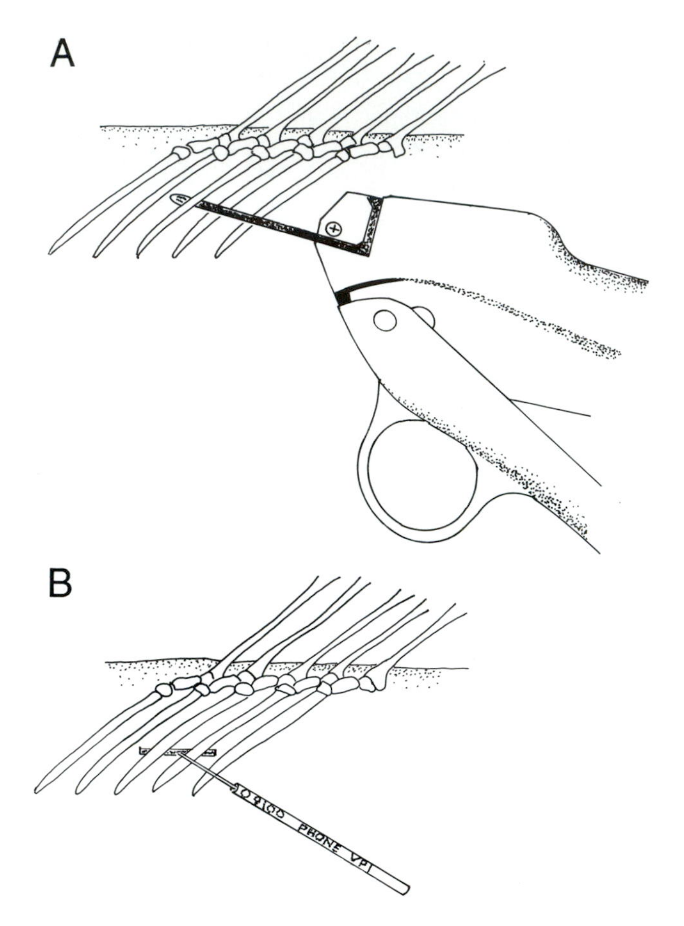

Figure 12.6 Placement of a T-bar tagging gun needle between pterygiophores (**A**), and position of a T-bar anchor tag behind the pterygiophores (**B**).

identification code, reward value, and an address for returning tags. The tags are available in a wide variety of sizes, colors, and styles and therefore can be used for a multitude of tagging studies.

Tag loss is the primary disadvantage of anchor tags. Tag loss is highly variable and depends on the type of fish being tagged, the duration of the tagging study, and the experience of the tagger. Ebener and Copes (1982) estimated T-bar tag loss for lake whitefish was 11.1% in the first year and 18.9% in the third year. Muoneke (1992) found that single-tagged white bass lost 24.8% of their tags per year.

Published studies of the effects of dart and T-bar anchor tags on fish have been contradictory. For example, Scheirer and Coble (1991) documented a decrease in growth of northern pike tagged with T-bar tags and Manire and Gruber (1991) found decreased growth of lemon sharks tagged with dart tags. Conversely, Tranquilli and Childers (1982) found no difference in growth between tagged and untagged largemouth bass. Dart and T-bar tags can cause irritation and open sores from constant rubbing of the tag against the body. Although these wounds cannot heal (Buckley and Blankenship 1990), proper insertion and use of small tags relative to fish size can reduce the amount of injury and tag loss.

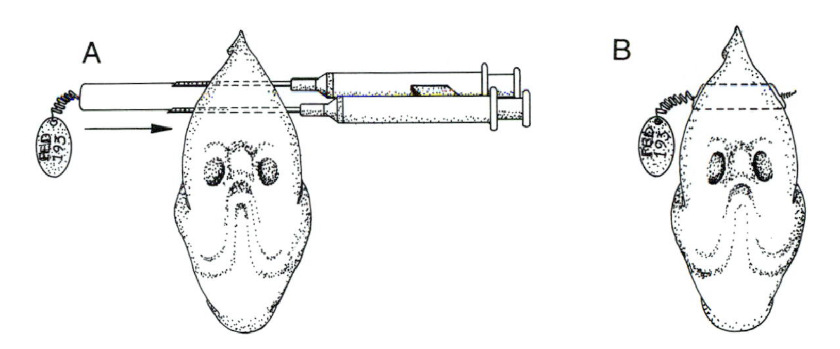

Figure 12.7 Application of Carlin tag by use of hypodermic needles to thread wires. (**A**) The hypodermic needles are inserted into the fish, and the wires are threaded through the needles. (**B**) The hypodermic needles are removed, leaving the tag in place (reprinted from Wydoski and Emery 1983).

12.3.3 Transbody and Transstructural Tags

Transbody tags traverse a fish's body, and they are the oldest form of the modern-day external tag. Several types of transbody tags have evolved through the years, but only a few are commonly used.

The Peterson disc tag (Figure 12.1) is the oldest and most widely used transbody tag (McFarlane et al. 1990). The Peterson tag consists of two discs, one on either side of the dorsal body, attached by a wire that is passed through the muscle of a fish. Identification information, reward values, and other information can be printed on the discs. Long retention time is the primary advantage of the Peterson tag. Its disadvantage is that it restricts growth in body thickness. Therefore, Peterson tags should not be used in long-term tagging studies.

Peterson tags are applied under the dorsal fin of the fish (Figure 12.1). One disc, holed in the center, is threaded onto the wire and the wire is pushed through the fish musculature. Once the wire is through the opposite side, the other disc is threaded and the wire is twisted so the discs are close to the skin surface. Excess wire is cut away.

The Carlin tag is the most popular of the loosely defined group of dangler tags (Figures 12.1, 12.7); the hydrostatic tag is similar (Figure 12.1). These tags are attached to the fish by U-shaped stainless steel wire. The open ends of the U-shaped wire are inserted into hypodermic needles or cannulas passed through the body of the fish; when the needles are withdrawn, the wire is pulled snugly against the body and the ends are crimped together. At the "bottom" of the U, at the fish side opposite the crimping, is a plate with information and instructions. As with all transbody tags, the Carlin tag has a long retention time. The Carlin tag is susceptible to entanglement because it is tied in loops. In general, the Carlin tag does not reduce the growth of fish; however, McAllister et al. (1992) found that yearling rainbow trout in raceways had significantly slower growth when tagged with Carlin tags than with T-bar anchor tags.

The spaghetti tag is a loop of thin vinyl tubing that is passed through the body of a fish and tied in a knot (Figure 12.1). Identification information is printed on the tubing. This tag was developed by Wilson (1953) for yellowfin tuna and albacore, but the tag has been used on many fish species since its invention. The tag is inexpensive

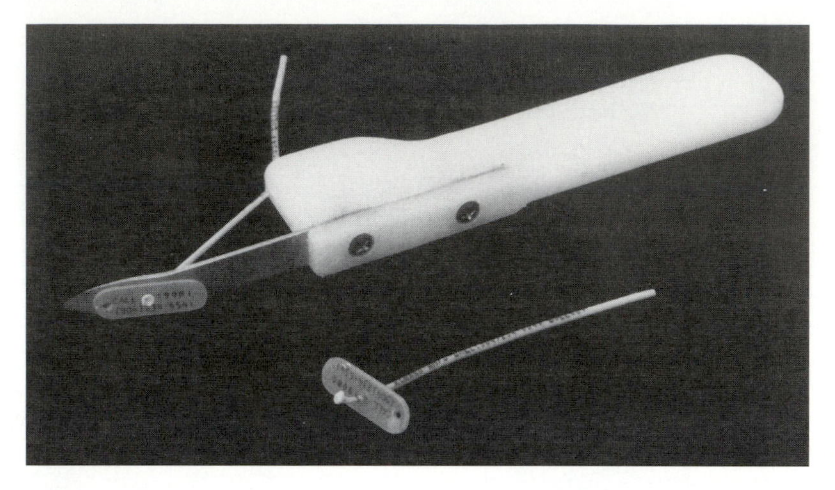

Figure 12.8 A modified scalpel used for inserting internal anchor tags (reprinted from Gutherz et al. 1990).

and has a long retention time. However, it is susceptible to entanglement because of its loop configuration. The lengthy application time is another disadvantage of the spaghetti tag.

The spaghetti tag is attached by threading the vinyl tubing through the dorsal musculature of the fish. As with the Carlin tag, a hypodermic needle or cannula is used to get the tubing through the fish. The tubing is knotted, proximal to its ends, such that the knot lies along the middle axis of the body.

Transstructural tags are used on opercula and jaws. Bachelor button and strap tags (Figure 12.1) were once commonly used on the opercula of fish and strap tags on their jaws. These tags are not commonly used now because they can substantially alter normal fish behavior.

12.3.4 Internal Anchor Tag

McFarlane et al. (1990) stated that the internal anchor tag (Figure 12.1) was the prototype for all anchor tags. The internal anchor tag is similar to the T-bar and dart tags, but it is most often inserted into the body cavity instead of the dorsal musculature. The anchor of the tag lays flush against the body wall, and the vinyl tubing, with all the pertinent information, extends outside the body. The internal anchor tag has been widely used in marine tagging programs.

The internal anchor tag is inserted from the ventral side through a small incision in the body wall. A sample specimen should be dissected to determine tagging location in order to avoid injury to internal organs. After the tagging location is determined by dissection, an external reference point should be chosen. The incision should be perpendicular to the body axis and should penetrate the muscle but not the peritoneum surrounding the body cavity (Nielsen 1992). The anchor should be pushed through the incision parallel to the body axis and towards the fish's head so the anchor will slide along the body cavity. Once the anchor is inserted it should be pulled back so it is centered under the incision. A special scalpel has been developed that can hold the internal anchor tag while making the incision (Figure 12.8).

The primary advantage of the internal anchor tag is that it has a high retention rate if it is applied correctly on the appropriate fishes. The tag is usually effective on large, wide-bodied fishes and can be inappropriate for small fishes. Fable (1990) found that the internal anchor tag yielded a higher return rate for king mackerel than did single-barbed dart tags. Weathers et al. (1990) documented no tag loss after 90 d for largemouth bass 131–568 mm in total length.

Like other anchor tags, the internal anchor tag can cause abrasions both internally and externally. Mattson et al. (1990b) observed the anterior edge of the anchor protruding through the body wall of striped bass. They also noted adhesion of scar tissue on the anchor for fish with and without anchor protrusion. Another disadvantage of the internal anchor tag is that the tagging procedure is difficult, requires experience, and is time-consuming. Extra care should be taken so that internal organs are not damaged during tag insertion; tagging is particularly risky before and during spawning because of enlarged reproductive organs.

12.3.5 Branding

Hot and cold branding have been used to produce a recognizable scar on the surface of the fish. Branding causes the tissue to scar in the shape of the branding symbol, and pigment is either concentrated or displaced at the branding site (McFarlane et al. 1990). Cold branding is preferred and is usually more successful than hot branding (Fujihara and Nakatani 1967; Smith 1973; McFarlane et al. 1990). Branding works best on fine-scaled and scaleless fishes.

Several types of coolants have been used for cold branding; the most common are liquid nitrogen (N_2) and pressurized carbon dioxide (CO_2). See Bryant and Walkotten (1980) and Knight (1990) for details on the construction of N_2 and CO_2 freeze-branding apparatuses. Various branding techniques have been used for hot branding, such as boiling water, propane torch, soldering iron, and lasers.

Typically, fish are anesthetized before they are branded so they do not move during the procedure (see Summerfelt and Smith 1990 and Chapter 5 for anesthetic techniques). The branding location should be on the lightest part of the body because the brand usually is dark colored (Knight 1990). Most brands are applied on the side of the fish in the midbody region (Nielsen 1992).

The fish is removed from water and the branding area is blotted dry. The brand is applied for 1–2 s with gentle pressure. If the branding iron sticks to the fish, it needs to be recooled or reheated. The applicator should be cleaned if the mark becomes distorted.

Branding is dangerous, whether cold gases or hot materials are used. Branders should wear insulated leather gloves and safety goggles. For cold branding, Knight (1990) suggested that CO_2 was safer to work with and yielded the same results as N_2.

Branding is typically a rapid process for marking many fish. For example, Bryant et al. (1990) branded 2,000 juvenile salmon in 2 h. Another advantage of branding is that the body surface is not penetrated; mortality rates are usually low (Bryant et al. 1990), and the growth of a branded fish is not impeded. Branding can be used on fish of various sizes because the branding apparatus can be adjusted. It takes several days for the brands to be visible after marking; the length of time varies with water temperature (Knight 1990).

The primary disadvantage of branding is that it is a short-term mark. As the fish grows, the mark becomes less legible. Like fin clipping, only a few groups of fish can be identified. Symbols that are closed (e.g., B, R, and D) are harder to distinguish because they lose identity faster than do open symbols (e.g., C, T, and I). Bryant et al. (1990) recommended straight-line letters, such as T, I, V, and X. They suggested not using letters U and V together because as the fish grows, U and V could be confused. More complex symbols can be used on larger brands because there is little chance of incomplete marking (Bryant et al. 1990).

12.3.6 Pigment Marks

"Pigments" used for marking include dyes, stains, inks, paints, and microscopic plastic chips. They are variously applied by immersion, spraying, injection (needle and needleless), and tattooing. Because these inert materials are imbedded in or under the epidermis, they are sometimes considered to form internal marks. In some respects, pigment marks were a precursor to the visible implant tag.

Pigment marking techniques were first employed during the 1920s, and their use increased during the 1960s and 1970s (McFarlane et al. 1990). Their popularity declined with the invention and development of new tags. Pigment marks have most commonly been used in mark–recapture and behavioral studies.

The most critical factor for durable marks is the proper placement of the pigment. Shallow injections in areas of lessened skin pigmentation provide longer-lasting marks. The cheek pad and ventral side of the body and head are potential target areas (Kelly 1967). Choice of material is also an important factor to consider.

When marks are applied with pressurized spraying devices, the granule size, amount of pressure, and distance from the surface of the target area are all important considerations (Nielson 1990). Pressure and distance are also important to consider with needleless inoculators, and granule size must be small enough not to plug the inoculator.

Pigment marks are generally very visible for the short term (months). Longer-term retention depends upon the material and application methods. Fluorescent granules, which are embedded under the skin by means of compressed air, are reported to remain up to 12 years (Nielson 1990), but detection requires careful examination with ultraviolet light by experienced personnel in a darkened area. The amount of fish growth after marking affects the longevity of the mark because formation of new, unpigmented cells in the area of marking causes fading.

The advantages of pigment marks are that they are relatively simple and inexpensive to apply. The materials can be injected with handheld syringes (Kelly 1967) or needleless injectors (Laufle et al. 1990). Fluorescent granules can be applied to thousands of fish by means of powered sprayers (Phinney et al. 1967).

A disadvantage of pigments is the limited number of available colors, which restricts the diversity of identification codes that can be used. This limitation can be overcome with tattoos, which can create numerous characters for codes, and microtaggants, which are microscopic, multicolored and multilayered plastic chips that have unlimited capacity for coding (Thompson et al. 1986).

12.4 INTERNAL TAGS AND MARKS

Internal tags and marks are imbedded within, or implanted beneath, the epidermis of the animal. Small internal tags have become increasingly popular in the last few decades. Developers of new tags have attempted to capitalize on the benign characteristics associated with internal tags. They have also made tags smaller so the tags are less obtrusive and more readily used with smaller animals.

Internal tags do not require mutilation or removal of body parts and they do not protrude from the body. If they can be used alone (but see below), they thus avoid adverse effects from abnormal behavior, predation, entanglements, and other factors. Another advantage is their high probability of retention over time. After some low but measurable chance of loss while the epidermis heals from the intrusion of tag placement, the tag likely will remain for the life of the animal. Tag retention varies to some degree depending upon the type and size of the tag, proper placement of the tag, species, and implant site. Tag retention should always be monitored and reported. The advantages of internal tags are realized only if proper tagging techniques are followed—tags should be made of biocompatible materials, placed in nonobtrusive locations, and be small in relation to the host and tagging location. These conditions require strict attention and cannot be taken for granted, especially when one works with previously untested species.

A disadvantage of all but recent internal tags is that they are not visible, so trained personnel equipped with detection devices may be necessary. Often, the presence of an internal tag bearing information about the fish must be signaled by an external fin clip or tag.

12.4.1 Subcutaneous and Body Cavity Tags

Subcutaneous and body cavity tags were the first internal tags. They were typically flat, elongated strips of celluloid, plastic, or magnetic material several centimeters long, and they were inserted subcutaneously between the epidermis and musculature or in the body cavity (Jones 1979). Because these tags were not visible, fin marks or (if appropriate) magnets were frequently used to indicate their presence. The use of subcutaneous and body cavity tags declined after development of other internal tags.

12.4.2 Visible Implant Tags

The desire to combine the positive attributes of external tags (visual identification of live fish) with internal tags (low biological effects) led to the development of visible implant tags (Bergman et al. 1992). In this unique concept, transparent tissue serves as a "window" for an internal tag. The visible implant tag is a patented concept and actually involves an assortment of different types of tags.

The original visible implant tag described by Haw et al. (1990) is alphanumerically coded (i.e., the tag code consists of numbers and letters) and made of polyester and diazo film (Figure 12.9). The flat, rectangular tag has three alphanumeric characters and is available in seven colors, the combination of which provides several thousand codes. The tag is available in two sizes: 2.5 mm long × 1.0 mm wide and 3.5 mm long × 1.5 mm wide.

The fluorescent elastomer tag is another in the family of visible implant tags. This tag is a polymer of biocompatible materials that is injected as a liquid into transparent tissue; within hours it cures into a pliable solid that makes a cohesive, well-defined mark. Coding presently is limited to six colors, but machine readability of different wavelengths is expected to increase the possible number of codes considerably. The colors are readily seen in ordinary light, but visibility is greatly enhanced by ultraviolet light in reduced natural light.

Suitable transparent tissues for tagging are more common than one might expect without careful examination, but some taxa do lack appropriate unpigmented tissue. A commonly used site in salmonid species is the transparent tissue posterior to the eye. Other taxa have clear tissue overlying the jaw area. The membrane between the soft fin rays of some fishes also appears to offer potential.

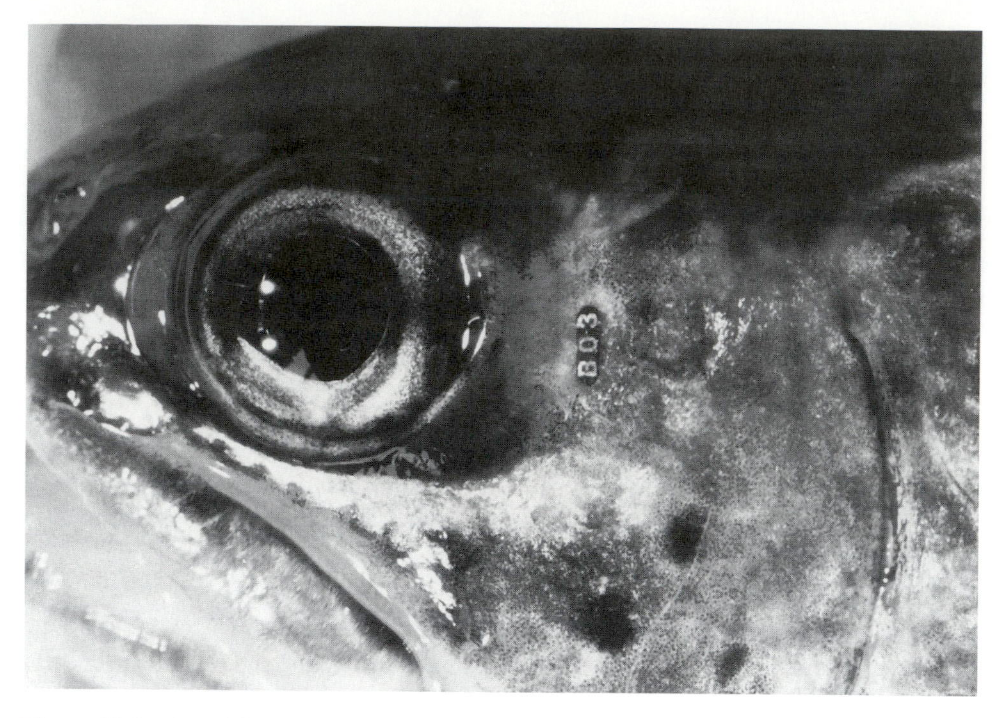

Figure 12.9 An alphanumerically coded visible implant tag placed in transparent tissues around the eye of a 185-mm rainbow trout (reprinted from Haw et al. 1990).

Elastomer tags can be delivered with very small needles (30 gauge) and can be used on smaller specimens than can alphanumeric tags. The alphanumeric visible implant tag is at least 1.0 mm wide and is injected with a flattened needle adapted to hold the tag. Until literature becomes available, researchers must examine and test for appropriate injection sites on their species.

The principal advantage of visible implant tags is that the codes can be read in live fish. In addition, the tags and applicators are relatively inexpensive and can be used proficiently with little experience. Their biggest disadvantage (in 1996) lies in the newness of the concept, the evolution of the new tags, and a corresponding lack of published results. However, two factors are emerging from published and personal communications: fish must be of a minimum size for good retention and retention varies among species regardless of size (Krouse and Calkins 1992; Blankenship and Tipping 1993; Bryan and Ney 1994).

Visible implant tags are being used for situations in which sacrificing the animal is not desired, visual identification is preferred, and relatively inexpensive tags are required. Such situations include broodstock operations and capture-and-release experiments. Visible implant elastomer tags can also be used to denote the presence of internal coded wire tags.

12.4.3 Coded Wire Tags

The coded wire tag system revolutionized Pacific salmon management after its invention in the early 1960s (Jefferts et al. 1963). Its use expanded quickly with salmonids, and by 1990 over 40 million juvenile salmonids were being tagged annually (Johnson 1990). Today it is the most frequently applied tag in the world. It

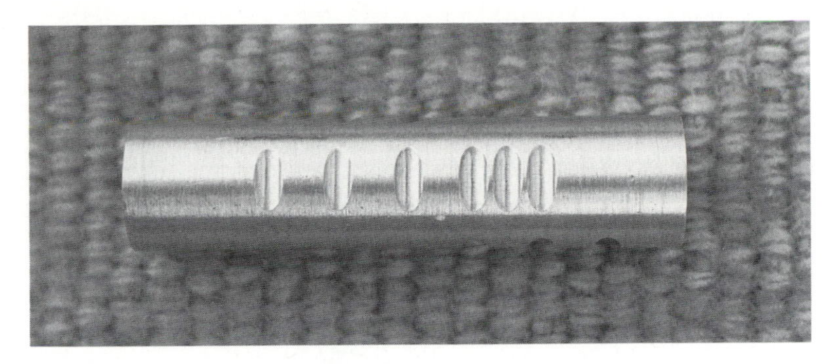

Figure 12.10 A coded wire tag with binary coding (reprinted from Nielsen 1992).

has been successfully used with more than 20 genera of fishes (Buckley and Blankenship 1990) and with various invertebrates (Krouse and Nutting 1990).

The coded wire tag is a very small piece of magnetized stainless steel wire (usually 1.1 mm long × 0.25 mm in diameter) that was originally coded with up to six longitudinal color stripes. Ambiguity associated with distinguishing different colors and problems with peeling of color stripes led to the present binary coding system of notches (Figure 12.10). The present coding system provides over 250,000 different codes. Typically, a unique code is used to identify a group of fish sharing a common trait (batch code). However, identification of an individual is now possible because coded wire tag wire can be obtained with continuous sequential codes.

Depending upon the number of animals to be tagged, handheld syringes, semiautomated injectors, or fully automated injectors can be used. A large-scale tagging and recovery operation is depicted in Figure 12.11. The body location for tagging is critical for high retention rates. In salmonids, the tag is inserted in the forward portion of the head or snout. Custom-made head molds for different species and sizes are used to accurately position the fish to receive the tag, which is delivered via a hollow needle that is injected into the target area. Snout tissue is not an acceptable site for species with large sinuses or other characteristics that may allow movement of tags (Fletcher et al. 1987); cheek muscle is a common target site for fish with these characteristics. Muscle tissue in other locations would be adequate. It is important, however, to place the tag parallel with the muscle striations or high tag loss can result (Dunning et al. 1990). An examination of the target site is necessary prior to tagging. New tagging sites, such as transparent postocular tissue, dorsal fins, and adipose fins, have also been successfully used (Heinricher Oven and Blankenship 1993) and offer the possibility of benignly recovering tags from live specimens.

The advantages of the coded wire tag are related to its very small size and biocompatibility. External and internal tissue damage from the injection process, and the tag itself, is minor and heals rapidly (Bergman et al. 1968; Fletcher et al. 1987; Buckley and Blankenship 1990).

Fish as small as 0.25 g have been successfully tagged with coded wires (Thrower and Smoker 1984). However, extra precautions must be taken with very small fish to avoid damage to sensitive tissues adjacent to the target area (Fletcher et al. 1987; Morrison and Zajac 1987).

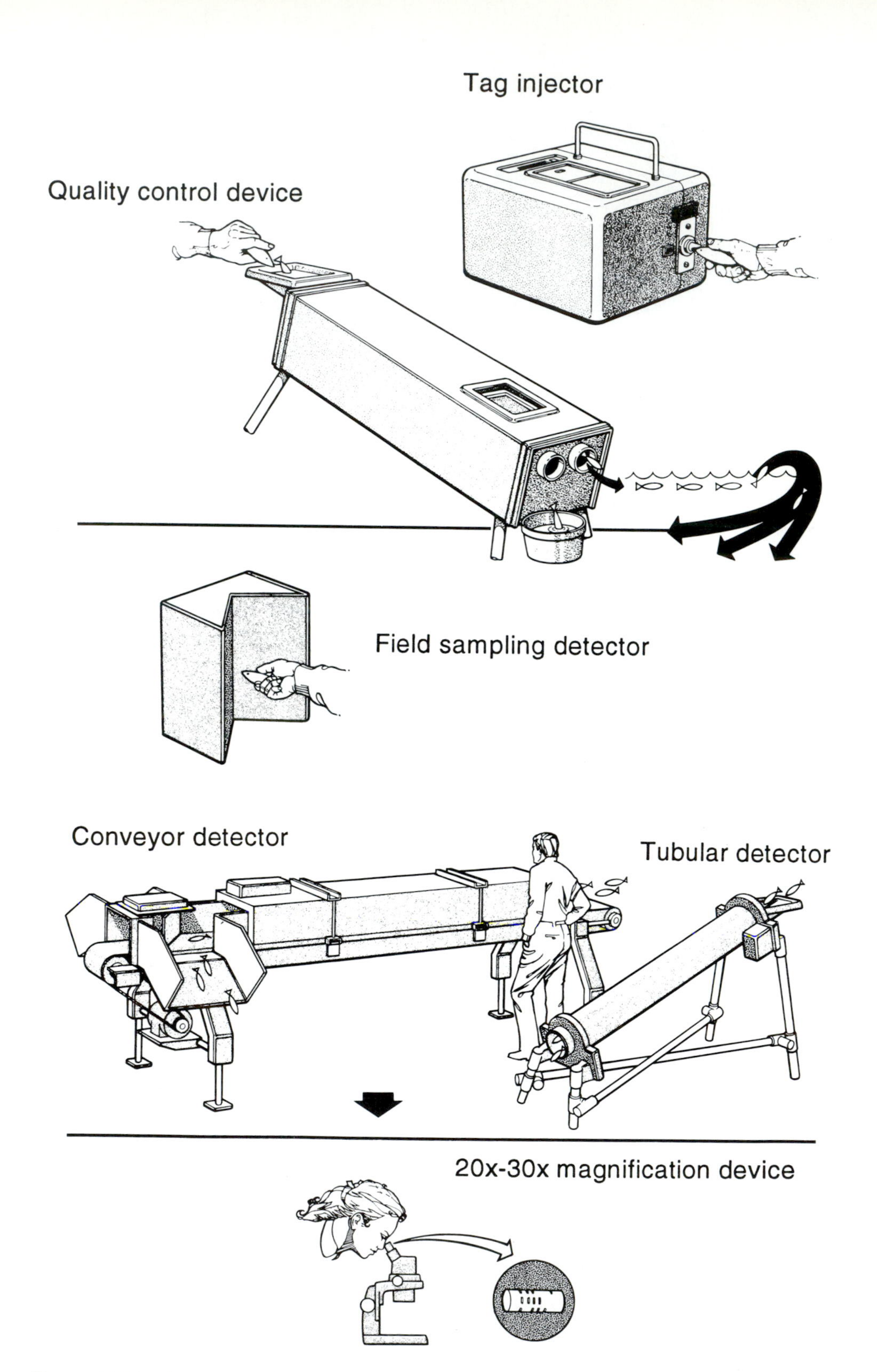

Figure 12.11 A binary-coded wire-tagging operation. The tagging system includes a tag injector and a quality control device, which are used before fish are released, and detection and sorting devices for recovered fish. Tags are decoded by use of a microscope. Illustration courtesy of Northwest Marine Technology (reprinted from Nielsen 1992).

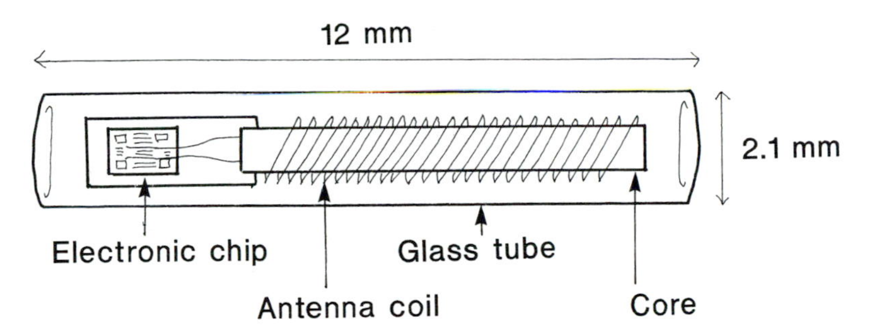

Figure 12.12 Schematic of a passive integrated transponder tag (reprinted from Nielsen 1992).

Tag retention rates are typically high, and losses that occur usually happen within the first 3–4 weeks after tagging (Blankenship 1990). High tag loss can result from improper tagging technique or improper tagging location site (Bailey and Dufour 1987).

In relative terms, application of coded wire tags is inexpensive. In 1994, tags cost US$0.05 to 0.10 a piece, depending upon quantity and delivery time. Major operations require large capital expenditures but, amortized over time, add little to the total cost of tagging. As with most tagging procedures, labor to apply the tags accounts for the largest proportion of the cost. Accounting for all application costs, including labor and amortization of equipment, the Pacific States Marine Fisheries Commission (1992) estimated the average cost of applying a coded wire tag in conjunction with an adipose fin clip was $0.11 per fish.

The primary disadvantage of the coded wire tag is that the fish need to be sacrificed to recover the tag. Another disadvantage is that tagging and detection of coded wire tags require advanced technological equipment. However, most of the equipment is rather simple to operate. Handheld and stationary field detectors are available to detect the presence of the magnetic tag. Typically, at least with salmonids, the adipose fin is clipped to denote the presence of a coded wire tag. Electronic detection by itself is feasible (Mattson et al. 1990a); however, equipment failure or improper technique can result in missed tags. Testing with magnetized standards at regular intervals is recommended (Morrison 1990). After the tag is detected and recovered, it needs to be interpreted by microscopy at 20–30× magnification.

12.4.4 Passive Integrated Transponder Tags

Invented in the 1980s, the passive integrated transponder (PIT) tag is a sophisticated electronic identification system. The tag is a computer chip and antenna encapsulated in a glass tube. Originally a larger tag, it shrank with computer chip technology. Although larger than a coded wire tag, it is still relatively small and measures 12.0 mm long × 2.1 mm in diameter (Figure 12.12). The tag requires an external energy source (i.e., the reading device) to be activated. Once energized, the tag instantaneously relays a unique code to an interrogation system (Prentice et al. 1990a). The tag is activated only for the amount of time it is exposed to the reader.

Passive integrated transponder tags can be injected with a handheld, modified, 12-gauge hypodermic needle or semiautomatic tag injector. Due to its high cost, passive integrated transponder tagging has been largely restricted to projects in which few fish need to be identified or they are very valuable (e.g., in behavioral studies or for captive broodstock). An exception to this is the extensive passive integrated transponder tagging of Pacific salmonids in the Columbia River. Large amounts of mitigative funds are used to monitor survival and migration of salmonids between numerous hydroelectric dams.

The two principal body locations where passive integrated transponder tags have been injected are in the body cavity and subcutaneously in the musculature. Prentice and Park (1984) found the body cavity preferable for Pacific salmon because the tag would be in a nonedible portion of the fish. Jenkins and Smith (1990) chose an intramusculature site posterior of the dorsal fin for captive red drum and striped bass broodfish. Both groups had high initial tag retention.

The advantage of the passive integrated transponder tag is that individual codes can be read from live fish. Over 34 billion codes are available. As would be expected from a small internal tag, no biological (growth and survival) or performance (swimming speed and stamina) effects have been found (Prentice et al. 1990a; Jenkins and Smith 1990).

The primary disadvantage of the passive integrated transponder tag is cost. Individual tags cost $4.75–6.00 (1994 U.S. dollars), depending upon quantity. Small-scale projects require about $2,000 for tagging and monitoring equipment, and large-scale operations can cost $20,000 or more (Nielsen 1992). Trained personnel are required for operating both the tagging and recovery systems. Detection ranges for passive integrated transponder tags are limited. Large detection systems under optimal conditions provide a maximum range of 18 cm; handheld detectors have a range of about 7.6 cm (Prentice et al. 1990b). Tag loss during spawning for tags placed in the body cavity has occurred (Prentice et al. 1990a); further investigation is needed.

12.5 CHEMICAL MARKS

Chemical marks are constant differences in the chemical composition of animal body tissues that can be recognized by humans. The marks may be natural ones that have been incorporated through variations in the natural environment; however, the chemical marks discussed in this section are those caused by biologists. Chemical marks can be artificially induced by means of immersion, injection, or ingestion. These marks are incorporated physiologically within the animal and require specialized equipment for detection. Some authors group pigment marks (e.g., dyes, stains, and inks) with chemical marks (Muncy et al. 1990). This chapter makes the distinction between marks that are visible externally (pigments) and marks that are incorporated within the tissues of the animal through natural and chemical processes. Calcified tissues (otoliths, bones, and scales) are the most common tissues used for chemical marks. These tissues are preferred because they incorporate elements or chemicals permanently and in a form that can provide a time line. Soft tissues lose chemical marks relatively quickly (weeks to months). Retention appears best in otoliths, next best in bones, and poorest in scales. Different elements, however, are incorporated at different rates in different tissues. For instance, strontium may be incorporated better in otoliths than in bones, but rare earth (lanthanide series) elements, such as europium and terbium, may be incorporated better in scales than in otoliths (Muncy et al. 1990).

Chemical marking includes two promising types of marking: elemental and fluorescent. Elemental marks require sophisticated techniques to measure the presence and amount of various metallic elements. Most often, the elements used are from the alkaline and rare earth groups. The most commonly used analytical techniques include atomic absorption spectroscopy, inductively coupled plasma mass spectrometry, X-ray-fluorescence spectrometry, and neutron activation analysis (Behrens Yamada and Mulligan 1990). These techniques can detect metallic elements at parts-per-million concentrations and less.

The most commonly used fluorescent compounds are tetracycline and calcein (Tsukamoto 1985; Beckman et al. 1990). If taken up by calcified structures in sufficient quantities, these compounds can be seen as yellow or orange rings under ultraviolet light. Visual detection of fluorescent compounds with ultraviolet light does not require the expensive equipment or high level of expertise needed for elemental analysis. Identification of marks in the field rather than in a distant laboratory is also more probable. However, the concentrations of fluorescent compounds required for visible detection arc magnitudes higher than elemental concentrations that can be measured by the more costly techniques. Consistent and reliable visual marks can be hard to obtain. Variables such as fish size, dosage, uptake method, and water chemistry can all play important roles in obtaining visual marks (Beckman et al. 1990). Fluorescent compounds can be detected nonvisually by other (more expensive) techniques, however.

The expenses associated with recovery of chemically marked fish are magnified if marked fish mingle with unmarked fish, because the unmarked animals must undergo the same expensive processing (e.g., otolith analysis) as marked fish. To avoid this, chemically marked fish can be given a secondary external mark, but this would add to the cost and time of marking the fish.

More common usage of chemical marks in the future will likely depend upon improvements of existing, or development of new, detection techniques that are less expensive and more user-friendly than current methods. Exciting work in chemical detection continues with electron microprobe analysis lasers (Coutant 1990) and in visual detection with a scanning electron microscope that uses back-scattered electron images.

The advantages of chemical marks relate mostly to their application: large numbers of small animals can be marked easily, quickly, and inexpensively by taking up markers from solution or ingesting them in feed. Newly hatched fry and even unhatched embryos can be chemically marked by immersion with little or no handling or anesthesia. Manual injection of chemicals is time-consuming and hence costly; however, the chemicals used for marking are relatively inexpensive, and such low concentrations are required that the marking aspect is very cheap. (If natural chemical marks can be used for some purpose, such as identifying the origins of individuals in a mixed population, marking costs are nil.) Finally, because chemical marks are typically incorporated into calcified tissues, the marks are long lasting.

Most of the disadvantages of chemical marks have to do with recovery or identification of marks. Foremost is the expensive laboratory equipment and technical expertise required for chemical analysis. Tissue preparation of otoliths and bones can be time-consuming and expensive.

12.6 NATURAL MARKS

Natural marks are derived from natural processes. These natural processes can be inherited (genetic) or obtained from the environment (e.g., indicative parasites). Without too much stretch, we include marks "naturally" acquired in unnatural environments, such as the eroded fins on fish raised in raceways and the widely spaced circuli on scales of hatchery-reared (well-fed, fast-growing) fish. Natural marks allow one to distinguish among groups of fish. For example, hatchery fish have been differentiated from wild stocks based on natural marks (Humphreys et al. 1990). The patterns in otolith or scale growth, body size, and color are the most commonly used natural marks (McFarlane et al. 1990). Natural marks are not intrusive, unlike the previously discussed marks and tags. Two major assumptions are associated with natural marks. First, natural marks should be present and stable throughout a study. Second, natural marks should provide a sure way to identify animals to group. Because natural marks vary through time and on individuals, these assumptions are usually not met. However, fisheries biologists have been successful in using natural marks to identify fish stocks. Because most natural marks are highly variable, statistical analysis becomes a useful tool in the identification process (Schweigert 1990).

12.6.1 Morphological Marks

Morphological marks are of two types: morphometric and meristic. Morphometric marks are based on body shape, size, and color. Bachman (1984), for example, identified individual brown trout based on their spotting patterns. Meristic marks are based on counts of serial physical features, such as the number of fin rays, myomeres, or lateral line scales. Because morphometric and meristic marks can be difficult to use, taxonomists, experienced fisheries biologists, and biostatisticians should be consulted when suitable features for morphological marks are sought.

The advantage of morphological marks is that nothing intrusive is done to the fish. The disadvantage is that marks can vary with environmental, genetic, and physiological factors. Thus, the permanence of a mark needs to be verified for each situation.

12.6.2 Scale and Otolith Marks

Scale and otolith size, shape, and circulus or other incremental patterns have been used for many years as naturally occurring marks. These marks are developed through genetic variation among populations, life history patterns, or environmental variation among habitats (Nielsen 1992). Ross and Pickard (1990) determined the contribution of hatchery-reared striped bass to a fishery based on scale shape and circulus spacing. Humphreys et al. (1990) found that hatchery-reared striped bass had thick, widely spaced circuli near the focus, which corresponded to fast growth in the hatchery. The fast growth was followed by a "growth check" resulting from handling, tagging, and adaptation to the natural environment. Conversely, wild striped bass had uniform circuli and no growth check. See Chapter 16 for detailed discussion on scale and otolith structure, terminology, and location.

As with morphological marks and genetic marks, the advantage of scales and otoliths as fish identifiers is that they are naturally produced. Except for otolith marking (see below), the fish mark themselves, thus reducing handling injury and the physiological stress of carrying a tag. Another advantage is that nearly all fish in a similarly affected group carry a mark.

Scales and otoliths have to be removed from the fish for analysis, which can be a

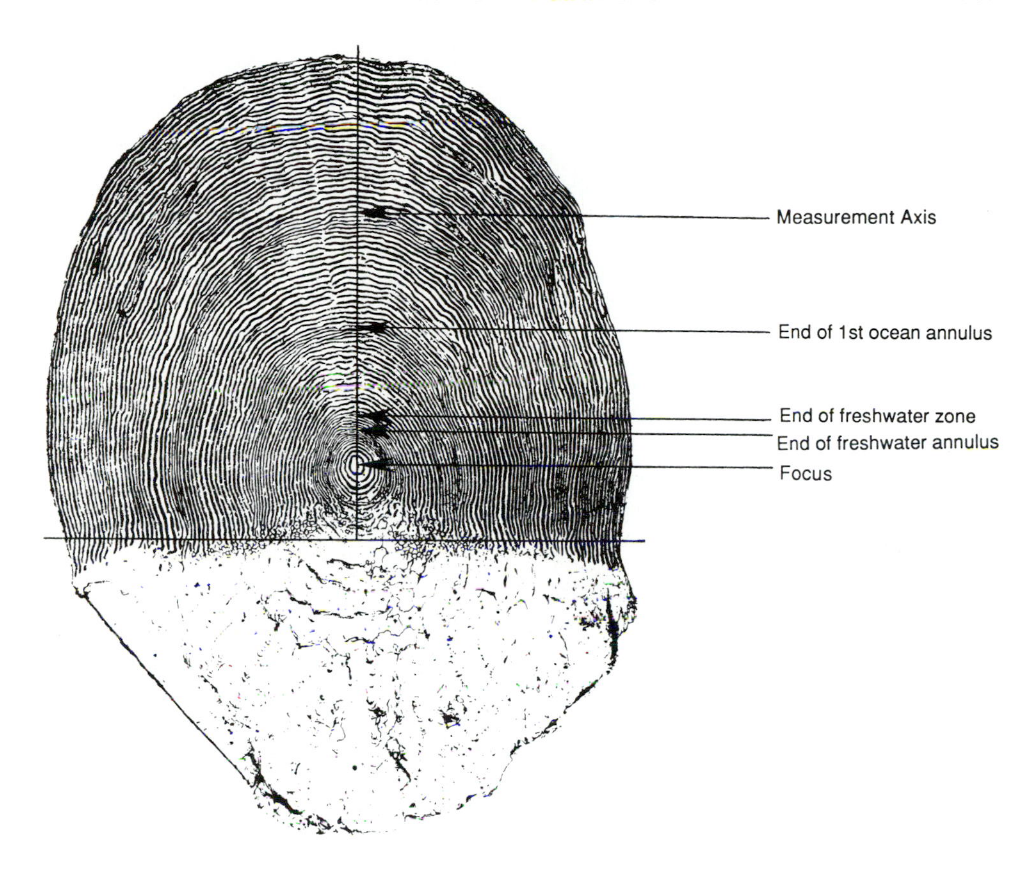

Figure 12.13 Cycloid scale from an age-1 chinook salmon, showing scale features, axis for measurement, and life history zones (reprinted from Davis et al. 1990).

disadvantage. Removal of otoliths requires that the fish be sacrificed (see Chapter 16). Removal of scales usually does not cause a problem (see Chapter 16), but it does break the skin surface and disturbs the mucous coating, which increases the risk of infection. Other disadvantages of using scale patterns as markers are that lost scales are regenerated and that scale edges are resorbed in times of fish stress, both of which can eliminate or obscure a mark.

Measuring the total number of circuli and the average spacing between circuli are common techniques used for distinguishing between groups of fish (Davis et al. 1990; Ross and Pickard 1990). To measure these features consistently from all scales, a standardized line from the scale's central point (focus) to the scale edge should be established. For cycloid scales, a reference line can be anywhere in the posterior quadrant (Figure 12.13). For ctenoid scales, Ross and Pickard (1990) recommended that measurements be taken along a reference line between the widely and closely spaced circuli (Figure 12.14).

Nielsen (1992) described a three-step process for using marks on scales and otoliths. First, collect and measure the structures from fish of known origin (baseline data) and determine if a difference in structures exists. If a difference does exist, set up discriminating criteria for the baseline groups. Second, collect and measure structures from fish of known origin and evaluate the accuracy of the discriminating

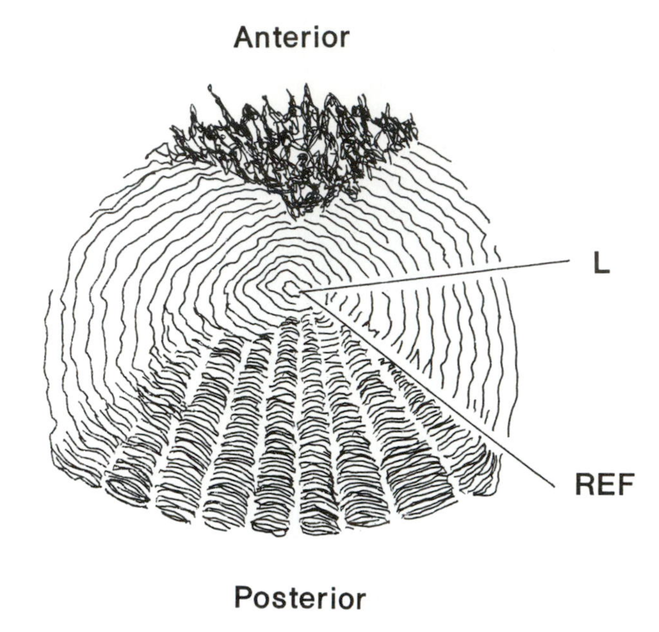

Figure 12.14 Ctenoid scale showing reference radius (REF) and luminance extraction line (L) for measuring scale features (reprinted from Nielsen 1992).

criteria. Third, use the discriminating criteria to assign fish of unknown origin to baseline groups based on structures and measurements.

Inducing marks on otoliths can be achieved by controlling the environment of captive fish. Controlling the environment alters otolith development and produces recognizable patterns on the otoliths. Temperature is the primary method used to mark otoliths, and altering water temperatures can produce a unique banding pattern (Volk et al. 1990). Chemicals have also been used to mark otoliths (see section 12.5).

Because otoliths are formed early, marking can be induced prior to hatching. Otolith marks cannot be shed, eroded, or resorbed as scale marks can be. Otolith marking is usually inexpensive because large groups of small fish can be marked concurrently.

Otolith marking does not allow for marking individuals. Another disadvantage is that otoliths often require extensive preparation before they can be analyzed (see Chapter 16). Once otoliths are prepared, distinguishing the patterns takes an experienced otolith reader.

12.6.3 Genetic Marks

Identifying fish stocks by genetic marks is complex. A fisheries biologist needs to have some understanding of genetics and biochemistry in order to use genetic marks. Results from genetic studies can be difficult to interpret and often require advanced statistical procedures. For these reasons, one should consult a geneticist and statistician before undertaking a genetic marking program.

The specific procedures used to analyze genetic marks are beyond the scope of this text. Therefore, fisheries biologists planning to use genetic marks are referred to Aebersold et al. (1987), Ryman and Utter (1987), Allendorf and Ferguson (1990),

Billington and Hebert (1990), Gharrett and Seeb (1990), Utter and Seeb (1990), Whitmore (1990), and Nielsen (1992). See Leary and Booke (1990) and Nielsen (1992) for ways to conduct protein electrophoresis and mitochondrial (mt) DNA analysis.

The primary advantage of genetic marks is that the marking method is typically natural. Each fish carries information contained in its genetic material. This information is often decoded by analyzing the biochemical phenotype for each individual. The number of individual fish needed for genetic identification is usually less than that needed for other marking techniques because every individual is marked (Nielsen 1992). However, naturally occurring genetic marks are rarely if ever completely diagnostic at the population or stock level. Rather, stocks are typically distinguished on the basis of differences in the frequencies of various genetic characters. Thus, genetic stock identification is a statistical process that usually requires that many individual fish be sampled and analyzed. The analysis yields a probability-based determination of stock origin, not an absolute identification as is possible with most artificial marks. Most marks are limited to one generation; however, genetic marks are passed from one generation to the next during reproduction and are typically expressed throughout all life stages. Finally, many fish species exhibit protein and mtDNA polymorphisms, thus providing the heterogeneity needed to differentiate fish stocks.

Several disadvantages are associated with genetic marks. Collecting genetic information can be complex and requires that the fish be sacrificed. However, many improvements have occurred, and many tests can now be done with small nonlethal tissue samples from fins, skin, or muscle (Nielsen 1992; Chapter 5). As stated earlier, data from genetic marks can be difficult to interpret. Therefore, many of the procedures require considerable expertise and experience.

12.6.4 Other Natural Marks

Some fish parasites have restricted distributions, and their occurrences in fish sometimes can reveal the origins of those fish or at least the environments through which the fish have passed in their life cycles (e.g., Moes et al. 1990). Chemical marks associated with unique chemistries of watersheds or water masses in which fish might have lived were mentioned in section 12.5. Interpretation of natural parasitic or chemical occurrences requires collaboration with specialists, but such information can yield rewarding information about fish origins or histories.

12.7 CHOICE OF A MARKING TECHNIQUE

Several authors have described the "perfect" mark or tag, but no one has been able to develop one. No one may ever accomplish such a feat because different studies have different goals that likely demand different characteristics of a mark or tag. The elusive search for the perfect marking system nevertheless continues to be pursued with ingenuity and expertise from within fisheries and from other fields of science. This search continues to provide an ever-increasing array of choices or tools from which to choose. Choosing an appropriate marking system is very important but can be difficult.

Because there is no perfect marking system or mark (Box 12.2), review of the literature and consultation with experienced professionals are essential in choosing the best technique for a particular study. Several factors must be considered, which will likely differ from one study to the next.

Box 12.2 The Perfect Mark (reprinted from Nielsen 1992)

The pursuit of the perfect mark has produced many lists of characteristics and criteria. Rounsefell and Kask (1945), writing at a time when most tags were buttons, straps, or body cavity plates, listed these criteria for selection of the proper tag:

(1) length of time the tag remains on the fish (the longer the better);
(2) ease of application (the easier and more uniform, the better);
(3) likelihood of being seen when the fish is recaptured.

Kelly (1967) described characteristics of an ideal dye (injected as a mark). His test is applicable, however, to most types of marks:

(1) long-lasting (preferably for the fish's lifetime);
(2) permanent on every fish;
(3) nontoxic and nonirritating;
(4) no effect on growth;
(5) requires little or no extra formulation (that is, little special preparation);
(6) allows rapid marking;
(7) inexpensive;
(8) nonencumbering;
(9) easily visible to an untrained observer;
(10) provides several different mark combinations;
(11) requires little specialized equipment.

Stott (1971) described the more general characteristics of an ideal mark:

(1) permanent on any fish;
(2) unmistakably recognizable (preferably individually) to anyone:
(3) no effect on growth, mortality, or behavior;
(4) no effect on predation rate or vulnerability to fishing gear.

Everhart and Youngs (1981), expanding on earlier work with Rounsefell, brought the characteristics of an ideal mark to an even dozen:

(1) remains unaltered during the fish's lifetime;
(2) no effect on behavior or vulnerability to predators;
(3) does not tangle with weeds or nets;
(4) inexpensive and easily obtained;
(5) fits any size fish with little alteration;
(6) easy to apply without anesthetic and with little or no stress on fish;
(7) identifies fish at least to group;
(8) creates no health hazard;
(9) does not harm food or aesthetic value of fish;
(10) easy to detect in the field by untrained individuals;
(11) causes no confusion in reporting;
(12) remains unaffected by preservation.

Objectives. The first and foremost step in choosing a marking system is to clearly define the objectives of the study and the information needed to meet those objectives. Care must be taken to choose a tag or mark that does not introduce biases or compromise the information.

Behavior and biological functions. It was previously stated, but is appropriate to reemphasize, that interpretation of tagging data normally requires assumptions that tagging has not distorted normal growth, maturation, mortality, and behavior of the target fish. Certain external tags might be appropriate when migration patterns with angler recovery is the objective but not when growth studies, in which tags might introduce a bias, are planned.

Mark retention. Different marks are retained physically or visibly at different rates. Mark retention can differ among taxa or sizes at which fish are marked. A pigment mark might be adequate for a short-term mark–recapture study but not for one in which the recaptures may occur a year later.

Informational capacity. Fishery management objectives often involve concerns about stocks or groups of fish. Examples are concerns about exploitation rates, survival, diet, and size at release from hatcheries. In these cases, tagging many fish all at once (batch tagging) is adequate. Sometimes, however, researchers need individual codes, such as for behavioral or genetic studies. A marking system must be capable of having enough codes to meet the needs of the study.

Tagging requirements. Before a tagging study begins, biologists must know the number of fish to be tagged and the particular characteristics of the animals being tagged. Costs and available resources are the primary constraints on the number of fish to be tagged. Costs and resources include the tag, application, and detection costs (equipment), the length of time and level of expertise required for marking, and the number of persons required. Fish characteristics to consider before tagging begins include the size, shape, and structure of the animal and whether handling stress and anesthesia pose a problem.

Recovery requirements. When mark–recapture studies are conducted, one must determine whether dead or live animals will be recaptured and whether multiple recaptures are necessary. The study may dictate whether trained personnel or commercial and recreational anglers can be used to identify marked fish and interpret tag codes. Mark visibility and cost influence these decisions. Coordination of recoveries with other agencies (local, national, international) may be required if fish move across political jurisdictions.

12.8 DESIGN OF A MARKING PROGRAM

Millions of fish throughout the world are tagged each year. In the Pacific Northwest of North America alone, over 40 million Pacific salmon are given coded wire tags each year (Johnson 1990). Regardless of the number of fish being tagged, the key to any successful tagging operation is careful planning and organization. Fisheries biologist should keep in mind the goal, objectives, and final report when they plan and organize a tagging program. A successful marking program requires that fisheries biologists understand the assumptions, advantages, and disadvantages of the tag or mark being used. These factors become extremely important during data analysis.

12.8.1 Planning

Planning begins with an expressed need for a tagging study. For example, an agency biologist might need to know the contribution of hatchery fish to a fishery or the percentage of a fish population that is harvested by anglers.

Once the need is defined, the goal and objectives of the project should be stated. The goal is usually an all-encompassing statement, such as "determine exploitation of white crappie in Thunderbird Reservoir, Oklahoma." Objectives are more precisely stated and are measurable. For example, "estimate angler harvest rates for age-2 and older white crappie by means of tag returns."

The next step in planning is to develop the methodology for meeting each objective. The best approach is to list several alternative strategies without considering their merits—a brain-storming approach. Once the list has been developed, each alternative is evaluated closely and the best one is chosen. For any alternative, the key considerations are required sample size; costs associated with capture; marking and recapturing; the type of analysis; and the time span of the project. A statistician should be consulted to make sure the design will yield valid results. Finally, it is always important to choose the tags that will cause the least possible distress to the fish (see Chapter 5).

12.8.2 Data Management and Analysis

Because large marking studies can produce large amounts of data, Johnson (1990) recommended that data should be handled by a centralized group. The group should be responsible for developing standardized forms and codes for species, selecting computer programs, and designing the overall sampling program. Similarly, the group should be responsible for quality control. A project leader should perform these tasks for small tagging studies.

Data analysis techniques for tagging and marking data vary a lot. Regardless of the technique used, expert help is usually needed (Nielsen 1992). As stated earlier, many of the techniques involve complicated statistical procedures. Thus, a statistician is usually needed to help analyze tagging and marking data. Computer programs have become popular for analyzing large data sets (Rexstad et al. 1990).

12.8.3 Public Relations

Many tagging studies require that sport and commercial fisheries report tagged fish. Therefore, it is imperative that tagging experiments be publicized through news releases. Local outdoor writers are usually willing to publish information pertaining to fisheries management and research activities. One should assure that news releases describe the tag (color, shape, size, and tagging location) and agency contacts if a tagged fish is caught. In addition to news releases, one should distribute posters around study areas and to local sporting goods stores. Posters and tag drop-off points should be easily seen and located near high-traffic areas. It is also helpful to provide return-postage-paid envelopes to anglers—doing so may increase tag return rates. Finally, it is always a good idea to contact anglers who returned a tag. Often, they are interested in matters such as fish size at time of tagging, sex, distance traveled, or other such information. Positive feedback such as this can result in continued cooperation in the future.

12.8.3.1 Reward Systems

Cash rewards are often used to help increase tag return rates. Haas (1990) found that a reward system increased tag returns by more than 50%. Obviously, this improved return could have a substantial influence on estimating parameters such as fish exploitation rates. Small individual rewards are most frequently used ($5–10). However, some programs give small rewards and put all the tag returns in an end-of-season lottery. In recent years, hats and t-shirts have been used as popular tag return incentives.

It is important that tags are clearly marked "reward" and that adequate information is provided so that people can redeem their rewards. Reward systems do not eliminate underreporting, but they can reduce it.

12.9 FURTHER INFORMATION

Numerous papers and books have been published on the various aspects of tagging aquatic organisms. However, *Marking and Tagging of Aquatic Animals: An Indexed Bibliography* (Emery and Wydoski 1987), *Fish-Marking Techniques* (Parker et al. 1990), and *Methods of Marking Fish and Shellfish* (Nielsen 1992) are fundamental resources for fisheries scientists interested in tagging or marking aquatic organisms.

12.10 REFERENCES

Aebersold, P. B., G. A. Winans, D. J. Teel, G. B. Milner, and F. M. Utter. 1987. Manual for starch gel electrophoresis: a method for the detection of genetic variation. NOAA (National Oceanographic and Atmospheric Administration) Technical Report NMFS (National Marine Fisheries Service) 61.

Allendorf, F. W., and M. M. Ferguson. 1990. Genetics. Pages 35–63 *in* C. B. Schreck and P. B. Moyle, editors. Methods for fish biology. American Fisheries Society, Bethesda, Maryland.

Bachman, R. A. 1984. Foraging behavior of free-ranging wild and hatchery brown trout in a stream. Transactions of the American Fisheries Society 113:1–32.

Bailey, R. F. J., and R. Dufour. 1987. Field use of an injected ferromagnetic tag on the snow crab (*Chionoecetes apilis o. fab.*). Journal du Conseil International pour l' Exploration de la Mer 43:237–244.

Beckman, D. W., C. A. Wilson, F. Lorica, and J. M. Dean. 1990. Variability in incorporation of calcein as a fluorescent marker in fish otoliths. American Fisheries Society Symposium 7:547–549.

Behrens Yamada, S., and T. J. Mulligan. 1990. Screening of elements for the chemical marking of hatchery salmon. American Fisheries Society Symposium 7:550–561.

Bergman, P. K., K. B. Jefferts, H. F. Fiscus, and R. Hager. 1968. A preliminary evaluation of an implanted coded-wire fish tag. Washington Department of Fisheries Fisheries Research Paper 3(1):63–84.

Bergman, P. K., F. Haw, H. L. Blankenship, and R. M. Buckley. 1992. Perspectives on design, use, and misuse of fish tags. Fisheries 17(4):20–25.

Billington, N., and P. D. N. Hebert. 1990. Technique for determining mitochondrial DNA markers in blood samples from walleyes. American Fisheries Society Symposium 7:492–498.

Blankenship, H. L. 1990. Effects of time and fish size on coded-wire tag loss from chinook and coho salmon. American Fisheries Society Symposium 7:237–243.

Blankenship, H. L., and J. M. Tipping. 1993. Evaluation of visible implant and sequentially coded-wire tags in sea-run cutthroat trout. North American Journal of Fisheries Management 13:391–394.

Bryan, R. D., and J. J. Ney. 1994. Visible implant tag retention by and effects on condition of a stream population of brook trout. North American Journal of Fisheries Management 14:216–219.

Bryant, M. D., C. A. Dolloff, P. E. Porter, and B. E. Wright. 1990. Freeze branding with CO_2: an effective and easy-to-use field method to mark fish. American Fisheries Society Symposium 7:30–35.

Bryant, M. D., and W. J. Walkotten. 1980. Carbon dioxide freeze branding device for use on juvenile salmonids. Progressive Fish-Culturist 42:55–56.

Buckley, R. M., and H. L. Blankenship. 1990. Internal extrinsic identification systems: overview of implanted wire tags, otolith marks, and parasites. American Fisheries Society Symposium 7:173–182.

Carline, R. F., and O. M. Brynildson. 1972. Effects of the Floy anchor tag on the growth and survival of brook trout (*Salvelinus fontinalis*). Journal of the Fisheries Research Board of Canada 29:458–460.

Coutant, C. C. 1990. Microchemical analysis of fish hard parts for reconstructing habitat use: practice and promise. American Fisheries Society Symposium 7:574–580.

Davis, N. D., K. W. Myers, R. V. Walker, and C. K. Harris. 1990. The fisheries research institute's high-seas salmonid tagging program and methodology for scale pattern analysis. American Fisheries Society Symposium 7:863–879.

Dell, M. B. 1968. A new fish tag and rapid, cartridge-fed applicator. Transactions of the American Fisheries Society 97:57–59.

Dunning, D. J., Q. E. Ross, B. R. Freidmann, and K. L. Marcellus. 1990. Coded-wire tag retention by, and tagging mortality of, striped bass reared at the Hudson River hatchery. American Fisheries Society Symposium 7:262–266.

Dunning, D. J., O. E. Ross, J. R. Waldman, and M. T. Mattson. 1987. Tag retention by and tagging mortality of Hudson River striped bass. North American Journal of Fisheries Management 7:535–538.

Eames, M. J., and M. K. Hino. 1983. An evaluation of four tags suitable for marking juvenile chinook salmon. Transactions of the American Fisheries Society 112:464–468.

Ebener, M. P., and F. A. Copes. 1982. Loss of Floy anchor tags from lake whitefish. North American Journal of Fisheries Management 2:90–93.

Eipper, A., and J. Forney. 1965. Evaluation of partial fin-clips for marking largemouth bass, walleyes, and rainbow trout. New York Fish and Game Journal 12:233–240.

Emery, L., and R. Wydoski. 1987. Marking and tagging of aquatic animals: an indexed bibliography. U.S. Fish and Wildlife Service Resource Publication 165.

Everhart, W. H., and W. D. Youngs. 1981. Principles of fishery science, 2nd edition. Cornell University Press, Ithaca, New York.

Fable, W. A., Jr. 1990. Summary of king mackerel tagging in the southeastern USA: mark–recapture techniques and factors influencing tag returns. American Fisheries Society Symposium 7:161–167.

Fletcher, D. H., F. Haw, and P. K. Bergman. 1987. Retention of coded-wire tags implanted into cheek musculature of largemouth bass. North American Journal of Fisheries Management 7:436–439.

Franzin, W. G., and G. A. McFarlane. 1987. Comparison of Floy anchor tags and fingerling tags for tagging white suckers. North American Journal of Fisheries Management 7:307–309.

Fujihara, M. P., and R. E. Nakatani. 1967. Cold and mild heat marking of fish. Progressive Fish-Culturist 29:172–174.

Gharrett, A. J., and J. E. Seeb. 1990. Practical and theoretical guidelines for genetically marking fish populations. American Fisheries Society Symposium 7:407–417.

Green, A. W., G. C. Matlock, and J. E. Weaver. 1983. A method for directly estimating the tag reporting rate of anglers. Transactions of the American Fisheries Society 112:412–415.

Gunn, J. M., J. M. Ridgeway, P. J. Rubec, and S. U. Qadri. 1979. Growth curtailment of brown bullheads tagged during the spring spawning period. Progressive Fish-Culturist 41:216–217.

Gutherz, E. J., B. A. Rohr, and R. V. Minton. 1990. Use of hydroscopic molded nylon dart tags and internal anchor tags on red drum. American Fisheries Society Symposium 7:152–160.

Haas, R. C. 1990. Effects of monetary rewards and jaw-tag placement on angler reporting rates for walleyes and smallmouth bass. American Fisheries Society Symposium 7:655–659.

Haegele, C. W. 1990. Anchor tag return rates for Pacific herring in British Columbia. American Fisheries Society Symposium 7:127–133.

Haw, F., P. K. Bergman, R. D. Fralick, R. M. Buckley, and H. L. Blankenship. 1990. Visible implanted fish tag. American Fisheries Society Symposium 7:311–315.

Heinricher Oven, J., and H. L. Blankenship. 1993. Benign recovery of coded-wire tags from rainbow trout. North American Journal of Fisheries Management 13:852–855.

Humphreys, M., R. E. Park, J. J. Reichle, M. T. Mattson, D. J. Dunning, and Q. E. Ross. 1990. Stocking checks on scales as marks for identifying hatchery striped bass in the Hudson River. American Fisheries Society Symposium 7:78–83.

Jefferts, K. B., P. K. Bergman, and H. F. Fiscus. 1963. A coded-wire identification system for macro-organisms. Nature (London) 198:460–462.

Jenkins, W. E., and T. I. J. Smith. 1990. Use of PIT tags to individually identify striped bass and red drum brood stocks. American Fisheries Society Symposium 7:341–345.

Jensen, A. C. 1967. Effects of tagging on the growth of cod. Transactions of the American Fisheries Society 96:37–41.

Johnson, J. K. 1990. Regional overview of coded wire tagging of anadromous salmon and steelhead in northwest America. American Fisheries Society Symposium 7:782–816.

Jones, R. 1979. Materials and methods used in marking experiments in fishery research. FAO (Food and Agriculture Organization of the United Nations) Fisheries Technical Paper 190.

Kelly, W. H. 1967. Marking freshwater and a marine fish by injected dyes. Transactions of the American Fisheries Society 96:163–175.

Knight, A. E. 1990. Cold-branding techniques for estimating Atlantic salmon parr densities. American Fisheries Society Symposium 7:36–37.

Krouse, H. L., and G. T. Calkins. 1992. Retention of visible implant tags in lake trout and Atlantic salmon. Progressive Fish-Culturist 54:163–170.

Krouse, J. S., and G. E. Nutting. 1990. Effectiveness of the Australian western rock lobster tag for marking juvenile American lobsters along the Maine coast. American Fisheries Society Symposium 7:94–100.

Laufle, J. C., L. Johnson, and C. L. Monk. 1990. Tattoo-ink marking method for batch-identification of fish. American Fisheries Society Symposium 7:38–41.

Leary, R. F., and H. E. Booke. 1990. Starch gel electrophoresis and species distinctions. Pages 141–170 *in* C. B. Schreck and P. B. Moyle, editors. Methods for fish biology. American Fisheries Society, Bethesda, Maryland.

Manire, C. A., and S. H. Gruber. 1991. Effect of M-type dart tags on field growth of juvenile lemon sharks. Transactions of the American Fisheries Society 120:776–780.

Matlock, G. C. 1981. Nonreporting of recaptured tagged fish by saltwater recreational boat anglers in Texas. Transactions of the American Fisheries Society 110:90–92.

Matthews, K. R., and R. H. Reavis. 1990. Underwater tagging and visual recapture as a technique for studying movement patterns of rockfish. American Fisheries Society Symposium 7:168–172.

Mattson, M. T., B. R. Friedman, D. J. Dunning, and Q. E. Ross. 1990a. Magnetic tag detection efficiency for Hudson River striped bass. American Fisheries Society Symposium 7:267–271.

Mattson, M. T., J. R. Waldman, D. J. Dunning, and Q. E. Ross. 1990b. Abrasion and protrusion of internal anchor tags in Hudson River striped bass. American Fisheries Society Symposium 7:121–126.

McAllister, K. W., P. E. McAllister, R. C. Simon, and J. K. Werner. 1992. Performance of nine external tags on hatchery-reared rainbow trout. Transactions of the American Fisheries Society 121:192–198.

McFarlane, G. A., and R. J. Beamish. 1990. Effect of an external tag on growth of sablefish (*Anoplopoma fimbria*), and consequences to mortality and age at maturity. Canadian Journal of Fisheries and Aquatic Sciences 47:1551–1557.

McFarlane, G. A., R. S. Wydoski, and E. D. Prince. 1990. Historical review of the development of external tags and marks. American Fisheries Society Symposium 7:9–29.

Moles, A., P. Rounds, and C. Kondzela. 1990. Use of the brain parasite *Myxobolus neurobius* in separating mixed stocks of sockeye salmon. American Fisheries Society Symposium 7:224–231.

Moring, J. R. 1990. Marking and tagging intertidal fishes: review of techniques. American Fisheries Society Symposium 7:109–116.

Morrison, J. A. 1990. Insertion and detection of magnetic microwire tags in Atlantic herring. American Fisheries Society Symposium 7:272–280.

Morrison, J., and D. Zajac. 1987. Histologic effect of coded-wire tagging in chum salmon. North American Journal of Fisheries Management 7:439–441.

Muncy, R. J., N. C. Parker, and H. A. Poston. 1990. Inorganic chemical marks induced in fish. American Fisheries Society Symposium 7:541–546.

Muoneke, M. I. 1992. Loss of Floy anchor tags from white bass. North American Journal of Fisheries Management 12:819–824.

Nielsen, L. A. 1992. Methods of marking fish and shellfish. American Fisheries Society Special Publication 23.

Nielson, B. R. 1990. Twelve-year overview of fluorescent grit marking of cutthroat trout in Ben Lake, Utah–Idaho. American Fisheries Society Symposium 7:42–46.

Pacific States Marine Fisheries Commission. 1992. Mass marking anadromous salmonids: techniques, options, and compatibility with the coded-wire tag system. Pacific States Marine Fisheries Commission (PSMFC), Portland, Oregon.

Parker, N. C., A. E. Giorgi, R. C. Heidinger, D. B. Jester, Jr., E. D. Prince, and G. A. Winans, editors. 1990. Fish-marking techniques. American Fisheries Society Symposium 7.

Phinney, D. E., D. M. Miller, and M. L. Dahlberg. 1967. Mass-marking young salmonids with fluorescent pigment. Transactions of the American Fisheries Society 96:157–162.

Prentice, E. F., T. A. Flagg, and C. S. McCutcheon. 1990a. Feasibility of using implantable passive integrated transponder (PIT) tags in salmonids. American Fisheries Society Symposium 7:317–322.

Prentice, E. F., T. A. Flagg, C. S. McCutcheon, and D. F. Brastow. 1990b. PIT-tag monitoring systems for hydroelectric dams and fish hatcheries. American Fisheries Society Symposium 7:323–334.

Prentice, E. F., and D. L. Park. 1984. A study to determine the biological feasibility of a fish tagging system. Annual Report (contract DE-A179-83BP11982, Project 83-19) to Bonneville Power Administration, Portland, Oregon.

Radcliffe, R. W. 1950. The effect of fin-clipping on the cruising speed of goldfish and coho salmon fry. Journal of the Fisheries Research Board of Canada 8:67–73.

Rexstad, E., K. P. Burnham, and D. R. Anderson. 1990. Design of survival experiments with marked animals: a case study. American Fisheries Society Symposium 7:581–587.

Ross, W. R., and A. Pickard. 1990. Use of scale patterns and shape as discriminators between wild and hatchery striped bass stocks in California. American Fisheries Society Symposium 7:71–77.

Rounsefell, G. A., and J. L. Kask. 1945. How to mark fish. Transactions of the American Fisheries Society 73:320–365.

Ryman, N., and F. Utter, editors. 1987. Population genetics and fishery management. University of Washington Press, Seattle.

Scheirer, J. W., and D. W. Coble. 1991. Effect of Floy FD-67 anchor tags on growth and condition of northern pike. North American Journal of Fisheries Management 11:369–373.

Schweigert, J. F. 1990. Comparison of morphometric and meristic data against truss network for describing Pacific herring stocks. American Fisheries Society Symposium 7:47–62.

Smith, J. R. 1973. Branding chinook, coho, and sockeye salmon fry with hot and cold metal tools. Progressive Fish-Culturist 35:94–96.

Stott, B. 1971. Marking and tagging. Pages 82–97 *in* W. E. Ricker, editor. Methods for the assessment of fish production in fresh waters, 2nd edition. Blackwell Scientific Publications, Oxford, UK.

Summerfelt, R. C., and L. S. Smith. 1990. Anesthesia, surgery, and related techniques. Pages 213–271 *in* C. B. Schreck and P. B. Moyle, editors. Methods for fish biology. American Fisheries Society, Bethesda, Maryland.

Thompson, K. W., L. A. Knight, Jr., and N. C. Parker. 1986. Color-coded fluorescent plastic chips for marking small fishes. Copeia 2:544–546.

Thrower, F. P., and W. W. Smoker. 1984. First adult return of pink salmon tagged as emergents with binary coded-wires. Transactions of the American Fisheries Society 113:803–804.

Tranquilli, J. A., and W. F. Childers. 1982. Growth and survival of largemouth bass tagged with Floy anchor tags. North American Journal of Fisheries Management 2:184–187.

Tsukamoto, K. 1985. Mass-marking of ayu eggs and larvae by tetracycline-tagging of otoliths. Bulletin of the Japanese Society of Scientific Fisheries 51:903–911.

Utter, F. M., and J. E. Seeb. 1990. Genetic marking of fishes: overview focusing on protein variation. American Fisheries Society Symposium 7:426–438.

Van Den Avyle, M. J. 1993. Dynamics of exploited fish populations. Pages 105–135 *in* C. C. Kohler and W. A. Hubert, editors. Inland fisheries management in North America. American Fisheries Society, Bethesda, Maryland.

Volk, E. C., S. L. Schroder, and K. L. Fresh. 1990. Inducement of unique otolith banding patterns as a practical means to mass-mark juvenile Pacific salmon. American Fisheries Society Symposium 7:203–215.

Waldman, J. R., D. J. Dunning, and M. T. Mattson. 1990. A morphological explanation for size-dependent anchor tag loss from striped bass. Transactions of the American Fisheries Society 119:920–923.

Weathers, K. C., S. L. Morse, M. B. Bain, and W. D. Davies. 1990. Effects of abdominally implanted internal anchor tags on largemouth bass. American Fisheries Society Symposium 7:117–120.

Whitmore, D. H., editor. 1990. Electrophoresis and isoelectric focusing techniques in fisheries management. CRC Press, Boca Raton, Flordia.

Wilson, R. C. 1953. Tuna marking, a progress report. California Fish and Game 39:429–442.

Wydoski, R., and L. Emery. 1983. Tagging and marking. Pages 215–237 *in* L. A. Nielsen and D. L. Johnson, editors. Fisheries techniques. American Fisheries Society, Bethesda, Maryland.

Chapter 13

Acoustic Assessment of Fish Abundance and Distribution

STEPHEN B. BRANDT

13.1 INTRODUCTION

The purpose of this chapter is to provide a basic introduction to the use of underwater acoustics for measuring fish abundances and distributions. It is written for someone new to acoustic technology and interested in exploring its potential. The references cited are largely limited to those in the peer-reviewed literature and have been selected to provide the reader with greater detail and examples of applications. General reviews, mathematical treatment, and additional references on the application of underwater acoustics to fish can be found in Forbes and Nakken (1972), Clay and Medwin (1977), Thorne (1983b, 1983c), Johannesson and Mitson (1983), Mitson (1984), Stanton and Clay (1986), MacLennan (1990), MacLennan and Simmonds (1992), and Smith et al. (1992). Commonly used terms are defined in Box 13.1.

13.1.1 Definition of Fisheries Acoustics

Fisheries acoustics is the use of transmitted sound to detect fish. Sound travels quickly and efficiently through water and reflects from fish and other organisms in the water. The returning echoes contain information on fish sizes, distributions, and abundances. Fish reflect sound well because various components of the fish's body (e.g., swim bladder and muscle) have densities quite different from that of the surrounding water.

The application of underwater acoustics to fish detection has been referred to as fisheries acoustics, hydroacoustics, underwater acoustics, and echo sounding. These terms have been used interchangeably. Other terms such as bioacoustics, acoustical oceanography, and sonar have broader meanings that reflect the wider use of sound in aquatic environments. Bioacoustics includes underwater communication and echolocation by animals and the study of animal sounds used in reproduction, feeding, and predator–prey interactions by fishes and mammals (e.g., Tavolga et al. 1981; Hawkins and Myrberg 1983; Purves and Pilleri 1983). Acoustical oceanography is a broad term that includes bottom detection and mapping, seismic subbottom profiling, and the study of properties of sound transmission and noise in the sea (Clay and Medwin 1977; Stanton and Clay 1986). Sonar was originally an acronym that included all *so*und *na*vigation and *r*anging applications, but its use in fisheries acoustics often refers to sideward-facing transducers (device from which the sound is transmitted). Echo sounding is a sonar application that typically refers to downward-facing transducers.

Box 13.1 Definitions of Acoustic Terms

Absorption: The loss of sound energy as sound travels through water, caused by friction and, in marine environments, molecular relaxation of certain compounds. Absorption increases with water salinity and sound frequency.

Acoustic axis: The center axis of the transmitted acoustic beam. Sound intensity is highest along the acoustic axis.

Acoustic pulse or ping: The burst of sound transmitted into the water by the transducer. The pulse has a specific frequency and duration determined by the transducer and echo sounder.

Acoustic scatterer or target: Objects, primarily fish, that reflect sound in water. When sound encounters an object in the water with a density different from that of water, a portion of the sound will be reflected back to the transducer as an echo.

Acoustic transect: The path of a ship that is collecting acoustic data continuously from one location to another.

Backscattering cross section (σ_{bs}): A measure of the reflectivity of an acoustic target; the ratio of the sound intensity reflected (I_r) from a target to the sound intensity incident (I_i) to the target at a distance, R, from the target; $\sigma_{bs} = R^2(I_r/I_i)$ (unit, square meters).

Bandwidth: The range of frequencies transmitted by an echo sounder (unit, Hz).

Beam angle: The full angle (in degrees) from the acoustic axis of the transducer at which the sound intensity is one-half (−3 decibels) that on the acoustic axis.

Beam directivity pattern: The pattern of sound intensities transmitted from or received by a transducer. The transducer is most sensitive along the acoustic axis.

Decibel (dB): A dimensionless unit used for expressing ratios of sound intensities. Decibels are defined as 10 times the logarithm of the ratio of a measured sound intensity (I_M) to a reference sound intensity (I_R): $10 \cdot \log_{10}(I_M/I_R)$.

Doppler effect: The change in frequency of reflected sound caused by the relative movement between the transducer and the acoustic target.

Dual-beam transducer: A transducer that has both a wide beam and a narrow beam. Sound is transmitted on the narrow beam and received on both the narrow and wide beams. The ratio of the size of the two received echoes allows a calculation of radial location of a target in the acoustic beam.

Echo: Sound reflected from an acoustic scatterer.

Echo counting: A signal-processing technique that counts the number of echoes received from individual targets.

Echogram: A qualitative graphic representation of echo voltages across time.

Echo sounder: An instrument used to transmit and receive electrical signals from the transducer.

Echo-squared integration: A signal-processing technique used to measure the total amount of acoustic energy reflected back to the transducer. The energy reflected is proportional to total backscattering cross section of all targets.

Box 13.1 Continued.

Frequency: The number of sinusoidal sound waves per unit time expressed in kilohertz, or 1,000 cycles per second.

Gain: The amount of amplification of an acoustic signal (unit, decibels, dB).

Intensity: The power per unit area of a sound wave (unit, W/m^2).

Pressure: A unit used to describe the force per unit area of sound transmission (units, Pa/m^2 or N/m^2).

Pulse duration: The duration in time from start to end of an acoustic pulse (unit, s).

Pulse length: The length of an acoustic pulse (unit, m).

Pulse transmission rate: The number of acoustic pulses transmitted per unit time (unit, pulses/s).

Signal-to-noise ratio: The ratio of the strength of a signal to the background noise. The higher the ratio, the better the signal.

Sonar: A general term for all *so*und *na*vigation and *r*anging hardware. Most often used to refer to horizontally facing transducers.

Sound speed (*c*): The speed of sound in water (unit, m/s).

Split-beam transducer: A four-quadrant transducer that measures the differential arrival times of echoes in order to define the location of a target in the acoustic beam.

Standard target: A target with a known target strength that is used to calibrate acoustic hardware.

Target strength: A measure of the proportion of sound that is reflected from an acoustic scatterer back to the transducer. It is expressed in decibels and is equivalent to 10 times the $\log_{10}$ of the backscattering cross section.

Time-varied gain (TVG): The amplification of an acoustical signal. It increases with time and is used to compensate for sound spreading and absorption in the water.

Transducer: A pressure-sensitive device that converts electrical energy into sound energy for sound transmission and sound energy into electrical energy during sound reception.

Transmission loss: The attenuation of sound intensity as sound travels through water. The reduction in sound intensity is caused by the spherical spreading of the sound wave and by sound absorption.

Trigger signal: Electrical signal that initiates the transmission of the acoustic pulse.

Wavelength: The length of one cycle of sound (unit, m).

13.1.2 History and Current Status

Echo sounding was developed largely during the first World War to detect submarines. One of the first published applications of echo sounding to fisheries research was that of Sund (1935), who discovered that midwater fish (presumably Atlantic cod) were confined to a narrow layer of the water column. Early applications of underwater acoustics used sound to record presence or absence of fishes and to locate aggregations of fishes (Sund 1935; Balls 1948). Only qualitative comparisons of the relative abundances of fishes were possible (Forbes and Nakken 1972). These applications were also limited to ideal, open-water situations.

Quantitative estimates of fish abundances were made possible in the late 1960s and early 1970s with the development of both echo-counting and echo-squared integration procedures (e.g., Dragesund and Olsen 1965; Craig and Forbes 1969; Forbes and Nakken 1972). These procedures provided two ways to convert echoes into measures of fish abundances. As acoustic hardware and data analyses continued to improve through the 1980s, the use of underwater acoustics began to gain wider acceptance and more frequent use among fisheries scientists. The development of multiple-beam transducers (Burczynski and Johnson 1986; Foote et al. 1986) and statistical procedures (Clay 1983; Lindem 1983; Stanton and Clay 1986) provided the means to estimate acoustic sampling volume and the sizes of fishes and improved the accuracy of acoustic assessment of fish stocks. During the past decade, the development of stable, scientific echo sounders, multifrequency applications, new transducer deployment techniques, standardized calibration procedures, and more realistic models of the sound-scattering properties of biological targets have improved the accuracy, precision, and reliability of acoustic measurements and have led to a variety of new types of applications. Advances in digital electronics and computer power have led to better ways to record, analyze, and display data and to make full use of the acoustic information. Overall, acoustic techniques are now accepted and routinely used for fish stock assessment and ecological research. The field of fisheries acoustics continues to evolve at a rapid pace.

13.1.3 Applications

The primary and traditional application of underwater acoustics is for fish stock assessment, particularly in marine environments (MacLennan and Simmonds 1992). Fish biomass, numerical abundances, and mean sizes have been measured successfully in aquatic environments ranging in size from small lakes (Thorne 1979, 1983a; Burczynski et al. 1987; Unger and Brandt 1989; Jacobson et al. 1990) to larger lakes (Chapman 1976; Brandt et al. 1991; Argyle 1992; Walline et al. 1992) to estuaries and coastal areas (Thorne 1977; Francis 1985; Luo and Brandt 1993) and to the open ocean (e.g., Cushing 1968; Bailey and Simmonds 1990; MacLennan and Simmonds 1992). Sophisticated acoustic techniques are also used by commercial fishers to find concentrations of fishes and to aim their fishing gear (MacLennan and Simmonds 1992). More recent applications of acoustics have made better use of the rich spatial and temporal information inherent in acoustic data to study the behavior and ecology of fish and zooplankton. Such studies have evaluated fish distributions in relation to features of the environment, assessed trophic interactions, monitored the effects of artificial structures on fish movements, and estimated fish production (see section 13.6.1). Other aquatic organisms also scatter sound, and acoustics have been used to study the distribution and biology of small zooplankton and larger invertebrates, such as Antarctic krill (see section 13.6.3).

13.1.4 Advantages and Limitations

The use of underwater acoustics for measuring abundances and distributions of pelagic fishes has several advantages over standard sampling techniques. The ability to "see" and count what is under the surface of the water without disturbing the environment is one key advantage of underwater acoustics. Because sound travels in water at approximately 1,500 m/s, the entire water column can be sampled quickly, and detailed maps of fish densities and mean sizes can be obtained over large bodies of water. The ability to use sound to measure fish abundances throughout the water column continuously alleviates many of the sampling problems created by the spatial patchiness of fish distributions. Acoustic techniques are thus particularly well suited for assessment of midwater fishes. When the transducer is mounted permanently in one location, fish abundances and movements across a particular volume of water can be monitored continuously (see section 13.6.2). The high rate of sampling makes acoustic techniques very cost effective and contributes to low variance. Acoustic techniques are unobtrusive in that fish are not harmed or interfered with when sampled. There is also little avoidance of the acoustic signal by fishes.

The use of acoustics has several limitations. Perhaps the most severe limitation of acoustic techniques is that fish species cannot be directly identified. Although substantial progress is being made on acoustic identification of targets (see section 13.5.7), traditional fish sampling programs are still required to verify species identity and to obtain detailed information on the biology (size frequency, age, sex, and diet) of fishes. Acoustic techniques cannot easily sample all parts of the aquatic environment. Fish that are near the surface or within about 0.5 m of the bottom of the water column cannot easily be detected. Thus, the proportion of the water column that can be sampled decreases in shallow-water environments (e.g., Unger and Brandt 1989). The maximum depth at which a fish can be detected is also limited because sound loses energy as it travels in water. Finally, trained personnel are required to operate acoustic hardware and evaluate acoustic data. Such training and experience are rarely available at academic institutions.

13.2 COMPONENTS OF UNDERWATER ACOUSTICS

13.2.1 Sound Transmission

In essence, the principle of acoustic detection of fishes is rather straightforward. Sound is transmitted into the water as a pulse. As the sound pulse travels through the water it encounters targets, such as fish, that reflect sound back toward the sound source. These echoes provide information on fish size, location, and abundance. The basic components of the acoustic hardware (Figure 13.1) function to transmit the sound, receive, record, and analyze the echoes. These components can be housed within one instrument package or can be separate.

Sound is transmitted into the water by sending pulses (termed pings) of sound pressures in a directed beam through the water column. The echo sounder uses a trigger signal to electronically generate the pulse that is transmitted to the transducer. Scientific echo sounders are equipped with several options that control the duration and rate of sound transmissions. The transducer, which is submerged underwater, converts the electrical pulse into sound pressure and transmits the sound into the water. Sound is transmitted as a wave of pressure that spreads outward from the transducer in a spherical pattern.

Transducers are designed to focus greater amounts of energy directly below the transducer, along the acoustic axis, and less energy away from this perpendicular direction. Normally, the same transducer is used for sound transmission and echo reception. The transmitted sound has a specific frequency (number of cycles per second, or hertz [Hz]) that differs among different types of acoustic hardware. The frequencies used for fishes generally range between 12 and 420 kHz, but frequencies of 1 kHz to 1 MHz have been used for special applications. Sound is normally transmitted as a short burst, or acoustic pulse, of a set duration, usually 0.1 to 1.0 ms. The pulse transmission rate is generally high (e.g., 1 to 10 times per second) to ensure

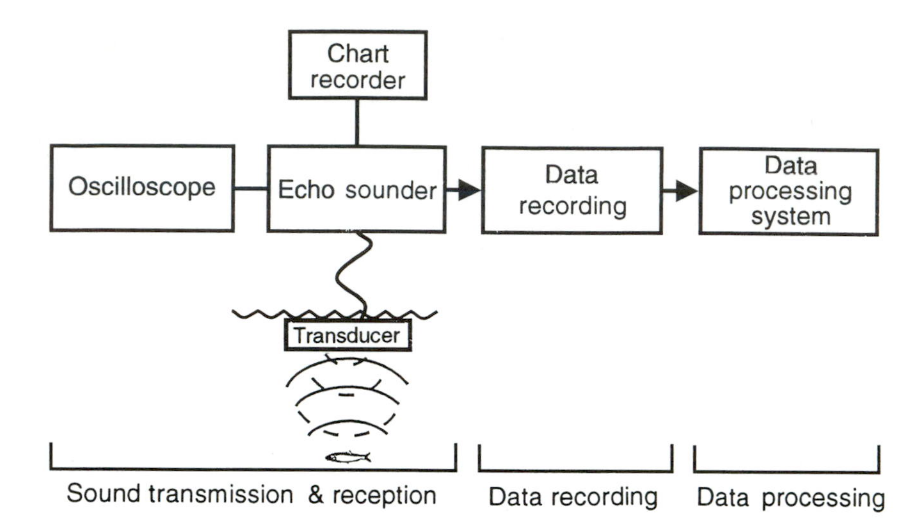

Figure 13.1 The basic components of acoustic hardware. An echo sounder with transducer is used to produce and receive sound. Sound waves are produced by the transducer and transmitted into the water (dashed lines). Sound reflects off a fish target (solid lines) and returns to the transducer as an echo. Echoes are viewed on an oscilloscope and a paper chart recorder. Data can be recorded and replayed for signal processing or can be processed directly.

continuous coverage of the water volume. The sound frequency, pulse duration, and pulse transmission rate determine the spatial resolution of the acoustic data (discussed in section 13.4.2).

13.2.2 Echo Production and Sound Reception

Sound travels as a pressure wave that results in a periodic expansion and contraction of the water. The pressure wave propagates at a speed of approximately 1,500 m/s in salt water. When the sound wave encounters a fish, or any other acoustic scatterer, an echo is reflected radially outward from the target. Immediately after the sound pulse is transmitted, the acoustic hardware begins to "listen" for echoes or sound pressures at the same frequency at which the pulse was transmitted. Echoes are produced by any object in the water having a density different from that of water. Fish are good acoustic targets because their swim bladders have a high density contrast with water. Swim bladders, filled with gas, contribute a large (90% or more) portion of the echo (Jones and Pearce 1957; Clay and Medwin 1977; Foote 1980b; Foote and Traynor 1988; MacLennan and Simmonds 1992).

13.2.3 Data Display and Analyses

Various methods have been developed to convert echoes into measures of fish locations, sizes, and abundances. Echoes received by the pressure-sensitive transducer are converted into electrical voltages, which are normally digitized for data recording and analyses. The time between sound transmission and echo reception and the size of the echo (voltage) are the basic pieces of information provided by acoustic hardware. The length of time between sound transmission and echo reception is determined by the distance of the acoustic target from the transducer and the speed of sound in water. Time can thus be converted into target depth (see section 13.5.1). The sizes and number of echoes provide the information on fish sizes and abundances. These are complex relationships.

Figure 13.2 shows how acoustic signals would appear as time-dependent variations in voltage levels for a single sound transmission. A high voltage occurs at the surface of the water column because a short residual reverberation of the transducer remains immediately after sound transmission. The bottom also produces a strong echo. Echoes between the bottom and surface are due to acoustic scatterers in the water column, presumably fish. These types of voltages are produced for each acoustic transmission. Echo voltages can be monitored on an oscilloscope and displayed graphically on a chart recorder. The chart recorder provides a good way to monitor the operation of the echo sounder and to get a qualitative view of fish distributions. The chart recorder works by producing marks (shades of gray or color) on a chart paper or computer screen that correspond to the size of the echo and the time it was received. Thus, for each ping (see Figure 13.2), there are large marks at the surface and at the bottom and smaller marks in the water column that represent echoes from fish. If the survey vessel is traveling in one direction at a constant speed, the composite of a series of consecutive pings produces a cross-sectional map of fish distributions. This map is called an echogram (Figure 13.3). Care must be taken in interpreting echograms because sampling volume increases with range from the transducer (see section 13.5.1) and is not corrected for in most displays (see Jacobson 1990). Modern echo sounders use color to display echoes because color has a greater dynamic range than do shades of gray.

Various methods are used to extract information on fish abundances and sizes from the acoustic signals (see section 13.5.1). The specific type of analyses that can be done depends on the type of acoustic hardware that is used and on the fish distributions. In essence, signal processing is focused on summing the total amount of echo energy received to estimate fish abundances and on relating the size of an individual echo to the size of the target from which the echo originated (see section 13.5.3). Signal processing can be done in the field or can be completed subsequently in the laboratory.

13.3 THE SONAR EQUATION

Various sonar equations are used to mathematically describe sound transmission and reflectance in water. Here, a simple equation is developed that describes factors that affect the size of an echo reflected off an individual target, such as a fish, and received at the transducer. Sonar equations can be expressed in terms of sound pressure or voltage amplitude or in a logarithmic form. The equations are often expressed in logarithmic form because typical ranges of sound intensity (power per unit area) encountered in underwater acoustics are quite large. For example, the ratio of the sound intensity reaching a fish to the sound intensity reflected from the fish typically differs by factors of 10^3 to 10^6 for large and small fish, respectively. The decibel (dB) is the unit used to express logarithmic differences in sound intensity. Decibels are a dimensionless unit based on a ratio of sound intensities. A decibel is specifically defined as $10 \cdot \log_{10}(I_A/I_B)$, where I_A and I_B are two sound intensities. In the example above, the decibel expression of the ratio of sound reflected from a fish to that encountering the fish target would be -30 dB ($10 \cdot \log_{10}[0.001]$) and -60 dB ($10 \cdot \log_{10}[0.000001]$) for a large and small fish, respectively. An easy way to think in decibels is to realize that a change in sound intensity of 3 dB is approximately a

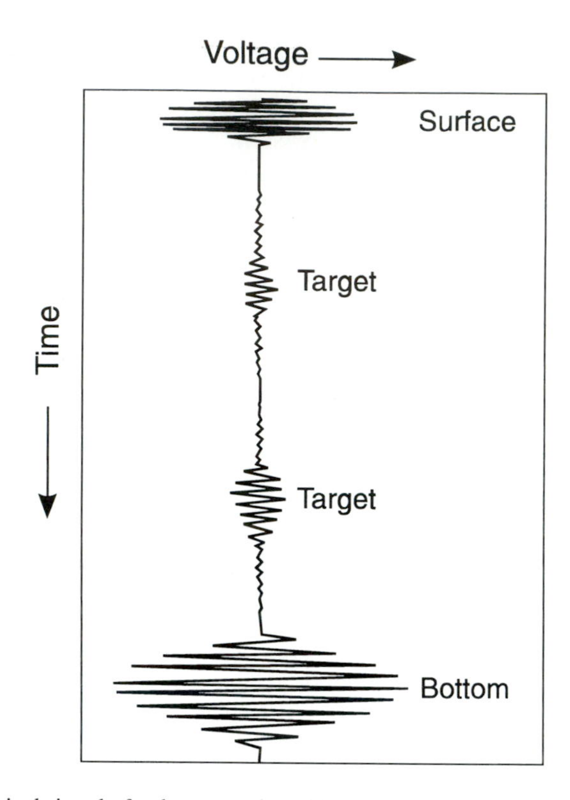

Figure 13.2 A typical signal of voltage produced by a single acoustic transmission versus time. Surface scatter is high because of the residual vibration of the transducer after the sound pulse is initiated. The size of the fish echoes differs with target size. The bottom normally produces a large echo.

doubling in sound intensity ($10 \cdot \log_{10}[2] = 3.01$), and a change of 10 dB is equivalent to an order-of-magnitude change in intensity ($10 \cdot \log_{10}[10] = 10.0$).

For an individual target, the size of the echo returning to the transducer (echo level, EL) depends on the amount of sound reaching the target, the proportion of the incident sound that is reflected from the target, and the amount of the reflected echo that returns to the transducer. The amount of sound that reaches a target depends on the amount of sound transmitted (the source level, SL) and the location of the target relative to the transducer. For decibel expression, any sound must be compared to a reference sound level. The source level is usually expressed as the sound intensity at (or "referred to") a range of 1 m from the transducer, and echoes received by the transducer are compared to this standardization source level. Different (sometimes arbitrary) reference levels may be used to characterize particular aspects of sound propagation and pattern as noted in the next paragraph.

Sound is not transmitted uniformly in all directions from the transducer surface. Most of the sound is transmitted directly perpendicular to the transducer face along what is called the acoustic axis (Figure 13.4). Sound intensity generally drops off rapidly at increasing angles from the acoustic axis. Thus, the amount of sound reaching a target depends on the angle location of the target relative to the transducer. The directivity pattern ($B[\theta]$) for a circular transducer describes the proportion of sound transmitted at different angles from the transducer (Figure

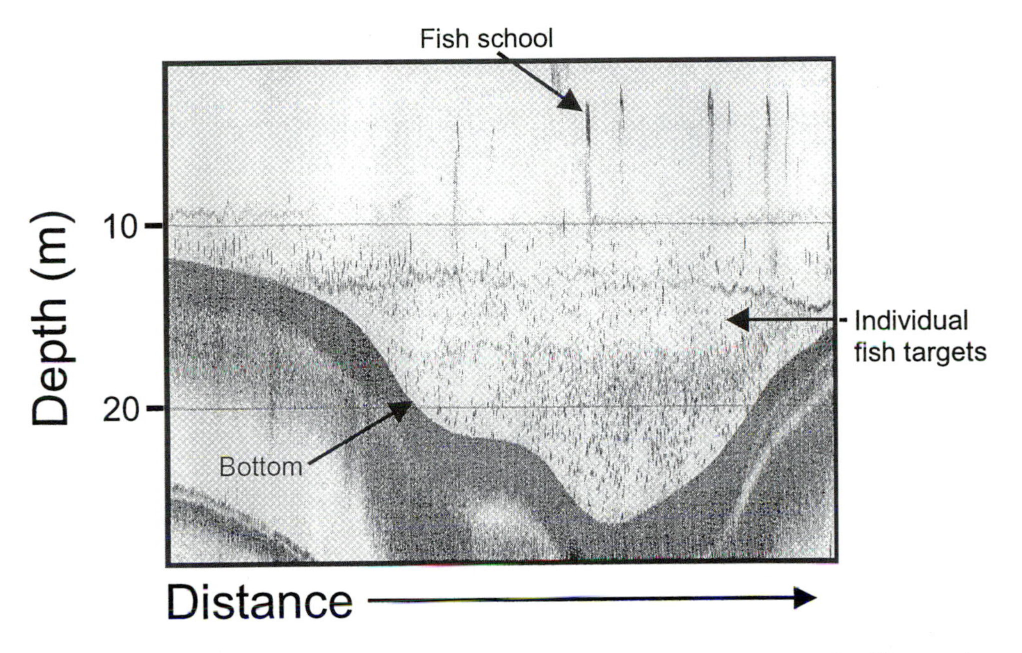

Figure 13.3 Example of an echogram, this one made along a transect across the Chesapeake Bay. The water surface is at the top of the echogram. The bottom is marked by the upper edge of the thick dark band undulating across the middle of the echogram. The lowermost heavy band is a "double echo" of the bottom, produced by sound waves that were reflected from the bottom, then from the water surface, and then from the bottom again before reaching the transducer. Light horizontal lines (at 10m and 20m) are reference marks automatically placed on the chart paper as it was pulled past the marking stylus. The marks between the surface and bottom are acoustic targets, mostly fish, in the water column. Discrete schools of fish near the surface, horizontal "bands" of fish in midwater, and loose aggregations or individual fish in the bay's deep channel have been detected.

13.4). By convention, $B(\theta) = 0.0$ dB at every point along the acoustic axis and is negative (lower intensity) away from the axis. The transducer also has the same directional sensitivity when receiving an echo.

Sound levels are weakened as the sound travels through water. This transmission loss (TL) occurs on the way to the target as well as on the returning echo. When the sound wave hits a target, a portion of the sound is reflected back in the direction of the transducer. Sound scatters from any surface having a density different from that of water (e.g., the bottom, bubbles, fish, or plankton). The amount of sound reflected is largely a function of target size and the difference in density between the target and water. The reflectivity of a target is called the backscattering cross section, or, in decibels, the target strength (TS). The target strength is expressed as the logarithm of the ratio of the sound intensity reflected from a target (I_r) to the sound intensity incident to a target (I_i) or $10 \cdot \log_{10}(I_r/I_i)$. Because the reflected sound intensity is always less than the incident sound intensity, target strength has negative values.

When the above terms are put together, the factors affecting the echo level received at the transducer from an individual target can be expressed in decibels by the following equation:

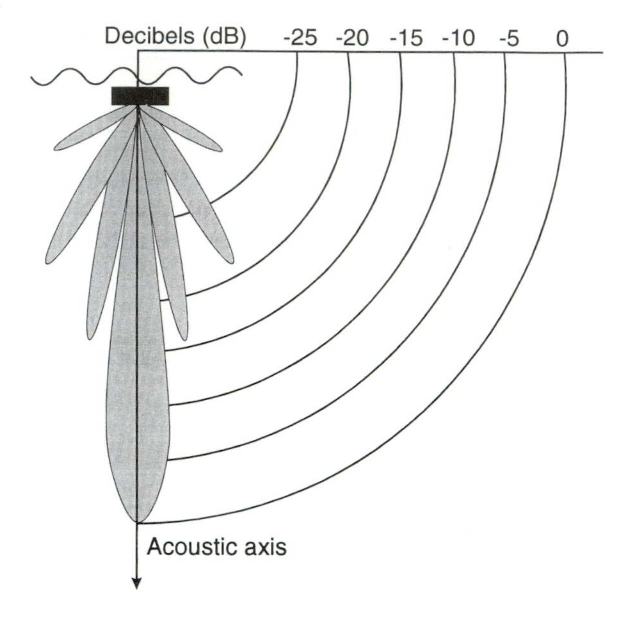

Figure 13.4 Two-dimensional directivity pattern depicting signal strength of an acoustic beam. The signal transmission is maximum along the acoustic axis (where the reference sound level is set at 0.0 dB by convention) and declines with greater angles. Low-intensity side lobes are commonly produced at wide angles from the acoustic axis.

$$\text{EL} = \underset{[\text{SL} + B(\theta) - \text{TL}]}{\begin{matrix}\text{sound}\\ \text{reaching}\\ \text{target}\end{matrix}} + \underset{[\text{TS}]}{\begin{matrix}\text{sound}\\ \text{reflected}\\ \text{by target}\end{matrix}} + \underset{[B(\theta) - \text{TL}]}{\begin{matrix}\text{sound}\\ \text{returning}\\ \text{to transducer}\end{matrix}}. \tag{13.1}$$

Combining terms, equation (13.1) becomes

$$\text{EL} = \text{SL} - 2\text{TL} + \text{TS} + 2B(\theta). \tag{13.2}$$

Thus, the echo level depends on the amount of sound transmitted into the water, the dissipation of sound as it travels through the water, the angular direction of the target from the transducer, and the proportion of the echo reflected by the target back towards the transducer.

Sound transmission loss is caused by two processes, both of which progressively increase with distance or range (R) from the transducer. Sound intensity, which is expressed on a per-unit-area basis, is reduced as sound spreads over a larger area during travel. A two-dimensional analogy is the spreading and dissipation of water rings radiating from the point at which a rock is dropped into a pond. Because sound spreads spherically in water, the loss of sound intensity is proportional to the increase in the surface area of a sphere with a radius equal to the distance the sound has traveled from the transducer. The surface area of a sphere is proportional to the square of its radius. Thus, at a range R from the transducer, the sound intensity has been reduced by a factor of R^2. Sound also spreads spherically from the target back to the transducer and is thereby reduced by an additional factor of R^2. From transmission to reception, therefore, spherical spreading reduces sound intensity overall by a factor of R^4. In decibel terms, relative to the initial sound intensity at 1 m from the transducer (the source level), the echo intensity is reduced by a factor of R^4 to $10 \cdot \log_{10}(1/R^4) = -40 \cdot \log_{10}(R)$ dB.

Sound energy is dissipated through absorption. Sound absorption is largely a frictional loss due to the conversion of sound energy into heat (Fisher and Simmons 1977). The sound absorption coefficient, α, is a function of sound frequency and water temperature and salinity. Sound absorption is lower in freshwater than in salt water and increases at higher frequencies. Clay and Medwin (1977) provided detailed equations for calculating absorption losses. Generally, sound absorption is expressed in decibels per meter, and total absorption is directly proportional (α is the proportionality constant) to the distance the sound travels ($2R$).

Substituting both transmission loss terms into equation (13.2), we obtain a simple form of a sonar equation:

$$\mathrm{EL} = \mathrm{SL} - 40 \cdot \log_{10} R - 2\alpha R + \mathrm{TS} + 2B(\theta). \tag{13.3}$$

Equation (13.3) can be expressed in sound intensity as

$$I_{\mathrm{EL}} = \frac{I_{\mathrm{SL}} \cdot \sigma_{\mathrm{bs}} \cdot B^2(\theta)}{R^4} (10^{-\alpha R/5}), \tag{13.4}$$

where I_{EL} and I_{SL} are the sound intensities received at the transducer and transmitted by the transducer (referred to a range of 1 m), respectively. The term σ_{bs}, backscattering cross section, is a measurement of the reflectivity of a target and is defined in equation (13.10). Sound intensity is proportional to sound pressure squared. Thus, the sonar equation for sound pressure (P) is

$$(P_{\mathrm{EL}})^2 = \frac{(P_{\mathrm{SL}})^2 \cdot \sigma_{\mathrm{bs}} \cdot B^2(\theta)}{R^4} (10^{-\alpha R/5}), \tag{13.5}$$

or

$$P_{\mathrm{EL}} = \frac{P_{\mathrm{SL}} \cdot \sqrt{\sigma_{\mathrm{bs}}} \cdot B(\theta)}{R^2} (10^{-\alpha R/10}). \tag{13.6}$$

The square root of the backscattering cross section has been referred to as the backscattering length (L_{bs}) with units of meters (Clay and Medwin 1977).

The source level and beam directivity are measured during acoustic calibrations. The transmission losses can be corrected by an appropriate time-variable amplification of the signals. The two unknowns left in equation (13.3) are target strength (TS) and angular location (θ) of the target in the beam. Most recent developments in acoustic techniques have been focused on improving our ability to measure or model these two unknowns (see section 13.5.2).

13.4 PREPARATIONS FOR FISH STOCK ASSESSMENT

Fish stock assessment is the most common application of underwater acoustics. Outlined in sections 13.4 and 13.5 are general procedures and guidelines for conducting an acoustic assessment of a fish population. These general procedures and methods of data analyses can be applied to most other types of applications of acoustics (see section 13.6).

13.4.1 Evaluation of Objectives

The first step in applying acoustics to fish stock assessment is to evaluate whether or not underwater acoustics is an appropriate tool for the question being asked and for the particular species and body of water of interest. Acoustic techniques will not be suitable for all species nor for all types of aquatic environments. It is very well suited for species that live in midwater. Some prior knowledge of the expected spatial distribution, sizes, and behavior of the target species will be needed to make this determination and then to select the appropriate acoustic hardware and design the survey strategy.

One must also determine the type of data required. Choices of acoustic instrumentation, method of data analyses, and design of the survey depend on whether one needs absolute measures of fish abundances (fish density per unit area or volume) or relative measures (changes in abundances between surveys), whether samples are required over a large area or over a fixed location across time, and whether information on species distribution is needed.

13.4.2 Selection of Acoustic Hardware

All quantitative acoustic analyses require that measurements are made with a scientific-quality echo sounder that has stable electronics and low noise levels and can be calibrated easily. The specific choice of acoustic hardware depends on several factors, including the type of questions being asked, whether absolute or relative density measurements of abundances are needed, the size and distribution of the fish being studied, the type of transducer being deployed, and the physical characteristics (e.g., salinity and depth range) of the aquatic environment being sampled. The specifications of the echo sounder and transducer will in turn set the sampling resolution and the ability to resolve individual targets, the depth range at which targets can be detected, the acoustic sampling volume, the types and sizes of organisms that can be detected, and the types of data analyses that can be done. Many technical factors, such as sound frequency, wavelength, and absorption, are interrelated, and there will be trade-offs in the selection of the acoustic hardware. Some primary considerations are discussed below.

Frequency. The frequency of the transmitted sound is the number of cycles per second (Figure 13.5). Typical echo sounders used for fisheries research have fixed frequencies of 12,000–420,000 cycles per second (12–420 kHz). The frequency of the sound largely sets the minimum size at which an individual acoustic target can be detected. Targets smaller than one wavelength reflect echoes that are disproportionately weaker than those reflected by targets a wavelength in size or larger. Sound wavelength is defined as the length (in meters) of one cycle of sound. Wavelength (λ) is the speed of sound in water (c) divided by sound frequency (f):

$$\lambda \text{ (m/cycle)} = c \text{ (m/s)}/f \text{ (cycles/s)}. \quad (13.7)$$

Typical values of wavelength at a sound speed of 1,500 m/s are 39.5 mm for a frequency of 38 kHz, 12.5 mm for 120 kHz, and 3.6 mm for 420 kHz. Wavelength is inversely proportional to sound frequency; thus, smaller targets can be detected at higher frequencies. But there are trade-offs in the choice of frequency. Higher-frequency sound is absorbed more readily in water than is lower-frequency sound, particularly in marine systems. Thus, the range at which targets of a given size can be detected is less for a higher-frequency echo sounder.

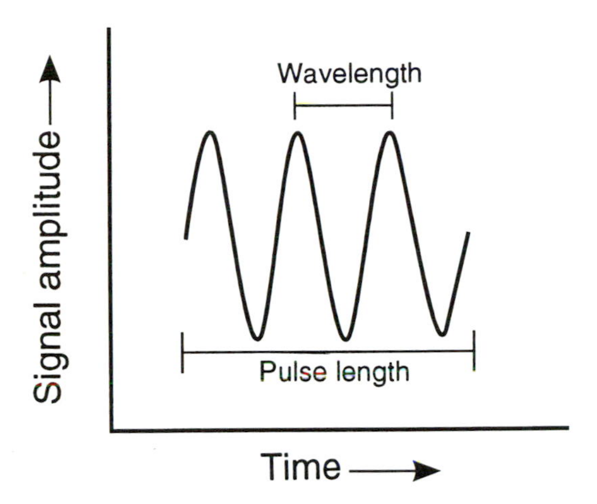

Figure 13.5 Signal amplitude plotted as a function of time to demonstrate the relationship between acoustic frequency and pulse length of an acoustic transmission. A pulse of sound is transmitted into the water over a short period (pulse duration or length). The sound is characterized by a fixed frequency (number of cycles per second). Its wavelength (length of one cycle of sound) is determined by that frequency and the speed of sound in water.

If the fish species of interest is large and deep in the water column, or if small zooplankton are abundant, then a lower-frequency transducer (e.g., 38 kHz) might be the best choice because it would induce relatively few masking echoes from small targets. A high-frequency system (e.g., 420 kHz) might be preferred for assessing small fishes or large zooplankton, particularly in fresh water. Clay and Horne (1994) provided more precise criteria for the choice of sound frequency based on fish scattering properties. Echo sounders that operate at more than one frequency at a time are useful for fish stock assessment in complex aquatic environments that contain a wide range of target sizes.

Spatial resolution. Echoes from individual targets often must be resolved to make quantitative measurements of fish abundances and size (see section 13.5.2). The pulse length is one factor that affects spatial resolution. Pulse length (in meters) is controlled by manipulating pulse duration (in seconds; Figure 13.5); time is multiplied by the speed of sound to calculate length. Pulse length sets the minimum difference in range that can be detected between echoes from individual targets (Figure 13.6). Targets separated by less than one-half pulse length in range from the transducer generate overlapping echoes. Scientific echo sounders can often transmit pulses at user-defined, digitally controlled rates. Pulse durations of 0.1 ms (0.075-m range resolution) to 1.0 ms (0.75-m range resolution) are commonly used. The minimum pulse length that can be used is set by the sound frequency because a pulse must be at least one wavelength long. Higher-frequency echo sounders have shorter pulse lengths and thus finer depth resolution than do lower-frequency echo sounders. Short pulse lengths, however, may be susceptible to higher noise levels.

The directionality or directivity of the sound beam pattern ($B[\theta]$) affects the beam's horizontal resolution (ability to distinguish targets at the same range). Because sound spreads as it travels, the effective sampling volume of a transducer increases with range from the transducer (Figure 13.6), but the horizontal spatial

Transducer resolution

High resolution
- Narrow beam
- Short pulse

Low resolution
- Wide beam
- Long pulse

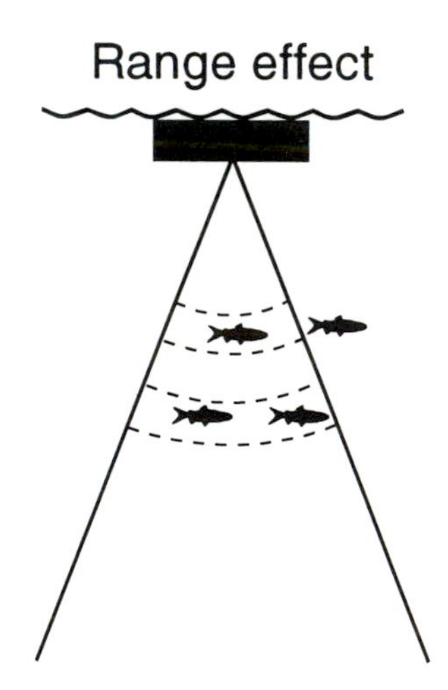

Figure 13.6 Geometry of vertical and horizontal target resolution within the acoustic beam in relation to differences in pulse length, beam angle, and range. The distance between paired broken curves is one pulse width.

resolution decreases. The transducer beam angle provides a good benchmark to characterize and compare sound directivity among different transducers. The beam angle is nominally defined by the cone at which the level of the transmitted sound intensity is equal to one-half of the intensity transmitted directly along the acoustic axis (Figure 13.4). This angle is referred to as the −3 dB point because −3 dB is equivalent to one-half loss in sound intensity ($10 \cdot \log_{10}[0.5] = -3$). The angle from the acoustic axis to the −3-dB point is one-half of the full beam angle. The beam angle is only an indicator of sound directivity and does not define the absolute boundary of any cone transmitted into the water nor the angular limits to sound transmission or target detection (see section 13.5.1). Targets located at wide angles from the acoustic axis produce small echoes that may not contribute much to the echoes received at the transducer.

Typical beam angles for fisheries applications range from about 5° to 30°. Transducers with small beam angles resolve individual targets better than do transducers with larger beam angles because less water is sampled during each sound transmission. There are trade-offs in the choice of beam angle (see the discussion by Clay and Horne 1994). Although wider beam angles have reduced ability to resolve individual targets, they sample more water, an advantage when fish are rare and widely dispersed. Narrow beams and short-pulse lengths are best when fish are very abundant or highly aggregated.

Pulse transmission rate. Pulse transmission rate is the number of pulses transmitted into the water per second. The depth of the water determines the maximum pulse transmission rate because the bottom echo must be received at the transducer before the next pulse is transmitted, otherwise signals interfere. Scientific echo sounders generally provide a range of pulse transmission rates (e.g., 1–10 times per second). High pulse transmission rates are useful for repeatedly measuring individual fish and for target tracking (e.g., Thorne and Johnson 1993) but add to the amount of data collected.

Other electronic equipment needs. In addition to a scientific echo sounder, transducer, and signal-processing hardware (e.g., computer) and software, other electronic equipment is often required. Acoustic data generally need to be recorded. It is best to use digital recorders (digital audiotape recorders and computer storage on hard or optical disk drives) for most purposes. An oscilloscope is generally required for calibrating and monitoring signal strengths. Paper chart recorders or color video screens are commonly used to display echo returns. A chart recorder is often part of the echo sounder.

13.4.3 Transducer Type and Deployment

The transducer is a pressure-sensitive device that generates a voltage when pressure is applied to it or generates sound when a voltage is applied to it. Transducers come in different sizes and can be constructed of different types of materials (see further discussions by Coates 1989 and MacLennan and Simmonds 1992). Selection of the transducer and echo sounder normally go hand in hand. In addition to transducer frequency and beam angle, one must decide whether the transducer will have a single-beam, dual-beam, split-beam, or some other type of configuration. The transducer configuration limits the ways in which fish sizes and abundances can be determined (see section 13.5.2).

Transducers can be deployed in the water in many ways (Figure 13.7). They can be either fixed in one place or towed through the water facing downward, upward, or off to the side. The type of equipment and signal processing are usually independent of how the transducer is deployed. Mobile acoustic surveys are commonly used for fish stock assessment because samples can be taken over large geographic areas. For mobile surveys, a downward- or sideward-facing transducer is towed through the water along a fixed line at a constant speed. Data are collected continuously along this transect line.

The transducer can be mounted through the hull of the assessment vessel or towed in a specially designed vehicle. Hull-mounted transducers limit surveys to one vessel,

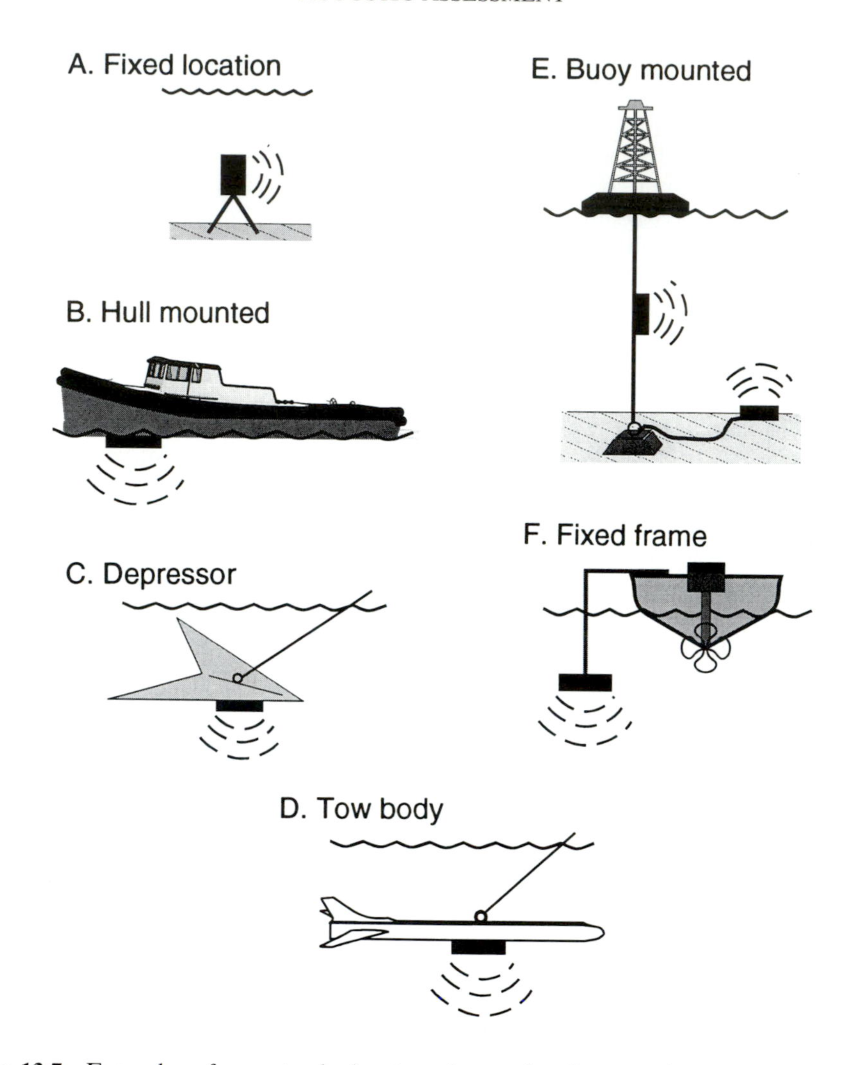

Figure 13.7 Examples of ways to deploy transducers. In all cases, the transducer must be under the water.

are subjected to the pitch and roll of the vessel, and may be more susceptible to bubble interference. Various methods can be used to tow the transducer. The vehicles in which transducers are mounted and towed are designed to minimize transducer movement from a horizontal plane and afford greater flexibility in the use of different survey vessels. The best location for towing a transducer is just beneath the water surface alongside or in front of the ship and outside the ship's wake, propeller noise, and bubble production. Vehicles that are towed deep in the water can be used to get closer to small targets or to locate fish that are deep or near the bottom. Towing speed should be selected to minimize noise caused by the vessel. There may be a trade-off in towing more slowly to minimize such noise levels and towing more quickly to maximize the survey area that can be sampled in the available time.

13.4.4 Survey Design

As in other types of sampling programs, the objective of a survey design is to sample a representative part of a population that can be used to estimate the entire population abundance. The survey design must, therefore, cover the geographic extent of the population and take into account the behavior and distribution of the fish. The peculiarities and limitations of acoustic stock assessment must be accommodated when sampling times and places are chosen. For example, restricted sampling volumes require that surveys be conducted when fish are away from the surface and bottom. Acoustic species identification is difficult, so acoustic sampling is most reliable when and where the fish of interest are most isolated from other species. As a final example, line transect methodologies require different statistical analyses than those used for isolated point samples.

Mobile acoustic surveys are generally conducted at a constant speed (e.g., 1.5–3 m/s) along fixed-line transects; geographic location is logged at regular intervals along the transect. Modern acoustic hardware allows information (from global position systems or loran) on the ship's location to be input directly into the data logging system at regular intervals along the transect. It is important to know the geographic location of the survey transects to map fish distributions and to use the data to make population estimates.

The design and timing of the acoustic survey depend on the spatial distribution and behavior of the target species. Acoustic surveys are generally more effective if done at a time when the fish are in the middle of the water column, dispersed, and relatively isolated from other species. For example, acoustic surveys are often done at night when schooling fishes have dispersed and may have migrated from the bottom into the water column (e.g., Brandt et al. 1991; Argyle 1992). Prior knowledge of the expected diel and seasonal distribution of the fish species helps define the most appropriate times to sample as well as the geographic coverage of the survey. A literature search and preliminary sampling (feasibility study) are highly recommended to define the most appropriate sampling times and locations.

Once the most appropriate sampling time(s) has been selected, survey transects must be laid out. The number of survey transects is a trade-off between maximizing the geographic coverage of a survey area and minimizing costs or time of ship charter. Various types of survey patterns have been used for fish stock assessment, and there are advantages and disadvantages for each type of survey design (Shotton 1981; Shotton and Bazigos 1984; MacLennan and Simmonds 1992). Commonly used survey designs include randomized, parallel, zigzag, and box transects (Figure 13.8).

Zigzag surveys collect data continuously along a track that spans the survey area. This type of survey minimizes survey time and maximizes the spatial coverage. However, care must be taken in data interpretation because data taken near the ends of transects are more spatially correlated than those taken near the centers of the transects (Williamson 1982). Equally spaced parallel transects and box designs (Figure 13.8) avoid this unequal spatial correlation and simplify interpolation of data between transects. Parallel and box designs also provide good spatial coverage so that the geographic distribution of the fish can be mapped.

Other types of survey designs include some element of randomness. For a stratified random design, the habitat is first subdivided into several strata based on environmental criteria that might be important to the fish (Jolly and Hampton 1990a, 1990b). Transects are then located at random locations (either run parallel to one another or

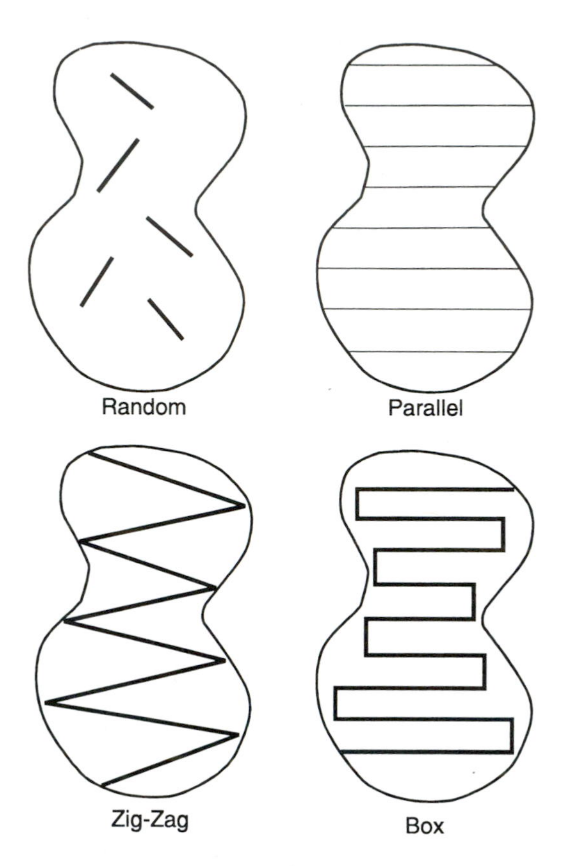

Figure 13.8 Various types of survey transect designs used for acoustic stock assessment of fishes.

oriented in a random direction) within each stratum. This type of survey design has the advantage of simplifying the assumptions used to estimate sampling variance, but it does not take full advantage of the ability of acoustic sampling to sample spatial heterogeneity in the environment. Randomized surveys require more survey time per unit sample. A partial random design locates transects at a random location within a regularly spaced grid (e.g., Jolly and Hampton 1990a, 1990b). Such randomized designs can place more effort in areas in which higher concentrations of fish are expected. Adaptive survey designs allow some changes to be made in the transect layout during the survey (e.g., to map fish concentrations).

The choice of survey design affects the statistical analyses of data. The strengths and weaknesses of various survey designs were discussed by MacLennan and Simmonds (1992). Replicate sampling along transects provides information on variance within transects. Adding some transects that are oriented perpendicular to the survey transects provides information on the gradients in fish distributions between transects.

13.4.5 Additional Sampling Requirements

Additional types of measurements are routinely collected during acoustic surveys, and time must be allocated to these endeavors when the survey cruise plan is designed. Other activities include physical and chemical measurements across the sample area, biological sampling, acoustic calibration (see section 13.4.6) and measurements of fish target strength (see section 13.5.2). Measurements that define the physical and chemical habitat of the fishes (e.g., temperature, salinity, oxygen levels, and water depth) should be taken routinely along the acoustic transect lines. These data provide criteria for extrapolating the acoustic measures of fish density to the rest of the aquatic environment and can also help in identifying species (section 13.5.7).

Biological samples (usually obtained with trawls or gill nets) should also be taken during acoustic surveys. Biological samples are needed to help identify acoustic targets, to corroborate acoustic estimates of fish sizes, and to collect other life history and ecological data on the fishes present in the sampling region. Biological samples should be large enough and taken with sufficient frequency to provide good estimates of fish sizes and relative species composition (as well as any other type of life history information desired). Similarity in the sizes and abundances of fishes estimated acoustically with those estimated by direct sampling helps validate both techniques (Mulligan and Kieser 1986; Costello et al. 1989; Bailey and Simmonds 1990; Parkinson et al. 1994). Biological sampling often can be targeted at concentrations of fishes or done when there is a marked change in the echo patterns of fish distributions.

13.4.6 Calibration

Calibration of the acoustic instruments is critical for quantitative measures of absolute densities and sizes of targets. Three procedures are used for a complete calibration (Robinson 1983; Foote 1990; Simmonds 1990). First, transducers require regular measurements of sound source levels and directivity patterns. These measurements are made in specialized tanks, where the exact locations of the transducer and a calibrated hydrophone can be determined. Second, standard targets are used to measure hardware performance in the field. A standard target with known acoustic target strength is placed beneath the transducer at various depths, and measured target strengths are compared with known values. The type and size of the standard target should be matched to the sound frequency and have a target strength similar to that of the fishes being assessed (Foote 1990). Precision-machined spheres made of copper or tungsten carbide have generally replaced the use of Ping-Pong balls for standard targets because the target strengths of metallic materials are largely independent of water temperature and acoustic frequency (Foote and MacLennan 1984). To ensure that the echo sounder is working consistently throughout a survey, calibration against standard targets should be done at least at the beginning and end of a short acoustic survey and at regular intervals during a longer survey.

A third method of calibration is used to measure recording levels and echo sounder amplification. A reference voltage signal is input to the echo sounder. The calibration signals are recorded at regular intervals (e.g., at the beginning of each recording tape). Recorded values are compared with the reference voltages to standardize signals. Most scientific echo sounders are equipped with such calibration tones.

13.5 APPLICATION OF ACOUSTICS TO FISH STOCK ASSESSMENT

13.5.1 Fish Abundance Estimation

Two basic techniques for measuring fish abundances with acoustic data are echo counting and echo integration. Measuring relative fish density is far simpler than measuring absolute fish densities. Measures of absolute density require an estimate of acoustic sampling volume. Before these analyses can be done, time must be converted to target depth, and echo voltages must be corrected for sound attenuation.

Measurement of target depth. Range detection of echoes is straightforward and is usually done automatically within the acoustic hardware or software. Range (or depth) of a target from the transducer is calculated by multiplying the time (t) between sound transmission and echo reception by the speed of sound in water (c) and then dividing by 2 to account for the two-way travel distance. Thus, target range (R) is

$$R = 0.5\,(t \cdot c). \tag{13.8}$$

Sound travels quickly in water, so time increments must be measured at a fine scale. For example, at a sound speed of 1,500 m/s, an echo received 0.1 s after sound transmission implies a target approximately 75 m from the transducer. The speed of sound in water increases slightly with increases in water temperature, salinity, and pressure (Leroy 1969; Del Grosso and Mader 1972; MacKenzie 1981). Clay and Medwin (1977, equation 3.2.20) provided a simplified, empirical formula for estimating sound speed from temperature (T, °C), salinity (S, ‰), and depth (Z, m):

$$c = 1449.2 + 4.6T - 0.055T^2 + 0.00029T^3$$
$$+ (1.34 - 0.010T)(S - 35) + 0.016Z; \tag{13.9}$$

Thus, for water with a salinity of 35‰, a temperature of 15°C, and a depth of 100 m, the speed of sound is 1,508 m/s. For a particular study, speed of sound is normally assumed to be constant throughout the water column despite minor differences associated with vertical gradients of temperature and salinity.

Time-varied gain. Sound loses intensity as it spreads through water and is absorbed. The farther the distance that the sound travels, the greater the loss of sound. Scientific echo sounders are generally equipped with a time-varied gain (TVG) that amplifies the sound signal with time (time being an electronic proxy for distance; signal amplification increases with elapsed time as a measure of distance) to correct for these sound losses (MacLennan 1987). The TVG correction for absorption differs between marine and freshwater environments and among transmission frequencies. Once TVG has been applied, the signal strength of echoes originating from different depths are comparable. Quantitative analyses require that an echo sounder have a stable and accurate TVG.

Echo counting. Echo counting is a relatively simple approach to measuring fish abundances (Craig and Forbes 1969; Forbes and Nakken 1972; Nickerson and Dowd 1977). An echo count is registered if an echo voltage from an individual target (see Figure 13.2) exceeds a predetermined threshold level. If fish are sufficiently dispersed and each echo represents an individual fish, then the number of echoes counted per unit time or distance traveled provides an estimate of the number of fish in the sampled water volume (i.e., fish density). Before echo counting can begin, echoes must be adjusted for sound attenuation so that echoes received from different depths have equal echo amplitudes. The appropriate TVG to apply is given in equation (13.3) as $40 \cdot \log_{10}(R) + 2\alpha R$. Echo-counting procedures require that threshold levels not be set so high that ambient noise is counted as fish or so low that fish targets are missed (Weimer and Ehrenberg 1975).

In order to compare fish abundances at different depths, counts of individual echoes must be corrected for the change in sampling volume with range from the transducer. The increase in sampling volume is proportional to the square of the range (R^2) from the transducer due to the spherical spreading of sound (section 13.3). If all fish targets are of equal size and the variability in echo strength and corresponding limits of detectability for a single fish can be ignored, then the number of echo counts per acoustic pulse from different depths can be compared by dividing the number of echoes by R^2. This method provides a quantitative, although relative, measure of fish density. A variation of echo counting has been used to count the number and estimate the size of fish schools. Fish abundance is then calculated by multiplying total school volume by an estimate of the fish density within the schools (Smith 1970, 1979; see discussion by MacLennan and Simmonds 1992).

A measure of actual water volume sampled by the acoustic beam is needed to estimate absolute fish densities (fish/m^3). Sampling volume is not simply the volume of the cone defined by the transducer beam angle. It is far more complex. For a given transducer, the effective sampling volume differs with the size and orientation of the fish. Larger fish can generally be detected at wider beam angles and at farther ranges than can smaller fish. Fish at the same range but different angles from the transducer produce echoes of different strength. A small fish near the acoustic beam axis could produce a stronger echo than might a larger fish off the axis (Figure 13.9).

Four basic approaches have been used to estimate acoustic sampling volume. First, volume can be calculated directly for a calibrated echo sounder if all the parameters and variables in the sonar equation (equation 13.3) are known. Second, sampling volume and mean target strength can be estimated from data on the size distribution of echo amplitudes. This approach is based on a model of beam directivity and models of the expected variations in target strengths of an individual fish (see section 13.5.2; Peterson et al. 1976; Clay 1983). Third, effective sampling volume can be measured directly by calculating the mean length of time targets at different depths remain in the acoustic beam as the transducer passes overhead at a fixed speed (Nunallee 1980, 1990; Kieser and Mulligan 1984; Crittenden et al. 1988; Thorne 1988). When this duration-in-beam technique is used in mobile ship surveys, fish are assumed to be stationary relative to transducer movement. This assumption is tenuous, and the approach may be most valid for fixed-transducer deployments (see section 13.6.2). Fourth, multibeam transducers (section 13.5.2) can directly measure the angle at which targets are detected and allow geometric calculations of sampling volume.

Echo-squared integration. Echo-squared integration is the most commonly used approach to measuring fish abundances. It is required in cases in which fish densities are too great for echo counting, and echoes reaching the receiver are composites of echoes derived from more than one individual. During echo integration, the total

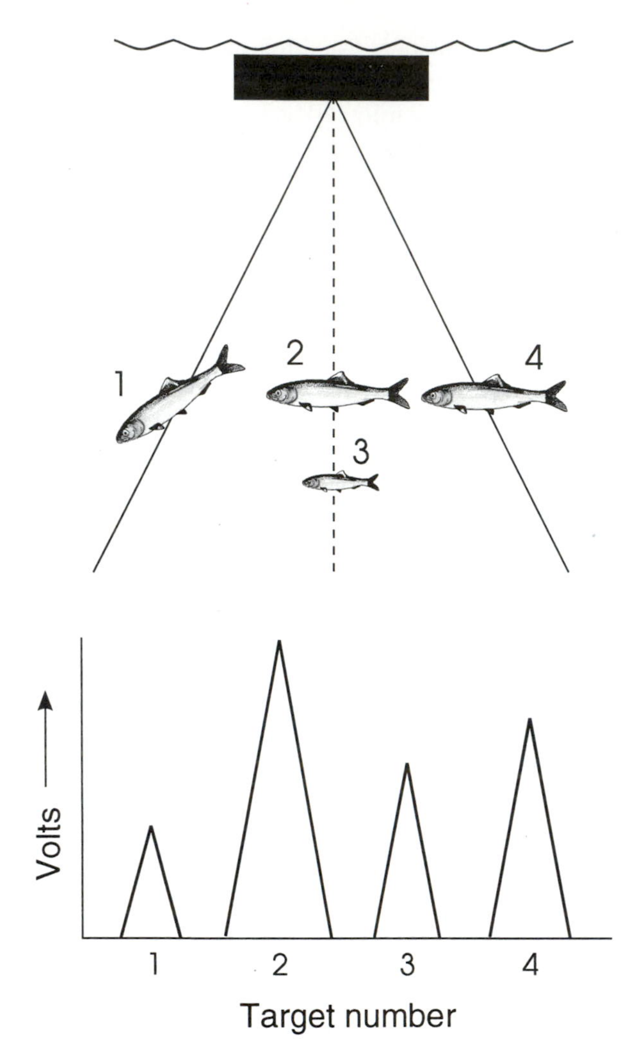

Figure 13.9 Relationship between the size (voltage amplitude) of a fish echo returned to the transducer and target size, target location in the acoustic beam, and target orientation. At the same beam angle and orientation, larger targets should have larger echoes than smaller targets (compare targets 2 and 3). The same target has a larger echo along the beam axis than off the axis (targets 2 and 4). Echo size should increase as the fish's length axis becomes more perpendicular to the line of sound propagation (compare targets 1 and 4). Differences in swim bladder morphology, however, affect target strength and may mask some orientational principles.

reflected energy, or sound intensity, is summed. Sound intensity is proportional to voltage squared; therefore, the signals must first be squared before they are summed. The total reflected energy is assumed to be directly proportional to total backscattering cross section of all targets in a volume of water (see equation 13.5). If all fish are the same size, numerical fish density is calculated by dividing the total echo-squared voltage received by the mean backscattering cross section of individual fish targets (Thorne 1971; Thorne et al. 1971; Clay and Medwin 1977; Foote 1983).

Fish target strength or mean backscattering cross section, is thus a critical parameter for estimating fish density (see section 13.5.2).

Signals must first be corrected for sound attenuation and depth-related changes in sampling volume before echo integration can be done. The TVG to correct for sound absorption ($2\alpha R$) is the same as that for echo counting (equation 13.3). Echo-squared integration sums across the entire acoustic sampling volume, which increases with range by a factor of R^2. This increase directly compensates for the decrease in sound intensity reaching a target at range R. Because these two effects are equal and opposite, they cancel. The total echo reflected from all the fish at a given range will spread spherically on its way back to the transducer. Echo intensity will again be reduced by a factor of R^2, but this time it is not compensated by volume integration. Thus, the TVG to adjust for signal spreading for echo integration (in decibels) is $20 \cdot \log_{10}(R)$.

13.5.2 Estimation of Fish Target Strength

The target strength of a fish is an acoustic measurement that is required for scaling echo-counting and echo-squared integration analyses and for estimating fish sizes. When the sound wave encounters a target in the water, sound is scattered in all directions in a complex manner. Only a portion of the sound will be reflected back in the direction from which the sound source originated. This amount is called the backscattering cross section. At a radial distance R from the target, the backscattering cross section (σ_{bs}) is defined by

$$\sigma_{\text{bs}} = R^2(I_r/I_i), \tag{13.10}$$

where I_r is the sound intensity scattered back from the target and I_i is the sound intensity incident to the target. The target strength is defined as the decibel equivalent of the sound reflected back towards the sound source at a distance of 1 m from the target, or $10 \cdot \log_{10}(\sigma_{\text{bs}}) = 10 \cdot \log_{10}(I_r/I_i)$. Fish target strengths generally range from -25 to -65 dB.

A tremendous amount of research has gone into measuring and modeling fish target strengths. The target strength of a fish for any given echo is complex, and the amount of sound scattered back toward the transducer depends on fish size, swim bladder volume and orientation, fish behavior, and acoustic frequency. The target strength of a fish can be considered analogous to fish length in that it is a given parameter for a given fish. But an individual fish actually has a distribution of target strength values when they are measured in the field. Small differences in the angle of the incident sound relative to the fish's swim bladder orientation and the shape of the fish's body or its orientation at the time the sound reaches it cause large variations in the measured target strength of an individual fish (e.g., Clay and Horne 1994). Normally, mean target strength calculated from a number of echoes is a good estimator of mean fish size. Once the acoustic target strength has been determined, it can be converted to fish size (see section 13.5.3).

Estimates of fish target strength made from acoustic data must be based on echoes originating from individual fish targets. Various criteria based on the shape and duration of an echo have been developed to distinguish an echo from an individual fish from an echo that is a composite of several closely spaced fish. It is often assumed that isolated fish are the same size as nearby fish that have more concentrated distributions. Once an individual echo is recognized, it must be corrected for any transmission losses in the signal strength ($40 \cdot \log_{10} R + 2\alpha R$). The position in the acoustic beam must also be determined so that corrections can be made for the beam directivity (see equation 13.3). The two common techniques used to determine the position of fish targets in the acoustic beam are the use of multibeam (dual-beam and split-beam) transducers and the use of statistical approaches and models.

Dual-beam and split-beam approaches. Multibeam transducers can be used to measure fish target strength and acoustic sampling volume directly. With multibeam transducers, one can use beam geometry to locate the angular position of a target in the beam and then remove the effects of the beam directivity pattern on echo strength. Two basic types of multibeam transducers have been developed. These are referred to as dual-beam and split-beam echo sounders (Traynor and Ehrenberg 1990). Both devices rely on some initial screening of the acoustic data to identify single targets. A measure of target strength is then obtained from each echo resulting from each acoustic pulse. However, given the wide variation in measures of target strength from the same individual fish, the mean target strength of several echoes provides the best estimate of target strength for a fish population. Various target-tracking, or echo-tracking, techniques have been developed to get a more precise measure of target strength. Using such an approach, echoes from an individual target are followed from ping to consecutive ping so that an ensemble of echoes can be obtained from each individual fish.

Dual-beam transducers were developed to determine the radial position of a target in the acoustic beam. A dual-beam echo sounder has overlapping wide and narrow-beam transducers (Figure 13.10). Sound is transmitted on the narrow-beam transducer. Echoes are received on both the narrow- and wide-beam transducers. The ratio of the amplitude of the echoes received by the two transducers is used to calculate the angular location of a target relative to the acoustic axis of the transducer. Dual-beam echo sounders have been applied to a variety of aquatic environments and types of organisms (e.g., Ehrenberg 1979; Ehrenberg and Traynor 1979; Traynor and Ehrenberg 1979; Traynor and Williamson 1983; Williamson and Traynor 1984; Dickie and Boudreau 1987; Ponton and Meng 1990; Wiebe et al. 1990; Brandt et al. 1991).

The split-beam system uses four independent transducer quadrants for sound reception (Foote et al. 1986; MacLennan and Svellingen 1989; Degnbol and Lewy 1990). A comparison of the differences in arrival times of an echo to each of the quadrants is used to determine both the angle and direction of the target relative to the transducer (Figure 13.11). Once the location of the echo in the acoustic beam has been determined, the acoustic sampling volume and target strength can be estimated. By pinpointing an individual target in three-dimensional space, echo tracking with a split-beam echo sounder may also be useful for tracking individual fish movements and estimating fish swimming speed and direction.

Statistical procedures. Statistical or modeling approaches can also be used to estimate the mean target strength of a population of fish. This approach recognizes that the expected size distribution of echoes received at the transducer is due to two effects: the various locations of the fish in the acoustic beam pattern and the variations in scattering from an individual target due to small changes in fish orientation and behavior (Peterson et al. 1976; Clay and Medwin 1977; Clay 1983). If fish are randomly distributed in the acoustic beam (have equal probability of being detected in any part of the acoustic beam), the expected size distribution of echoes

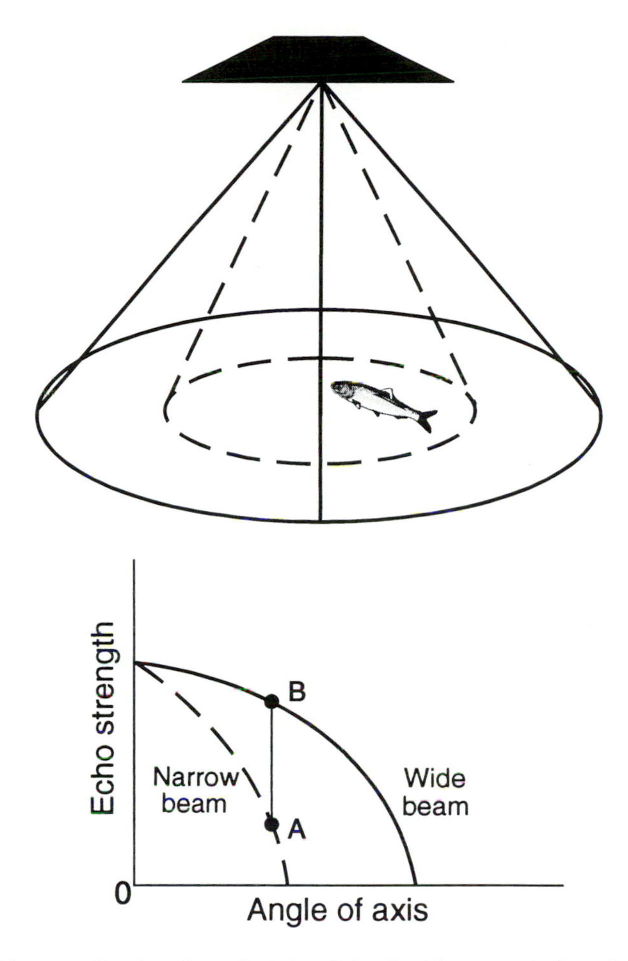

Figure 13.10 Diagram showing the principle of the dual-beam technique for locating fish in the beam pattern. Sound is transmitted on the narrow beam and received on both narrow- and wide-beam transducers. The ratio of the echo strengths (A:B) received by the narrow and wide beams uniquely determines the angular location of the target from the transducer.

due to beam directivity can be calculated directly. Targets located to the side of the beam axis will have a small echo. Large echoes will be produced only by targets on or near the beam axis. The combination of all possibilities produces an expected probability distribution of target amplitudes caused by differences in target locations in the acoustic beam.

The size of the echo at any location in the acoustic beam depends on the orientation of the target relative to the horizontal plane of the transducer and on the shape (e.g., flexed or straight body) of the target, which can vary as the fish swims. For a given size fish at a fixed location in the acoustic beam, the echoes reflected from repeated measurements of the fish vary. A given size fish will thus produce a distribution of target strengths, and therefore there is not a one-to-one correspondence between target strength distributions and fish size distributions. Theoretical predictions (Peterson et al. 1976; Huang and Clay 1980; Clay 1983; Palumbo et al. 1993) and subsequent experimental corroboration (Huang 1977; Penrose and Kaye

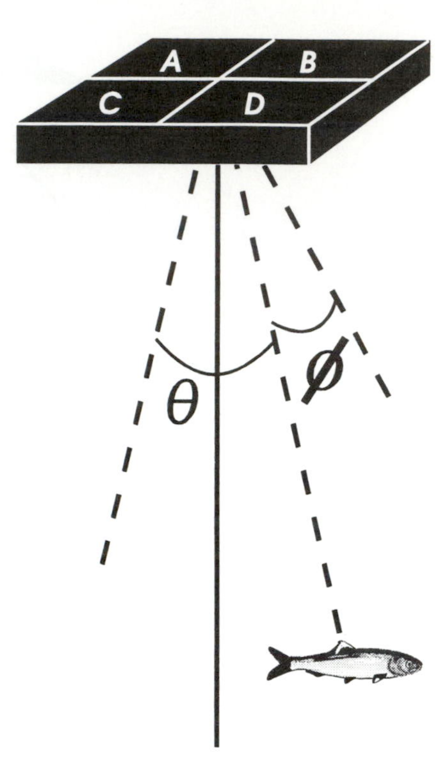

Figure 13.11 Diagram showing the principle of the split-beam technique for locating fish in the beam pattern. The echo from a target will reach each of the four segments (A, B, C, and D) of the transducer at slightly different times. Differential arrival times can be used to locate targets in three-dimensional space. Angles θ and ϕ represent two of the four beam quadrants.

1979; Clay and Horne 1994) have shown that the echoes from targets of equal size have a size frequency that can be characterized by Rayleigh (Rayleigh 1945; Peterson et al. 1976) or Ricean (Rice 1954; Clay and Heist 1984; Stanton and Clay 1986) probability distributions.

When both effects are combined, the resultant size distribution of echo amplitudes can be predicted. If a match can be found between this prediction and the echo amplitudes taken from observed survey data, then mean target strengths and fish density can be calculated directly because the total volume of the acoustic beam is known (Clay 1983). This technique requires a large number of echoes to obtain a good representation of the size and distribution of echo amplitudes. This statistical approach has been applied successfully in several studies (e.g., Craig and Forbes 1969; Peterson et al. 1976; Ehrenberg et al. 1981; Clay 1983; Ehrenberg 1983; Lindem 1983; Clay and Heist 1984; Stanton 1985a, 1985b; Stanton and Clay 1986; Rudstam et al. 1987, 1988; Bjerkeng et al. 1991; Walline et al. 1992; Parkinson et al. 1994).

13.5.3 Measurements of Fish Size and Biomass

Target strength is an acoustic measurement. To measure fish size or biomass density (g/m^3), it is necessary to define the relationship between the acoustic size of a target and the actual size (length or biomass) of the target. Much effort has been spent on defining the relationship between target strength and fish size and anatomy, swim bladder volume, fish behavior, and fish orientation to the incident sound wave (Haslett 1969; Love 1969, 1971a, 1971b, 1977; McCartney and Stubbs 1971; Nakken and Olsen 1977; Olsen 1990; Foote and Nakken 1978; Foote 1979, 1987; Penrose and

Kaye 1979; Fedotova and Shatoba 1983; Sun et al. 1985; Buerkle 1987; Rose and Leggett 1988; Arnaya et al. 1989; Ona 1990; Do and Surti 1990; Thiebaux et al. 1991; Clay and Horne 1994). The target strength of a fish depends on characteristics of the fish's body, primarily the size and orientation of the swim bladder (this is typically related to body length or mass) and, secondarily, on the body shape (Foote 1980a, 1980b; Midttun 1984; Saenger 1988; Blaxter and Batty 1990; Clay and Horne 1994). If fish length can be estimated from acoustic target strength, biomass densities of fish can then be computed from acoustic measures of fish numerical density and fish weight–length regressions. Standard weight–length relationships developed to date for various fish species are summarized in Chapter 15.

Several techniques have been used to define the relationship between fish length and target strength. One approach is to directly measure the target strength of a fish in the laboratory or with experimental devices designed for use at sea (see MacLennan and Simmonds 1992). These types of measurements have been made on live fish as well as dead or preserved specimens (e.g., Love 1971b, 1977; Nakken and Olsen 1977; Edwards and Armstrong 1984; Foote 1991). Target strengths are usually measured from the dorsal view of the fish and with the sound source originating directly above the fish. Target strength measurements taken from the dorsal view on a rigid fish body are generally higher than those measured on a free-swimming fish in the field. Measures of target strength taken from a sagittal (side) view are more appropriate if horizontal-facing transducers are used in acoustic surveys. Because most of the echo is caused by the swim bladder (Foote 1980b) and backscattering varies with body shape as the fish swims (Clay and Huang 1980), target strength measurements should be done on live, free-swimming (or caged or tethered) fishes with inflated swim bladders. Target strength measurements should be made over a range of fish sizes so that a species-specific relationship between fish size and acoustic target strength can be defined.

If it is not possible to do individual experiments on the species of fish being surveyed, one can use regression equations (see Box 13.2) that have been established for different types of fishes and are available in the literature (Love 1971b, 1977; Nakken and Olsen 1977; Foote 1987; Bjerkeng et al. 1991). For these equations, target strength (TS) is normally regressed against fish length (L, measured in centimeters or meters) as follows:

$$\text{TS} = a \cdot \log(L) + d; \tag{13.11}$$

a and d are empirically derived constants. Target strength and d are expressed in decibels. MacLennan and Simmonds (1992) provided a list of equations more comprehensive than those given in Box 13.2. Perhaps the best known equations are those derived by Love (1971b), which were based on measurements made from the dorsal view of several sizes and species of fish. A degree of caution should be taken if regressions from the literature are used. Experimental techniques differ among investigators, and target strengths differ with fish behavior, tilt angle, and physiological condition (Foote 1980a; MacLennan at al. 1990b; Ona 1990). Physiological condition may vary from habitat to habitat. It is preferable to directly measure the target strength of the species of fish in the aquatic environment being surveyed.

Fish sizes can also be estimated from models of the acoustic scattering properties of targets. Such models have evolved from consideration of fishes or swim bladders as ellipsoids and spheres (Anderson 1950; Haslett 1962; Furusawa 1988) or elon-

Box 13.2 Equations relating fish target strengths (TS, in decibels) to fish length (L)

Fish length is given in meters (L_m) or centimeters (L_{cm}), and acoustic wavelength (λ) is given in meters. Measurements for equations 13.3–13.7 were made at 38 kHz.

Equation	Target	TS of 25-cm fish	Reference
1. $TS = 19.1 \cdot \log_{10}(L_m) + 0.9 \cdot \log_{10}(\lambda) - 23.9$	Many species	−36.7	Love 1971a
2. $TS = 24.5 \cdot \log_{10}(L_m) - 4.5 \cdot \log_{10}(\lambda) - 26.4$	Many species	−34.8	McCartney and Stubbs 1971
3. $TS = 24.6 \cdot \log_{10}(L_{cm}) - 66.6$	Atlantic cod	−32.2	Nakken and Olsen 1977
4. $TS = 17.2 \cdot \log_{10}(L_{cm}) - 60.8$	Sprat	−36.8	Nakken and Olsen 1977
5. $TS = 13.6 \cdot \log_{10}(L_{cm}) - 56.8$	Atlantic herring	−37.8	Nakken and Olsen 1977
6. $TS = 20 \cdot \log_{10}(L_{cm}) - 71.9$	Clupeids	−43.9	Foote 1987
7. $TS = 20 \cdot \log_{10}(L_{cm}) - 67.4$	Gadoids	−39.4	Foote 1987
8. $TS = 20 \cdot \log_{10}(L_{cm}) - 65.3$	Capelin	−37.3	Rose and Leggett 1988

gated cylinders (e.g., Clay 1991, 1992) to more complex shapes, realistic shapes that have a fixed, gas-filled component (Foote and Traynor 1988; Do and Surti 1990; Clay and Horne 1994) and a weaker echo from a bent, fluid cylinder that can change form in a manner that simulates a fish's body as the fish swims (Clay and Heist 1984; Clay 1991). These models can be used to directly predict the size distributions of fish target strengths from different size fishes (e.g., Foote 1985; Furusawa 1988; Clay and Horne 1994).

Another approach that has been used to establish a relationship between target strength and fish size is to compare mean target strength to the mean size of fish caught with more traditional sampling gears such as gill nets (Rudstam et al. 1987) or trawls (Argyle 1992). This technique assumes that the fish sampling method is unbiased with respect to fish size (e.g., Wardle 1983). Several comparisons of acoustic data and biological samples can be used to establish a regression between mean target strength and mean length or weight of the fish caught in the net. To apply this approach, it is important to ensure that the acoustic techniques and nets sample the same water volume (i.e., that acoustic data are restricted to match data from the water sampled with other sampling gear).

As mentioned before, the target strength of an individual echo is a poor estimator of the size of that target (Clay and Huang 1980; Dawson and Karp 1990). Likewise, the size distribution of fish derived by converting each target strength measurement into fish size will not match the size distribution of fish in the environment. In general, the size distribution of fish estimated with acoustics will be much broader than that of the actual fish population. Acoustic measures of mean fish size are normally good approximations of mean fish sizes in a population and are sufficient for most population estimates. Groups of fish (e.g., juveniles and adults) must differ widely in size (e.g., 50–100 mm) if they are to be separated in survey data taken from single echo sounders.

13.5.4 Population Abundance Estimations

Once the acoustic survey is completed, the abundance of the total population must be estimated. This involves extrapolating measures of fish density to the entire aquatic environment. Acoustic surveys are generally done by running a series of transect lines across the surface of a body of water (see section 13.4.4). Measures of fish density are generally made across a set number of consecutive pings (Williamson 1982) or at regular fixed distances along a transect. Data are also summed across regular depth intervals (e.g., every 1, 5, or 10 m) or throughout the entire water column. These measures of fish density can be stratified by geographic location, water depth, bottom depth, or some other environmental factor such as water temperature, and densities of fishes in areas that were not sampled can be estimated by extrapolation. Fish densities are often estimated between sampled areas by means of various data interpolation routines such as kridging.

13.5.5 Sampling Variance

It is difficult to estimate the true variance of an acoustic population estimate (see Chapter 2) because several variables contribute to the calculation of fish density, each of which contributes to the sample variance. This is not different from other types of sampling techniques. For example, when trawls are used to sample fishes, variance can be caused by variations in catchability of target fish at different times of the day, visibility of the trawl to fish, efficiency of the trawl operation, boat speed, trawl sample volume, and fish distributions and behavior. Similarly, in acoustic sampling, variance may be caused by variations in hardware operation, noise levels, fish orientation and physiological state, air content of the swim bladder, and swimming behavior. These factors are assumed to be small relative to the errors in identifying species (see 13.5.7) and estimating fish target strength.

As in all population estimation techniques, sample variance arises from subsampling fishes that have a nonuniform distribution (Cochran 1977). Survey design also affects the way in which variance of population abundance is estimated (see section 13.4.4). MacLennan and Simmonds (1992) and Foote and Stefansson (1993) discussed the relationships between variance and survey design. References on acoustic sampling precision and error estimation include Moose and Ehrenberg (1971), Ehrenberg (1974), Bodholt (1977), Smith (1979), Williamson (1982), Aglen (1983), Robinson and Hood (1983), Efron and Tibshirani (1986), MacLennan and MacKenzie (1988), Godo (1990), Jolly and Hampton (1990b), Robotham and Castillo (1990), Simmonds (1990), Foote et al. (1991), Godo and Wepestad (1993), Hansson (1993), Smith and Gavaris (1993), and Johnson et al. (1994).

13.5.6 Bias and Noise in the Acoustic Data

Potential sources of error or bias in acoustic data are calibration errors, errors in the use of conversion factors (e.g., regressions of target strength to fish length; speed of sound), missed portions of the fish population (those near the surface, near the bottom, or in tight schools), shading effects, ship avoidance by fish, and noise. Shading occurs when high-density schools of fish reflect or absorb sufficient sound energy to reduce the amount of sound reaching the deeper parts of the fish school (Thoreson 1991; Appenzeller and Leggett 1992; Foote et al. 1992; Furusawa et al. 1992). Shading is minimized if sampling is done when the fish population is dispersed. Fish avoidance may be a problem if the research vessel is particularly noisy (Olsen et al. 1983; Gerlotto and Freon 1992). Sideward- or forward-facing transducers may be useful if fish are avoiding the research vessel. Fish detection and avoidance of the acoustic signal itself should not be a problem for most species because acoustic frequencies are well above the sound detection capability of fish (Popper and Fay 1973; Hawkins 1986). Clupeid fish have been observed to avoid strong acoustic pulses with a frequency near 100 kHz (Dunning et al. 1992; Nestler et al. 1992; Ross et al. 1993), but the effect is limited in range, and the mechanisms are not understood.

Another potential source of bias in acoustic signal processing is noise. If noise is counted or integrated as fish, the assessment will be biased. Noise received through the transducer is amplified through the time-varied-gain of the echo sounder and thus may appear to be greater in deeper water. Noise has physical, biological, and electronic origins (Wenz 1972). Physical noise is caused by turbulence, breaking waves, propeller wash or bubbles produced by the research vessel, wave action, and methane bubbles rising from bottom sediments (e.g., Dalen and Lovik 1981; Rudstam and Johnson 1992). Biological noise is the production of echoes by biological scatterers that are not being assessed. Electronic noise, generated by internal or external electronic interference, is generally evident as event noise and continuous noise. Event noise lasts for a short time and usually seems to be present throughout the water column; it may occur if the transducer comes out of the water in rough seas or if a sporadic surge occurs in the electrical supply, and it is recognized on an echogram as a sharp line extending throughout the water column. Continuous noise is evident as sinusoidal waves or dots on an echogram and is usually caused by interference from electronic sources such as other echo sounders used in ship navigation. These markings are readily identifiable on a chart record but cannot be electronically differentiated from signals generated by biological scatterers. Estimates of resident noise levels can be made by measuring signal levels when the echo sounder is not pinging. This measured noise level can be used to set a threshold level to remove the noise from echo-squared integration analyses.

13.5.7 Species Identification

One of the most severe limitations of acoustic fish stock assessment is the difficulty of directly identifying the species that produced the acoustic scattering. This is not a problem if acoustic surveys are done in an aquatic environment dominated by one species or if species can be separated by size alone. Problems with species identification can be minimized by conducting the survey in a part of the environment or at a time of day or year when the target species dominates a region or is spatially separated from other species. When abundance estimates of a single species of fish inhabiting a multispecies environment are needed, a measure of relative species composition is required.

One approach to allocating acoustic measures of total fish biomass among species is to estimate the proportion of each species in the environment with direct biological sampling during the survey. The differential weighting of the acoustic measurement should be done on a biomass or acoustic backscattering basis, not on a numerical basis. The precision and accuracy of a biomass estimate of a species' abundance is directly proportional to the precision and accuracy of the sampling procedure used to identify the fishes. For example, if trawling suggests that the fish composition in an area is 20% species A and 80% species B but is actually 10% species A and 90% species B, the abundance estimate of species A would be 100% too high and the abundance estimate of species B would be 11% too low. Variance about these estimates also depends on the variance about the estimate of relative species composition. Most net-sampling techniques are not equally effective at catching different species or sizes of fishes (see Chapters 6 and 7) and also have a high variance because the underlying fish distribution is unknown. Ambiguity in species composition can be minimized by aiming the sampling net at specific layers or concentration of fishes and by using additional species identification clues (e.g., target size) provided by the acoustic data (e.g., Holliday 1977).

Another way to estimate the relative abundance of species is to allocate acoustic biomass measurements to species based on the type of habitat from which the echoes originated. Species often have preferred habitats. Information on the basic ecology, life history, and habitat preferences of the target species may help in identifying acoustic echoes. Water temperature is a good example of a habitat feature that correlates well with fish distributions. Fish have preferred temperatures and often occupy narrow thermal ranges (Fry 1947; Ferguson 1958; Coutant 1977; Magnuson et al. 1979). Sympatric species may live in habitats with different water temperatures (e.g., Brandt et al. 1980). If this is the case, acoustic measurements may be allocated to different species on the basis of water temperature. Biological sampling based on habitat criteria such as preferred temperatures may reduce the amount of sampling needed to measure relative species abundances. Other aspects of fish behavior, such as diel differences in vertical distribution or schooling (e.g., Rudstam and Johnson 1992) or measurements of fish size or swimming speed, may also aid in species identification.

A third method to help identify acoustic targets is to use the information contained in each echo. Classification of echo data from individual targets or fish schools has recently proven successful in identifying fish species (e.g., Deuser et al. 1979; Rose and Leggett 1988, 1990; Richards et al. 1991; Goyke 1995). Echoes can be categorized on the basis of attributes of the acoustic signal (echo shape and size), attributes of an ensemble of signals (patchiness, degree of aggregation, and schooling patterns), attributes of the environment from which the signal originated (e.g., water temperature), or a combination of the above. Multivariate statistical approaches (e.g., principal component analyses or neural networks) and discriminate analyses can then be used to group targets with similar attributes to help identify targets or separate zooplankton from fishes (Stanton and Clay 1986; Rose and Leggett 1988; Kjaergaard et al. 1990; Cochrane et al. 1991; Richards et al. 1991; Ramani and Patrick 1992).

13.6 ADDITIONAL APPLICATIONS OF ACOUSTICS

13.6.1 Fish Ecology

One of the most recent advances in the field of acoustics is the recognition that acoustic data contain rich spatial and temporal information on fish distributions and abundances. Fish and zooplankton distributions are characteristically patchy in time and space because of animal behavior (e.g., migrations, schooling, dispersal, and predator–prey interactions) or physical complexity (e.g., thermal structure, currents, and advection) in the environment. During the past decade, acoustic techniques have been used extensively to study fish ecology, behavior, distribution, predator–prey interactions, and responses to environmental factors. The ability of acoustics to measure fish distributions continuously across space and time makes acoustics an ideal tool for examining fish distributions at a wide range of spatial and temporal scales.

The ways in which acoustics have been applied to fish ecology are diverse. Studies have been done on vertical migrations and diel differences in fish distributions (Isaacs et al. 1974; Janssen and Brandt 1980; Nash et al. 1989; Weston and Andrews 1990b; Levy 1991; Luecke and Wurtsbaugh 1993); school dispersal, size, and geometry (Smith 1970, 1977; Spigarelli et al. 1973; Holliday and Larson 1979; Misund et al. 1992, 1995; Reid and Simmonds 1993; Scalabrin and Masse 1993); fish distribution in space and time (Heist and Swenson 1983; Boudreau 1992; Stables and Thomas 1992); fish distribution in response to specific environmental factors such as water temperature (e.g., Spigarelli et al. 1973; Kaye 1979b; Nash et al. 1989; Levy et al. 1991), light (Blaxter and Currie 1967; Sameoto et al. 1985), internal propagating waves (Chindonova and Shulepov 1965; Andreyeva and Makshtas 1977; Smith 1977; Kaye 1979a) and tides (Thorne et al. 1990); food web interactions among trophic levels (Janssen and Brandt 1980; Sprong et al. 1990; Levy 1991); the spatial scales of predator–prey interactions (Schneider 1989; Rose and Leggett 1990; Brandt and Mason 1994; Horne and Schneider 1994); spawning behavior (Eckmann 1991); the size structure of pelagic food webs (Robinson 1983; Sprules et al. 1991; Stanton and Holliday 1991); fish migration in rivers (Thorne and Johnson 1993); and the distribution and behavior of fishes in the vicinity of reefs or artificial structures, such as power plant intakes (Johnston and Hopelain 1990; Johnson et al. 1992, 1994; Skalski et al. 1993). Acoustic techniques are also useful for examining spatial patchiness of other aquatic organisms (Greenlaw and Pearcy 1985; Nash et al. 1987; Nero and Magnuson 1989a, 1989b; Nero et al. 1990; Watkins et al. 1990; Greene et al. 1994).

Recent applications have integrated acoustic data with ecological and bioenergetic models to map how well fish should grow in different parts of the aquatic environment and to predict the production potential of aquatic environments (Brandt et al. 1991, 1992; Brandt and Kirsch 1993; Luo and Brandt 1993; Rudstam et al. 1993). Acoustic data have also been linked with virtual population analyses to estimate fish mortality (Sparholt 1990). Finally, acoustic techniques might have potential for mapping certain components of a fish's habitat, such as the bottom type and the distribution of aquatic plants (e.g., Thomas et al. 1990).

13.6.2 Fixed-Location Transducer Deployment

Fixed-location transducer deployments (Figure 13.7) are often used to measure passage or migration of fishes in rivers or to monitor changes in fish abundance at a particular location across time. Transducers can face horizontally across rivers, be aimed at structures such as hydroelectric intakes, or be attached to moored or free-drifting buoys at which data can be recorded or transmitted to a shore-based receiving station (e.g., Wiebe et al. 1995). Both echo-counting (e.g., Ransom and Steig 1994) and echo-integration signal processing have been used to analyze data collected with fixed-location transducers. Examples of the types of applications of stationary acoustics include monitoring fish migrations in rivers (Gaudet 1990; Johnston and Hopelain 1990; Mesiar et al. 1990; Thorne and Johnson 1993), monitoring fish distributions at artificial reefs, assessing the effectiveness of bypass routes, diversion screens, and spillways for fish passage at hydroelectric facilities

(Johnson et al. 1992, 1994; Skalski et al. 1993; Ransom and Steig 1994), and studying changes in fish distributions over time (Thorne et al. 1990; Stables and Thomas 1992).

Fixed-location transducers are commonly used in rivers. Rivers are relatively complex environments for applying acoustic techniques and require careful placement of transducers and great care in calibration. Transducers are normally deployed facing sideways in a manner to minimize reflections from the surface and bottom of the water and to determine fish direction and movement rates. Riverbeds can be modified to funnel fish past the acoustic beam. High-resolution echo sounders (narrow beam angles and short pulse length) are commonly used in rivers. Echo or target tracking of individual fish is often used during fixed-transducer applications to determine the direction of fish movement (Thorne and Johnson 1993; Johnson et al. 1994). The direction of the fish migration can be determined by directing the beam obliquely down- or upstream and then evaluating the changes in range of a target. Another way to measure the direction of fish movement is to place a series of transducers along the river and watch the order in which the targets are detected by the respective transducers. Split-beam transducers (whereby the location of the target can be tracked in three-dimensional space) have recently proven to be very effective in providing directional information and distinguishing fish from drifting debris.

13.6.3 Invertebrate Assessment

Acoustic techniques are commonly used for assessing invertebrates (Smith et al. 1992). One of the most common applications has been the use of acoustics to monitor krill abundances in the Antarctic Ocean (Falk-Petersen and Kristensen 1985; Dalen and Kristensen 1990; Everson et al. 1990; Greene et al. 1991a; Hewitt and Demer 1991; McClatchie et al. 1994). Acoustics have also been used to study squid (Jefferts et al. 1987), amphipods (Melnik et al. 1993), and small zooplankton (McNaught 1969; Beamish 1971; Greenlaw 1977, 1979; Pieper 1979; Holliday and Pieper 1980; Sameoto 1982; Kristensen and Dalen 1986; Holliday et al. 1989; Greene and Wiebe 1990; Haney et al. 1990; Stanton 1990; Melnik et al. 1993). The data analyses used for invertebrates are largely the same as those used for fish, as long as individual organisms can be resolved (Greene et al. 1989; Smith et al. 1992).

Zooplankton studies require the use of higher-frequency sound (200–1,000 kHz; e.g., Holliday and Pieper 1980; Richter 1985; Chu et al. 1992; Madureira et al. 1993) than that typically used for fish (38–420 kHz). Higher-frequency echo sounders have a more restricted depth range. The transducer can be moved closer to the zooplankton to improve the ability of the echo sounder to resolve individual zooplankton. This has been done successfully by mounting the transducer on deep-towed vehicles or submersibles (e.g., Pieper and Holliday 1984; Holliday et al. 1989; Greene et al. 1991b).

Zooplankton (and fish) sizes can also be determined with multifrequency acoustics (McNaught 1969; Holliday and Pieper 1980; Greenlaw and Johnson 1983; Pieper and Holliday 1984; Costello et al. 1989; Holliday et al. 1989; Pieper et al. 1990; Cochrane et al. 1991). The size of an individual target that can be detected depends on the acoustic frequency. Higher frequencies can better detect smaller zooplankton (Chu et al. 1992). When several frequencies are used at the same time, abundances can be estimated for zooplankton of various sizes. As many as 21 frequencies have been used to study zooplankton (Pieper et al. 1990). The absence of swim bladders in zooplankton makes the sound-scattering characteristics of zooplankton quite different from those of fishes (Greenlaw and Johnson 1982; Foote et al. 1990; Stanton 1990; Stanton et al. 1993a, 1994a, 1994b). Different models have been developed to relate acoustic target strength to zooplankton size. Scattering models for zooplankton have evolved (e.g., from spheres to cylinders to bent cylinders) and have improved the theoretical predictions of zooplankton target strengths (Stanton 1988, 1989; Chu et al. 1992, 1993; Stanton et al. 1993a, 1993b). Sizes and abundances of zooplankton estimated with multifrequency echo sounders generally agree well with those estimated from pump samples (Costello et al. 1989; Pieper et al. 1990).

13.7 DEVELOPING TECHNOLOGIES

Developments in acoustic data analyses and sensor technology are continuing. In the field of acoustic data analyses, one of the most promising developments is the use of geostatistical approaches to map fish distributions, interpolate fish abundances between transect lines, and calculate the variances of fish population estimates (Guillard et al. 1990; Hansson 1993; Petitgas 1993a, 1993b). These methods recognize the continuous spatial structure of acoustic data and the inherent differences between acoustic data and random point samples (e.g., Cressie 1993). The linking of acoustic data with geographic information system software and ecological modeling has also helped to make more use of the ecological and spatial information in acoustic data (e.g., Greene and Wiebe 1990; Brandt and Kirsch 1993; Goyke and Brandt 1993; Luo and Brandt 1993). These methods also facilitate graphic visualization of acoustic data (Greene and Wiebe 1990; Brandt et al. 1992; Greene et al. 1994).

Transducer deployment strategies also are evolving. Transducers have been towed in deep water to assess fish abundances near the bottom (Coombs and Cordue 1995) and right at the surface to assess fish abundances in very shallow water (Unger and Brandt 1989). Transducers have been mounted on submersibles to allow the transducers to get closer to the targets (Greene et al. 1991b) and to anchored or free-drifting buoys so that long-term data sets can be collected (e.g., Wiebe et al. 1995). Side-scan sonar is the use of a towed transducer that faces horizontally (Hewitt et al. 1976; Kubecka et al. 1994). This technique has potential for detecting fish in shallow water, near the surface or bottom of the water column, or to the front and side of the survey vessel to assess fish avoidance (Misund 1993).

Other types of acoustic sensor technologies are being applied to fishes and invertebrates. The use of multifrequency echo sounders that emit sound at two to three fixed frequencies at the same time is becoming routine. Because target strengths are a function of sound frequency, multifrequency systems are used to simultaneously assess the abundances of different-sized targets or to separate fishes from zooplankton (Holliday et al. 1989; Nero et al. 1990; Pieper et al. 1990). Each frequency also provides a unique set of echoes that may be used for direct species identification. Wideband, broadband, or chirp acoustic systems (Simmonds and Copeland 1986; Kjaergaard et al. 1990; Simmonds and Armstrong 1990; Zakharia 1990) use information in closely contiguous frequencies. Very-low-frequency (e.g., 2 kHz) sound has been used to detect fish at long ranges (>50 km) from the transducer (Revie et al. 1990). Low-frequency sound has also been used to assess fish sizes. The swim bladder of a fish resonates at a particular frequency, and its resonance frequency is correlated with the size of the swim bladder (Holliday 1972; Hawkins 1977; Love 1978, 1993).

Sector-scanning or multibeam sonars emit a wide-beam acoustic signal and receive echoes at several elements within the transducer that are quickly scanned. This gear samples a large volume of water but can discriminate targets at a high angular resolution (see Arnold et al. 1990). Three-dimensional, multibeam echo sounders are being developed (Jaffe et al. 1995; McGehee and Jaffe 1996) to define patch structure of fish and zooplankton distributions and to help evaluate predator–prey interactions. Acoustic Doppler current profilers (ACDPs) were originally designed to measure ocean currents, but they have been modified to measure zooplankton abundances (Flagg and Smith 1989; Plueddemann and Pinkel 1989; Heywood et al. 1991; Smith et al. 1992; Roe and Griffiths 1993; Ashjian et al. 1994; Cochrane et al. 1994). The ACDP is based on the principal (Doppler shift) that sound reflected from a moving object shifts frequency slightly in proportion to the speed at which the object is moving. The ACDP is a high-frequency acoustic system that measures the Doppler shift in sound frequency caused by the advection of small planktonic particles to determine current speed and direction.

13.8 TRAINING

Specialized training is required to use acoustic techniques (Thorne 1992). The most effective applications of acoustic techniques have been done by individuals trained in both fisheries and underwater acoustics or by teams composed of experts from both disciplines. Short courses can provide a potential user with the tools to operate equipment, understand the basic concepts, and make decisions on the appropriateness of this technology for a particular objective. However, in-depth training and experience will be required for data survey design, analyses, and interpretation.

13.9 REFERENCES

Aglen, A. 1983. Random errors of acoustic fish abundance estimates in relation to the survey grid density applied. FAO (Food and Agriculture Organization of the United Nations) Fisheries Report 300:293–298.

Anderson, V. C. 1950. Sound scattering from a fluid sphere. Journal of the Acoustical Society of America 22:426–431.

Andreyeva, I. B., and Y. P. Makshtas. 1977. Internal waves and sound scattering layers in the thermocline. Oceanology 17:287–289.

Appenzeller, A. R., and W. C. Leggett. 1992. Bias in hydroacoustic estimates of fish abundance due to acoustic shadowing: evidence from day–night surveys of vertically migrating fish. Canadian Journal of Fisheries and Aquatic Sciences 49:2179–2189.

Argyle, R. L. 1992. Acoustics as a tool for the assessment of Great Lakes forage fisheries. Fisheries Research 14:179–196.

Arnaya, I. N., N. Sano, and K. Iida. 1989. Studies on acoustic target strengths of squid II: effect of behavior on averaged dorsal aspect target strength. Bulletin of the Faculty of Fisheries Hokkaido University 40:83–99.

Arnold, G. P., M. G. Walker, and B. H. Holford. 1990. Fish behavior achievements and potential of high-resolution sector-scanning sonar. Rapports et Procés-Verbaux des Réunions Conseil International pour l'Exploration de la Mer 189:112–122.

Ashjian, C. J., S. L. Smith, C. N. Flagg, A. J. Mariano, W. J. Behrens, and P. V. Z. Lane. 1994. The influence of a Gulf Stream meander on the distribution of zooplankton biomass in the Slop Water, the Gulf Stream, and the Sargasso Sea, described using a shipboard acoustic Doppler current profiler. Deep-Sea Research 41:23–50.

Bailey, R. S., and E. J. Simmonds. 1990. The use of acoustic surveys in the assessment of the North Sea herring stock and a comparison with other methods. Rapports et Procés-Verbaux des Réunions Conseil International pour l'Exploration de la Mer 189:9–17.

Balls, R. 1948. Herring fishing with the echometer. Journal du Conseil International pour l'Exploration de la Mer 15:193–206.

Beamish, P. 1971. Quantitative measurements of acoustic scattering from zooplankton organisms. Deep-Sea Research 18:811–822.

Bjerkeng, B., R. Borgstrom, A. Braband, and B. Faafeng. 1991. Fish size and distribution and total fish biomass estimated by hydroacoustical methods: a statistical approach. Fisheries Research 11:41–73.

Blaxter, J. H. S., and R. S. Batty. 1990. Swimbladders 'behavior' and target strength. Rapports et Procés-Verbaux des Réunions Conseil International pour l'Exploration de la Mer 189:233–244.

Blaxter, J. H. S., and R. I. Currie. 1967. The effect of artificial lights on acoustic scattering layers in the ocean. Symposia of the Zoological Society of London 19:1–14.

Bodholt, H. 1977. Variance error in echo integrator output. Rapports et Procés-Verbaux des Réunions Conseil International pour l'Exploration de la Mer 170:196–204.

Boudreau, P. R. 1992. Acoustic observations of patterns and aggregation in haddock (*Melanogrammus aeglefinus*) and their significance to production and catch. Canadian Journal of Fisheries and Aquatic Sciences 49:23–31.

Brandt, S. B., and J. Kirsch. 1993. Spatially-explicit models of striped bass growth potential in Chesapeake Bay. Transactions of the American Fisheries Society 122:845–869.

Brandt, S. B., J. J. Magnuson, and L. B. Crowder. 1980. Thermal habitat partitioning by fishes in Lake Michigan. Canadian Journal of Fisheries and Aquatic Sciences 37:1557–1564.

Brandt S. B., and D. M. Mason. 1994. Landscape approaches for assessing spatial patterns in fish foraging and growth. Pages 211–238 *in* K. Fresh, editor. Theory and application in fish feeding. University of South Carolina Press, Columbia.

Brandt, S. B., D. M. Mason, and E. V. Patrick. 1992. Spatially-explicit models of fish growth rate. Fisheries 17(2):23–25.

Brandt, S. B., and six coauthors. 1991. Acoustics measures of the abundance and size of pelagic planktivores in Lake Michigan. Canadian Journal of Fisheries and Aquatic Sciences 48:894–908.

Buerkle, U. 1987. Estimation of fish length from acoustic target strengths. Canadian Journal of Fisheries and Aquatic Sciences 44:1782–1785.

Burczynski, J. J., and R. L. Johnson. 1986. Application of dual-beam acoustic survey techniques to limnetic populations of juvenile sockeye salmon (*Oncorhynchus nerka*). Canadian Journal of Fisheries and Aquatic Sciences 43:1776–1788.

Burczynski, J. J., P. H. Michaletz, and G. M. Marrone. 1987. Hydroacoustic assessment of the abundance and distribution of rainbow smelt in Lake Oahe. North American Journal of Fisheries Management 7:106–116.

Chapman, D. W. 1976. Acoustic estimates of pelagic ichthyomass in Lake Tanganyka with an inexpensive echosounder. Transactions of the American Fisheries Society 105:581–587.

Chindonova, Y. G., and V. A. Shulepov. 1965. Sound-scattering layers as indicators of internal waves in the ocean. Oceanology 5:78–81.

Chu, D., K. G. Foote, and T. K. Stanton. 1993. Further analysis of target strength measurements of Antarctic krill at 38 and 120 kHz: comparison with deformed cylinder model and inference of orientation distribution. Journal of the Acoustical Society of America 93:2985–2988.

Chu, D., T. K. Stanton, and P. H. Wiebe. 1992. Frequency dependence of sound backscattering from live individual zooplankton. International Council for the Exploration of the Sea Journal of Marine Science 49:97–106.

Clay, C. S. 1983. Deconvolution of the fish scattering PDF from the echo PDF for a single transducer sonar. Journal of the Acoustical Society of America 73:1989–1994.

Clay, C. S. 1991. Low-resolution acoustic scattering models: fluid-filled cylinders and fish with swim bladders. Journal of the Acoustical Society of America 89:2168–2179.

Clay, C. S. 1992. Composite ray-mode approximations for backscattered sound from gas-filled cylinders and swimbladders. Journal of the Acoustical Society of America 92:2173–2180.

Clay, C. S., and B. G. Heist. 1984. Acoustic scattering by fish-acoustic models and a two-parameter fit. Journal of the Acoustical Society of America 75:1077–1083.

Clay, C. S., and J. K. Horne. 1994. Acoustic models of fish: the Atlantic cod (*Gadus morhua*). Journal of the Acoustical Society of America 96:1661–1668.

Clay, C. S., and K. Huang. 1980. Backscattering cross-sections of live fish: PDF and aspect. Journal of the Acoustical Society of America 67:795–802.

Clay, C. S., and H. Medwin. 1977. Acoustical oceanography: principles and applications. Wiley, New York.

Coates, R. F. W. 1989. Underwater acoustic systems. Wiley, New York.

Cochran, W. G. 1977. Sampling techniques, 3rd edition. Wiley, New York.

Cochrane, N. A., D. Sameoto, and D. J. Belliveau. 1994. Temporal variability of euphausiid concentrations in a Nova Scotia shelf basin using a bottom-mounted acoustic Doppler current profiler. Marine Ecology Progress Series 107:55–66.

Cochrane, N. A., D. Sameoto, A. H. Herman, and J. Neilson. 1991. Multiple-frequency acoustic backscattering and zooplankton aggregations in the inner Scotian Shelf basins. Canadian Journal of Fisheries and Aquatic Sciences 48:340–355.

Coombs, R. F., and P. L. Cordue. 1995. Evolution of a stock assessment tool: acoustic surveys of spawning hoki (*Macruronus novaezelandiae*) off the west coast of South Island, New Zealand 1985–91. New Zealand Journal of Marine and Freshwater Research 29(2):175–194.

Costello, J. H., R. E. Pieper, and D. V. Holliday. 1989. Comparison of acoustic and pump sampling techniques for the analysis of zooplankton distributions. Journal of Plankton Research 11:703–709.

Coutant, C. C. 1977. Compilation of temperature preference data. Journal of the Fisheries Research Board of Canada 34:739–745.

Craig, R. E., and S. T. Forbes. 1969. A sonar for fish counting. Fiskeridirektoratets Skrifter Serie Havundersokelser 15:210–219.

Cressie, N. A. C. 1993. Series in probability and mathematical statistics: statistics for spatial data, 2nd edition. Wiley, New York.

Crittenden, R. N., G. L. Thomas, D. A. Marino, and R. E. Thorne. 1988. A weighted duration-in-beam estimator for the volume sampled by a quantitative echo sounder. Canadian Journal of Fisheries and Aquatic Sciences 45:1249–1256.

Cushing, D. H. 1968. Direct estimation of a fish population acoustically. Journal of the Fisheries Research Board of Canada 25:2349–2364.

Dalen, J., and K. E. Kristensen. 1990. Comparative studies of theoretical and empirical target-strength models of euphausiids (krill) in relation to field-experiment data. Rapports et Procés-Verbaux des Réunions Conseil International pour l'Exploration de la Mer 189:336–344.

Dalen, J., and A. Lovik. 1981. The influences of wind-induced bubbles on echo integration surveys. Journal of the Acoustical Society of America 69:1653–1659.

Dawson, J. J., and W. A. Karp. 1990. In situ measures of target-strength variability of individual fish. Rapports et Proces-Verbaux des Réunions Conseil International pour l'Exploration de la Mer 189:264–273.

Degnbol, P., and P. Lewy. 1990. Interpretation of target-strength information from split beam data. Rapports et Procés-Verbaux des Réunions Conseil International pour l'Exploration de la Mer 189:274–282.

Del Grosso, V. A., and C. W. Mader. 1972. Speed of sound in pure water. Journal of the Acoustical Society of America 52:1442–1446.

Deuser, L. M., D. Middleton, T. D. Plemons, and J. K. Vaughan. 1979. On the classification of underwater acoustic signals II. Experimental applications involving fish. Journal of the Acoustical Society of America 65:44–45.

Dickie, L. M., and P. R. Boudreau. 1987. Comparison of acoustic reflections from spherical objects and fish using a dual-beam echosounder. Canadian Journal of Fisheries and Aquatic Sciences 44:1915–1921.

Do, M. A., and A. M. Surti. 1990. Estimation of dorsal aspect target strength of deep-water fish using a simple model of swimbladder backscattering. Journal of the Acoustical Society of America 87:1588–1596.

Dragesund, O., and S. Olsen. 1965. On the possibility of estimating year-class strength by measuring echo-abundance of O-group fish. Fiskeridirektoratets Skrifter Serie Havundersokelser 13:47–75.

Dunning, D. J., Q. E. Ross, P. Geoghegan, J. J. Reichle, J. K. Menezes, and J. K. Watson. 1992. Alewives avoid high-frequency sound. North American Journal of Fisheries Management 12:407–416.

Eckmann, R. 1991. A hydroacoustic study of the pelagic spawning behavior of whitefish (*Coregonus lavaretus*) in Lake Constance. Canadian Journal of Fisheries and Aquatic Sciences 48:995–1002.

Edwards, J. I., and F. Armstrong. 1984. Target strength experiments on caged fish. Scottish Fisheries Bulletin 48:12–20.

Efron, B., and R. Tibshirani. 1986. Bootstrap methods for standard errors, confidence intervals and other measures of statistical accuracy. Statistical Science 2:54–77.

Ehrenberg, J. E. 1974. Recursive algorithm for estimating the spatial density of acoustic point scatterers. Journal of the Acoustical Society of America 56:542–547.

Ehrenberg, J. E. 1979. A comparative analysis of in-situ methods for directly measuring the acoustic target strength of individual fish. IEEE (Institute of Electrical and Electronics Engineers) Journal of Oceanics Engineering 4:141–152.

Ehrenberg, J. E. 1983. A review of in-situ target strength estimation techniques. FAO (Food and Agriculture Organization of the United Nations) Fisheries Report 300:85–90.

Ehrenberg, J. E., T. J. Carlson, J. J. Traynor, and N. J. Williamson. 1981. Indirect measurement of the mean acoustic backscattering cross-section of fish. Journal of the Acoustical Society of America 69:955–962.

Ehrenberg, J. E., and J. J. Traynor. 1979. Evaluation of the dual-beam acoustic fish target strength measurement method. Journal of the Fisheries Research Board of Canada 36:1065–1071.

Everson, I., J. L. Watkins, D. G. Bone, and K. G. Foote. 1990. Implications of a new acoustic target strength for abundance estimates of Antarctic krill. Nature 345:338–340.

Falk-Petersen, S., and A. Kristensen. 1985. Acoustic assessment of krill stocks in Ullsfjorden (North Norway). Sarsia 70:83–90.

Fedotova, T. A., and O. E. Shatoba. 1983. Acoustic backscattering cross-section of cod averaged by sizes and inclination of fish. FAO (Food and Agriculture Organization of the United Nations) Fisheries Report 300:63–68.

Ferguson, R. G. 1958. The preferred temperatures of fish and their midsummer distribution in temperate lakes and streams. Journal of the Fisheries Research Board of Canada 15:607–624.

Fisher, F. H., and V. P. Simmons. 1977. Sound absorption in sea water. Journal of the Acoustical Society of America 62:558–564.

Flagg, C. N., and S. L. Smith. 1989. On the use of the acoustic Doppler current profile to measure zooplankton abundance. Deep-Sea Research 36:455–474.

Foote, K. G. 1979. On representing the length dependence of acoustic target strengths of fish. Journal of the Fisheries Research Board of Canada 36:1490–1496.

Foote, K. G. 1980a. Effect of fish behavior on echo energy: the need for measurements of orientation distributions. Journal du Conseil International pour l'Exploration de la Mer 39:193–201.

Foote, K. G. 1980b. Importance of the swimbladder in acoustic scattering by fish: a comparison of gadoid and mackerel target strengths. Journal of the Acoustical Society of America 67:2084–2089.

Foote, K. G. 1983. Linearity of fisheries acoustics, with addition theorems. Journal of the Acoustical Society of America 73:1932–1940.

Foote, K. G. 1985. Rather-high-frequency sound scattering by swimbladdered fish. Journal of the Acoustical Society of America 78:688–700.

Foote, K. G. 1987. Fish target strengths for use in echo integrator surveys. Journal of the Acoustical Society of America 82:981–987.

Foote, K. G. 1990. Spheres for calibrating an eleven-frequency acoustic measurement system. Journal du Conseil International pour l'Exploration de la Mer 46:284–286.

Foote, K. G. 1991. Summary of methods for determining fish target strength at ultrasonic frequencies. Journal du Conseil International pour l'Exploration de la Mer 48:211–217.

Foote, K. G., A. Aglen, and O. Nakken. 1986. Measurement of fish target strength with a split-beam echo sounder. Journal of the Acoustical Society of America 80:612–622.

Foote, K. G., I. Everson, J. L. Watkins, and D. G. Bone. 1990. Target strengths of Antarctic krill (*Euphausia superba*) at 38 and 120 kHz. Journal of the Acoustical Society of America 87:16–24.

Foote, K. G., H. P. Knudsen, R. J. Kornliussen, P. E. Nordbo, and K. Roang. 1991. Postprocessing system for echo sounder data. Journal of the Acoustical Society of America 90:37–47.

Foote, K. G., and D. N. MacLennan. 1984. Comparison of copper and tungsten carbide spheres. Journal of the Acoustical Society of America 75:612–616.

Foote, K. G., and O. Nakken. 1978. Dorsal-aspect target-strength functions of six fishes at two ultrasonic frequencies. Fiskeridirektoratets Skrifter Serie Havundersokelser 1978(3):1–95.

Foote, K. G., E. Ona, and R. Thoreson. 1992. Determining the extinction cross section of aggregating fish. Journal of the Acoustical Society of America 91:1983–1989.

Foote, K. G., and G. Stefansson. 1993. Definition of the problem of estimating fish abundance over an area from acoustic line-transect measurement of density. International Council for the Exploration of the Sea Journal of Marine Science 50:369–381.

Foote, K. G., and J. J. Traynor. 1988. Comparison of walleye pollock target strength estimates determined from *in situ* measurements and calculations based on swimbladder form. Journal of the Acoustical Society of America 83:9–17.

Forbes, S. T., and O. Nakken, editors. 1972. Manual of methods for fisheries resource survey and appraisal, part 2: the use of acoustic instruments for fish detection and abundance estimation. FAO (Food and Agriculture Organization of the United Nations) Management of Fisheries and Science 5.

Francis, R. I. C. C. 1985. Two acoustic surveys of pelagic fish in Hawke Bay, New Zealand, 1980. New Zealand Journal of Marine and Freshwater Research 19:375–389.

Fry, F. E. J. 1947. Effects of the environment on animal activity. University of Toronto Studies Biology Series 55.

Furusawa, M. 1988. Prolate spheroidal models for predicting general trends of fish target-strength. Journal of the Acoustical Society (Japan) (E) 9:13–24.

Furusawa, M., K. Ishii, and Y. Miyanohana. 1992. Attenuation of sound by schooling fish. Journal of the Acoustical Society of America 92:987–994.

Gaudet, D. M. 1990. Enumeration of migrating salmon populations using fixed-location sonar counters. Rapports et Proces-Verbaux des Réunions Conseil International pour l'Exploration de la Mer 189:197–209.

Gerlotto, F., and P. Freon. 1992. Some elements on vertical avoidance of fish schools to a vessel during acoustic surveys. Fisheries Research 14:251–259.

Godo, O. R. 1990. Factors affecting accuracy and precision in abundance estimates of gadoids from scientific surveys. Doctoral dissertation. University of Bergen, Bergen, Norway.

Godo, O. R., and V. G. Wepestad. 1993. Monitoring changes in abundance of gadoids with varying availability to trawl and acoustic surveys. International Council for the Exploration of the Sea Journal of Marine Science 50:39–51.

Goyke, A. P. 1995. Acoustic estimates of fish abundance and distribution in Lake Ontario. Doctoral dissertation. University of Maryland, College Park.

Goyke, A., and S. B. Brandt. 1993. Spatial models of salmonid growth rates in Lake Ontario. Transactions of the American Fisheries Society 122:870–883.

Greene, C. H., T. K. Stanton, P. H. Wiebe, and S. McClay. 1991a. Acoustic estimates of Antarctic krill. Nature 349:110.

Greene, C. H., and P. H. Wiebe. 1990. Bioacoustical oceanography: new tools for zooplankton and micronekton research in the 1990s. Oceanography 3:12–17.

Greene, C. H., P. H. Wiebe, and J. Burczynski. 1989. Analyzing zooplankton size distributions using high-frequency sound. Limnology and Oceanography 34:129–139.

Greene, C. H., P. H. Wiebe, R. T. Miyamoto, and J. Burczynski. 1991b. Probing the fine structure of ocean sound-scattering layers with ROVERSE technology. Limnology and Oceanography 36:193–204.

Greene, C. H., P. H. Wiebe, and J. E. Zamon. 1994. Acoustic visualization of patch dynamics in oceanic ecosystems. Oceanography 7(1):4–12.

Greenlaw, C. F. 1977. Backscattering spectra of preserved zooplankton. Journal of the Acoustical Society of America 62:44–52.

Greenlaw, C. F. 1979. Acoustical estimation of zooplankton populations. Limnology and Oceanography 24:226–242.

Greenlaw, C. F., and R. K. Johnson. 1982. Physical and acoustical properties of zooplankton. Journal of the Acoustical Society of America 72:1706–1710.

Greenlaw, C. F., and R. K. Johnson. 1983. Multiple-frequency acoustical estimation. Biological Oceanography 2:3–4.

Greenlaw, C. F., and W. G. Pearcy. 1985. Acoustical patchiness of mesopelagic micronekton. Journal of Marine Research 43:163–178.

Guillard, J. D, D. Gerdeaux, and J. M. Chautru. 1990. The use of geostatistics for abundance estimation in lakes: the example of Lake Annecy. Journal du Conseil International pour l'Exploration de la Mer 189:410–414.

Haney, J. F., A. Craggy, K. Kimball, and F. Weeks. 1990. Light control of evening vertical migrations by *Chaoborus punctipennis* larvae. Limnology and Oceanography 35:1068–1078.

Hansson, S. 1993. Variation in hydroacoustic abundance of pelagic fish. Fisheries Research 16:203–222.

Haslett, R. W. G. 1962. Determination of the acoustic back-scattering patterns and cross sections of fish models and ellipsoids. British Journal of Applied Physics 13:349–357.

Haslett, R. W. G. 1969. The target strengths of fish. Journal of Sound Vibration 9:181–191.

Hawkins, A. D. 1977. Fish sizing by means of swimbladder resonance. Rapports et Procés-Verbaux des Réunions Conseil International pour l'Exploration de la Mer 170:122–129.

Hawkins, A. D. 1986. Underwater sound and fish behavior. Pages 114–115 *in* T. J. Pitcher, editor. The behavior of teleost fishes. Croom Helm, London.

Hawkins, A. D., and A. A. Myrberg, Jr. 1983. Hearing and sound communication under water. Pages 347–405 *in* B. Lewis, editor. Bioacoustics, a comparative approach. Academic Press, London.

Heist, B. G., and W. A. Swenson. 1983. Distribution and abundance of rainbow smelt in western Lake Superior as determined from acoustic sampling. Journal of Great Lakes Research 9:343–353.

Hewitt, R. P., and D. A. Demer. 1991. Krill abundance. Nature 353:310.

Hewitt, R. P., P. E. Smith, and J. C. Brown. 1976. Development and use of sonar mapping for pelagic stock assessment in the California Current area. U.S. National Marine Fisheries Service Fishery Bulletin 74:281–300.

Heywood, K. J., S. Scrope-Howe, and E. D. Barton. 1991. Estimation of zooplankton abundance from shipborne ADCP backscatter. Deep-Sea Research 38:677–691.

Holliday, D. V. 1972. Resonance structure in echoes from schooled pelagic fish. Journal of the Acoustical Society of America 51:1322–1333.

Holliday, D. V. 1977. Two applications of the Doppler effect in the study of fish schools. Rapports et Procés-Verbaux des Réunions Conseil International pour l'Exploration de la Mer 170:21–30.

Holliday, D. V., and H. L. Larson. 1979. Thickness and depth distributions of some epipelagic fish schools off southern California. U.S. National Marine Fisheries Service Fishery Bulletin 77:489–494.

Holliday, D. V., and R. E. Pieper. 1980. Volume scattering strengths and zooplankton distributions at acoustic frequencies between 0.5 and 3 Mhz. Journal of the Acoustical Society of America 67:135–146.

Holliday, D. V., R. E. Pieper, and G. S. Kleppel. 1989. Determination of zooplankton size and distribution with multifrequency acoustic technology. Journal du Conseil International pour l'Exploration de la Mer 46:52–61.

Horne, J. K., and D. C. Schneider. 1994. Lack of spatial coherence of predators with prey: a bioenergetic explanation for Atlantic cod feeding on capelin. Journal of Fish Biology 45(Supplement A):191–207.

Huang, K. 1977. PDF of backscattered sound from live fish. Master's thesis. University of Wisconsin-Madison, Madison.

Huang, K., and C. S. Clay. 1980. Backscattering cross sections of life fish: PDF and aspect. Journal of the Acoustical Society of America 67:795–802.

Isaacs, J. D., S. A. Tont, and G. L Wick. 1974. Deep scattering layers: vertical migration as a tactic for finding food. Deep-Sea Research 21:651–656.

Jacobson, P. T. 1990. Pattern and process in the distribution of cisco, *Coregonus atredii*, in Trout Lake, Wisconsin. Doctoral dissertation. University of Wisconsin-Madison, Madison.

Jacobson, P. T., C. S. Clay, and J. J. Magnuson. 1990. Size, distribution, and abundance by deconvolution of single-beam acoustic data. Rapports et Procés-Verbaux des Réunions Conseil International pour l'Exploration de la Mer 189:304–316.

Jaffe, J. S., E. Reuss, D. McGehee, and G. Chandran. 1995. FTV, a sonar for tracking macrozooplankton in three dimensions. Deep-Sea Research 45:1495–1512.

Janssen, J., and S. B. Brandt. 1980. Feeding ecology and vertical migration of adult alewife (*Alosa pseudoharengus*) in Lake Michigan. Canadian Journal of Fisheries and Aquatic Sciences 37:177–184.

Jefferts, K., J. J. Burczynski, and W. G. Pearcy. 1987. Acoustical assessment of squid *Loligo opalescens* off the General Oregon coast. Canadian Journal of Fisheries and Aquatic Sciences 44:1261–1267.

Johannesson, K. A., and R. B. Mitson. 1983. Fisheries acoustics: a practical manual for biomass estimation. FAO (Food and Agriculture Organization of the United Nations) Fisheries Technical Paper 240.

Johnson, G. E., J. R. Skalski, and D. J. Degan. 1994. Statistical precision of hydroacoustic sampling of fish entrainment at hydroelectric facilities. North American Journal of Fisheries Management 14:323–333.

Johnson, G. E., C. M. Sullivan, and M. W. Erho. 1992. Hydroacoustic studies for developing a smolt bypass system at Wells Dam. Fisheries Research 14:221–237.

Johnston, S. V., and J. S. Hopelain. 1990. The application of dual-beam target tracking and Doppler-shifted echo processing to assess upstream salmonid migration in the Klamath River, California. Rapports et Procés-Verbaux des Réunions Conseil International pour l'Exploration de la Mer 189:210–222.

Jolly, G. M., and I. Hampton. 1990a. A stratified random transect design for acoustic surveys of fish stocks. Canadian Journal of Fisheries and Aquatic Sciences 47:1282–1291.

Jolly, G. M., and I. Hampton. 1990b. Some problems in the statistical design and analysis of acoustic surveys to assess fish biomass. Rapports et Procés-Verbaux des Réunions Conseil International pour l'Exploration de la Mer 189:415–420.

Jones, F. R. H., and G. Pearce. 1957. Acoustic reflection experiments with perch (*Perca fluviatilis*) Linn. to determine the proportion of the echo returned by the swimbladder. Journal of Experimental Biology 35:437–450.

Kaye, G. T. 1979a. Acoustic remote sensing of high-frequency internal waves. Journal of Geophysical Research 84:7017–7022.

Kaye, G. T. 1979b. Correlation between acoustic scatterers and temperature gradients. Journal of Marine Research 37:319–326.

Kieser, R., and T. J. Mulligan. 1984. Analysis of echo counting data: a model. Canadian Journal of Fisheries and Aquatic Sciences 41:451–458.

Kjaergaard, N., L. Bjorno, E. Kjaergaard, and H. Lassen. 1990. Broadband analysis of acoustic scattering by individual fish. Rapports et Procés-Verbaux des Réunions Conseil International pour l'Exploration de la Mer 189:370–380.

Kristensen, A., and J. Dalen. 1986. Acoustic estimation of size distribution and abundance of zooplankton. Journal of the Acoustical Society of America 80:601–611.

Kubecka, J. A., A. Duncan, W. M. Duncan, D. Sinclair, and A. J. Armstrong. 1994. Brown trout populations of three Scottish lochs estimated by horizontal sonar and multimesh gill nets. Fisheries Research 20:29–48.

Leroy, C. C. 1969. Development of simple equations for accurate and more realistic calculations of the speed of sound in sea water. Journal of the Acoustical Society of America 46:216–226.

Levy, D. A. 1991. Acoustic analysis of diel vertical migration behavior of *Mysis relecta* and kokanee (*Oncorhynchus nerka*) within Okanagan Lake, British Columbia. Canadian Journal of Fisheries and Aquatic Sciences 48:67–72.

Levy, D. A., R. L. Johnson, and J. M. Hume. 1991. Shifts in fish vertical distribution in response to an internal seiche in a stratified lake. Limnology and Oceanography 36:187–192.

Lindem, T. 1983. Successes with conventional in situ determination of fish target strength. FAO (Food and Agriculture Organization of the United Nations) Fisheries Report 300:104–111.

Love, R. H. 1969. Maximum side aspect target strength of an individual fish. Journal of the Acoustical Society of America 46:746–752.

Love, R. H. 1971a. Measurements of fish target strength: a review. U.S. National Marine Fisheries Service Fishery Bulletin 69:703–715.

Love, R. H. 1971b. Dorsal-aspect target strength of an individual fish. Journal of the Acoustical Society of America 49:816–823.

Love, R. H. 1977. Target strength of an individual fish at any aspect. Journal of the Acoustical Society of America 62:1397–1403.

Love, R. H. 1978. Resonant scattering by swimbladder bearing fish. Journal of the Acoustical Society of America 64:571–580.

Love, R. H. 1993. A comparison of volume scattering strength data with model calculations based on quasisynoptically collected fishery data. Journal of the Acoustical Society of America 94:2255–2268.

Luecke, C., and W. A. Wurtsbaugh. 1993. Effects of moonlight and daylight on hydroacoustic estimates of pelagic fish abundance. Transactions of the American Fisheries Society 122:112–120.

Luo, J., and S. B. Brandt. 1993. Bay anchovy production and consumption in mid-Chesapeake Bay based on a bioenergetics model and acoustic measure of fish abundances. Marine Ecology Progress Series 98:223–236.

MacKenzie, K. V. 1981. Nine-term equation for sound speed in the oceans. Journal of the Acoustical Society of America 70:807–812.

MacLennan, D. N. 1987. Time-varied-gain functions for pulsed sonars. Journal of Sound Vibration 110:511–522.

MacLennan, D. N. 1990. Acoustical measurement of fish abundance. Journal of the Acoustical Society of America 87:1–15.

MacLennan, D. N., and I. G. MacKenzie. 1988. Precision of acoustic fish stock estimates. Canadian Journal of Fisheries and Aquatic Sciences 45:605–616.

MacLennan, D. N., A. E. Magurran, T. J. Pitcher, and C. E. Hollingworth. 1990b. Behavioral determinants of fish target strength. Rapports et Procés-Verbaux des Réunions Conseil International pour l'Exploration de la Mer 189:245–253.

MacLennan, D. N., and E. J. Simmonds. 1992. Fisheries acoustics. Chapman and Hall, London.

MacLennan, D. N., and I. Svellingen. 1989. Simple calibration technique for the split-beam echosounder. Fiskeridirektoratets Skrifter, Serie Havundersokelser 18:365–379.

Madureira, L. S. P., P. Ward, and A. Atkinson. 1993. Difference in backscattering strength determined at 120 and 38 kHz for three species of Antarctic macroplankton. Marine Ecology Progress Series 93:17–24.

Magnuson, J. J., L. B. Crowder, and P. A. Medvick. 1979. Temperature as an ecological resource. American Zoologist 19:331–343.

McCartney, B. S., and A. R. Stubbs. 1971. Measurements of the acoustic target strength of fish in dorsal aspect, including swimbladder resonance. Journal of Sound Vibration 15:397–420.

McClatchie, S., C. H. Greene, M. C. Macaulay, and D. R. M. Sturley. 1994. Spatial and temporal variability of Antarctic krill: implications for stock assessment. International Council for the Exploration of the Sea Journal of Marine Research 51:11–18.

McGehee, D., and J. S. Jaffe. 1996. Three-dimensional swimming behavior of individual zooplanktors: observations using the acoustical imaging system FishTV. International Council for the Exploration of the Sea Journal of Marine Research 53:363–369.

McNaught, D. C. 1969. Acoustical determination of zooplankton distributions. Proceedings Conference on Great Lakes Research 11:76–84.

Melnik, N. G., O. A. Timoshkin, and V. G. Sideleva. 1993. Hydroacoustic measurement of the density of the Baikal macrozooplanter *Macrohectopus branickii*. Limnology and Oceanography 38:425–434.

Mesiar, D. C., D. M. Eggars, and D. M. Gaudet. 1990. Development of techniques for the application of hydroacoustics to counting migratory fish in large rivers. Rapports et Proces-Verbaux des Réunions Conseil International pour l'Exploration de la Mer 189:223–232.

Midttun, L. 1984. Fish and other organisms as acoustic targets. Rapports et Procés-Verbaux des Réunions Conseil International pour l'Exploration de la Mer 184:25–33.

Misund, O. A. 1993. Dynamics of moving masses: variability in packing density, shape, and size among herring, sprat and saithe schools. International Council for the Exploration of the Sea Journal of Marine Sciences 50:145–160.

Misund, O. A., A. Aglen, A. K. Beltestad, and J. Dalen. 1992. Relationships between the geometric dimensions and biomass of schools. International Council for the Exploration of the Sea Journal of Marine Science 49:305–315.

Misund, O. A., A. Aglen, and E. Fronaes. 1995. Mapping the shape, size, and density of fish schools by echo integration and a high-resolution sonar. International Council for the Exploration of the Sea Journal of Marine Science 52:11–20.

Mitson, R. B. 1984. Fisheries sonar. Fishing News Books, London.

Moose, P. H., and J. E. Ehrenberg. 1971. An expression for the variance of abundance estimates using a fish echo integrator. Journal of the Fisheries Research Board of Canada 28:1293–1301.

Mulligan, T. J., and R. Kieser. 1986. Comparison of acoustic population estimates of salmon in a lake with a weir count. Canadian Journal of Fisheries and Aquatic Sciences 43:1373–1385.

Nakken, O., and K. Olsen. 1977. Target strength measurements of fish. Rapports et Procés-Verbaux des Réunions Conseil International pour l'Exploration de la Mer 170:52–69.

Nash, R. D. M., J. J. Magnuson, C. S. Clay, and T. K. Stanton. 1987. A synoptic view of the Gulf Stream from with 70 kHz sonar: taking advantage of a closer look. Canadian Journal of Fisheries and Aquatic Sciences 44:2022–2024.

Nash, R. D. M., J. J. Magnuson, T. K. Stanton, and C. S. Clay. 1989. Distribution of peaks of 70 kHz acoustic scattering in relation to depth and temperature during day and night at the edge of the Gulf Stream—Echofront '83. Deep-Sea Research 36:587–596.

Nero, R. W., and J. J. Magnuson. 1989a. Characterization of patches along the transects using higher resolution 70 kHz integrated echo data. Canadian Journal of Fisheries and Aquatic Sciences 46:2056–2064.

Nero, R. W., and J. J. Magnuson. 1989b. Effects of changing spatial scale on acoustic observations of patchiness in the Gulf Stream. Landscape Ecology 6:279–291.

Nero, R. W., J. J. Magnuson, S. B. Brandt, T. K. Stanton, and J. M. Jech. 1990. Finescale biological patchiness of 70 kHz acoustic scattering at the edge of the Gulf Stream—Echofront '85. Deep-Sea Research 37:999–1016.

Nestler, J. M., G. R. Ploskey, J. Pickens, J. Menezes, and C. Schilt. 1992. Responses of blueback herring to high-frequency sound and implications for reducing entrainment at hydropower dams. North American Journal of Fisheries Management 12:667–683.

Nickerson, T. B., and R. G. Dowd. 1977. Design and operation of survey patterns for demersal fishes using the computerized echo counting system. Rapports et Procés-Verbaux des Réunions Conseil International pour l'Exploration de la Mer 170:232–236.

Nunallee, E. P., Jr. 1980. Application of an empirically scaled digital echo integrator for assessment of juvenile sockeye salmon, (*Oncorhynchus nerka* Walbaum) populations. Doctoral dissertation. University of Washington, Seattle.

Nunallee, E. P. 1990. An alternative to thresholding during echo-integration data collection. Rapports et Procés-Verbaux des Réunions Conseil International pour l'Exploration de la Mer 189:92–94.

Olsen, K. 1990. Fish behavior and acoustic sampling. Rapports et Procés-Verbaux des Réunions Conseil International pour l'Exploration de la Mer 189:147–158.

Olsen, K., J. Angell, F. Pettersen, and A. Lovik. 1983. Observed fish reactions to a surveying vessel with special reference to herring, cod, capelin and polar cod. FAO (Food and Agriculture Organization of the United Nations) Fisheries Report 300:131–138.

Ona, E. 1990. Physiological factors causing natural variations in acoustic target strength of fish. Journal of the Marine Biological Association of the United Kingdom 70:107–127.

Palumbo, D., J. D. Penrose, and B. A. White. 1993. Target strength estimation from echo ensembles. Journal of the Acoustical Society of America 94:2766–2775.

Parkinson, E. A., B. E. Rieman, and L. G. Rudstam. 1994. Comparison of acoustic and travel methods for estimating density and age composition of kokanee. Transactions of the American Fisheries Society 123:841–854.

Penrose, J. D., and G. T. Kaye. 1979. Acoustic target strengths of marine organisms. Journal of the Acoustical Society of America 65:374–380.

Peterson, M. L., C. S. Clay, and S. B. Brandt. 1976. Acoustic estimates of fish density and scattering function. Journal of the Acoustical Society of America 60:618–622.

Petitgas, P. 1993a. Geostatistics for fish stock assessment: a review and an acoustic application. International Council for the Exploration of the Sea Journal of Marine Science 50:285–298.

Petitgas, P. 1993b. Use of a disjunctive kriging to model areas of high pelagic fish density in acoustic fisheries surveys. Aquatic Living Resources 6:201–209.

Pieper, R. E. 1979. Euphausiid distribution and biomass determined acoustically at 102 kHz. Deep-Sea Research 26:687–702.

Pieper, R. E., and D. V. Holliday. 1984. Acoustic measurements of zooplankton distributions in the sea. Journal du Conseil International pour l'Exploration de la Mer 41:226–238.

Pieper, R. E., D. V. Holliday, and G. S. Keppel. 1990. Quantitative zooplankton distributions from multifrequency acoustics. Journal of Plankton Research 12:433–441.

Plueddemann, A. J., and R. Pinkel. 1989. Characterization of the patterns of diel migration using a Doppler sonar. Deep-Sea Research 36:509–530.

Ponton, D., and H. J. Meng. 1990. Use of dual-beam acoustic technique for detecting young whitefish, *Coregonus so.*, juveniles: first experiments in an enclosure. Journal of Fish Biology 36:741–750.

Popper, A. N., and R. R. Fay. 1973. Sound detection and processing by teleost fishes: a critical review. Journal of the Acoustical Society of America 53:1515–1529.

Purves, P. E., and G. E. Pilleri. 1983. Echolocation in whales and dolphins. Academic Press, London.

Ramani, N., and P. H. Patrick. 1992. Fish detection and identification using neural networks—some laboratory results. Institute of Electrical and Electronic Engineers Journal of Oceanics Engineering 17:364–368.

Ransom, B. H., and T. W. Steig. 1994. Using hydroacoustics to monitor fish at hydropower dams. Lake and Reservoir Management 9(1):163–169.

Rayleigh, J. W. S. 1945. The theory of sound, 2nd edition 1986. Reprinted by Dover, New York.

Reid, D. G., and E. J. Simmonds. 1993. Image analysis techniques for the study of fish school structure from acoustic survey data. Canadian Journal of Fisheries and Aquatic Sciences 50:886–893.

Revie, J., D. E. Weston, F. R. Harden Jones, and G. P. Fox. 1990. Identification of fish echoes located at 65 km range by shore-based sonar. Journal du Conseil International pour l'Exploration de la Mer 46:313–324.

Rice, O. R. 1954. Mathematical analysis of random noise. Pages 133–294 *in* N. Wax, editor. Selected papers on noise and stochastic. Dover, New York.

Richards, L. J., R. Kieser, T. J. Mulligan, and J. R. Candy. 1991. Classification of fish assemblages based on echo integration surveys. Canadian Journal of Fisheries and Aquatic Sciences 48:1264–1272.

Richter, K. E. 1985. Acoustic scattering at 1.2 Mhz from individual zooplankters and copepod populations. Deep-Sea Research 32:149–161.

Robinson, B. J. 1983. In-situ measurements on the target strength of pelagic fishes. FAO (Food and Agriculture Organization of the United Nations) Fisheries Report 300:99–111.

Robinson, B. J., and C. Hood. 1983. A procedure for calibrating acoustic survey systems with estimates of obtainable precision and accuracy. FAO (Food and Agriculture Organization of the United Nations) Fisheries Report 300:59–62.

Robotham, V. H., and J. Castillo. 1990. The bootstrap method: an alternative for estimating confidence intervals of resources surveyed by means by hydroacoustic techniques. Rapports et Procés-Verbaux des Réunions Conseil International pour l'Exploration de la Mer 189:421–424.

Roe, H. S. J., and G. Griffiths. 1993. Biological information from an Acoustic Doppler Current Profiler. Marine Biology 115:339–346.

Rose, G. A., and W. C. Leggett. 1988. Hydroacoustic signal classification of fish schools by species. Canadian Journal of Fisheries and Aquatic Sciences 45:597–604.

Rose, G. A., and W. C. Leggett. 1990. The importance of scale to predator–prey spatial correlations: an example of Atlantic fishes. Ecology 71:33–43.

Ross, Q. E., D. J. Dunning, R. Thorne, J. K. Menezes, G. W. Tiller, and J. K. Watson. 1993. Response of alewives to high-frequency sound at a power plant intake in Lake Ontario. North American Journal of Fisheries Management 13:291–303.

Rudstam, L. G., C. S. Clay, and J. J. Magnuson. 1987. Density and size estimates of cisco (*Coregonus artedii*) using analysis of echo peak PDF from a single-transducer sonar. Canadian Journal of Fisheries and Aquatic Sciences 44:811–821.

Rudstam, L. G., and B. M. Johnson. 1992. Development, evaluation, and transfer of new technology. Pages 507–524 *in* J. F. Kitchell, editor. Food web management: a case study of Lake Mendota. Springer Verlag, New York.

Rudstam, L. G., R. C. Lathrop, and S. R. Carpenter. 1993. The rise and fall of a dominant planktivore: direct and indirect effects on zooplankton. Ecology 74:303–319.

Rudstam, L. G., T. Lindem, and S. Hansson. 1988. Density and in-situ target strength of herring and sprat: a comparison between two methods of analyzing single-beam sonar data. Fisheries Research 6:305–315.

Saenger, R. A. 1988. Swimbladder size variability in mesopelagic fish and bioacoustical modeling. Journal of the Acoustical Society of America 84:1007–1017.

Sameoto, D. 1982. Zooplankton and micronekton abundance in acoustic scattering layers on the Nova Scotian slope. Canadian Journal of Fisheries and Aquatic Sciences 39:760–777.

Sameoto, D., N. A. Cochrane, and A. W. Herman. 1985. Response of biological acoustic backscattering to ships' lights. Canadian Journal of Fisheries and Aquatic Sciences 42:1535–1543.

Scalabrin, C., and J. Masse. 1993. Acoustic detection of the spatial and temporal distribution of fish schools in the Bay of Biscay. Aquatic Living Resource 6:255–267.

Schneider, D. C. 1989. Identifying the spatial scale of density-dependent interaction of predators with schooling fish in the southern Labrador current. Journal of Fish Biology 35:109–115.

Shotton, R. 1981. Acoustic survey design. Pages 629–688 *in* J. B. Suomala, editor. Meeting on hydroacoustical methods for the estimation of marine fish populations, volume 2. The Charles Stark Draper Laboratory, Cambridge, Massachusetts.

Shotton, R., and G. P. Bazigos. 1984. Techniques and considerations in the design of acoustic surveys. Rapports et Procés-Verbaux des Réunions Conseil International pour l'Exploration de la Mer 180:34–57.

Simmonds, E. J. 1990. Very accurate calibration of a vertical echo sounder: a five-year assessment of performance and accuracy. Rapports et Procés-Verbaux des Réunions Conseil International pour l'Exploration de la Mer 189:183–191.

Simmonds, E. J., and F. Armstrong. 1990. A wideband echo sounder: measurements on cod, saithe, herring, and mackerel from 27 to 54 kHz. Rapports et Procés-Verbaux des Réunions Conseil International pour l'Exploration de la Mer 189:381–387.

Simmonds, E. J., and P. J. Copeland. 1986. A wide band constant beamwidth echosounder for fish abundance estimation. Proceedings of the Institute of Acoustics 11:54–60.

Skalski, J. R., A. Hoffman, B. H. Ransom, and T. W. Steig. 1993. Fixed-location hydroacoustic monitoring designs for estimating fish passage using stratified random and systematic sampling. Canadian Journal of Fisheries and Aquatic Sciences 50:1208–1221.

Smith, P. E. 1970. The horizontal dimensions and abundance of fish schools in the upper mixed layer as measured by sonar. Pages 563–591 *in* G. B. Farquhar, editor. Proceedings of the international symposium on biological sound scattering in the ocean. U.S. Department of the Navy, Maury Center for Ocean Science, Washington, DC.

Smith, P. E. 1977. The effects of internal waves on fish school mapping with sonar in the California Current area. Rapports et Procés-Verbaux des Réunions Conseil International pour l'Exploration de la Mer 170:223–231.

Smith, P. E. 1979. Precision of sonar mapping for pelagic fish assessment in the California Current. Journal du Conseil International pour l'Exploration de la Mer 38:33–40.

Smith, S. L., and S. Gavaris. 1993. Improving the precision of abundance estimates of eastern Scotian Shelf Atlantic cod from bottom trawl surveys. North American Journal of Fisheries Management 13:35–47.

Smith, S. L., and seven coauthors. 1992. Acoustic techniques for the in situ observation of zooplankton. Archives fuer Hydrobiologie 36:23–43.

Sparholt, H. 1990. A stochastic integrated VPA for herring in the Baltic Sea using acoustic estimates as auxiliary information for estimating natural mortality. Journal du Conseil International pour l'Exploration de la Mer 46:325–332.

Spigarelli, S. A., G. P. Romberg, and R. E. Thorne. 1973. A technique for simultaneous echo location of fish and thermal plume. Transactions of the American Fisheries Society 102:462–466.

Sprong, I., B. R. Kuipers, and H. Witte. 1990. Acoustic phenomena related to an enriched benthic zone in the North Sea. Journal of Plankton Research 12(6):1251–1261.

Sprules, W. G., S. B. Brandt, E. H. Jin, M. Munawar, D. J. Stewart, and J. Love. 1991. Patterns in the biomass size structure of the total Lake Michigan pelagic community. Canadian Journal of Fisheries and Aquatic Sciences 48:105–115.

Stables, T. B., and G. L. Thomas. 1992. Acoustic measurement of trout distributions in Spada Lake, Washington, using stationary transducers. Journal of Fish Biology 40:191–203.

Stanton, T. K. 1985a. Volume scattering: echo peak PDF. Journal of the Acoustical Society of America 77:1358–1366.

Stanton, T. K. 1985b. Density estimate of biological sound scatterers using sonar echo peak PDFs. Journal of the Acoustical Society of America 78:1868–1873.

Stanton, T. K. 1988. Sound scattering by cylinders of finite length, 2: elastic cylinders. Journal of the Acoustical Society of America 83:64–67.

Stanton, T. K. 1989. Sound scattering by cylinders of finite length, 3: deformed cylinders. Journal of the Acoustical Society of America 83:691–705.

Stanton, T. K. 1990. Sound scattering by zooplankton. Rapports et Procés-Verbaux des Réunions Conseil International pour l'Exploration de la Mer 189:353–362.

Stanton, T. K., D. Chu, P. H. Wiebe, and C. S. Clay. 1993a. Average echoes from randomly oriented random-length finite cylinders: zooplankton models. Journal of the Acoustical Society of America 94:6463–3472.

Stanton, T. K., and C. S. Clay. 1986. Sonar echo statistics as a remote sensing tool: volume and seafloor. IEEE (Institute of Electrical and Electronics Engineers) Journal of Oceanics Engineering 11:79–99.

Stanton, T. K., C. S. Clay, and D. Chu. 1993b. Ray representation of sound scattering by weakly scattering deformed fluid cylinders: simple physics and application to zooplankton. Journal of the Acoustical Society of America 94:3454–3462.

Stanton, T. K., and D. V. Holliday. 1991. Using sonar to examine the oceanic food chain. Journal of the Acoustical Society of America 89:465–466.

Stanton, T. K., and six coauthors. 1994a. On acoustic estimates of zooplankton biomass. International Council for the Exploration of the Sea Journal of Marine Science 51:505–512.

Stanton, T. K., P. H. Wiebe, D. Chu, and L. Goodman. 1994b. Acoustic characterization and discrimination of marine zooplankton and turbulence. International Council for the Exploration of the Sea Journal of Marine Science 51:469–479.

Sun, Y., R. D. M. Nash, and C. S. Clay. 1995. Acoustical measurements of the anatomy of fish at 220 kHz. Journal of the Acoustical Society of America 78:1772–1776.

Sund, O. 1935. Echo sounding in fishery research. Nature 135:953.

Tavolga, W. N., A. N. Popper, and R. R. Fay, editors. 1981. Hearing and sound communication in fishes. Springer-Verlag, New York.

Thiebaux, M. L., P. R. Boudreau, and L. M. Dickie. 1991. An analytical model of acoustic fish refection for estimation of maximum dorsal aspect target strength. Canadian Journal of Fisheries and Aquatic Sciences 48:1772–1782.

Thomas, G. L., S. Thiesfeld, S. Bonar, G. Pauly, and R. W. Crittenden. 1990. Estimation of submergent plant biovolume using acoustic range information. Canadian Journal of Fisheries and Aquatic Sciences 47:805–812.

Thoreson, R. 1991. Absorption of acoustic energy in dense herring schools studies by the attenuation in the bottom echo signal. Fisheries Research 10:317–327.

Thorne, R. E. 1971. Investigations into the relation between integrated echo voltage and fish density. Journal of the Fisheries Research Board of Canada 27:269–273.

Thorne, R. E. 1977. Acoustic assessment of Pacific hake and herring stocks in Puget Sound, Washington and southeastern Alaska. Procés-Verbaux des Réunions Conseil International pour l'Exploration de la Mer 170:265–278.

Thorne, R. E. 1979. Hydroacoustic estimates of adult sockeye salmon (*Oncorhynchus nerka*) in Lake Washington 1972–1975. Journal of the Fisheries Research Board of Canada 36:1145–1149.

Thorne, R. E. 1983a. Application of hydroacoustic assessment techniques to three lakes with contrasting fish distributions. FAO (Food and Agricultural Organization of the United Nations) Fisheries Report 300:269–277.

Thorne, R. E. 1983b. Assessment of population abundance by hydroacoustics. Biological Oceanography 2:252–262.

Thorne, R. E. 1983c. Hydroacoustics. Pages 239–259 *in* L. A. Nielsen and D. L. Johnson, editors. Fisheries techniques. American Fisheries Society, Bethesda, Maryland.

Thorne, R. E. 1988. An empirical evaluation of the duration-in-beam technique for hydroacoustic estimation. Canadian Journal of Fisheries and Aquatic Sciences 45:1244–1248.

Thorne, R. E. 1992. Current status of training and education in fisheries acoustics. Fisheries Research 14:135–141.

Thorne, R. E., J. Hedgepeth, and J. Campos. 1990. The use of stationary hydroacoustic transducers to study diel and tidal influences on fish behavior. Rapports et Procés-Verbaux des Réunions Conseil International pour l'Exploration de la Mer 189:167–175.

Thorne, R. E., and G. E. Johnson. 1993. A review of hydroacoustic studies for estimation of salmonid downriver migration past hydroelectric facilities on the Columbia and Snake rivers in the 1980s. Reviews in Fisheries Science 1:27–56.

Thorne, R. E., J. E. Reeves, and A. E. Millikan. 1971. Estimation of the Pacific hake (*Merluccius productus*) population in Port Susan, Washington, using an echo integrator. Journal of the Fisheries Research Board of Canada 28:1275–1284.

Traynor, J. D., and J. E. Ehrenberg. 1990. Fish and standard-sphere target-strength measurements obtained with a dual-beam and split-beam echo-sounding system. Rapports et Procés-Verbaux des Réunions Conseil International pour l'Exploration de la Mer 189:325–335.

Traynor, J. D., and N. J. Williamson. 1983. Target strength measurements of walleye pollock (*Theraga chalcogramma*) and a simulation study of the dual beam method. FAO (Food and Agricultural Organization of the United Nations) Fisheries Report 300:112–124.

Traynor, J. J., and J. E. Ehrenberg. 1979. Evaluation of the dual beam acoustic fish target strength measurement method. Journal of the Fisheries Research Board of Canada 36:1065–1071.

Unger, P. A., and S. B. Brandt. 1989. Seasonal and diel changes in sampling conditions for acoustic surveys of fish abundance in small lakes. Fisheries Research 7:353–366.

Walline, P. D., S. Pisanty, and T. Lindem. 1992. Acoustic assessment of the number of pelagic fish in Lake Kinneret, Israel. Hydrobiologia 231:153–163.

Wardle, C. S. 1983. Fish reactions to towed gears. Pages 167–195 *in* A. MacDonald and I. G. Priede, editors. Marine biology at sea. Academic Press, London.

Watkins, J. L., D. J. Morris, C. Ricketts, and A. W. A. Murray. 1990. Sampling biological characteristics of krill: effect of heterogeneous nature of swarms. Marine Biology 107:409–415.

Weimer, R. T., and J. E. Ehrenberg. 1975. Analysis of threshold-induced bias inherent in acoustic scattering cross-section estimates of individual fish. Journal of the Fisheries Research Board of Canada 32:2547–2551.

Wenz, G. M. 1972. Review of underwater acoustic research: noise. Journal of the Acoustical Society of America 51:1010–1024.

Weston, D. E., and H. W. Andrews. 1990a. Monthly estimates of fish numbers using long-range sonar. Journal du Conseil International pour l'Exploration de la Mer 47:104–111.

Weston, D. E., and H. W. Andrews. 1990b. Seasonal sonar observations of the diurnal shoaling times of fish. Journal of the Acoustical Society of America 87:673–680.

Wiebe, P. H., C. H. Greene, T. K. Stanton, and J. Burczynski. 1990. Sound scattering by live zooplankton and micronekton: empirical studies with dual-beam acoustical system. Journal of the Acoustical Society of America 88:2346–2360.

Wiebe, P. H., K. E. Prada, T. C. Austin, T. K. Stanton, and J. J. Dawson. 1995. New tool for bioacoustical oceanography: free-drifting/moored autonomous acoustic platform for long-term measurement of biomass. Sea Technology 37(2):10–14.

Williamson, N. J. 1982. Cluster sampling estimation of the variance of abundance estimates derived from quantitative echo sounder surveys. Canadian Journal of Fisheries and Aquatic Sciences 39:229–231.

Williamson, N. J., and J. J. Traynor. 1984. In-situ target-strength estimation of Pacific whiting (*Merluccius productus*) using a dual-beam transducer. Journal du Conseil International pour l'Exploration de la Mer 41:285–292.

Zakharia, M. E. 1990. A prototype wideband sonar for fisheries in lakes and rivers. Rapports et Procés-Verbaux des Réunions Conseil International pour l'Exploration de la Mer 189:394–397.

Chapter 14

Field Examination of Fishes

RICHARD J. STRANGE

14.1 INTRODUCTION

Fisheries scientists expend considerable effort collecting fishes by a variety of methods (see Chapters 6, 7, 8, 10, 20, and 21). Once in hand, fishes can be examined in a number of ways that depend on the objectives of the study. Routine examination nearly always includes identification of species and determination of length and weight. Further, fishes may be sexed, scales taken, various organs measured, and stomach contents preserved. These observations, upon further analysis, reveal population structure, age and growth relationships, relative health, and diet: data that provide the foundation for fisheries management decisions. Sometimes fisheries scientists are called to assist in emergency situations in which dead and dying fishes threaten an economic loss or indicate an environmental problem. In these cases, examination focuses on signs of ill health.

14.2 ROUTINE EXAMINATION

14.2.1 Basic Observations

In a routine collection, fishes are first sorted by species. Fishes to be returned alive should be handled as little as possible and kept in adequate water (see Chapter 5). Species of interest are weighed and measured, scales may be taken for determination of age and growth (see Chapter 16), and if fish can be sexed externally, sex may be recorded. Some, but not most, fishes can be sexed reliably by external examination if they are sexually mature; virtually no immature fishes can be sexed externally. External evidence of sex includes brighter coloration in males (e.g., bluegill), difference in the genital opening (channel catfish), and difference in the shape of the head (trout and salmon). If the sex is of interest and the fish cannot be sexed externally, then dissection can be used to identify the paired gonads. In immature fish, the gonads are small, and testes can be recognized as light-colored, opaque, fine-textured organs whereas ovaries are granular and translucent. In mature fish, the gonads are much enlarged. The testes are milky white; in the ovaries developing eggs, which may be pink, yellow, green, or black, are clearly evident.

If analysis of diet will be conducted, stomach contents must be preserved (see Chapter 17). A nonlethal method of obtaining stomach contents by washing them out of the fish (lavage) has been used on a number of species. Techniques vary, but essentially a stream of water is directed down the fish's esophagus while the fish is turned head downward; the stomach contents are collected as they wash from the mouth. In small fishes, the stomach contents can be saved by preserving the entire specimen in 10% formalin, but an incision in the body cavity must be made close to the stomach to ensure that the formalin quickly penetrates the gut. With larger fishes, the stomach is dissected out of the fish in the field and preserved separately in formalin.

14.2.2 Necropsy-Based Fish Health Assessment

Condition and organ indices. Thorough, quantitative examination can yield useful information on the health and condition of fishes. Methods of obtaining standardized data from gross necropsy have been developed to facilitate comparison of fishes from different localities observed by different people. Caution must be used with these methods when comparing fish of different species, different sexes, or even the same species and sex if fish were collected at different times of the year. Condition and organ indices are simply ratios of one measurement to another. The most common is condition factor, K or C, is the ratio of weight to the cube of length. This yields a number near unity useful for comparing robustness; that is, the higher the number, the plumper the fish (see Chapter 15). Organosomatic indices are the ratio (times 100 would yield a percentage) of an organ to the whole body. The gonadosomatic index (GSI),

$$\text{GSI} = \text{ovary weight/body weight},$$

describes the relative size of the ovaries or testes and is useful in indicating time of spawning. The hepatosomatic (liver) index, similarly calculated, may decrease in starved fish or increase in fish subjected to toxins. Other indices (e.g., splenosomatic and viscerosomatic) are occasionally used as well.

A problem with simple indices is that they vary depending on the individual weight of the fish (i.e., a plump fish and a thin fish with the same length and ovary weight will have different GSIs). This problem led Legler (1977) to develop relative gonad weight G_r:

$$G_r = (\text{ovary weight} \times 100)/W_s,$$

where W_s is the length-specific weight (see Chapter 15). The hepatosomatic index can be treated similarly and yield a relative liver weight (L_r). The amount of visceral fat present in a fish is indicative of prior feeding success and can be measured alone or as part of a systematic condition assessment.

Systematic condition assessment. Recently, a method of systematically recording a variety of characteristics that can be observed during gross necropsy has gained acceptance as a way to evaluate the relative health of fish (Goede and Barton 1990). The strength of this "autopsy-based condition assessment" is that a number of specific observations are made on each animal and compared with previous findings that represent a reference or norm for the species. Standard observations include easily measured blood constituents (hematocrit, leukocrit, and plasma protein), damage to extremities, and state of maturity, as well as the condition of gills, pseudobranch, mesenteric fat, spleen, hind gut, kidney, liver, and bile. Although this approach may miss effects that might be detected by more sophisticated analyses (histological, biochemical, or toxicological), its intention is to provide an easily used method for detecting trends in the health and condition of fish populations. It should not be used to attempt a definitive diagnosis, but it is a valuable assessment technique as are routine postmortem examinations in human and veterinary medicine.

Box 14.1 Some Examples of Pathogen-Induced Fish Diseases

Disease	Causative agent	Typical host
Parasites		
ich or white spot	*Ichthyoptherius multifilis*	freshwater fishes
Epistylus infection	*Epistylus* spp.	freshwater fishes
Trichodina infection	*Trichodina* spp.	freshwater fishes
whirling disease	*Myxosoma cerebralis*	trout
gyro	*Gyrodactylus* spp.	freshwater fishes
yellow "grub" of muscle	*Clinostonum marginatum*	freshwater fishes
tapeworm	*Ligula intestinalis*	minnows, suckers
fish "louse"	*Argulus* spp.	freshwater fishes
anchor parasite	*Lernea* spp.	freshwater fishes
fungus	*Saprolegnia* spp.	freshwater fishes
Bacteria		
furunculosis	*Aeromonas salmonicida*	trout
motile *Aeromonas* septicemia	*Aeromonas hydrophila*	freshwater fishes
columnaris disease	*Flexibacter* spp.	freshwater fishes
vibriosis	*Vibrio anguillarum*	saltwater fishes
enteric redmouth	*Yersinia ruckeri*	trout
Edwardsiella septicemia	*Edwardsiella tarda*	catfishes
bacterial kidney disease	*Renibacterium salmoninarum*	trout
Viruses		
infectious pancreatic necrosis	ether stable virus	young trout
infectious hemopoietic necrosis	rhabdovirus	salmon
viral erythrocytic necrosis	blood cell virus	saltwater fishes
channel catfish virus	herpesvirus	catfishes

14.3 EMERGENCY EXAMINATION

14.3.1 Definition of Health and Illness in Wild Fishes

Healthy fish are relatively free from pathogens and environmental conditions that increase mortality or reduce growth and fecundity. Healthy fish may, however, harbor pathogens and probably have encountered unfavorable environmental conditions. Disease organisms and environmental stressors (see Chapter 5) are normal in the lives of fish, and fish usually have the ability to withstand occasional exposures to these challenges without suffering ill effects. There are times when the pathogens become so numerous or the environment becomes sufficiently stressful as to threaten the lives of the fish or, short of that, slow their growth and inhibit their reproduction. When this happens, we term the fish diseased. Fisheries managers, researchers, and aquaculturists must be capable of detecting overt diseases in fishes because serious health problems in fish populations can affect the success of the fisheries biologists' work. Box 14.1 lists some examples of fish pathogens (parasites, bacteria, and viruses) that fisheries workers may encounter.

Pathogens are not the only causes of fish disease. In fact, adverse environmental conditions are often more important causes of ill health in fishes than are pathogens. Environmental conditions that can harm fish include degraded water quality (low dissolved oxygen, high ammonia, and excessive siltation), toxic pollutants (chlorine, pesticides, and heavy metals), and poor nutrition. Other factors, such as overcrowding and excessive competition, can reduce the ability of fish to obtain sufficient food and reproduce normally. Even when pathogens are directly responsible for disease, the infections are often secondary to degraded water quality or other environmental stressors; that is, adverse environmental conditions often allow infectious agents to become pathogenic.

Widespread pathogenic disease is rarely detected in wild fishes. Disease organisms are, of course, present in the wild, and wild fishes certainly die of infections, but the majority of mass fish kills are caused by adverse environmental conditions. In fish culture, however, pathogens often cause substantial mortalities because fish in culture ponds and raceways are generally crowded. Crowding encourages the transmission of pathogens, and poor water quality as a result of crowding stresses the fish.

14.3.2 Limitations of Field Diagnosis

Fish health cannot be completely assessed through the interpretation of clinical signs. Obvious indications of disease can be detected during field examination, but without thorough laboratory analysis even a specialist in fish health cannot diagnose disease in a fish. The causative agent of disease, be it a pathogen, toxicant, or adverse environmental situation, must be identified before a positive diagnosis can be made. The goal, then, is either to rule out or confirm specific, suspected disease agents. A biologist without fish health training, even with a microscope and other diagnostic equipment, can only be expected to recognize obvious indications of ill health and to be able to take, package, and send samples correctly for further analysis. In the following sections, references to clinical signs and specific disease organisms are intended only as examples, not as a key to definitive diagnosis.

14.3.3 Investigation of Fish Kills

Unfortunately, fisheries workers are sometimes called upon to evaluate mass die-offs of fish. Investigation of a fish kill can involve a variety of agencies. At the federal level, the U.S. National Marine Fisheries Service, U.S. Fish and Wildlife Service, U.S. Coast Guard, and U.S. Environmental Protection Agency may act depending on the situation. The state fish and wildlife agency and the state water quality agency are nearly always involved. Expertise in biology, chemistry, and statistics is needed. Although examination of individual fish is included in the assessment of a fish kill, it is important to determine the extent of the affected area and estimate the number of fish killed.

Generally, an interagency pollution response team will be activated upon report of a fish kill. This group should have predetermined guidelines (AFS 1992) to follow regarding the initial notification, field procedures, public relations, and report preparation. Typically, a fish kill is short lived and requires immediate response. Water samples for later testing for possible toxicants should be taken inside and outside the kill zone. Basic water quality characteristics, such as temperature, dissolved oxygen, pH, and alkalinity, should be measured on site. Individual fish should be examined for signs of ill health and frozen for future analysis. Often the cause of the die-off cannot be immediately determined.

Data collected during a fish kill are potential evidence in a legal proceeding, and care should be taken to insure that all observations are well documented and not easily lost or altered. It is best to enter written documentation in a bound notebook, to use indelible ink, and to identify the person responsible for each measurement (see Chapter 1). Notes should be duplicated and stored separately from the original. Photographs and videotapes can be extremely valuable, but care must be taken to identify the date, time, exact location, and personnel involved. Any samples collected must be uniquely numbered and labeled. Further, a "chain of custody" must be established, meaning that there must be a complete list of persons who have had responsibility for assuring the security and integrity of each sample through time, including dated signatures of each party when the sample moves from one person to another.

The primary role of the fishery biologist in assessing the fish kill is to determine the number, species, and size of the fish killed, the areal extent of the kill, and ultimately the monetary value of the fish killed. Accurately counting dead fish in a large kill is a very difficult task. Not all areas of the lake or stream may be accessible, not all dead fish will be visible to investigators in accessible areas, and the area of die-off may simply be too large for the personnel to cover completely. Because of these problems, a statistical sampling design is usually employed (AFS 1992). There are various sampling protocols for different situations, and the persons responsible for the count should know how to apply these prior to notification of a kill.

Once an estimate of the fish kill is made, the information needs to be used to determine the monetary and economic value of the loss. This economic information is needed by courts and regulatory agencies in setting levels of mitigation, if responsible parties can be identified. The simplest and most widely accepted method of valuation of fish is replacement cost. The "cost to raise" has been established for many species, even those not routinely reared in hatcheries (AFS 1992). This method of valuation can be employed whether or not the area of the kill is actually restocked. Although replacement cost has legal precedent and broad acceptance, it is an underestimate of the value of the dead fish because it neglects the value to the user of the resource, primarily the angler. There are methods of assessing the economic value of recreational opportunity that may be lost for some time following a fish kill. The two most important are the travel cost method and the contingent ("bidding") method (Weithman 1993). Analysis of all expenditures that anglers make to go fishing is the basis of the travel cost method, whereas the contingent method relies on anglers' responses to hypothetical questions about the value of the fishery.

14.3.4 Behavioral Signs

Even before a fish is in hand, we can observe behavioral indications of environmental stress or ill health. In early morning during the summer, it is not unusual to see fish gulping at the surface, or "piping," in highly eutrophic lakes and ponds. This is an adaptive response to low dissolved oxygen; by pulling the surface film over their gills, the fish are respiring the water with the highest oxygen content. Fish can tolerate occasional, short-term oxygen deficiencies without ill effect, but piping late in the day or piping by fishes tolerant of low dissolved oxygen, such as black bullhead or common carp, is indicative of a dangerous situation.

A behavioral indication of external parasites, more often observed in hatchery raceways than in the wild, is the fish rubbing itself against the bottom. This is termed "flashing" because when you look down into the water, the silvery side of the fish turns up. Behavior of stream fishes feeding on aquatic insect drift can be mistaken for flashing. Fish near death often abandon normal behavior and may be seen swimming aimlessly near the surface or finning quietly in shallow water without their normal wariness. Such individuals can usually be dipnetted without difficulty. On rarer occasions, fish may be seen in convulsions (sometimes as a result of pesticide poisoning) or whirling (a sign of certain infectious diseases).

14.3.5 External Signs

Body conformation and color. If at all possible, select living fish for examination. Fish undergo rapid postmortem changes that complicate evaluation of both external and internal signs. It is helpful first to observe fish alive and swimming. To do this, place the living fish in a clear plastic bag of water. Signs of ill health include excessive mucus production on body and gills (often the result of toxicant irritation or ectoparasites), fins clamped close to the body and shimmying (a general sign of illness), and faded or blotchy coloration (often a sign of stress and some bacterial infections) (Francis-Floyd 1988).

For further examination, kill the fish. Any time fish are handled, humane procedures should be followed, and an approved institutional animal care and use protocol should be on file (see Chapter 5). Does the fish's body conform to what is considered normal for the species? The body should be fairly robust with a head proportional to the body and eyes proportional to the head. An emaciated fish with a large head and eyes is malnourished. Such fish may be found in overcrowded conditions in which food is scarce. Protruding eyeballs, "pop-eye" or exophthalmia, may be the result of parasites or gas bubble disease, which is caused by water supersaturated with nitrogen.

Fins. The fin membranes should be intact to the end of the rays, free of slime or cottony fungus (e.g., *Saprolegnia* spp.), and without hemorrhagic areas . In the spring, it is normal to see occasionally the lower lobe of the caudal fin frayed in centrarchids because of nesting activity. Otherwise, frayed fins (Figure 14.1) may indicate the attack of bacteria such as *Flexibacter* spp., especially when slime or hemorrhage is evident. In advanced cases of fin "rot," fins may be entirely eroded. Also inspect fin membranes for encysted parasites that appear as small white (e.g., *Ichthyoptherius multifilis*) or black (e.g., *Neascus* spp.) spots; these may appear elsewhere on the body as well.

Skin, scales, and mucus. Scales should lie flat and be firmly attached. A thin, clear, evenly distributed mucus film should cover the fish, and the surface of the fish should be free of reddened areas, bloody sores (hemorrhagic lesions), nodular growths, and fungus (Figure 14.1). Reddened areas and lesions are evidence of systemic (widespread, internal) infections of bacteria (e.g., *Aeromonas* spp.) or superficial bacterial infections (e.g., *Flexibacter columnaris*). Skin lesions may be complicated by parasite infestations (e.g., *Epistylus* spp.) or fungus. Skin lesions in wild fish are seen most often during the early spring when rising water temperatures encourage bacterial growth at a time when fish are least resistant to it. An increased prevalence of skin lesions also has been associated with fish from water with high organic load, such as below a sewage outfall, and a correspondingly high bacterial community. Nodular growths are a typical host reaction to some parasites; such growths may also be tumors caused by chemical pollution. Large ectoparasites of the skin (e.g., *Lernea* spp.) can be seen without magnification.

Gills. Pull back or cut off the operculum and carefully inspect the gills. In a freshly killed fish, the gills should be bright red and without a thick mucus covering. The gills are frequently a site of ectoparasites; they are also a sensitive area constantly exposed to the water and, as such, are often the first tissue to show an

Figure 14.1 An apparently healthy bluegill (upper) compared with one showing a number of overt signs of disease (lower).

adverse reaction to unfavorable water quality. Nearly all cultured fish and many wild fish display some gill damage. Though a microscope is necessary to verify a healthy gill, the unaided eye can easily detect pale color, excess mucus, and erosion of the posterior edge of the respiratory tissue (holobranch).

14.3.6 Internal Signs

Technique for opening the fish. After carefully examining the outside of the fish, begin the internal examination, or necropsy, by laying the fish on its side and making an incision just above the vent, along the top of the rib cage, and forward through the pectoral girdle and fifth nonrespiratory gill arch, which marks the rear of the opercle cavity. Scissors work best for fish up to 0.5 kg. Take a shallow bite with the scissors so that the internal organs are not damaged. Pull the flap downward to open the body cavity, freeing the forward edge and any clinging mesenteries. The body wall can now be cut along the bottom and removed to expose the internal organs fully (Figure 14.2). This method of opening a fish is useful in general examination and

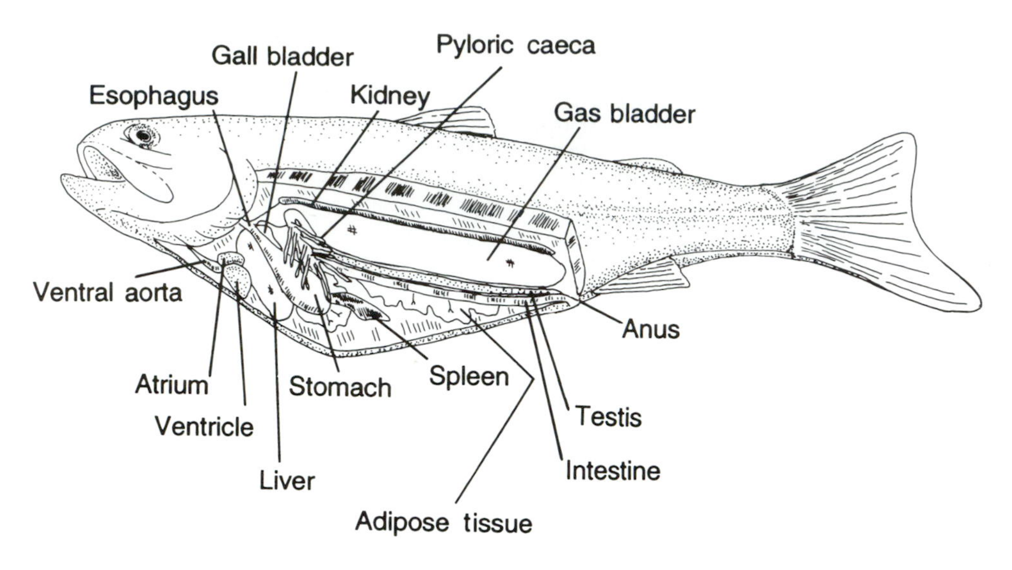

Figure 14.2 A trout dissected to show normal internal organs.

inspection for parasites. An aseptic technique for opening a fish to obtain material for bacterial isolation is described in section 14.4.3.

Digestive tract. Carnivorous fish, which include most game species, have a gut that is a short, simple S-shaped tube. Beginning anteriorly, the esophagus extends from the oral cavity to a muscular, elastic stomach. The stomach empties into either a straight or simply convoluted intestine that terminates at the vent or anus (Figure 14.2). The liver, which among other vital functions produces bile, lies around the anterior portion of the stomach. It is connected to the gut by the gall bladder and bile duct. Pyloric caeca, fingerlike out-pocketings of the digestive tract in the region of the pylorus, are present in many fishes (e.g., trout and centrarchid bass) but not in all (e.g., catfishes). Pyloric caeca often appear to be draped over the front of the stomach because the digestive tract's S-shaped curve places the pylorus anterior to the stomach. Although not part of the digestive tract, the spleen is easily recognized as a small, darkly colored organ located near the posterior portion of the stomach.

Although normal livers usually have a rich, reddish brown color, they may appear somewhat pale because of species and dietary differences (e.g., glycogen buildup in cultured fish). Such color variation does not necessarily reflect ill health. The liver, however, should be firm, uniformly colored, and free of spots. Bacterial diseases sometimes produce abscesses in the liver that appear as light-colored areas; small tumors also may have a similar appearance. Larvae of trematode parasites (small white "grubs") can be seen in the liver without magnification. Enlargement or nodules in the spleen are considered pathologic (diseased) conditions. The amount of mesenteric fat deposits along the gut (particularly the pyloric caeca) indicates the previous feeding success of the fish. Little fat signifies a fish with low energy reserves whereas heavy fat deposits are often seen in cultured fish. Inflammation in or around the gut or anywhere in the peritoneal cavity is a frequent sign of a systemic bacterial infection. Large amounts of mucus in an otherwise empty intestine are associated with some viral diseases. Large parasites (e.g., nematodes and cestodes) can be seen easily when present inside the gut, but many parasites, especially protozoa, are microscopic. A stomach and intestine containing food items means the fish has fed recently, usually a sign of good health. Absence of food in the gut, however, may only mean that the fish has not eaten recently.

Kidney. The kidney lies along the backbone above the gas bladder. In systemic infection, bacteria tend to accumulate in the kidney. This explains why the kidney often shows signs of bacterial infections and why the kidney is the best organ from which to isolate pathogens (section 14.4.3). A healthy kidney is uniformly dark in color and has a flat or slightly concave surface. In contrast, a kidney with white spots and puffiness is indicative of advanced systemic infection or toxicosis.

Muscle. Slices through the white muscle of fish may reveal hemorrhagic areas and either free or encysted larvae of trematodes and cestodes, which can be seen readily without magnification. Cavity-like lesions in the muscle occasionally are caused by chronic, systemic bacterial infections. Although most fish pathogens cannot be transmitted to humans, and the few exceptions are killed by cooking, obviously parasitized fish are usually rejected by anglers for aesthetic reasons. There are cases, then, where parasite occurrences not severe enough to be harmful to the fish population are harmful to the fisheries resource.

14.4 SAMPLING FOR DISEASE ORGANISMS

14.4.1 Diagnostic Expertise

As stated in section 14.3.2, a biologist without extensive fish health training can only be expected to spot gross indications of ill health and to take, package, and send samples correctly for further analysis. The previous sections outlined methods of detecting obvious disease. The following sections discuss the proper way to obtain and handle samples.

Where can samples be sent for definitive identification or pathogens or toxic residue analyses? Because there is a different arrangement for handling such samples in each state, and the arrangements change with fluctuations in funding and staffing, a universal list would be cumbersome and soon outdated. Although state fish and wildlife agencies usually do not have their own complete diagnostic facilities for fish health, they almost always have a formal or informal arrangement with a university, federal installation, or private practitioner to provide definitive identifications of pathogens when necessary. Similarly, fish and wildlife agencies usually depend on other groups for toxic residue analyses after the more obvious environmental causes of fish kills (e.g., dissolved oxygen and temperature) have been ruled out. Often, pollutant analyses are conducted by the state water quality agency with which the fish and wildlife agency cooperates in investigating fish kills (section 14.3.3). Individuals with fish health problems can find diagnostic expertise through their state fish and wildlife agency or allied federal organizations in the U.S. Departments of the Interior and Agriculture. Also, many fish health specialists are active in the Fish Health Section of the American Fisheries Society, which is an excellent source of information and expertise.

14.4.2 Sampling for Parasites

Selection and care of specimens. Though field examination of fish is best begun with living fish, live fish are mandatory for parasite examination because many external parasites leave fish within a few minutes after the host's death. Many of the larger parasites, both internal and external, can be viewed with the unaided eye or with the help of a hand lens. An external examination and necropsy conducted in the field, therefore, may provide some useful information. However, the basic parasite examination requires at least a good dissection microscope. A thorough examination, which leads to a definitive inventory of parasites, requires a compound microscope and a person with fish health training. It is often necessary, therefore, to transport living fish suspected of having a parasite problem to a laboratory where they can be more fully examined.

Basic parasite examination. Although fish health training is necessary to make a complete parasitological report on a fish, a studious biologist who routinely examines fish for parasites can soon learn to recognize the important ones and make judgements regarding the severity of the infection. A basic parasite examination (Hoffman 1967) is begun by killing the fish. If the fish is small, submerge the fish in a petri dish of water and examine the fish under a dissection microscope. If the fish is large, keep the body surface moist, remove portions of the fins, and examine those portions under the dissection microscope. Next, take mucus scrapings from the body of the fish, place them in a drop of water between a slide and cover slip, and examine them under the compound microscope.

Cut away a gill arch and examine it under a dissection microscope. Then place a bit of gill tissue in a drop of water between a slide and cover slip and examine it under the compound microscope. Open the fish as described in section 14.3.6, remove the viscera, and examine the organs submerged in a petri dish of water (or, preferably, physiological saline) under the dissection microscope. While examining the viscera, tease apart the internal organs with forceps and scrape the inside of the gut to reveal parasites. Tease apart and examine some of the body musculature and check the eyes and brain under a dissection microscope as well.

14.4.3 Sampling for Bacteria

Selection and care of specimens. It should be recognized that a number of fish must be screened for bacterial pathogens in order to make a valid conclusion about the health of a population. Whereas methods of determining the actual sample size necessary are outside the scope of this chapter, it should be obvious that a larger sample of fish is necessary to rule out an active pathogen in a larger population of fish or when incidence of disease is low (McDaniel 1979). Usually, when fish are transported to fish health diagnostic laboratories for isolation of the vector (pathogen causing the disease), the first step is bacterial identification. There is no reason, however, that bacterial cultures cannot be begun in the field, as long as an aseptic technique is followed carefully. In fact, field cultures are preferable (when done properly) because of rapid postmortem decomposition and resulting contamination of the tissues from normal gut and skin bacteria. All fish must be alive when collected and, ideally, freshly killed when bacterial isolation is attempted. If bacterial isolation will not be done in the field and fish cannot be kept alive, they may be enclosed individually in sealed plastic bags and stored on wet ice for preferably less than 8 h and absolutely not more than 48 h.

Bacterial isolation. Although identification of bacterial pathogens requires specialized training as well as access to a variety of stains, media, and laboratory equipment, taking an initial culture is fairly simple and requires only a few basic

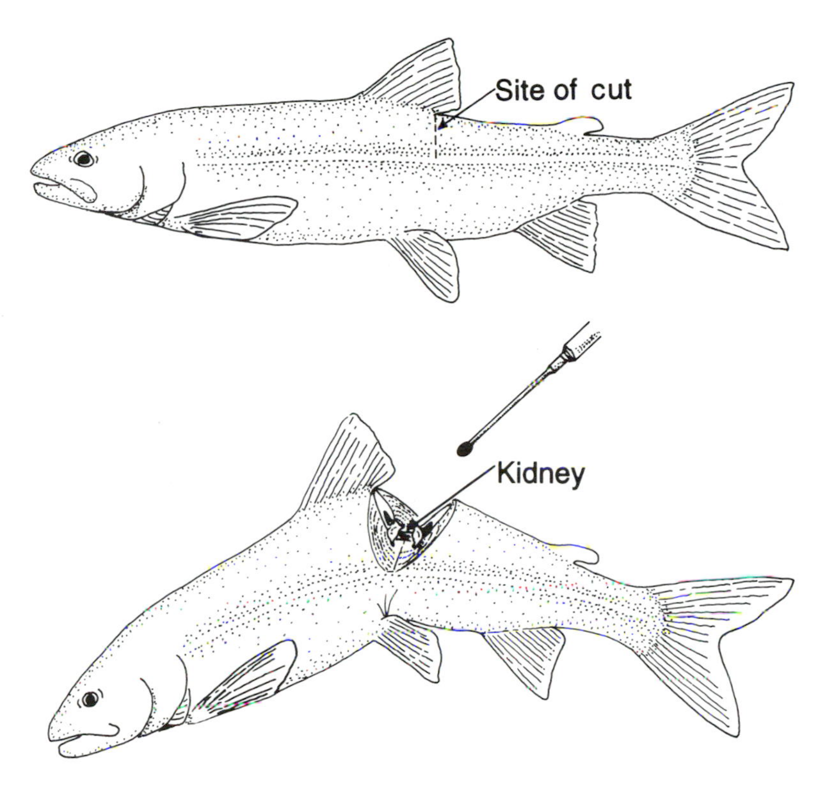

Figure 14.3 Aseptic technique for obtaining kidney tissue for bacterial culture.

tools. After conducting the external examination and killing the fish (section 14.3.5), sear the area immediately behind the dorsal fin with a glowing red spatula to kill any bacteria on the surface. Dip a pair of scissors in high-proof alcohol and heat them in a flame. Cut down through the seared area, severing the spinal cord but not going deeper. Bend the head and tail together to expose the kidney, a dark red mass below the spinal cord (Figure 14.3). Insert a flamed inoculating loop down into the kidney, extracting a loop of tissue. Streak the tissue on about a quarter of a sterile nutrient agar plate (blood agar and trypticase soy agar are good for isolation). Flame the loop and streak across the plate at right angles from the edge of the initial streaks. Flame the loop again. Streak the tissue across the plate a third time from the edge of the second streaks. This procedure will spread out the bacteria, permitting the growth of individual colonies. Isolations from fish may be incubated at room temperature (20–25°C), although faster growth of colonies will occur under incubation at 37°C.

14.4.4 Sampling for Viruses

Diagnostic work with viruses requires specialized procedures and equipment. Select fish suspected of viral disease and transport them to a fish health diagnostic center as described in the section for bacterial sampling (14.4.3). A biologist who routinely transports samples for viral analysis may want to learn to make the initial tissue homogenates, which, if buffered, may be stored for up to a week under refrigeration (McDaniel 1979).

14.5 SAMPLING BLOOD AND TISSUE

14.5.1 Rationale for Collection of Blood and Tissue Samples

As mentioned in section 14.3.1, adverse environmental conditions are more often the cause of widespread ill health in wild fishes than are pathogens. Often the specific stressor can be readily identified in the environment, as in the case of an acutely depressed dissolved oxygen level. At other times, however, overt signs of disease and mortality may occur in fish populations when no obvious adverse environmental conditions appear and pathogens are not suspected. In these instances, it is often useful to obtain blood and tissue samples from fish for pathology studies and residue analysis. Such analyses also may be useful in evaluating sublethal stress (Wedemeyer and Yasutake 1977) and often are employed in toxicological and physiological research on fishes. In addition, there is strong interest in the genetic composition of fish stocks, and sampling tissue for genetic analysis is becoming widespread (see Chapter 5).

14.5.2 Sampling Blood

Techniques of obtaining blood. Although there are a number of ways to obtain blood samples from fish, perhaps the easiest and most consistent method involves tapping the blood vessels that run in the hemal arches of vertebrae between the anal fin and caudal peduncle. Fish less than 15 cm long must be sacrificed to obtain even a minimal blood sample for microanalysis (0.1 mL). Place the fish under deep anesthesia by means of tricaine methanesulfonate (see Chapter 5), wrap the fish in a paper towel for ease of handling, with a razor blade cut the tail off just above the caudal peduncle, and immediately place a heparinized (or otherwise anticoagulant-treated) capillary tube, with a volume of about 0.25 mL, against the severed caudal vessel. The tube will fill with blood, and additional tubes may be collected if possible and necessary, though they must be filled without delay because fish blood coagulates quickly.

Blood samples from fish longer than 15 cm may be obtained without killing the fish. Anesthetize the fish to immobility (50 to 100 mg/L of tricaine methanesulfonate) and prepare a heparinized syringe or Vacutainer® blood sampler. Syringes may be heparinized by rinsing with a heparin solution (commercial, injectable preparations of 10,000 units/mL work well). Vacutainers are available with a variety of anticoagulants. While one person holds the fish ventral side up, another inserts the needle (3.81 cm; 21 gauge for small fish, 18 gauge for large fish) at the midline between the anal fin and the caudal peduncle until the needle stops against the vertebral column (Figure 14.4). Apply slight suction with the syringe or break the seal on the Vacutainer; if blood flow does not begin, slight twisting and movement of the needle may initiate it. When an adequate sample is obtained, withdraw the syringe, remove the needle, and expel the blood into a microcentrifuge tube (1.5 mL).

Preservation of samples. Depending on the analysis to be performed, whole blood may be stored up to several hours on ice. Hematocrit (packed red blood cell volume), however, should be run within minutes after blood collection because red blood cells swell when blood is exposed to the air. Most blood characteristics are determined on plasma, the fluid portion of the blood without the cells. It is best, therefore, to centrifuge blood immediately after collection, separating the plasma from the cells, and to freeze the plasma sample. This can be accomplished in the field by using a portable generator for the centrifuge and a cooler with dry ice.

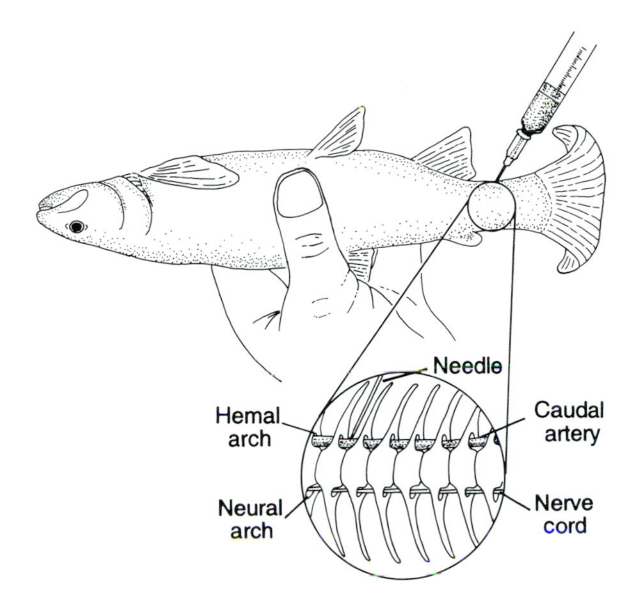

Figure 14.4 Technique for obtaining a blood sample from a large fish without killing the fish.

14.5.3 Sampling for Histology

Postmortem changes in histological features (microscopic tissue structures) occur rapidly, making it mandatory that histological samples be taken from fish collected alive. Immediately after killing the fish, remove the tissues of interest. Make small, thin slices and drop them into labeled vials of fixative. The necessity of taking thin (2 mm) slices deserves emphasis because thin slices promote rapid penetration by the fixative and minimize later concern about artifacts caused by poor fixation. Bouin's fluid (Galigher and Kozloff 1971) is an excellent fixative for fish tissue, although buffered, 10% formalin is nearly as good and has the advantage of being compatible with routine procedures for slide preparation in large, automated histopathology laboratories. The volume of fixative in the vial should be at least 10 times the volume of tissue to be fixed.

Although most fish tissues have some diagnostic value in a histopathology study (Roberts 1978), gill, liver, and kidney tissues are almost always included. Kidney tissue is difficult to dissect from fish because it adheres to the body wall and is soft. In small fish, thin slices can be taken through the vertebral column to include the kidney.

14.5.4 Sampling for Residue Analysis

Tissue for most residue analyses may be taken from dead fish as long as the tissue is not putrid. Wrap tissue for heavy metal analysis in plastic and wrap tissue to be analyzed for organic compounds in aluminum foil. Keep wrapped tissues on ice until they can be frozen. It is important to sample white muscle because it constitutes the edible portion; take white muscle tissue from the fish's back, in front of the dorsal fin. Analyze organ tissues as well because they tend to accumulate toxins to a greater degree than does muscle. Blood plasma and brain tissue are important because they reflect physiologically available residues and do not include an inactive, bound component that may occur in other tissues. Quantitative analysis of tissue for toxicants is an expensive procedure. This limits the number of samples that can be tested. In order to sample broadly for a toxicant believed to be at low incidence of occurrence, a number of fish may be pooled into a single sample for analysis. In that way, 5 or 10 fish can be tested for the price of 1 fish.

14.6 REFERENCES

AFS (American Fisheries Society). 1992. Investigation and valuation of fish kills. American Fisheries Society Special Publication 24.

Francis-Floyd, R. 1988. Behavioral diagnosis. Pages 305–316 *in* M. K. Stoskopf, editor. Tropical fish medicine. The veterinary clinics of North America: small animal practice, volume 18, number 2. Saunders, Philadelphia.

Galigher, A. E., and E. N. Kozloff. 1971. Essentials of practical microtechnique. Lea & Febiger, Philadelphia.

Goede, R. W., and B. A. Barton. 1990. Organismic indices and an autopsy based assessment as indicators of health and condition of fish. American Fisheries Society Symposium 8:93–108.

Hoffman, G. L. 1967. Parasites of North American freshwater fishes. University of California Press, Berkeley.

Legler, R. E. 1977. New indices of well-being for bluegills. Master's thesis. University of Missouri, Columbia.

McDaniel, D., editor. 1979. Procedure for the detection and identification of certain fish pathogens. American Fisheries Society, Fish Health Section, Bethesda, Maryland.

Roberts, R. J., editor. 1978. Fish pathology. Baliere Lindall, London, England.

Wedemeyer, G. A., and W. J. Yasutake. 1977. Clinical methods for the assessment of the effects of environmental stress on fish health. U.S. Fish and Wildlife Service Technical Paper 89.

Weithman, A. S. 1993. Socioeconomic benefits of fisheries. Pages 159–175 *in* C. C. Kohler and W. A. Hubert, editors. Inland fisheries management in North America. American Fisheries Society, Bethesda, Maryland.

Chapter 15

Length, Weight, and Associated Structural Indices

RICHARD O. ANDERSON AND
ROBERT M. NEUMANN

15.1 INTRODUCTION

Length and weight data provide statistics that are cornerstones in the foundation of fishery research and management. The objective of this chapter is to present information on methods of measurement and the calculation and some potential interpretations of structural indices from such measurements.

The numbers and sizes of fish in a population determine its potential to provide benefits for commercial and recreational fisheries. Length and weight data also provide the basis for estimating growth, standing crop, and production (tissue growth) of fish in natural waters as well as in hatcheries and laboratories. Annual production is the generation of tissue weight per unit area—often expressed as kilograms per hectare—in the course of a year. It is determined by the rate of reproduction (number of viable offspring), the rate of growth (change in weight of individuals), and the rate of mortality (loss of individuals in an age-group) (Box 15.1). These functional rates determine population dynamics over time, as well as structural attributes such as density (number per given area), biomass (weight per given area), and length frequency (section 15.6) at any point in time.

Population "structure" is the number (or proportion) of individuals in each age- or size-group of the population. One challenge for a fishery manager is to identify problems and opportunities presented by existing population structures. Effective adjustments of functional attributes, such as altering mortality rate with length limit regulations, may result in a population structure that better supports management objectives for that fish population and community.

15.1.1 Uses of Length Measurements

Fish length is important to recreational anglers. In many fisheries, length is used to define legal size for harvest. Weithman and Anderson (1978) and Weithman and Katti (1979) developed a fish quality index that describes the value of a captured fish in terms of the world record length for that species. Gabelhouse (1984a) used this fish quality index to define, for several recreationally important species, length categories that can be used to evaluate length frequencies of fish samples in terms of management objectives. Fishery managers thereby can use length-frequency data to assess fish populations and to monitor them over time in response to management strategies.

Box 15.1 The Production Process

The production process is best illustrated by a graph (Allen 1951) representing the ages and sizes of fish in a population. In the model population of largemouth bass illustrated here, the number and mean weight of each age-group are assumed to be consistent year after year. The number of fish and average fish weight of each age-group at annulus formation in the spring is illustrated in the figure (modified from Anderson 1974). The straight line connecting each age-group incorporates all mortality during the growing season. The area ABEF represents the total weight of 100 age-2 fish that die each year; the area ABCD represents the annual production accomplished by this cohort. The total area under the curve is annual production for the population.

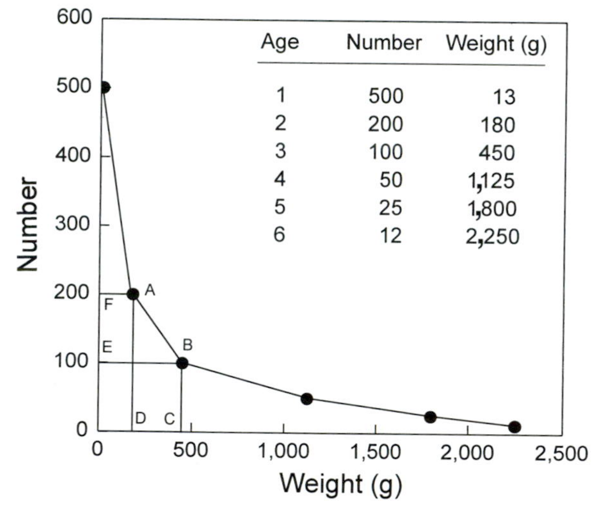

Based on information contained in this figure, answer the following questions.

1. What is the total number and biomass of fish in this population in the spring?
2. What is the annual production of the population?
3. What is the ratio between production and biomass?
4. If 180 g is assumed to be minimum harvestable size and there was no natural mortality, what number and weight harvested could be sustained each year?

After finishing this chapter, develop answers to these questions.

5. If it is assumed that the average individual in each cohort is in good condition, can you estimate values for the size structure indices of proportional stock density (PSD) and relative stock density (RSD)?
6. What assumptions would you make about the size and productivity of the lake for this population and other components of the fish community?
7. Would you like to go fishing there?

Answers to these questions can be found in Appendix 15.1.

15.1.2 Uses of Weight Measurements

Weights of individuals and aggregate biomasses are also key attributes of populations. The production process results in the creation of tissue by individuals within populations. Although production can be expressed as calories or as weights of carbon, protein, or dried tissue, all these measures are normally based on a measurement of wet or live weight. From a fishery perspective, total weight or weight per unit area is the statistic normally reported for harvest or standing stock. Weight is the common basis for reporting catches, whether made by anglers or by commercial fishers. Weight at age and annual weight increments are statistics that describe the growth process. Annual weight increments best reflect how fish of various sizes gain in value to the fishery; annual weight gains and appropriate growth efficiency values can be used to estimate consumption of prey. Length and weight data can also be used to calculate indices of condition (section 15.4).

15.2 SAMPLING AND MEASURING CONSIDERATIONS

All methods of sampling fish populations have some inherent bias. The sampling bias for or against certain sizes or particular gender of fish is related to an individual's vulnerability to the gear—that is, to the probability that a fish will be caught by a gear. Factors influencing vulnerability are the times and places the gear is used (fish may not be present or, if they are, they may not be accessible to the gear because of their behavior or habitat preferences) and the selectivity of the gear (the sizes and types of fishes the gear can retain when fish encounter it). To the extent they can be recognized, sampling biases should be acknowledged and corrected for (Chapters 6–8, 10).

The number of fish to be measured and weighed is influenced by sampling objectives, the variability among individuals and populations studied, and the number and sizes of fish collected (Chapter 2). Often only subsamples can or need to be measured or weighed. Research often requires larger sample sizes than are necessary for management surveys. In general, it is desirable to measure more fish for a length-frequency distribution (section 15.6) than to weigh for a weight–length relationship (section 15.3). For many species, sexes should be distinguished because males and females may differ in their morphologies. The best measures of variability and confidence are determined from the length and weight of individuals rather than from the total and average weights for a group within a size-class. In the field, errors in weight measurements are more common than are errors in length measurements, and care should be taken to obtain precise measurements of weight (Gutreuter and Krzoska 1994).

In fishery management and stock assessment studies, it is desirable to follow standard practices. The justification is not that one measure is inherently superior to another but that conventional techniques facilitate data comparability and communication of results. For example, there are two conventions for designating a length-group in length-frequency histograms: a 10-cm length-group may include fish from 10.0 to 10.99 cm or from 9.5 to 10.49 cm. This difference can lead to problems in communication and data comparison. The first convention is advantageous because length limits, when they are established, are based on whole numbers, as are the defined stock and quality lengths (Anderson 1976) used for determination of length-frequency indices (section 15.7). Many scientific and working groups have adopted this convention (Holden and Raitt 1974).

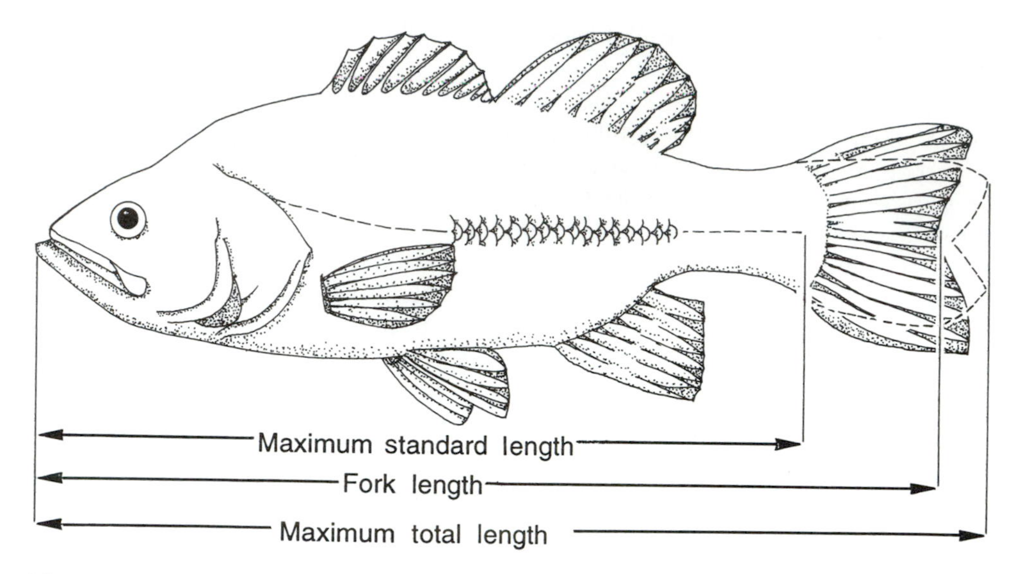

Figure 15.1 Common measurements of fish length—maximum standard, fork, and maximum total.

15.2.1 Length Measurements

Two approaches are used to measure fish length: measurement of the whole body and measurement of body parts. Whole-body measurements are most commonly used in fishery investigations, although partial measurements are useful in food studies or for fish that have been beheaded or have damaged tails or heads.

Three common whole-body measures of fish are total, fork, and standard lengths (Figure 15.1). Total length is defined as the length from the anterior-most part of the fish to the tip of the longest caudal fin rays. Total length is measured by two conventions: when the lobes of the caudal fin are compressed dorsoventrally and when the caudal fin is left spread in a "natural" position. Europeans favor the "natural" orientation; North Americans typically measure with the caudal fins compressed. In Canada during the 1920s "total length" meant fork length.

Fork length is the length from the most anterior part of the fish to the tip of the median caudal fin rays. Fork length is commonly used in fishery studies in Canada and for certain species such as tunas. It has been favored for species with "forked" caudal fins (those whose dorsal and ventral rays are longer than median rays) because the longest rays are often eroded by contact with rocks, debris, or hatchery walls.

Standard length is the length from the tip of the upper jaw to the posterior end of the hypural plate. The hypural plate, originating from the posterior end of the vertebral column, is an array of altered vertebral elements that support the rays of the caudal fin. In practice, standard length may be measured to some external feature such as the position of the last lateral line scale, the end of the fleshy caudal peduncle, or the midline of a crease that forms when the tail is bent sharply. Standard length is most commonly used in taxonomic studies because it is unaffected by caudal fin anomalies. Maximum standard length, used by some investigators, is measured from the most anterior part of the fish; this is longer than standard length for fish that have protruding lower jaws. Standard length is a less convenient measure for fishery workers than either fork or total length.

Less common measuring techniques are used for special situations. Measures of specific body parts are used when intact fish are not obtainable. Pectoral length, the distance from the posterior insertion of the pectoral fin to the posterior margin of the longest caudal fin rays, is used when fish have been beheaded (Laevastu 1965). Head length is also used when body damage impedes other measurements. For spawning or recently dead Pacific salmon, which often have frayed tails and enlarged or damaged jaws, measurements are made from the eye to the hypural bone. In the United States, the measurement typically is made from the middle of the orbit (midorbital–hypural distance); in Canada, it is made from the hind margin of the orbit (postorbital–hypural distance). Length measurements for paddlefish are usually reported as the length from the anterior edge of the eye to the fork of the tail (Ruelle and Hudson 1977).

Because different measurements of length are used, it is sometimes necessary to convert one to another to make comparisons. Such conversion functions are usually linear and often reduce to a constant of proportionality. These functions may be estimated for any sample by measuring any combination of total, fork, standard, or other lengths and applying regression techniques. Geometric mean (GM) regressions (Ricker 1984) give much better results than do ordinary regressions when the regression is projected over a wide range or used for a different population (Chapter 2). Any new sample, even from the same population, is likely to include a different range or distribution of lengths and will require a different line. If the complete range of lengths in a population is used, and the lengths are more or less evenly distributed by length-classes, the GM and ordinary regressions are practically the same. In a great majority of cases, a simple proportion is adequate. An alternative is to use published conversion factors for the species concerned (e.g., Carlander 1969, 1977).

Measuring devices. Various devices used to measure fish length include measuring boards, tape measures, calipers, and plastic or foil sheets. The choice depends in part on the budget and the sizes and species of fish measured. Often one person makes the measurements and a second records them. Data can be entered into a portable computer or data logger while measurements are being made. Electronic measuring boards have also been developed. Audiotape recorders are sometimes used to record data if a worker is alone.

Measuring boards consist of a linear scale on a board with a rigid headpiece. Finfish are usually measured with the fish's mouth closed, slight pressure exerted against the headpiece, and the fish's body positioned on its right side so that its head is facing the observer's left. Herke (1977) described a plexiglass measuring board that is easily fabricated. Aluminum rulers have an advantage of durability. When pectoral lengths are measured, the headpiece is undesirable. Measuring boards should be maintained in good condition. Warped or loose headpieces and worn or faded markings can be a cause of error. Semiautomated measuring boards with computer interfaces are becoming more common (Morizur 1995; Sigler 1995).

Tape measures are desirable for measuring large marine fish. Calipers may be used for small fishes; although precise, they are more time consuming and cumbersome to use than are other devices.

Plastic or foil sheets are useful when lengths are the only data recorded for individual fish (Holden and Raitt 1974). Removable sheets are fixed across the measuring board. Instead of recording fish length, a pin hole is made at the terminus of the measurement. Many fish can be measured on the same sheet and length-frequency distributions then can be easily determined.

Accuracy and precision of length measurements. Accuracy refers to the difference between any measurement and the actual value. Precision refers to the reproducibility of a measurement (Chapter 2). Accuracy of length measurements is determined by the physical state of fish, ability of the investigator, and technique used. Fish should be measured fresh if possible. Rigor mortis, drying, and preservation result in shrinkage. If rigor mortis has occurred, fish should be flexed several times to relax the musculature. Accuracy can be influenced by any numerical bias for even, odd, or other specific divisions on a scale read by an investigator. The consistency with which fish are pressed against the headpiece and with which caudal lobes are compressed affects precision. Fish size and morphology also influence precision. The accuracy and precision of length and weight data and the speed of measurement are influenced by working conditions. Whatever is practical should be done to maintain good working conditions; adequate lighting and comfort of workers are of primary concern.

15.2.2 Weight Measurements

Weighing fish is more difficult and time consuming than is measuring their length. Weighing fish under field conditions presents special challenges. Spring-loaded scales are commonly used for weighing individual fish in the field. However, only fish that weigh at least 10% of scale capacity can be weighed with reasonable precision (Gutreuter and Krzoska 1994); therefore, a set of scales with different capacities should be used to cover a useful range of weights. Jennings (1989) reported that for weighing fish less than 1 kg, small handheld spring-loaded scales were similar in precision to triple-beam balances; these small spring-loaded scales were easy to use and to carry into the field. Scales with dashpot temperature compensators are available. Electronic scales with digital readout are available for field use. Large individuals or bulk lots of fish are usually weighed on hanging or platform scales. Small live fish are often weighed in a tared volume of water. In the field, portable, battery-powered electronic balances are often used for weighing small fish. A wide array of scales and electronic balances offering excellent accuracy are available for laboratory use.

Because most fish maintain near-neutral buoyancy in water throughout their lives, their specific gravity is close to 1.0. Body volume is proportional to weight and can provide a substitute measure. Volumetric measure of water displaced can prove useful for weighing large numbers of live fish as a lot—a need that often arises in fish hatcheries.

Accuracy and precision of weight measurements. The accuracy associated with weight measurements is determined by the accuracy of the weighing device, the amount of moisture on fish, and changes in the fish caused by death or preservation. Of these variables, the accuracy of the weighing device is easiest to control. Accuracy of a scale can be determined with a set of standard weights. Periodic checks of measuring equipment are recommended.

All fish carry with them a quantity of water on the body surface and in the buccal cavity. The error caused by surface water is inversely related to fish size. Small fish have a greater surface area for their volume or weight than do large fish. Because the quantity of surface water is variable among fish, it results in a loss of precision. To increase precision, allow water to drip from the fish or blot the fish before they are

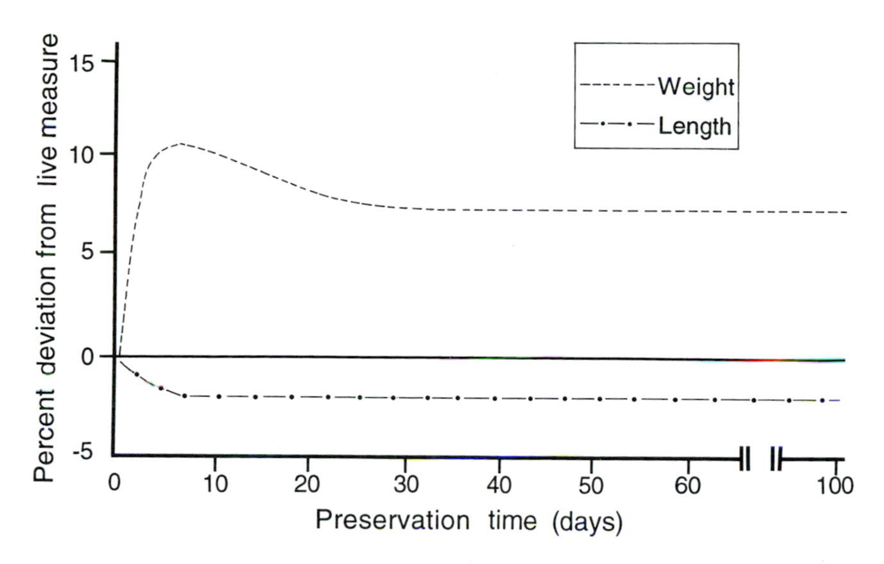

Figure 15.2 Typical patterns of effect on length and weight that might be observed for fishes preserved in 10% freshwater formalin.

weighed. Parker (1963) reported that blotting of fish before they were weighed improved the accuracy but increased the standard deviation of weights slightly as compared with drip drying. Damp chamois or cloth can be used to blot much of the surface moisture from fish without causing damage. When wet fish are being weighed in the field, water inevitably accumulates on the scale. Ensure proper drainage or compensate for water accumulation by adjusting the tare as often as necessary. When platform scales are used, take care to avoid drops of water on the balance beam.

Some measurements of body weight have been made after removal of the stomach contents. In laboratory studies, fish are often fasted before they are weighed. Variations in the degree of hydration of sex products also can influence body weight. Usually body weight should include the weight of sex organs because the energy and materials of these organs are part of fish production.

Weighing devices are sensitive to motion from the wind, the boat, and the fish. Wind problems can be avoided or reduced with a suitable windbreak. The problem of boat movement can be circumvented by weighing fish on shore. If it is necessary to weigh fish in a boat, wind problems can be reduced if the boat is allowed to drift freely with the waves rather than being anchored. The problem of fish movements can be alleviated by either sacrificing or anesthetizing fish. Anesthesia also helps reduce handling stress (Chapter 5).

15.2.3 Preservation Effects

The effects of preservation on length, weight, and indices of well-being deserve special attention. The method and duration of preservation affects both length and weight (Chapter 5). The effects of preservation in formalin have been studied most frequently; the pattern of effects on length and weight that has emerged from these studies is depicted in Figure 15.2. To estimate live weight, appropriate conversion factors are needed to convert length and weight data of dead or preserved fish.

15.3 WEIGHT–LENGTH RELATIONSHIPS

Le Cren (1951:202) stated "The analysis of length–weight data has usually been directed towards two rather different objects. First, towards describing mathematically the relationship between length and weight, primarily so that one may be converted to the other. Secondly, to measure the variation from the expected weight for length of individual fish or relevant group of individuals as indications of fatness, general 'well-being,' gonad development, etc." The term condition was applied to analyses of the second type. Condition (section 15.4) could be eminently useful as an inexpensive, easily measured surrogate if it could be demonstrated that under given circumstances it is a robust predictor of fecundity, reproduction, growth, or mortality rates.

A variety of useful concepts, centering on the body shape of individual fish, arise from the consideration of combined weight–length data. The power function

$$W = aL^b, \qquad (15.1)$$

where W is weight, L is length, and a and b are parameters, has proven to be a useful model for weight as a function of length (Figure 15.3). In general, b less than 3.0 represents fish that become less rotund as length increases and b greater than 3.0 represents fish that become more rotund as length increases. For most species and populations, b is greater than 3.0. If b equals 3.0, growth may be isometric, meaning that the shape does not change as fish grow.

The parameters a and b in equation 15.1 can be estimated by linear regression of logarithmically transformed weight–length data (Chapter 2). When weight–length data are transformed, the curvilinear relation between weight and length becomes "straightened," which allows for estimation of a and b by means of linear regression procedures (Figure 15.3). Weight–length relationships based on transformed data are usually reported in the form

$$\log_{10}(W) = a' + b \cdot \log_{10}(L), \qquad (15.2)$$

where a' is $\log_{10}a$ and is the y-axis intercept and b is the slope of the equation. The intercept (a) in equation 15.1 is estimated by taking the antilogarithm of a' in equation 15.2; b is the same in both equations 15.1 and 15.2.

When data are collected for weight–length relationships, individual fish should be measured to the nearest millimeter and weighed properly to achieve precision and accuracy (section 15.2.2). Five fish per length interval (e.g., 1 cm) usually make an adequate sample. Measure more fish if males and females have different weight–length relationships. Use appropriate equipment and techniques to collect weight–length data for small fish. Do not include small fish in weight–length analyses if either accuracy or precision is low.

15.4 INDICES OF CONDITION

Length is the primary determinant of weight of fishes. However, there can be a wide variation in weight between fish of the same length both within and between populations (Figure 15.3). Indices of condition, or well-being, are more easily interpreted and compared than are a and b in weight–length relationships. There are three basic variations of indices of condition for whole fish—Fulton condition factor, relative condition factor, and relative weight.

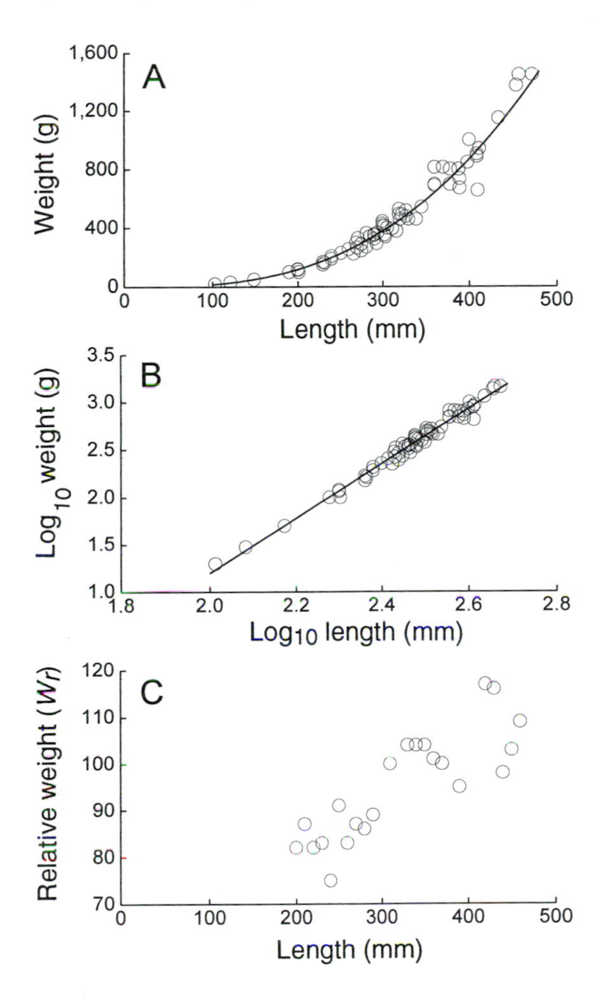

Figure 15.3 Weight–length relationships and relative weight (W_r) plotted as a function of length for largemouth bass collected from Murdo Lake, South Dakota. **(A)** The curvilinear relationship between weight (W) and length (L) ($W = aL^b$); **(B)** the weight–length relationship for the data in (A) logarithmically transformed ($\log_{10}[W] = a' + b \cdot \log_{10}[L]$); (**(C)** mean W_r per centimeter length-group for 20-cm and longer largemouth bass collected by spring electrofishing (W_r data from Lindgren 1991).

15.4.1 Fulton Condition Factors (*K* and *C*)

Fulton-type condition factors are of the form

$$K = (W/L^3) \times 100{,}000 \tag{15.3}$$

when metric units (millimeters, grams) are used and

$$C = (W/L^3) \times 10{,}000 \tag{15.4}$$

when English units (inches, pounds) are used. The constants 100,000 and 10,000 used in each equation are simply scaling constants to convert small decimals to mixed numbers so that the numbers can be more easily comprehended. A standard convention of subscript abbreviations is used to designate if K or C was calculated

Box 15.2 Comparisons of Relative Weight (W_r) and the Fulton-Type Index of Condition (C)

Bennett (1970) suggested a range of values for C (inches, pounds) for largemouth bass and bluegills in "poor," "average," or "normal" condition, and "very fat" or of "good plumpness." Note how average, or normal, C varies with shape of the two species. Weight of a fish for each species at each selected length was calculated for the high and low value of C; these calculated weights for each length were then used to determine W_r (see section 15.4.3). The descriptive values of C are in general agreement with values of W_r for 300-mm largemouth bass and 150-mm bluegill but are inconsistent for other lengths. The variable W_r values for different lengths of fish at a constant C demonstrate a problem of Fulton-type condition factors—many species change their shape as they grow. With appropriate standard weight (W_s) equations (see section 15.4.3), fish of all lengths in good condition have a W_r of about 100.

Bennett's classification	Range of C	W_r for total lengths (mm) of:			
		125	200	300	380
Largemouth bass					
"Poor" condition	3.5	80	73	67	64
	4.5	102	93	86	83
"Average" condition	4.6	104	96	88	85
	5.5	125	114	106	101
"Very fat"	5.6	127	116	108	103
	6.5	148	135	125	120
		125	150	175	200
Bluegill					
"Poor" condition	6.0	85	80	77	73
	7.0	99	94	89	86
"Normal" condition	7.1	101	95	91	87
	8.0	114	107	102	98
"Good plumpness"	8.1+	115	109	103	99

based on measurements of maximum total length (K_{TL} or C_{TL}), fork length (K_{FL} or C_{FL}), or standard length (K_{SL} or C_{SL}). When no subscript is used, maximum total length is the normal measurement.

A practical problem exists because Fulton condition factors vary for the same fish, depending on whether metric or English units are used. Also, because K and C increase with length for fish with b greater than 3, comparisons should be limited to fish of similar lengths. Comparison of K and C between species is usually impossible because different fishes have different shapes. For example, Bennett (1970) suggested that largemouth bass and bluegills are in "normal" or "average" condition when C is 4.6 to 5.5 for largemouth bass and 7.1 to 8.0 for bluegills (Box 15.2).

15.4.2 Relative Condition Factor (K_n)

Relative condition factor (K_n) compensates for allometric growth, that is, when shape changes as fish grow (Le Cren 1951). It is calculated for each fish as

$$K_n = (W/W'), \tag{15.5}$$

where W is weight of the individual and W' is the length-specific mean weight for a fish in the population under study as predicted by a weight–length equation calculated for that population.

Relative condition was used to advantage by Le Cren (1951) to compare male and female Eurasian perch collected in different seasons within one population. The concept of K_n was expanded by Swingle (1965) and Swingle and Shell (1971) by establishing state average weight–length relationships for several fishes in Alabama. A practical advantage of K_n is that average fish of all lengths and species have a value of 1.0, regardless of the species or units of measurement. Disadvantages of K_n are that averages may not describe fish in good condition and that averages can vary from one geographic location to another. If agencies use different weight–length equations, comparison and communication will be difficult.

15.4.3 Relative Weight (W_r)

Relative weight (W_r) represents a refinement of the K_n concept (Wege and Anderson 1978); W_r is given by the equation

$$W_r = (W/W_s) \times 100, \tag{15.6}$$

where W is the weight of an individual and W_s is a length-specific standard weight predicted by a weight–length regression constructed to represent the species. The form of the W_s equation is

$$\log_{10}(W_s) = a' + b \cdot \log_{10}(L), \tag{15.7}$$

where a' is the intercept value and b is the slope of the $\log_{10}$(weight)–$\log_{10}$(length) regression equation and L is the maximum total length of the fish. Note that the form of a W_s equation is the same as that for a typical weight–length equation (equation 15.2). A basic concept of W_r is that the standard should describe the inherent shape of a fish in good condition. When W_r values are well below 100 for an individual or a size-group, problems may exist in food or feeding conditions; when W_r values are well above 100, fish may not be making the best use of a surplus of prey.

Several approaches have been used to estimate the a' and b of the W_s equation (Wege and Anderson 1978; Hillman 1982; Anderson and Gutreuter 1983; Murphy et al. 1991). The first W_s equation was proposed for largemouth bass. Wege and Anderson (1978) used the summary of largemouth bass weight–length data compiled by Carlander (1977) to develop the equation. A curve was fitted to the 75th-percentile weights to develop the W_s equation. For certain species, such as white crappie and black crappie (Neumann and Murphy 1991), smallmouth bass (Kolander et al. 1993), and redear sunfish (Pope et al. 1995), this technique produced W_s equations that were length biased. A length-biased W_s equation leads to W_r values that consistently increase or decrease with fish length in a large sample of populations. For example, W_r increased with fish length in 52 of 56 black crappie population samples and 92 of 104 white crappie population samples when W_r was based on W_s equations developed by the Wege and Anderson (1978) technique.

Murphy et al. (1990) developed a W_s equation for walleye based on a method similar in premise to the original Wege and Anderson (1978) technique, in that it used 75th-percentile weights. However, Murphy et al. developed regression equations for each population included in their study, whereas the Carlander (1977) summary used by Wege and Anderson was based on pooled weight–length data. The method of Murphy et al. (1990) was termed the regression-line–percentile (RLP) technique. When it was applied to white and black crappies, it provided W_r values that did not consistently increase or decrease with fish length (Neumann and Murphy 1991). Developers of W_s equations recommend minimum fish lengths for which the W_s equations are valid. Minimum lengths are designated because weight measurements of small fish tend to be quite variable (low precision and accuracy). It is important that W_s equations be evaluated for comparability across lengths and species (Box 15.3).

Standard weight equations have been proposed for a variety of fishes (Table 15.1). The procedure for obtaining a W_s value for a given length of a particular fish species is outlined in Box 15.4. Alternatively, tables of W_s values can be prepared or W_r and W_s can be calculated with computer programs. A consistency in the use of any standard promotes communication and understanding (Box 15.5).

In concept, a mean W_r of 100 for a broad range of size-groups may reflect ecological and physiological optimality for populations. In laboratory experiments, McComish (1971) fed midge larvae to bluegills at a range of daily rations. The highest values of growth efficiency were estimated to be when final W_r was near 100 (Figure 15.4).

The difficulty in defining good condition for any species has led to the suggestion that W_s, the reference weight in the W_r index, should be thought of simply as a benchmark for comparison of samples and populations, not as any reliable measure of optimality (Murphy et al. 1990). Murphy et al. (1991) suggested that it remains appropriate to model W_s to represent better-than-average populations of a given species, but targets for specific application of the W_r index may need to be adjusted based on management objectives.

Trends or patterns in W_r can be evaluated by plotting individual W_r values by fish length or mean W_r values for length-groups (Figure 15.3). Calculation of mean W_r for an entire sample can mask important length-related trends in fish condition (Murphy et al. 1991). The length-groups defined by the five-cell model of Gabelhouse (1984a; Table 15.2) provide a convenient basis for determination of W_r values. Low W_r for a length-group should provide evidence of competition as a factor influencing growth. Fish of the same length may be considered ecological equivalents even if they differ in age (Gutreuter 1987). Larkin et al. (1956) showed that body size can provide a more interpretable basis than age for expression of growth rates.

Relative weight as a predictor. The processes of ingestion, digestion, and metabolism are influenced by numerous physical and biological factors. The rates of these processes determine rates of growth and change in body composition and condition. A description of fish body composition and methods of analyzing fish proximate composition can be found in Busacker et al. (1990). Research studies have demonstrated relations between condition, or W_r, and proximate composition of fish (McComish et al. 1974; Rose 1989; Brown and Murphy 1991b; Neumann and Murphy 1992). Fat content can be directly related to W_r (Box 15.6). Regression models of Brown (1989) indicated that when visceral fat reserves were depleted in

Box 15.3 Development and Evaluation of Standard Weight (W_s) Equations

The regression-line–percentile (RLP) technique (Murphy et al. 1991) is the most promising method developed to date to estimate W_s equations. The appropriate sample size needed to develop an RLP W_s equation can be determined using the method suggested by Brown and Murphy (1996). However, any equation should be evaluated for comparability across lengths and species. Standards proposed for a few species have a slope less than 3.0, which is unlikely to approximate normal or good form (Murphy et al. 1991). One test for comparability across lengths has been a lack of bias wherein slope deviations were about equally distributed on either size of the standard; however, in some cases the samples compared to the standard were the same ones used to develop the standard (R. S. Cone in Springer et al. 1990). The logic of equally distributed slope deviations may not be valid if depleted stocks of game fish, such as largemouth bass, are common and the relative weight (W_r) of preferred to memorable lengths is higher than it is for stock to quality lengths (see section 15.7.1 for definition of stock, quality, preferred, and memorable lengths). One possible way to test comparability across length-groups would be to analyze proximate composition of fish in good condition. Percentages of ash, water, fat, and protein may be misleading because the sum of percentages should all add to about 100. A more sensitive basis of comparison should be absolute weight of ash, protein, and fat as a function of length. Absolute weight of ash as a function of length was highly correlated for bluegill ($r = 0.997$, McComish et al. 1974). Grams of protein as a function of length should reflect good condition at a high percentile (75–90%) of variability. Given a constant change in form, the ratio of grams of protein to grams of fat for each length-group should be similar.

Comparability across species might be tested by the percentage of mean population W_r values that fall within a W_r range of 85–104 for an adequate number of populations over a range of habitats and locations. These percentages for northern pike, sauger, walleye, white bass, white bass × striped bass (Willis et al. 1991), and white and black crappies (Neumann and Murphy 1991) ranged from 74 to 80%; in contrast to this range is the 37% for yellow perch (Willis et al. 1991). A 54% value for striped bass might be expected because this species has been stocked in many reservoirs with unsatisfactory summer water temperature and dissolved oxygen conditions. Given comparability across species, the species and size-group with the lowest W_r might be recognized as an important problem or opportunity to enhance the fish community.

juvenile striped bass and hybrid striped bass, expected W_r values declined to about 78 and 73, respectively. These research results are based on laboratory experiments and need to be tested for populations in nature.

In temperate fishes that spawn in the early spring, lipid reserves can be translocated to developing eggs (Hayes and Taylor 1994). Bevier (1988) collected adult white crappie from two large Missouri reservoirs in the fall and in the following spring prior to spawning. In one reservoir, mean W_r values in the fall and spring were both 94; in the other reservoir, mean W_r values in the fall and spring were 84 and 85,

Box 15.4 Calculation of Relative Weight (W_r)

To calculate W_r for an individual fish one must determine the standard weight (W_s) for a fish of that length from the W_s equation for that species (Table 15.1). For example, to calculate W_s and W_r for three walleyes that were each 450 mm long and weighed 800, 963, and 1,090 g, the W_s equation for walleye is used:

$$\log_{10}(W_s) - 5.453 + 3.180 \cdot \log_{10}(L),$$

L being the maximum total length of the walleyes (450 mm). The standard weight is calculated as

$$\begin{aligned}\log_{10}(W_s) &= -5 .453 + 3.180 \cdot \log_{10}(450)\\ &= -5.453 + 3.180\ (2.653)\\ &= 2.984.\end{aligned}$$

The antilog of 2.984 gives the nontransformed W_s of a 450-mm walleye as 963 g.

Relative weight for the three walleyes collected is obtained by dividing the actual weight by the W_s for a 450-mm walleye (963 g) and multiplying by 100:

$$W_r = 100 \times (W/W_s);$$

$W_r = 100 \times (800/963)$ $= 83$ $\quad$ $W_r = 100 \times (963/963)$ $= 100$ $\quad$ $W_r = 100\ (1{,}090/963)$ $= 113$

Higher values of W_r indicate greater plumpness.

respectively. In the spring samples, ova size frequencies, the proportion of mature eggs, and relative gonad weight (G_r; Chapter 14) were directly related to W_r. Reproductive success in experimental ponds was better in ponds stocked with fish with higher W_r. Neumann and Murphy (1992) also found a direct relation between the proportion of mature eggs and W_r for white crappie in a Texas reservoir. Willis (1987) demonstrated a correlation between abundance of young gizzard shad and W_r of the parental stock the previous fall. Relative weight and proximate composition of adults may influence not only the quantity of eggs but also their quality and the survival and growth of newly hatched fish (Chambers et al. 1989; Brown and Taylor 1992).

Relative weight may be a more robust predictor of fecundity than of growth. Among largemouth bass that ranged from 250 to 260 mm at capture in small impoundments during late summer and fall, Wege and Anderson (1978) found a significant relationship ($r^2 = 0.81$) between W_r and estimated annual length increment . Estimated increments ranged from 11 mm at a W_r of 78 to 136 mm at a W_r of 110. Gutreuter and Childress (1990) developed correlation models for annual length increments of largemouth bass collected in November from large reservoirs in

Box 15.5 Comparison of Standard or Average Weights for Largemouth Bass

Two standard weight (W_s) equations have been developed for largemouth bass—Wege and Anderson (1978) and Henson (1991). Even though the regression-line–percentile technique used by Henson is an improved technique for development of a W_s equation, the editors have elected to recognize the original W_s equation as the proposed standard (Table 15.1). The original standard has had widespread adoption and long-term use by many organizations and individuals. Also, there is a relatively small difference in the values of W_r determined with the two equations. Biologists in Alabama have used state average weight–length tables to determine relative condition (K_n) of largemouth bass for many years (Swingle and Shell 1971). Prentice (1987) calculated an average weight–length relationship for largemouth bass in Texas. Note the variation between these different standards. A consistency in the use of any standard promotes communication and understanding.

	Condition index by the method of:			
Largemouth bass length (mm)	Wege and Anderson	Henson	Swingle and Shell	Prentice
200	100	95	1.01	0.92
300	100	98	0.96	0.94
380	100	100	0.93	0.95
510	100	102	0.89	0.95

Texas; average r^2 values for 2 years were 0.25 and 0.34. Relative weight values ranged from 55 to 144 with a mean of 85. The data did not produce precise models for indirect assessment of growth rates. Similar conclusions were made for data from populations of white crappie. However, Gabelhouse (1991), working at six Kansas reservoirs over 3 years, found high degrees of correlation ($r = 0.71$ to 0.87) between mean W_r and mean length at capture of white crappies aged 1 through 4. Willis et al. (1991) found that mean population W_r values for yellow perch were significantly correlated with growth, but correlation coefficients were generally low. Higher correlation coefficients were observed between W_r and proportional stock density (PSD; section 15.7.1) of yellow perch (Willis et al. 1991) and northern pike populations (Willis and Scalet 1989). Predicted mean W_r values for northern pike at PSD values of 40 and 70 were 97 and 103, respectively; populations with high density or slow growth had an expected mean W_r of 87.

Wege and Anderson (1978) found that the standard error of mean W_r was negatively correlated ($r = -0.75$) with density of 20- to 30-cm largemouth bass in small impoundments; variability of W_r was a better predictor of density than was mean W_r. When density was high, variability was low and vice versa.

Seasonal changes in consumption and proximate composition or condition can be quite substantial. The mean fat content of yellow perch in two Michigan lakes increased from about 5% in June to 17% in August (Hayes and Taylor 1994). Mean W_r of yellow perch in a South Dakota lake increased from about 103 in June to 118

Table 15.1 Intercept (a') and slope (b) parameters for standard weight (W_s) equations that have been proposed for various fish species and minimum total lengths (mm) recommended for application. The standard equation format is $\log_{10}(W_s) = a' + b \cdot \log_{10}$ (total length). Metric (M) equations are in millimeters and grams; English (E) equations are in inches and pounds. (Updated from Murphy et al. 1991.)

Species[a]	Intercept (a') M	Intercept (a') E	Slope (b)	Minimum total length	Source
Black crappie	−5.618	−3.576	3.345	100	Neumann and Murphy (1991)
Bluegill*	−5.374	−3.371	3.316	80	Hillman (1982)
Brook trout*	−5.085	−3.467	3.043	130	Whelan and Taylor (1984)
Brown trout (lotic)	−4.867	−3.366	2.960	140	Milewski and Brown (1994)
Burbot	−4.868	−3.454	2.898	200	Fisher et al. (1996)
Chain pickerel	−5.824	−3.293	3.243	150	R. M. Neumann (University of Connecticut, unpublished); M. K. Flammang (Texas Parks and Wildlife Department, unpublished)
Channel catfish	−5.800	−3.829	3.294	70	Brown et al. (1995)
Chinook salmon	−4.661	−3.243	2.901	200	Halseth et al. (1990)
Cutthroat trout					C. G. Kruse and W. A. Hubert (University of Wyoming, unpublished)
Lentic	−5.192	−3.514	3.086	130	
Lotic	−5.189	−3.492	3.099	130	
Freshwater drum	−5.419	−3.575	3.204	100	Blackwell et al. (1995)
Gizzard shad*	−5.376	−3.580	3.170	180	Anderson and Gutreuter (1983)
Golden shiner	−5.593	−3.611	3.302	50	Liao et al. (1995)
Lake trout	−5.681	−3.778	3.246	280	Piccolo et al. (1993)
Largemouth bass*	−5.316	−3.490	3.191	150	Wege and Anderson (1978)
Mountain whitefish	−5.086	−3.478	3.036	140	Rogers et al. (1996)
Muskellunge	−6.066	−4.052	3.325	380	Neumann and Willis (1994)
Northern pike	−5.437	−3.745	3.096	100	D. W. Willis (South Dakota State University, unpublished)
Northern pike × muskellunge	−6.126	−4.095	3.337	240	K. B. Rogers and K. Koupal (Colorado State University, unpublished)
Paddlefish (overall)	−5.027	−3.340	3.092	280	Brown and Murphy (1993)
Males	−4.494	−3.063	2.910	280	
Females	−4.073	−2.822	2.782	280	
Pumpkinseed	−5.179	−3.289	3.237	50	Liao et al. (1995)
Rainbow trout					D. G. Simpkins and W. A. Hubert (University of Wyoming, unpublished)
Lentic	−4.898	−3.354	2.990	120	
Lotic	−5.023	−3.432	3.024	120	
Redear sunfish	−4.968	−3.263	3.119	70	Pope et al. (1995)
Rock bass*	−4.883	−3.209	3.083	100	Marteney (1983)
Sauger	−5.492	−3.671	3.187	70	C. S. Guy (Kansas State University, unpublished)
Smallmouth bass	−5.329	−3.491	3.200	150	Kolander et al. (1993)
Spotted bass	−5.392	−3.533	3.215	100	Wiens et al. (in press)
Striped bass	−4.924	−3.358	3.007	150	Brown and Murphy (1991b)
Walleye	−5.453	−3.642	3.180	150	Murphy et al. (1990)
Walleye × sauger	−5.692	−3.760	3.266	170	Flammang et al. (1993)
White bass	−5.066	−3.394	3.081	115	Brown and Murphy (1991b)
White bass × striped bass	−5.201	−3.448	3.139	115	Brown and Murphy (1991b)
White crappie	−5.642	−3.618	3.332	100	Neumann and Murphy (1991)
Yellow perch	−5.386	−3.506	3.230	100	Willis et al. (1991)

[a]Equations were developed by the RLP technique unless asterisked.

in August and declined to 84 in October (Guy and Willis 1991). Relative weight is most likely to be a predictor of growth rate when data are collected during the period of most rapid growth. Legler (1977) evaluated condition of bluegills in four Missouri lakes—two each with satisfactory and unsatisfactory growth. Condition increased to late June and early July in lakes with satisfactory growth but declined in the other

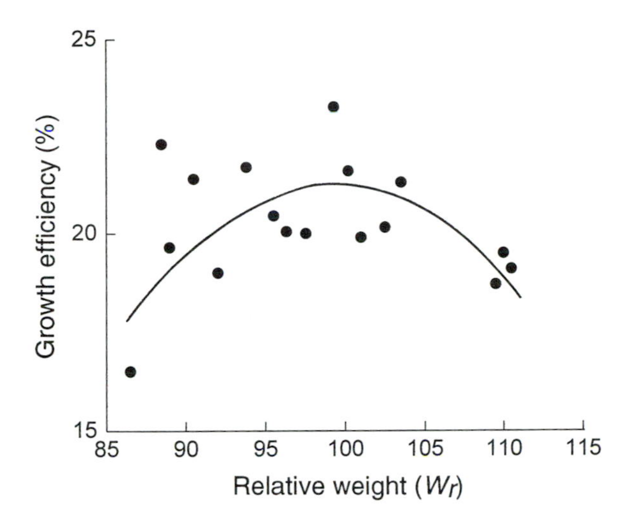

Figure 15.4 Live-weight growth efficiency of bluegills fed midge larvae as a function of relative weight at the end of the experiment (data from experiment D by McComish 1971).

two. Mean liver weight index values were also higher during this period in lakes with satisfactory growth. No consistent differences in index values between populations were apparent in late spring or fall. An abundant food source such as midge larvae would be needed to sustain or improve condition of bluegills during a season of frequent spawning and high growth potential (Hillman 1982).

Because of the potential variability of W_r as a function of season and size of fish, it may or may not be an accurate predictor of growth rate. However, high or low values at any size or season should indicate problems and possibly management opportunities. Mitzner (1990) recommended that stocking density or frequency for channel catfish in Iowa lakes should be decreased if mean W_r is less than 80 and increased if mean W_r is greater than 100.

Low values of W_r, that is, low energy reserves (fat and protein), should be expected to be related to relatively high rates of mortality (Newsome and Leduc 1975). Low fat content or condition in the fall has been associated with high winter mortality of young Colorado squawfish (Thompson et al. 1991) and adult alewives (O'Gorman et al. 1990). Malnutrition of any organism probably increases vulnerability to pathogens or other environmental stresses and increases the probability of natural mortality. Angler exploitation of walleye in Oneida Lake, New York, was inversely related to annual growth increments, leading to the conclusion that well-fed, rapidly growing walleyes were harder to catch than were hungry walleyes (Forney 1980).

Murphy et al. (1990) reviewed the characteristics of a desirable condition index. From a management perspective, an index should be easy to interpret and consistent in meaning across length-classes and species. Given appropriate W_s equations, the W_r index offers these advantageous characteristics.

15.5 WEIGHT MODELS

Small impoundments are convenient experimental microcosms for basic research on ecological concepts and principles as well as for applied research on fish population dynamics and fisheries management. State and federal programs stimu-

Table 15.2 Length categories that have been proposed for various fish species. Measurements are minimum total lengths for each category, except paddlefish body length is anterior edge of eye to fork of tail. English units (E) are in inches, metric (M) in centimeters (updated from Willis et al. 1993).

Species	Stock		Quality		Preferred		Memorable		Trophy		Source
	E	M	E	M	E	M	E	M	E	M	
Black bullhead	6	15	9	23	12	30	15	38	18	46	Gabelhouse (1984a)
Black crappie	5	13	8	20	10	25	12	30	15	38	Gabelhouse (1984a)
Blue catfish	12	30	20	51	30	76	35	89	45	114	Gabelhouse (1984a)
Bluegill	3	8	6	15	8	20	10	25	12	30	Gabelhouse (1984a)
Brook trout											
Lentic	8	20	13	33							Anderson (1980)
Lotic	5	13	8	20							
Brown trout (lotic)	6	15	9	23	12	30	15	38	18	46	Milewski and Brown (1994)
Burbot	8	20	15	38	21	53	26	67	32	82	Fisher et al. (1996)
Chain pickerel	10	25	15	38	20	51	25	63	30	76	Gabelhouse (1984a)
Channel catfish	11	28	16	41	24	61	28	71	36	91	Gabelhouse (1984a)
Chinook salmon (landlocked)	11	28	18	46	24	61	30	76	37	94	Hill and Duffy (1993)
Common carp	11	28	16	41	21	53	26	66	33	84	Gabelhouse (1984a)
Cutthroat trout	8	20	14	35	18	45	24	60	30	75	C. G. Kruse and W. A. Hubert (unpublished)
Flathead catfish	14	35	20	51	28	71	34	86	40	102	Quinn (1991)
Freshwater drum	8	20	12	30	15	38	20	51	25	63	Gabelhouse (1984a)
Gizzard shad	7	18	11	28							Anderson and Gutreuter (1983)
Green sunfish	3	8	6	15	8	20	10	25	12	30	Gabelhouse (1984a)
Lake trout	12	30	20	50	26	65	31	80	39	100	Hubert et al. (1994)
Largemouth bass	8	20	12	30	15	38	20	51	25	63	Gabelhouse (1984a)
Muskellunge	20	51	30	76	38	97	42	107	50	127	Gabelhouse (1984a)
Northern pike	14	35	21	53	28	71	34	86	44	112	Gabelhouse (1984a)
Paddlefish	16	41	26	66	33	84	41	104	51	130	Brown and Murphy (1993)
Pumpkinseed	3	8	6	15	8	20	10	25	12	30	Gabelhouse (1984a)
Rainbow trout	10	25	16	40	20	50	26	65	31	80	D. G. Simpkins and W. A. Hubert (unpublished)
Redear sunfish	4	10	7	18	9	23	11	28	13	33	Gabelhouse (1984a)
Rock bass	4	10	7	18	9	23	11	28	13	33	Gabelhouse (1984a)
Sauger	8	20	12	30	15	38	20	51	25	63	Gabelhouse (1984a)
Smallmouth bass	7	18	11	28	14	35	17	43	20	51	Gabelhouse (1984a)
Spotted bass	7	18	11	28	14	35	17	43	20	51	Gabelhouse (1984a)
Striped bass (landlocked)	12	30	20	51	30	76	35	89	45	114	Gabelhouse (1984a)
Walleye	10	25	15	38	20	51	25	63	30	76	Gabelhouse (1984a)
Walleye × sauger	9	23	14	35	18	46	22	56	27	69	Flammang et al. (1993)
Warmouth	3	8	6	15	8	20	10	25	12	30	Gabelhouse (1984a)
White perch	5	13	8	20	10	25	12	30	15	38	Gabelhouse (1984a)
White bass × striped bass	8	20	12	30	15	38	20	51	25	63	Gabelhouse (1984a)
White bass	6	15	9	23	12	30	15	38	18	46	Gabelhouse (1984a)
White crappie	5	13	8	20	10	25	12	30	15	38	Gabelhouse (1984a)
Yellow perch	5	13	8	20	10	25	12	30	15	38	Gabelhouse (1984a)
Yellow bass	4	10	7	18	9	23	11	28	13	33	Anderson and Gutreuter (1983)
Yellow bullhead	6	15	9	23							Anderson (1980)

Box 15.6 Estimated Fat Contents of Bluegills

In laboratory experiments, McComish et al. (1974) fed bluegills of different sizes a range of daily rations. Proximate composition analyses were made and regression models were developed. Note the direct relation between fat content and relative weight (W_r) for 150-mm bluegills. Note also the relatively constant fat content for fish of various lengths but all with a W_r of 100. Further research is needed to test for similar relationships for fish from natural environments and for other species.

150-mm bluegills						
W_r	70	80	90	100	110	120
Fat (%)	3.32	4.57	5.75	7.00	8.24	9.40

Bluegills with W_r = 100						
Length (cm)	10	12	14	16	18	20
Fat (%)	7.00	6.98	7.02	6.98	7.04	7.06

lated the construction of many small impoundments in the southern United States in the 1940s and beyond for purposes of soil conservation or water for irrigation or wildlife needs. An entomologist, H. S. Swingle, was concerned about these waters as habitats for mosquitos that could cause the spread of malaria. His early observations led to the conclusion that ponds with fish had few or no mosquitos. This led to further studies of how these waters might best be used to produce fish as food for a rural population. The results of management efforts in experimental ponds were often evaluated by census of the fish populations after the ponds were drained.

A good or balanced fish population was defined by Swingle (1950) as a population that can sustain a harvest of good-sized fish in proportion to the productivity of the water. Structural characteristics based on biomass values were developed empirically for fish populations in small impoundments. Dominant species in these communities were usually largemouth bass and bluegill.

Swingle (1950) characterized fishes that often feed on fish as carnivores, or *C* species, and those that often feed on invertebrates and serve as food for carnivores as forage, or *F* species. The *F/C* ratio is the weight of *F* species divided by weight of *C* species. The most desirable range of *F/C* is 3.0 to 6.0.

The *Y/C* ratio was developed because some individuals in the *F* group are too large to be prey. The *Y* value is the total weight of fish in the *F* group small enough to be eaten by the average-sized adult in the *C* group. The *Y/C* ratio is the total weight of *Y* divided by the total weight of *C*. The desirable range of *Y/C* is 1.0 to 3.0.

Another of Swingle's models is A_T, the percentage of total weight of a fish population composed of harvestable-size fish. The most desirable range of A_T for largemouth bass and bluegill populations is 60 to 85%. Swingle's *E* value for a species or group is the percentage of weight of the entire fish community composed of that species or group. The desirable range of *E* for largemouth bass in small impoundments is 14 to 25%.

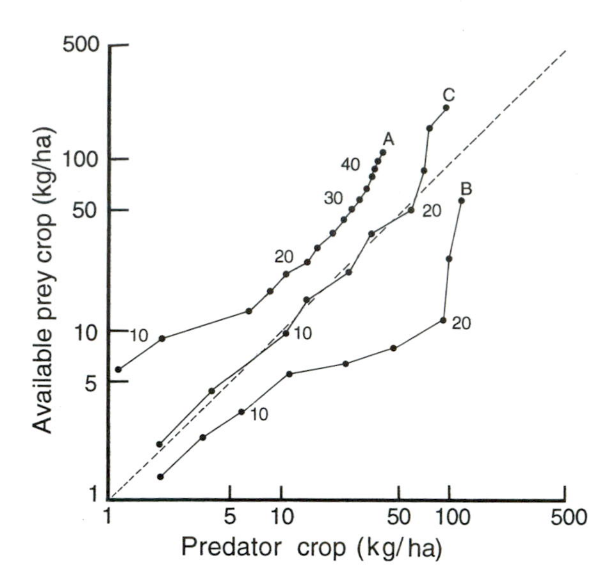

Figure 15.5 Logarithmic plots of available prey:predator (*AP*/*P*) for three general conditions: (A) excess prey for all lengths of predators; (B) prey deficiencies for all lengths of predators; and (C) prey adequacy for small predators but excess for large predators (>20 cm). Diagonal dashed line indicates the minimum desirable *AP*/*P* ratio (from Noble 1981). Numbers along the curves are predator lengths (cm), and points represent 2.5-cm length increments.

Jenkins and Morais (1978) developed the "available prey:predator" ratio (*AP*/*P*), which is similar in concept to Swingle's *Y*/*C* ratio. For the determination of *AP*/*P*, predators of each size are equated to an appropriate size of largemouth bass on the basis of maximum size of prey that each can consume. Biomass of prey small enough to be eaten by a particular size of predator is plotted on log–log scales as a function of the cumulative biomass of predators (Figure 15.5). The *AP*/*P* ratio has been successful for documenting shortages and surpluses of available prey and for documenting changes in the seasonal availability of prey (Jenkins 1979; Timmons et al. 1980; Stephen 1986).

15.6 LENGTH-FREQUENCY HISTOGRAMS

Length-frequency distributions reflect an interaction of rates of reproduction, recruitment, growth, and mortality of the age-groups present. These distributions and changes in distributions with time can help in understanding the dynamics of populations and in identifying problems such as year-class failures or low recruitment, slow growth, or excessive annual mortality.

For histograms and general stock assessment purposes, sample at least 100 adult or stock-length fish (see section 15.7.1 for definition of stock length). The width of length-groups for the histogram depends on maximum fish lengths; a 1.0-cm interval is commonly used for species that reach 30 cm, a 2.0-cm interval for 60-cm species, and a 5.0-cm interval for 150-cm species. Length-frequency histograms, such as the one in Figure 15.6, can provide important information about the status of a population if the histogram represents the size structure of the population. Fre-

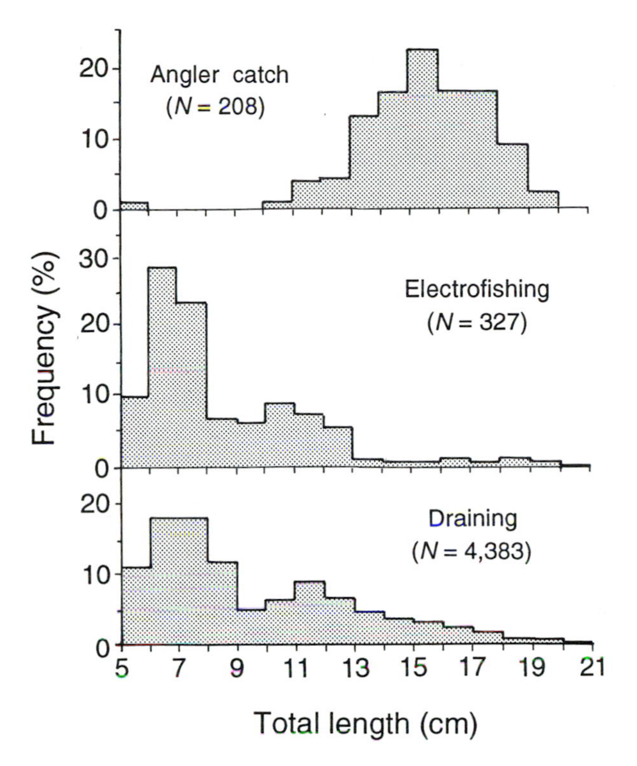

Figure 15.6 Length-frequency distributions of age-1 and older bluegills (≥5 cm) sampled from angler catch, electrofishing, and lake draining at Ridge Lake, Illinois, in fall 1985 (from Santucci and Wahl 1991). Estimated proportional stock density (PSD) values for these three sampling methods were 63 for angling, 16 for electrofishing, and 18 for lake draining.

quently, however, the interpretation of data must recognize gear-related influences on the sizes and numbers of fish that were captured (Chapters 6, 7, 8, and 10).

Length frequencies can be plotted as percentage or proportional distributions of individuals among length-classes, as in Figure 15.6, or as absolute numbers of individuals in each length-class. For comparative purposes, absolute numbers are often standardized to catch per unit of sampling effort (CPUE), such as the number caught per hour of electrofishing or per trap-net night. Standardization adjusts for differences in sampling effort between sampling times or locations. Size distributions with standardized fish numbers are better indicators of relative fish abundances than are percentage distributions. However, CPUE comparisons are valid only if the sampling methodology (gear, procedures) has been consistent.

15.7 LENGTH-FREQUENCY INDICES

Problems with weight models (section 15.5) are that data collection is expensive and many fish are killed in the process. In the 1950s, Swingle was unable to recognize unbalanced fish populations and communities by unfavorable growth or mortality rates because he could not determine fish age. He used reproductive success of largemouth bass and bluegill as a practical indication of a balanced or unbalanced fish community (Swingle 1956).

Michaelson (1970) advanced the understanding of the dynamics of balanced and unbalanced bluegill populations by analyses of structure and function of two balanced and two unbalanced populations in Missouri ponds. This experimental approach was expanded to a study of 38 fish communities in the Midwest; the results of that study strengthened the foundation of the concept and application of structural indices based on weight–length and length-frequency data (Anderson 1978).

15.7.1 Stock Density Indices

Proportional stock density (PSD) and relative stock density (RSD) are numerical descriptors of length-frequency data. Given representative samples of a population, stock density indices are easily calculated and can provide insight or predictive ability about population dynamics.

Proportional stock density. Proportional stock density (Anderson 1976) is calculated as

$$\text{PSD} = \frac{\text{number of fish} \geq \text{minimum quality length}}{\text{number of fish} \geq \text{minimum stock length}} \times 100. \tag{15.8}$$

Values of PSD range from 0 to 100. All expressions of PSD should be rounded to the nearest whole number and should be reported without the percent symbol; decimals represent unfounded accuracy (Box 15.7).

Gabelhouse (1984a) summarized the development of PSD and other stock density indices. Stock length has been variously defined as the approximate length at maturity, minimum length effectively sampled by traditional fisheries gears, and the minimum length of fish that provide recreational value. Quality length was defined by Anderson (1978) as the minimum size of fish most anglers like to catch. For largemouth bass, minimum stock length is 20 cm and quality length is 30 cm. The PSD for a largemouth bass sample is the percentage of 20-cm and longer fish that are also longer than 30 cm; balanced largemouth bass populations may have PSD values of 40–70 (Table 15.3).

Anderson and Weithman (1978) defined stock and quality lengths as percentages of world record lengths; for example, stock length was defined as 20–26% of record length. They recommended that this approach be applied to coolwater fishes such as yellow perch, walleye, smallmouth bass, northern pike, and muskellunge. Anderson (1980) summarized suggested stock and quality lengths for 26 species of fish.

Table 15.2 lists both English and metric units for designation of length categories. Stock length for white crappie is either 5 in or 13 cm (not 12.5 cm). Small differences in index values are due to measurement units. Willis et al. (1993) encouraged fisheries biologists to use the established values in either English or metric units, rather than converting from English to metric units.

Fisheries biologists usually collect as many fish as possible within the allotted worker-hour and time constraints. Procedures have been developed to determine whether population PSD is within a given interval (Weithman et al. 1980) to determine what sample sizes are needed to achieve selected levels of precision (Miranda 1993), and to calculate confidence intervals for PSD (Box 15.8; Gustafson 1988).

Relative stock density. Relative stock density (Wege and Anderson 1978) is the percentage of fish of any designated length-group in a sample and is calculated as

$$\text{RSD} = \frac{\text{number of fish} \geq \text{specified length}}{\text{number of fish} \geq \text{minimum stock length}} \times 100. \tag{15.9}$$

Like PSD, RSD should be rounded to the nearest whole number and should be reported without the percent symbol (Box 15.7).

Relative stock density was first used for largemouth bass; the specified length was 15 in (38 cm), and the percentage of stock-length fish that were also 15 in long came to be known as RSD-15. Gabelhouse (1984a) noted the need for more than a two-cell (stock and quality lengths) model for size structure analysis. His example involved two bluegill populations. Both populations had a PSD of 60, meaning that 60% of stock-length bluegills (8 cm) also were longer than quality length (15 cm). However, one population contained no bluegills over 18 cm, whereas the other contained numerous bluegills over 20 cm and a few that even exceeded 25 cm. Thus, using Weithman's (1978) fish quality index, Gabelhouse developed a five-cell length categorization system based on percentages of world record length.

Gabelhouse (1984a) defined the length ranges from which stock (S), quality (Q), preferred (P), memorable (M), and trophy (T) lengths should be chosen for 70 species of fish (the 70 include two hybrids and one family). Minimum lengths for Q, P, M, and T categories corresponded to near 36–41, 45–55, 59–64, and 74–80% of world record lengths, respectively. Gabelhouse then recommended minimum S, Q, P, M, and T lengths for 27 species (including one hybrid) of warm and coolwater fishes. Proposed standards are presented for 39 taxa in Table 15.2. Adoption of these five standard length categories would facilitate communication within the fisheries profession (Willis et al. 1993).

Gabelhouse (1984a) discussed two systems of RSD calculation: traditional (as described above) and incremental. Traditional RSD values are calculated as the percentage of stock-length fish that are also longer than the defined minimum length for the size categories of quality (RSD-Q, or PSD), preferred (RSD-P), memorable (RSD-M), and trophy (RSD-T). Incremental RSD values are calculated as the percentage of stock-length fish consisting of individuals between the minimum lengths for the size categories (Box 15.7). Thus, incremental RSD values are relative stock density of stock to quality length (RSD S–Q), quality to preferred length (RSD Q–P), preferred to memorable length (RSD P–M), and memorable to trophy length (RSD M–T) fish plus trophy length fish (RSD-T). Thus, RSD S–Q identifies the percentage of stock-length fish that are from stock to quality length. The sum of incremental values is 100.

Gabelhouse (1984a) suggested that traditional RSD calculations may be best suited for among-lake comparisons, where lessening the effects of variable year-class strength (dominant or less-dominant age-groups) would be helpful. Traditional RSD calculations may be more useful for one-time or first-time assessments of a population. Traditional RSD calculations are the most commonly used and hence are easiest to communicate.

15.7.2 Young–Adult Ratio

Reynolds and Babb (1978) proposed the young–adult ratio (YAR) as a means to provide a relative measure of reproductive success and population structure for largemouth bass. The ratio of numbers in particular length-groups can be used to

Box 15.7 Calculation of Stock Density Indices

Below are examples of how to calculate proportional stock density (PSD) as well as traditional and incremental relative stock density (RSD) values based on length-frequency data from a sample of largemouth bass captured by electrofishing.

Step 1: Know the formulas

$$\text{PSD} = \frac{\text{number of fish} \geq \text{minimum quality length}}{\text{number of fish} \geq \text{minimum stock length}} \times 100.$$

$$\text{RSD (traditional)} = \frac{\text{number of fish} \geq \text{specified length}}{\text{number of fish} \geq \text{minimum stock length}} \times 100.$$

$$\text{RSD (incremental)} = \frac{\text{number of fish in a length-class}}{\text{number of fish} \geq \text{minimum stock length}} \times 100.$$

Step 2: Know the length category designations

Length categories for largemouth bass		Length-frequency sample, largemouth bass: Length-class	Number
Minimum stock (S) length	= 20 cm	20–29.9 cm	50
Minimum quality (Q) length	= 30 cm	30–37.9 cm	30
Minimum preferred (P) length	= 38 cm	38–50.9 cm	10
Minimum memorable (M) length	= 51 cm	51–62.9 cm	7
Minimum trophy (T) length	= 63 cm	63+ cm	3

Step 3: Do the calculations

(1) PSD

$$\begin{aligned} \text{PSD} &= (\text{number} \geq 30\text{ cm})/(\text{number} \geq 20\text{ cm}) \\ &= [(30 + 10 + 7 + 3)\ /(50 + 30 + 10 + 7 + 3)] \times 100 \\ &= (50/100) \times 100 = 50 \end{aligned}$$

(2) Traditional RSD

$$\begin{aligned} \text{RSD-Q} &= \text{PSD} = 50 \\ \text{RSD-P} &= (\text{number} \geq 38\text{ cm})/(\text{number} \geq 20\text{ cm}) \\ &= [(10 + 7 + 3)/100] \times 100 = 20 \\ \text{RSD-M} &= (\text{number} \geq 51\text{ cm})/(\text{number} \geq 20\text{ cm}) \\ &= [(7 + 3)/100] \times 100 = 10 \\ \text{RSD-T} &= (\text{number} \geq 63\text{ cm})/(\text{number} \geq 20\text{ cm}) = (3/100) \times 100 = 3 \end{aligned}$$

(3) Incremental RSD

$$\begin{aligned} \text{RSD S–Q} &= (\text{number 20–29.9 cm})/(\text{number} \geq 20\text{ cm}) \\ &= (50/100) \times 100 = 50 \\ \text{RSD Q–P} &= (\text{number 30–37.9 cm})/(\text{number} \geq 20\text{ cm}) = (30/100) \times 100 \\ &= 30 \\ \text{RSD P–M} &= (\text{number 38–50.9 cm})/(\text{number} \geq 20\text{ cm}) \\ &= (10/100) \times 100 = 10 \end{aligned}$$

Box 15.7 Continued.

$$
\begin{aligned}
\text{RSD M–T} &= (\text{number } 51\text{–}62.9 \text{ cm})/(\text{number} \geq 20 \text{ cm}) \\
&= (\ 7 / 100) \times 100 = 7 \\
\text{RSD-T} &= (\text{number} \geq 63 \text{ cm})/(\text{number} \geq 20 \text{ cm}) \\
&= (\ 3 / 100) \times 100 = 3
\end{aligned}
$$

Note:

$$(\text{RSD Q–P}) + (\text{RSD P–M}) + (\text{RSD M–T}) + (\text{RSD-T}) = 30 + 10 + 7 + 3 = 50 = \text{PSD}$$

$$(\text{RSD Q–P}) + (\text{RSD P–M}) + (\text{RSD M–T}) + (\text{RSD-T}) + (\text{RSD S–Q}) = 30 + 10 + 7 + 3 + 50 = 100$$

calculate the index. In samples of largemouth bass collected in late summer or fall, YAR was defined as the number of fish 15.0 cm or less divided by the number 30.0 cm or greater. When adult density was low (<25/ha) in small, midwestern impoundments, the index ranged from less than 1 to greater than 60; at moderate adult densities (50–75/ha) the expected index ranged from 1 to 10 (Reynolds and Babb 1978). These ratios suggest that overharvest or depleted stocks of largemouth bass can result in reproduction that is too low (YAR < 1) or excessive (YAR > 60).

15.7.3 Stock Density Indices: Population and Community Models

Given representative samples of a population, PSD and RSD values can provide insight or predictive ability about population dynamics. Both high and low numbers and wide variation over time are evident in populations with functional problems (i.e., unsatisfactory recruitment, growth, mortality; section 15.1). Novinger and Legler (1978) demonstrated important relationships between bluegill population structure and dynamics in small impoundments. Regression analyses illustrated relationships between PSD and mean length at age 3, annual length increments, and mortality of age-1 fish. Populations with PSD near 0 had maximum densities of stock- to quality-length bluegills. Highly variable year-class strength of bluegill was evident in populations with low PSD (Anderson 1973; Price 1977). Both population density and biomass of largemouth bass and brook trout have been related to values of PSD (Figure 15.7).

Generally accepted objective ranges for stock density index values have been developed for fisheries managers wishing to maintain or create a balanced fish

Table 15.3 Generally accepted stock density index ranges for balanced fish populations (from Willis et al. 1993). Given are proportional stock density (PSD); relative stock density of preferred length fish (RSD-P); and relative stock density of memorable length fish (RSD-M). Indices for crappies are based on fish from midwestern ponds.

Species	PSD	RSD-P	RSD-M	Source
Bluegill	20–60	5–20	0–10	Anderson (1985)
Crappies	30–60	>10		Gabelhouse (1984b)
Largemouth bass	40–70	10–40	0–10	Gabelhouse (1984a)
Northern pike	30–60			Anderson and Weithman (1978)
Walleye	30–60			Anderson and Weithman (1978)
Yellow perch	30–60			Anderson and Weithman (1978)

Box 15.8 Sequential Sampling and Estimating Precision for Proportional Stock Density

When fishery biologists collect length data to estimate proportional stock density (PSD), they can use sequential sampling methods to eliminate wasted effort. Using sequential sampling, a biologist can continuously monitor how many stock-length and quality-length fish are being captured during each sampling effort unit (e.g., each 15 min of electrofishing) and end sampling once the cumulative sample provides a reliable estimate of PSD. Thus, the total number of stock-length and quality-length fish are constantly compared, and sampling continues until the population can be classified according to size structure.

The table below, developed by Weithman et al. (1980), indicates the number of quality-length fish needed in a sample of stock-length fish to categorize PSD within one of three intervals: 0–39, 40–60, or 61–100. Weithman et al. (1980) also provided a table in which PSD can be categorized within five PSD ranges. Based on the table below, sequential sampling ends as soon as the number of quality-length largemouth bass is less than or equal to the lower limit in the 0–39 column, greater than or equal to the upper limit in the 61–100 column, or within the range in the 40–60 column.

	Number of quality-length by PSD range		
Number of stock-length fish sampled	PSD 0 to 39	PSD 40 to 60	PSD 61 to 100
10	0		10
15	2		13
20	4		16
25	7		18
30	9		21
35	11		24
40	13		27
45	15		30
50	17	25	33
55	19	27–28	36
60	21	29–31	39
65	23	32–33	42
70	26	34–36	44
75	28	36–39	47
80	30	38–42	50
85	32	40–45	53
90	34	42–48	56
95	36	44–51	59
100	38	46–54	62

Suppose that a biologist is sampling largemouth bass by means of electrofishing. The biologist collects fish for 15-min intervals and keeps a record of the cumulative number of stock-length and quality-length largemouth bass. During the first electrofishing period, the biologist collects 15 stock-length bass, of which 5 are quality length. Because the number of quality-length largemouth

Box 15.8 Continued.

bass among the 15 stock-length bass sample is not less than or equal to the lower limit in the 0–39 column (i.e., 2), greater than or equal to the upper limit in the 61–100 column (13), or within the range in the 40–60 column, the biologist continues to sample. During the next electrofishing run, the biologist collects 20 more stock-length largemouth bass, of which only 1 is quality length. Now the biologist has a cumulative total of 35 stock-length largemouth bass, of which 6 are quality length. Because the number of quality-length largemouth bass is less than the lower limit (11) for a sample of 35 stock-length fish in the 0–39 column, the PSD can be reliably categorized as being in the range of 0–39. Weithman et al. (1980) noted that biologists never need to sample more than 100 fish to get a reliable estimate of PSD; in the example above, only 35 stock-length bass were needed.

Miranda (1993) developed a method that biologists can use to approximate the sample size (N) required for estimating PSD before fish collection begins. In this case, the biologist must hypothesize the PSD of the population beforehand. If a biologist has a good idea of the PSD before sampling, this method is more advantageous than the method of Weithman et al. (1980) because it is known beforehand how many fish are needed for a reliable sample. However, the method of Miranda (1993) may be disadvantageous because PSD needs to be hypothesized; if a biologist does not have an idea of PSD, it is usually set at 50, which requires the largest sample size.

The methods of Miranda (1993) and Weithman et al. (1980) are based on predetermined levels of confidence. For example, based on the method of Weithman et al. (1980), the biologist is 90% confident that PSD is within a certain interval. Oftentimes, biologists collect as many fish as possible in the time that they can allocate to a particular sampling effort. Gustafson (1988) developed easy-to-use tables for biologists to approximate confidence intervals for PSD. These tables allow biologists to obtain a measure of precision for the PSD estimate after the collection is completed. For example, if a biologist collected 80 largemouth bass and PSD was 20, Gustafson's tables show that the biologist is 95% certain that the PSD is within the range of 10 to 30. However, fewer fish may have been required to categorize PSD if a sequential sampling approach had been used.

population (Table 15.3). A balanced fish population is one that is intermediate between the extremes of a large number of small fish and a small number of large fish and indicates that the rates of recruitment, growth, and mortality may be satisfactory (Anderson and Weithman 1978). In most cases, objective ranges were developed from simple models that used recruitment, growth, and mortality rates to predict population size structure (Willis et al. 1993).

In special cases, fisheries managers desire to manage largemouth bass–bluegill communities in small impoundments for stock density index ranges that are higher or lower than the objective range of balanced populations for each species. Objective ranges for largemouth bass–bluegill communities managed under various options, such as "big bass" and "panfish" options, are summarized in Table 15.4. For

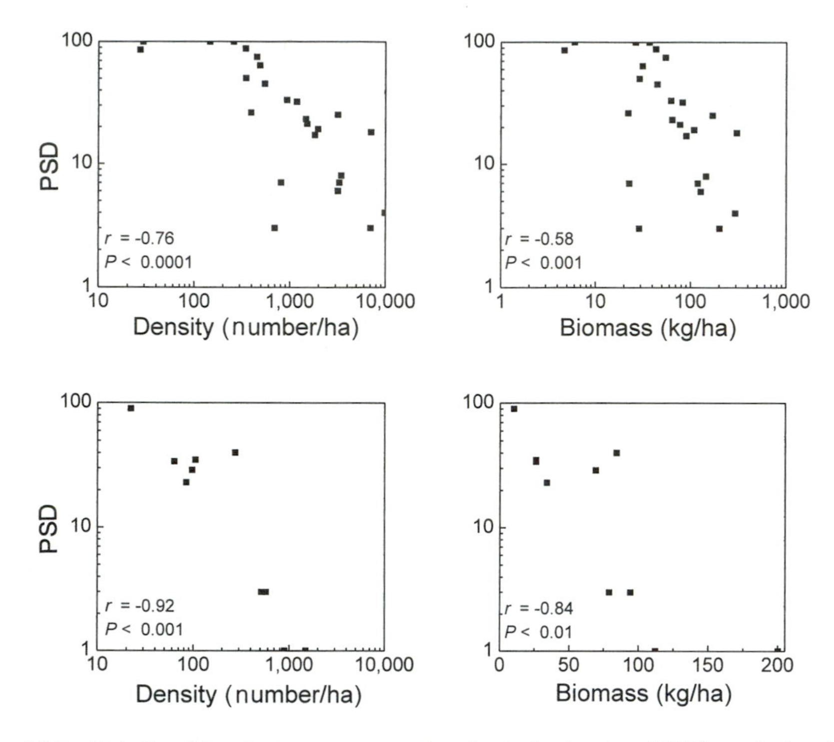

Figure 15.7 Relationships between proportional stock density (PSD) and density and between PSD and biomass of brook trout in Wyoming beaver ponds (top; from Johnson et al. 1992) and largemouth bass in small South Dakota impoundments (bottom; from Hill and Willis 1993).

example, the largemouth bass PSD objective range of 50–80 for the "big bass" option is higher than the range of 40–70 for balanced populations.

Relative weight can be used in conjunction with stock density indices to assess fish populations. Changes in W_r and PSD for a fish population over time can be visualized by plotting mean W_r and PSD over a time period of several years (Figure 15.8).

Wege and Anderson (1978) demonstrated, for small impoundments in late summer and fall, that largemouth bass 200–300 mm long and 300–350 mm long were likely to have mean W_r values near 100 in balanced populations with a PSD of 60

Table 15.4 Stock density index objective ranges for largemouth bass and bluegill under three different management strategies (from Willis et al. 1993). Given are proportional stock density (PSD); relative stock density of preferred length fish (RSD-P); and relative stock density of memorable length fish (RSD-M).

Management strategy	Largemouth bass			Bluegill	
	PSD	RSD-P	RSD-M	PSD	RSD-P
Panfish	20–40	0–10		50–80	10–30
Balance	40–70	10–40	0–10	20–60	5–20
Big bass	50–80	30–60	10–25	10–50	0–10

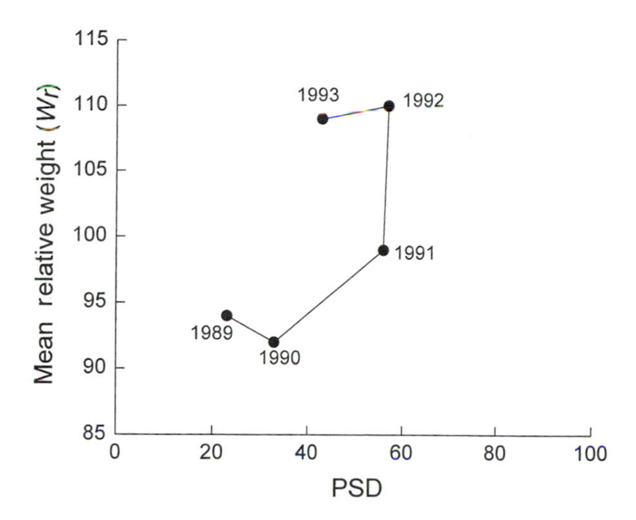

Figure 15.8 Phase plane graph showing changes in mean relative weight (W_r) and proportional stock density (PSD) of largemouth bass collected from 1989 to 1993 in Murdo Lake, South Dakota, after implementation of a 300–380-mm slot length limit in 1989 (from Neumann et al. 1994).

(RSD S–Q = 40; RSD Q–P = 40; and RSD P–M = 20). Davies (1987) determined that K_n of largemouth bass and bluegill was likely to be near 1.0 when PSD of largemouth bass was 40–65. However, biologists need to be aware that both PSD and W_r can vary seasonally (Pope and Willis 1996).

Fish community structure and dynamics can be visualized by plotting PSD of panfish as a function of PSD of game fish in both simple and multispecies communities. Weighted average values of PSD for panfish and game fish were calculated to assess changes in community structure in two Wisconsin lakes by combining northern pike and largemouth bass as game fish and bluegill, yellow perch, and pumpkinseed as panfish (Goedde and Coble 1981; Figure 15.9). Community structures in the lower left of the figure reflect mediocre quality of both game fishes and panfish. Low PSD of panfish populations reflects low density of game fish, ineffective predation associated with habitat problems, or both. A community in the lower right of the graph reflects low density game fish stocks and low recruitment to stock length. A community in the upper left reflects high game fish density and relatively low recruitment of panfish to quality size. Community structures in the upper right may reflect a "balance of nature," with game fish at or near carrying capacity and panfish or food fish at or near 50% of carrying capacity for these species.

15.8 A NEW ERA

The current outlook on natural resource conservation includes biodiversity and ecosystem-based management. These broader responsibilities are an advancement in philosophy from past traditional management goals in fisheries, such as maximum sustained harvest and other models based on single species. The objective of preservation (no change) in biodiversity of native resident species has placed heavy demands for time and funds for programs and actions to protect rare and endangered

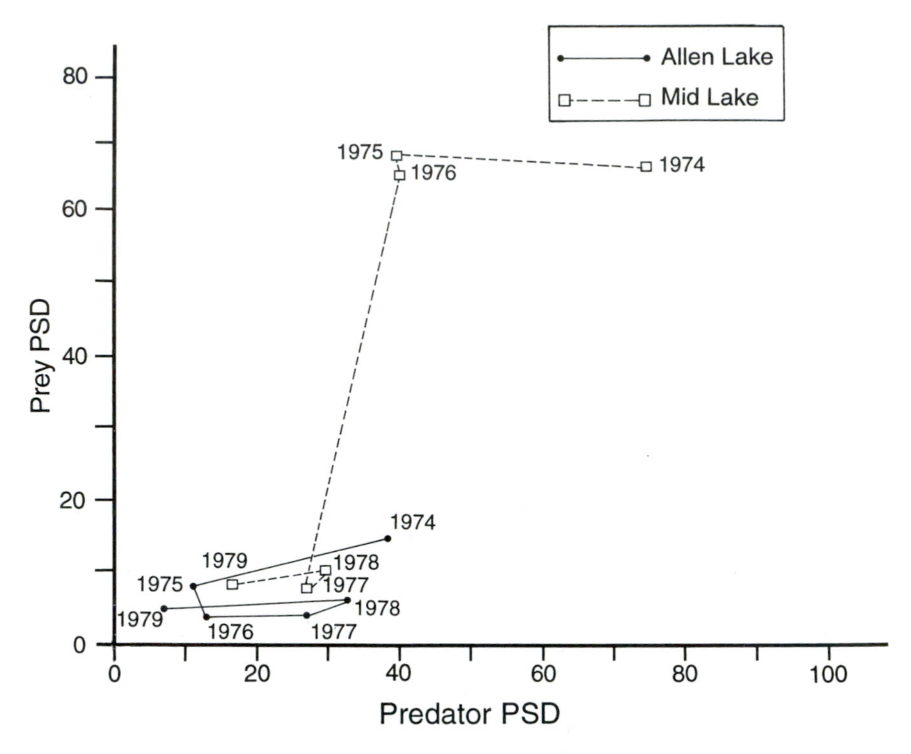

Figure 15.9 Phase plane graph showing the time trajectory of weighted proportional stock density (PSD) for game fishes (largemouth bass and northern pike) and panfishes (bluegill, yellow perch, and pumpkinseed) in two Wisconsin lakes. Mid Lake was opened to angling in 1976 after having been closed for 20 years. Allen Lake had been continuously exploited (from Goedde and Coble 1981).

species. There are few models that identify priorities in or progress toward ecosystem-based management. In contrast with natural systems, many fisheries are managed within artificial systems such as impoundments. These artificial systems commonly contain introduced species because no fish communities are native to these newly created (from an evolutionary perspective) waters. To further compound the problem, many riverine species have declined in abundance after impoundments have been constructed on a riverine system. Thus, the relation of ecosystem-based management and biodiversity to these systems are unclear, depending on the water type being managed.

Our perspective is that the structural indices based on length and weight of key species in aquatic communities will help managers recognize the problems in fish populations and community structure and dynamics. These problems provide opportunities for management of the resource and the level of public benefits provided. The need and opportunities for research to help meet this need are extensive. The concept of balance in fish populations and communities is based on information from lentic environments and populations of warmwater and coolwater species. However, density-dependent, predator–prey, and environmental interactions are the determinants of performance and production in all aquatic ecosystems.

15.9 REFERENCES

Allen, K. R. 1951. The Horokiwi Stream, a study of a trout population. New Zealand Marine Department, Fisheries Bulletin 10, Wallington.

Anderson, R. O. 1973. Application of theory and research to management of warmwater fish populations. Transactions of the American Fisheries Society 102:164–171.

Anderson, R. O. 1974. Influence of mortality rate on production and potential sustained harvest of largemouth bass populations. Pages 18–28 *in* J. L. Funk, editor. Symposium on overharvest and management of largemouth bass in small impoundments. American Fisheries Society, North Central Division, Special Publication 3, Bethesda, Maryland.

Anderson, R. O. 1976. Management of small warm water impoundments. Fisheries 1(6):5–7, 26–28.

Anderson, R. O. 1978. New approaches to recreational fishery management. Pages 73–78 *in* G. D. Novinger and J. G. Dillard, editors. New approaches to the management of small impoundments. American Fisheries Society, North Central Division, Special Publication 5, Bethesda, Maryland.

Anderson, R. O. 1980. Proportional stock density (PSD) and relative weight (W_r): interpretive indices for fish populations and communities. Pages 27–33 *in* S. Gloss and B. Shupp, editors. Practical fisheries management: more with less in the 1980's. Proceedings of the 1st Annual Workshop of the New York Chapter American Fisheries Society. (Available from New York Cooperative Fishery Research Unit, Ithaca, New York).

Anderson, R. O. 1985. Managing ponds for good fishing. University of Missouri Extension Division, Agricultural Guide 9410, Columbia.

Anderson, R. O., and S. J. Gutreuter. 1983. Length, weight, and associated structural indices. Pages 283–300 *in* L. A. Nielsen and D. L. Johnson, editors. Fisheries techniques. American Fisheries Society, Bethesda, Maryland.

Anderson, R. O., and A. S. Weithman. 1978. The concept of balance for coolwater fish populations. American Fisheries Society Special Publication 11:371–381.

Bennett, G. W. 1970. Management of lakes and ponds. Van Nostrand Reinhold, New York.

Bevier, P. W. C. 1988. Density dependent interrelations of white crappie populations in experimental ponds. Master's thesis. University of Missouri, Columbia.

Blackwell, B. G., M. L. Brown, and B. R. Murphy. 1995. Development and evaluation of a standard weight (W_s) equation for freshwater drum. Prairie Naturalist 27:51–61.

Brown, M. L. 1989. Standard weight (W_s) development for striped bass and hybrid striped bass, and the relationship of relative weight (W_r) to proximate composition. Master's thesis. Texas A&M University, College Station.

Brown, M. L., F. Jaramillo, Jr., D. M. Gatlin, III, and B. R. Murphy. 1995. A revised standard weight (Ws) equation for channel catfish. Journal of Freshwater Ecology 10:295–302.

Brown, M. L., and B. R. Murphy. 1991a. Relationship of relative weight (W_r) to proximate composition of juvenile striped bass and hybrid striped bass. Transactions of the American Fisheries Society 120:509–518.

Brown, M. L., and B. R. Murphy. 1991b. Standard weight (W_s) development for striped bass, white bass, and hybrid striped bass. North American Journal of Fisheries Management 11:451–467.

Brown, M. L., and B. R. Murphy. 1993. Management evaluation of body condition and population size structure for paddlefish: a unique case. Prairie Naturalist 25:93–108.

Brown, M. L., and B. R. Murphy. 1996. Selection of minimum sample size for application of the regression-line–percentile technique. North American Journal of Fisheries Management 16:427–432.

Brown, R. W., and W. W. Taylor. 1992. Effects of egg composition and prey density on the larval growth and survival of lake whitefish (*Coregonus clupeaformis* Mitchell). Journal of Fish Biology 40:381–394.

Busacker, G. P, I. R. Adelman, and E. M. Goolish. 1990. Growth. Pages 363–387 *in* C. B. Schreck and P. B. Moyle, editors. Methods for fish biology. American Fisheries Society, Bethesda, Maryland.

Carlander, K. D. 1969. Handbook of freshwater fishery biology, volume 1. Iowa State University Press, Ames.

Carlander, K. D. 1977. Handbook of freshwater fishery biology, volume 2. Iowa State University Press, Ames.

Chambers, R. C., W. C. Leggett, and J. A. Brown. 1989. Egg size, female effects, and the correlations between early life history traits of capelin, *Mallotus villosus*: an appraisal at the individual level. U.S. National Marine Fisheries Service Fishery Bulletin 87:515–523.

Davies, W. D. 1987. Developing appropriate harvest strategies for warmwater fish populations in southeastern United States. Pages 197–206 *in* R. A. Barnhart and T. D. Roeloff, editors. Catch and release fishing: a decade of experience. Humbolt State University, Arcata, California.

Fisher, S. J., D. W. Willis, and K. L. Pope. 1996. An assessment of burbot (*Lota lota*) weight–length data from North American populations. Canadian Journal of Zoology 74:570–575.

Flammang, M. K., D. W. Willis, and B. R. Murphy. 1993. Development of condition and length-categorization standards for saugeye. Journal of Freshwater Ecology 8:199–207.

Forney, J. L. 1980. Evolution of a management strategy for the walleye in Oneida Lake, New York. New York Fish and Game Journal 27:105–141.

Gabelhouse, D. W., Jr. 1984a. A length-categorization system to assess fish stocks. North American Journal of Fisheries Management 4:273–285.

Gabelhouse, D. W., Jr. 1984b. An assessment of crappie stocks in small midwestern private impoundments. North American Journal of Fisheries Management 4:371–384.

Gabelhouse, D. W., Jr. 1991. Seasonal changes in body condition of white crappies and relations to length and growth in Melvern Reservoir, Kansas. North American Journal of Fisheries Management 11:50–56.

Goedde, L. E., and D. W. Coble. 1981. Effects of angling on a previously fished and an unfished warmwater fish community in two Wisconsin lakes. Transactions of the American Fisheries Society 110:594–603.

Gustafson, K. A. 1988. Approximating confidence intervals for indices of fish population size structure. North American Journal of Fisheries Management 8:139–141.

Gutreuter, S. 1987. Considerations for estimation and interpretation of annual growth rates. Pages 115–126 *in* R. C. Summerfelt and G. E. Hall, editors. Age and growth of fish. Iowa State University Press, Ames.

Gutreuter, S., and W. M. Childress. 1990. Evaluation of condition indices for prediction of growth of largemouth bass and white crappie. North American Journal of Fisheries Management 10:434–441.

Gutreuter, S., and D. J. Krzoska. 1994. Quantifying precision of in situ length and weight measurements of fish. North American Journal of Fisheries Management 14:318–322.

Guy, C. S., and D. W. Willis. 1991. Seasonal variation in catch rate and body condition for four fish species in a South Dakota natural lake. Journal of Freshwater Ecology 6:281–292.

Halseth, R. A., D. W. Willis, and B. R. Murphy. 1990. A proposed standard weight (*Ws*) equation for inland chinook salmon. South Dakota Department of Game, Fish and Parks Fisheries Report 90-7, Pierre.

Hayes, D. B., and W. W. Taylor. 1994. Changes in the composition of somatic and gonadal tissues of yellow perch following white sucker removal. Transactions of the American Fisheries Society 123:204–216.

Henson, J. C. 1991. Quantitative description and development of a species-specific growth form for largemouth bass, with application to the relative weight index. Master's thesis. Texas A&M University, College Station.

Herke, W. H. 1977. An easily made, durable fish-measuring board. Progressive Fish-Culturist 39:47–48.

Hill, T. D., and W. G. Duffy. 1993. Proposed minimum lengths for size categories of landlocked chinook salmon. Prairie Naturalist 25:261–262.

Hill, T. D., and D. W. Willis. 1993. Largemouth bass biomass, density and size structure in small South Dakota impoundments. Proceedings of the South Dakota Academy of Science 72:31–39.

Hillman, W. P. 1982. Structure and dynamics of unique bluegill populations. Master's thesis. University of Missouri, Columbia.

Holden, M. J., and D. F. S. Raitt. 1974. Manual of fishery science, part 2: methods of resource investigation and their application. FAO (Food and Agriculture Organization of the United Nations) Fisheries Technical Paper 115.

Hubert, W. A., R. D. Gipson, and R. A. Whaley. 1994. Interpreting relative weights of lake trout stocks. North American Journal of Fisheries Management 14:212–215.

Jenkins, R. M. 1979. Predator–prey relations in reservoirs. Pages 123–134 *in* H. Clepper, editor. Predator–prey systems in fisheries management. Sport Fishing Institute, Washington, DC.

Jenkins, R. M., and D. I. Morais. 1978. Prey–predator relations in the predator-stocking-evaluation reservoirs. Proceedings of the Annual Conference Southeastern Association of Fish and Wildlife Agencies 30(1976):141–157.

Jennings, M. R. 1989. Use of spring scales for weighing live fish in the field. North American Journal of Fisheries Management 9:509–511.

Johnson, S. L., F. J. Rahel, and W. A. Hubert. 1992. Factors influencing the size structure of brook trout in beaver ponds in Wyoming. North American Journal of Fisheries Management 12:118–124.

Kolander, T. D., D. W. Willis, and B. R. Murphy. 1993. Proposed revision of the standard weight (W_s) equation for smallmouth bass. North American Journal of Fisheries Management 13:398–400.

Laevastu, T. 1965. Manual of methods in fisheries biology. FAO (Food and Agriculture Organization of the United Nations) Manual in Fishery Science 1, Rome.

Larkin, P. A., J. G. Terpenning, and R. R. Parker. 1956. Size as a determinant of growth rate in rainbow trout *Salmo gairdneri*. Transactions of the American Fisheries Society 86:84–96.

Le Cren, E. D. 1951. The length–weight relationship and seasonal cycle in gonad weight and condition in the perch *Perca fluviatilis*. Journal of Animal Ecology 20:201–219.

Legler, R. E. 1977. New indices of well-being for bluegills. Master's thesis. University of Missouri, Columbia.

Liao, H., C. L. Pierce, D. H. Wahl, J. B. Rasmussen, and W. C. Leggett. 1995. Relative weight (W_r) as a field assessment tool: relationships with growth, prey biomass, and environmental conditions. Transactions of the American Fisheries Society 124:387–400.

Lindgren, J. P. 1991. Evaluation of largemouth bass harvest regulations for South Dakota waters. Master's thesis. South Dakota State University, Brookings.

Marteney, R. E. 1983. Rock bass populations of the Ozark National Scenic Riverways. Master's thesis. University of Missouri, Columbia.

McComish, T. S. 1971. Laboratory experiments on growth and food conversion by the bluegill. Doctoral dissertation. University of Missouri, Columbia.

McComish, T. S., R. O. Anderson, and F. G. Goff. 1974. Estimation of bluegill (*Lepomis macrochirus*) proximate composition with regression models. Journal of the Fisheries Research Board of Canada 31:1250–1254.

Michaelson, S. M. 1970. Dynamics of balanced and unbalanced bass–bluegill populations in Boone County, Missouri. Master's thesis. University of Missouri, Columbia.

Milewski, C. L., and M. L. Brown. 1994. Proposed standard weight (*Ws*) equation and length category standards for stream-dwelling brown trout. Journal of Freshwater Ecology 9:111–116.

Miranda, L. E. 1993. Sample sizes for estimating and comparing proportion-based indices. North American Journal of Fisheries Management 13:383–386.

Mitzner, L. 1990. Assessment of maintenance stocked channel catfish in Iowa lakes. Iowa Department of Natural Resources, Federal Aid in Fish Restoration Completion Report, Reservoir Investigations Project F-94-R 6, Des Moines.

Morizur, Y. 1995. Comment: bar-coded measuring systems. Transactions of the American Fisheries Society 124:640–641.

Murphy, B. R., M. L. Brown, and T. A. Springer. 1990. Evaluation of the relative weight (W_r) index, with new applications to walleye. North American Journal of Fisheries Management 10:85–97.

Murphy, B. R., D. W. Willis, and T. A. Springer. 1991. The relative weight index in fisheries management: status and needs. Fisheries 16(2):30–38.

Neumann, R. M., and B. R. Murphy. 1991. Evaluation of the relative weight (W_r) index for assessment of white crappie and black crappie populations. North American Journal of Fisheries Management 11:543–555.

Neumann, R. M., and B. R. Murphy. 1992. Seasonal relationships of relative weight to body composition in white crappie, *Pomoxis annularis* Rafinesque. Aquaculture and Fisheries Management 23:243–251.

Neumann, R. M., and D. W. Willis. 1994. Relative weight as a condition index for muskellunge. Journal of Freshwater Ecology 9:13–18.

Neumann, R. M., D. W. Willis, and D. D. Mann. 1994. Evaluation of largemouth bass slot length limits in two small South Dakota impoundments. Prairie Naturalist 26:15–32.

Newsome, G. E., and G. Leduc. 1975. Seasonal changes of fat content in the yellow perch (*Perca flavescens*) of two Laurentian lakes. Journal of the Fisheries Research Board of Canada 32:2214–2221.

Noble, R. L. 1981. Management of forage fishes in impoundments of the southern United States. Transactions of the American Fisheries Society 110:738–750.

Novinger, G. D., and R. E. Legler. 1978. Bluegill population structure and dynamics. Pages 37–49 *in* G. D. Novinger and J. G. Dillard, editors. New approaches to the management of small impoundments. American Fisheries Society, North Central Division, Special Publication 5, Bethesda, Maryland.

O'Gorman, R., R. A. Berqstedt, and C. P. Schneider. 1990. A cold winter will trigger a major mortality of alewives in Lake Ontario. U.S. Fish and Wildlife Service Research Information Bulletin 90-12, Washington, DC.

Parker, R. R. 1963. Effects of formalin on length and weight of fishes. Journal of the Fisheries Research Board of Canada 20:1441–1455.

Piccolo, J. J., W. A. Hubert, and R. A. Whaley. 1993. Standard weight equation for lake trout. North American Journal of Fisheries Management 13:401–404.

Pope, K. L., M. L. Brown, and D. W. Willis. 1995. Proposed revision of the standard weight (*Ws*) equation for redear sunfish. Journal of Freshwater Ecology 10:129–134.

Pope, K. L., and D. W. Willis. 1996. Seasonal influences on freshwater fisheries sampling data. Reviews in Fisheries Science 4:57–73.

Prentice, J. A. 1987. Length–weight relationships and average growth rates of fishes in Texas. Texas Parks and Wildlife Department, Inland Fisheries Data Series 6, Austin.

Price, R. 1977. Age and growth of an unbalanced bluegill population in Michigan. Transactions of the American Fisheries Society 106:331–333.

Quinn, S. P. 1991. Evaluation of a length-categorization system for flathead catfish. Proceedings of the Annual Conference Southeastern Association of Fish and Wildlife Agencies 43(1989):146–152.

Reynolds, J. B., and L. R. Babb. 1978. Structure and dynamics of largemouth bass populations. Pages 50–61 *in* G. D. Novinger and J. G. Dillard, editors. New approaches to the management of small impoundments. American Fisheries Society, North Central Division, Special Publication 5, Bethesda, Maryland.

Ricker, W. E. 1984. Computation and use of central trend lines. Canadian Journal of Zoology 62:1897–1905.

Rogers, R. B., L. C. Bergsted, and E. P Bergersen. 1996. Standard weight equation for mountain whitefish. North American Journal of Fisheries Management 16:207–209.

Rose, C. J. 1989. Relationship between relative weight (*Wr*) and body composition in immature walleye. Master's thesis. Texas A&M University, College Station.

Ruelle, R., and P. L. Hudson. 1977. Paddlefish (*Polyodon spathula*): growth and food of young-of-the-year and a suggested technique for measuring length. Transactions of the American Fisheries Society 106:609–613.

Santucci, V. J., and D. H. Wahl. 1991. Use of creel census and electrofishing to assess centrarchid populations. American Fisheries Society Symposium 12:481–491.

Sigler, M. F. 1995. Bar-coded measuring systems: response to comment. Transactions of the American Fisheries Society 124:641–642.

Springer, T. A, and six coauthors. 1990. Properties of relative weight and other condition indices. Transactions of the American Fisheries Society 119:1048–1058.

Stephen, J. L. 1986. Effects of commercial harvest on the fish community of Lovewell Reservoir, Kansas. Pages 211–217 *in* G. E. Hall and M. J. Van Den Avyle, editors. Reservoir fisheries management: strategies for the 80's. American Fisheries Society, Southern Division, Reservoir Committee, Bethesda, Maryland.

Swingle, H. S. 1950. Relationships and dynamics of balanced and unbalanced fish populations. Alabama Agricultural Experiment Station, Auburn University, Bulletin 274.

Swingle, H. S. 1956. Appraisal of methods of fish population study, part 4. Determination of balance in farm fish ponds. Transactions of the North American Wildlife and Natural Resources Conference 21:299–318.

Swingle, H. S. 1965. Length–weight relationships of Alabama fishes. Auburn University, Department of Fisheries and Allied Aquacultures Series 1, Auburn, Alabama.

Swingle, W. E., and E. W. Shell. 1971. Tables for computing relative conditions of some common freshwater fishes. Alabama Agricultural Experiment Station, Auburn University, Circular 183.

Thompson, J. M., E. P Bergersen, C. A. Carlson, and L. R. Kaeding. 1991. Role of size, condition, and lipid content in the overwinter survival of age-0 Colorado squawfish. Transactions of the American Fisheries Society 120:346–353.

Timmons, T. J., W. L. Shelton, and W. D. Davies. 1980. Differential growth of largemouth bass in West Point Reservoir, Alabama–Georgia. Transactions of the American Fisheries Society 109:176–186.

Wege, G. J., and R. O. Anderson. 1978. Relative weight (W_r): a new index of condition for largemouth bass. Pages 79–91 *in* G. D. Novinger and J. G. Dillard, editors. New approaches to the management of small impoundments. American Fisheries Society, North Central Division, Special Publication 5, Bethesda, Maryland.

Weithman, A. S. 1978. A method of evaluating fishing quality—development, testing, and application. Doctoral dissertation. University of Missouri, Columbia.

Weithman, A. S., and R. O. Anderson. 1978. A method of evaluating fishing quality. Fisheries 3(3):6–10.

Weithman, A. S., and S. K. Katti. 1979. Testing of fishing quality indices. Transactions of the American Fisheries Society 108:320–325.

Weithman, A. S., J. B. Reynolds, and D. E. Simpson. 1980. Assessment of structure of largemouth bass stocks by sequential sampling. Proceedings of the Annual Conference Southeastern Association of Fish and Wildlife Agencies 33(1979):415–424.

Whelan, G. E., and W. W. Taylor. 1984. Fisheries report. ELF Communications System Ecological Monitoring Program. Annual report for ecosystems—tasks 5.8, 5.9, 5.10 for ITT Research Institute, Chicago, Illinois. U.S. Navy Electronics Systems Command Technical Report E06548-8, Washington, DC.

Wiens, J. R., C. S. Guy, and M. L. Brown. In press. A revised standard weight (*Ws*) equation for spotted bass. North American Journal of Fisheries Management.

Willis, D. W. 1987. Reproduction and recruitment of gizzard shad in Kansas reservoirs. North American Journal of Fisheries Management 7:71–80.

Willis, D. W., C. S. Guy, and B. R. Murphy. 1991. Development and evaluation of a standard weight (W_s) equation for yellow perch. North American Journal of Fisheries Management 11:374–380.

Willis, D. W., B. R. Murphy, and C. S. Guy. 1993. Stock density indices: development, use, and limitations. Reviews in Fisheries Science 1:203–222.

Willis, D. W., and C. G. Scalet. 1989. Relations between proportional stock density and growth and condition of northern pike populations. North American Journal of Fisheries Management 9:488–492.

APPENDIX 15.1: Answers to Questions in Box 15.1

1. Number = 887; biomass = 216 kg.

2. Production = 190 kg. For example, age-1 to age-2 production = mean number [(500 + 200)/2] times weight gain (180 − 13 g) = 58,450 g. No estimate of tissue growth of age-0 fish that die is possible without data on number hatched. Apparent production (weight of survivors) of age-1 fish was 6.5 kg (500 × 13 g).

3. Production : biomass = 0.88

4. Number = 200; weight = 161 kg. For example age-2 mortality = number that die (100) times mean weight [(180 + 450 g)/2] = 31.5 kg; age-6 mortality = 12 × 2,250 g = 27 kg.

5. If good condition is W_r = 100 (Wege and Anderson 1978), then length at age and assumed length ranges and distributions for the largemouth bass population could be as follows.

			Number of fish of:			
Age	Length (mm)	Length range (mm)	Stock to quality length	Quality to perferred length	Preferred to memorable length	Memorable length
2	235	200–290	200			
3	314	280–370	20	80		
4	419	368–468		5	45	
5	485	433–533			22	3
6	521	467–567			6	6

Proportional stock density (PSD) = (167/387 · 100 = 43; relative stock density of preferred length bass = 82/387 · 100 = 21.

6. Given the numbers and growth rate in this example, good productivity and relatively small area (54 kg/ha, 4 ha) might be assumed. Because largemouth bass of all sizes were in good condition and the population exhibited a stable age distribution, the fish community is probably well balanced with an F/C ratio of about 3 and a PSD for panfish species of 40–60.

7. Given low to moderate fishing effort and regulated harvest to sustain population structures, this should be an enjoyable place to fish year after year.

Chapter 16

Determination of Age and Growth

DENNIS R. DEVRIES AND RICHARD V. FRIE

16.1 INTRODUCTION AND OBJECTIVES

Although the phrase "age and growth" is commonly used throughout fisheries, we rarely stop to consider what it means. Age refers to some quantitative description of the length of time that an organism has lived, whereas growth is a measure of the change in body or body part size between two points in time, and growth rate is a measure of change in some metric of fish size as a function of time. Both age and growth can be measured across intervals that span a wide temporal range, from hours to days to years. Although age and growth describe different aspects of a fish, they are closely related and typically used together in the field of fisheries.

Age and growth information is extremely important in almost every aspect of fisheries. Because growth provides an integrated assessment of the environmental (e.g., temperature [Fry 1971; Brett 1979; Magnuson et al. 1979] and water chemistry [Sadler and Lynam 1986]) and endogenous conditions (e.g., genetic factors; Gjerde 1986) affecting a fish (Wootton 1990), growth is a useful metric with which to evaluate habitat suitability, prey availability, or the influence of management activities on a target species. In addition, growth is a relatively easy variable to measure, particularly when compared with obtaining information on prey availability or predator diets.

16.1.1 Value of Age and Growth Information

Why is age and growth information of use to fisheries professionals? A multitude of questions and problems exist that can (and often must) be addressed by age and growth information. Suppose, for example, that one needs to assess whether a management strategy has accomplished its desired objective for the target species. More specifically, suppose one needs to assess whether the protection of a largemouth bass population by means of a size regulation on harvest has improved a "stunted" bluegill population. How would such an evaluation be conducted? As management strategies are implemented, some metric must be chosen to determine the ultimate outcome of those strategies on the target species. Managers can collect a sample of the target species before the management strategy is implemented and another sample after the strategy is put in place. After all of the fish are aged (as detailed in this chapter), the mean length achieved by fish of each age can be calculated and compared from before and after the management manipulation to determine whether any changes in length at age (and hence growth) of the target species have simultaneously occurred with the manipulation.

In this chapter, we first describe approaches and techniques used to quantify fish age. Then we discuss ways in which fish growth can be quantified and thus used to address research and management questions in fisheries.

16.2 APPROACHES TO AGING FISH

Although several approaches to aging fish exist, we restrict our discussions to the three most common: (1) direct observation of individuals, (2) use of length-frequency analyses, and (3) use and analysis of hard parts of fish. For a description of the way in which we designate ages, see Box 16.1.

16.2.1 Observation of Individuals

Observations of individuals is the most direct and accurate method for quantifying fish age, containing essentially no error. This technique can involve the direct observation of individuals held in confinement. Confinement is sometimes carried out in aquacultural settings, when fish are spawned and their larvae are maintained in laboratory aquaria. In such a case, the number of days since hatching is known. In a more natural setting, fish of known age (e.g., from a hatchery setting as just described) can be marked and released into natural systems and recaptured at a later date; the period of time between capture and recapture can be quantified to determine fish age. Although it is the most accurate method, however, direct observation is typically the most labor-intensive, time-consuming, and costly way to determine fish ages. Because of this, it is infrequently used in fisheries; its real value may lie in its use as a validation technique to evaluate fish ages as determined by other methods.

16.2.2 Length-Frequency Analysis

Length-frequency analysis can sometimes be used to separate age-groups; an underlying assumption is that fish length within each age-group is normally (or at least unimodally) distributed around the peak value (see Chapter 2 for a discussion of the normal distribution). For species that cannot be aged reliably with hard parts (e.g., tropical fishes; Tesch 1971), length-frequency analysis may be the only approach available. In its simplest form, called the Petersen method (Isaac 1990), a large sample of fish is collected by some method that does not yield biased distributions of fish age or size (see Chapters 6 and 7). The frequency of individuals is plotted as a function of fish length (Figure 16.1), and distinct peaks are identified in the resulting length-frequency distribution. Each peak is assigned an age, beginning with the youngest age present (see Box 16.1) for the mode of smallest fish. For example, in the spring, before fish have spawned, the peak representing the smallest fish in the length-frequency distribution would be assigned to age-1 fish. Later in the summer, after fish have spawned, the peak of smallest fish in the length-frequency distribution would be assigned to age-0 fish.

Although this procedure appears relatively simple in principle, it is generally not simple in practice. Because older fish typically grow in length more slowly than younger fish, variation among individuals within age-groups tends to increase relative to differences between age-groups as fish grow older, obscuring differences in fish length among ages (Figure 16.1). As such, this technique is not generally useful for older fish. Further, variation in hatching time and growth rates among individuals

Box 16.1 Conventions for Age Designations

A great deal of confusion centers around the nomenclature used to designate fish ages. Although both Roman and Arabic numerals are used to designate fish age, we use Arabic numerals to designate the current year of life of a fish, as recommended by the American Fisheries Society in its guides for authors each year. Because the time at which fish lay down an annual mark differs among species and regions, we consider fish to have completed their current year of life on 1 January in the northern hemisphere and 1 July in the southern hemisphere (see figure below; Hile 1950; Jearld 1983). We do not use a plus sign to designate growth between the last annulus and the edge of the hard part (for fish aged with hard parts), as has previously been done. For example, a northern hemisphere fish that is spawned and hatched in spring of one year is designated as age 1 for an entire year beginning on 1 January of the next year (see figure below). Similarly, a northern hemisphere fish is considered to be age 4 during the period between 1 January of its fourth year of life and 1 January of its fifth year of life (see figure below). Finally, a fish is considered to be age 0 (or young of year) during the time from hatching until the next 1 January.

Conventions for age designations

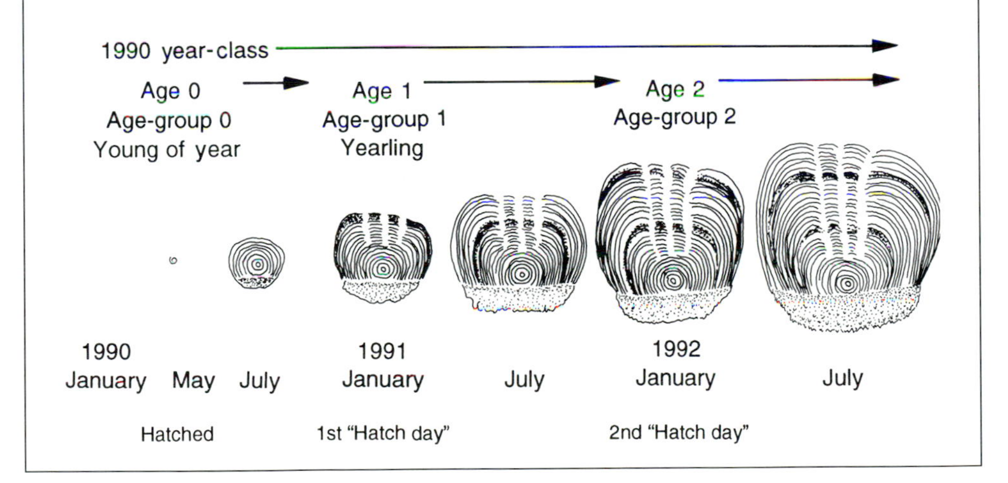

within an age-group can lead to size-groups that do not necessarily correspond to different age-groups. For example, Crumpton et al. (1988) found six to eight age-groups when aging black crappie from otoliths (as in section 16.2.3.2) but could distinguish only two age-groups with length-frequency analysis. Factors that enhance the ability to separate age-groups when using the Petersen method include (1) a large sample that provides an accurate representation of the lengths of fish present in the population to be aged, (2) a relatively short spawning period, and (3) relatively rapid growth that is uniform among individuals within age-groups.

Despite these limitations, substantial progress has been made in graphical and statistical approaches to separate age-groups from length-frequency data (Mac-

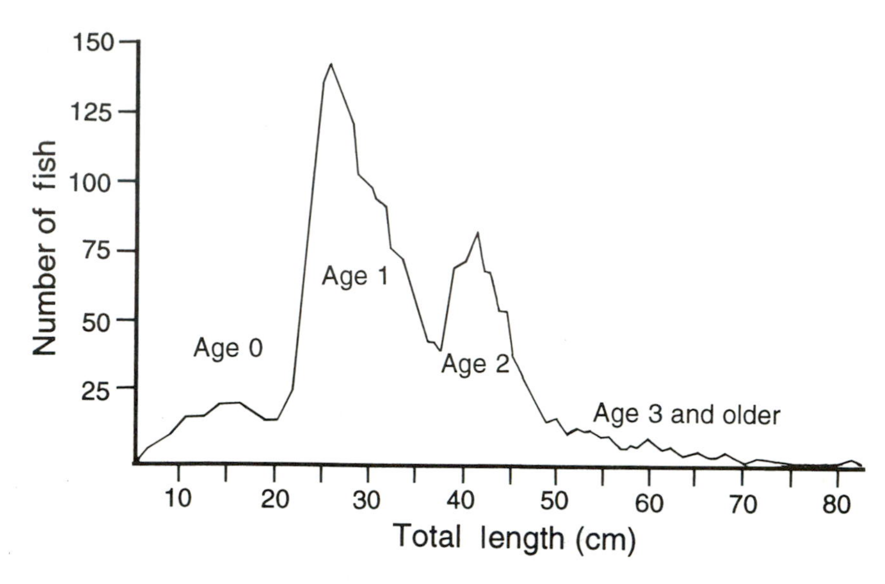

Figure 16.1 Length-frequency distribution of catch of haddock, indicating the length-groups of fish caught and how ages were assigned. Note the difficulty in assigning ages to fish that are older than age 3 (from Lux 1971).

Donald 1987; Isaac 1990). Early techniques were based on visual separation of modes and later techniques involved distinguishing age-groups by manually plotting length-frequency distributions on probability paper (Cassie 1954; Bhattacharya 1967). Given the difficulties that can be associated with visually separating modes in length-frequency data (discussed above), statistical methods (e.g., maximum likelihood estimation; McNew and Summerfelt 1978) have been used. A variety of statistical methods exist that are beyond the scope of this text; these have been recently reviewed by MacDonald (1987) and Isaac (1990). It is likely that this approach to aging fish will continue to evolve and progress, given the continual improvement in statistical capabilities and computer technology. As such, length-frequency analysis provides a low-cost technique to age fish, although the circumstances under which it can be applied are somewhat limited.

16.2.3 Use of Hard Parts

Examination of hard parts is the most frequently used method for aging fishes. Parts that have been used to age fishes include scales, otoliths, fin spines, fin rays, cleithra, vertebrae, opercular bones, and dentary bones. The use of hard parts is based on the appearance of marks, which have been given a variety of names including annual marks, annual rings, or annuli (annulus is the singular) for annular growth, and daily rings or daily increments for daily growth. The mechanism for the formation of annual marks is thought to be related to changes in growth during alternating periods of relatively fast (e.g., spring and summer) and slow (e.g., fall and winter) growth. As such, because the contrast between seasons in tropical latitudes is less pronounced than in temperate regions, determination of age and growth of tropical fishes is difficult (Tesch 1971), and length-frequency analyses may be required for these fishes. Daily increments are formed by differential deposition of calcium carbonate and protein over a 24-h period, which may be due to an internal circadian rhythm (Mugiya et al. 1981; Campana and Neilson 1985).

16.2.3.1 Choice of Hard Part

Historically, scales have been the most often used hard part. The clear advantage of scales over most other hard parts is that scales can be nonlethally removed from a fish, whereas bony structures (i.e., otoliths, cleithra, vertebrae, opercular bones, and dentary bones) require sacrificing the fish. This nonlethality can be important, particularly when working with rare species or when removal of individuals may alter fish density, such as in an experimental setting. However, the possibility exists that scales may not form distinguishable annuli because of slow growth during periods of food limitation, and scales may actually be resorbed in cases of severe stress (the "Crichton effect"; Simkiss 1974), leading to the potential for aging errors when ages are based on scales that have been partly demineralized. In addition, because scales can be regenerated after damage or removal, ages based on partly regenerated scales will not be accurate.

In contrast, although otoliths require that fish be sacrificed, they continue to form annuli or increments (growth may be slow or nonexistent) and do not appear to be resorbed during periods of food deprivation or stress (Marshall and Parker 1982; Campana 1983a, 1983b; Geen et al. 1985; Neilson and Geen 1985). In addition, when comparisons between hard parts have been conducted, increment counts from otoliths are often more precise than those obtained from scales (e.g., Lowerre-Barbieri et al. 1994; see also references in the next paragraph). A further advantage of otoliths is that they can be used to quantify the daily age of age-0 fish (see recent reviews in Campana and Neilson 1985; Jones 1986, 1992). Because age 0 has historically been a difficult life stage with which to work, the ability to quantify daily age provides new directions for age and growth studies (Beamish and McFarlane 1987; Stevenson and Campana 1992).

A number of factors can affect the choice of hard part and should be considered whenever such factors can be identified and are relevant. For example, latitude may affect the choice of scales versus otoliths for aging black and white crappies. In three studies at southern North American latitudes, ages as determined from otoliths of white and black crappies were more precise than those determined from scales (in Florida, Schramm and Doerzbacher 1985; in Oklahoma, Boxrucker 1986; in Mississippi, Hammers and Miranda 1991). In northern latitudes, precision of ages determined by both hard parts were similar (Kruse et al. 1993). Clearly, in southern latitudes otoliths are the hard part of choice, whereas in northern systems scales may be the best choice, given that fish need not be sacrificed.

In the end, the choice of which hard part to use to obtain accurate age information will require validation of ring deposition in the hard part (Beamish and McFarlane 1983; see section 16.4) combined with comparative study of aging techniques across hard parts within a species (e.g., Brennan and Cailliet 1989; Welch et al. 1993; see also references in the preceding paragraph). Because a large number of validation studies have been conducted, a review of the literature will help to identify any studies that are available for the particular species with which you are working.

Regardless of the hard part to be used for aging, it is critical that the fish population be sampled with techniques that are unbiased relative to fish length, sex, and maturity (see Chapters 6, 7, 8, and 10 for a discussion of sampling techniques). For example, if fish are collected with a gear type that selectively samples larger fish,

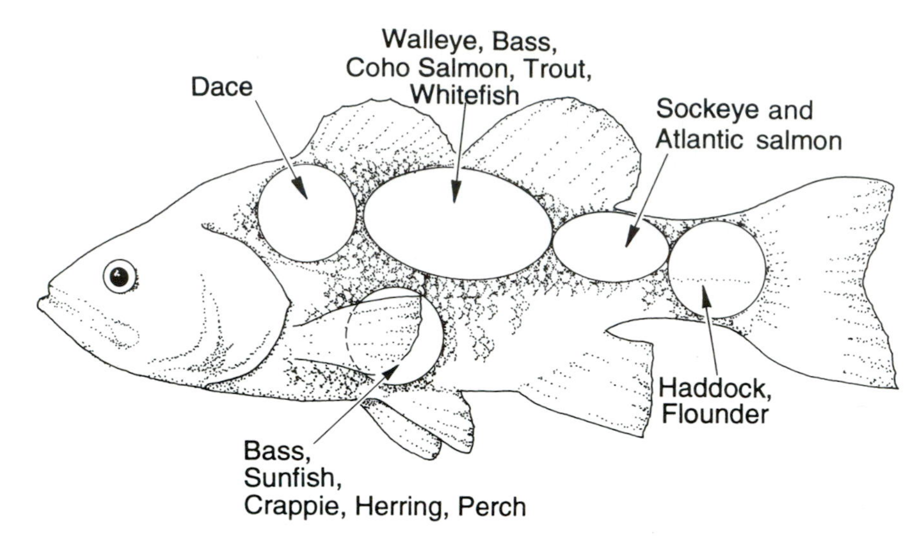

Figure 16.2 General body areas from which scales may be removed from fish of different taxa.

the calculated age structure and growth rates will not represent that of the entire population. In addition, if fish are sampled differentially among habitats that differ in their value to fishes, differences in age structure or growth rates may occur, and age and growth results will be in error. Though we will not further address sampling here, unbiased samples are essential for accurate determination of age and growth.

16.2.3.2 Techniques for Using Hard Parts

The techniques for collection, storage, and preparation of hard parts are organized into three sections: scales, otoliths, and other bony structures.

Scales. Because scales have traditionally been the structure most often used to age fish, we will begin our discussion with them. Scales have maintained their popularity because they are relatively easy to collect and do not require that fish be sacrificed. Scales for different fish taxa should be taken from different areas of the body (Figure 16.2), and within a species, the area from which scales are removed should be consistent. Once the area to be sampled is identified, the area should be wiped clean of mucus and dirt to keep the scales clean so that they are easier to work with later. Scales can be removed by scraping from the front of the fish toward the tail with the blade of a knife (scales will slide out) or by pulling individual scales off with forceps. Sampled scales should be placed directly into a coin envelope that is labeled with species, sample location, fish length and weight, date, and other pertinent information. Placing the knife blade or forceps tip into the envelope and pinching the envelope gently easily removes the scales from the implement to the inside of the envelope. The knife or forceps must be cleaned completely of all scales before the next fish is sampled. It is impossible to determine the quality of the scales in the field, but a sample of 5–10 scales should be adequate for most species. Scales may be stored dry in the labeled coin envelopes.

Scales often are thick and translucent and sometimes curl as they dry, making observation and measurement difficult. Thus, for viewing scales it is generally

Figure 16.3 A scale press (from Smith 1954).

preferable either to place the scales between glass slides that are taped together or to make impressions of the scales on acetate slides (Smith 1954). Impressions can be made by taping the scales to an acetate slide or by placing the scales between two or three acetate slides and pressing them through a scale press (Figure 16.3). When three acetate slides are used, the middle slide contains the scale impression and will exit the scale press uncurled; the outer two slides will be curled from pressing and can be discarded. If possible, use of an impression on an acetate slide is the best alternative because it allows a clear view of the surface of the scale without having to work with the scale itself. Scales or scale impressions either can be viewed directly under a dissecting microscope or can be projected with a microprojector or a microfiche projector to produce a larger image of the scale. The latter is preferred, given that image quality improves with size; in addition, the error associated with measuring an object (as required for back-calculation of growth; see section 16.3.5) decreases with size.

Although there are numerous scale types (e.g., ganoid and rhombic; Moyle 1993), two types, cycloid (typically found on soft-rayed fishes) and ctenoid (typically found on spiny-rayed fishes), are most common. Both cycloid and ctenoid scales share some features but differ in that ctenoid scales have small spines, called cteni, on the posterior field. Both scale types have a focus, which forms first, and radii that extend linearly from the focus to the anterior margin. Circuli (singular is circulus) form regularly and, when growth slows or ceases in the winter, these circuli become crowded and begin to form incompletely. When growth resumes in the spring, the circuli that form grow around or cut over the incomplete circuli, leading to the characteristic "cutting over" and the appearance of annuli (Figure 16.4; Lagler 1956; Everhart and Youngs 1981). Though this phenomenon is seen in both cycloid and

Figure 16.4 Magnified, projected image of a fish scale. The pencil point indicates a growth annulus (reprinted from Snow and Sand 1992, with permission).

ctenoid scales, it is typically more prevalent in ctenoid scales. The annulus is usually defined to be at the outer border of the closely spaced circuli (Figure 16.4). The definition of exactly where the annulus occurs is not necessarily critical for the determination of fish age (i.e., because one is interested in simply counting the annuli) but is extremely important for measurements required for back-calculation of lengths at previous ages (see section 16.3.5). See Lagler (1956) and Tesch (1971) for additional figures and photographs of scales.

Otoliths. Otoliths (or "earstones") require more effort to collect than do scales, given that the fish must be sacrificed, and they are more difficult to find and remove than are scales, particularly for small fish. Fish have three pairs of otoliths: the largest pair, the sagittae, is preferred for aging. Dissection of the head is required for otolith removal. This can be accomplished by cutting away the gill arches and removing the otoliths through the base of the skull or by cutting longitudinally through the top of the skull and across the head (Tesch 1971; Secor et al. 1991, 1992). Once exposed, the otoliths should be carefully removed with a pair of fine forceps to prevent breakage.

Upon removal, otoliths can be stored dry in labeled coin envelopes (as with scales) or in labeled vials containing water, ethanol, or a mixture of water and glycerine (50:50). In the latter, an antifungal additive (e.g., thymol) should be used to prevent fungal growth (Chilton and Beamish 1982). Otoliths should never be stored in formalin, as nonneutralized formalin will lead to decalcification, after which ages cannot be determined. Because the solution in which otoliths are stored can influence the otolith (e.g., water can lead to a chalky coating and glycerin can cause the otolith to clear), otoliths stored in any type of solution should be checked regularly for any changes in appearance. In addition, if otolith removal from the fish is not possible at the time of collection, fish can be frozen until otoliths can be removed. For a complete discussion of otolith selection, removal, and storage, see Secor et al. (1991, 1992).

The amount of preparation required to quantify fish age (in days or years) by use of otoliths differs greatly among species, among individuals within a species (especially among fish of different sizes), and even among otoliths within an individual, ranging from essentially no preparation (i.e., viewing the otolith directly after removal from the fish) to extensive preparation (e.g., soaking in glycerine, burning, dying, and sectioning). Because there is no single method that is best for viewing all types of otoliths, the investigator must assess various combinations of preparation, backgrounds (i.e., light versus dark), light sources (transmitted, incidental, or fiber optic), type of microscope (compound, dissecting, or scanning electron), and degree of magnification (Brothers 1987). The best combination of these factors differs among fish species, as well as between viewing annular and daily growth. If otoliths are opaque (as is often the case with thick otoliths), they may need to be soaked in oil (e.g., clove or cedar), alcohol, glycerine, xylol, or cresol. Given that clearing may continue to the point at which rings are no longer visible, otoliths need to be carefully monitored when soaking. In addition, caution must be used when otoliths are viewed whole because ring formation may not take place on the entire otolith and may be limited to one surface, particularly in older, slow-growing fish (Beamish and Chilton 1982). To assure that whole-view counts are accurate, it is always advisable to section a subsample of otoliths and count marks to compare with counts from the whole otolith prior to sectioning. Other techniques that may enhance otolith appearance (either whole otoliths or sections) include burning the otolith in a flame (Christensen 1964; Chilton and Beamish 1982), dying (Albrechtsen 1968), and acid etching (Haake et al. 1982; Campana and Neilson 1985). For a complete discussion of methods used to prepare otoliths (to enhance their ability to be read), see Campana and Neilson (1985) and Secor et al. (1991, 1992).

When sectioning an otolith is necessary (e.g., for clouded otoliths or for otoliths from old fish), the otolith must be sectioned through the nucleus (or kernel) to avoid missing the early marks (Figures 16.5, 16.6). The nucleus of the otolith can be recognized by its central groove (the sulcus acusticus), which extends from the anterior to the posterior end of the inner surface of the otolith (Figure 16.7). Otoliths to be sectioned can be mounted in thermoplastic cement or embedded in clear epoxy and cut transversely along the dorsoventral plane with a Buehler Isomet low-speed saw or with a jeweler's saw (Beamish 1979; Casselman 1983). An alternative, less expensive technique involves use of a variable-speed drill to grind otoliths to produce thin, readable sections (Maceina 1988). Sections can then be mounted on a microscope slide for viewing. After cutting, the section may require additional grinding and polishing with very fine-grade wet–dry sandpaper (400–600 grit carborundum). For more specific methods to be used for otolith preparation and sectioning, see Secor et al. (1991, 1992).

Relative to annular age and growth, otoliths have alternating hyaline and opaque

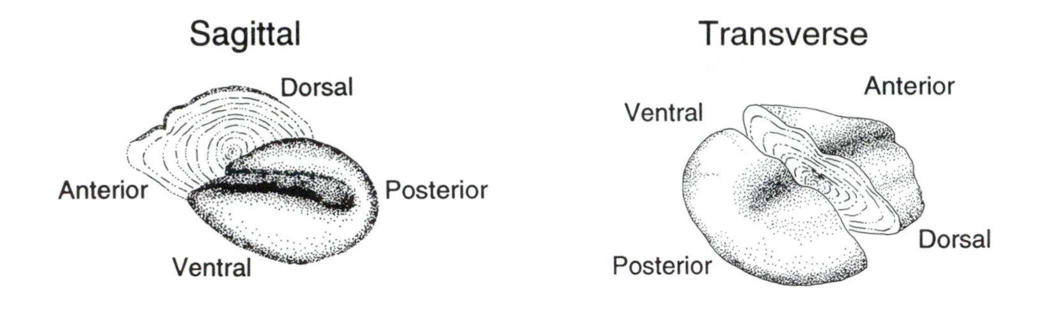

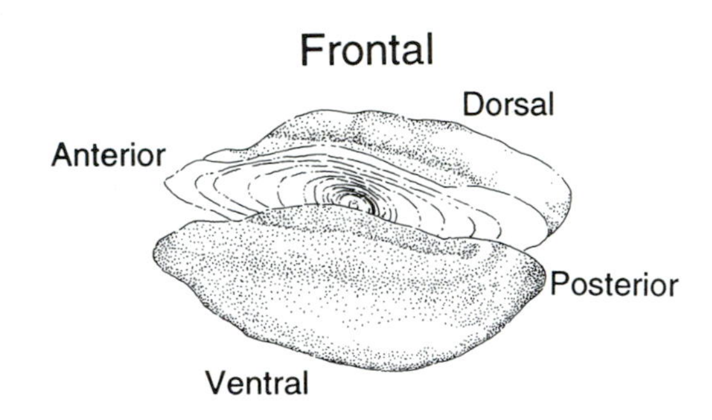

Figure 16.5 Examples of sectioned otoliths, showing three sectioning planes (from Secor et al. 1991).

bands, which represent periods of active growth (hyaline bands) and periods of slow growth (opaque bands). Taken together, one hyaline and adjacent opaque band compose one year of growth; opaque bands are typically counted to age the fish. The process is the same for daily age and growth, but the age at which increment formation begins must be determined independently, and this number added to the number of rings to determine the actual age and hatching date of the fish.

A complete description of counting procedures can be found in Beamish (1979) for annular aging and in Campana (1992) for daily aging. Counting rings on otoliths remains almost as much an art form as a science. In a study designed to quantify the

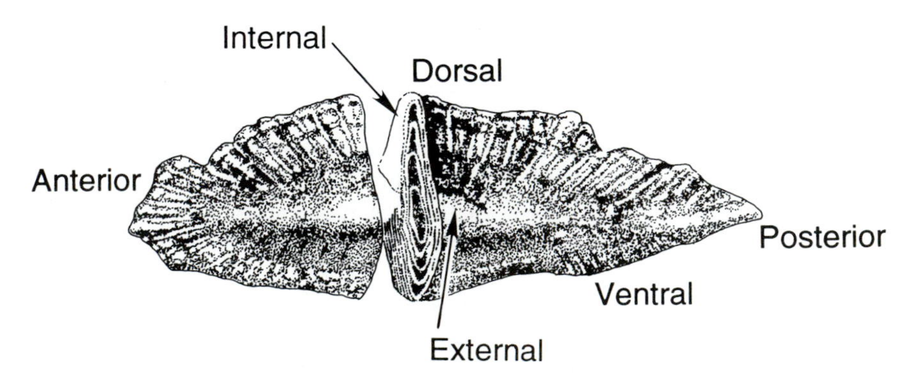

Figure 16.6 Otolith from Pacific hake, showing position of transverse section used for age determination (from Beamish 1979).

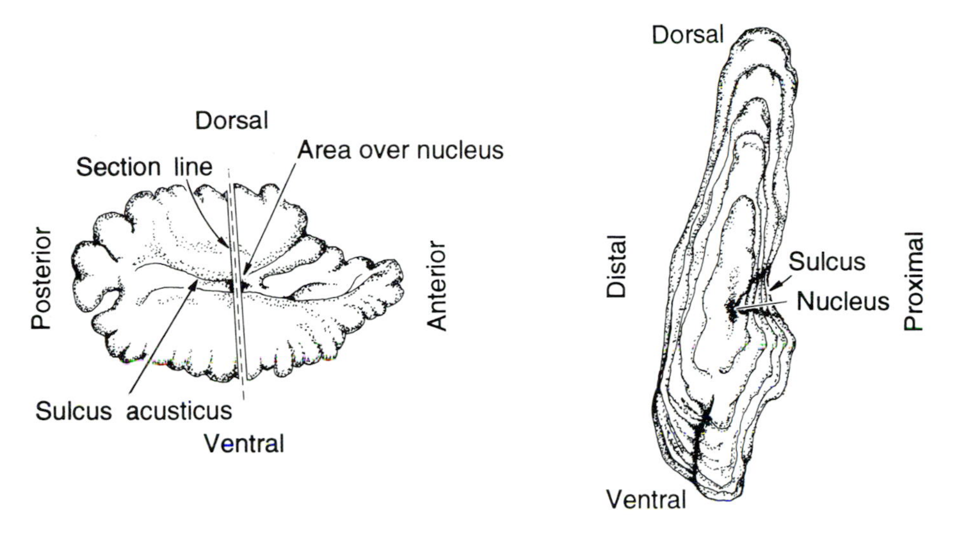

Figure 16.7 Otolith from a redfish, indicating the position of the sulcus acusticus and line along which to section the otolith (left) and illustrating the sectioned otolith (right).

accuracy and precision of age estimates, Campana and Moksness (1991) found that daily age estimates of larval herring by a number of expert readers underestimated the true fish age; there was no relationship between precision of counts and accuracy. Clearly, there is no substitute for experience when it comes to reading otoliths.

Other bony structures. Although removal of fin spines and rays does not require sacrificing the fish, removal of opercular bones, cleithra, dentary bones, and vertebrae obviously does. Rays and spines are easier to collect than other parts that require dissection of the fish.

Removal of pectoral spines (e.g., from a channel catfish) is relatively easy, requiring only the use of pliers to grasp the spine at the articulation, press it flat against the fish body, and rotate it counterclockwise until it is completely dislocated (Sneed 1951; Ashley and Garling 1980), although removal of the entire spine may not be required (Crumpton et al. 1987). Fin rays should be removed by cutting the ray below the articulation, just at the body surface. Remaining tissue should be removed from spines and rays (with bleach, if necessary), and the structures should be dried and stored dry in coin envelopes or vials.

The choice of internal bony structure (i.e., opercular bone, cleithrum, dentary bone, or vertebrae) to be used for age and growth information depends on characteristics of the species being studied. For example, because elasmobranchs lack calcareous otoliths and skeletal parts, traditional aging methods do not apply to them, and vertebral centra have been found useful (Calliet et al. 1983). In this case, the larger, more anterior centra should be used. The structure should be cleaned of any remaining tissue, either by soaking in distilled water and then drying (after which dry tissue can typically be easily removed) or by soaking in bleach followed by a distilled water rinse. For aging paddlefish, the dentary bone is the structure of choice

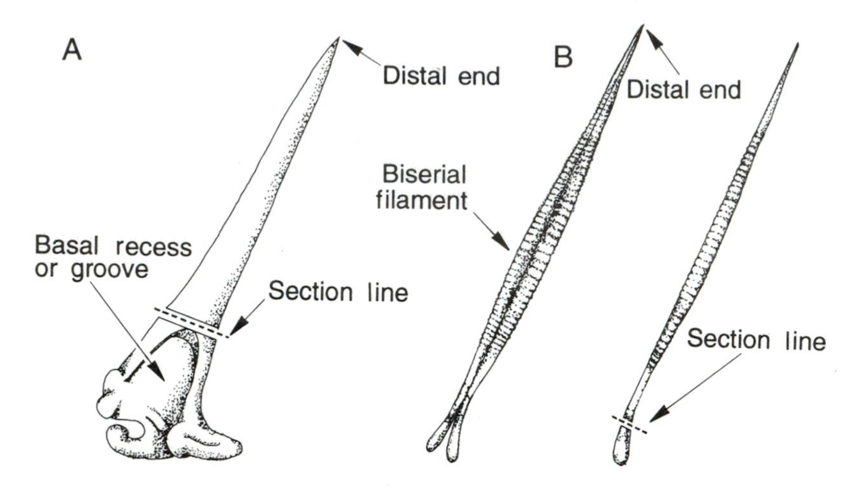

Figure 16.8 **(A)** Channel catfish spine indicating the location at which to section. **(B)** Diagram of a fin ray (biserial) showing the location for sectioning after filaments are separated.

(Adams 1942). The ideal portion to be used is either the lateral or mesial limb; the mesial limb is preferred due to its greater width. After removal, the dentary bone should be cleaned of any attached tissue and can be stored in alcohol or dry in labeled vials or coin envelopes.

Determining age from spines and rays requires that they be sectioned near the base in an exactly transverse plane (Figure 16.8A) to assure that all annuli are included. Spines, in particular, sometimes suffer from the disadvantage that early annuli may be obscured in older fish. The exact location of the section depends on the species. For example, channel catfish spines should be sectioned at the distal end of the basal groove (Sneed 1951), whereas for flathead catfish the articulating process of the pectoral spine is the best location for sectioning (Turner 1982). A dremmel saw or a fine jeweler's saw can be used to section spines (Witt 1961; Quinn 1989). The thickness of the section must be adjusted to assure that it is thin enough for annuli to be visible; 0.5–1.0 mm is typical. Sections may require soaking (e.g., in acetic acid; Carroll and Hall 1964) and may need to be ground and polished for annuli to be easily observed and quantified. Sections are best viewed in water and under a microscope, although they can be projected (e.g., with a microfiche projector as for scales) for counting and measuring rings.

Similarly, fin rays should be sectioned near the base (Figure 16.8B), which can also be done with a jeweler's saw. As with fin spines, section thickness can vary, and sections may require grinding and polishing. Additional details regarding the removal, preparation, and determination of age from fin rays can be found in Scidmore and Glass (1953), Pycha (1955), and Beamish (1981).

16.2.3.3 False and Missing Rings

The only requirement that must be met for obtaining reliable age information from any hard part is that the ring count (whether it be annuli or daily rings) correspond with fish age for individuals of known age. Because annuli are thought to form in hard parts when growth resumes following a period of slow or nonexistent growth, factors that affect fish growth also affect the formation of annuli on hard parts. Such factors include rapidly fluctuating temperatures (unrelated to annual changes in season), starvation, adverse environmental conditions, injury, and spawning activity (Van Oosten 1957; Schramm 1989). These factors can lead to the formation of extra rings (called false annuli, checks, or halo bands) or to missing rings (called missing annuli). Additional problems with the use of hard parts include fish not forming an annulus during a summer of slow growth and peripheral annuli being lost due to resorption (e.g., scale margins can be resorbed during spawning) or erosion. All of these biases must be considered in any study of fish age.

These problems, particularly the presence of extra rings, are also important in the daily aging of fish. Numerous subdaily rings are typically present, to the extent that experienced readers do not always agree on which rings represent true daily rings (Campana and Moksness 1991). As a consequence, validation studies (discussed in section 16.4) are extremely important to studies of daily age and growth, as much to provide a check on the interpretation of rings by the readers as a check on the frequency of the increments (Campana 1992).

16.2.4 Use of an Age–Length Key to Determine Population Age Structure

In practice, fisheries workers take length measurements from a large number of fish but estimate age from a relatively small, fixed subsample (Ricker 1975). For example, a field crew may select hard parts for aging from the first 10 fish within each predetermined length-group. Although random subsamples for aging (i.e., the number aged is proportional to the number of fish in each length-group) are superior (Kimura 1977), an age–length key (Box 16.2) from fixed stratified subsamples (i.e., samples that are stratified or segregated by length-group) has very little bias if length-group intervals are small (Westrheim and Ricker 1978, 1979; Reed and Wilson 1979).

Although fixed stratified subsampling reduces subsequent laboratory processing costs, it does not represent the distribution of lengths or ages in the total collected sample because the subsample is stratified by fish length. Serious error would result if one were to calculate the mean age of the sample from only the fish that were subsampled because the estimate of the mean would be from a uniform distribution of lengths (i.e., a flat-topped histogram), whereas the catch sample of lengths is probably normally distributed (bell-shaped histogram; Chapter 2).

An age–length key helps to correct this subsampling bias. It can be used to convert the distribution of ages and lengths from the fixed, stratified subsample into the catch sample distribution, such that estimates of mean age, mean length at age, and number of fish of each age represent the sampled population. One must multiply the total number of fish collected in each length-group by the proportions of each length-group that are a given age in the age–length key (Box 16.2). Counts are summed through length-groups to obtain the number in each age-group in the entire sample (Box 16.2).

16.2.5 Value of Age Information

The age composition of a representative sample of fish can be used to assess recruitment to the sampling gear, compare relative abundance of age-groups, and estimate total mortality from a catch curve (Van Den Avyle 1993). If more than 1 year of samples are available, the worker can use the age composition to track year-class abundance through time to assess environmental effects and the outcome of management actions. For example, abundance, growth, and mortality can be

Box 16.2 Computation of an Age-Frequency Distribution from an Age–Length Key

Below is an example of a table that can be used to calculate the age-frequency distribution for a fish sample, in this case, a sample of white crappies, based on the relationship between age and length for a subsample of the fish. The length distribution of fish in the entire sample is shown in the first two columns. Scales are taken from up to 10 fish per length-group, and ages are determined for these subsamples (column 3). The age percentages in each subsample are then used to allocate the entire length-group sample to age-groups (last five columns). For example, consider the 20 white crappies in the 21-cm length-group. Scales were aged for 10 of these fish, showing that 5 fish (50%) were age 2 and 5 (50%) were age 3. Thus, 10 fish in the full length-group sample (50% of 20) are assigned to age-group 2 and 10 fish to age-group 3. The age structure of the population is estimated by summing the numbers within each age-group column.

Length-group (cm)	Number in sample	Number (age) in subsample	Sample allocation per age-group				
			Age 1	Age 2	Age 3	Age 4	Age 5
13	64	10(1)	64				
14	28	10(1)	28				
15	11	10(1)	11				
16	8	4(1), 4(2)	4	4			
17	12	1(1), 9(2)	1	11			
18	36	10(2)		36			
19	51	10(2)		51			
20	32	10(2)		32			
21	20	5(2), 5(3)		10	10		
22	31	1(2), 9(3)		3	28		
23	38	10(3)			38		
24	23	1(2), 7(3), 2(4)		2	16	5	
25	16	4(3), 6(4)			6	10	
26	15	1(2), 9(4)			2	13	
27	23	10(4)				23	
28	18	8(4), 2(5)				14	4
29	13	2(4), 8(5)				3	10
30	8	1(4), 7(5)				1	7
31	2	2(5)					2
All	449		108	149	100	69	23

monitored through time to assess a minimum-length regulation on largemouth bass as year-classes recruit to vulnerable sizes and ages (Figure 16.9). In Figure 16.9, age-3 and age-4 fish underwent substantial mortality during 1979–1980 and age-2 fish were abundant and grew slowly during 1979–1980. In 1980, age-2 fish were extremely abundant.

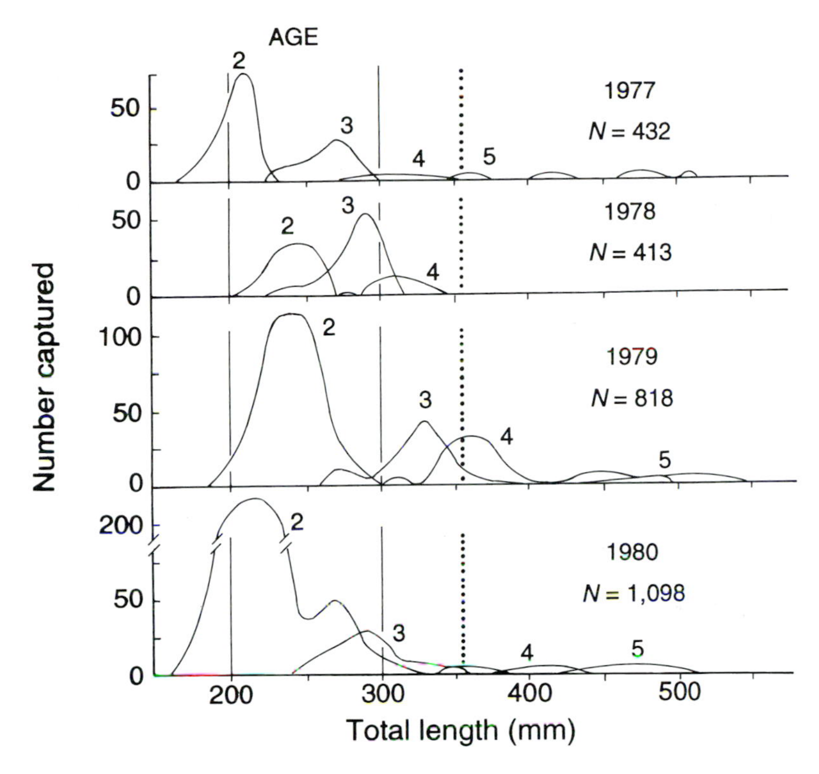

Figure 16.9 Example of tracking age-groups, growth, and mortality through time to evaluate a fishery. Figure contains length-frequency distributions of age-2 and older largemouth bass collected in Ross Lake, Ohio, in spring 1977–1980. Solid vertical lines denote minimum stock and quality lengths (see Chapter 15); dotted vertical lines indicate the 356-mm-minimum-length limit (adapted from Carline et al. 1984).

16.3 GROWTH DETERMINATION

Growth is one of the most important and reliable indicators of fish health, population production, and habitat quality. Faster growth results in more fish reaching desirable sizes for sport anglers and commercial fishers. The topic of growth determination is a vast field that could take volumes to cover (e.g., Weatherley and Gill 1987). However, we restrict ourselves to the area of growth that is directly related to the processes and techniques associated with aging fishes. For a detailed discussion of how to quantify fish growth, see Busacker et al. (1990). Further, although growth can be described in terms of both fish length and weight, we will restrict our discussion to growth in length. Because length and weight are related (Chapter 15), the techniques we describe for length can be applied to weight as well. As with fish age (section 16.2), there are a number of ways to determine fish growth, including direct observation, observation of changes in length-frequency distributions through time, and back-calculation of length and growth from hard parts. In this section we will briefly review techniques associated with direct observation and length-frequency analysis, followed by discussions of calculation of growth from an age-length key and use of historical growth data (via back-calculation).

16.3.1 Direct Observation

As with fish age, direct observation is the least variable technique for quantifying growth, the only variation would be caused by error associated with measuring the fish. Given the difficulty in identifying individual fish, this technique is difficult to use in practice and is generally restricted to aquacultural settings or laboratory experimental situations in which individual fish can be held, identified, and remeasured.

16.3.2 Length-Frequency Analysis

Length-frequency analysis can be used to estimate growth by following changes in the modal length for an age-group (determined as in section 16.2.2) through time. To use this procedure, one must assume that the modal length for an age-group represents the length for that age-group. Given the difficulties that can be associated with identifying age-groups (described in section 16.2.2), use of this technique is often limited to situations in which growth cannot be determined in any other way (e.g., tropical fishes).

16.3.3 Use of Mean Length at Age of Capture

Fish can be aged (as described in section 16.2), and mean lengths for all of the fish collected within each age-group can be calculated to estimate growth. Caution is advised because such length-at-age estimates can have unknown bias from differential mortality by size and from size-selective sampling. Because of the gear used, younger fish are often missed; if so, growth information about them will be missing. Size at capture should be interpreted with knowledge of date of capture. For example, an age-5 fish caught in September is likely to be longer than an age-5 fish caught in May. Also, growth history specific for individual year-classes is masked unless several years of samples are collected. For example, reporting the mean length of an age-3 largemouth bass to be 30 cm has limited value unless one knows whether the sampled fish from that year-class averaged 22 or 28 cm long at age 2.

We also wish to make it clear that there is a distinction between the amount of growth and the rate of growth. For example, a paired *t*-test (see Chapter 2) of the mean increments in length at each age can help determine the statistical validity of growth rate differences between two populations of the same species. In Figure 16.10, average differences between the paired increments do not differ statistically (P = 0.46, df = 5), indicating that the slopes of the lines connecting the mean lengths of fish in the two lakes are approximately parallel to one another. This means that average growth rate for the 6 years of life is about the same for fish from both lakes. However, fish from Lake A must have grown at a faster rate before age 1 than did fish from Lake B because age-1 fish from Lake A are about 40 mm longer than are age-1 fish from Lake B. This 40 mm difference is maintained thereafter. Median lengths for age-1 through age-6 fish from Lake A are greater than for fish from Lake B (Mann–Whitney *U*-test, P = 0.047; see Chapter 2). Thus, fish from Lake A are consistently longer than fish from Lake B, but growth rate is faster in Lake A only until age 1.

From a modeling, or theoretical, point of view, incremental growth is often assessed via the von Bertalanffy equation. For details of this approach, see Busacker et al. (1990).

16.3.4 Length at Age from an Age–Length Key

Mean length for an age-group can be calculated from an age-length key by multiplying the midpoint of the length-group interval by the number of fish of that

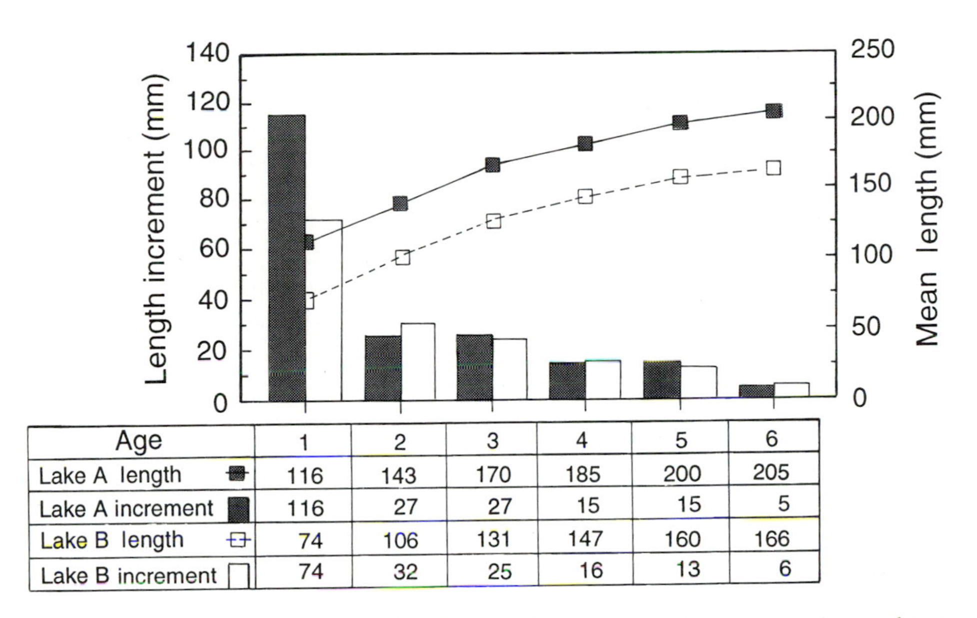

Age	1	2	3	4	5	6
Lake A length	116	143	170	185	200	205
Lake A increment	116	27	27	15	15	5
Lake B length	74	106	131	147	160	166
Lake B increment	74	32	25	16	13	6

Figure 16.10 Comparison of mean lengths at each age at time of capture and annual mean growth increment to each age for bluegills from Lake A and Lake B.

age within that length-group, summing these products for a given age-group, and then dividing the sum of the products by the total number in that age-group. Formulae for calculating the sample mean and variance of a frequency distribution (Remington and Schork 1970), specifically for the mean and variance in fish length for an age-group containing counts within length-groups, are

$$\bar{x} = \frac{\Sigma fx}{\Sigma f} \tag{16.1}$$

and

$$s^2 = \frac{(\Sigma f)(\Sigma fx^2) - (\Sigma fx)^2}{(\Sigma f)\,[(\Sigma f) - 1]}; \tag{16.2}$$

$\bar{x}$ = mean length,
s^2 = sample variance of length,
f = frequency, or number of fish of the selected age within the length-group, and
x = midpoint of the length-group interval.

For an example of these calculations, applied to the data in Box 16.2, see Box 16.3. A computer program, such as FishCalc (available from the Computer User Section of the American Fisheries Society), automates the calculation and application of an age–length key to catch samples.

16.3.5 Back-Calculation of Annular Growth

Information on the growth history of year-classes of fish from a single sample can be obtained with back-calculation. The ability to back-calculate is based on a

Box 16.3 Computation of Mean Length at Age, and Its Variance, from an Age–Length Key

Mean length at age can be calculated from an age–length key (as in Box 16.2) by multiplying the midpoint of the length-group interval by the number of fish of a given age within that length-group, summing these products across length-groups, and then dividing the sum of these products by the total number in that age-group. The variance can be calculated similarly, as given by equation (16.2) in text. Below is an example of this calculation for length at age 1 for the data presented in Box 16.2.

Midpoint length (x)	Number at age (f)	fx	fx^2
13 cm	64	832	10,816
14 cm	28	392	5,488
15 cm	11	165	2,475
16 cm	4	64	1,024
17 cm	1	17	289
Sum	108	1,470	20,092

$$\text{Mean length at age} = (\Sigma fx)/(\Sigma f) = 1{,}470/108 = 13.61 \text{ cm.}$$

$$\text{Variance} = \frac{(\Sigma f)(\Sigma fx^2) - (\Sigma fx)^2}{(\Sigma f)[(\Sigma f) - 1]} = \frac{(108)(20{,}092) - (1{,}470)^2}{(108)(107)} = 0.78.$$

Below are the mean length and variance values for fish of ages 1–5 from the data in Box 16.2.

Age	Mean length (cm)	Variance
1	13.61	0.78
2	19.01	1.79
3	22.86	1.25
4	26.64	1.82
5	29.30	0.77

repeatable relationship between the size of the hard part and fish length. If such a relationship exists, then one can back-calculate fish lengths at all previous marks on the hard part by translating the distance between annuli on the hard part into annual growth in fish length or weight (see Box 16.4). A subsample of fish of each age is necessary to estimate growth of groups of fish hatched in different years (i.e., all fish in a sample do not always need to be aged). The only data that are required are the length of fish at capture, the radius of the hard part at capture (measured from the nucleus to the margin), and the radius of the hard part to the outer edge of each of the increments (either annuli or daily rings). These measurements can be obtained with a microscope with an ocular micrometer, a digital capture system (Frie 1982), or a microprojector, as described previously for fish age (see section 16.2). The three methods most commonly used for back-calculation are the direct proportion method (Le Cren 1947), the Fraser–Lee method (Carlander 1982; Frie 1982; Busacker et al. 1990) and the Weisberg method (Weisberg and Frie 1987; Weisberg 1993a, 1993b).

16.3.5.1 Direct Proportion Method

The direct proportion method can be used when the relationship between body length and hard-part radius is linear and has an intercept that does not differ from the origin (Box 16.4). In this situation, hard-part growth is directly proportional to body growth. According to Le Cren (1947), length at previous ages is given by

$$L_i = \frac{S_i}{S_c} L_c; \tag{16.3}$$

L_i = back-calculated length of the fish when the ith increment was formed,
L_c = length of the fish at capture,
S_c = radius of hard part at capture, and
S_i = radius of the hard part at the ith increment.

The assumption that the regression between body length and hard-part radius has the origin as its intercept must be verified. If the intercept differs from the origin, errors in back-calculated growth can result, and another technique must be used.

16.3.5.2 Fraser–Lee Method

The Fraser–Lee method of back-calculation has been widely accepted for most of this century and is applicable when the intercept of the relationship between fish length and hard-part radius is not at the origin (Box 16.4). The formula for back-calculating lengths at previous annuli is

$$L_i = \frac{L_c - a}{S_c} S_i + a; \tag{16.4}$$

$\frac{L_c - a}{S_c}$ = slope of a two-point regression line to estimate L_i,

a = intercept parameter,

and L_i, L_c, S_c, and S_i are defined as above.

The slope of the Fraser–Lee back-calculation equation, $(L_c - a)/S_c$, is calculated for each fish as the slope of a line connecting two points: (S_c, L_c) and $(0, a)$. The value for a is calculated as the intercept of the regression of fish length at capture on hard-part radius at capture for a wide range of fish sizes of the species being examined. A standard a value for a species can be used (as suggested by Carlander 1982 to keep any bias caused by this parameter consistent among workers), or the a value can be determined for an individual sample of a species (because growth can vary substantially among populations of the same species). However, in the latter case, care must be taken to assure that an even distribution of fish across the entire range of fish lengths is included (this is particularly difficult for small fish) because the intercept can vary with fish size (Hudson and Bulow 1984). A recently published alternative approach is to generate a biologically determined intercept, which is defined as the fish length at which otolith length equals zero (Campana 1990).

Box 16.4 Back-Calculation of Past Growth

Back-calculation is a technique that allows fisheries biologists to obtain information on past growth of a fish based on the relationship between the radius of a hard part and fish length. If the relationship is directly proportional, then it can be graphically demonstrated as follows.

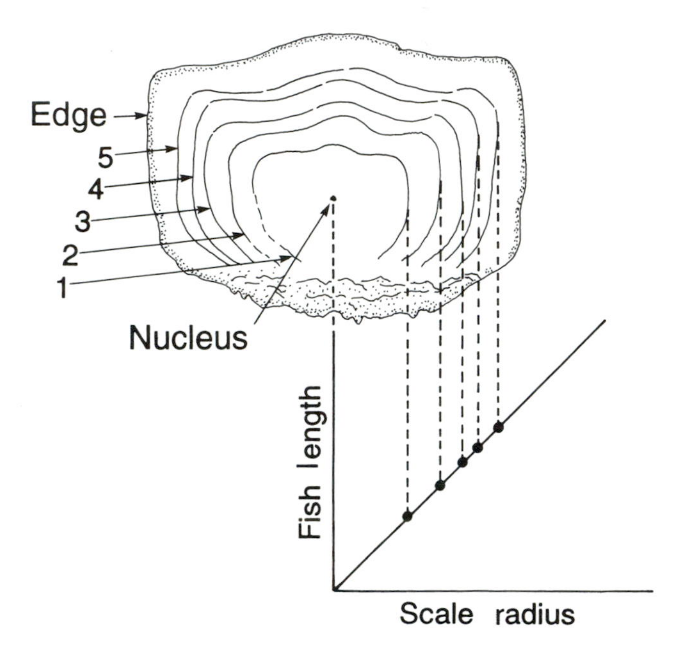

Changes in fish length are given by text equation (16.3). For example, if the length of a fish at capture is 171 mm, the scale radius at capture is 3.75 mm, and the scale radii (S_i) at previous annuli are as presented below, then the lengths at these annuli (L_i) are as follows.

Increment (i)	S_i	L_i
1	1.49	67.9
2	2.56	116.7
3	3.04	138.6
4	3.31	150.9
5	3.58	163.2
Capture	3.75	171

However, the relationship is typically not this simple. For example, fish hatch at lengths that are greater than 0 mm, and their hard parts, or increments on their hard parts, do not always begin to form upon hatching. Given this, the intercept for the relationship between fish size and hard-part radius is not always through the origin, as required for the previous calculations. In these cases, a more complex equation is used (see equation 16.4) in which the intercept can assume a nonzero value (e.g., see figure below). Further, the slope to use for an individual fish in equation (16.4) is calculated as the slope of the

Box 16.4 Continued.

line connecting two data points for that fish—(S_c, L_c) and (0, a). The slope is $(L_c - a)/S_c$, where S_c is the radius of the hard part at capture, L_c is the length of the fish at capture, and a is the intercept of the regression of fish length on hard-part radius.

As an example, if the length of a fish at capture is again 171 mm, the scale radius at capture is 3.75 mm, and the intercept (a) of the regression of fish length on otolith radius is 28.0, the slope ($[L_c - a]/S_c$) is 38.1. Given the following otolith radii (S_i) to previous annuli, the fish lengths at those previous annuli (L_i) are as follows.

Increment (i)	S_i	L_i
1	1.49	84.8
2	2.56	125.5
3	3.04	143.8
4	3.31	154.1
5	3.58	164.4
Capture	3.75	171

Below is an illustration of a linear regression of body lengths at capture on hard-part lengths at capture. The intercept parameter (a) is where the regression line crosses the vertical axis and is used in the Fraser–Lee formula for back-calculation. Data are from 950 walleyes from Lake of the Woods, Minnesota. Number of data points on the graph are represented as follows: * represents 1, + represents 10 or more, and all other numbers are actual number of points.

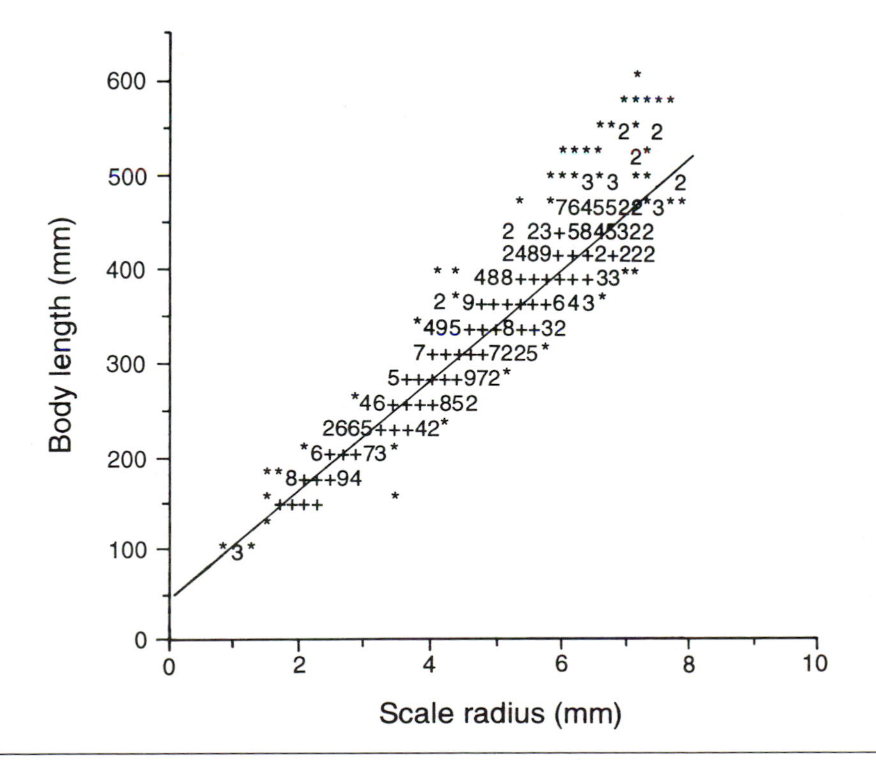

Table 16.1 Mean back-calculated lengths for each year-class showing Lee's phenomenon for bluegill. Younger fish in the sample appear to have better growth.

		Length at age					
Age-group	Year-class	1	2	3	4	5	6
1	1988	90.5					
2	1989	85.3	113.6				
3	1990	78.9	112.2	139.6			
4	1991	75.9	108.9	133.8	150.3		
5	1992	66.2	96.0	129.1	147.5	160.3	
6	1993	59.7	92.2	126.2	145.4	156.6	166.2

Note that *a* is used twice in the Fraser–Lee formula—in the slope of the two-point regression line for each fish and as the intercept (from the regression of fish length on hard-part radius). It has been suggested that *a* can be interpreted as the length of the fish when it first forms the hard part. However, *a* can at times be negative because it is estimated from a regression of length at capture on length of hard part at capture: $L_c = mS_c + a$, where *m* is the slope of the regression and *a* is the intercept (see second illustration Box 16.4). As such, the statistical interpretation of *a* (including negative values) is that it is the best value that will predict L_c accurately in that equation.

When back-calculated lengths at age are smaller for older fish than for younger fish in the sample (Table 16.1), Lee's phenomenon (Ricker 1975) might be occurring. Potential explanations for Lee's phenomenon include that (1) older fish had decreased vulnerability to predation or fishing mortality because they are the slower-growing survivors of their year-class, (2) the sampling gear is selective for faster-growing individuals of the youngest ages, (3) the aging technique is faulty, (4) bias exists in the intercept parameter, and (5) the ratio of fish length to hard-part radius varies systematically with fish growth rate. Reverse Lee's phenomenon (i.e., when back-calculated lengths at age are larger for older fish than for younger fish in the sample) could imply that faster-growing members of the year-class escaped predation or fishing mortality better than did slower-growing members or that the intercept parameter is biased.

Although the Fraser–Lee method is computerized for easy use (see Frie 1982 and DisBCal [from the Computer User Section of the American Fisheries Society]) and allows for a nonzero intercept of the regression of fish lengths on hard-part radius, it also has several problems that must be considered. First, in the Fraser–Lee method, the increments on the hard parts are not used to estimate the parameter *a*. That is, models fit to fish length and hard-part radius at capture are assumed to be applicable to earlier lengths at age. Second, the statistical comparison of means between two or more time periods or populations is unreliable because the estimate of back-calculated length is from a sample size of one fish on a two-point regression (equation 16.4; Figure 16.11). The sampling distribution or valid statistical measures of precision have not been published for the Fraser–Lee method. If between-fish differences in growth are presumed to exist, then it is difficult to identify whether differences in back-calculated lengths are caused by a biased intercept parameter (i.e., *a*), sampling error, environmental or management change, or the life history stages the fish went through (i.e., the influence of the age of an average fish when it grew).

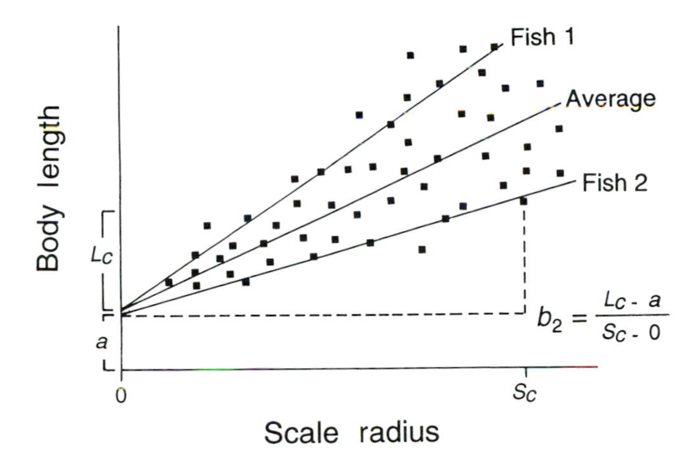

Figure 16.11 Illustration of how the Fraser–Lee formula back-calculates body length from distances between the focus and each annulus on a hard part for each fish, one at a time. Back-calculated lengths at age i are calculated for individual fish as described by text equation (16.4). Each fish has its own slope b in the Fraser–Lee equation, which is estimated from a two-point regression: one point is (S_c, L_c) and the other is (0, a); S_c is the radius of the hard part at capture, L_c is the length of fish at capture, and a is the intercept parameter.

16.3.5.3 Weisberg Method

The Weisberg method of back-calculation involves a linear modeling approach (Weisberg and Frie 1987; Weisberg 1993a, 1993b). With this technique, variability in growth due to environmental influence (i.e., when an average fish grew) can be separated from variability in age-dependent growth while an average fish grew. Thus, explicit statistical tests (e.g., analysis of variance, Dowdy and Wearden 1991) can be used to suggest whether a management action (or some other environmental effect) caused a change in growth from one time period to another.

With this technique, models are fit to the hard-part increments and the results translated into statements of length when needed (Weisberg and Frie 1987). A two-way analysis of variance (Dowdy and Wearden 1991) is used to partition the variation in growth increments into components due to age effects, and year and environmental effects. As such, all increments are used as replicate observations, and age and year and environment are the class or treatment variables (Table 16.2). A related multiple-regression approach, in which a number of variables can be assessed, is described by Maceina (1992).

The Weisberg method has several advantages over the Fraser–Lee method. The method is more precise when used with sparse samples than is the Fraser–Lee method because it uses all data from the hard parts instead of data from a single hard part one fish at a time (Table 16.2). The Weisberg method has been shown to have more desirable sampling and statistical assumptions (Wagner 1989) than the Fraser–Lee method because it calculates means and standard deviations instead of back-calculating individual fish lengths. Unlike the Fraser–Lee method, assumptions and mechanics of the method are statistically explicit. Statistically valid estimates of precision of the mean back-calculated increment are obtained. Finally, as with the Fraser–Lee method, the Weisberg method is in microcomputer form for easy application (Weisberg 1993a, 1993b).

Table 16.2 Mean measured increments (mm) between annuli on scales of walleye, typical of data used with the Weisberg method of back-calculation. Rows correspond to fish age and columns correspond to the calendar year during which growth occurred. Diagonals correspond to the mean increments of each year-class during each year and at every age. The Weisberg method uses variance across the rows to model environmental (year) effects on growth and variance down the columns to model the influence of age on growth.

	Year of growth								
Age	1986	1987	1988	1989	1990	1991	1992	1993	1994
1	1.10	1.36	1.28	1.23	1.18	1.40	1.48	1.08	1.74
2		1.20	1.11	1.22	1.21	1.16	1.30	0.97	1.70
3			1.14	1.00	1.04	0.99	1.24	0.84	1.29
4				0.93	0.81	0.82	1.04	0.72	0.96
5					0.66	0.66	0.72	0.58	0.82
6						0.56	0.65	0.53	0.66
7							0.51	0.45	0.52
8								0.43	0.44
9									0.39

16.3.6 False and Missing Rings

De Bont (1967) suggested five requirements for obtaining reliable age and growth information from any hard part, which remain relevant today. The first, correspondence of ring count with fish age for individuals of known age, applies to aging fish and was discussed in section 16.2.3.3. The other four requirements necessary for obtaining accurate growth information via back-calculation are (1) regular increases in body size with increasing ring count; (2) constancy of scale number (for use of scales) throughout life; (3) constant ratio between the increment in the radius of the hard part and the increment in body size; and (4) growth rings formed regularly and at the same time of year (for annular growth). Although these criteria have been addressed for a number of structures in a wide variety of fish species, investigators should assure that they have been met for the species being aged.

16.4 VALIDATION AND VERIFICATION OF MARKS ON HARD PARTS

Whenever hard parts are used to derive age and growth information from fishes, a necessary component of the study is an evaluation of the aging technique by both validation and verification. Verification refers to a measurement of the precision (Chapter 2) of age measurement from hard parts and is typically accomplished by having two or more people independently count the increments on the hard part. Verification is sometimes accomplished by a single reader conducting multiple independent counts on the same hard part. Although this will certainly lead to decreased between-count variation, relative to using more than one reader, it does not provide any check as to the accuracy (i.e., validation; see below) of the counts. Rather, it verifies only that the single reader is consistent in his or her reading technique across fish. Even when two or more readers (e.g., reader X and reader Y) read the same hard parts, systematic biases can influence the data that are produced and conclusions that are drawn. As such, counts by multiple readers can be directly compared using a number of techniques, including regression analysis of reader Y's counts as a function of reader X's counts, paired *t*-test of the two readers' counts, and a nonparametric Wilcoxon matched-pairs signed-ranks test. Campana et al. (1995) endorsed the use of an age bias graph, in which the mean age (± some estimate of precision) as determined by reader Y is graphed as a function of the mean age (± some estimate of precision) as determined by reader X. Such a graph makes visual determination of systematic biases among readers relatively easy to identify. Further, the readers must not know the length of the fish before determining the age. Aging fish without such accessory information is less precise, more time-consuming, and frustrating; however, the resulting age determination will be a more valid statistic because systematic bias is minimized.

Validating counts obtained from hard parts, which refers to the accuracy of the age measurement, is "the confirmation of the temporal meaning of an increment" (Wilson et al. 1983). Validation must be an integral part of every age and growth study. Typically, validation takes the form of counting the rings on fish of known age (e.g., fish whose hard parts have been marked at a previous known point in time or fish that have been released and recaptured at a later date) and comparing the count with the known age. Although comparison of results of age measurements made independently on a fish via a number of techniques can enhance the confidence in age estimates, this is not validation and should not be considered as such.

Relative to the validation of age estimates, validation of back-calculation as a technique to assess fish growth has seldom been conducted. Validation of back-calculation is difficult, requiring that the observed lengths of individual fish, or means of a group of fish, be compared with those generated by back-calculation (Francis 1990). Validation is typically attempted by comparing mean back-calculated lengths of a group of fish with mean lengths of another group. However, as Francis (1990) pointed out, because among-cohort variation can be great, comparison between cohorts is often not appropriate because the two cohorts may be growing under different conditions, leading to differences in growth rates. Thus, the preferred validation method is to compare measured and back-calculated lengths of individuals. Davies and Sloane (1986) compared observed and back-calculated lengths of stocked rainbow trout in a Tasmanian impoundment and found that back-calculated lengths at both ages 3 and 4 were within 5% of observed values. For both walleye and smallmouth bass, Klumb and Frie (unpublished data) found that back-calculated lengths agreed with observed lengths within 17%, for both the Fraser–Lee and Weisberg methods. For short-term growth (7–15-d intervals) of chinook salmon, Bradford and Geen (1987) found that observed growth rates were not correlated with growth rates as back-calculated from otolith increment widths. However, for longer-term growth (51-d intervals), the relationship was significant. They concluded that otolith growth may be more conservative than is somatic growth and emphasized that caution must be used when short-term growth estimates via back-calculation are to be used. Considering this same phenomenon, Campana (1990) noted that these same concerns apply to measurements of annuli on both scales and otoliths.

Validation studies have been conducted for a great number of taxa (Geffen 1992); though its use is increasing, validation is still not a standard practice in age and growth studies (Beamish and McFarlane 1983; Campana and Neilson 1985). Although validation is critical to the interpretation of the results of any age and growth study, validation may not need to be conducted as a component of a study as long as a validation study has been completed for the species of interest living under similar conditions. However, it is preferable that some form of validation can be included in any study of age and growth.

As an additional benefit, validating hard-part counts from specimens of known age can help readers determine exactly what an annual or daily ring is, as opposed to accessory marks, checks, and so on. This is particularly important in the study of daily otolith increments, where formation of subdaily rings is common and a great deal of experience is required to distinguish true daily rings (Campana 1992; Geffen 1992).

16.5 REFERENCES

Adams, L. A. 1942. Age determination and rate of growth in *Polyodon spathula*, by means of the growth rings of the otoliths and dentary bone. American Midland Naturalist 28:617–630.

Albrechtsen, K. 1968. A dying technique for otolith age reading. Journal of Conservation 32:278–280.

Ashley, K. W., and D. L. Garling, Jr. 1980. Improved method for preparing pectoral spine sections of channel catfish for age determination. Progressive Fish-Culturist 42:80–81.

Beamish, R. J. 1979. Differences in age of Pacific hake (*Merluccius productus*) using whole and sections of otoliths. Journal of the Fisheries Research Board of Canada 36:141–151.

Beamish, R. J. 1981. Use of fin-ray sections to age walleye pollock, Pacific cod, and albacore, and the importance of this method. Transactions of the American Fisheries Society 110:287–299.

Beamish, R. J., and D. E. Chilton. 1982. Preliminary evaluation of a method to determine the age of sablefish (*Anoplopoma fimbria*). Canadian Journal of Fisheries and Aquatic Sciences 39:277–287.

Beamish, R. J., and W. N. McFarlane. 1983. The forgotten requirement for age validation in fisheries biology. Transactions of the American Fisheries Society 112:735–743.

Beamish, R. J., and W. N. McFarlane. 1987. Current trends in age determination methodology. Pages 15–42 *in* R. C. Summerfelt and G. E. Hall, editors. Age and growth of fish. Iowa State University Press, Ames.

Bhattacharya, C. G. 1967. A simple method of resolution into Gaussian components. Biometrics 23:115–135.

Boxrucker, J. 1986. A comparison of the otolith and scale methods for aging white crappies in Oklahoma. North American Journal of Fisheries Management 6:122–125.

Bradford, M. J., and G. H. Geen. 1987. Size and growth of juvenile chinook salmon back-calculated from otolith growth increments. Pages 453–461 *in* R. C. Summerfelt and G. E. Hall, editors. Age and growth of fish. Iowa State University Press, Ames.

Brennan, J. S., and G. M. Cailliet. 1989. Comparative age-determination techniques for white sturgeon in California. Transactions of the American Fisheries Society 118:296–310.

Brett, J. R. 1979. Environmental factors and growth. Pages 599–675 *in* W. S. Hoar, D. J. Randall, and J. R. Brett, editors. Fish physiology, volume 8. Academic Press, New York.

Brothers, E. B. 1987. Methodological approaches to the examination of otoliths in aging studies. Pages 319–330 *in* R. C. Summerfelt and G. E. Hall, editors. Age and growth of fish. Iowa State University Press, Ames.

Busacker, G. P., I. R. Adelman, and E. M. Goolish. 1990. Growth. Pages 363–387 *in* C. B. Schreck and P. B. Moyle, editors. Methods for fish biology. American Fisheries Society, Bethesda, Maryland.

Calliet, G. M., L. K. Martin, D. Kusher, P. Wolf, and B. Welden. 1983. Techniques for enhancing vertebral bands in age estimation of California elasmobranchs. Pages 157–165 *in* E. D. Prince and L. M. Pulos, editors. Proceedings of the international workshop on age determination of oceanic pelagic fishes: tunas, billfishes, and sharks. NOAA (National Oceanic and Atmospheric Administration) Technical Report 8, Washington, DC.

Campana, S. E. 1983a. Feeding periodicity and the production of daily growth increments in the otoliths of steelhead trout (*Salmo gairdneri*) and starry flounder (*Platichthys stellatus*). Canadian Journal of Zoology 61:1591–1597.

Campana, S. E. 1983b. Calcium deposition and otolith check formation during periods of stress in coho salmon, *Oncorhynchus kisutch*. Comparative Biochemistry and Physiology 75A:215–220.

Campana, S. E. 1990. How reliable are growth back-calculations based on otoliths? Canadian Journal of Fisheries and Aquatic Sciences 47:2219–2227.

Campana, S. E. 1992. Measurement and interpretation of the microstructure of fish otoliths. Canadian Special Publication of Fisheries and Aquatic Sciences 117:59–71.

Campana, S. E., M. C. Annand, and J. I. McMillan. 1995. Graphical and statistical methods for determining the consistency of age determinations. Transactions of the American Fisheries Society 124:131–138.

Campana, S. E., and E. Moksness. 1991. Accuracy and precision of age and hatch date estimates from otolith microstructure examination. International Council for the Exploration of the Sea Journal of Marine Science 48:303–316.

Campana, S. E., and J. D. Neilson. 1985. Microstructure of fish otoliths. Canadian Journal of Fisheries and Aquatic Sciences 42:1014–1032.

Carlander, K. D. 1982. Standard intercepts for calculating length from scale measurements for some centrarchid and percid fishes. Transactions of the American Fisheries Society 111:332–336.

Carline, R. F., B. L. Johnson, and T. J. Hall. 1984. Estimation and interpretation of proportional stock density for fish populations in Ohio impoundments. North American Journal of Fisheries Management 4:139–154.

Carroll, B., and G. E. Hall. 1964. Growth of catfishes in Norris Reservoir, Tennessee. Journal of the Tennessee Academy of Science 39:86–91.

Casselman, J. M. 1983. Age and growth assessment of fish from their calcified structures—techniques and tools. Pages 1–18 *in* E. D. Prince and L. M. Pulos, editors. Proceedings of the international workshop on age determination of oceanic pelagic fishes: tunas, billfishes, and sharks. NOAA (National Oceanic and Atmospheric Administration) Technical Report 8, Washington, DC.

Cassie, R. M. 1954. Some uses of probability paper in analysis of size frequency distributions. Australian Journal of Marine and Freshwater Research 5:513–522.

Chilton, D. E., and R. J. Beamish. 1982. Age determination methods for fishes studied by the groundfish program at the Pacific Biological Station. Canadian Special Publication of Fisheries and Aquatic Sciences 60.

Christensen, J. M. 1964. Burning of otoliths, a technique for age determination of soles and other fish. Journal of Conservation 29:73–81.

Crumpton, J. E., M. M. Hale, and D. J. Renfro. 1987. Aging of three species of Florida catfish utilizing three pectoral spine sites and otoliths. Proceedings of the Annual Conference Southeastern Association of Fish and Wildlife Agencies 38(1984):335–341.

Crumpton, J. E., M. M. Hale, and D. J. Renfro. 1988. Bias from age-grouping black crappie by length-frequency as compared to otolith aging. Proceedings of the Annual Conference Southeastern Association of Fish and Wildlife Agencies 40(1986):65–71.

Davies, P. E., and R. D. Sloane. 1986. Validation of aging and length back-calculation in rainbow trout, *Salmo gairdneri* Rich., from Dee Lagoon, Tasmania. Australian Journal of Marine and Freshwater Research 37:289–295.

De Bont, A. F. 1967. Some aspects of age and growth of fish in temperate and tropical waters. Pages 67–88 *in* S. D. Gerking, editor. The biological basis of freshwater fish production. Wiley, New York.

Dowdy, S., and S. Wearden. 1991. Statistics for research. Wiley, New York.

Everhart, W. H., and W. D. Youngs. 1981. Principles of fishery science, 2nd edition. Cornell University Press, Ithaca, New York.

Francis, R. I. C. C. 1990. Back-calculation of fish length: a critical review. Journal of Fish Biology 36:883–902.

Frie, R. V. 1982. Measurement of fish scales and back-calculation of body-lengths using a digitizing pad and microcomputer. Fisheries 7(5):5–8.

Fry, F. E. J. 1971. The effects of environmental factors on the physiology of fish. Pages 1–98 *in* W. S. Hoar and D. J. Randall, editors. Fish physiology, volume 6. Academic Press, New York.

Geen, G. H., J. D. Neilson, and M. Bradford. 1985. Effects of pH on the early development and growth and otolith microstructure of chinook salmon, *Oncorhynchus tshawytscha*. Canadian Journal of Zoology 63:22–27.

Geffen, A. J. 1992. Validation of otolith increment deposition rate. Canadian Special Publication of Fisheries and Aquatic Sciences 117:101–113.

Gjerde, B. 1986. Growth and reproduction in fish and shellfish. Aquaculture 57:37–55.

Haake, P. W., C. A. Wilson, and J. M. Dean. 1982. A technique for the examination of otoliths by SEM with application to larval fishes. Pages 12–15 *in* C. F. Bryan, J. V. Conner, and F. M. Truesdale, editors. Proceedings of the fifth annual larval fish conference. Louisiana State University Press, Baton Rouge.

Hammers, B. E., and L. E. Miranda. 1991. Comparison of methods for estimating age, growth, and related population characteristics of white crappies. North American Journal of Fisheries Management 11:492–498.

Hile, R. 1950. A nomograph for the computation of the growth of fish from scale measurements. Transactions of the American Fisheries Society 78:156–162.

Hudson, W. F., and F. J. Bulow. 1984. Relationships between squamation chronology of the bluegill, *Lepomis macrochirus* Rafinesque, and age–growth methods. Journal of Fish Biology 24:459–469.

Isaac, V. J. 1990. The accuracy of some length-based methods for fish population studies. International Center for Living Aquatic Resources Management Technical Report 27, Manila, Phillippines.

Jearld, A., Jr. 1983. Age determination. Pages 301–324 *in* L. A. Nielsen and D. L. Johnson, editors. Fisheries techniques. American Fisheries Society, Bethesda, Maryland.

Jones, C. 1986. Determining age of larval fish with the otolith increment method. U.S. National Marine Fisheries Service Fishery Bulletin 84:91–103.

Jones, C. M. 1992. Development and application of the otolith increment technique. Canadian Special Publication of Fisheries and Aquatic Sciences 117:1–11.

Kimura, D. K. 1977. Statistical assessment of the age–length key. Journal of the Fisheries Research Board of Canada 34:317–324.

Kruse, C. G., C. S. Guy, and D. W. Willis. 1993. Comparison of otolith and scale age characteristics for black crappies collected from South Dakota waters. North American Journal of Fisheries Management 13:856–858.

Lagler, K. F. 1956. Freshwater fishery biology, 2nd edition. Brown, Dubuque, Iowa.

Le Cren, E. D. 1947. The determination of the age and growth of the perch (*Perca fluviatilis*) from the opercular bone. Journal of Animal Ecology 16:188–204.

Lowerre-Barbieri, S. K., M. E. Chittenden, Jr., and C. M. Jones. 1994. A comparison of a validated otolith method to age weakfish, *Cynoscion regalis*, with the traditional scale method. U.S. National Marine Fisheries Service Fishery Bulletin 92:555–568.

Lux, F. 1971. Age determination in fishes. U.S. Fish and Wildlife Service Fish and Wildlife Leaflet 637.

MacDonald, P. D. M. 1987. Analysis of length-frequency distributions. Pages 371–384 *in* R. C. Summerfelt and G. E. Hall, editors. Age and growth of fish. Iowa State University Press, Ames.

Maceina, M. J. 1988. Simple grinding procedure to section otoliths. North American Journal of Fisheries Management 8:141–143.

Maceina, M. J. 1992. A simple regression model to assess environmental effects on fish growth. Journal of Fish Biology 41:557–565.

Magnuson, J. J., L. B. Crowder, and P. A. Medvick. 1979. Temperature as an ecological resource. American Zoologist 19:331–343.

Marshall, S. L., and S. S. Parker. 1982. Pattern identification in the microstructure of sockeye salmon (*Oncorhynchus nerka*) otoliths. Canadian Journal of Fisheries and Aquatic Sciences 39:542–547.

McNew, R. W., and R. Summerfelt. 1978. Evaluation of a maximum-likelihood estimator for analysis of length-frequency distributions. Transactions of the American Fisheries Society 107:730–736.

Moyle, P. B. 1993. Fish: an enthusiast's guide. University of California Press, Berkeley.

Mugiya, Y., N. Watabe, J. Yamada, J. M. Dean., D. G. Dunkelberger, and M. Shimizu. 1981. Diurnal rhythm in otolith formation in the goldfish, *Carassius auratus*. Comparative Biochemistry and Physiology 68A:659–662.

Neilson, J. D., and G. H. Geen. 1985. Effects of feeding regimes and diel temperature cycles on otolith increment formation in juvenile chinook salmon (*Oncorhynchus tshawytscha*). U.S. National Marine Fisheries Service Fishery Bulletin 83:91–100.

Pycha, R. L. 1955. A quick method of preparing permanent fin-ray and spine sections. Progressive Fish-Culturist 17:192.

Quinn, S. P. 1989. Flathead catfish abundance and growth in the Flint River, Georgia. Proceedings of the Annual Conference Southeastern Association of Fish and Wildlife Agencies 42(1988):141–148.

Reed, W. J., and K. H. Wilson. 1979. Comment on "Bias in using an age-length key to estimate age-frequency distributions" by S. J. Westrheim and W. J. Ricker. Journal of the Fisheries Research Board of Canada 36:1159–1160.

Remington, R. D., and M. A. Schork. 1970. Statistics with application to the biological and health sciences. Prentice-Hall, Englewood Cliffs, New Jersey.

Ricker, W. E. 1975. Computation and interpretation of biological statistics of fish populations. Fisheries Research Board of Canada Bulletin 191.

Sadler, K., and S. Lynam. 1986. Some effects of low pH and calcium on the growth and tissue mineral content of yearling brown trout, *Salmo trutta*. Journal of Fish Biology 29:313–324.

Schramm, H. L., Jr. 1989. Formation of annuli in otoliths of bluegills. Transactions of the American Fisheries Society 118:546–555.

Schramm, H. L., Jr., and J. F. Doerzbacher. 1985. Use of otoliths to age black crappie from Florida. Proceedings of the Annual Conference Southeastern Association of Fish and Wildlife Agencies 36(1982):95–105.

Scidmore, W. J., and A. W. Glass. 1953. Use of pectoral fin rays to determine age of the white sucker. Progressive Fish-Culturist 15:114–115.

Secor, D. H., J. M. Dean, and E. H. Laban. 1991. Manual for otolith removal and preparation for microstructural examination. University of South Carolina, Baruch Institute for Marine Biology and Coastal Research Technical Report 91-1, Columbia.

Secor, D. H., J. M. Dean, and E. H. Laban. 1992. Otolith removal and preparation for microstructural examination. Canadian Special Publication of Fisheries and Aquatic Sciences 117:19–57.

Simkiss, K. 1974. Calcium metabolism of fish in relation to ageing. Pages 1–12 *in* T. B. Bagenal, editor. Ageing of fish. Unwin Brothers Limited, London.

Smith, S. H. 1954. Method of producing plastic impressions of fish scales without using heat. Progressive Fish-Culturist 16:75–78.

Sneed, K. E. 1951. A method for calculating the growth of channel catfish, *Ictalurus lacustris punctatus*. Transactions of the American Fisheries Society 80:174–183.

Snow, H. E., and C. J. Sand. 1992. Comparative growth of eight species of fish in fifty-five northwestern Wisconsin lakes. Wisconsin Department of Natural Resources Research Report 153, Madison.

Stevenson, D. K., and S. E. Campana, editors. 1992. Otolith microstructure examination and analysis. Canadian Special Publication of Fisheries and Aquatic Sciences 117.

Tesch, F. W. 1971. Age and growth. Pages 98–130 *in* W. E. Ricker, editor. Methods for assessment of fish production in fresh waters. Blackwell Scientific Publications, London.

Turner, P. R. 1982. Procedures for age determination and growth rate calculations of flathead catfish. Proceedings of the Annual Conference Southeastern Association of Fish and Wildlife Agencies 34(1980):253–262.

Van Den Avyle, M. J. 1993. Dynamics of exploited fish populations. Pages 105–135 *in* C. C. Kohler and W. A. Hubert, editors. Inland fisheries management in North America. American Fisheries Society, Bethesda, Maryland.

Van Oosten, J. 1957. The skin and scales. Pages 207–244 *in* M. E. Brown, editor. The physiology of fishes, volume 1. Academic Press, New York.

Wagner, B. 1989. Backcalculation of fish lengths: a comparison of methodologies. Master's thesis. University of Missouri, Columbia.

Weatherley, A. H., and H. S. Gill, editors. 1987. The biology of fish growth. Academic Press, New York.

Weisberg, S. 1993a. A computer program for using hard-part increment data to estimate age and environmental effects in fish populations. University of Minnesota, Minnesota Sea Grant College Program, St. Paul.

Weisberg, S. 1993b. Using hard-part increment data to estimate age and environmental effects. Canadian Journal of Fisheries and Aquatic Sciences 50:1229–1237.

Weisberg, S., and R. V. Frie. 1987. Linear models for the growth of fish. Pages 127–143 *in* R. C. Summerfelt and G. E. Hall, editors. Age and growth of fish. Iowa State University Press, Ames.

Welch, T. J., M. J. Van Den Avyle, R. K. Betsill, and E. M. Driebe. 1993. Precision and relative accuracy of striped bass age estimates from otoliths, scales, and anal fin rays and spines. North American Journal of Fisheries Management 13:616–620.

Westrheim, S. J., and W. E. Ricker. 1978. Bias in using an age–length key to estimate age-frequency distributions. Journal of the Fisheries Research Board of Canada 35:184–189.

Westrheim, S. J., and W. E. Ricker. 1979. Reply to comment on "Bias in using an age–length key to estimate age-frequency distributions" by W. J. Reed and K. H. Wilson. Journal of the Fisheries Research Board of Canada 36:1160–1161.

Wilson, C. A., E. B. Brothers, J. M. Casselman, C. L. Smith, and A. Wild. 1983. Glossary. Pages 207–208 *in* E. D. Prince and L. M. Pulos, editors. Proceedings of the international workshop on age determination of oceanic pelagic fishes: tunas, billfishes, and sharks. NOAA (National Oceanic and Atmospheric Administration) Technical Report 8, Washington, DC.

Witt, A., Jr. 1961. An improved instrument to section bones for age and growth determination of fish. Progressive Fish-Culturist 23:94–96.

Wootton, R. J. 1990. Ecology of teleost fishes. Chapman and Hall, New York.

Chapter 17

Quantitative Description of the Diet

STEPHEN H. BOWEN

17.1 INTRODUCTION

There are many reasons why biologists and managers want to know about the diets of fishes. Like all animals, fishes function to convert the organic energy and material they ingest into living biomass. Their productivity in this process both as individuals and as populations is influenced by the quality and quantity of the food they are able to obtain (Ney 1990; Bowen et al. 1995). Managers are often called upon to develop plans for stocking and harvest that match predator populations with the available forage base. In aquaculture, the need for supplemental feeding may depend on the adequacy of naturally available foods. Aquatic ecologists often choose fishes as models for research on specific trophic strategies; fish are usually larger than invertebrate consumers, making analysis of gut contents and study of digestive physiology more practical. Fish are often behaviorally sophisticated and have feeding behaviors that point to important factors affecting their nutrition and growth. Changing environmental conditions such as those resulting from eutrophication or warming affect the fortunes of fish species differently, in part because these changing conditions affect the availability of food types. Thus, an understanding of fish diet and its influence on growth can be essential to understanding the ecological role and the productive capacity of many populations.

There are also reasons why biologists and managers should approach the study of fish diets with caution. We already know in general what most fish species eat. To provide more detail for a specific population is usually expensive in terms of time and other resources. Thus, food habit studies are difficult to justify as a part of routine assessments. There should be some reason to suspect the quantity or quality of available food limits the population. Has growth declined in comparison to earlier years? Are the fish in poor condition? Has fecundity at length declined? Have large changes been observed in the availability of preferred prey? These are some indications that the diet may be affecting the population and should be investigated further. The purpose of this chapter is to help biologists choose and apply techniques for the study of diet that are appropriate to particular fish populations.

17.2 COLLECTION OF FISHES FOR STUDY OF DIET

Of the many techniques devised for collecting fishes, some are more useful than others for description of diet. Some capture techniques result in loss of information as a result of regurgitation. Capture techniques that stress fish severely, such as rotenone treatment, electroshocking, overnight gillnetting, and trawling at depth, are more prone than others to create problems. Seine, cast-net, and short-term gill- or trammel-net sets are generally less traumatic. In some cases, a very sudden shock may be useful. One biologist found that a single-pronged spear allowed fish to regurgitate, but a three-pronged spear stunned fish on impact, and regurgitation was not a problem (M. S. Christensen, Rhodes University, personal communication). Some fish species are more likely to regurgitate than others. Piscivorous predators have large, distensible esophagi that make regurgitation fairly easy, and loss of gut contents can be a problem with these species. If preliminary collections over a 24-h cycle yield mostly specimens with empty stomachs, different collection techniques should be compared to see if one yields a greater proportion of animals with food in the digestive tract than others.

Postcapture digestion may also result in loss of valuable dietary information. Gill nets and traps of various sorts may hold fish for long hours during which much of the diet can be digested beyond the point at which identification is possible. Consequently, these capture techniques are most useful when catch rates are high and specimens can be removed from the gear soon after capture. Some postcapture digestion may be tolerable if gill nets or traps are the only effective means by which the target species can be collected.

A third difficulty for a diet study arises when fish in traps feed on foods they would not normally eat. For example, some species of catfish that are not normally piscivorous may feed on small fishes when confined with these prey in a trap. Because of the problems of postcapture digestion and atypical feeding, traps are of limited use in diet studies.

17.3 SAMPLING STRATEGIES

Many factors may influence the amount and type of food found in the digestive tracts of fish. The effects of the diel cycle, seasonal changes, size and territoriality of fish, and differential digestion rates must be considered in the design of an efficient and accurate sampling program.

17.3.1 Diel Effects

The behavior of most fishes is highly structured within the diel cycle (Johnson and Dropkin 1993). Diel changes in habitat, feeding intensity, and the diet itself must be considered. During nonfeeding hours, fish often seek shelter in low-risk environments. Just prior to feeding, they move into habitats where suitable food is available. Some fish feed intensively during a single extended feeding period lasting 6 h or longer. Others may feed intensively for only 1 or 2 h, followed by low levels of feeding activity for several hours. Still other fish are known to feed intensively at dusk and dawn, and there is little evidence of their feeding at other hours. Although given patterns of diel feeding behavior are commonly associated with particular species, fishes can completely reverse their diel cycle in response to changes in food availability or predator distribution (Bowen and Allanson 1982). In order to maximize the amount of information gained per fish collected, fish must be collected when their stomachs are fullest. Therefore, it is usually a good idea to begin a diet study with a 24-h series of collections made at regular intervals of 2, 3, or 4 h. Data on diel variation in both catch per effort and stomach fullness help in the planning of an efficient sampling program.

Diet itself may vary with time of day. Salmonids in large streams may feed on invertebrate drift during the night and small fish during the day. I once found a population of golden shiners that fed on periphyton at night, on cladocerans at dusk and dawn, and on sessile invertebrates during the day. Comparison of stomach contents from a series of 24-h collections is necessary to determine if the diet varies consistently according to time of day. If no such variation is apparent, then gut fullness and catch per effort determine the best time of day to sample. If a consistent pattern of diel variation is detected, it adds to both the interest and the complexity of a study. For example, if one is interested in seasonal changes in diet, diel variation in diet must be measured in each season.

17.3.2 Seasonal Effects

Fishes are highly responsive to seasonal changes in food availability. Changes within trophic categories—from mayflies to chironomids for example—are common as invertebrate populations mature and emerge. Changes from one trophic category to another are also known to occur. Bluegills may switch from a diet of invertebrates to a diet of algae as invertebrate numbers decline toward the end of summer (Kitchell and Windell 1970). Amazon River fishes highly specialized for feeding on invertebrates in the rainy season turn to detritus as an interim food resource at the end of the dry season (Lowe-McConnell 1975). Thus, to adequately describe the trophic resources utilized by a fish population, one often must sample at frequent intervals throughout the year.

17.3.3 Effects of Fish Size and Territoriality

Diets selected by fish vary according to fish size and sex. As fish grow larger, they may switch from one prey type to another or select larger individuals of the same prey type (Werner and Hall 1974, 1976; Buijse and Houthuijzen 1992). The change may be abrupt and associated with a change in habitat, as when salmon smolts change from invertebrate diets in freshwater to fish diets at sea. In many cases, the transition is gradual, as with lake herring, which change slowly from a diet of copepod nauplii to a diet of copepod adults as they grow. If the change is abrupt, the goals of the study may be served by analysis of gut contents from fish divided into two size-groups. When the diet changes more gradually, 10 or more size-groups may be required to document the change, and it may be necessary to expand the range of habitats sampled in order to obtain statistically significant numbers of individuals for all size-groups.

Adult males of some species establish and defend breeding territories that they rarely leave (Bowen 1984). They may feed infrequently and must depend on food resources within their territories. As a result, their diets may be considerably different from those of the females and subadult males. This may be of no consequence to the goals of a study if one is interested in preadult stages or production of females only. On the other hand, a contrast between the diets and nutritional status of adult males and adult females may help identify important factors whereby diet affects fitness.

17.3.4 Differential Digestion Rates

Stomach contents may not accurately reflect the diet of a fish for two reasons. First, some prey, such as tubificids (Kennedy 1969) or protozoa, may be digested rapidly and leave little recognizable trace in the digestive tract. The importance of these prey may be underestimated or overlooked entirely. If such a bias is suspected, it may be useful to hold the fish in an aquarium, watch how they feed, and then collect samples of their food for comparison with gut contents. A second type of bias can result from differential rates of digestion for various prey. Slowly digested prey may accumulate relative to rapidly digested prey and, as a result, be overrepresented in the gut (Mann and Orr 1969; Gannon 1976). This effect can be minimized if fish are collected at the peak of daily feeding intensity when all food items in the anterior gut have been ingested recently. It is also possible to compensate for the effects of differential rates of digestion by determining the gut passage time for each type of food (Klumpp and Nichols 1983).

17.4 REMOVAL, FIXATION, AND PRESERVATION OF GUT CONTENTS

17.4.1 Removal of Gut Contents

To protect valuable fish, many workers have developed methods for removal of stomach contents from fish that are to be kept alive. One method employs a pump to flush stomach contents from the fish with one or more volumes of water (Giles 1980). Tests in which the stomachs were dissected after pumping showed this is a highly effective method of removing prey.

Another technique uses clear plastic (acrylic) tubes that are open on both ends. The tube is wetted and then inserted through the esophagus and into the stomach (Van Den Avyle and Roussel 1980). For each fish being sampled, the diameter of the tube is chosen so that it fully distends the esophagus and surrounds the contents of the stomach. This method works best with piscivorous predators that typically feed on one or a few large prey at a time. The esophagus can distend to a circular diameter equal to that of the stomach, and the prey fit inside the tube. A thumb or the palm of the hand is used to close the end of the tube so that a vacuum is produced as the tube is withdrawn. A systematic assessment of this technique showed that reproducibility was good across workers and that greater than 80% of the prey were recovered (Cailteux et al. 1991).

Collection of stomach contents from live animals works best with more robust fishes, such as perches, sunfishes, and the many catfishes. The technique has also been successful with trouts (Bechara et al. 1993). Delicate fishes such as the herrings and many carps and minnows are unlikely to survive the experience. Although these approaches require a special apparatus, valuable fish are conserved and, in many cases, removal of gut contents from live animals is faster than removal by dissection.

Nevertheless, it is often preferable to remove gut contents by dissection. Dissection allows the worker to examine the intestine for fullness and to be confident that all prey are removed from the stomach or anterior gut segment. Data on sex and gonad maturity may be collected at the same time. If data on intestinal contents are needed, dissection is essential.

Before dissection, fish are killed as humanely as possible with an overdose of anesthetic, a sharp blow to the head, or (for small fish) severing the spinal column with scissors or a knife (see Chapter 5). Open the coelom to expose the viscera. Using blunt scissors, sever the esophagus, the last few millimeters of the intestine, and the mesentery at its dorsal point of attachment. This excision allows the visceral mass to be lifted out of the coelom and placed into a dissection pan for more careful manipulation. Separate the digestive tract from the rest of the viscera and divide it into segments as dictated by the needs of the study. Open each segment by slitting it lengthwise with fine scissors. For piscivorous fishes, prey items can be lifted directly from the stomach. For fish that eat smaller prey, it is often useful to hold the slit segment open with forceps over a petri dish or small beaker and then to wash the food items from the segment with water from a wash bottle. If quantitative recovery of gut contents is not required, the food may be extruded by sliding a blunt probe along the length of the segment. This technique may, however, also extrude the gut mucosa. Do not mistake this tissue as part of the diet.

17.4.2 Fixation and Preservation of Gut Contents

Ten percent formalin (a 3.9% formaldehyde solution) is fully adequate as a fixative. Because it rapidly hardens prey tissues, partly digested prey are more likely to stay intact and, thus, are easier to identify. Before samples are examined in detail, remove excess formaldehyde by soaking the samples in several changes of water. Afterward, preserve the samples in a 45–70% aqueous solution of alcohol. Methanol, ethanol, and isopropanol are all satisfactory. Wear plastic gloves and work in a fume hood to minimize exposure to residual formaldehyde. Formaldehyde (as a gas or in a solution) is a suspected carcinogen and can cause permanent damage to fingernails and sinus tissues.

The availability of field time and facilities usually determines when gut contents are removed and fixed. It is best to remove and fix gut samples immediately after fish are captured. This minimizes postcapture digestion and avoids difficulties associated with dissection of hardened fish tissues (Borgeson 1963). However, when tens or hundreds of fish are captured at one time and all are needed for the study, it is often necessary to fix the gut contents in situ with the least investment of time possible. In warm climates, hold fish on ice if more than a few minutes will elapse before fixation. Fishes of 100 g live weight or less can be fixed whole. In order to halt postcapture digestion, slit the coelom to allow entry of formalin or inject formalin directly into the coelom (Emmett et al. 1982). For larger fishes, it is usually more efficient to fix only the digestive tract.

The goals of the study will determine the extent to which individual samples must be kept separate and identified. When gut contents are removed in the field, they may be stored in small numbered vials or plastic bags, and corresponding data (e.g., length, weight, and sex) may be recorded on data sheets with reference to the same sample number. When intact digestive tracts are fixed, they may be stored in separate containers. Several digestive tract samples may be stored in a single container if they are wrapped individually in cheese cloth and each has a paper label written in pencil. For some studies it may be adequate to pool gut samples according to size-groups of the fish. Although this results in loss of data on variation among individuals, it requires less handling and greatly simplifies subsequent analyses.

17.5 IDENTIFICATION OF DIET COMPONENTS

17.5.1 Identification of Partly Digested Prey

Identifying stomach contents is often made difficult by digestion. Even recently ingested food items may be ground by jaw or pharyngeal teeth to a point at which recognition of many individual fragments is difficult. As a result, it has become customary to identify food items by finding some characteristic part of the organism that is resistant to digestion. For invertebrates, various parts of the exoskeleton, such as eye capsules, head shields, and tarsal claws have been used (Ahlgren and Bowen 1992; Bechara et al. 1993). Fish prey can be identified by the characteristic form of their otoliths (Whitfield and Blaber 1978). Macrophytes may be identified by a characteristic shape or sculpturing along the edge of the leaves. In contrast, algal cells are usually found intact in the anterior digestive tract, and identification of these presents no special problems.

17.5.2 Level of Identification

Some studies have included identification of prey items to the species level. Although taxonomic identification is a useful tool, this level of precision may not be necessary. If the purpose of the investigation is to assess the effects of fish predation on the invertebrate community in a given habitat, then identification to species is required (Diel 1992). When studying how competing fishes divide the available prey base, broad categories defined at the order or family level can be fully adequate (Bergman and Greenberg 1994). In some instances, only the relative size of prey appears to be important in diet selection. Unless there is a specific need for identification to species, higher taxonomic levels of identification produce the most useful data per unit of time invested.

17.6 QUANTITATIVE DESCRIPTION OF DIET

There are three simple approaches to quantitative description of the diet: frequency of occurrence, percent composition by number, and percent composition by weight. Each approach provides distinctly different information and, consequently, it is necessary to discuss each one in some detail. Box 17.1 compares the approaches by means of a sample data set.

17.6.1 Frequency of Occurrence

Frequency of occurrence is the fastest approach to quantitative analysis of fish diets. When examining gut samples from individual fish, compile a cumulative list of foods found and record the presence or absence of each for each specimen. When all specimens have been examined, the proportion of the fish that contained one or more of a given food type is calculated as the frequency of occurrence for that food type. For example, if 18 out of a sample of 22 bluegills contained one or more chironomids, then the frequency of occurrence of chironomids in the diet would be 0.82 or 82%.

These results indicate the extent to which fish in the sample can be characterized as a single feeding unit. If nearly all fish contain the same set of prey types, this will be clearly documented with uniformly high frequencies of occurrence. Under some circumstances, or with some fish species, individuals appear to concentrate on finding and ingesting just one food type. In a single seine haul, I found juvenile Mozambique tilapia that had been feeding on ants, ostracods, or detritus, but few had eaten a mixture of these foods. This selective or opportunistic feeding behavior is reflected in lower frequencies of occurrence. Similarly, a group of omnivorous fish that all feed on periphyton but specialize individually in selection of invertebrate prey have high frequency-of-occurrence values for most algae types and low values for most invertebrate types.

Although frequency of occurrence often gives valuable insights, there are limits to the information it can provide. High frequency of occurrence does not mean that a given food type is of nutritional importance to the consumer. The food type may be consumed with great regularity by most members of a population but in very small quantities. For example, a fish population whose diet is made up of 99% benthic invertebrates may unavoidably ingest sedimented algae to the extent that algae make up 1% of the diet. Because these algal cells are small and tend to be evenly distributed across a substrate, the frequency of occurrence for many algal species may be 100%. The frequency of occurrence for individual invertebrate species may not exceed 50%. This does not mean that algae are more important than invertebrates in the consumer's diet. Frequency of occurrence describes the uniformity with which groups of fish select their diet but does not indicate the importance of the various types of food selected.

17.6.2 Percent Composition by Number

When percent composition by number is measured, the number of food items of each food type is determined for each fish examined. The metric is the number of items of a given food type expressed as a percentage of the total number of food items counted. After a preliminary examination of a few samples, it is often necessary to give some thought to the appropriate unit to be counted. If the diet consists of invertebrate prey that are largely disarticulated by the pharyngeal apparatus, some characteristic fragment must be counted. To minimize counting time, it is best to choose some fragment that is found only once per prey. Not all foods, however, are well suited to a counting approach. Detritus and higher plants that are ingested bit by bit are not found in discrete units of uniform size, and, thus, counts of these particles have little meaning.

For fish that feed on large prey (>2 mm maximum dimension), it is usually possible to count all food items in the stomach. However, most fish feed on small prey, and their stomachs often contain such great numbers that counting all items directly is impractical. Under these circumstances, subsampling is necessary. Using a beaker and a magnetic stirrer, suspend the gut contents in a known volume of water. When invertebrate prey are to be counted, a 3- or 4-mm-inside-diameter glass tube is used like a pipette to transfer a known volume of the suspension to a petri dish for counting at the required magnification. The tube can be calibrated against a volumetric cylinder. The same transfer can be made more conveniently by means of a rubber bulb and pipette that has been made to transfer a preset volume by cutting off the tip of the glass pipette barrel and recalibrating the apparatus. Adjust the volume of the suspension, the volume transferred, or both so that 50 to 100 items are counted for each prey type of interest. Subsampling errors become significant if fewer than 50 items are counted, but counts above 100 do little to reduce subsampling errors (Allanson and Kerrich 1961). Food items greater than 100 μm can usually be counted in a petri dish under a dissecting microscope. A piece of 1-mm graph paper fixed to the bottom of the dish helps one to keep track of which items have been counted.

When algal cells or other microscopic foods (<100 μm maximum dimension) are to be counted, the subsample usually has to be much smaller. This is because there can be thousands of cells in 1 mL of suspension, a number impractical to count. One simple and economic approach to counting small particles is the drop transect method (Edmonson 1974; APHA 1989:10200F.2c.4). The approach takes advantage of the repeatability with which a given pipette makes water drops that are virtually identical in volume. To determine what that volume is, count how many drops are needed to obtain 2 mL in a 5-mL graduated cylinder or weigh drops on a balance and

Box 17.1 Recording and Summarizing Dietary Data

Hypothetical data collected from a population of fathead minnows show how different quantitative measures support different kinds of interpretations. For each fish specimen, the characteristic measurement for each food item is recorded according to taxon. At the end of the examination, the number of measurements is determined to provide count data in the last column.

DATA SHEET

Fish specimen number: Place of capture:

Capture date: Analysis date:

Taxon	Measurements	Count
Plecoptera		
Ephemeroptera		
Baetis		
Ephemerella		
Leptophlebiidae		
Heptageniidae		
Tricoptera		
Psycoglypha subborealis		
Other Tricoptera		
Chironomidae		
Tanypodinae		
Orthocladiinae		
Tanytarsini		
Chironomini		
Chironomid pupae		
Other invertebrates		
Simuliidae		
Adult insects		
Algae		
Diatoms		
Green algae		
Blue-green algae		
Detritus		

Box 17.1 Continued.

When all specimens have been examined, the data are summarized in a table. The data typically are summarized as means, as shown below; for real data, means should be accompanied by some measure of variability, such as standard deviation or range. From the frequency-of-occurrence column, we conclude that the majority of fathead minnows consumed Simuliidae, Orthocladiinae, Tanytarsini, diatoms, and detritus. Other food items were ingested by just a few individuals in the group. Percent-by-number data show that of the five food types ingested by the majority of individuals, only detritus particles were eaten in large numbers. Among the invertebrate prey, Orthocladiinae were eaten in relatively large numbers; Simuliidae and Tanytarsini, although present in most diets, were eaten in small numbers. The percentage-by-weight data show that detritus was by far the most important source of organic matter in the diet. Among the invertebrate prey, Orthocladiinae was the most important source of biomass eaten, but the relatively rare but large *Psycoglypha subborealis* was also an important contributor.

Taxon	Frequency of occurrence	Mean percentage by number	Mean percentage by weight
Plecoptera	0.012	0.2	1.0
Ephemeroptera			
Baetis	0.011	1.3	3.1
Ephemerella	0.006	1.2	3.9
Leptophlebiidae	0.008	0.2	0.3
Heptageniidae	0.010	0.9	1.5
Tricoptera			
Psycoglypha subborealis	0.040	0.1	11.4
Other Tricoptera	0.044	2.3	4.9
Chironomidae			
Tanypodinae	0.034	1.7	4.8
Orthocladiinae	0.721	12.9	18.5
Tanytarsini	0.658	2.3	0.9
Chironomini	0.240	0.3	0.3
Chironomid pupae	0.187	2.5	1.2
Other invertebrates			
Simuliidae	0.962	0.8	5.1
Adult insects	0.004	1.3	0.9
Algae			
Diatoms	0.985	10.3	0.8
Green algae	0.221	0.3	0.03
Blue-green algae	0.110	1.1	0.07
Detritus	1.000	60.3	41.3
Total		100	100

convert weight to volume (1 g = 1 mL). Once the volume of a drop from the pipette is known, the pipette can be used to subsample the suspension of food items that includes small particles. A single water drop of the suspension is the subsample.

Put one drop containing the small diet particles in the center of a clean glass slide and cover it with a 22-mm × 22-mm coverslip. Surface tension draws the sample out under the coverslip. At magnifications less than 200 times, it is often practical to count all particles under the coverslip. Begin by positioning the top edge of the microscope field on what appears to be the upper left corner of the coverslip. Count particles as they are found during microscope sweeps from left to right across the top edge of the coverslip. Then move down one microscope field and count particles while sweeping from right to left. Continue this process until the entire slide has been covered. At higher magnifications, sweeping the entire coverslip takes too long to be practical, so a number of transects are counted beginning at arbitrarily chosen locations on the side edge of the coverslip. To determine what proportion of the total coverslip is swept by one transect, determine the diameter of the microscope field by use of a stage micrometer. Divide the counts by the proportion counted to estimate the total number of particles of each type under the coverslip.

More precise and accurate techniques for counting small and microscopic food items are available. They are typically more complex and require specialized equipment and quite a bit more time. Instructions for their use make up substantial sections of several reference volumes on limnological methods (Lind 1985; APHA 1989; Wetzel and Likens 1991). In most cases, the investment is unlikely to yield much improvement in the data because counting errors are small compared with sampling errors and the variation in diet selection by fish.

Because bacteria are so small and difficult to distinguish from other particles to which they are attached, an epifluorescence microscope is required for counting them. Suspend the diet in particle-free water and pass a small subsample through a 0.2-μm pore membrane filter. Stain the bacteria trapped on the filter with a fluorochrome; acridine orange is most commonly used. Under the microscope, the light beam from the illumination unit is focused on the membrane filter and causes the fluorochrome in stained bacteria to fluoresce. By means of this approach, bacteria can be distinguished quantitatively from detritus. For details of bacterial counting methods, see Zimmerman and Meyer-Reil (1974) and Zimmerman et al. (1978).

Unlike the frequency-of-occurrence approach, which provides a quantitative description for only the entire sample of fish, calculating percent composition by number describes the diet of an individual fish. To summarize results for the entire sample, calculate mean values. Remember that percentage data are not normally distributed and must be transformed (Chapter 2) for calculation of confidence intervals.

Percent-composition-by-number data are of little value on their own, but they are useful as a step toward other calculations. The raw data from which percent composition by number is derived can be used together with estimates of feeding rate to assess the effect of predators on prey population dynamics. More commonly, percent-composition-by-number figures are used with estimates of prey weight as described below.

17.6.3 Percent Composition by Weight

In the percent-composition-by-weight approach to quantitative analysis of the diet, the weight of items of each food type is expressed as a percentage of the total weight of ingested food for an individual fish. If prey are large enough to be handled individually, they may be weighed directly (Johnson and Dropkin 1993). Both wet and dry weights have been used. Wet weight is obtained by gently blotting surface fluid from the item and then weighing the item. Dry weight is obtained by drying the item until it attains a constant weight. Preweighed aluminum foil pans sold by most scientific supply houses are useful for drying. Because the pans do not absorb moisture and are light in weight, they add little to weighing error.

Drying is most convenient in an oven, and a wide range of drying temperatures has been recommended. At 105°C drying is rapid and bacterial action is essentially stopped, but changes in the real or apparent chemical composition may be undesirable artifacts. For example, some researchers have expressed concern that volatile organic components may be lost at 105°C. So far as I have been able to determine, this has not yet been established as a significant problem for any nutrient. Others have suggested that drying at 105°C and higher temperatures may result in physical and chemical modification of sample components to produce "artifact lignin," a tarlike substance that may interfere with subsequent chemical analyses. Drying at 105°C may also denature proteins (sometimes called the Milliard browning reaction) so that they are difficult to extract and quantify using standard procedures. In contrast, drying at lower temperatures (60°C or 80°C) may permit microbial activity to alter the sample's chemical composition substantially. Despite these potential limitations, many researchers have successfully used 105°C for drying gut contents after determining that this temperature had no significant effect on the specific nutrients of interest. Freeze drying is more expensive in equipment and time but avoids these potential artifacts so it is the best technique when one is interested in the chemical composition of the diet. Whether freeze dried or oven dried, dry weights are generally more precise than wet weights and provide information more representative of nutritional value.

Weights of prey items collected from the stomach or foregut are not exactly what the consumer ate: they show the weight ingested minus the weight digested. For some purposes, these data may be fully adequate. Johnson and Dropkin (1993) used dry weights of items collected from stomachs to compare the diets of fish species that make up the community in the Juniata River, Pennsylvania. An implicit assumption of their analysis is that any differences in the rate at which prey are digested would equally bias the diet description for each fish species. Because there were large differences between most fishes, any failure of this assumption apparently had little effect on the outcome of the analysis.

If goals of the study require estimates of the weight of prey actually ingested, it is necessary to estimate their weight at ingestion by reconstruction. In this procedure, some hard structure such as an operculum, vertebra, or head capsule is measured and compared to a graph or equation showing the relationship between hard-structure size and whole-prey weight (Pearre 1980; Litvak and Hansell 1990; Bechara et al. 1993). Litvak and Hansell (1990) and Bechara et al. (1993), among others, provide equations for estimation of prey weight from the dimensions of digestion-resistant parts for a variety of freshwater invertebrate prey. Depending on the level of precision needed, it may be preferable to derive these relationships for prey collected from the fish's feeding habitat.

Because many invertebrates and most algae are too small to weigh directly, weight must be estimated from measurements made through a microscope (Ahlgren and Bowen 1992; Bechara et al. 1993). Although this approach is potentially labor-intensive, a microcomputer combined with a digitizer can be used to reduce time and increase accuracy. A camera lucida (drawing tube) can be used to trace the outlines of food particles from several microscope fields onto a single sheet of paper. The paper is then put on the digitizer and the length (long axis) and width (perpendicular to the long axis) are recorded for each particle. The volume of the food item is estimated by calculation of the volume of a geometric solid of similar size and shape. The shape approximation need not be complex because error in estimation of a single particle's volume is small relative to subsampling error. Total volume of each food type in the gut is then calculated by extrapolating volume in the subsample to the whole sample. However, because volume is not a nutritionally meaningful measure, estimated wet volume should be converted to dry weight based on 1 mm^3 being equal to 0.1 mg. The exact conversion varies among taxa, but the given value is precise enough for many studies (Cummins and Wuycheck 1971). If greater precision is required, then the milligrams dry mass per cubed millimeter can be determined directly by measuring the mass and volume of a sample containing only one type of food item (Ahlgren and Bowen 1992).

Biologists have tended to ignore material in diets that they cannot recognize as part of an organism. This is clearly a mistake because some fish derive most or all of their nutritional needs from detritus (Bowen 1978; Bowen et al. 1984). To distinguish between detritus and inorganic sediment, combust the sample in a muffle furnace to constant weight at 550°C. Times required to reach constant weight range from 30 min to 12 h depending on the size and composition of the sample. Accuracy can be improved by cooling the samples in a desiccator prior to weighing. The weight loss on combustion provides a very close approximation of total organic weight (Dean and Gorham 1976). For the purpose of these calculations it may be reasonable to assume that organisms are 100% organic matter (with the exception of mollusks, for which special calculations must be made). Organic weight in the sample that cannot be accounted for as prey weight must be present in the form of detritus. The weight of detritus may also be estimated directly from measurements made through the microscope (Ahlgren 1990a, 1990b).

It is important to cross-check calculations for small-particle diets to see if the results from reconstructed weight and indirect weight estimates are realistic. Weigh the whole sample (dry weight, corrected for subsampling) and compare that weight to the sum of estimated weights of all prey. Unless extensive digestion is suspected, the sum of the individual prey weights should not exceed the total sample weight. In many cases when fish feed on benthic invertebrates, algae, or macrophytes, the sum of estimated prey weights may be much less than total sample weight. Inorganic sediment and organic detritus are usually responsible for the difference.

Percent composition by weight is the only one of the three widely used methods that begins to identify the food or foods important for a fish's nutrition. Although percent composition by number does describe the diet quantitatively, the weight differs for each food type, and, thus, no comparison of types is possible. Which is more important, 10,000 bacteria or 1 amphipod? Abundance alone does not provide an answer. Percent composition by weight quantifies food types in directly comparable weight units. With the exception of mollusks, food value is roughly proportional to weight. Thus, percent-composition-by-weight data do suggest the relative importance of individual food types in the nutrition of the fish.

17.7 DATA ANALYSIS AND INTERPRETATION

It is a common experience that once data on fish diets are compiled, biologists are not entirely sure what those data mean. To move from simple knowledge to useful understanding, the data must be analyzed and interpreted. Fish biologists use the full repertoire of standard statistical analyses applied in other fields of biology, as well as a few more specialized indices developed by ecologists for analysis of trophic relationships. The more common of these indices are introduced below.

17.7.1 Selectivity Indices

Do fish feed at random on all available prey or do they feed preferentially on a selected subset? Ivlev (1961) studied this question in an exhaustive series of aquarium studies and developed one of the first quantitative descriptions of diet selection, his selectivity index. Since then, several alternative methods for calculating selectivity have been proposed. All selectivity indices are based on a comparison of the relative abundance of a given prey type in the diet to the relative abundance of that prey type in the environment. Thus, a selectivity index value is calculated for each individual prey type indicating whether the prey type is selected for, selected against, or eaten with the same frequency that it occurs in the environment. Today the Manley–Chesson index (Chesson 1983) and the Strauss index (Strauss 1979) are generally preferred. The Strauss index (L) is calculated as

$$L = r_i - p_i,$$

where r_i is the relative abundance of prey type i in the diet (as a proportion of the total number of prey in the diet) and p_i is the relative abundance of prey type i in the environment. Possible values range from $+1$, which indicates perfect selection for a prey type, and -1, which indicates perfect selection against it. An advantage of this index is that Strauss provided procedures for calculation of confidence intervals and tests of statistical difference between selectivity index values for different prey.

Development of selectivity analyses requires data on the abundance of prey in the environment, and this requirement can create some interesting challenges. Although the details of strategies for sampling invertebrate prey in aquatic habitats are beyond the scope of this chapter (see Chapter 11), it is worth considering the general limitations to analysis and interpretation of data on prey abundance. Not all prey are equally vulnerable to a given sampling gear. Thus, differences in the abundance of prey in diets and in the environment can be due to sampling gear selectivity rather than to selective ingestion of prey by fish. Systematic, exhaustive sampling can overcome this limitation in some environments (Bechara et al. 1992). Once reliable data on abundance in the environment are obtained, interpretation of the basis for selection becomes an interesting challenge. Are some prey preferred because they are less cryptic, because they yield more benefit to the consumer per unit of effort spent feeding, or because they are less distasteful than alternative prey? Study of these questions leads to much greater understanding of the intriguing interactions between predator and prey.

17.7.2 Diet Overlap Indices

In any community of fishes, the diets of several species often are very similar. In comparing pairs of species, it would be interesting to know in quantitative terms just how similar their diets are. Diet overlap indices were developed to allow such comparisons.

One of the first overlap indices was proposed by Schoener (1971):

$$C_{xy} = 1 - 0.5\ (\Sigma\,|p_{xi} - p_{yi}|)$$

where C_{xy} is the index value, p_{xi} is the proportion of food type i used by species x, p_{yi} is the proportion of food type i used by species y, and the vertical bars mean absolute (positive) values of the difference. Other more sophisticated measures have been developed, but Schoener's index requires relatively few assumptions and is widely used (Crowder 1990). Diet overlap indices provide relative measures of the extent to which species use the same food resources but do not provide absolute measures of competition per se because the prey may or may not be a limiting resource for one or both species. An alternative approach to quantification of food resource use by a community of fishes was developed by Litvak and Hansell (1990) based on principal components analysis. For those having the necessary background in statistics, this approach takes advantage of more of the information in the data set and provides greater insight into the details of resource use.

17.7.3 Hybrid Diet Indices

Biologists have sometimes worried about the correct interpretation of frequency-of-occurrence, percentage-by-number, and percentage-by-weight data. For example, some authors have suggested that percentage by number overemphasizes the importance of smaller prey because they weigh so much less than larger prey; on the other hand, percentage by weight overemphasizes the importance of large prey because it takes more time to capture and eat a large number of small organisms (Hynes 1950; Pinkas et al. 1971; George and Hadley 1979; Hysop 1980).

Swynnerton and Worthington (1940) sought to reconcile these apparent conflicts by development of the "points method." Food items are sorted into categories, usually species, and each category is assigned a numerical score based on (1) the number of items present and (2) the average size of the items in the category. This can best be described as a hybrid numerical–volumetric index applied subjectively according to how the sample looks to an investigator. Hynes (1950) preferred this technique because it avoided the tedium of other methods. As reviewed by Hysop (1980), many modifications of the points method have been proposed, but all preserve the basic subjective scoring concept. There is considerable question whether or not this technique can be applied consistently by one individual over time. When points are reported in published papers, there is little basis for evaluating the data because the subjective basis for assigning points is not and cannot be described. Thus, the points method cannot meet one of the fundamental requirements for valid scientific techniques—repeatability.

An alternative approach has involved development of indices that combine frequency-of-occurrence, percentage-by-number, and percentage-by-weight data. Pinkas et al. (1971) proposed the "index of relative importance" (IRI), which they calculated as

$$\text{IRI} = (\%\text{ by number} + \%\text{ by volume}) \times (\%\text{ frequency of occurrence}).$$

George and Hadley (1979) proposed the use of a "relative importance index" (Ri_a), which is derived from the "absolute importance index" (Ai_a). For food item a,

$$Ai_a = \%\text{ frequency of occurrence} + \%\text{ total number} + \%\text{ total weight};$$

$$Ri_a = 100Ai_a/\Sigma Ai_a,$$

in which Ai_a is summed over all food types.

Although these indices may at first offer some intuitive appeal, the summing and multiplication of percentages that are dimensionless ratios produces numbers of no definable meaning. Hybrid indices that combine frequency-of-occurrence, percentage-by-number, and percentage-by-weight data obscure the real information and should not be used.

The fault lies not with frequency-of-occurrence, percentage-by-number, or percentage-by-weight data but with the investigators' failure to define the objective of their analysis. For example, if an investigator wants to determine the effect of a predator on the population dynamics of its several prey species, then percentage by number provides essential data for description of the diet. For such a study, it makes no difference how large or small a given prey may be. If the goal is to measure the contribution of prey to the predator's nutrition, then percentage by weight is fully adequate as a diet descriptor. A sample of 25 smelt stomachs that contains in total 5,000 cladocerans weighing 50 mg and 100 fish larvae weighing 950 mg indicates that 95% of the smelt population's gross energy comes from fish, in spite of the relative numbers of prey consumed. It may be that cladocerans are digested more rapidly than are fish larvae or that the few fish larvae are not representative of the true diet because the sample size was too small. Effects of differential digestion rates are discussed in section 17.3.4, and sample size is a fundamental issue treated in any introductory statistics text. Thus, these problems are not biases attributable to percentage by weight as a descriptor of diet composition, and application of hybrid indices would add only confusion.

17.8 FUTURE INVESTIGATIONS

Techniques discussed above are potentially useful in studying how diet affects the production and well-being of fish populations, but these techniques are unlikely to furnish much new insight on their own.

Associated with each food item are costs of acquiring the food and benefits once the food is eaten and digested. The cost-benefit analyses used to develop optimal foraging theory are built on diet descriptions like those discussed above and lead to substantial advances in our understanding of fish habitat selection and production. Several papers on this topic are available (Werner 1974, 1977; Werner and Hall 1974; Horn 1983; Mittelbach 1983). Although this approach is not used as extensively today, the perspectives that emerged from this and subsequent work will be valuable to many researchers in the future. As new questions are developed, the techniques of cost-benefit analyses may be used for further studies.

Fishes, like other consumers, have basic nutritional requirements for dietary protein and energy (Box 17.2). Getting these two nutritional requirements from their diets presents fish with different challenges for different feeding strategies (Bowen et al. 1995). Predators usually get plenty of protein because they feed on protein-rich animal prey. But the rate of energy intake is often limited by prey defenses and by low prey numbers. At the other extreme are the detritivores. Detritivores have food available to them in quantities that far exceed their capacity to consume, but the nutritional value of detritus is limited by its low level of digestible protein. Herbivores that feed on algae or aquatic vascular plants are limited to some extent by both

Box 17.2 Nutritional Considerations

As we consider the significance of diet for growth, how do we compare the nutritional roles of different types of food? Is 1 g of algae just as useful as 1 g of chironimids? If not, why not? The figure discussed here (modified from Bowen et al. 1995) summarizes work intended to answer these questions.

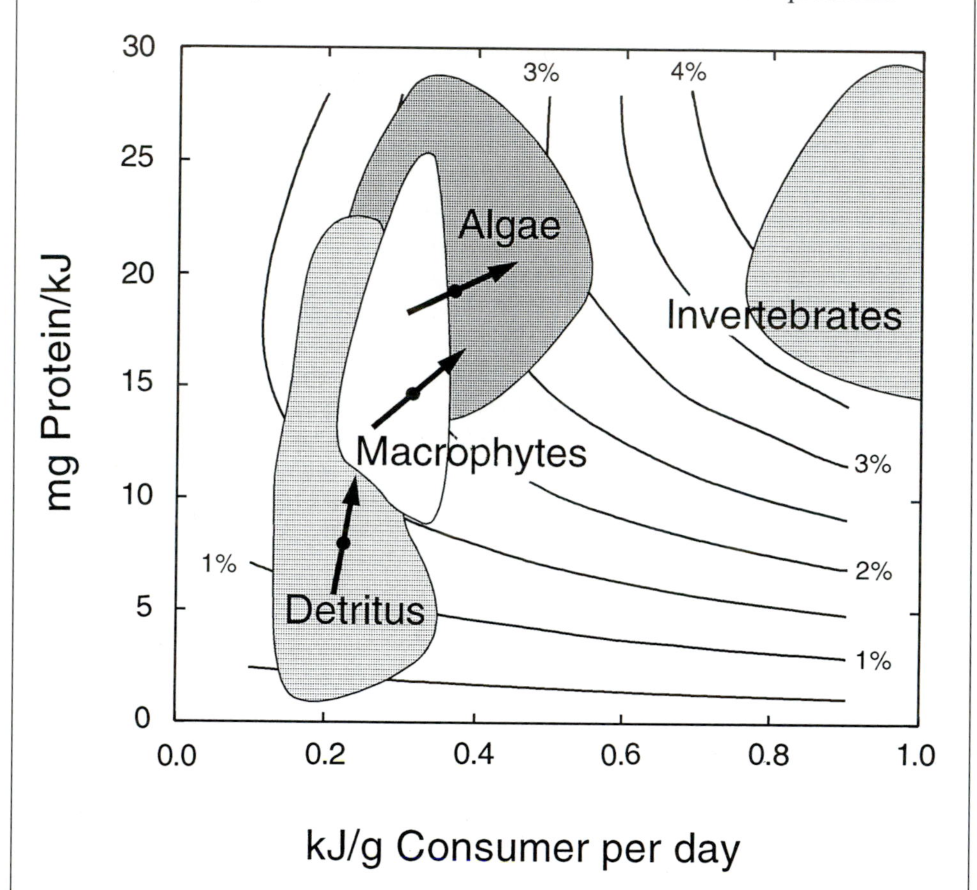

This figure summarizes the results of growth experiments in which the omnivorous blue tilapia was fed 16 artificial diets ad libitum in aquaria. Protein and energy levels in the diets spanned the ranges for detritus, algae, macrophytes, and invertebrate prey in aquatic habitats. Sixty-four groups of fish were fed daily for 41 d. Feeding rate was measured daily, and assimilation efficiencies were measured during the last week of the experiment. At the end of the experiment, specific growth rate (SGR) was calculated as

$$\text{SGR} = \log_e(W_{41} - W_0)/41,$$

where W_0 is the live weight at the start of the experiment and W_{41} is the live weight of fish at the end of the experiment. Values of SGR multiplied by 100 are equivalent to percent of body weight gain per day. A multiple-regression model of SGR as a function of energy assimilation and protein to energy ratio accounted for 94% of the variation in SGR.

Box 17.2 Continued.

In this figure, SGR isopleths labeled 1%, 2%, 3%, and 4% are plotted in terms of energy assimilation rate (*x*-axis) and the protein richness of the food assimilated (*y*-axis). The results show that energy content of the diet need not limit growth when protein levels are high because the fish can compensate for low energy levels by eating more food. There is no similar opportunity to compensate for low protein level, increased ingestion does not significantly increase growth. At high protein levels, protein is fully adequate, and only the energy assimilation rate affects growth. At intermediate protein levels, interaction between protein level and energy level affects growth.

The combined effects of dietary protein and energy levels help to identify constraints associated with different feeding strategies. Shaded areas show the range of protein and energy values relevant to diets of detritus, macrophytes, algae, and invertebrate prey. Arrows centered on mean protein and energy values for each food type show the direction of the maximum gradient on the surface, that is, the combination of change in protein and energy that maximizes the increase in growth relative to the mean values. For detritivores, we expect that low protein levels are the main obstacle to growth, and thus a successful detritivore should feed selectively on the most protein-rich detritus available. Algae are relatively rich in protein, and the main limitation to growth for fish on algal diets is assimilation of energy. Thus we expect fishes that feed extensively on algae to have adaptations that increase ingestion and digestion of this diet. Macrophytes lie between detritus and algae in protein value, and both selection for higher protein levels and adaptations for increased assimilation of energy are likely to be important for macrophyte consumers. The point for mean values corresponding to invertebrate prey lies outside the range of values plotted, but for fishes feeding on these protein-rich prey, the rate of energy assimilation most limits growth.

energy intake rate and protein content of their diets. Omnivores complement scarce but protein-rich animal prey with energy-rich but protein-poor detritus or plant foods. Thus, intake of dietary protein, energy, or both potentially limits the production of fish populations. Based on literature values for feeding rates and the digestibility of protein and energy in various types of food, it may be possible to estimate initially the relative roles of various diet components in supporting growth of a consumer by means of a figure like that in Box 17.2. Confirmation of these estimates requires direct experimentation with the fish and prey being studied.

17.9 REFERENCES

Ahlgren, M. O. 1990a. Diet selection and the contribution of detritus to the diet of the juvenile white sucker, (*Catostomus commersoni*). Canadian Journal of Fisheries and Aquatic Sciences 47:41–48.

Ahlgren, M. O. 1990b. Nutritional significance of facultative detritivory to the juvenile white sucker (*Catostomus commersoni*). Canadian Journal of Fisheries and Aquatic Sciences 47:49–54.

Ahlgren, M., and S. Bowen. 1992. Comparison of quantitative light microscopy techniques used in diet studies of detritus-consuming omnivores. Hydrobiologia 239:79–83.

Allanson, B. R., and J. E. Kerrich. 1961. A statistical method for estimating the number of animals found in field samples drawn from polluted rivers. Internationale Vereinigung für Theoretischen und Angewandte Limnologie Verhandlungen 14:491–494.

APHA (American Public Health Association), American Water Works Association, and Water Pollution Control Federation. 1989. Standard methods for the examination of water and wastewater, 17th edition. APHA, Washington, DC.

Bechara, J. A., G. Moreau, and L. Hare. 1993. The impact of brook trout *(Salvelinus fontinalis)* on an experimental stream benthic community: the role of spatioa and size refugia. Journal of Animal Ecology 62:451–464.

Bechara, J. A., G. Moreau, and D. Planas. 1992. Top-down effects of brook trout (*Salvelinus fontinalis*) in a boreal forest stream. Canadian Journal of Fisheries and Aquatic Sciences 49:2093–2103.

Bergman, E., and L. A. Greenberg. 1994. Competition between a planktivore, a benthivore and a species with ontogenetic diet shifts. Ecology 75:1233–1245.

Borgeson, D. P. 1963. A rapid method for food-habits studies. Transactions of the American Fisheries Society 92:434–435.

Bowen, S. H. 1984. Differential habitat utilization by sexes of *Sarotherodon mossambicus* in Lake Valencia: significance for fitness. Journal of Fish Biology 24:115–121.

Bowen, S. H. 1987. Composition and nutritional value of detritus. Pages 192–216 *in* D. J. W. Moriarty and R. S. V. Pullin, editors. Detritus and microbial ecology in aquaculture. Proceedings of the International Center for Living Aquatic Resources Management Conference. International Center for Living Aquatic Resources Management, Manila, Philippines.

Bowen, S. H., and B. R. Allanson. 1982. Behavioral and trophic plasticity of juvenile *Tilapia mossambica* in utilization of the unstable littoral habitat. Environmental Biology of Fishes 7:357–362.

Bowen, S. H., A. A. Bonetto, and M. O. Ahlgren. 1984. Microorganisms and detritus in the diet of a typical neotropical riverine detritivore, *Prochilodus platenses* (Pisces: Prochilodontidae). Limnology and Oceanography 29:1120–1122.

Bowen, S. H., E. V. Lutz, and M. O. Ahlgren. 1995. Dietary protein and energy as determinants of food quality: trophic strategies. Ecology 16:899–907.

Buijse, A. D., and R. P. Houthuijzen. 1992. Piscivory, growth and size-selective mortality of age 0 pikeperch (*Stizostedion lucioperca*). Canadian Journal of Fisheries and Aquatic Sciences 49:894–902.

Cailteux, R. L., W. F. Porak, and S. Crawford. 1991. Reevaluating the use of acrylic tibes for collection of largemouth bass stomach contents. Proceedings of the Annual Conference Southeastern Association of Fish and Wildlife Agencies 44(1990):126–132.

Chesson, J. 1983. The estimation and analysis of preference and its relationship to foraging models. Ecology 64:1297–1304.

Crowder, L. B. 1990. Community ecology. Pages 609–632 *in* C. B. Schreck and P. B. Moyle, editors. Methods for fish biology. American Fisheries Society, Bethesda, Maryland.

Cummins, K. W., and J. C. Wuycheck. 1971. Caloric equivalents for investigations in ecological energetics. International Association of Theoretical and Applied Limnology Special Communication 18:1–158.

Dean, W. E., and E. Gorham. 1976. Major chemical and mineral components of profundal surface sediments in Minnesota lakes. Limnology and Oceanography 21:259–284.

Diel, S. 1992. Fish predation and benthic community structure: the role of omnivory and habitat complexity. Ecology 75:1646–1661.

Edmonson, W. T. 1974. A simplified method for counting phytoplankton. IBP (International Biological Programme) Handbook 12:14–15.

Emmett, R. T., W. D. Muir, and R. D. Pettit. 1982. Device for injecting preservative into the stomach of fish. Progressive Fish-Culturist 44:107–108.

Gannon, J. E. 1976. The effects of differential digestion rates of zooplankton by alewife, *Alosa pseudoharengus*, on determinations of selective feeding. Transactions of the American Fisheries Society 105:89–95.

George, E. L., and W. F. Hadley. 1979. Food and habitat partitioning between rock bass (*Ambloplites rupestris*) and smallmouth bass (*Micropterus dolomieui*) young of the year. Transactions of the American Fisheries Society 108:253–261.

Giles, N. 1980. A stomach sampler for use on live fish. Journal of Fish Biology 16:441–444.

Horn, M. H. 1983. Optimal diets in complex environments: feeding strategies of two herbivorous fishes from a temperate rocky intertidal zone. Oecologia 58:345–350.

Hynes, H. B. N. 1950. The food of freshwater sticklebacks (*Gasterosteus aculeatus* and *Pygosteus pungitius*) with a review of methods used in studies of the food of fishes. Journal of Animal Ecology 19:36–58.

Hysop, E. J. 1980. Stomach contents analysis—a review of methods and their application. Journal of Fish Biology 17:411–429.

Ivlev, V. S. 1961. Experimental ecology of the feeding of fishes. Yale University Press, New Haven, Connecticut.

Johnson, J. H., and D. S. Dropkin. 1993. Diel variation in diet composition of a riverine fish community. Hydrobiologia 271:149–158.

Kennedy, C. R. 1969. Tubificid oligochaetes as food of dace. Journal of Fish Biology 1:11–15.

Kitchell, J. F., and J. T. Windell. 1970. Nutritional value of algae to bluegill sunfish, *Lepomis macrochirus*. Copeia 1:186–190.

Klumpp, D. W., and P. D. Nichols. 1983. Nutrition of the southern sea garfish *Hyporhamphus melanochir*: gut passage rate and daily consumption of two food types and assimilation of seagrass components. Marine Ecology Progress Series 12:207–216.

Lind, O. T. 1985. Handbook of common methods in limnology, 2nd edition. Kendall/Hunt, Dubuque, Iowa.

Litvak, M. K., and R. I. C. Hansell. 1990. Investigation of food habit and niche relationships in a cyprinid community. Canadian Journal of Zoology 68:1873–1879.

Lowe-McConnell, R. H. 1975. Fish communities in tropical freshwaters. Longman, London.

Mann, R. H. K., and D. R. Orr. 1969. A preliminary study of the feeding relationships of fish in a hard-water and a soft-water stream in southern England. Journal of Fish Biology 1:31–44.

Mittelbach, G. G. 1983. Optimal foraging and growth in bluegills. Oecologia 59:157–162.

Ney, J. J. 1990. Trophic economics in fisheries: assessment of demand-supply relationships between predators and prey. Reviews in Aquatic Sciences 2:55–81.

Pearre, S., Jr. 1980. The copepod width–weight relation and its utility in food chain research. Canadian Journal of Zoology 58:1884–1891.

Pinkas, L., M. S. Oliphant, and I. L. K. Iverson. 1971. Food habits of albacore, bluefin tuna and bonito in Californian waters. California Fish and Game 152:1–105.

Schoener, T. W. 1971. Theory of feeding strategies. Annual Review of Ecology and Systematics 2:369–404.

Strauss, R. E. 1979. Reliability estimates for Ivlev's electivity index, the forage ratio, and a proposed linear index of food selection. Transactions of the American Fisheries Society 108:344–352.

Swynnerton, G. H., and E. B. Worthington. 1940. Notes on the food of fish in Haweswater (Westmorland). Journal of Animal Ecology 9:183–187.

Van Den Avyle, M. J., and J. E. Roussel. 1980. Evaluation of a simple method for removing food items from live black bass. Progressive Fish-Culturist 42:222–223.

Werner, E. E. 1974. The fish size, prey size, handling time relation in several sunfishes and some implications. Journal of the Fisheries Research Board of Canada 31:1531–1536.

Werner, E. E. 1977. Species packing and niche complementarity in three sunfishes. American Naturalist 111:553–578.

Werner, E. E., and D. J. Hall. 1974. Optimal foraging and the size selection of prey by the bluegill sunfish (*Lepomis macrochirus*). Ecology 55:1042–1052.

Werner, E. E., and D. J. Hall. 1976. Niche shifts in sunfishes: experimental evidence and significance. Science 191:404–406.

Wetzel, R. G., and G. E. Likens. 1991. Limnological analyses, 2nd edition. Springer-Verlag, New York.

Whitfield, A. K., and S. J. M. Blaber. 1978. Food and feeding ecology of piscivorous fishes at Lake St. Lucia, Zululand. Journal of Fish Biology 13:675–691.

Zimmerman, R., R. Iturriaga, and J. Becker-Birck. 1978. Simultaneous determination of the total number of aquatic bacteria and the number thereof involved in respiration. Applied and Environmental Microbiology 36:926–935.

Zimmerman, R., and L. A. Meyer-Reil. 1974. A new method for fluorescence staining of bacterial populations on membrane filters. Kieler Meeresforschungen 30:24–27.

Chapter 18

Underwater Observation

ANDREW DOLLOFF, JEFFREY KERSHNER,
AND RUSSELL THUROW

18.1 INTRODUCTION

Underwater observation by divers or remote sensing equipment is among the most versatile, cost-effective techniques for obtaining accurate information on the composition, distribution, abundance, and behavior of aquatic organisms in their natural surroundings. Many types of studies, ranging from broad-based surveys or inventories to highly specialized observations of behavior (Drew et al. 1976), incorporate underwater methods. Underwater observation may be most appropriate when other sampling methods, such as capture by electroshocking, seines, or gill nets, are not effective because of conditions such as extreme conductivity (either too low or too high), excessive depth, or habitat complexity. Underwater methods also can be used to help describe the distribution, abundance, and life history of threatened, endangered, or sensitive species that cannot be sampled by potentially destructive methods.

Although underwater observation methods have a place in the repertoire of nearly all fisheries professionals, these methods are not without unique shortcomings. Whether by divers or cameras, underwater observations are possible only in waters of adequate visibility. In many streams, rivers, lakes, and estuaries, it is not possible to observe aquatic organisms directly because of high turbidity. In addition, the computation of many vital fisheries statistics, such as growth and mortality rates and production, is not possible without accurate measurements of fish length, weight, age, and other characteristics. If, for example, an investigator needs to know how many smallmouth bass are using newly installed habitat structures in a shallow, clearwater stream, underwater methods are among the logical choices for sampling. If the investigator also needs to know the state of maturity, biomass, and diet of those fish, then some other method either in addition to or in place of underwater methods will be required.

As is true with any specialized sampling method, there are technical skills to acquire and safety precautions to follow. In this chapter we describe procedures for the safe application of underwater observation. We refer to all practitioners of underwater observation as divers, regardless of the specific equipment used. Although we describe both skin and scuba (self-contained underwater breathing apparatus) diving techniques, this chapter is not intended as a substitute for diver certification programs sponsored by the National Association of Underwater Instructors or the Professional Association of Diving Instructors.

Figure 18.1 Snorkelers counting fish in a river.

18.2 UNDERWATER OBSERVATION TECHNIQUES AND EQUIPMENT

18.2.1 Snorkel

Snorkeling requires the least equipment of all underwater observation techniques and is one of the simplest ways to observe organisms under water. Typical snorkeling equipment includes a mask, snorkel, wet or dry suit, and swim fins or wading boots, depending on the environment to be investigated (Box 18.1). Snorkeling can be used in remote locations where it may be difficult to use other sampling apparatus. Snorkeling is especially useful for observing activities such as spawning, behavioral interactions, use of favored feeding and resting positions, movements, and for estimating the numbers and size structures of populations. Snorkeling is effective in a variety of environments. Small streams and rivers are normally well suited for snorkel observations provided underwater visibility is adequate (Figure 18.1). Small ponds and impoundments are also typically good environments for snorkeling, particularly in the littoral zone.

Snorkeling techniques will vary depending on the study objectives and environments to be surveyed. A generalized protocol for streams follows. Divers typically enter the stream either upstream or downstream from the area to be sampled. A short resting period usually follows to allow divers to become acclimated and to allow any organisms disturbed by the divers' approach to resume normal behavior. Divers in deep water typically proceed in a downstream direction by floating, whereas in shallow water, divers move upstream by pulling themselves along the bottom. In lentic or marine environments a similar protocol is followed except that snorkelers use swim fins to propel themselves through the water.

The type and consistency of data collected by snorkeling depends greatly on light conditions and the time of day (Spyker and Van Den Berghe 1995; Thurow and Schill

Box 18.1 Equipment Checklist for Visual Underwater Surveys

Wet or dry suit
Neoprene hood, gloves, and booties
Knee pads
Mask and snorkel
Wading boots or swim fins
Underwater data recorder (e.g., slate or cuff)
Clipboard and data forms
Calibrated thermometer
Flagging
Tape measure
Hand tally counter
Stadia rod
Underwater dive light and spare batteries
Flashlight or headlamp
Fish silhouette for measuring visibility
First aid kit that includes swimmer's ear drops
Knife
Mask defogger
Neoprene or dry suit repair kit
Camera and film
Drinking water
High-energy food
Dive watch

Additional items for scuba diving

Tanks and backpack
Regulator
Pressure gauge
Depth gauge
Buoyancy compensator vest
Dive tables or computer
Weight belt

Additional items for hookah

Air compressor or air bottles
Umbilical hoses and filters
Regulator
Harness
Masks
Pony bottle and regulator (emergency air)
Depth gauge
Weight belt
Communication equipment (e.g., microphones and cables)

1996). Investigators usually establish protocols for daytime sampling that specify certain hours with optimum light conditions (e.g., 1000–1700 hours). Observations conducted at night or during twilight hours require handheld or fixed position underwater lights or chemical (cyalume) light sticks. Disturbance or displacement of organisms can be minimized by not shining light directly on the animals, by directing lights to the underside of the water surface (Contor 1989), or by using color filters (Riehle and Griffith 1993).

Snorkeling may require special considerations when conditions are less than ideal. For example, in waters with dark substrate, benthic fishes and other benthic organisms may be difficult to observe by snorkeling. Snorkelers may fail to detect or incorrectly identify target organisms, count them more than once, or incorrectly estimate size (Griffith et al. 1984). Counting organisms accurately in a dense population can be difficult (Heggenes et al. 1990). Some species and sizes of fish are more difficult to see than others, especially species that remain near the substrate (Hillman et al. 1992) or concealed by cover (Rodgers et al. 1992). Differences in fish behavior during different times of the day or year may influence observability (Campbell and Neuner 1985; Rodgers et al. 1992; Spyker and Van Den Berghe 1995). In large rivers, multiple divers are needed to estimate populations, which increases observer bias. Snorkeling also has limited application in deep areas of lakes, ponds, rivers, and marine environments.

18.2.2 Scuba

More specialized equipment is used for scuba diving (Box 18.1). Divers wear tanks filled with compressed air, which is delivered via a mouthpiece that regulates air flow.

Figure 18.2 Diver wearing scuba equipment.

Depth and pressure gauges, a buoyancy compensator, watch, and weight belt are frequently used in addition to wet or dry suits, masks, and fins. Scuba is typically limited to locations where large amounts of equipment can be transported.

Scuba divers may remain submerged for much longer periods of time than may snorkelers (Figure 18.2). This is advantageous in large rivers, lakes, ponds, and many marine environments in which organisms live at greater depths than snorkelers can access. Generalized sampling protocols for scuba divers are similar to those used by snorkelers. Longer resting periods may be needed to acclimate divers. Scuba diving is noisier than snorkeling, and the bubble trail emitted by divers may initially frighten fish or other aquatic organisms.

As with snorkeling, light conditions and underwater visibility are important for successful scuba diving. Scuba divers are able to work at depths where light penetration may be limited. In many situations underwater lights will be necessary, even for daytime conditions.

Scuba diving is frequently unnecessary in small streams or rivers where divers do not need to remain fully submerged for long periods. In rivers with strong currents, scuba equipment may be unsafe.

18.2.3 Hookah

Hookah diving (surface-supplied air) is popular for collecting aquatic organisms (Ahlstedt and McDonough 1993), ship and oil rig maintenance, and suction dredging (Figure 18.3). Commercial mussel divers typically use hookah rigs, which allow them to remain underwater for long periods relatively unburdened by cumbersome

Figure 18.3 Diver entering the water with a hookah diving rig.

equipment. In hookah diving, air is delivered to a diver via an umbilical hose from an air compressor or compressed air bottles that are mounted on a boat or on land. Although a diver's mobility and range may be limited by the umbilical, the major advantage of hookah is that air supply is almost unlimited, allowing for maximum time under water. Another major feature of hookah rigs is that clear voice communication with surface support is made possible by running a telephone cable in conjunction with the air supply hose.

Hookah diving, like scuba, may be impractical in small streams or small shallow rivers that can be sampled by snorkeling. Hookah is most useful in larger rivers, lakes, and ponds where divers must remain submerged for relatively long intervals and where obstructions are few and the chance of umbilical fouling is minimal.

18.2.4 Alternative Methods

Although visual observations by divers generally provide accurate estimates of fish abundance (Greene and Alevizon 1989), various remote sensing methods have been developed for specialized underwater sampling where conditions such as extreme cold or excessive depths preclude or inhibit direct observation. The use of underwater cameras has become popular as a means to conduct surveys. Waterproof cameras (still or video) may be placed in blinds (camouflaged areas in a fixed location) and remotely operated to take pictures at predetermined frequencies. Remote cameras and lighting systems allow investigators to obtain information on populations of organisms at any time of the day or night. Remote systems can be set to operate continuously or at predetermined intervals. More sophisticated systems have triggering mechanisms that can be activated by movements such as the passing of a school of fish, making these systems particularly useful for monitoring migrations. Cameras may also be mounted on a remotely operated vehicle or towed behind a boat to conduct photo surveys along fixed transect lines or in some pattern (Bergestedt and Anderson 1990; Beauchamp et al. 1992, 1994). Remotely operated vehicles are often the most desirable alternative in deep offshore areas where diver access is impractical.

Remote cameras have several disadvantages. Cameras, light sources, underwater housings, and remote sensing equipment are expensive to buy and maintain. Divers are generally required to install and service equipment on a periodic basis. Remote cameras are limited by the same conditions affecting underwater observers: low-light or turbid water may render cameras ineffective or result in poor sampling efficiency. We recommend that remote cameras be used with other methods to get the best results.

18.2.5 Record Keeping

Underwater observation requires special considerations for record keeping. Fish counts and other data can be recorded directly by the diver or communicated to an assistant. Waterproof slates or cuffs are probably the most popular methods for divers to record data. Erasable slates are made from thin sheets of plastic, opaque plexiglas (Jacobson 1994), or formica board (Gardiner 1984). A short (30 cm) piece of surgical tubing can be attached through a hole in the slate and a pencil inserted into the free end of the tubing. The slate can be carried or attached to the diver; some divers carry slates or small clipboards in specially designed pockets or simply in the partially unzipped top of their wet suits. A permanent marker can be used to create lines or a grid on the slate. The diver records information on the slate and then transfers it to a paper copy or to an assistant.

Because a slate may be awkward, particularly when a diver must use both hands, many divers prefer to record data on a waterproof cuff made of white, polyvinyl chloride (PVC) pipe (Figure 18.4). A cuff leaves a diver's hands free, thereby reducing the possibility of losing valuable data. Cuffs are manufactured from thin-walled, 10-cm-diameter pipe; the length of the cuff is adjusted to fit the diver's forearm. Thurow (1994) modified this basic design by cutting a length of pipe in half lengthwise and drilling holes at the corners of each half. Surgical tubing is threaded through each pair of holes and used to secure the half-round pipe to the diver's arm. Pencils are attached to the ends of the surgical tubing. One drawback to slates and cuffs is that the writing surface is limited and the diver must frequently stop and transfer data to other media.

An underwater scroll is another alternative for recording data. A scroll (Ogden 1977) consists of a long strip of translucent, matte-surface polyester drafting film that is fed between two rollers across a flat sheet of stiff plastic. A scroll eliminates the need for single sheets of paper and provides a large, continuous writing surface. Tabular data forms, maps, or photographs can be attached to the flat plastic surface and data can be recorded on the film. The film is periodically advanced to provide a clean writing surface. Scrolls can be awkward to operate, however, particularly by a diver wearing gloves or mitts.

Various sizes of waterproof paper are available for recording dive survey information (Graham 1992). Data forms can be copied and attached to a clipboard to provide a hard writing surface. Large rubber bands are used to fix the paper to the

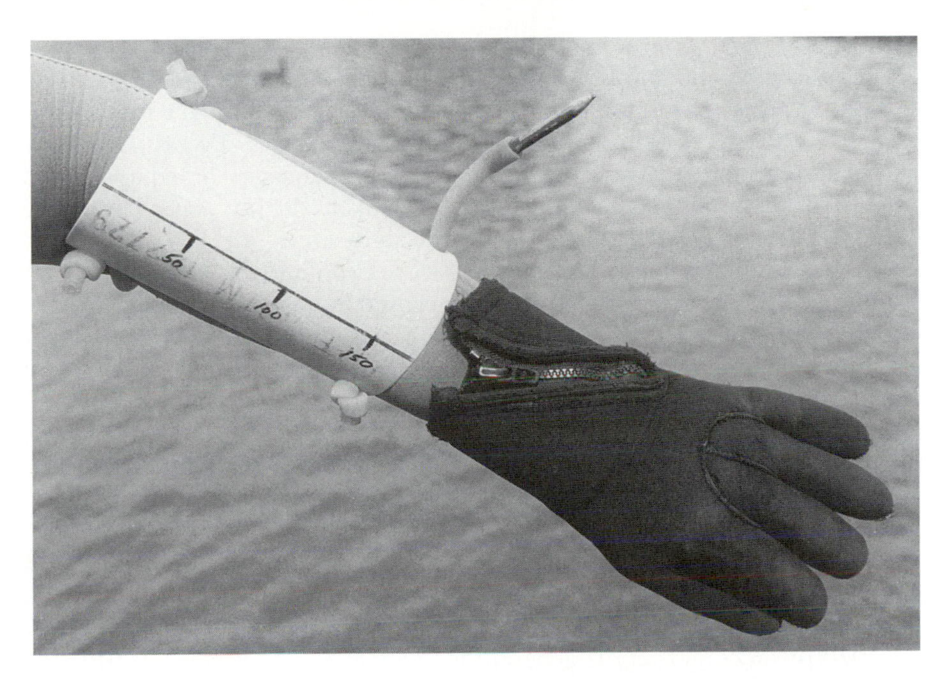

Figure 18.4 Diving cuff for recording data underwater. A half-round length of PVC pipe is held on the forearm by surgical tubing, a free end of which holds a pencil.

clipboard to prevent the paper from moving freely. Taxonomic keys or sample protocols also can be copied onto waterproof paper and attached or laminated to the clipboard. Pencils can be inserted into a length of surgical tubing and attached to the clipboard. Data are recorded on the field forms and later transcribed to other media. The clipboard has disadvantages like those of the waterproof slate, and turning pages with gloves is particularly difficult. Waterproof paper has a tendency to smear, so caution must be used when data are recorded and transcribed.

There are several alternatives to diver-recorded data. Divers can use sign language (Lee 1967) or verbal communication to transfer information to an assistant. The assistant records the data, and the diver is free to continue the survey. The assistant also provides the diver with information on survey boundaries, sample protocols, and potential hazards. Sophisticated electronic data recording devices also are available, including underwater radios for diver-to-diver and diver-to-surface communication, tape recorders, and diver-held cameras. Such devices are less frequently used because of high cost.

18.3 SAFETY AND TRAINING

18.3.1 Hazards

As with any field activity, underwater safety is of paramount importance. Fast-moving water, cold water temperatures, poor visibility, physical obstructions, and other environmental factors, including contaminants and potentially dangerous organisms, are among the conditions divers may encounter (Somers 1976; U.S. Navy 1985; Flemming and Max 1990). Never dive alone: divers must always have a partner, either in the water or outside on the bank or in a boat. Always assess the potential hazards of a site before entering the water. Divers in regulated rivers face special risks; always check with dam operators for scheduled water release times.

Never attach ropes or lines to divers in streams, lakes, or marine environments that are subject to strong currents, wave action, or tidal changes. Divers should avoid areas of extreme water velocity and turbulence; appendages or equipment may become wedged or pinned against rocks or sunken debris. Always move into low-velocity areas when entering or leaving the study area.

One of the most common hazards divers face is hypothermia, a potentially lethal condition of below-normal body temperature caused by exposure to cold (Box 18.2). Divers in all conditions are subject to hypothermia, although divers conducting winter and night surveys are probably at greatest risk. Divers submerged for prolonged periods may be subject to hypothermia even when water temperatures exceed 20°C. To prevent hypothermia, divers must wear appropriate protective clothing, eat high-energy foods, drink plenty of liquids, and take periodic breaks to restore body temperature. Divers and their partners must have first aid training so they are able to recognize and treat hypothermia.

Although less common than hypothermia, divers in warm waters (>40°C) may experience hyperthermia, a condition characterized by abnormally high body temperature (Flemming and Max 1990). Divers who experience hyperthermia feel lightheaded or dizzy and report general muscular weakness accompanied by faint trembling sensations. Because even momentary unconsciousness can have disastrous consequences under water, divers must avoid overheating and dehydration by taking frequent breaks, being appropriately outfitted (e.g., a loose t-shirt, neck covering, and copious amounts of water-resistant sun block), and drinking plenty of fluids.

Turbid water is another hazard frequently encountered by divers. For most purposes, the utility and safety of underwater methods decreases with decreasing visibility. Most waters are subject to varying reductions in visibility resulting from turbidity. Typical situations include highly turbid nearshore zones of marine environments, lakes, and ponds after storms or high winds and streams or rivers after heavy rain or downstream from dams. In otherwise clear waters, divers may experience temporary reductions in underwater visibility as a result of these natural or human-induced disturbances and should be prepared to deal with reduced visibility.

Underwater obstructions including debris jams, submerged logs, boulders, bedrock outcrops, vegetation, reefs, and human-created structures including riprap, dams, and diversions have the potential to entrap divers. Extreme caution should be exercised when surveying near underwater obstructions. Additional hazards may include coils of barbed wire, monofilament fishing line and nets, and broken glass. We recommend that all divers carry a knife or other tool suitable for cutting, sawing, or prying.

Divers should be aware of potential hazardous materials including chemical and microbial contaminants (Losonsky 1991) and take appropriate precautions (Masterson 1991). Before conducting a survey, divers should become thoroughly familiar with potential sources of contaminants, such as mines, sewage outfalls, agricultural runoff, and industrial discharges. In general, diving for visual observations in contaminated water poses unacceptable risks to a diver. Divers should avoid contaminated water unless they are thoroughly acquainted with the risks and the use of protective clothing systems (Barsky 1991).

One of the most troublesome organisms is also one of the smallest: *Giardia*

Box 18.2 Hypothermia

Tolerance for cold varies greatly among individuals and depends on factors such as age, physical condition, experience, and specific tasks performed. Other factors, such as duration and depth of dive and water conditions (current, turbidity, and salinity), also influence a diver's susceptibility to hypothermia. Some of the most important but often overlooked stress times occur immediately before and after a dive, when the diver may be unprotected from exposure to wind, rain, and snow.

Professional divers eat properly (balanced diet high in carbohydrates), exercise regularly, and wear protective suits appropriate for the conditions under which they will be working. A properly fitting wet suit usually protects a diver for about 30 min at temperatures down to 5–6°C. For dives of longer duration or in colder water, a dry suit is necessary and may be used in conjunction with either passive (insulated underwear) or active (electric underwear) heating accessories. In truly extreme conditions, such as under ice or in arctic diving, specially designed hot-water or other liquid-filled suits are available. These specialty suits severely limit a diver's mobility and freedom, however, and are extremely expensive to purchase, use, and maintain.

Regardless of experience, conditioning, and equipment, no diver can afford to ignore the warning signs of hypothermia. A diver undergoing hypothermia first notices a loss in manual dexterity, followed by general weakness and clumsiness in arm and leg movements. As cooling progresses, judgement becomes impaired, and the diver responds to otherwise normal situations with confusion or indifference. A decreased core temperature of 0.5–0.8°C may result in a temporary loss of 10–20% in mental capacity and 40% in memory. Divers working long hours in deep, cold water may experience a synergistic interaction with nitrogen narcosis. Support personnel must take steps to remove a diver from the water at the first signs of erratic or irrational behavior. Do not trust the diver's sensation of sufficient warmth. When under the stress of hypothermia, divers may feel that they are sufficiently warm or even hot.

Rewarming may take several hours or overnight. Use plenty of blankets or warm (<40°C), not hot, water. Handle severely hypothermic individuals carefully, because stress can cause cardiac fibrillation and arrest. Emphasize warming of the body core to avoid sending cold blood into the core from the extremities. Be prepared to administer cardiopulmonary resuscitation if necessary.

lamblia, a protozoan that causes giardiasis when ingested, is ubiquitous in freshwater environments throughout the world. Symptoms of giardiasis are highly variable and include diarrhea, fatigue, loss of weight, dehydration, and nausea. All water, including pristine springs in wilderness, should be considered potential sources of *G. lamblia*. Avoid ingesting any water that has not been filtered, boiled, or chemically treated. During underwater surveys, divers should use care to avoid swallowing water. If symptoms of giardiasis are experienced, consult a physician immediately.

Divers should be familiar with potentially dangerous organisms before attempting a survey. Divers in marine environments may encounter a variety of organisms

Box 18.3 Safety Checklist

1. Be in good physical condition as certified by a physician within the last 12 months.
2. Never dive alone: diving teams consist of at least two persons. For scuba surveys the buddy system is mandatory, and divers must always be accompanied by observers on the bank or in a boat to assist in case of an accident.
3. Be certified in cardiopulmonary resuscitation and basic first aid. Be sure that personnel can recognize and treat hypothermia.
4. Carry a first aid kit at all times, including a device for extracting venom and medication to prevent anaphylactic shock (resulting from insect stings), as prescribed or recommended by a physician.
5. Be certain that persons not in the field know the schedule. Carry two-way radios or cellular telephones for rapid communication with emergency personnel. Carry telephone numbers and addresses of nearest medical facilities, fire or rescue stations, and police departments.
6. Develop and follow a safety plan for all underwater operations. Address the following as appropriate:
   a. water temperature and maximum time of submersion;
   b. for scuba, dive submersion tables and telephone number for the closest decompression chamber;
   c. escape routes and rescue protocols for potentially dangerous situations;
   d. equipment requirements and condition; and
   e. potential risks of the environment and recommended prevention and treatment.

including predators (e.g., sharks and barracudas, large marine mammals, and poisonous fish, reptiles, and invertebrates) (U.S. Navy 1985). Divers in freshwater environments may encounter potentially dangerous animals such as poisonous snakes, large mammals, venomous insects, poisonous plants, and microorganisms both in the water and along riparian areas of lakes and rivers.

It is important to have a safety plan for any underwater observation. A safety checklist is the first step for any diving operation (Box 18.3).

18.3.2 Training

Training is essential for successful application of underwater surveys. An initial investment in crew training will help ensure diver safety and collection of accurate information. More than most techniques employed in fisheries investigations, diving is physically and mentally exhausting, particularly during cold weather. It is important to identify individuals with an interest and aptitude for underwater observation during crew selection and early training.

Training should address safety, equipment, observation techniques, and data collection and recording (Thurow 1994). Training usually has two stages. The first stage enables crew members to become familiar with surroundings and equipment. If divers do not have experience it is important to let them practice underwater and

Figure 18.5 Flotarium used to photograph fish for use in a species identification booklet. The fish's common name and species code can be imprinted directly onto the photo.

become accustomed to using a mask, snorkel, and other gear. For scuba, divers must be recently certified. Certified divers who have not used scuba for at least 3 years should schedule an open-water dive review with a certified instructor. Even individuals who dive regularly may have completely different diving experiences from situations they will encounter in their current job. Recreational scuba experience in the tropics is not adequate preparation for diving in temperate climates with a wet or dry suit. It is important that all divers have a basic understanding of the dive environment and of their equipment and how to use it. Training in survival and rescue procedures is also useful. During training, emphasize that divers must not harass or unnecessarily disturb aquatic organisms. Divers have special responsibility not to exploit for personal gain information gathered under the auspices of a scientific diving program, nor should they reveal sensitive information, such as the location of game fish, on which others might capitalize.

The second stage of training is the specific training divers receive for the survey they are expected to perform. Both experienced and novice divers need periodic training in species identification and life history characteristics of the species they are to observe. Whereas adult specimens may be easily recognized, juveniles are often difficult to identify. Several methods are available to train divers in identification. Divers can view and compare color photographs with live specimens at different life stages, or they capture and display specimens in small aquaria, either above or below the water surface (Figure 18.5). Viewing videotapes of organisms in natural environments, visiting aquaria, and working with experienced observers are effective ways to learn species identification (Thurow 1994).

Divers often need to estimate the size as well as the species of organisms. Many species flee, hide, or attack when approached closely by a diver, so size must be estimated from a distance. Several methods have been developed to estimate organism size. One method is to view the animal and estimate its size relative to fixed points. The diver then swims to the reference points and measures the distance with a measuring device (Cunjak and Power 1986; Baltz et al. 1987). Swenson et al. (1988) developed a calibrated bar that attaches to the divers mask. The diver observes length on the bar and measures distance to the organism to estimate its length. Divers can also practice estimating fish size by viewing wooden dowels or fish silhouettes of known lengths underwater. Accuracy of size estimate improves with training (Griffith 1981).

Accurately estimating abundance requires a well-designed survey protocol that is consistently implemented. Divers should be trained in situations similar to the ones they will encounter in the field. During training, it is useful to have experienced divers sample an area and compare results with trainees. Sampling protocols should be thoroughly tested in field environments before implementation. Abundance estimates derived by diving can be compared with other methods, such as electrofishing. Graham (1992) tested several underwater observation techniques before a suitable method was developed to estimate fish density around artificial structures in the low-visibility waters of a Virginia reservoir. He found that counts of fish made immediately after descent differed significantly from counts made 3 and 5 min after arriving at the bottom.

18.4 ENVIRONMENTAL INFLUENCES

The accuracy of underwater surveys is influenced by the behavior and life stages of individual species and by environmental features of the sampling unit, such as depth, temperature, clarity of the water, and the type and abundance of cover.

18.4.1 Depth

The survey area must have sufficient depth to enable the diver to submerge a mask. Shallow water may limit the diver's ability to view organisms that hide beneath or behind obstructions. Divers may count fish in water that is deep enough to submerge a mask but too shallow to float, provided the observer is able to crawl through the area. Dive teams in shallow water may have difficulty maintaining an organized line while floating downstream (Schill and Griffith 1984). Excessive depths impose a number of limitations including low light penetration and decreased time for conducting a survey because of the need for diver decompression to avoid nitrogen narcosis.

18.4.2 Temperature

Water temperature influences organism behavior and may result in biased underwater counts. Observers should carry a calibrated thermometer and measure water temperatures before sampling and periodically as sampling proceeds. Organism behavior may change seasonally in response to changing temperature. For example, as temperatures decline in the fall, many stream-dwelling fishes adopt one of two strategies: migration or concealment (Bustard and Narver 1975).

Salmonids may migrate from summer habitat into other portions of the watershed as temperatures decline below 10°C (Bjornn 1971). Movement into concealment cover at reduced water temperatures is well documented for a variety of resident and anadromous salmonids, including juvenile chinook salmon (Edmundson et al. 1968; Hillman et al. 1987), juvenile steelhead (Edmundson et al. 1968; Everest and

Chapman 1972; Bustard and Narver 1975), cutthroat trout (Bustard and Narver 1975), and rainbow trout (Campbell and Neuner 1985). The accuracy of underwater counts of juvenile salmonids declines with decreased water temperatures (Shepard et al. 1982; Contor 1989; Hillman et al. 1992; Riehle and Griffith 1993); at water temperatures below 9°C, most juvenile salmonids were hidden during the daytime, and counts underestimated the true population. Accuracy of counts improved as temperature increased.

18.4.3 Visibility

Water clarity can severely limit a diver's ability to survey organisms. Although no consensus has been reached, researchers agree that the minimum acceptable visibility is dependent on the species to be counted and the nature of the physical habitat. Palmer and Greybill (1986) noted a significant positive correlation between visibility and numbers of fish observed as visibility increased beyond 2 m. Water clarity must be sufficient to enable divers to see the bottom in the deepest sampling units, identify species, and see organisms that are fleeing.

Divers should not assume that visibility is adequate without measuring it. Water that appears clear from above the surface may look different underwater. Silt, algae, and other particulate material such as flecks of mica can scatter light and reduce visibility. Divers should periodically measure the visibility of a standard object in water to be surveyed. Suitable objects include silhouettes of target organisms or local substrate. Estimate visibility by averaging three measurements of the maximum distance at which the standard object can be distinguished.

Storms and other events may temporarily reduce underwater visibility. When this occurs, terminate the survey and return after conditions improve. Divers also may encounter low visibility as a result of turbulence in isolated portions of a sampling unit. Divers should survey all areas surrounding the turbulence before attempting to survey turbulent areas. Fish frequently maintain positions outside areas of extreme turbulence although they may seek cover in turbulence when disturbed.

Surveys of some organisms, such as sessile mollusks, fish eggs, and algae, may be accomplished even when visibility is near zero by divers probing into the sediment (Nalepa and Gauvin 1988) or by operating an apparatus such as a pump to remove sediment within a proscribed area (Flath and Dorr 1984).

18.4.4 Cover

The type and abundance of cover can limit the application of underwater surveys. Surveys in simple habitats lacking cover may be more accurate than those in complex habitats with abundant cover (Rodgers et al. 1992). Organisms may be more visible where cover consists of surface turbulence and depth as compared with areas where cover consists of vegetation, woody debris, or other structures. We recommend that divers describe and quantify the cover in their sampling sites to relate physical features of the sampling area with survey results.

18.5 APPLICATIONS

18.5.1 Precision and Accuracy

The statistical precision (Chapter 2) of underwater estimates of fish abundance is derived by replicating counts. Counts may be replicated temporally within the same unit (Colton and Alevizon 1981; Sale and Douglas 1981; Slaney and Martin 1987;

Box 18.4 Bounded Counts

Divers frequently must assume that some fish escape observation and that their estimates of abundance by direct count are lower than the actual abundance. To account for this bias, Hicks and Watson (1985) used the bounded counts method (Regier and Robson 1967) to estimate the seasonal abundance of rainbow and brown trout in the Rangitikei River. The estimated number of fish present, $\hat{N}$ is calculated according to the formula

$$\hat{N} = 2N_m - N_{m-1},$$

where N_m is the largest and N_{m-1} the second largest count in a series of passes through the sample unit. Data for the rainbow trout are presented below.

	Number of fish that were:			
Pass	Large	Medium	Small	Total fish
1	15	35	29	79
2	7	43	41	91
3	14	34	39	87
$\hat{N}$	16	51	43	95

Spyker and Van Den Berghe 1995) or spatially across multiple units within a stratum (Hankin and Reeves 1988). In the simplest example, observers make three counts in the same unit, calculate the mean and variance, and place confidence limits around the mean. Replicate counts require independence and may be completed by individual divers or teams of divers. Hankin and Reeves (1988) replicated counts of age-1 steelhead in 30 pools of a small stream (2–16 m wide). Mean counts ranged from 1 to 60 steelhead. Of the replicate counts, 87% were within 15% of the mean count. Stream size does not appear to influence the repeatability of counts by trained divers. Schill and Griffith (1984) made 28 replicate counts in 10 reaches of a large stream (77–99 m wide); 93% of the replicate counts were within 15% of the average count.

Although variation in replicate counts is typically small (Schill and Griffith 1984; Hankin and Reeves 1988) the accuracy of underwater estimates has been difficult to assess because the true population density is usually unknown (Hillman et al. 1992). Diver counts should be calibrated with other methods of estimating population size because the accuracy of diver estimates varies among aquatic environments (Hankin and Reeves 1988; Rodgers et al. 1992). The accuracy of underwater estimates has been assessed by comparing diver counts with abundance estimates derived from electrofishing (Griffith 1981; Hankin and Reeves 1988), seining (Goldstein 1978), and piscicides (Northcote and Wilke 1963; Dibble 1991; Hillman et al. 1992). Slaney and Martin (1987) and Zubik and Fraley (1988) reported a technique that combines diver counts and mark–recapture estimates and can be used to calibrate diver counts in remote streams. A method known as bounded counts, which accounts for direct count bias, is presented in Box 18.4.

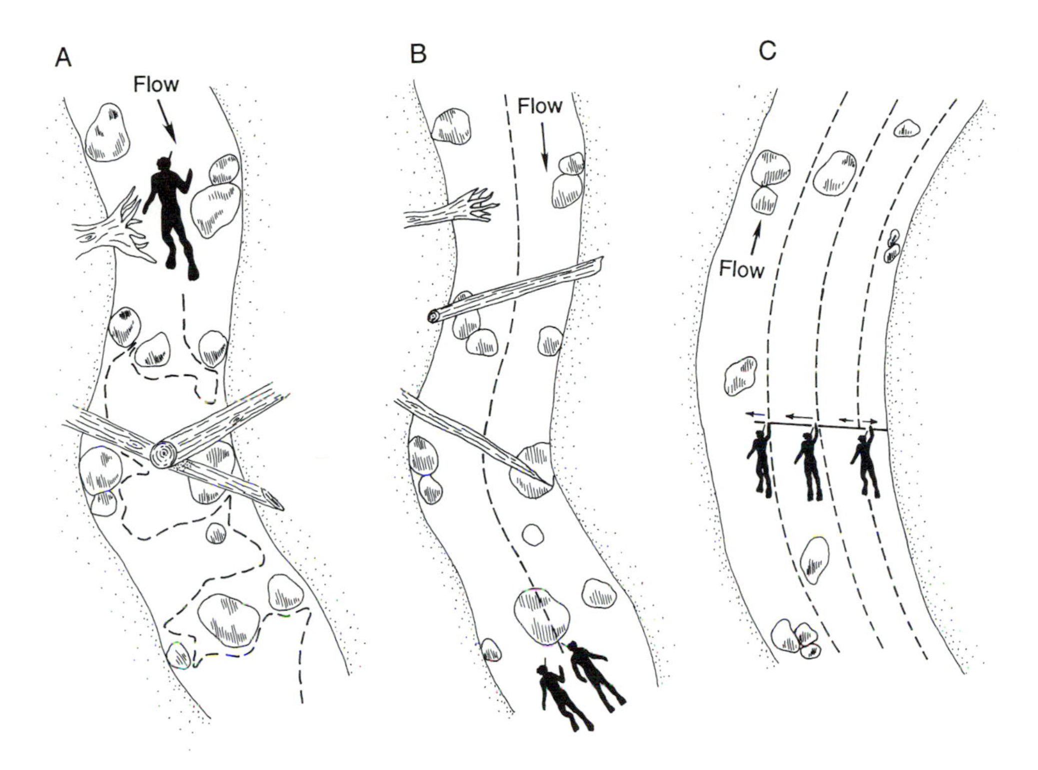

Figure 18.6 Routes (dashed lines) of diver(s) surveying a stream or river: **(A)** single diver zigzagging upstream in a small stream; **(B)** two divers surveying equal lanes in a larger river; and **(C)** three (or more) divers drifting downstream in a deep river and holding onto a pipe to maintain uniform counting lanes.

Diver observation is an efficient method to estimate fish populations over large areas. Hankin and Reeves (1988) found that for the same cost, basinwide visual technique estimates (see section 18.5.2.3) were 1.7 to 3.3 times more accurate than were electrofishing estimates. Although fish population estimates derived by diver counts are often less accurate than estimates derived by other methods, divers often can examine more habitat units at a lower overall cost.

18.5.2 Underwater Survey Procedures

In flowing waters, divers moving upstream are less likely to startle fish and cause them to flee or change their behavior because most stream-dwelling fish orient facing into the current. Whenever conditions permit, divers should enter streams downstream from the unit to be sampled and proceed slowly upstream, avoiding sudden movements (Figure 18.6); a diver who moves slowly can nearly touch fish before the fish are frightened (Heggenes et al. 1990). When it is impractical to move upstream, divers may enter the water upstream from the sampling unit and float downstream with the current, remaining as motionless as possible. Size and complexity of the sampled unit, underwater visibility, and the type of estimate determine the number of observers needed to complete a particular survey.

Regardless of the type of habitat surveyed, divers or other crew members can measure habitat features such as surface area, depth, substrate, and discharge after fish are observed and counted. Fish counts are typically converted to density estimates, expressed as the number of fish per 100 m^2 or per hectare, to facilitate spatial or temporal comparison of populations. The following section describes various types of survey procedures and considerations for the number of observers required.

18.5.2.1 Direct Enumeration

Direct enumeration is an accurate method of counting many species such as stream-dwelling salmonids (Northcote and Wilkie 1963; Slaney and Martin 1987), northern pike in shallow margins of lakes (Turner and Mackay 1985), and fishes associated with structure in marine habitats (Thresher and Gunn 1986). This method assumes that all organisms in a sample unit have equal probability of being seen and counted. Typically, one or more divers count all organisms in a single pass through a sample unit. Precision can be evaluated by making multiple passes through a sample unit (Keast and Harker 1977; Hicks and Watson 1985).

In small streams, where a diver can see from bank to bank from a single point underwater, fish can be counted using a variety of techniques. Depending on the characteristics of the unit, the diver typically proceeds from downstream up, zigzagging between banks and taking care to conduct a thorough search of the stream margins and all cover components, such as undercut banks, substrate interstices, and accumulations of woody debris (Figure 18.6A). Where water depth, turbulence, or clarity limit the ability to see and identify fish accurately, the diver first moves up one bank of the unit and counts all fish to the limits of visibility and then repeats the procedure for the opposite bank. Divers may allow the current to carry them downstream in units that are too deep or swift to permit them to move upstream. Although water clarity may allow one diver to see the entire channel width, additional divers may be needed to count concealed or less conspicuous organisms. Very shallow water habitats such as riffles typically require more observers than do deep water habitats.

Multiple divers are usually necessary to obtain accurate fish counts in larger rivers and lakes. For relatively homogeneous habitats, sampling units can be divided into equal lanes and divers move slowly upstream, counting all fish within an assigned lane (Figure 18.6B). If the unit is too deep, turbulent, or complex, divers can use natural features such as a line of boulders to partition the unit. In water too deep to move upstream, two divers can each hold one end of a baton or multiple divers can position themselves along a length of plastic pipe to maintain a uniform counting lane (Figure 18.6C). Fish are counted as the divers drift downstream through the unit (Schill and Griffith 1984). In all situations, the distance between divers should always be less than the maximum underwater visibility. Divers must start and stop at the same time, remain in their assigned lanes, and move at the same speed. To avoid double-counting fish, divers must not count fish that move among lanes. When it is not feasible to count all fish from bank to bank, divers may count fish within a subunit of the stream channel. The area surveyed is estimated by multiplying the length of the unit by the width of the sampled corridor.

18.5.2.2 Expansion Estimates

The expansion method (Box 18.5) can be used to estimate the total population of fish in locations such as individual habitat units of large rivers, lakes, and marine environments where total enumeration is not feasible. Divers partition the sample area into relatively homogeneous strata. In rivers, divers are randomly assigned lanes

Box 18.5 Expansion Estimates

To estimate the number of Flathead River cutthroat trout that were greater than 254 mm in total length, Zubik and Fraley (1988) divided a 45.1-m-wide, 2.2-km-long section of river into four lanes. Divers were randomly assigned to lanes and instructed to count fish within a 4.0-m-wide visual corridor. Because only 36% of the actual river corridor was sampled, the first step in calculating the number of fish per kilometer was to develop an expansion factor (EF):

EF = 45.1 (mean river width)/4 (counting lanes)/4 (visual distance) = 2.82.

The average total number of fish seen (44) in three passes was then multiplied by 2.82 to yield an expanded total population estimate of 124 fish, which was then divided by 2.2 km to yield a density of 56 fish/km.

Zubik and Fraley also were able to calculate confidence intervals about their estimate.

	Number of fish in lane:				
Pass or statistic	1	2	3	4	All
1	13	13	8	13	47
2	11	5	11	15	42
3	20	9	5	10	44
Mean	15	9	8	13	44
SE	2.73	2.31	1.73	1.45	1.45

The standard error of the total was similarly expanded (1.45 [2.82] = 4.09) and adjusted to a kilometer basis (4.09/2.2 = 1.86). The 95% confidence interval was $N \pm t(\text{SE})$ with 2 df or 56 ± 4.303 (1.86) = 56 ± 8.

to count fish (Zubik and Fraley 1988). If all fish within a particular stratum have an equal probability of being seen and counted, the total number of fish can be estimated by dividing the number of fish counted in the counting lanes by the percent of the stratum that was sampled. Average density, variance, and confidence intervals can be calculated by replicating counts within individual strata and lanes (Slaney and Martin 1987).

18.5.2.3 Basinwide Estimates

Hankin and Reeves (1988) developed procedures for estimating fish numbers within entire watersheds. The basic premise of the basinwide method is that if there is a consistent relationship between diver counts and the true numbers of fish, then it is possible to calculate a calibration ratio and correct for the bias associated with diver counts. Basinwide sampling consists of two distinct phases. During the first phase, divers count fish in a large sample selected from the total number of habitat units of a particular type. During the second phase, a crew determines the true number of fish (typically by electrofishing) in a subsample of at least 10 habitat units of each type sampled by divers in the first phase. The second-phase sample provides a means to evaluate the relationship between diver counts and estimates of the "true" numbers of fish. Equations to calculate the mean, variance, and total number of fish can be found in Hankin and Reeves (1988).

Later versions of the basinwide methodology revised the variance estimator for total numbers of fish and outlined a method to optimally allocate first- and second-phase sampling effort (Dolloff et al. 1993). The optimum allocation procedure provides a means to minimize overall sampling variance based on the relative cost of first-phase (diver estimated) and second-phase (electrofishing) sampling.

18.5.2.4 Mark–Recapture Estimates

Underwater observation also can be used with other techniques to derive mark–recapture population estimates (Van Den Avyle 1993). Researchers have captured fish by angling or some other method and marked them with tags that are visible underwater (Helfman 1981; Slaney and Martin 1987; Zubik and Fraley 1988; Vore 1993). Different colored tags or marks can be used to mark each size-class of fish (Nielsen 1992). After the marked fish redistribute themselves in the sampling unit, divers record the number of marked and unmarked fish by species and size-class. A total population estimate can be derived by pooling the estimates for each size-class, provided that the recovery efficiencies (ratio of marked fish observed to marked fish) are constant over all size-classes in the sample.

18.5.2.5 Line Transect Estimates

Most underwater estimates of fish abundance assume constant probability of sighting out to the limits of a diver's visibility. In practice, this means that all fish within a diver's field of vision (lane) are seen with certainty. This approach works well in terrestrial environments (Burnham et al. 1980; Seber 1982) and in aquatic environments for highly visible or sessile species (Lyons 1987). The probability of sighting many other species, however, is not constant, and hence this approach cannot be used to estimate abundance. Many benthic organisms are cryptically colored and frequently occupy interstitial spaces or other inconspicuous locations, whereas others may flee at the approach of a diver. For these organisms a method that assumes a decreasing probability of sighting with increasing distance from an established line may be more appropriate (Ensign et al. 1995).

When conducting line transect surveys, divers must travel along a well-defined line to ensure that measurements of distance from the line to observed targets are both precise and accurate. Ensign et al. (1995) advised investigators to mark lines of travel with lengths of colored twine oriented parallel to the current and anchored at the upstream and downstream ends of the sampled units. When multiple lines are established, the spacing between lines should be greater than the maximum underwater visibility. A diver identifies and counts fish on both sides of the line. The positions of all target fish are marked with weighted, color-coded markers attached to floats. After completing counts, the diver measures the distance from each marker to the survey line and records the species.

The specific model used to estimate density depends on the amount and quality of the available data and the degree to which assumptions about the data can be met. Ensign et al. (1995) strongly recommended models that correct estimates of the total number of fish counted for unequal probability of sighting.

18.5.2.6 Habitat Use Estimates

Direct underwater observation has become increasingly popular for observing organisms in their natural environments. Because many species apparently do not modify their behavior in the presence of divers, underwater observations provide relatively unbiased information on habitat use. For example, because salmonids tend to occupy specific locations in the water column, known as focal points, many researchers have used underwater methods to study habitat use of different life stages (Fausch and White 1981; Rimmer et al. 1984; Cunjak and Power 1986; Cunjak 1988). Upon encountering a fish, the diver carefully notes the species and location and estimates the fish's size. A weight and float can be used to mark the organism's location. If organisms do not modify their behavior when approached, divers can measure or estimate various macro- and microhabitat features. By comparing habitat characteristics at the locations of fish with measurements of available habitat, investigators are then able to develop estimates of habitat use.

More accurate estimates of size, weight, or food habits can be obtained by collecting organisms by means of several techniques. Lethal capture methods include explosive charges (Everest 1978) and spear guns. Nonlethal capture methods include slurp guns (Morantz et al. 1987), nets (Bonneau et al. 1995), diver-operated electrofishing probes (James et al. 1987), and various types of pumps (Flath and Dorr 1984) and suction dredges (Koch 1992).

18.6 REFERENCES

Ahlstedt, S. A., and T. A, McDonough. 1993. Quantitative evaluation of commercial mussel populations in the Tennessee River portion of Wheeler Reservoir, Alabama. Pages 38–49 *in* K. S. Cummings, A. C. Buchanan, and L. M. Koch, editors. Proceedings of a symposium: conservation and management of freshwater mussels. Upper Mississippi River Conservation Committee, Rock Island, Illinois.

Baltz, D. M., B. Vondracek, L. R. Brown, and P. B. Moyle. 1987. Influence of temperature on microhabitat choice by fishes in a California stream. Transactions of the American Fisheries Society 116:12–20.

Barsky, S. M. 1991. Practical systems for contaminated water diving. Undersea Biomedical Research 18:253–258.

Beauchamp, D. A., B. C. Allen, R. C. Richards, W. A. Wurtsbaugh, and C. R. Goldman. 1992. Lake trout spawning in Lake Tahoe: egg incubation in deepwater macrophyte beds. North American Journal of Fisheries Management 12:422–499.

Beauchamp, D. A., P. E. Budy, B. C. Allen, and J. M. Godfrey. 1994. Timing, distribution, and abundance of kokanees spawning in a Lake Tahoe tributary. Great Basin Naturalist 54(2):130–141.

Bergestedt, R. A., and D. R. Anderson. 1990. Evaluation of line transect sampling based on remotely sensed data from underwater video. Transactions of the American Fisheries Society 119:86–91.

Bjornn, T. C. 1971. Trout and salmon movements in two Idaho streams as related to temperature, food, stream flow, cover, and population density. Transactions of the American Fisheries Society. 100:423–438.

Bonneau, J. L., R. F. Thurow, and D. L. Scarnecchia. 1995. Capture, marking, and enumeration of juvenile bull trout and cutthroat trout in small, low-conductivity streams. North American Journal of Fisheries Management 15:563–568.

Burnham, K. D., D. A. Anderson, and J. L. Laake. 1980. Estimation of the density of animals from line transect sampling of biological populations. Wildlife Monographs 72:1–202.

Bustard, D. R., and D. W. Narver. 1975. Aspects of the winter ecology of juvenile coho salmon (*Oncorhynchus kisutch*). Canadian Journal of Fisheries and Aquatic Sciences 32:667–680.

Campbell, R. F., and J. H. Neuner. 1985. Seasonal and diurnal shifts in habitat utilization by resident rainbow trout in western Washington Cascade Mountain streams. Pages 39–48 *in* F. W. Olson, R. G. White, and R. H. Hamre, editors. Symposium on small hydropower and fisheries. American Fisheries Society, Western Division and Bioengineering Section, Bethesda, Maryland.

Colton, D. E., and W. S. Alevizon. 1981. Diurnal variability in a fish assemblage of a Bahamian coral reef. Environmental Biology of Fishes 3:341–345.

Contor, C. R. 1989. Diurnal and nocturnal winter habitat utilization by juvenile rainbow trout in the Henry's Fork of the Snake River, Idaho. Master's thesis. Idaho State University, Pocatello.

Cunjak, R. A. 1988. Physiological consequences of overwintering: the cost of acclimatization? Canadian Journal of Fisheries and Aquatic Sciences 45:443–452.

Cunjak, R. A., and G. Power. 1986. Winter habitat utilization by stream resident brook trout (*Salvelinus fontinalis*) and brown trout (*Salmo trutta*). Canadian Journal of Fisheries and Aquatic Sciences 43:1970–1981.

Dibble, E. D. 1991. A comparison of diving and rotenone methods for determining relative abundance of fish. Transactions of the American Fisheries Society 120:663–666.

Dolloff, C. A., D. G. Hankin, and G. H. Reeves. 1993. Basinwide estimation of habitat and fish populations in streams. U.S. Forest Service General Technical Report SE-83.

Drew, E. A., J. N. Lythgoe, and J. D. Woods. 1976. Underwater research. Academic Press, New York.

Edmundson, E., F. H. Everest, and D. W. Chapman. 1968. Permanence of station in juvenile chinook salmon and steelhead trout. Journal of the Fisheries Research Board of Canada 25:1453–1464.

Ensign, W. E., P. L. Angermeier, and C. A. Dolloff. 1995. Use of line transect methods to estimate abundance of benthic stream fishes. Canadian Journal of Fisheries and Aquatic Sciences 52:213–222.

Everest, F. H. 1978. Diver-operated device for immobilizing fish with a small explosive charge. Progressive Fish-Culturist 49:121–122.

Everest, F. H., and D. W. Chapman. 1972. Habitat selection and spatial interaction by juvenile chinook salmon and steelhead trout in two Idaho streams. Journal of the Fisheries Research Board of Canada 29:91–100.

Fausch, K. D., and R. J. White. 1981. Competition between brook trout (*Salvelinus fontinalis*) and brown trout (*Salmo trutta*) for positions in a Michigan stream. Canadian Journal of Fisheries and Aquatic Sciences 38:1220–1227.

Flath, L. E., and J. A. Dorr, III. 1984. A portable, diver-operated, underwater pumping device. Progressive Fish-Culturist 46:219–220.

Flemming, N. C., and M. D. Max. 1990. Scientific diving: a general code of practice. UNESCO (United Nations Educational Scientific and Cultural Organization) Paris, France, and University of Florida Sea Grant College Program, Gainesville.

Gardiner, W. R. 1984. Estimating population densities of salmonids in deep water in streams. Journal of Fish Biology 24:41–49.

Goldstein, R. M. 1978. Quantitative comparison of seining and underwater observation for stream fishery surveys. Progressive Fish-Culturist 40:108–111.

Graham, R. J. 1992. Visually estimating fish density at artificial structures in Lake Anna, Virginia. North American Journal of Fisheries Management 12:204–212.

Greene, L. E., and W. E. Alevizon. 1989. Comparative accuracies of visual assessment methods for coral reef fishes. Bulletin of Marine Science 44:899–912.

Griffith, J. S. 1981. Estimation of the age-frequency distribution of stream-dwelling trout by underwater observation. Progressive Fish-Culturist 43:51–53.

Griffith, J. S., D. J. Schill, and R. E. Gresswell. 1984. Underwater observation as a technique for assessing fish abundance in large western rivers. Proceedings of the Annual Conference Western Association of Fish and Wildlife Agencies 63:143–149.

Hankin, D. J., and G. H. Reeves. 1988. Estimating total fish abundance and total habitat area in small stream based on visual estimation methods. Canadian Journal of Fisheries and Aquatic Sciences 45:834–844.

Heggenes, J., A. Brabrand, and S. J. Saltveit. 1990. Comparison of three methods for studies of stream habitat use by young brown trout and Atlantic salmon. Transactions of the American Fisheries Society 119:101–111.

Helfman, G. S. 1981. Twilight activities and temporal structure in a freshwater fish community. Canadian Journal of Fisheries and Aquatic Sciences 38:1405–1420.

Hicks, B. J., and N. R. N. Watson. 1985. Seasonal changes in abundance of brown trout (*Salmo trutta*) and rainbow trout (*S. gairdneri*) assessed by drift diving in the Rangitikei River, New Zealand. New Zealand Journal of Marine and Freshwater Research 19:1–10.

Hillman, T. C., J. S. Griffith, and W. S. Platts. 1987. Summer and winter habitat selection by juvenile chinook salmon in a highly sedimented Idaho stream. Transactions of the American Fisheries Society 116:185–195.

Hillman, T. W., J. W. Mullan, and J. S. Griffith. 1992. Accuracy of underwater counts of juvenile chinook salmon, coho salmon, and steelhead. North American Journal of Fisheries Management 12:598–603.

Jacobson, L. K. 1994. Factors limiting outmigration of cutthroat trout from St. Charles Creek to Bear Lake. Master's thesis. Utah State University, Logan.

James, P. W., S. C. Leon, V. Z. Alexander, and O. E. Maughan. 1987. Diver-operated electrofishing device. North American Journal of Fisheries Management 7:597–598.

Keast, A., and J. Harker. 1977. Strip counts as a means of determining densities and habitat utilization patterns in lake fishes. Environmental Biology of Fishes 1:181–188.

Koch, L. M. 1992. Qualitative mussel survey in pool 24 of the upper Mississippi River at Mississippi River mile 292 in June 1989. Missouri Department of Conservation, Jefferson City.

Lee, O. 1967. The complete illustrated guide to snorkel and deep diving. Doubleday, New York.

Losonsky, G. 1991. Infections associated with swimming and diving. Undersea Biomedical Research 18:181–186.

Lyons, J. 1987. Distribution, abundance, and mortality of small littoral-zone fishes in Sparkling Lake, Wisconsin. Environmental Biology of Fishes 18:93–107.

Masterson, B. F. 1991. Protection of recreational divers against water-borne microbial hazards. Undersea Biomedical Research 18:197–204.

Morantz, D. L., R. K. Sweeney, C. S. Shirvell, and D. A. Longard. 1987. Selection of microhabitat in summer by juvenile Atlantic salmon (*Salmo salar*). Canadian Journal of Fisheries and Aquatic Sciences 44:120–129.

Nalepa, T. F., and J. M. Gauvin. 1988. Distribution, abundance, and biomass of freshwater mussels (Bivalvia: Unionidae) in Lake St. Clair. Journal of Great Lakes Research 14:411–419.

Nielsen, L. A. 1992. Methods of marking fish and shellfish. American Fisheries Society Special Publication 23.

Northcote, T. G., and D. W. Wilkie. 1963. Underwater census of stream fish populations. Transactions of the American Fisheries Society 92:146–151.

Ogden, J. C. 1977. A scroll apparatus for the recording of notes and observations underwater. Marine Technology Society Journal 11:13–14.

Palmer, K., and J. Greybill. 1986. More observations on drift diving. Freshwater Catch 30:22–23.

Regier, H. A., and D. S. Robson. 1967. Estimating population number and mortality rates. Pages 31–66 *in* S. D. Gerking, editor. The biological basis of freshwater fish production. Blackwell Scientific Publications, Oxford, UK.

Riehle, M. D., and J. S. Griffith. 1993. Changes in habitat use and feeding chronology of juvenile rainbow trout in fall and the onset of winter in Silver Creek, Idaho. Canadian Journal of Fisheries and Aquatic Sciences 50:2119–2128.

Rimmer, D. M., U. Paim, and R. L. Saunders. 1984. Changes in the selection of microhabitat by juvenile Atlantic salmon (*Salmo salar*) at the summer–autumn transition in a small river. Canadian Journal of Fisheries and Aquatic Sciences 41:469–475.

Rodgers, J. D., M. F. Solazzi, S. L. Johnson, and M. A. Buckman. 1992. Comparison of three techniques to estimate juvenile coho salmon populations in small streams. North American Journal of Fisheries Management 12:79–86.

Sale, P. F., and W. A. Douglas. 1981. Precision and accuracy of visual census technique for fish assemblages on coral patch reefs. Environmental Biology of Fishes 6:333–339.

Schill, D. J., and J. S. Griffith. 1984. Use of underwater observations to estimate cutthroat trout abundance in the Yellowstone River. North American Journal of Fisheries Management 4:479–487.

Seber, G. A. F. 1982. The estimation of animal abundance and related parameters. Hafner Press, New York.

Shepard, B. B., J. J. Fraley, T. M. Weaver, and P. Graham. 1982. Life histories of westslope cutthroat and bull trout in the upper Flathead River Basin, Montana. Montana Department of Fish, Wildlife and Parks, Kalispell.

Slaney, P. A., and A. D. Martin. 1987. Accuracy of underwater census of trout populations in a large stream in British Columbia. North American Journal of Fisheries Management 7:117–122.

Somers, L. H. 1976. Research diver's manual. University of Michigan Press, Ann Arbor.

Spyker, K. A., and E. P. Van Den Berghe. 1995. Diurnal abundance patterns of Mediterranean fishes assessed on fixed transects by scuba divers. Transactions of the American Fisheries Society 124:216–224.

Swenson, W. A., W. P. Gobin, and T. D. Simonson. 1988. Calibrated mask-bar for underwater measurement of fish. North American Journal of Fisheries Management 8:382–385.

Thresher, R. E., and J. S. Gunn. 1986. Comparative analysis of visual census techniques for highly mobile, reef-associated piscivores (Carangidae). Environmental Biology of Fishes 17:93–116.

Thurow, R. F. 1994. Guide to underwater methods for study of salmonids in the intermountain west. U.S. Forest Service General Technical Report INT-GTR-307.

Thurow, R. F., and D. J. Schill. 1996. Comparison of day snorkeling, night snorkeling, and electrofishing to estimate bull trout abundance and size structure in a second-order Idaho stream. North American Journal of Fisheries Management 16:314–323.

Turner, L. J., and W. C. Mackay. 1985. Use of visual census for estimating population size in northern pike (*Esox lucius*). Canadian Journal of Fisheries and Aquatic Sciences 42:1835–1840.

U.S. Navy. 1985. U.S. Navy diving manual. NAVSEA 0994-LP-001-9010. U.S. Department of the Navy, Washington, DC.

Van Den Avyle, M. 1993. Dynamics of exploited fish populations. Pages 105–134 *in* C. C. Kohler and W. A. Hubert, editors. Inland fisheries management in North America. American Fisheries Society, Bethesda, Maryland.

Vore, D. W. 1993. Size, abundance, and seasonal habitat utilization of an unfished trout population and their response to catch and release fishing. Master's thesis. Montana State University, Bozeman.

Zubik, R. J., and J. J. Fraley. 1988. Comparison of snorkel and mark–recapture estimates for trout populations in large streams. North American Journal of Fisheries Management 8:58–62.

Chapter 19

Advances in Underwater Biotelemetry

JIM WINTER

19.1 INTRODUCTION

Advances in the electronics industry have enabled scientists to develop sophisticated telemetry methods to monitor the locations, behavior, and physiology of free-ranging aquatic animals. Underwater biotelemetry involves attaching to an aquatic organism a device that relays biological information. The information is relayed via ultrasonic or radio signals to a remote receiving system (Figure 19.1). Telemetry provides a means to monitor the biology of animals not readily visible, to collect data with a minimal influence on an animal's behavior and health, to collect more data than are normally gathered by techniques such as mark and recapture, and to compare physiological and behavioral data collected in the laboratory and in natural systems. In addition, telemetry often provides the only means to solve some biological problems. This chapter describes telemetry systems, methods of attaching transmitters, methods of tracking free-ranging aquatic animals, and means of sampling and processing data.

If the device attached to the animal emits a signal, it is called a transmitter. If the transmitter returns a signal in response to one sent to it, it is called a transponder. Many transponders have their own power source and are called active transponders. Others, which are termed passive transponders, have a current induced in them by the interrogating system. Passive integrated transponders, or PIT tags, are minitransponders that have recently been introduced to tag and identify fingerling fish (Chapter 12).

Transmitters have an electronic oscillator circuit that produces a signal by inducing high-frequency vibrations in the water or air. A hertz is a measure of the frequency of the signal and is equal to one cycle per second. Signals usually are ultrasonic, in the 20–300 kHz frequency range, or are radio signals, in the 27–300 MHz range. For comparison, the standard amplitude modulation (AM) radio uses frequencies from 550 to 1,600 kHz and the standard frequency modulation (FM) radio uses 88 to 108 MHz.

Ultrasonic signals are received by a microphone or hydrophone submerged in the water. Radio signals, on the other hand, are received by a variety of antenna types located above the water. Both receiving systems pass the signals to a receiver that converts the signals to a form that is audible via headphones or that can be electronically processed and automatically recorded. The transmitted signal allows the investigator to locate the animal by means of a wide variety of methods: by boat,

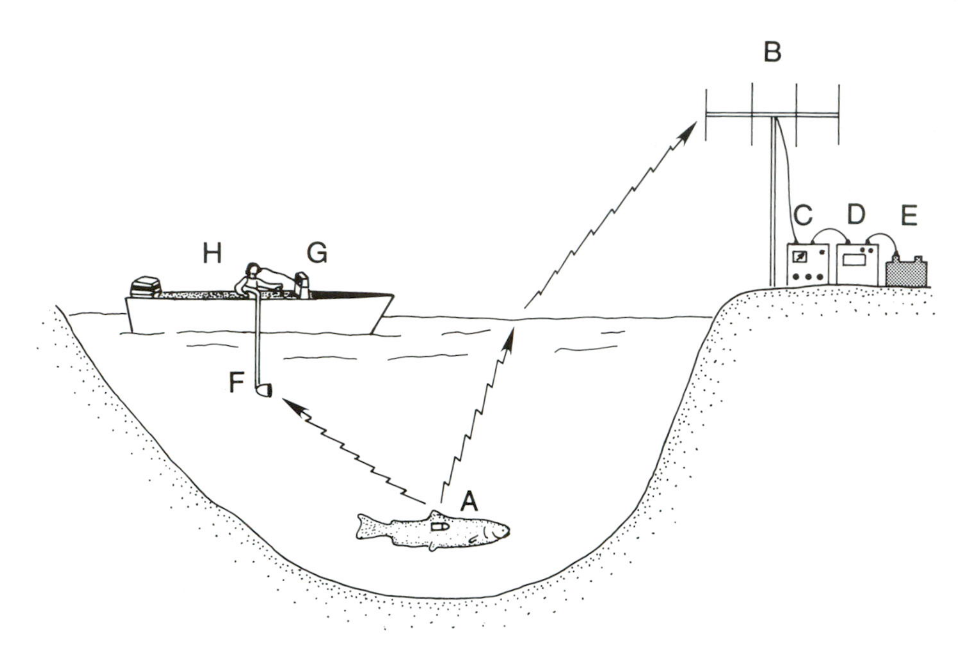

Figure 19.1 Generalized scheme of two of the many types of underwater biotelemetry systems. On the left is an ultrasonic tracking system used to locate a fish equipped with a transmitter. On the right is a fixed radio receiving station used to record environmental or physiological data (e.g., temperature) from the transmitter. Features are **(A)** a transmitter, which emits ultrasonic or radio signals (usually only one type of signal); **(B)** a radio receiving antenna (Yagi type); **(C)** a radio receiver; **(D)** a data logger or data collection computer, which contains a signal decoder and a microprocessor to store data; **(E)** a battery; **(F)** an ultrasonic hydrophone; **(G)** an ultrasonic receiver; and **(H)** headphones, a recorder, and manual data collection materials.

on foot, by truck, from a fixed receiving station, by airplane, or by satellite. If the transmitter has an electronic sensor, it can also relay information about the animal's environment (e.g., water temperature) or about the animal's physiology (e.g., heart rate).

There are many factors to consider before beginning a telemetry project. The first step is to compile simple yes or no questions that can be readily tested. Too many researchers have proceeded on telemetry projects without forming testable hypotheses and thus have produced a collection of maps of movements that cannot be meaningfully analyzed. Another important consideration is whether the problem can be solved by means other than telemetry. There is a tendency to adopt techniques because they are novel and flashy, not because they are the best means to answer a question. For example, gill netting probably would be more efficient than telemetry to determine whether a species is found at a particular depth, especially if no other information is desired, such as diel vertical movement patterns or temperature selection. Because telemetry equipment is expensive and tracking is labor-intensive, the cost per unit of information is high unless one collects large amounts of data. In addition, some problems might be more appropriate to study in the laboratory because it is too difficult to obtain sufficient observations or to account for the many variables in natural systems.

19.2 TELEMETRY SYSTEMS

Underwater biotelemetry had its origins in the late 1950s. Initially, ultrasonic transmitters were developed because their low-frequency, long-wave signals travel well through water and are minimally affected by water conductivity. While ultrasonic transmitters were being developed for aquatic animals, radio transmitters were being developed for terrestrial use. In the late 1960s, radio transmitters were modified to work underwater.

The early developments in underwater biotelemetry have been documented by Stasko (1975), Stasko and Pincock (1977), Winter et al. (1978), and Winter (1983). Other reviews contain detailed information on ultrasonic (Ireland and Kanwisher 1978; Mitson 1978; Nelson 1978; Hawkins and Urquhart 1983) and radio (Cochran 1980; Mech 1983; Kenward 1987; Samuel and Fuller 1994) telemetry equipment and techniques. Many excellent papers are regularly published in proceedings of international telemetry conferences (Long 1977, 1979; Amlaner and Macdonald 1980; Amlaner 1989; Priede and Swift 1992). This section on telemetry systems describes features common to ultrasonic and radio systems, the advantages of each system, the types of ultrasonic and radio equipment, sensing transmitters, and the selection of an equipment supplier.

19.2.1 Features Common to Ultrasonic and Radio Systems

Transmitter signals. Ultrasonic and radio transmitters (tags) have many common components and features. Both may emit continuous-wave or pulsing signals. Continuous (whistling) signals are distinguished more easily from background noise and are detected at slightly greater distances than are pulsing signals (beeps). In addition, animal activity is more easily detected and recorded with continuous signals than pulsing signals. By changing the orientation of the transmitter, an active animal causes the intensity of the signal to waver. In contrast, pulsing signals use less energy and thus increase transmitter life. The pulse length ("on" time), pulse rate (number of pulses per minute), and pulse interval (time between pulses) can be used to identify individual transmitters and to transmit environmental measurements in sensing tags (see section 19.2.5). If the pulse interval is much longer than 1 s, it is difficult to determine signal direction. Furthermore, pulse lengths less than 15 ms become difficult to detect by ear (Potgeiter 1989). The ratio of transmitter on time to total time (duty cycle) is usually 2 to 4%. Because pulsing signals conserve battery life and can be easily used to transmit biological parameters, pulsing signals rather than continuous signals are most commonly used.

Transmitter encapsulation. Simple, expendable tags are encapsulated with epoxy, wax, urethane, silicone, or dental acrylic. They are encapsulated by pouring (potting) the material over the components in a mold or by dipping the components into the material. For ultrasonic tags, the density of the potting material should be close to that of water for best signal transmission. Although potting transmitters is simple and produces a minimum package size, it does not produce minimum weight or allow easy replacement of components. Tubes that are filled with oil (e.g., dehydrated castor oil) or air are often used to encase complex, expensive, and recoverable tags. Compared with potted transmitters, oil- or air-filled tubes allow easier replacement of components and better transmission of ultrasonic signals between the signal transducer and water. Tubes are larger than potted transmitters and do not weigh much less, unless they are filled with air. Although most commercial firms supply tags encapsulated, someone who knows how to encapsulate tags can customize an attachment method to fit a unique animal.

Transmitters are usually turned on by soldering exposed wires or by activating magnetic-reed switches. After wires are soldered, they must be sealed well with epoxy or another encapsulating material to prevent water from leaking into the transmitter. Magnetic-reed switches are activated by removing a magnet taped over the switch of the transmitter. They are widely used because they are convenient and so small that they do not increase the size of most transmitters. If transmitters with magnetic-reed switches are stored close together, the magnets may cancel each other, and the batteries will drain. Magnetic-reed switches also occasionally malfunction.

Batteries. Because the battery generally represents more than 50% of the volume and up to 80% of the weight of the transmitter, choice of battery is critical. Choice of batteries for transmitters is largely determined by battery energy per unit weight or volume (V/g or V/mL). Other considerations are battery cost (minor), shelf life, initial voltage per cell, voltage drop during discharge, performance under environmental conditions, and available sizes and shapes (Nelson 1978). Trade-offs among transmitter life, size, and range are necessary. Larger batteries have greater energy capacities and permit longer transmitter life but increase the transmitter size. Although increasing the current drain may produce a slight increase in signal range, the transmitter life is greatly decreased. Transmitter life in hours can be estimated by dividing the manufacturer's milliampere-hour (mAh) rating for a battery by the average current drain (mA) of the transmitter. However, transmitter life can be considerably less than estimated depending on battery freshness, variability in battery production, and environmental factors. All batteries lose some of their energy from self-discharge. If a battery is on the shelf a long time, a large percent of its total capacity is lost. Telemetry manufacturers can reduce shelf life problems by using certified cells that are manufactured more carefully and are fresher. Certified cells are expensive, must be ordered in large volumes, and are not available in many sizes. The use of certified cells is decreasing because new cells are being developed that have good shelf life and performance.

Five types of batteries are used in transmitters and receivers: lithium, mercury oxide, silver oxide, alkaline, and rechargeable nickel–cadmium (Kenward 1987). Mercuric oxide cells are now rarely used because of possible environmental contamination. Lithium batteries are widely used in transmitters because they produce the highest voltage (approximately 3.04 V) per unit weight and volume. Furthermore, lithium batteries have excellent low-temperature performance, good high-temperature tolerance, long shelf life, good efficiency over a wide range of voltages, and low cost. The sizes and shapes of lithium cells that can be used in transmitters are increasing greatly because of demand for new sizes in other electronic products.

Receivers. A biotelemetry receiver filters input signals, amplifies them, and converts them to a form that is audible to an investigator or is processed by an electronic signal detector. Whether a signal is detected depends largely on the transmitted signal level, noise level, receiver sensitivity, receiver bandwidth, electronic detector, and human hearing. The most important limiting factors are the transmitted signal level and the noise level. To detect a faint transmitter signal, a receiver must have good sensitivity (the minimum level of signal that can be detected above the receiver's internal noise). Most receivers are close to optimum sensitivity.

A receiver must have a narrow frequency bandwidth to exclude as much ambient noise as practical. Bandwidth is the range of frequencies that will be passed through the narrowest filter to the listener or electronic detector. The transmitter signal must be within the bandwidth to be detected. If receiver bandwidths are too narrow, there is an increased chance of not detecting a signal when the tuning dial is off center frequency (Nelson 1978). Furthermore, the frequencies of some transmitters change slightly (drift) with different water temperatures and may not be detected if the bandwidth is very narrow. Because most telemetry systems rely on human hearing, the real system bandwidth is the human ear, or about 50 Hz. Although the human ear may seem to be a crude instrument compared with electronic detectors, the human ear is a near-optimal receiver with high sensitivity (Urick 1975). For example, with moderate pulse lengths, detection levels by the ear are often 20–25 dB lower than detection levels by electronic detectors. In addition, the ear and brain are good signal filters and processors. Therefore, human hearing is usually used to search for faint signals. Electronic detectors can have narrower bandwidths than the ear and eliminate much noise. They are often used in recording data from transmitters with strong signals. They replace much labor and eliminate errors caused by human fatigue. If electronic detectors have very narrow bandwidths, they are vulnerable to slight drifts in transmitter frequency and imprecise tuning of the receiver. In reality, electronic detectors require a wider bandwidth than does the ear, thus limiting their sensitivity (Voegeli 1989).

Other characteristics are also important in choosing receivers. First, the frequency selector should be accurate, and the receiver should be capable of separating many transmitter frequencies in the same locale. Second, a receiver should permit quick searching of frequencies, especially when radio-tracking from an airplane. Specific frequencies can be programmed into most receivers; thus, the frequencies can be manually or automatically scanned for quick searches or for recording data electronically. Although you may not require portable receivers initially, consider the possible need for portable receivers in future projects before making purchases. Some data-recording receivers are too tiring to carry in the field and are too susceptible to weather and accidents. If the receiver may be used in the field eventually, it may be good to have a receiver that is separate from the data recorder. A hand-carried receiver should be small, light, and sturdy. In addition, a portable receiver should use rechargeable batteries and get long use per charge. Field receivers also should be capable of working over a wide range of temperatures (−25 to 50°C). Because fish tracking is often conducted in rain and fog, a receiver should have waterproof switches and be moisture resistant. A receiver that has good moisture resistance will have a tightly sealed receiver box or the circuit board coated to prevent corrosion.

19.2.2 Advantages and Disadvantages of Ultrasonic and Radio Systems

The advantages and disadvantages of ultrasonic and radio telemetry systems must be weighed with respect to characteristics of the study area and of the animal. The most important habitat characteristics to consider are water depth, water conductivity (dissolved ions), current speed, habitat size, plant densities, and temperature gradients. One needs to consider whether the animal is very mobile, pelagic, or benthic. In addition, one must estimate the amount of time that may lapse between locations because of bad weather and limited labor. As the mobility of the animal and the size of the study area increases, the tracking system must be more mobile.

Ultrasonic telemetry is suited well for studies in salt water, fresh water with high conductivity, and deep water because these habitats cause little reduction in signal strength. Animals can be located very accurately (<3–4 m) with ultrasonic systems from a boat. Because the hydrophone or receiving device in ultrasonic systems must be submerged in the water, searching for highly mobile animals over great distances or through holes in the ice is difficult and time-consuming. Ultrasonic signals are adversely affected by macrophytes, algae, thermoclines, water turbulence, raindrops, and boat motors. Dense macrophytes, high concentrations of particulate matter, suspended algae, or algae attached to hydrophones can reduce signal range from several hundred meters to a few meters. Thermoclines or temperature gradients reduce range because ultrasonic signals are refracted downward from the warmer water (see section 19.2.3). Ultrasonic telemetry is not suitable for use in turbulent water near dams or rapids because trapped air bubbles reduce the signals. Because ultrasonic transmitters are usually built without crystals that minimize frequency changes, fewer individuals can be distinguished by different frequencies than with radio telemetry. Individual ultrasonic transmitters are coded by differences in pulse rates, which are measured by a stopwatch or an electronic pulse counter, by unique pulse sequences, or by unique signals (see section 19.2.5) on one of several channels in a sensing transmitter.

Radiotelemetry, on the other hand, is suited well for shallow, low-conductivity fresh water and for turbulent water. Because radio receiving antennas do not require contact with water, radio telemetry can be used to search large areas to find highly mobile species (e.g., salmon) or for shore-based data recording. Antennas can be mounted on airplanes, boats, trucks, snowmobiles, and portable towers or hand carried along streams (Winter et al. 1978). Radio signals are little affected by vegetation, algae, or thermoclines. Signals can be received through the ice, but they can be considerably reduced under slushy conditions. Because radio transmitters use crystals to minimize frequency changes, each tag can transmit on a different frequency and provide easy identification of many individuals at the same location and time.

The disadvantages of radiotelemetry are that it cannot be used in salt water, except if the animal surfaces periodically, and that signals are reduced by increasing depth and water conductivity (see section 19.2.4). If water depth is usually less than 5 m and the species is not highly migratory (e.g., largemouth bass), high conductivity may not be a problem. However, with depths greater than 5 m and conductivities greater than 400 μS/cm (μmhos/cm), radiotelemetry may not work well for highly mobile species (e.g., walleye). Radio signals are also deflected by metal objects, terrestrial vegetation, and terrain. However, experienced trackers recognize the inconsistent bearings and decreased signal range caused by deflections. Thus, they take bearings from additional locations or relocate the antennas.

Combined acoustic (ultrasonic) and radio transmitters (CARTs) have been built (Solomon and Potter 1988). Because CARTs have the advantages of both telemetry systems, they can be used for fish (e.g., Atlantic salmon) that migrate between salt water and fresh water. To conserve power, the ultrasonic stage turns off after a predetermined time period and allows the radio transmitter to have a longer life. The predetermined time period corresponds to the approximate time that a fish needs to travel through the estuary to the river. A CART is larger than a separate ultrasonic or radio transmitter and has slightly less life.

Some features are similar for ultrasonic and radio systems. When used in environments for which they are suited, both types of transmitters have about the same battery life and signal ranges. Ultrasonic and radio transmitters can be built in similar sizes and to sense the same variables. In addition, both receiving systems may receive interference from power lines, unshielded ignition systems, lightning, and citizens band radio. Finally, the cost of the equipment is about the same. The cost of a manual receiver in 1996 was about US$700 to $1,500 and a programmable scanning receiver was $2,100 to $3,200. Transmitter prices begin at about $140 each and become more costly for environmental sensing or other capabilities. A Yagi antenna or a hydrophone costs $100 to $400.

19.2.3 Ultrasonic Telemetry Systems

Transmitters. Ultrasonic transmitters usually operate from 20 to 300 kHz, with 50 to 100 kHz being the most common range of operation. An oscillator circuit is used to induce vibrations in a transducer that sends ultrasonic vibrations through the water. Transducers are devices that transform energy from one form to another; for example, at signal levels, electrical energy is changed into mechanical or acoustic energy and vice versa. The frequency of the vibrations is determined by the transducer's inherent mechanical ability to vibrate or resonate. Transducers are usually radially vibrating, ceramic cylinders because this type is inexpensive and approximately omnidirectional at resonance (Stasko and Pincock 1977). After batteries, transducers are often the largest transmitter component.

Hydrophones. The most important component of an ultrasonic system is the underwater microphone or hydrophone (Figure 19.2). Ultrasonic signals from a transmitter produce vibrations in transducers within the hydrophone. The vibrations are converted to electrical impulses that are sent to the receiver. When purchasing equipment, obtain hydrophones and receivers with compatible electronic specifications. If they are purchased from the same manufacturer, they probably will be matched. Better quality hydrophones often can be obtained from other sources, but hydrophone and receiver specifications must match or the units must be tested together.

Hydrophones have different angles (beamwidth) of signal reception and receiving patterns. Omnidirectional (360°) hydrophones are useful for searching and for fixed receiving stations. In quiet environments, they have almost the same signal-to-noise ratios as directional hydrophones. However, in noisy environments you can obtain a better signal-to-noise ratio by using directional hydrophones that block noise from most directions. Smaller beamwidths allow more accurate bearings, but one must be more careful in orienting the hydrophone to detect faint signals. Directional hydrophones have a linear arrangement (array) of transducers or a single transducer surrounded by a cone that directs signals to the transducer. Linear array hydrophones have good sensitivities and receive vertically flattened, fan-shaped patterns. They are more streamlined than are cone-shaped (parabolic) hydrophones and thus less subject to noise induced by boat movement or water currents. Cone hydrophones, on the other hand, have a very low acoustic-to-electrical conversion efficiency, receive a cone-shaped signal pattern, and are inexpensive. To reduce ambient electrical interference, good hydrophones have a shielded, twisted-pair lead cable instead of a coaxial cable and a low-noise preamplifier near the hydrophone to boost the signal through the cable (Stasko and Pincock 1977).

Directional hydrophones are usually mounted on rotatable, vertical pipes that are suspended in the water. If the hydrophone is mounted on a boat, the housing should

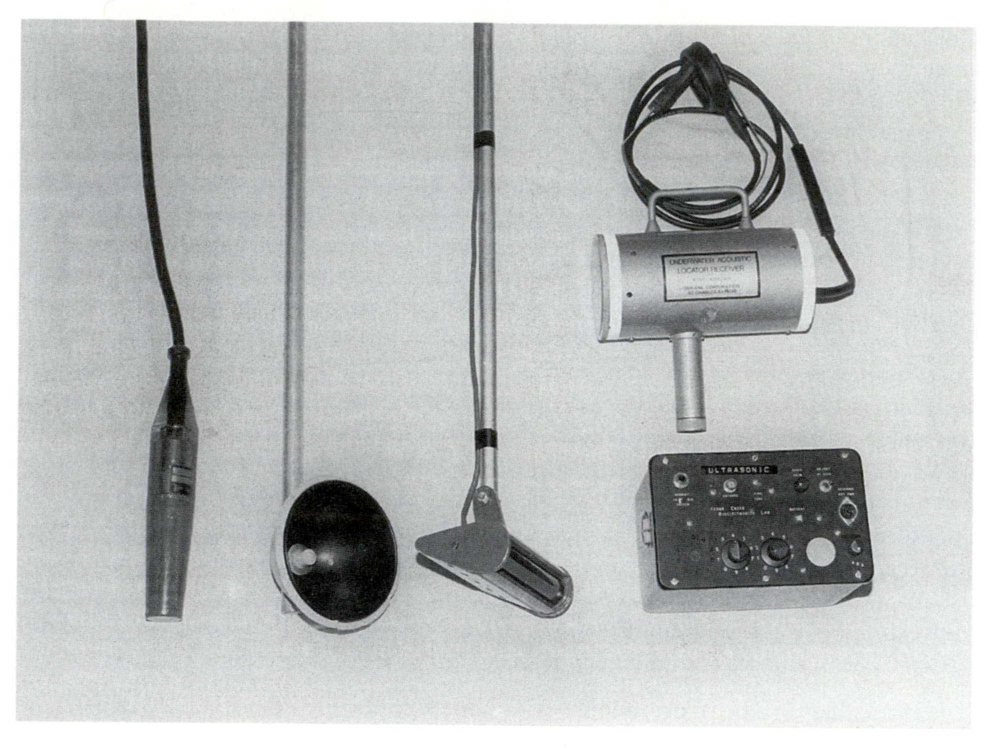

Figure 19.2 Ultrasonic telemetry equipment. Left (from left to right): omnidirectional, cone, and linear array hydrophones. Bottom right: a standard ultrasonic receiver. Top right: an ultrasonic receiver and headset for underwater use.

be streamlined to reduce bubbles and permit monitoring at slow speeds. Hydrophone mounting systems that can withstand moderate boat speeds have been described by Stasko and Polar (1973).

Signal transmission. The choice of frequency depends on the size of the transducer, which in turn affects the package size that an animal must carry and the acoustic output of the transducer. Small transducers produce high frequencies and, therefore, require more power to produce a given signal range; large transducers produce low frequencies and require less power output (Stasko and Pincock 1977). Because ultrasonic transmission characteristics are well known in salt water (Urick 1975), transmitter range can be mathematically estimated for deep water environments (Nelson 1978). Transmission characteristics are not well known for fresh water, but transmitter ranges are greater in fresh water than in salt water.

Ultrasonic signals are affected by spreading, absorption, noise, and other factors. As sound radiates outward from a source, its energy is spread over an increasing area, and the signal strength decreases. In deep water, spreading is approximately spherical, and the sound level decreases about 6 dB each time the distance doubles (Nelson 1978). Sound spreading is independent of frequency, temperature, and salinity. Some of the sound energy is also absorbed by water and converted to heat. For typical signals from ultrasonic transmitters, absorption increases with increasing frequency, increasing salinity, and decreasing temperatures. Spreading losses are much more significant than the frequency-dependent losses to adsorption.

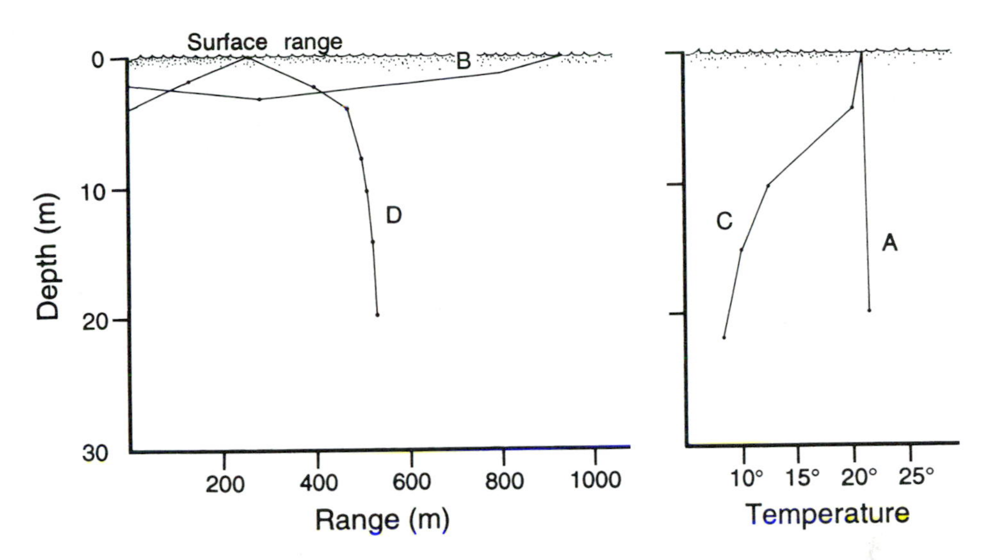

Figure 19.3 Effect of thermal stratification on the signal range of ultrasonic transmitters. Sound travels faster in warm than in cold water, and thus bends away from warmer water. When the water column is isothermal (curve a), sound is bent toward the surface (curve b) because of pressure effects on sound velocity, and the signal range is excellent. If the water body is thermally stratified (curve c), sound is bent downward by the warmer water more than it is bent upward by pressure effects (curve d), and the range is considerably reduced (adapted from Brumbaugh 1980, with permission).

Noise can interfere with signal recognition even though the signal level is strong. Noise can be caused by wind, wave action, boat engines, boat movement, rain drops, ice cracking, and biological sources such as snapping shrimp. Other factors that reflect, scatter, or refract signals cause signal losses that are difficult to calculate. Signal reflections from the bottom and vegetation result in less transmitter range in shallow water than over deep water. Scattering of the signal can be caused by small suspended particles, plankton, fish, and air bubbles from waves. Warm surface water, thermoclines, and temperature gradients cause downward refraction of the signal that can result in the signal not being detected at the surface (Figure 19.3; Brumbaugh 1980). The refracted signal sometimes can be detected by submerging the hydrophone to a deeper level or placing it on the bottom. In addition, submerged objects block signals, causing a sound shadow on one side of the object. Because sources of noise and reflections can render ultrasonic systems ineffective, consider or monitor these factors before beginning ultrasonic telemetry projects.

19.2.4 Radio Telemetry Systems

Transmitters. Radio telemetry systems usually operate from 27 to 300 MHz. In underwater radio transmitters, an oscillator circuit produces electromagnetic vibrations at a frequency determined by a crystal. Signals are transmitted through either circular (loop) or straight wire (whip) antennas. Transmitters with loop antennas are simpler to make because they work in air or water with no additional tuning. Whip antennas, on the other hand, require that the prototype transmitters be tuned in the water during construction so that the proper capacitor can be determined for subsequent production of transmitters. Whip antennas can also be tuned by trimming

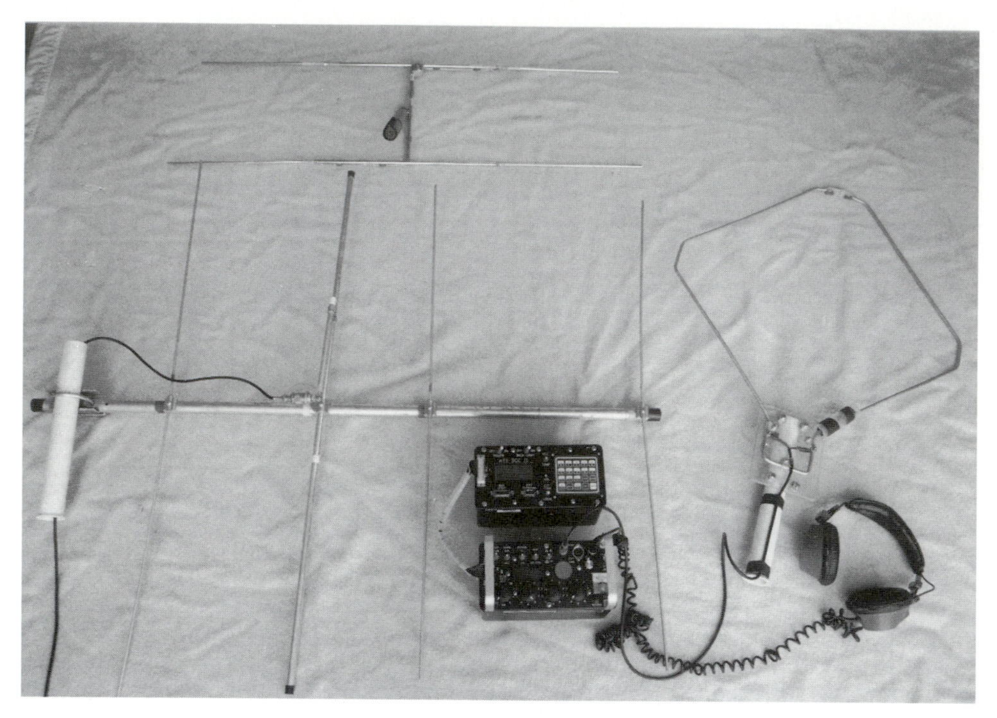

Figure 19.4 A scanning radio receiver, headsets, data collection computer, and three types of antenna. Left, a 4-element, high frequency (151 mHz) Yagi antenna; top, an H-antenna; and right, a loop antenna (49–54 mHz).

the antenna length until the maximum signal is produced. Because the signal transmission pattern of a loop antenna is much more directional than that of a whip antenna, signal levels from loop antennas on moving animals fluctuate greatly and make location of animals more difficult. The transmitter whip antenna should be straight for maximum efficiency and range. Flexible antennas that are implanted in the body cavity or extended out the body cavity bend and are theoretically less efficient. For a given size battery, whip antennas result in a longer, narrower package than do loop antennas and are better suited for implanting in narrow body cavities.

Receiving antennas. Receiving antennas are usually loop, Yagi, H-Adcock, or omnidirectional antennas. Each antenna has special characteristics. All have greater signal detection range as their height increases above the water. Loop antennas are usually small, circular or diamond-shaped metal tubes with a handle for hand carrying or mounting on an airplane (Figure 19.4). A loop antenna is tuned in the field by turning a screw on a variable capacitor until the antenna resonates at the correct frequency and produces the strongest signal. Loop antennas are bidirectional, and maximum signals occur when either edge is pointed at the transmitter. When the plane of the antenna is perpendicular to the transmitter direction, a minimum or no (null) signal occurs. Because the null is easier to distinguish than the maximum signal, listeners generally use the null to take bearings. The signal detection range of a loop antenna is usually less than 1 km. The low range makes it useful for maneuvering close to an animal or recovering lost transmitters. A loop antenna normally has lower sensitivity and optimal performance over a smaller range of frequencies than other receiving antennas.

Figure 19.5 A 53-MHz Yagi radio antenna mounted on a boat.

Yagi antennas consist of a series of small diameter metal tubes (elements) attached perpendicularly to a long metal tube (boom) (Figures 19.1, 19.4, and 19.5). The required lengths of the elements and boom increase with decreasing frequency and increasing wavelength of signals. As you increase the number of elements on Yagi antennas, the accuracy and sensitivity of the Yagi also increases. Yagi antennas are widely used because they have excellent directionalities, sensitivities, and ranges. They are also strong and perform well for long periods without adjustment. Yagi antennas can be mounted on towers, trucks, snowmobiles, and boats (Winter et al. 1978). Because high-frequency (>100 MHz) Yagis are small, they can be hand carried or mounted on airplanes.

The orientation of the Yagi is important. To maximize the reception of radio waves, Yagi antennas should be mounted with the elements vertical because radio waves from underwater transmitters are usually vertically oriented (polarized). Yagis are unidirectional. Usually one Yagi is mounted on a mast, and the direction in which the antenna is pointing is read on a compass disk near the base. Yagi antennas receive a peak signal when the end of the boom with the smallest element is pointed at the transmitter. There also is a null (no signal) point on each side of the direction to the peak signal. Because the nulls are sharper than the peak, bearings are determined by taking the midpoint between the two nulls.

Yagi antennas also can be mounted in pairs; the two Yagis are placed parallel, spaced one or two wavelengths apart, and pointed in the same direction (Mech 1983). This is called a null-peak system. A switch box is used to select whether the signals from both antennas are fed to the receiver in phase or out of phase. When the signals from both antennas are received in phase, there is a slight increase in sensitivity (gain) over a single Yagi. When both antennas are pointed at the transmitter and the signals are received out of phase, there is very sharp null because of signal cancellation. Therefore, null-peak systems have much better accuracy and repeatability of bearings than does a single Yagi. Precision on bearings to a transmitter are within ±3° for a single Yagi and within ±1° for null-peak systems; however, the error in distance estimates increases with distance from the transmitter.

The H and H-Adcock antennas are very popular because they are easily hand carried and some are collapsible. The H-Adcock design has two 1.5-m cross elements attached to a 1.5-m boom (Kenward 1987). Signals from the elements have a 180° phase difference that produces a sharp null when the elements are equidistant from the transmitter. The antenna has good sensitivity and accuracy (±3°); however, the reception is bidirectional. An H-antenna has one of its elements shorter than the other, which makes it unidirectional (Figure 19.4). When the shorter element is pointed toward the transmitter, a maximum signal is received. Based on their size and sensitivity, H and H-Adcock antennas are probably most ideal for signals around 100 MHz (Kenward 1987).

Omnidirectional antennas are usually ground plane antennas or straight wire whips. A ground plane antenna consists of a short, vertical, quarter-wavelength element that has four slightly longer elements placed perpendicular to the base of the vertical element and spaced equally along it (Amlaner 1980). A whip antenna can be a straight, stiff wire, or coaxial cable. These whip antennas are designed for use over metal surfaces, typically a motor vehicle roof, where the metal acts as the ground for the signal. Omnidirectional antennas are used for quickly finding a transmitter signal or for recording telemetry data where null points must be avoided. They have relatively low ranges. Short coaxial whip antennas can be used to pinpoint transmitters, locate fish along tailraces of dams, and record fish passage through fish ladders (Stuehrenberg et al. 1990).

Signal transmission. Transmission of radio signals depends on transmitter depth, water conductivity, signal frequency, reflection, and noise. As the depth of the transmitter increases, transmitter range decreases almost exponentially. Only those signals that travel almost vertically through the water will emerge and travel through air. At the water–air interface, signals from a typical horizontally oriented underwater antenna are changed to vertically oriented radio waves. Because transmitter range also decreases with increasing conductivity, radiotelemetry is suitable for conductivities above 400–600 μS/cm only if the animal remains close to the surface. The effects of increased depth and conductivity (Figure 19.6) are greater for frequencies higher than 100 MHz, which are commonly used in terrestrial telemetry, than for frequencies of 30–60 MHz, which are commonly used in underwater telemetry (Eiler 1990). If the water has low conductivity (<100 μS/cm) and shallow depth, one can use high frequencies (>100 MHz) and a high-frequency, hand-carried Yagi antenna, which has greater range than a loop antenna.

Losses in radio signal strength and errors in determining signal direction also occur when signals are reflected off vegetation, terrain, and buildings. Furthermore, noise from power lines, ignition systems, and citizens band radios can interfere with signal detection. Therefore, one should use a receiver to monitor the study area for noises before beginning a radiotelemetry project and should avoid using transmitters with the same frequencies as the noises. In addition, antennas should be installed in an open area if possible and oriented with the directional end away from sources of noise.

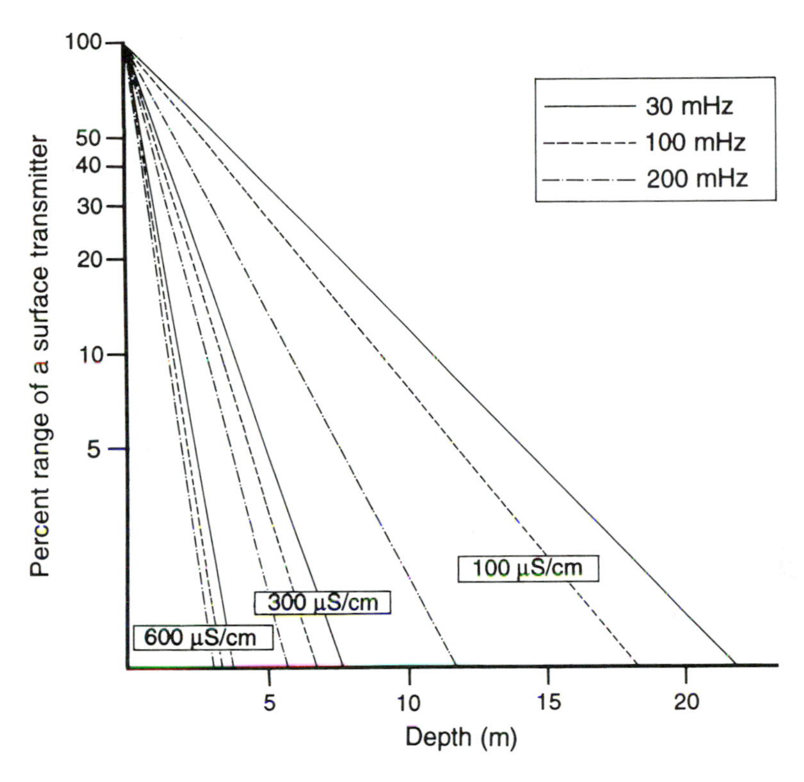

Figure 19.6 Attenuation of different frequencies of radio signals with increasing water depth and conductivity. The curves were generated by a plane wave equation from theoretical data on the properties of water (adapted from V. B. Kuechle, unpublished data).

19.2.5 Coded and Sensing Transmitters

Signals from ultrasonic and radio transmitters can relay information in addition to the animal's location and are called coded signals. Coded signals can provide information on an animal's identity and, from sensors in the transmitter, on physiological, behavioral, or environmental variables. Signals can be coded by differences in pulse rate, pulse interval, signal amplitude, frequency, and so on.

Signal coding is often used to identify tagged individuals. Ultrasonic transmitters have historically used differences in pulse rates, unique pulse sequences, or unique signals on one of several channels to identify transmitters. Because many studies today are tracking hundreds of fish, such as salmon migrating upstream, there is not sufficient time to search many unique frequencies for the presence of individuals or to collect much data from each fish. To increase monitoring time, some studies use fewer frequencies but each with many transmitters that are coded by differences in pulse rates at the same frequency (Eiler 1990; Stuehrenberg et al. 1990). Furthermore, this method can be used with receivers that cannot separate huge numbers of closely spaced frequencies. Studies with multiple codes per frequency depend on only a limited number of transmitters being at the same location at the same time. If the pulses from different tags overlap, the identity or data transmitted during the overlap is lost.

Transmitters that sense environmental and physiological variables provide valuable tools for understanding animal adaptations and requirements. Generally, ultrasonic and radio transmitters can be built to measure the same variables. Transmitters that measure temperature and motion are readily available from commercial sources. Many other types have been built, but their designs and production methods usually need improvement (Williams and White 1990). Sensing transmitters (except temperature) and recording equipment seem expensive; however, the cost per unit of information obtained often is lower than that of location telemetry.

Monitoring water temperature is valuable because the data can be used to relate animal movements to temperature and to reveal animal depth distribution in stratified waters during the summer. A temperature-sensing transmitter can be produced by replacing a fixed resistor in a location tag with a thermistor whose resistance changes with temperature (Nelson 1978). Temperature is usually coded by changes in pulse rate or interval between pulses. Because the circuitry is simple, temperature tags can be produced reliably and at a cost slightly greater than location tags.

Before attachment, each transmitter must be calibrated at known temperatures in a water bath. Although the temperature–pulse relationship is nonlinear, regression curves or equations can be developed to convert each tag's pulse rate or interval to temperature. Temperature transmitters can detect changes of ±0.1°C; however, the accuracy of the tag in measuring temperature is often ±1.0°C. Accuracy is largely determined by how much the tag will drift from the calibration curve with time. Water temperatures can be measured by an external transmitter (Ross and Siniff 1982) or by an implanted transmitter with a thermistor extending through the body wall (Rochelle 1974). In addition, the temperatures of specific body locations can be determined by inserting probes (Standora 1977; Prepejchal et al. 1980; Standora et al. 1984).

Motion-sensing transmitters are readily available commercially and are being used in fish studies (Eiler 1990). These transmitters have mercury switches that detect motion and alter the pulse rate. In a mortality-sensing tag, the transmitter pulses at the normal rate when the animal is active. If the fish does not move for a specified period of time (e.g., 6 h), a timing device is not reset, and the transmitter pulse rate doubles. In an activity-sensing tag, the transmitter produces a slow pulse rate when the animal is inactive and faster pulse rates when the animal is active. Biologists and engineers need to do more studies to be able to correlate specific behaviors with pulse rates.

Transmitters that measure water pressure give a direct indication of an animal's depth (Luke et al. 1973; Ross et al. 1979). In pressure-sensing transmitters, a strain gauge transducer responds to changes in water pressure and correspondingly alters circuit resistance and pulse characteristics. Pressure can be converted to depth because with each meter of depth, water pressure increases about 0.10 kg/cm^2. Each tag must be calibrated in a pressure chamber. Because the pressure transducer is expensive and the circuits are complex, pressure tags are quite expensive. Pressure tags can now be built small and with lives approaching those of location tags.

Many other variables have been or can be monitored from free-ranging aquatic animals. Some of these are salinity (Mani and Priede 1980), light (Nelson 1978), swimming speed (Nelson 1978), tail beat frequency (Ross et al. 1981), swimming activity (electromyograms) from sensors in muscles (Luke et al. 1979; Rogers et al. 1984), swimming direction (Nelson 1978), and heart beat and electrical activity (Priede and Young 1977; Scharold and Gruber 1991).

Sensing transmitters can be built to transmit several variables, that is, they are multichannel systems (Nelson 1978). The most common method is to transmit the information from each sensor (channel) sequentially by use of one transmitter frequency. One channel is used for a reference code. Switching of channels (multiplexing) can be rapid or slow. In rapid multiplexing, each sensor transmits for one pulse interval. In slow multiplexing, each sensor transmits for a fixed time interval, for example 5 s. Multichannel transmitters have been described by Ferrel et al. (1974) and Prepejchal et al. (1980). The simplest but most time-consuming method for recording data from sensing transmitters is to time a certain number of pulses or pulse intervals with a stopwatch or electronic pulse counter. These data also can be recorded automatically (see section 19.4.7).

A recent innovation is the use of microcontrollers, or microprocessors, to control transmitter on and off time, provide customized identification codes, and provide other time functions. Most microcontroller transmitters use a watch-type quartz crystal that can control timing very accurately. Because these transmitters contain a microcontroller, they can be programmed for specific on and off cycles. By turning the transmitter off regularly, such as at night or over the winter, battery life can be greatly conserved. The microcontroller can also be programmed to provide unique identification pulse codes or to duplicate codes specified by a regulatory agency. In addition, the transmitters can be easily programmed for the time functions necessary in activity- or mortality-sensing tags. Because the transmitter options can now be changed by the software instead of choosing among tags with fixed options, options are more numerous and less expensive. Presently, microcontroller transmitters weigh as little as 1.5 g. A key limitation on tag size is that existing microcontrollers require at least 1.5 V. Therefore, more than one silver oxide battery must be used.

19.2.6 Selection of a Supplier

Choosing a supplier of telemetry equipment is a crucial decision because equipment, performance, and service varies greatly among telemetry manufacturers. The first step is to review the literature for names of researchers who frequently use telemetry. They can recommend reliable equipment and manufacturers.

Researchers sometimes consider whether to make their equipment or to buy it from an established telemetry company. Although commercial equipment seems expensive, an investigator's time is better spent collecting data than building equipment. Avoid contracting equipment development to companies, friends, colleagues, or students not engaged in the business of animal tracking. This has too often led to reinventing the wheel and usually does not pay off in reliability or service. If the desired equipment does not exist, most commercial telemetry companies are interested in contracting to develop it. One must also be careful of companies that adapt equipment developed for some other purpose (e.g., marking underwater salvage objects or tracking terrestrial wildlife) with little modification to aquatic animals. Reliable firms usually build a large number of transmitters specifically for aquatic animals and consider aquatic animal tracking an important part of their business.

There are several things to evaluate about a telemetry company based on their reputation and what they promise to do. A firm should fill orders quickly, repair the equipment quickly, and loan replacement equipment. Because technical problems arise during planning or conducting a project, a good company will provide advice and visit the site to solve a problem. Good firms also should be willing to adapt or build equipment for the project rather than change your project to fit their equipment. Finally, the supplier should provide instructions and circuit diagrams with the equipment. Most problems that arise in the field are simple and can be solved with basic information. Some electronic problems can be easily solved by a local radio shop if its personnel have circuit diagrams to follow.

19.3 METHODS OF ATTACHING TRANSMITTERS

The transmitter attachment method depends on the morphology and behavior of the species, the nature of the aquatic ecosystem, and the objectives of the project. Many techniques have been tried (Winter 1983), and many species have been tagged (Table 19.1). Dart tags are useful for big fishes and whales that cannot be handled. Harnesses and epoxy are used on crabs and lobsters. Transmitters are easily attached to turtles with epoxy or screws. Seals can be tagged with tail collars and with transmitters glued to the fur. Collars and head mounts are used on alligators and crocodiles. Many species have been tagged with buoyant transmitters that are towed at the surface or below it. In addition, externally mounted or towed transmitters have been devised to drop off animals after the tracking is completed. The release is accomplished by corrodible metal links in salt water and biodegradable twine in fresh water. Deep-sea fishes can be fed transmitters that are embedded in bait.

The most widely used attachment methods for fish are external attachment, stomach insertion, and surgical implantation. Investigators have designed these methods to have minimal effects on fish and have observed tagged fish in tanks; however, few studies have experimentally and quantitatively tested for effects of transmitters on fishes (Lewis and Muntz 1984; Mellas and Haynes 1985). Before attaching transmitters to experimental animals, one should practice attachment methods on dead and captive animals. This practice should include dissection and necropsy to understand how tag placement relates to the anatomy. Fish generally should not be equipped with transmitters that weigh more than 1.25% in water or 2% in air of the fish's weight out of water.

19.3.1 External Transmitters

External attachment is quicker and easier than surgical implantations and can be used for fish that are spawning and feeding. Because external attachment allows the organism to recover quickly after tagging, it is useful if animals must be released immediately or data must be collected over a short period. External transmitters are also necessary for sensing some environmental variables such as water temperature. Because of the higher center of gravity caused by dorsal external transmitters, animals must make greater compensation for balance than when other methods are used. In addition, external transmitters cause increased drag on swimming organisms. Transmitter drag usually does not affect basic movement patterns but could affect swimming speed or energy expenditure. The balance and drag problems can be alleviated by reducing the size of the transmitter. Balance problems also can be alleviated if the transmitter is constructed in two equal-sized packages that can be mounted opposite to each other on the fish's back (Lewis and Muntz 1984). Finally, external transmitters can cause abrasions on the animal, and they can snag on vegetation, monofilament fishing line, and other objects.

The external attachment method with the widest application attaches the transmitter alongside the dorsal fin by means of two stainless steel or teflon-coated

Table 19.1 Methods of attaching transmitters to aquatic animals.

Method	Taxon	Reference
	External	
Along dorsal fin (pins and wires)	Walleye, bass, perch	Winter et al. (1978)
	Perch, northern pike	Ross and Siniff (1982)
	White sturgeon	Haynes et al. (1978)
+ Corrodible release	Shortnose sturgeon	Buckley and Kynard (1985)
	Porpoise, dolphin, whale	Leatherwood and Evans (1979)
Straps and pins	Salmon	Hallock et al. (1970)
Saddle	Rainbow trout	Winter (1976)
Hook and leader	White bass	Henderson et al. (1966)
Radio towed on the surface	Basking shark, loggerhead turtle, manatee, sperm whale	Priede and French (1991)
Dart	Angel shark	Standora and Nelson (1977)
	Whale	Leatherwood and Evans (1979)
Alligator clips (dorsal fin)	Cutthroat trout	McCleave and Horrall (1970)
+ Biodegradable twine	Lake trout	Lee (1984)
Harness	Red king crab	Rusanowski et al. (1990)
	American lobster	Lund and Lockwood (1970)
+ Corrodible release	Green sea turtles	Ireland (1980)
	Leatherback turtle	Keinath and Musick (1993)
Screws	Turtle	Ireland and Kanwisher (1978)
Nylon cable ties	Leatherback turtle	Standora et al. (1984)
Epoxy, glue	Turtle	Ireland and Kanwisher (1978)
	Lobster	Collins and Jensen (1992)
To the fur	Seals	Fedak et al. (1983)
	Harp and hooded seals	Folkow and Blix (1992)
Sutures	Weddell seal	Siniff et al. (1971)
Collar		
Flipper	Weddell seal	Siniff (1970)
Neck	American alligator	Standora (1977)
Head	Estuarine crocodile	Yerbury (1980)
	Stomach	
Ultrasonic transmitter	White bass	Henderson et al. (1966)
Ingested in bait	Grenadiers	Armstrong et al. (1992)
Radio whip antenna		
Out gill cavity	Atlantic salmon	McCleave et al. (1978)
Anchored in mouth	Salmon	Monan et al. (1975)
Out behind maxillary	Salmon	Gray and Haynes (1979)
	Implantation	
Ultrasonic transmitter	Flathead catfish	Hart and Summerfelt (1975)
External thermistor	Striped bass	Coutant and Carroll (1980)
Radio whip antenna		
Out of body cavity	Northern pike, suckers	Ross and Kleiner (1982)
In body cavity	Walleye	Einhouse (1981)
Radio loop antenna		
In body cavity	Frog	Seitz et al. (1992)
	Water snake	Tiebout and Cary (1987)
Radio transmitter		
Under the skin	Weddell seal	Siniff (1970)

electrical wires (Figure 19.7). A surgical needle can be used to pass the wires through the adipose tissue beneath the dorsal fin and between the pterygiophore bones. An alternative method is to insert large hypodermic needles through the fish from one side and thread the wires through the needles from the other side. After removing the hypodermic or surgical needles, the transmitter is pulled snug against the fish, and the wires are threaded through holes in a soft plastic plate (see Chapter 12).

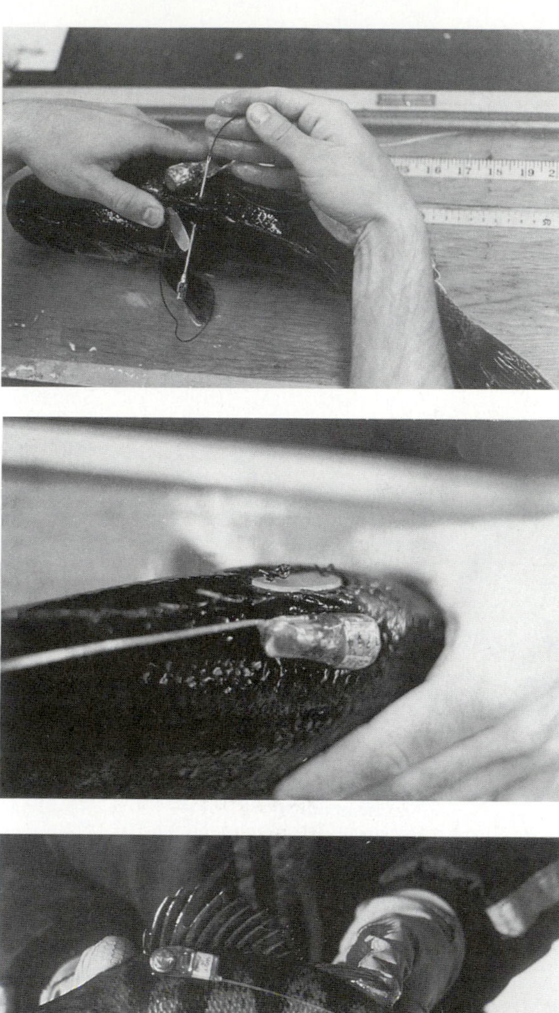

Figure 19.7 Procedures for attaching an external transmitter next to the dorsal fin of a fish. **(Top)** Threading an attachment wire through a hypodermic needle that passes through the fish's body beneath the fin. **(Middle)** Dorsal view of a transmitter and its plastic anchoring plate. **(Bottom)** Lateral view of an external radio and its whip antenna on a yellow perch.

When the transmitter and plate are positioned against the fish, knots are tied on the wires to prevent the wires from pulling through the fish. A neoprene pad placed between the plate and the fish will prevent abrasion. Furthermore, one should attach a radio transmitter so that the whip antenna does not extend beyond the caudal fin.

19.3.2 Stomach-Inserted Transmitters

Internal transmitters do not cause drag, cannot become snagged, and are less likely to cause abrasion. Because an internal transmitter is below a fish's center of gravity, one can use a heavier package internally than externally without creating balance problems. Stomach insertion can be done quickly and probably requires the shortest habituation time by fish. The disadvantages of stomach insertion are that transmitters may be difficult to get into the fish's mouth or past the pharyngeal food-crushing structure, may be regurgitated by many species, and may rupture the esophagus or stomach (Winter 1983; Nielsen 1992). In addition, external variables are difficult to monitor with stomach tags.

Stomach tags are inserted through the mouth into the stomach by means of tubes or ingested bait. One can use a plunger system consisting of a tube within a tube or a veterinarian's bolling gun. These tubes should be coated with glycerin so that they slide through the gullet smoothly, but one must be careful not to force the tube. Radio whip antennas are extended out the gullet and anchored with a barb in the roof of the mouth, passed through the tissue behind the maxillary, or passed out the gill cavity (Table 19.1).

19.3.3 Surgically Implanted Transmitters

Surgical implantation into the body cavity is excellent for physiological telemetry (e.g., heart rate) and is the best method for long-term attachment. However, external factors are also difficult to monitor unless a sensor is passed through the body wall. Implantation takes longer to perform and requires a longer recovery period. Usually, it is best to hold animals a day before surgery to ensure that the capture method has not caused much stress and also a day after surgery to ensure that they are recovering properly before release. In some species and at remote locations, releasing the fish quickly may be preferable to transporting or confining them. Surgical implants are more likely to cause infection than are other methods, especially in warm water, and incisions can be slow to heal in cold water. Although eggs can be removed easily from an anesthetized fish to create space for a transmitter, implants may not be suitable for some gravid females. Mortality of fish from implantation is higher after spawning than during other times. Therefore, the best times to implant in most fish are in the spring and early fall when the water is cool and after spawning when fish have recovered their body condition.

Surgical implantation (Figure 19.8) is not difficult and does not require elaborate equipment; however, it does require some planning (Hart and Summerfelt 1975). Experiment before beginning the field work to find the dosage of anesthetic (Chapter 5) that will keep an animal anesthetized for about 15 min and allow recovery in 15 min or less. Few anesthetics are approved by the U.S. Food and Drug Administration for use with fish, and one should consult experts to determine what is the best method currently available. Second, learn how to suture (e.g., from a veterinarian) and then practice on a piece of cloth stretched over the open end of a can. Poor suturing technique is time-consuming and often the cause of lost transmitters. Use of surgical staples is quicker and possibly as reliable as sutures (Mortensen 1990), whereas use of adhesives may not be as reliable (Petering and Johnson 1991).

Fish can be placed in a V-shaped tagging trough and water passed over the gills via a siphon system or squeeze bottle. An incision is made slightly to one side of the midventral line to avoid blood vessels and is made just large enough to admit the transmitter. The incision location varies among species but should be in the area

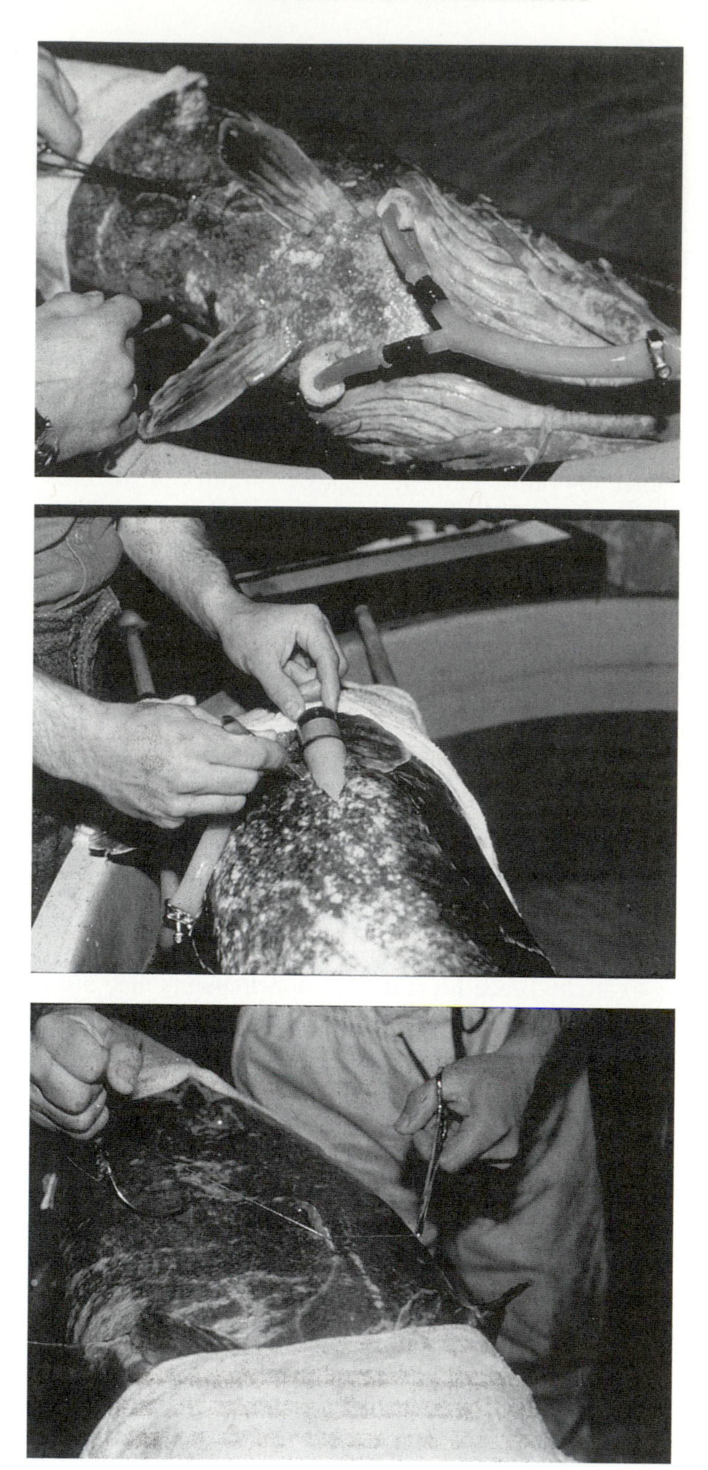

Figure 19.8 Surgical implantation of a transmitter into the body cavity of an Antarctic toothfish. **(Top)** Siphon for bathing gill cavities with an anesthetic or water. **(Middle)** Inserting the transmitter into the body cavity. **(Bottom)** Suturing the incision with monofilament line.

where the body is the deepest and not be near vital organs such as the heart or liver. Keep the instruments sterile by soaking them in Zephiran chloride, alcohol, or another antiseptic and keep water out of the incision. One can apply antibiotics into the body cavity, but this is not necessary if reasonably sterile techniques are used. After making an incision, an ultrasonic transmitter or a radio transmitter with a loop antenna is placed into the body cavity. Whip antennas or sensing probes can be passed through the body wall and trailed (Ross and Kleiner 1982). If a fish is large, a flexible whip antenna can be placed inside the body cavity by means of a hemostat forceps before the transmitter package is implanted. It is questionable whether extending a whip antenna through the body wall increases the signal range much because a radio signal is only slightly attenuated by the fish's body and the external whip is rarely straight. After implanting the transmitter, the incision is sutured. Although monofilament suture material is available from medical supply houses, 3-kg monofilament fishing line works as well. Incisions can be closed by one continuous suture for speed, or each suture can be tied off separately with square knots for greater reliability. For most fish, the muscle and skin are usually sutured together; for large fish, the muscle should be sutured first and then the skin with finer material. The needle must pass completely through the muscle so that the sutures will not pull the muscle through the incision.

Implanted transmitters can be lost through transintestinal expulsion. In this process, the transmitter is encapsulated by tissues, absorbed through the intestinal wall, and expelled out the anus. The encapsulation process is a response to a foreign body or an irritant (Marty and Summerfelt 1988). Transintestinal expulsion has been documented in channel catfish (Summerfelt and Mosier 1984) and rainbow trout (Chisholm and Hubert 1985; Helm and Tyus 1992). However, we do not know how many species exhibit this phenomenon. Transintestinal expulsion has not been a problem in long-term studies of species such as walleye (Einhouse 1981), Colorado squawfish, and razorback sucker (Tyus 1988). Large transmitters and transmitters coated with certain substances (e.g., paraffin) are more likely to be expelled. To minimize the likelihood of expulsion, a transmitter should be smooth and relatively small and not have a sticky coating (e.g., poorly cured epoxy). The potting substance should be inert, such as high-grade medical epoxy or silicone, beeswax (Helm and Tyus 1992), or ABS tubes. The problem of transintestinal expulsion will not eliminate the usefulness of implanted transmitters in most fish species. However, before using implanted transmitters on a particular species, ask other researchers if expulsion is a problem in that species, conduct a preliminary experiment, or order extra transmitters to compensate for possible tag loss.

19.4 METHODS OF TRACKING

Aquatic animals can be located by boat, airplane, triangulation, appearance at fixed stations, automatic tracking systems, and satellites. Data can be recorded manually or automatically. The choice of the tracking system depends on the habitat characteristics and size, animal behavior, animal mobility, number of specimens needed, and the amount of information required.

19.4.1 Boat

Tracking by boat usually involves stopping at specific locations and listening for transmitter signals. The engine must be at low revolutions per minute or off because of engine electrical interference; inboard engines usually have better noise suppression than outboard motors. Once a signal is located, the boat is often positioned above the animal. This provides a very accurate location, depending on transmitter depth, and allows limnological measurements at the site. As a tracker in a boat approaches an animal, the receiver volume and sensitivity (gain) must be continually adjusted downward to obtain nulls. The hydrophone or Yagi antenna will go from direction ahead while approaching the animal to an omnidirectional signal above the animal to direction astern after the boat passes over the animal. Hydrophones are more accurate at close distances (<3–4 m) than is a Yagi (<15 m); however, accurate (<8 m) radio locations can be made by switching to a low-gain coaxial whip antenna. In shallow water, data can be biased because the boat can disturb the animal. Disturbance will not alter home range or migration patterns if location procedures are not used very frequently. However, if movements are being followed continuously, then the animal's direction of movement or behavior could be affected. Some investigators reduce disturbance by attempting to stay a certain distance from the fish as estimated by signal strength, but this can result in considerable location error.

Once you know the animal's location, you must locate this site on a map. The animal's location is usually determined from landmarks by visual sightings, compass readings, sextant angles, and radar readings. Loran C radio readings on terrestrial radio beacons and satellite global positioning systems (GPSs) are now available (see section 19.4.6). Sextants, radar, loran C, and GPSs are recommended because they are the most accurate methods and least affected by waves and weather. Sextants are the least expensive ($150) and are used to measure the angles between prominent landmarks very accurately and precisely (±0.5°). The GPS (or NAVSTAR) is a satellite navigational system that also can be used to determine animal locations (Degler and Tomkiewicz 1993; see section 19.4.6). Unlike the Argos system (section 19.4.6), the satellite is the transmitter and sends a signal to a GPS receiver on the ground. The GPS receiver computes its location by measuring the time delay in arrival of synchronous signals from four satellites. The system was developed in 1973 and consists of 24 satellites that orbit the earth every 12 h. A given point on earth has continuous coverage. The GPS allows calculation of latitude, longitude, and altitude of the receiver. The GPS can be used to determine the location of an animal if the GPS receiver is on a boat, airplane, or person at the animal's location. The accuracy is about ±20–30 m and should improve greatly very soon. Portable GPS units are available commercially for $500–1,000 (1996 costs).

19.4.2 Airplane

Tracking from an airplane is useful for locating highly mobile radio-tagged animals (Mech 1983). Airplanes provide the greatest detection range for radio signals, usually double or triple that of a Yagi on the ground, but the cost is high. Because airplanes travel rapidly, a tracker can miss an animal while checking many frequencies. In addition, there may not be enough time to detect a weak signal from a deep transmitter. To minimize these problems, a receiver in which specific frequencies can be programmed and quickly scanned should be used. Approval is necessary to attach antennas to an airplane and can be obtained at the airport from a mechanic certified (in the United States) by the Federal Aviation Administration. Gilmer et al. (1981) recommended side orientation for high-frequency Yagis and either side orientation or a combination of side and front orientation for low-frequency loop antennas. To detect weak signals, use a plane with the ignition noise suppressed. In addition, use padded monophonic headsets that fit tightly around the ears to reduce interference from noises in the plane. Transmitter locations are determined by switching between antennas and listening for the maximum signal as the plane passes over the transmitter. In addition, circles of decreasing circumference can be flown until the receiver volume and gain are turned as low as possible (Mech 1983). Locations can be determined to within 100 m.

19.4.3 Triangulation

Animals can be located from a distance by taking bearings from two or more locations (triangulation). Bearings can be made by moving the receiving unit between locations or by using several stationary hydrophones or directional radio antennas. Bearings cannot be plotted accurately on a map unless the sites from which the bearings were taken are mapped accurately. Because the most accurate locations occur when bearings intersect at 90°, move mobile antennas to achieve bearings as close to this angle as possible. Fixed stations should be located so that it is not possible for the animal to be near the imaginary line between the stations but very likely for the animal to be near the imaginary perpendicular bisector (most accurate). Moving animals can cause great error in triangulation; thus, the most accurate triangulations are made if observers communicate and take bearings simultaneously from separate locations.

19.4.4 Appearance at Fixed Stations

Receiving systems and recording units can be set up at certain locations, such as along rivers (Potter et al. 1992; Eiler 1995) or dams (Stuehrenberg et al. 1990). The basic system records the presence or absence of a transmitter frequency near the hydrophone or antenna. In streams, narrow-beam hydrophones or directional Yagis can be pointed across the streambed. Where the animal can be in any direction around the station, omnidirectional antennas or hydrophones are used. Omnidirectional hydrophones and antennas can be arranged in a grid or transect; the animals can then be located according to which hydrophone or antenna receives the strongest signal (Zinnel 1980). Hydrophones can be moored on the ocean bottom and linked by cables to a receiving station on land (Urquhart and Smith 1992). Ultrasonic signals also can be detected by a series of sonar buoys, each containing a hydrophone, ultrasonic receiver, interfacing unit, and radio transmitter. The ultrasonic signals are relayed by radio signals to a receiving station on land (Solomon and Potter 1988). For deep ocean bottoms, a fixed receiving station has been developed that contains twin hydrophones, receiving and recording telemetry equipment, a video camera, and a video recorder (Armstrong et al. 1992). The system floats to the surface after a magnesium–iron link decays and releases the ballast.

19.4.5 Automatic Tracking Systems

Ultrasonic transmitters can be located automatically by measuring the time that it takes the signal to travel through water. A time delay system measures the differences in time that it takes the transmitter signal to reach each hydrophone in an array of three or more fixed hydrophones (Urquhart and Smith 1992). This system is very accurate. However, it depends on having a strong signal to trigger the time counter and having animals whose home ranges are similar to the range that hydrophones can detect signals. A similar method can determine the distance to the animal by measuring the time that it takes to send a signal to a tag and receive a signal back from that tag (transponder). If the direction to the transponder is also known and the transponder returns information on the animal's depth, the animal can be located accurately in three dimensions. Transponders theoretically save battery life because they transmit information only when needed; however, the receiving portion of the transponder, except in high-power-output transponders, often uses sufficient current to cancel the savings.

Automatic radio-tracking systems have been developed that use bearings from Yagi antennas, grids of antennas, signal time delays, and Doppler shifts. Bearings can be measured from continuously rotating Yagi antennas on towers (Cochran et al. 1965; Deat et al. 1980). The oldest system used two large (>21 m) towers with rotating antennas. There were 52 pairs of receivers, one receiver for each animal and tower. The bearings from each tower were initially recorded by lights on continuously moving film and transcribed by hand. Later a computer was added to record bearings. Another system involved establishing a grid of wire antennas over the study area and using a microprocessor-controlled switching system to measure signal strength at each antenna junction (Zinnel 1992). Radio transmitters can be located by the differences in time of signal arrival at different receiving stations. Yerbury's (1980) transmitters on crocodiles sent 10–20-mW signals to the receivers. The accuracy was ±50 m at 5 km. A similar system has been used on large mammals by Lemnell et al. (1983). These transmitters are large (140 g) and have high power outputs. Finally, the Doppler frequency shift of the transmitter signal among antennas can be measured to determine the location of the animal (Burchard 1989; Angerbjorn and Becker 1992). A Doppler frequency shift occurs when the frequency of sound or radio waves from a given source increases or decreases because the source and the observer are in rapid motion with respect to one another. Doppler tracking systems sequentially switch rapidly among four to eight quarter- to half-wavelength whip antennas that are spaced quarter wavelengths apart. This creates or accentuates the frequency shift.

Automatic tracking systems are extremely expensive and most are not easily relocated. The number of individuals and species within the range of the antennas or hydrophones is small; thus, the systems are limited to very economically important or well-funded problems. Because the systems are largely experimental, much effort is needed to make them operational (Kenward 1987; Samuel and Fuller 1994).

19.4.6 Satellite Telemetry

One of the most exciting recent developments has been in techniques to monitor animals by satellite (Taillade 1992). One system uses two Argos–TIROS satellites that are in polar orbits around the earth. Transmitters that communicate with these satellites are called platform transmitter terminals (PTTs). For animals, PTTs are as small as 60–150 g and are usually powered by lithium batteries or solar cells. All PTTs use the same frequency (401.650 MHz), but individual PTTs are identified by a code in their transmitted signal. The signal also can contain information from at least eight sensors connected to the PTT. Depending on the latitude of the tagged animal, these satellites will make 6–28 passes over a given transmitter each day. A PTT transmits messages every 90 s, and a satellite pass over a position lasts about 10–12 min. To locate an animal, the satellite must receive at least two messages. The signal is then transmitted to processing equipment on the ground. Location of a PTT is made by calculating the shift in the frequency of the transmitted signal (Doppler effect) caused by the movement of the satellite with respect to the PTT. The accuracy of locations ranges from ±150 m to several kilometers depending on animal behavior, environmental conditions, and data processing (Samuel and Fuller 1994). The cost of an animal PTT is about $2,000–3,000 each (1996 costs).

Satellites have been used to track movements of remote, wide-ranging animals and to receive data on animal behavior or the environment. Because of the size of the PTTs and their high radio frequencies, aquatic animals must be large and periodically above the surface of the water, such as dolphins (Jennings and Gandy 1980). Whales, basking sharks, and sea turtles have been tracked by PTTs towed on the surface of the water (Priede and French 1991). Other species, such as sea turtles, penguins, and Arctic seals, can be located by satellite when they surface or come out of the water to rest or nest (Ancel et al. 1992; Folkow and Blix 1992; Hays 1992; Keinath and Musick 1993). Battery current in a PTT can be conserved by having a clock or a saltwater switch turn on the PTT when the animal is above the surface.

Satellites have been used to monitor salmon movements in rivers where weather and mountains make airplane tracking too difficult (Eiler 1995). Signals from radio transmitters on salmon are recorded at a series of fixed recording stations as the salmon move upstream (Figure 19.9). The transmitter frequency and code, date and time, hydrological data, and meteorological conditions are then relayed from the recording station to a satellite. The information is downloaded at one of two main global receiving centers (in this case, Camp Springs, Maryland) where it can be accessed by modem from the investigator's distant laboratory.

Commercial GPSs are available that automatically track large animals that are at or above the water surface (see section 19.4.1). The animal carries a package containing a GPS receiver, sensors, a data storage system, a radio location transmitter, and a radio modem to relay data. The package can store GPS locations and data from sensors for up to a year. A tracker can use a portable command module containing a computer, modem, and GPS unit to download data from the animal's package. The package on the animal weighs about 1.8 kg. Satellite telemetry will be used more frequently by aquatic biologists as the technology improves and because it is more cost effective for remote, inhospitable areas than are other methods.

19.4.7 Automatic Data Recording

Signals from transmitters can be automatically decoded and recorded. The most basic system records the presence or absence of a signal on a strip chart (Siniff et al. 1971; Samuel and Fuller 1994) or tape recorder (Lucas et al. 1992). If a manual receiver is used, only one animal can be monitored, but the record can be continuous. Timers also can be used to turn on the recorder and receiver at specific time intervals. If a scanning receiver is used, then signals from many animals can be recorded at specific time intervals. Coded signals can be deciphered by electronic counters or decoders. Counters simply measure the number of pulses in a time interval. Decoders not only count pulses, but they can measure pulse intervals, pulse widths, and signal amplitude. These variables are converted from digital to analog (chart) form for recording. Although these systems reduce labor in data collecting, they require considerable manual transcribing of data for analysis.

A more elaborate method of recording and decoding data involves electronic data loggers, data collection computers, and microcomputers (Pincock 1980; Kuechle et al. 1989). These systems decode the signal and store the data in a microprocessor along with the date, time, and other data not obtained from the tag (e.g., water temperature and antenna identification). Simple data loggers do not allow the

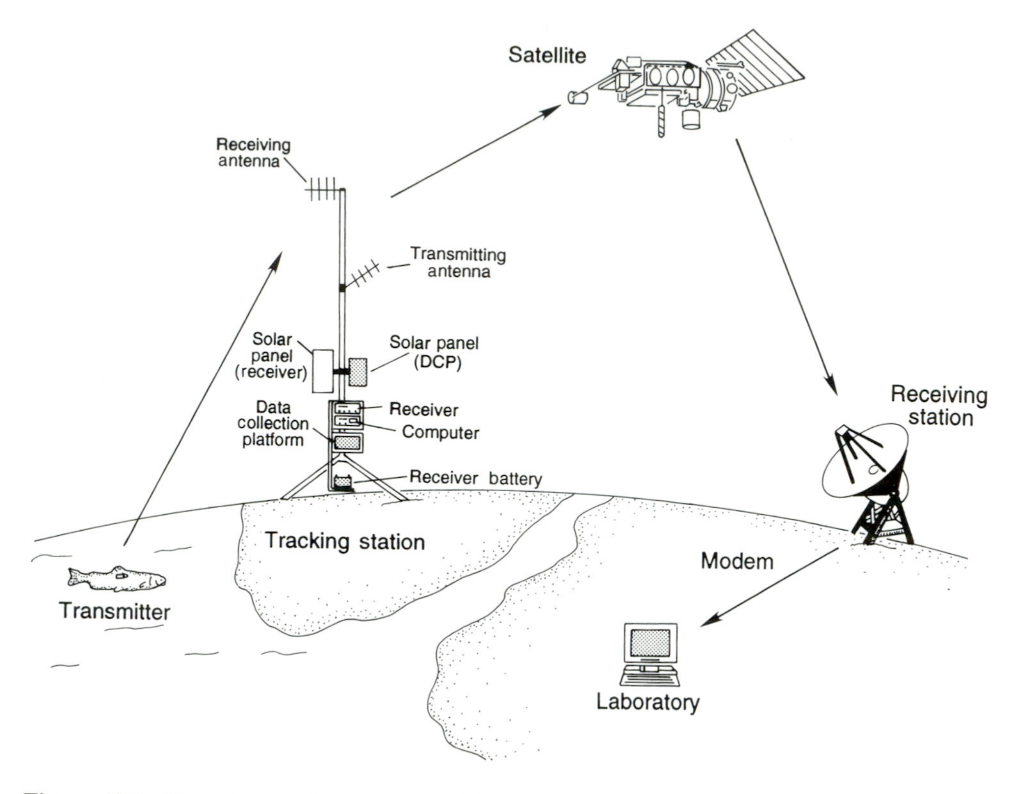

Figure 19.9 Remote tracking station (left) used to detect and record salmon tagged with radio transmitters. Fish data received by the antenna and processed by the receiver and data collection computer are transferred to the data collection platform (DCP). The DCP transmits the data to a satellite (right), which relays them to a receiving station. From a distant laboratory, an investigator retrieves the data with a modem (modified from Eiler 1995).

investigator to reprogram the decoder and microprocessor in the field. Data collection computers are similar to data loggers except that they can interpret data, control the receiver, and differentiate among multiple antennas (Figure 19.4). They have a simple calculator-type keyboard that allows the investigator to pass new instructions to the decoder; however, these decoders have little ability to analyze data while data are being collected. Data loggers or data collection computers interface with personal computers, and the data can be downloaded into standard computer spreadsheets or statistical programs. Data loggers and data computers come ready for use because they are specifically made for telemetry work. They are available from telemetry companies as units built into receivers or as accessories that connect to a scanning receiver. Software is available from the manufacturers to process and transfer data. Built-in data loggers require fewer external cables and connectors than do external data loggers coupled to conventional receivers. If the receiver must also be used in the field, then a conventional receiver is more advantageous because it is lighter, less expensive, and less susceptible to inclement weather and accidents. Commercial data loggers or data computers cost $3,000–4,000 (1996 costs). Finally, a conventional microcomputer can be used to control the receiver and sampling, decode signals, interpret the coded data, display data, reject noise, identify bad data, and store the data (Kuechle et al. 1989; Stuart and Friend 1989; Urquhart and Smith 1992). The computer software can be easily modified in the field, and data can be periodically analyzed while data collection is still in progress. Microcomputers usually require some modification to be used with telemetry systems. Data can be stored on floppy disks, computer tapes, and compact discs.

19.5 SAMPLING AND DATA PROCESSING

19.5.1 Field Considerations

There are many precautions to take when using telemetry equipment. A good rule for telemetry studies is to have a spare of every piece of equipment to avoid data loss because of accidental breakages or failures. Transmitters should be stored in padded boxes because crystals and transducers are very sensitive to jolts. If transmitters are not to be used for weeks, they should be kept in a refrigerator to keep the batteries fresher. It is a good idea to activate long-lived transmitters several days before testing or tagging. Because many batteries develop a coating on their electrodes, this coating must be burned off by use before the battery will deliver full voltage. In addition, if any transmitters are defective, the problem is likely to appear during the first days of activation. Test each transmitter for signal range before attaching it to an animal. This will reduce the chance of attaching a poor transmitter. Because a receiver's frequency selector seldom matches exactly the theoretical transmitter frequency, determine the correct receiver reading for each transmitter before releasing the animals. The correct receiver reading can be determined by disconnecting the antenna and reducing the receiver volume and gain controls until the transmitter is detected at one frequency. Finally, it is useful to carry a deactivated transmitter during tracking to test the receiving system if problems are suspected. However, do not carry an activated transmitter. Its signal may be so strong that it interferes with reception. Furthermore, it may be received at other frequencies and cause a mistake in recognizing an animal.

There are several precautions to take with receivers. Moisture reduces the sensitivity of a receiver, makes the gain control noisier, and shorts out the receiver. Cover receivers with plastic bags to keep moisture out. After use in wet conditions, receivers should be opened and placed in a warm area to dry. Lighter fluid or contact cleaner can be applied to noisy gain control switches. If the receiver is likely to be jarred, a padded carrying case may prevent damage. In the winter, one can place hand warmers in the carrying bag with the receiver to maintain battery and sensitivity performance. Finally, the nicad (nickel–cadmium) batteries in a receiver should be discharged to a low level before recharging, otherwise they will not deliver their full capacity in subsequent use; they develop a "memory" of how much they usually deliver.

Antennas and, to a lesser extent, hydrophones seldom malfunction. When they do, the most likely problem is the connecting cable. If the signal is discontinuous, check the connections. Because antenna cables get bent or twisted, the wire inside the cable may be broken. Replace cables periodically. Receiving systems can be checked in the field in the absence of a test transmitter. If the antenna or hydrophone is working, an increase in noise level should be heard when the cable is connected to the receiver. Running a finger over the hydrophone also should produce noise.

19.5.2 Sample Size Considerations

The number of tagged animals necessary in telemetry studies depends on cost, labor, availability of animals, type of data desired, and methods of statistical analysis. Studies of home range and migration patterns probably benefit more from long-term tracking of individuals than from close daily monitoring. If animals are so widely distributed that they must be located by boat, truck, or airplane, 20–40 individuals are usually the maximum number that can be monitored concurrently. The number can be larger if individuals are located less frequently or if they can be located from fixed receiving stations. Researchers who are interested in the fate of fish, such as where salmon spawn, often use 50–300 radio-transmittered fish. Studies on fish physiology or behavioral responses to environmental factors usually monitor less than 20 individuals concurrently, especially if the investigators use mobile tracking systems, because these studies require very frequent monitoring of individuals.

Before beginning a project, calculate the number of individuals needed for the statistical tests that will likely be used (Chapter 2). Furthermore, order more transmitters than needed for statistical tests because a third or more of the subjects will die, lose the transmitter, get caught by anglers, or suffer transmitter failure. Because sample sizes in underwater biotelemetry are often small, investigators must be careful not to scatter their tagging efforts over many sites, different species, and different size-classes. It is better to answer one question well than many questions poorly.

The number of data points or locations collected depends on the type of study, the difficulty of locating individuals, the choice of whether or not to record data automatically, and the transmitter life, which is less for many types of sensing transmitters than for location tags. Studies on home range and habitat use require that individuals be monitored throughout a season, and each individual is usually located every 6–48 h. If the time between successive locations of an animal is too short (e.g., less than 4–6 h), locations will be dependent on previous locations, or autocorrelated (Cresswell and Smith 1992; Samuel and Fuller 1994). Autocorrelation can bias data analyses. By taking locations at greater time intervals, autocorrelation will be avoided and sampling costs reduced. Subsampling sets of data also can reduce dependence of data points because the time interval between locations is increased. Procedures are available to determine whether a sufficient number of locations have been established to estimate the area of a home range (Winter 1977; Winter and Ross 1982).

The number of data points collected is often very large in studies of fish physiology and behavioral responses or in studies using sensing transmitters. Studies on orientation of animals to physical, chemical, or celestial factors often monitor individuals for a few days or less, and locations are made minutes to an hour apart. Experiments using sensing transmitters and studies on activity patterns usually record data at evenly spaced intervals of an hour or less. If data from sensing transmitters are automatically recorded, the number of data points can be in the thousands.

19.5.3 Methods of Searching

Appropriate methods of searching for animals equipped with transmitters are important in reducing cost, labor, and bias in data. Some trackers begin searches in the place where they last located the animal because this usually saves much time in searching. This method is not good for animals that are very mobile, may occupy deep water, or have unpredictable movement patterns (e.g., walleye) because you may miss detecting them. In addition, there will be considerable bias in determining habitat use because all habitats were not monitored equally.

The best method of searching for fish is to use equally spaced transects and to vary the starting point each day. The distance between transects and monitoring points along the transect should be less than the typical transmitter signal range to reduce the chance of not detecting a transmitter. By varying the starting point, one will avoid finding the same fish at the same time of the day and avoid the misconception that they are always behaving the same way.

19.5.4 Sampling Time

The time of the day when monitoring is done is important. If investigators sample during the same hours every day, they will get a biased view of fish habitat preferences, activity periods, and behavioral preferences. Ideally, all hours should be sampled equally in most studies. A good, practical method is to divide the day into equal sampling periods and try to arrange work schedules so that at the end of 1 or 2 weeks approximately equal time has been allocated to each period. The least biased approach is to choose the days and the hours sampled on a given day from a random numbers table. In practice, this is extremely difficult to do because investigators' schedules are not sufficiently flexible. However, another good method is to combine periodic close monitoring of a few individuals over 24-h periods with the sampling system normally used for all individuals. This will verify if the usual sampling periods are representative of the animal's typical activity. If investigators have good background information on when a behavior occurs, they may allocate all the monitoring time to a specific period (e.g., feeding) or behavior (e.g., spawning). This approach is very thorough and cost effective.

19.5.5 Data Plots

To process movement data quantitatively, bearings are often drawn on a plot board. Plot boards have a Cartesian coordinate system (x- and y-axes) constructed on a map that has an accurate representation of the shoreline. From a plot board, each animal location is given x- and y-coordinate values. Alternatively, a computer program can be used to calculate x- and y-coordinates by use of angular bearings from known receiver locations. Commercial and public domain programs are available to perform this conversion (Larkin and Halkin 1994). Coordinates can be entered into a computer file or spreadsheet along with information on the fish's identification number, date, time, weather conditions, water characteristics, habitat type, and environmental or physiological variables monitored by the transmitter on the animal. Popular computer graphic and statistical programs can plot scattergrams (x-y data) that can be overlaid on a lake map. Many home range analysis programs also have plotting capabilities (see section 19.5.6). The scale of the axes must be the same as that of the map to overlay plots of animal movements on the map. Sometimes, however, it is difficult to get the scales to match. In this case, one can determine the coordinates for many points along the shoreline and use the computer plotting program to draw a map of the lake to the correct scale. Computer programs developed for geographic information systems can be useful in preparing habitat maps to overlay on maps of movement patterns (White and Garrott 1990; Samuel and Fuller 1994).

19.5.6 Data Analysis

Because telemetry can generate large amounts of data, analysis of data is usually done with a computer. The comprehensive book by White and Garrott (1990) is very helpful for project design and data analysis. One of the most basic measures in movement studies is the distance between locations. A computer program to determine distance between points on a Cartesian coordinate system can be written easily from a formula found in most algebra books. The program can calculate distances from the previous location to determine swimming speed and activity or from fixed locations such as a stream mouth, underwater structure, or industrial discharge site to determine behavioral responses. Behavioral interactions between individuals can be analyzed by calculating the distances between individuals at specific times. Studies on animal orientation usually calculate the angle of movement with respect to the previous locations or some stimulus. Batschelet (1965) summarized statistical analysis of data on orientation and biological rhythms.

Home range areas and habitat use are determined by convex polygons, grid cell methods, probability models, harmonic means, Fourier transformations, Kernel methods, and other methods (Macdonald et al. 1980; Samuel and Fuller 1994). The simplest and oldest method involves drawing a convexed polygon by connecting the peripheral locations. The area of the polygon is determined by a planimeter (see Chapter 4) or by a computer. Home range size also can be determined from a gridded map by summing the areas of squares that contain plotted locations (Winter 1977; Winter and Ross 1982). Habitat use can be determined from the number of locations in grid squares containing each habitat type. Probability models and harmonic means are used because their area estimates do not change greatly with increasing sample size compared with convexed polygons and grid cell methods. There are many cost-free programs available for microcomputers that analyze home range sizes, rates of movement, and habitat use (White and Garrott 1990; Larkin and Halkin 1994).

There are some important statistical considerations before movement data or data from sensing transmitters are analyzed. To apply statistics properly, one must define the population of interest, the experimental unit, and the hypotheses to be tested. Because huge numbers of locations or data points can be collected on an individual by telemetry, people sometimes wrongly conclude that the data points are the samples and sample sizes are larger than they really are. The experimental units are the animals, not the number of locations or the environmental measurements. The animals are the units that are grouped (e.g., by sex or age) for hypothesis testing or to receive treatments. Pseudoreplication is another problem that can occur. This is the use of inferential statistics to test for treatment effects on data from experiments in which the treatments are not replicated or replicates are not statistically independent (Hurlbert 1984). Many components of a study can be replicated, but the experimental unit is the level at which replication is obligatory. Because automatic-recording studies can have huge data sets, autocorrelated data points, and missing data points, one may need to subsample the data set to obtain a more manageable, better-quality data set.

Another statistical problem is that certain variables such as home range size, distance moved upstream, or mean depth selection may not be normally distributed. One should plot the data and test for skewness in the data. If the data are skewed, examine some other measure of central tendency, such as the median, or transform the data. In addition, use nonparametric statistics to make comparisons when data are skewed. If the variable measured for the experimental units or animal is normally distributed and the sample size is large, then parametric statistics can be used.

Another consideration is that the data must be presented in units of measure appropriate to the way data were collected. For example, investigators often determine locations on animals hours apart, yet some report the animal's rate of movement in centimeters per second or meters per minute for comparison with data measured more exactly in the laboratory. Lastly, statistical techniques are used to decode sensing transmitters. Computer regression programs are used to make calibration curves and convert pulse interval times to the environmental variable.

19.5.7 Outlook

The underwater biotelemetry field has made some outstanding advances over the last 10 years. Advances should continue rapidly in the future as technology becomes available from the consumer electronics market, the biomedical field, the space industry, and the military. In addition, important ecological research areas such as environmental disasters, endangered species, and problems in economically important populations will provide some stimuli and funds to develop new techniques. However, many applications of telemetry currently are limited only by the imaginations of the investigators on how to study a problem.

19.6 REFERENCES

Amlaner, C. J., Jr. 1980. The design of antennas for use in radio telemetry. Pages 251–261 *in* Amlaner and Macdonald (1980).

Amlaner, C. J., Jr., editor. 1989. Biotelemetry 10: proceedings of the tenth international symposium on biotelemetry. University of Arkansas Press, Fayetteville.

Amlaner, C. J., Jr., and D. W. Macdonald, editors. 1980. A handbook on biotelemetry and radio tracking. Pergamon Press, Elmsford, New York.

Ancel, A., J. Gender, J. Lignon, P. Jouventin, and Y. LeMaho. 1992. Satellite radio-tracking of emperor penguins walking on sea-ice to refeed at sea. Pages 201–202 *in* Priede and Swift (1992).

Angerbjorn, A., and D. Becker. 1992. An automatic location system for wildlife telemetry. Pages 68–75 *in* Priede and Swift (1992).

Armstrong, J. D., P. M. Bagley, and I. G. Priede. 1992. Tracking deep-sea fish using ingestible transmitters and an autonomous sea-floor instrument package. Pages 376–386 *in* Priede and Swift (1992).

Batschelet, E. 1965. Statistical methods for the analysis of problems in animal orientation and certain biological rhythms. American Institute of Biological Sciences, Washington, DC.

Brumbaugh, D. 1980. Effects of thermal stratification on range of ultrasonic tags. Underwater Telemetry Newsletter 10(2):1–4.

Buckley, J., and B. Kynard. 1985. Yearly movements of shortnose sturgeons in the Connecticut River. Transactions of the American Fisheries Society 114:813–820.

Burchard, D. 1989. Direction finding in wildlife research by Doppler effect. Pages 169–177 *in* Amlaner (1989).

Chisholm, I. M., and W. A. Hubert. 1985. Expulsion of dummy transmitters by rainbow trout. Transactions of the American Fisheries Society 114:766–767.

Cochran, W. W. 1980. Wildlife telemetry. Pages 507–520 *in* S. D. Schemnitz, editor. Wildlife management techniques, 4th edition. The Wildlife Society, Washington, DC.

Cochran, W. W., D. W. Warner, J. R. Tester, and V. B. Kuechle. 1965. Automatic radio tracking system for monitoring animal movements. BioScience 15:98–100.

Collins, K. J., and A. C. Jensen. 1992. Acoustic tagging of lobsters on the Poole Bay artificial reef. Pages 354–358 *in* Priede and Swift (1992).

Coutant, C. C., and D. S. Carroll. 1980. Temperatures occupied by ten ultrasonic-tagged striped bass in freshwater lakes. Transactions of the American Fisheries Society 109:195–202.

Cresswell, W. J., and G. C. Smith. 1992. The effects of temporally autocorrelated data on methods of home range analysis. Pages 272–284 *in* Priede and Swift (1992).

Deat, A., C. Mauget, R. Mauget, D. Maurel, and A. Sempere. 1980. The automatic, continuous and fixed radio tracking system of the Chize Forest: theoretical and practical analysis. Pages 439–451 *in* Amlaner and Macdonald (1980).

Degler, R., and S. Tomkiewicz. 1993. GPS update. Telonics Quarterly 6(2):2–3. Telonics, Inc., Mesa, Arizona.

Eiler, J. H. 1990. Radio transmitters used to study salmon in glacial rivers. American Fisheries Society Symposium 7:364–369.

Eiler, J. H. 1995. A remote satellite-linked tracking system for studying Pacific salmon with radio telemetry. Transactions of the American Fisheries Society 124:184–193.

Einhouse, D. W. 1981. Summer–fall movements, habitat utilization, diel activity and feeding behavior of walleyes in Chautauqua Lake, New York. Master's thesis. State University College, Fredonia, New York.

Fedak, M. A., S. S. Anderson, and M. G. Curry. 1983. Attachment of a radio tag to the fur of seals. Journal of Zoology (London) 200:298–300.

Ferrel, D. W., D. R. Nelson, T. C. Sciarrotta, E. A. Standora, and H. C. Carter. 1974. A multichannel ultrasonic biotelemetry system for monitoring marine animal behavior at sea. ISA (Instrument Society of America) Transactions 13:120–131.

Folkow, L. P., and A. S. Blix. 1992. Satellite tracking of harp and hooded seals. Pages 214–218 *in* Priede and Swift (1992).

Gilmer, D. S., L. M. Cowardin, R. L. Duval, L. M. Mechlin, C. W. Shaiffer, and V. B. Kuechle. 1981. Procedures for the use of aircraft in wildlife biotelemetry studies. U.S. Fish and Wildlife Service Resource Publication 140.

Gray, R. H., and J. M. Haynes. 1979. Spawning migration of adult chinook salmon (*Oncorhynchus tshawytscha*) carrying external and internal radio transmitters. Journal of the Fisheries Research Board of Canada 36:1060–1064.

Hallock, R. J., R. F. Elwell, and D. H. Fry, Jr. 1970. Migrations of adult king salmon, *Oncorhynchus tshawytscha*, in the San Joaquin delta. California Department of Fish and Game Bulletin 151.

Hart, L. G., and R. C. Summerfelt. 1975. Surgical procedures for implanting ultrasonic transmitters into flathead catfish (*Plyodictis olivaris*). Transactions of the American Fisheries Society 104:56–59.

Hawkins, A. D., and G. G. Urquhart. 1983. Tracking fish at sea. Pages 103–166 *in* A. G. MacDonald and I. G. Priede, editors. Experimental biology at sea. Academic Press, London.

Haynes, J. M., R. H. Gray, and J. C. Montgomery. 1978. Seasonal movements of white sturgeon (*Acipenser transmontanus*) in the mid-Columbia River. Transactions of the American Fisheries Society 107:275–280.

Hays, G. C. 1992. Assessing the nesting beach fidelity and clutch frequency of sea turtles by satellite tracking. Pages 203–213 *in* Priede and Swift (1992).

Helm, W. T., and H. M. Tyus. 1992. Influence of coating type on retention of dummy transmitters implanted in rainbow trout. North American Journal of Fisheries Management 12:257–259.

Henderson, H. F., A. D. Hasler, and G. G. Chipman. 1966. An ultrasonic transmitter for use in studies of movements of fishes. Transactions of the American Fisheries Society 95:350–356.

Hurlbert, S. H. 1984. Pseudoreplication and the design of ecological field experiments. Ecological Monographs 54:187–211.

Ireland, L. C. 1980. Homing behavior of juvenile green turtles. Pages 761–764 *in* Amlaner and Macdonald (1980).

Ireland, L. C., and J. S. Kanwisher. 1978. Underwater acoustic biotelemetry: procedures for obtaining information on the behavior and physiology of free-swimming aquatic animals in their natural environments. Pages 341–379 *in* D. I. Mostofsky, editor. The behavior of fish and other aquatic animals. Academic Press, New York.

Jennings, J. G., and W. F. Gandy. 1980. Tracking pelagic dolphins by satellite. Pages 753–755 *in* Amlaner and Macdonald (1980).

Keinath, J. A., and J. A. Musick. 1993. Movements and diving behavior of a leatherback turtle, *Dermochelys coriacea*. Copeia 1993:1010–1017.

Kenward, R. 1987. Wildlife radio tagging: equipment, field techniques and data analysis. Academic Press, New York.

Kuechle, V. B., J. M. Haynes, and R. A. Reichle. 1989. Use of small computers as telemetry data collectors. Pages 695–699 *in* Amlaner (1989).

Larkin, R. P., and D. Halkin. 1994. A review of software packages for estimating animal home ranges. Wildlife Society Bulletin 22:274–287.

Leatherwood, S., and W. E. Evans. 1979. Some recent uses and potentials of radiotelemetry in field studies of cetaceans. Pages 1–31 *in* H. E. Winn and B. L. Olla, editors. Behavior of marine animals. Cetaceans, volume 3. Plenum, New York.

Lee, W. C. 1984. Hooking mortality of lake trout (*Salvelinus namaycush*) in high elevation Colorado reservoirs. Master's thesis. Colorado State University, Fort Collins.

Lemnell, P. A., G. Johnson, H. Helmerson, O. Holmstrand, and L. Norling. 1983. An automatic radio-telemetry system for position determination and data acquisition. Pages 76–93 *in* D. G. Pincock, editor. Proceedings of the fourth international conference on wildlife biotelemetry. Technical University of Nova Scotia, Halifax.

Lewis, A. E., and W. R. A. Muntz. 1984. The effects of external ultrasonic tagging on the swimming performance of rainbow trout, *Salmo gairdneri* Richardson. Journal of Fish Biology 25:577–585.

Long, F. M., editor. 1977. Proceedings of the first international conference on wildlife biotelemetry. University of Wyoming, Laramie.

Long, F. M., editor. 1979. Proceedings of the second international conference on wildlife biotelemetry. University of Wyoming, Laramie.

Lucas, M. C., G. G. Urquhart, A. D. F. Johnstone, and T. J. Carter. 1992. A portable 24-hour recording system for field telemetry: first results from continuous monitoring of heart rate of adult Atlantic salmon (*Salmo salar* L.) during the spawning migration. Pages 400–409 *in* Priede and Swift (1992).

Luke, D. M., D. G. Pincock, P. D. Sayre, and A. L. Weatherly. 1979. A system for the telemetry of activity-related information from free swimming fish. Pages 77–85 *in* F. M. Long, editor. Proceedings of the second international conference on wildlife biotelemetry. University of Wyoming, Laramie.

Luke, D. M., D. G. Pincock, and A. B. Stasko. 1973. Pressure-sensing ultrasonic transmitter for tracking aquatic animals. Journal of the Fisheries Research Board of Canada 30:1402–1404.

Lund, W. A., Jr., and R. C. Lockwood, Jr. 1970. Sonic tag for large decapod crustaceans. Journal of the Fisheries Research Board of Canada 27:1147–1151.

Macdonald, D. W., F. G. Ball, and N. G. Hough. 1980. The evaluation of home range size and configuration using radio tracking data. Pages 405–424 *in* Amlaner and Macdonald (1980).

Mani, R., and I. G. Priede. 1980. Salinity telemetry from estuarine fish. Pages 617–621 *in* Amlaner and Macdonald (1980).

Marty, G. D., and R. C. Summerfelt. 1988. Inflammatory response of channel catfish to abdominal implants: a histological and ultrastructural study. Transactions of the American Fisheries Society 117:401–416.

McCleave, J. D., and R. M. Horrall. 1970. Ultrasonic tracking of homing cutthroat trout (*Salmo clarki*) in Yellowstone Lake. Journal of the Fisheries Research Board of Canada 27:715–730.

McCleave, J. D., J. H. Power, and S. A. Rommel, Jr. 1978. Use of radio telemetry for studying upriver migration of adult Atlantic salmon (*Salmo salar*). Journal of Fish Biology 12:549–558.

Mech, L. D. 1983. A handbook of animal radio-tracking. University of Minnesota Press, Minneapolis.

Mellas, E. J., and J. M. Haynes. 1985. Swimming performance and behavior of rainbow trout (*Salmo gairdneri*) and white perch (*Morone americana*): effects of attaching telemetry transmitters. Canadian Journal of Fisheries and Aquatic Sciences 42:488–493.

Mitson, R. B. 1978. A review of biotelemetry techniques using acoustic tags. Pages 269–283 *in* J. E. Thorpe, editor. Rhythmic activities of fishes. Academic Press, New York.

Monan, G. E., J. H. Johnson, and G. F. Esterberg. 1975. Electronic tags and related tracking techniques aid in study of migrating salmon and steelhead trout in the Columbia River Basin. U.S. National Marine Fisheries Service Marine Fisheries Review 37(2):9–15.

Mortensen, D. G. 1990. Use of staple sutures to close surgical incisions for transmitter implants. American Fisheries Society Symposium 7:380–383.

Nelson, D. R. 1978. Telemetering techniques for the study of free-ranging sharks. Pages 419–482 *in* E. S. Hodgson and R. F. Mathewson, editors. Sensory biology of sharks, skates, and rays. U.S. Department of the Navy, Office of Naval Research, Arlington, Virginia.

Nielsen, L. A. 1992. Methods of marking fish and shellfish. American Fisheries Society Special Publication 23, Bethesda, Maryland.

Petering, R. W., and D. L. Johnson. 1991. Suitability of a cyanoacrylate adhesive to close incisions in black crappies used in telemetry studies. Transactions of the American Fisheries Society 120:535–537.

Pincock, D. 1980. Automation of data collection in ultrasonic biotelemetry. Pages 471–476 *in* Amlaner and Macdonald (1980).

Potgeiter, J. 1989. Audibility of narrow pulses from animal trackings under field conditions. Pages 715–718 *in* Amlaner (1989).

Potter, E. C. E., D. J. Solomon, and A. A. Buckley. 1992. Estuarine movements of adult Atlantic salmon (*Salmo salar* L.) in Christchurch Harbour, southern England. Pages 400–409 *in* Priede and Swift (1992).

Prepejchal, W., M. M. Thommes, S. A. Spigarelli, J. R. Haumann, and P. E. Hess. 1980. An automatic underwater radiotelemetry system to monitor temperature responses of fish in a fresh-water environment. Argonne National Laboratory, Ecological Sciences 108, Argonne, Illinois. (Available from National Technical Information Service, United States Department of Commerce, 5285 Port Royal Road, Springfield, Virginia 22161.)

Priede, I. G., and J. French. 1991. Tracking of marine animals by satellite. International Journal of Remote Sensing 12:667–680.

Priede, I. G., and S. M. Swift. 1992. Wildlife telemetry: remote monitoring and tracking of animals. Ellis Horwood, New York.

Priede, I. G., and A. H. Young. 1977. The ultrasonic telemetry of cardiac rhythms of wild brown trout (*Salmo trutta* L.) as an indicator of bioenergetics and behaviour. Journal of Fish Biology 10:299–319.

Rochelle, J. M. 1974. Design of gateable transmitter for acoustic telemetering tags. IEEE (Institute of Electrical and Electronics Engineers) Transactions on Biomedical Engineering 21:63–66.

Rogers, S. C., D. W. Church, A. H. Weatherley, and D. G. Pincock. 1984. An automated ultrasonic telemetry system for the assessment of locomotor activity in free-ranging rainbow trout, *Salmo gairdneri* Richardson. Journal of Fish Biology 25:697–710.

Ross, L. G., W. Watts, and A. H. Young. 1981. An ultrasonic biotelemetry system for the continuous monitoring of tail-bear rate from free-swimming fish. Journal of Fish Biology 18:479–490.

Ross, M. J., and C. F. Kleiner. 1982. Shielded-needle technique for surgically implanting radio-frequency transmitters in fish. Progressive Fish-Culturist 44:41–43.

Ross, M. J., V. B. Kuechle, R. A. Reichle, and D. B. Siniff. 1979. Automatic radio telemetry recording of fish temperature and depth. Pages 238–247 *in* F. M. Long, editor. Proceedings of the second international conference on wildlife biotelemetry. University of Wyoming, Laramie.

Ross, M. J., and D. B. Siniff. 1982. Temperatures selected in a power plant thermal effluent by adult yellow perch (*Perca flavescens*) in winter. Canadian Journal of Fisheries and Aquatic Sciences 39:346–349.

Rusanowski, P. C., E. L. Smith, and M. Cochran. 1990. Monitoring the nearshore movement of red king crabs under sea ice with ultrasonic tags. American Fisheries Society Symposium 7:384–389.

Samuel, M. D., and M. R. Fuller. 1994. Wildlife radiotelemetry. Pages 370–418 *in* T. A. Bookhout, editor. Research and management techniques for wildlife and habitats, 5th edition. The Wildlife Society, Bethesda, Maryland.

Scharold, J., and S. H. Gruber. 1991. Telemetered heart rate as a measure of metabolic rate in the lemon shark, *Negaprion brevirostris*. Copeia 1991:942–953.

Seitz, A., U. Faller-Doepner, and W. Reh. 1992. Radio-tracking of the common frog (*Rana temporaria*). Pages 484–489 *in* Priede and Swift (1992).

Siniff, D. B. 1970. Studies at McMurdo station of the population dynamics of Antarctic seals. Antarctic Journal of United States 5(4):129–130.

Siniff, D. B., J. R. Tester, and V. B. Kuechle. 1971. Some observations on the activity patterns of Weddell seals as recorded by telemetry. Antarctic Pinnipedia 18:173–180.

Solomon, D. J., and E. C. E. Potter. 1988. First results with a new estuarine fish tracking system. Journal of Fish Biology 33(Supplement A):127–132.

Standora, E. A. 1977. An eight-channel radio telemetry system to monitor alligator body temperatures in a heated reservoir. Pages 70–78 *in* F. M. Long, editor. Proceedings of the first international conference on wildlife biotelemetry. University of Wyoming, Laramie.

Standora, E. A., and D. R. Nelson. 1977. A telemetric study of the behavior of freeswimming Pacific angel sharks, *Squatina californica*. Bulletin Southern California Academy of Sciences 76:193–201.

Standora, E. A., J. R. Spotila, J. A. Keinath, and C. R. Shoop. 1984. Body temperatures, diving cycles, and movement of a subadult leatherback turtle, *Dermochelys coriacea*. Herpetologica 40(2):169–176.

Stasko, A. B. 1975. Underwater biotelemetry, an annotated bibliography. Canada Fisheries and Marine Service Technical Report 534. (Available from Micromedia Limited, 144 Front Street West, Toronto, Ontario M5J 2L7, Canada.)

Stasko, A. B., and D. G. Pincock. 1977. Review of underwater biotelemetry with emphasis on ultrasonic techniques. Journal of the Fisheries Research Board of Canada 34:1261–1285.

Stasko A. B., and S. M. Polar. 1973. Hydrophone and bow-mount for tracking fish by ultrasonic telemetry. Journal of the Fisheries Research Board of Canada 30:119–121.

Stuart, J. L., and T. Friend. 1989. DATACOL: an automated telemetry data collection system for physiological studies of swine farrowing. Pages 406–410 *in* Amlaner (1989).

Stuehrenberg, L., A. Giorgi, and C. Bartlett. 1990. Pulse-coded radio tags for fish identification. American Fisheries Society Symposium 7:370–374.

Summerfelt, R. C., and D. Mosier. 1984. Transintestinal expulsion of surgically implanted dummy transmitters by channel catfish. Transactions of the American Fisheries Society 113:760–766.

Taillade, M. 1992. Animal tracking by satellite. Pages 149–160 *in* Priede and Swift (1992).

Tiebout, H. M., and J. R. Cary. 1987. Dynamic spatial ecology of the water snake, *Nerodia sipedon*. Copeia 1987:1–18.

Tyus, H. M. 1988. Long-term retention of implanted transmitters in Colorado squawfish and razorback sucker. North American Journal of Fisheries Management 8:264–267.

Urick, R. J. 1975. Principles of underwater sound, 2nd edition. McGraw-Hill, New York.

Urquhart, G. G., and G. W. Smith. 1992. Recent developments of a fixed hydrophone array system for monitoring movements of aquatic animals. Pages 342–353 *in* Priede and Swift (1992).

White, G. C., and R. A. Garrott. 1990. Analysis of wildlife radio-tracking data. Academic Press, New York.

Williams, T. H., and R. G. White. 1990. Evaluation of pressure-sensitive radio transmitters used for monitoring depth selection by trout in lotic systems. American Fisheries Society Symposium 7:390–394.

Winter, J. D. 1976. Movements and behavior of largemouth bass (*Micropterus salmoides*) and steelhead (*Salmo gairdneri*) determined by radio telemetry. Doctoral dissertation. University of Minnesota, Minneapolis.

Winter, J. D. 1977. Summer home range movements and habitat use by four largemouth bass in Mary Lake, Minnesota. Transactions of the American Fisheries Society 106:323–330.

Winter, J. D. 1983. Underwater biotelemetry. Pages 371–395 *in* L. A. Nielsen and D. L. Johnson, editors. Fisheries techniques. American Fisheries Society, Bethesda, Maryland.

Winter, J. D., V. B. Kuechle, D. B. Siniff, and J. R. Tester. 1978. Equipment and methods for radio tracking freshwater fish. University of Minnesota Agricultural Extension Service Miscellaneous Report 152.

Winter, J. D., and M. J. Ross. 1982. Methods in analyzing fish habitat utilization from telemetry data. Pages 273–279 *in* N. B. Armantrout, editor. Proceedings of the symposium on the acquisition and utilization of aquatic habitat inventory information. American Fisheries Society, Western Division, Bethesda, Maryland.

Yerbury, M. J. 1980. Long range tracking of *Crocodylus porosus* in Arnhem Land, northern Australia. Pages 765–776 *in* Amlaner and Macdonald (1980).

Voegeli, F. M. 1989. Ultrasonic tracking, position monitoring, and data telemetry systems. Pages 279–284 *in* Amlaner (1989).

Zinnel, K. C. 1980. Behavior of walleye pike in experimental channels as monitored by a microcomputer utilizing radio telemetry. Master's thesis. University of Minnesota, Minneapolis.

Zinnel, K. C. 1992. Behavior of free-ranging pocket gophers (*Geomys bursariu s*). Doctoral dissertation. University of Minnesota, Minneapolis.

Chapter 20

Sampling the Recreational Creel

STEPHEN P. MALVESTUTO

20.1 INTRODUCTION

Management of a fishery resource requires information on the environment in which the fish live, on the other organisms with which the fish interact, on the biology of fish species to be managed, and on the people who use the fish stocks for food and recreation. Information on people using fisheries resources (e.g., anglers) and on the biology of fish species harvested can be collected using a creel survey.

Traditionally, a creel is the woven basket in which harvested fish are stored, and a creel survey involves counting anglers and sampling anglers' creels at particular recreational sites. Nowadays, anglers store harvested fish in many ways, such as in wire baskets, coolers, live wells, and on stringers, but the basic objectives of the creel survey remain the same: to estimate the amount of angling activity and the harvest of different kinds of fishes in number and weight.

These objectives frequently are efficiently accomplished by collecting information on site. In the field, creel survey biologists are responsible for correctly applying statistical sampling designs, properly conducting field protocol, obtaining accurate counts of anglers (by roving areas by boat, car, or foot or at access points), and conducting interviews of anglers (and perhaps other recreational users). Working outdoors at recreational fishing sites and interacting with anglers can be very rewarding. Normally, anglers are very friendly and want to talk about their fishing experiences, and there are opportunities to meet interesting people and learn about the fish community by observing fish and taking biological data. On-water work can be particularly challenging, such as when anglers are interviewed in strong currents or inclement weather and fish must be passed from boat to boat to obtain harvest information.

Creel surveys are extremely important because they are the only commonly applied sampling technique during which management agency personnel are required to interact with their clientele (anglers) through a formal interview process. This interaction provides a tremendous opportunity not only to obtain data necessary for sound fishery management but also to gain public support for agency activities and to educate anglers concerning ecology, resource conservation, and fishing. Creel surveys thus are an integral part of any management program designed to collect information for improvement of a fishery resource and to enhance benefits to those who are engaged in recreational fishing.

The traditional goal of fisheries management has been to maximize the number or weight of certain fish species harvested annually by humans for commercial or recreational purposes. This goal is the core philosophy of maximum sustained yield. For recreational fisheries management, this goal normally is to improve fishing success (e.g, number of fish caught per hour) for certain species. Classical examples of the mathematical underpinnings of maximum sustained yield can be found in fisheries texts by Beverton and Holt (1957) and Ricker (1975).

Given that yield from a recreational fishery encompasses more than just fishing success, the concept of optimum yield, which incorporates economic and social benefits associated with fishing, certainly is more applicable to the recreational setting (Anderson 1975; Bailey et al. 1986; American Fisheries Society 1987; Malvestuto 1989). It is becoming increasingly important to coalesce data collection efforts across disciplines to provide cost-effective means of understanding recreational fisheries for implementation of better management plans. When anglers are interviewed on site, or later by mail or telephone, human dimension questions can be incorporated into the data-gathering process. When integrated with traditional catch and effort data, human dimension data (Chapter 22) can enhance understanding of the clientele and guide management toward more socially equitable decisions.

Several terms are used frequently in this chapter. Fishing activity is referred to as fishing effort or fishing pressure. The basic unit of recreational fishing effort is the angler-hour, which is 1 h of fishing by a single angler. In the past, the terms catch and harvest have been used synonymously; however, given that recreational benefit may be derived from catching fish and returning them to the water (catch and release), it is more appropriate to treat harvest as a component of catch, where catch is defined as fish harvested plus fish released. Catch per unit of fishing effort (CPUE), the number of fish caught per angler-hour, for example, and harvest per unit of effort (HPUE) typically are used as indices of harvestable stock density (Ricker 1975) and fishing success (Malvestuto 1983). For simplicity here, the generic terms CPUE and catch will be used unless the discussion specifically refers to obtaining harvest data only (see Box 6.1 on CPUE). Also, the commonly used expression creel census will not be used in this chapter. A census implies total enumeration of a defined angler population, which normally is impossible for large recreational fisheries. A survey refers to a sample of anglers from the population and is the correct terminology.

Since the first edition of this book, there have been two major contributions to the recreational fishery survey literature. The first is the proceedings of American Fisheries Society Symposium 12, *Creel and Angler Surveys in Fisheries Management* (Guthrie et al. 1991), and the second is the follow-up techniques manual, *Angler Survey Methods and Their Applications in Fisheries Management* (Pollock et al. 1994). Much of the new material presented in this chapter stems from these important contributions to the literature.

20.2 THEORETICAL FRAMEWORK: SAMPLING THE ANGLING POPULATION

There are two basic considerations concerning collection of recreational fishery information: (1) what statistical survey design provides the best quantitative estimates of the fishery characteristics of interest; and (2) how can anglers be contacted to obtain the needed information? This section will address commonly used survey sampling designs and the following section (20.3) will treat angler contact methods.

Box 20.1 shows a four-step process for planning fishery surveys (Malvestuto 1989,

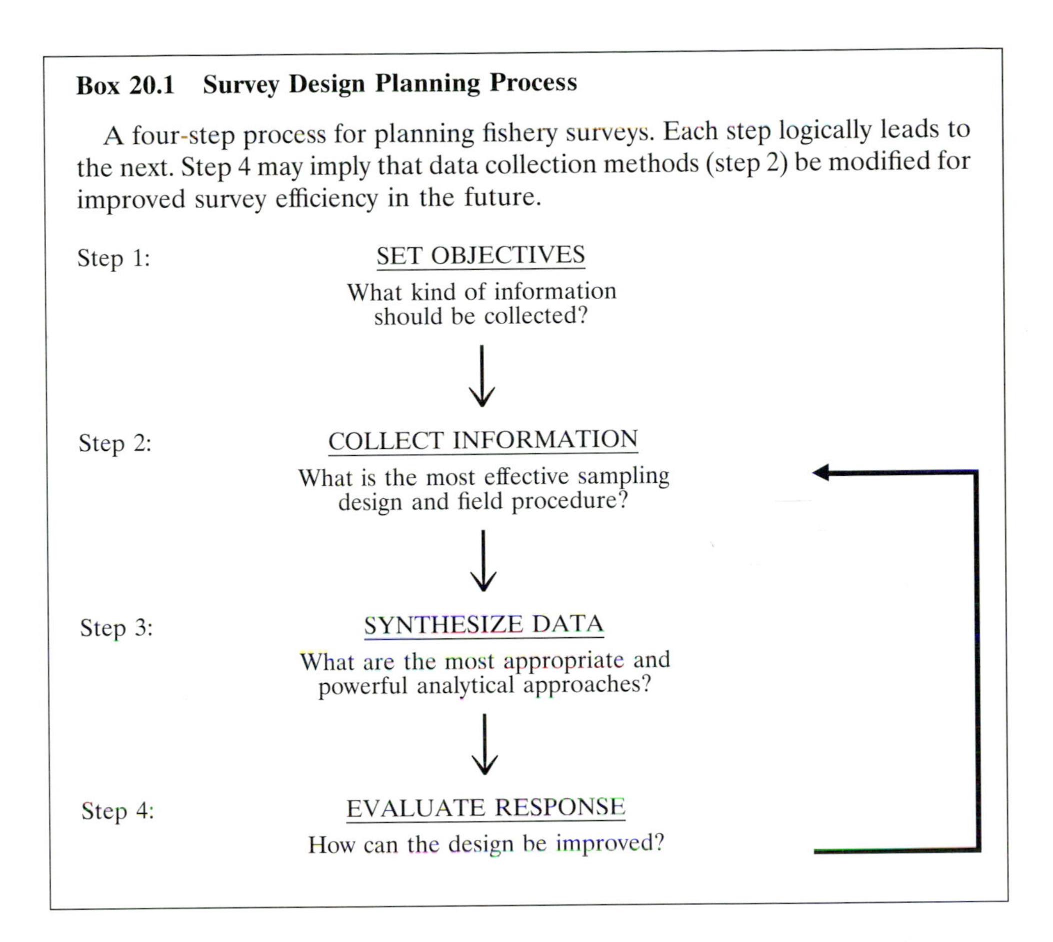

1991). Before deciding on the sampling design and other data collection techniques, you must clearly state the objectives of the survey (step 1). The objectives should identify the anglers of concern (target population), define the temporal and spatial dimensions of sampling, and specify the kind of information to be collected. Preferably, all variables to be measured should be listed, and the magnitude of change in key variables to be statistically documented over time and space should be specified. In general, the smaller the change to be statistically documented, the more intensive the sampling effort will need to be. For example, documenting a 25% change in fishing effort annually at the 90% level of confidence, all other factors being equal, will require about four times as much sampling time as would documenting a 50% change (Malvestuto 1991). Also see statistical considerations in Chapter 2 of this text.

The survey design must be customized to meet survey objectives and fit spatial and temporal characteristics of the physical site and angling activities (Malvestuto 1991; Pollock et al. 1994). The sampling procedures must provide for efficient collection of on-site or post-trip data that truly describe the target population. This is step 2 in the planning process. Methods cannot be sufficiently detailed if the survey objectives are not clearly and specifically defined.

To choose anglers in a representative manner on site requires that the fisheries biologist divide the time or space dimensions (or both) of fishing into sampling units

(SUs) that can be chosen at random. Statistically valid estimates require that all anglers in the target population be given some probability of being sampled. Although anglers cannot be chosen directly at random, the SUs within which they fish can be (see Chapter 1). Sampling design concepts are covered in detail in survey technique books such as Cochran (1977) and Jessen (1978); Pollock et al. (1994) provided a succinct overview of statistical concepts. Bazigos (1974), Malvestuto et al. (1980), Caddy and Bazigos (1985), Bayley and Petrere (1989), and Malvestuto and Meredith (1989) presented concepts and examples specifically for surveys of small-scale commercial and subsistence fisheries in developing countries. Malvestuto (1983), Van Den Avyle (1986), Guthrie et al. (1991), and Pollock et al. (1994) treated the topic for recreational fisheries in the United States.

Data, once collected, must be synthesized quantitatively according to statistical requirements imposed by the sampling design in order to meet survey objectives. This is step 3 of the planning process. Quantitative procedures should provide the most unbiased and precise estimates possible (see Chapter 2). Results generated from the survey can be used not only to fulfill study objectives but also to evaluate statistical effectiveness of the sampling design (step 4). Inefficient design features should be discarded, and modifications are in order if objectives were not met. An inefficient design requires redefinition of the data collection methodology for improved data quality in the future.

The sampling designs outlined in this section are simple random sampling, stratified random sampling, and stratified two-stage probability sampling.

20.2.1 Simple Random Sampling

In simple random sampling, the spatial and temporal framework is divided into nonoverlapping SUs, a given number of which are then chosen for the sample randomly and with equal probability. Box 20.2 outlines a simple random sampling process used to choose sampling days over a 59-d period. The specific objective of such sampling could be to describe, over the period indicated, a fishery on a particular body of water that can be canvassed entirely during one fishing day (for example, 6 AM to 6 PM). Monetary restrictions allow sampling to occur on only 10 of the 59 possible days.

The random sampling process begins by numbering SUs (fishing days) from 1 to 59. Next, 10 random numbers are chosen from a random numbers table (or by means of a random number generator on a calculator or computer) by looking at two-digit random numbers and choosing the first 10 different numbers between 01 and 59. Sampling days are chosen without replacement, so once a particular day has been chosen, it cannot be sampled again; therefore, its number is ignored if it appears again in the random draw.

20.2.2 Stratified Random Sampling

Stratification is the division of a population into subpopulations (strata) that are then subjected to simple random sampling. Stratification may be necessary for logistical or administrative reasons (water body crosses district or state lines), comparisons among defined subgroups, or reduction of sampling variance by dividing a heterogeneous population into more homogeneous groups (Cochran 1977; Jessen 1978; Pollock et al. 1994). To reduce variability associated with estimates of fishing effort, for example, days within the survey period are often grouped into weekdays and weekend days. The usual situation is that weekend days receive consistently higher levels of fishing effort than do weekdays. A simple random sample of days

Box 20.2 Simple Random Sampling

Outline of a simple random sampling procedure in which 10 sampling units (SUs, fishing days) were chosen from 59 possible units over a 2-month survey period. Randomly chosen days are circled on the calendars.

SAMPLING UNIT DEFINITIONS

SU	Date
1	1 Feb
2	2 Feb
3	3 Feb
4	4 Feb
↓	↓
57	29 Mar
58	30 Mar
59	31 Mar

RANDOM CHOICE

SU	Date
51	23 Mar
24	24 Feb
09	09 Feb
57	29 Mar
30	02 Mar
39	11 Mar
21	21 Feb
36	08 Mar
37	09 Mar
26	26 Feb

FEBRUARY

S	M	T	W	T	F	S
	1	2	3	4	5	6
7	8	(9)	10	11	12	13
14	15	16	17	18	19	20
(21)	22	23	(24)	25	(26)	27
28						

MARCH

S	M	T	W	T	F	S
	1	(2)	3	4	5	6
7	(8)	(9)	10	(11)	12	13
14	15	16	17	18	19	20
21	22	(23)	24	25	26	27
28	(29)	30	31			

would probably include both day types and would give a highly variable estimate of daily fishing effort. Taking a simple random sample of weekdays and another simple random sample of weekend days, however, would provide two relatively precise estimates of fishing effort that then could be combined to estimate effort for the entire study period.

A stratified random sampling plan based on the same 2-month survey period identified in the simple random sampling example is presented in Box 20.3. Days within the weekday stratum are enclosed by the dark rectangles on the calendar. The stratified random sampling plan requires that the sampling units for each stratum be chosen separately. The weekday stratum contains 43 SUs, and the weekend stratum contains 16; weekdays thus are consecutively numbered from 01 to 43 and weekend days from 01 to 16 so that two separate random samples can be chosen. In the example, 10 sampling days are allocated equally between strata, 5 d each.

Equal allocation of SUs among strata, as is done in Box 20.3, often will not be the best design. Cochran (1977) gives the general rule that more samples should be taken within a stratum if (1) the stratum is larger than others being sampled (more fishing days or more fishing effort); (2) the characteristic being measured is more variable within the stratum; and (3) the stratum costs less to sample. These three consider-

Box 20.3 Stratified Random Sampling

Outline of a stratified random sampling procedure in which five sampling units (SUs, fishing days) were chosen from each of two strata, one composed of weekdays, the other of weekend days. The strata are separated by the outlines on the calendars; randomly chosen days are circled.

SAMPLING UNIT DEFINITIONS

WEEKDAY STRATUM

SU	Date
1	1 Feb
2	2 Feb
3	3 Feb
4	4 Feb
↓	↓
41	29 Mar
42	30 Mar
43	31 Mar

WEEKEND STRATUM

SU	Date
1	6 Feb
2	7 Feb
3	13 Feb
4	14 Feb
↓	↓
14	21 Mar
15	27 Mar
16	28 Mar

RANDOM CHOICES

WEEKDAY STRATUM

SU	Date
43	31 Mar
25	05 Mar
01	01 Feb
22	02 Mar
40	26 Mar

WEEKEND STRATUM

SU	Date
05	20 Feb
13	20 Mar
03	13 Feb
14	21 Mar
12	14 Mar

FEBRUARY

S	M	T	W	T	F	S
	(1)	2	3	4	5	6
7	8	9	10	11	12	(13)
14	15	16	17	18	19	(20)
21	22	23	24	25	26	27
28						

MARCH

S	M	T	W	T	F	S
	1	(2)	3	4	(5)	6
7	8	9	10	11	12	13
(14)	15	16	17	18	19	(20)
(21)	22	23	24	25	(26)	27
28	29	30	(31)			

ations can be interrelated mathematically to provide optimum allocation of sampling units among strata (Cochran 1977: section 5.5; Malvestuto and Knight 1991; Pollock et al. 1994: section 3.3.4) .

Apart from potential gains in precision through stratification, this technique must be used if independent estimates are desired for defined subsets of the total population for descriptive or comparative purposes. Common subsets are geographical regions, habitat types, months or seasons of the year, and fishing locations (bank versus boat, for example). Sometimes, multiple stratification is desirable (season × day type × habitat), but given that each stratum must be independently sampled, the number of SUs required, and thus the expense, quickly can become too large for practical purposes.

20.2.3 Stratified Two-Stage Probability Sampling

Because of time, cost, and logistical constraints, fisheries may be divided into smaller units for on-site sampling purposes. The stratified random sampling scheme presented earlier, for example, requires the creel clerk to cover the entire body of water and to remain on the water for the entire fishing day. Frequently this is impossible. A solution is to subdivide each fishing day into secondary, or subsampling, units. The stratified sampling procedure then is conducted in two stages: (1) fishing days or primary sampling units (PSUs) are chosen; and (2) within each randomly chosen PSU, one or more secondary sampling units (SSUs) are randomly chosen.

Box 20.4 shows a subsampling design in which each fishing day is divided into six SSUs (two time periods × three lake sections), and one of these SSUs is chosen from each PSU. The box shows a random sample of SSUs (marked with an X) from the five weekdays and five weekend days previously chosen using stratified random sampling (Box 20.3). Note that the time periods are labeled AM and PM and may realistically represent a morning sampling period of 6 h (6 AM to 12 noon) and an afternoon sampling period of 6 h (12 noon to 6 PM) in a 12-h fishing day.

The two-stage probability sampling process requires that decisions be made about choosing PSUs and SSUs based on equal (uniform) or unequal (nonuniform) probabilities. If the design stratifies days into weekday and weekend strata, as per the scheme depicted in Box 20.3, then PSUs (fishing days) typically are given equal probabilities within strata for random sampling purposes. This makes sense unless it is known that certain weekdays or weekend days should have higher or lower probabilities than others, based on consistently higher or lower use on those days. For example, if factories let out on Wednesday afternoons such that fishing effort and harvest are typically higher on Wednesdays relative to the other four weekdays, then Wednesdays should be given a higher sampling probability among days within the weekday stratum.

For the two-stage sampling process, it is more typical that SSUs are chosen based on unequal probabilities. In Box 20.4, PSUs are chosen with equal probabilities, but SSUs are chosen with nonuniform probabilities, a desirable approach when SSUs have consistently different levels of fishing activity (or other characteristics of interest). The AM and PM time periods are given sampling probabilities of 0.4 and 0.6, respectively, because previous observation indicated that 40% of the fishing effort would occur in the morning and 60% in the afternoon. Unequal probabilities associated with lake sections are based on the same rationale.

Sampling probabilities associated with the six SSUs are calculated by multiplying the individual time period probabilities by the lake section probabilities. (Note that in Box 20.4 the sum of the probabilities equals 1.0 to represent the entire lake for an entire day.) In order for the nonuniform probabilities to be reflected in the random draw, the number range from 00 to 99 is divided into six unequal intervals proportional to the unequal probabilities associated with the six time period-lake section combinations. Thus, the first SSU (AM-1), with a sampling probability of

Box 20.4 Stratified Two-Stage Sampling

Outline of a stratified two-stage probability sampling procedure in which each primary sampling unit (PSU, fishing day) was divided into six secondary sampling units (SSUs, time period and lake section categories), one of which was chosen with nonuniform probability sampling from each primary unit. Chosen secondary units are marked with an "X." Time periods are morning (AM) and afternoon (PM).

Time period	probabilities
AM	0.40
PM	0.60
	1.00

Lake section	probabilities
1	0.50
2	0.25
3	0.25
	1.00

SSU	Probabilities	Number Ranges
AM-1	0.20	00–19
AM-2	0.10	20–29
AM-3	0.10	30–39
PM-1	0.30	40–69
PM-2	0.15	70–84
PM-3	0.15	85–99

RANDOM CHOICES

WEEKDAY STRATUM

PSU	Number chosen	SSU
1 Feb	83	PM-2
2 Mar	39	AM-3
5 Mar	09	AM-1
26 Mar	16	AM-1
31 Mar	62	PM-1

WEEKEND STRATUM

PSU	Number chosen	SSU
13 Feb	44	PM-1
20 Feb	86	PM-3
14 Mar	07	AM-1
20 Mar	50	PM-1
21 Mar	74	PM-2

Weekdays

	1 Feb AM	1 Feb PM	2 Mar AM	2 Mar PM	5 Mar AM	5 Mar PM	26 Mar AM	26 Mar PM	31 Mar AM	31 Mar PM
Lake Section 1					X		X			X
Lake Section 2		X								
Lake Section 3			X							

Weekends

	13 Feb AM	13 Feb PM	20 Feb AM	20 Feb PM	14 Mar AM	14 Mar PM	20 Mar AM	20 Mar PM	21 Mar AM	21 Mar PM
Lake Section 1		X			X			X		
Lake Section 2										X
Lake Section 3				X						

0.20, gets 20 digits (00–19) out of the entire number range; the second SSU (AM-2), with a sampling probability of 0.10, gets 10 digits (20–29) out of the entire number range, and so on through the remaining SSUs.

This stratified, two-stage sampling design is similar to that discussed in detail by Malvestuto et al. (1978). The advantage associated with this approach is that, on average, those sampling units receiving the most fishing effort will occur more often in the sample; more information will accrue and precision will increase, particularly if the probabilities assigned to the units accurately reflect a true distribution. Fishing effort is typically used to establish probabilities because it is more easily measured (party or boat counts, for example) than are other characteristics; however, if information is available on other characteristics that are more closely allied to the objectives of the study (harvest or HPUE), this information should be used to establish sampling probabilities (Bayley et al. 1991). Additionally, a single set of sampling probabilities (as for SSUs in the current example) may not hold for all strata; a separate set may be desirable for weekdays and weekends, and these, in turn, may change seasonally.

The more accurate the probabilities, the better the gain in precision from this type of design. Stanovick and Nielsen (1991) discussed strategies for allocating sampling effort and showed that expert opinion of fisheries managers could be used to derive accurate stratum weights and sampling probabilities for access points along the James River in Virginia. They also provided a method for deriving sampling probabilities for multiple-use recreational surveys. Hayne (1991) provided an excellent overview of nonuniform probability sampling.

The example in Box 20.4 is based on a two-stage subsampling design; however, the subsampling process can be extended to three or more stages (multistage probability sampling). Hayne (1991) warned that it becomes statistically complicated to apply nonuniform probability sampling to more than one stage of a multistage design. Thus, days typically should be chosen with uniform probabilities, then subsamples within days can be chosen with nonuniform probabilities. If days also are chosen with nonuniform probabilities, then only one day can be chosen per temporal block, and variances must be estimated differently (see Hayne 1991 for details). There are several applications of two-stage, nonuniform probability sampling in Guthrie et al. (1991).

20.2.4 Effectiveness of Survey Design Alternatives

There has been little work directed at assessing the effectiveness of various survey design alternatives. Although randomization is required for statistical validity, even random samples can have undesirable characteristics. For example, the simple random sample of days chosen from the months of February and March in Box 20.2 shows there were no days sampled until the 9th of February; thus, early February would not be represented in the data. Also, there is a large gap in March, the 12th–22nd, when no samples would be taken.

A better representation of the entire 2-month period might be established if sample days were more evenly distributed across the temporal dimension. This can be accomplished by taking a systematic random sample of days. The systematic sample is generated by randomly choosing a start day and then taking days that follow based on a fixed time interval. The fixed time interval is calculated by dividing the number of days in the survey period by the number of days to be sampled (sample size). Thus, for the simple random sample in Box 20.2, 59 total days divided by 10 sample days gives a fixed interval of about 6 d to cover the survey period evenly. The start day is chosen at random from within the first 6 d of February, and then every sixth day from that point on would be included in the sample to give 10 d total. The same process could be applied within weekday and weekend strata separately (Box 20.3). See Cochran (1977) and Pollock et al. (1994) for details of systematic random sampling.

The potential gain in precision due to stratification is based on the premise that the subpopulations used to define strata are sufficiently homogeneous within themselves, and contrast enough among themselves, for the strata to "explain" or "control" a significant amount of the total variability associated with important variables. Malvestuto and Knight (1991) evaluated the benefits of stratification of days within months into weekdays and weekend days, as in the example in Box 20.3. They found that stratification into day types significantly improved the precision of estimates of fishing effort but not of estimates of HPUE or harvest on several reservoirs and a tailwater. Sztramko (1991) found that removal of day type stratification increased the variability of estimates of fishing effort and harvest by about 10%; he also found that by isolating the 9-d opening period of the fishing season as a stratum, variability of fishing effort was reduced by 27% and that of harvest by 19% for the Long Point Bay fishery in Lake Erie.

Malvestuto and Knight (1991) showed that generally most of the variance associated with estimates of HPUE resided within sample days rather than between days, suggesting that for estimation of HPUE, more within-day sampling should be conducted, even at the expense of reducing the total number of days sampled. This finding was corroborated by Lester et al. (1991) who, based on historical data from 47 Ontario lakes, found that almost all (90%) of the variance associated with CPUE resided within days rather than among days. More recently, based on analysis of 4 years of data on Lake Guntersville, Alabama, Crow and Malvestuto (1996) found that choosing two SSUs (3–4-h sampling periods) at random within each PSU (fishing day) rather than just one reduced the sampling error of annual estimates of effort by an average of 15% and of HPUE by an average of 23%.

For river fisheries, Meredith and Malvestuto (1991) showed that relatively complex stratification schemes in time and space did little to improve precision of estimates of fishing effort, HPUE, or harvest. Precision of HPUE estimates improved by about 20% when temporal strata were defined as months rather than as seasons. Dent and Wagner (1991) recommended the use of monthly rather than seasonal strata to improve precision. The smaller the time blocks within the survey period, the more the temporal variability in fishery estimates will be controlled by the survey design (Hayne 1991).

With respect to more optimally allocating sampling effort over months of the year, Malvestuto et al. (1979), using months as strata, showed that variability of fishing effort, and to a lesser degree of HPUE and harvest, could be modeled as a function of monthly air temperature and rainfall patterns. These relationships provided a way to more optimally allocate monthly sampling effort without having prior survey data (but having prior air temperature and rainfall data). Application of this method produced no improvement in the precision of annual estimates of effort, harvest, or HPUE, however (Knight and Malvestuto 1991). Palsson (1991) observed no improvement in catch-per-boat estimates of Pacific cod when sampling days were more optimally allocated across temporal strata.

ROVING CREEL INTERVIEW SCHEDULE

Date _________ Time _______ Lake Section _________ Sample # ______ # in Party ____

Fishing From: Bank _____ Boat ______ With ___ # rods Sex: M ____ F____

Location: Open Water _____ Tree Shelter _____ Rip-Rap _____ Pier _____ Bridge ______

"Good morning (good afternoon). My name is _________________ and I am conducting an angler survey for the (affiliation). We are collecting information that will be used to help manage this resource. Do you mind if I ask you a few questions about your fishing trip today?"

"What county and state did your fishing trip today originate from?"

County __________________ State _________

"What do you estimate that you will spend on the following items for today's fishing trip?"

Gas $_______ Food $_______ Bait $_______ Lodging $_______

If fishing from boat: "Which landing did you use to launch your boat?"_____________________

"What time did you begin fishing today?" AM _________ PM _________

"What time do you think that you will finish fishing today?" AM _________ PM _________

"Now I would like to ask you some questions about your catch."

"What kind of fish are you fishing for?"____

"How many have you caught and released?"____

"How would you rate your fishing success today on a scale of poor, fair, good, or excellent?"

Poor_____Fair_____Good_____Excellent____

"Would you mind if I record the number and sizes of fish that you have harvested?"

SPECIES CAUGHT		LENGTH-CLASS (SPECIFY)							TOTAL
	No.								
	Wt.								
	No.								
	Wt.								
	No.								
	Wt.								
	No.								
	Wt.								
	No.								
	Wt.								

"That completes the interview. Thank you very much for your time. Do you have any comments that you would like to make about the management of this resource?"____________________

____

Figure 20.1 Example of a roving creel survey interview form. The example is meant to show logical questionnaire construction, not to provide an exhaustive list of questions.

20.3 ANGLER CONTACT METHODS

Once sampling units have been defined and randomly chosen, creel clerks must contact anglers to collect information necessary to meet the stated objectives of the survey. Anglers can be contacted on site at the time of their fishing trips or, alternatively, after their fishing trips by mail, telephone, or door-to-door (household) surveys. Anglers thus are either interviewed face-to-face or receive a questionnaire to fill out and return to the survey agency. This interaction with anglers provides a means of collecting a variety of information that may be deemed pertinent to management decision making. Figure 20.1 shows an example of a scripted interview for an on-site creel survey, and section 20.4 addresses on-site interviewing techniques. Chapter 22 gives examples of various types of questioning for mail, telephone, and direct interview questionnaires.

The remainder of this section summarizes angler contact methods frequently applied to recreational fishery surveys. On-site survey methods are emphasized, and summaries of roving, access point, and aerial techniques are provided. Mail and telephone surveys are described in a secondary manner (see Chapter 22 for additional details). Angler contact methods can be combined into "complemented surveys," which are discussed briefly in closing.

20.3.1 On-Site Angler Contact Methods

On-site contact, or intercept, methods maximize response rates (percentage of the sample responding to the interview) because anglers are contacted in person during their fishing trips. Memory recall biases are minimized because questions generally pertain only to the trip in progress, and creel clerks can identify species of fish harvested and obtain numbers, lengths, and weights of fish by direct observation. Disadvantages of on-site methods relative to mail and telephone surveys are high cost per interview, difficulty of relating survey results to the population at large, and logistical problems of contacting a representative sample of anglers over expansive geographical areas (Deuel 1980a).

20.3.1.1 Roving Surveys

The roving survey technique is an on-site intercept method by which the creel clerk contacts anglers as he or she moves through the fishing area along a predetermined route. Creel clerks usually travel by boat, but it is possible to rove among access points by means of a vehicle (see description of the "bus route" method in section 20.3.1.2). In certain situations, shore anglers must be contacted on foot.

The original statistical formulation of this approach (Robson 1961) dictates that (1) the route completely covers the survey area, (2) the clerk begins the route at a randomly chosen point of departure, (3) the clerk randomly chooses one of the two alternative directions of travel, and (4) the clerk travels at a constant speed. In practice, item (2) may be logistically difficult to accomplish, although boat launching sites can be chosen at random. In certain instances, item (3) may be impossible because of strong flows in rivers or regulations that restrict direction of boat travel.

It is critical that the clerk make a complete circuit of the survey area and stay on schedule (see section 20.5.1 for a discussion of proper field procedures for taking roving counts of anglers). When anglers are too numerous for all to be interviewed, the clerk should systematically skip parties in an objective manner (every 2nd, or 3rd, or 10th group, for example) to allow for a complete circuit. This coverage insures that interviews of anglers are spread over the entire sample area in proportion to the density of angling activity. If anglers are too few in number to occupy the full time of the clerk, then the clerk should slow down enough to complete a full circuit in the allotted time period.

The primary weakness of the roving survey is that catch and effort information is based on uncompleted rather than on completed fishing trips; that is, anglers are contacted while they are still fishing. Uncompleted fishing trips are measured from the time anglers begin to fish until the time of interview; completed fishing trips are measured from the time anglers begin to fish until the time they finish.

Obtaining unbiased estimates of catch and harvest rates based on uncompleted trip interviews requires that rate estimators (number of fish caught per hour) not be dependent on the length of time anglers fished before being interviewed. The literature contains conflicting viewpoints on the validity of this assumption. Earlier published information in which data on uncompleted versus completed trips were actually compared (Carlander et al. 1958; Von Geldern 1972; Malvestuto et al. 1978) suggested that the assumption was reasonable. More recent studies have shown that the assumption is not always valid. For example, MacKenzie (1991) found that during winter, uncompleted trips provided unbiased estimates of catch rates for northern pike in Lake Hortonia, Vermont, but that harvest rates were negatively biased relative to estimates from completed trips. Phippen and Bergersen (1991) found over a 129-d season that harvest rate was generally positively biased when based on uncompleted trip data relative to completed trip data from Parvin Lake, Colorado. This difference did not cause a meaningful difference in the final estimates of harvest from the two kinds of interviews. Dent and Wagner (1991) found no statistical difference ($P > 0.50$) in catch rates from uncompleted versus completed trips on Pomme de Terre Lake, Missouri, for either weekdays or weekend days. The relationship between catch or harvest rate and fishing time appears to be fishery specific, not only with respect to the site but also with respect to fish species.

Kokel et al. (1991) found that estimates of other angling descriptors were statistically different ($P < 0.05$) when based on data gathered from uncompleted versus completed fishing trips. These descriptors included certain trip expenditures, particularly gas cost for boat anglers, and distance traveled for bank and boat anglers fishing on the James River, Virginia.

Another problem with roving surveys occurs because the probability of contacting an angler is proportional to trip length. Therefore, creel clerks will tend to interview anglers who spend more time on the water, and thus clerks will overestimate the mean length of a fishing trip (length-of-stay bias). Malvestuto et al. (1978) found that the arithmetic mean of trip length from a roving survey overestimated actual trip length on West Point Lake, Alabama–Georgia, but the harmonic mean compensated for this positive bias (Box 20.5). Length-of-stay bias might affect estimates of other descriptors when anglers who fish longer, and thus are more likely to be intercepted with the roving technique, actually represent a different population of anglers than those who would normally be intercepted at access points. Kokel et al. (1991) believed that anglers who fished longer on the James River in Virginia likely were those who traveled farther and spent more money to reach the site, thus explaining relative differences in estimates of trip expenditures and travel distances derived from uncompleted trip interviews (roving) versus completed trip interviews (access points). Other disadvantages associated with the roving survey are that night surveys are generally impossible and, because anglers are interrupted while fishing, lengthy interviews generally should be avoided because public relation problems can occur (see section 20.4.1 on behavioral protocol).

The primary advantages of roving surveys as outlined by Von Geldern (1972) are that (1) contact of anglers is more time efficient if multiple access points are present, that is, waiting time between interviews is limited to travel time between anglers; (2)

Box 20.5 Arithmetic and Harmonic Means

Calculation of arithmetic and harmonic mean fishing time (hours) from a hypothetical set of seven interviews from a roving creel survey. Each interview provides a value for time spent fishing based on anglers' recall of when they started fishing that day and projected time of completion (see Figure 20.1 for example questionnaire).

Sample Data

Fishing time (t_i) for each (ith) interview measured in hours is converted to its reciprocal ($1/t_i$) for calculation of the harmonic mean. The first step is to sum (Σ) fishing time and its reciprocal over the total number of interviews (n) in the sample.

Interview number	Fishing time (t_i)	Reciprocal fishing time ($1/t_i$)
1	2.0	0.50
2	6.0	0.17
3	4.2	0.24
4	8.5	0.12
5	3.7	0.27
6	4.5	0.22
7	5.8	0.17
$n = 7$	$\sum_{i=1}^{n} t_i = 34.7$	$\sum_{i=1}^{n} 1/t_i = 1.69$

Arithmetic Mean Fishing Time ($\bar{t}_A$)

$$\bar{t}_A = \sum_{i=1}^{n} t_i/n.$$

1. Sum fishing hours (t_i) over the number of interviews (n):

$$\text{Total fishing hours} = \sum_{i=1}^{n} t_i = 34.7.$$

2. Calculate the arithmetic mean fishing time ($\bar{t}_A$) by dividing total fishing hours by the number of interviews (n):

$$\text{Arithmetic mean fishing time} = \bar{t}_A = \left(\sum_{i=1}^{n} t_i\right)/n = 34.7/7 = 4.96 \text{ h}.$$

Harmonic Mean Fishing Time ($\bar{t}_H$)

$$\bar{t}_H = n/\sum_{i=1}^{n} 1/t_i.$$

Box 20.5 Continued.

1. Sum reciprocals of fishing hours ($1/t_i$) over the number of interviews (n):

$$\text{Total reciprocal hours} = \sum_{i=1}^{n} 1/t_i = 1.69.$$

2. Calculate a harmonic mean fishing time ($\bar{t}_H$) by dividing the number of interviews (n) by total reciprocal fishing hours:

$$\text{Harmonic mean fishing time} = \bar{t}_H = n/\sum_{i=1}^{n} 1/t_i = 7/1.69 = 4.14 \text{ h}.$$

The harmonic mean gives a smaller estimate of average fishing time than does the arithmetic mean, possibly correcting for length-of-stay bias associated with roving creel surveys.

all angler types (rental and private boat, shore, and public and private pier) can be contacted in proportion to their actual abundance; and (3) interviews can be combined with angler counts over large areas. With respect to the last item, the creel clerk either makes separate count and interview circuits, randomly choosing which comes first, or combines counts and interviews into a single circuit ("count-as-you-go"); the latter is more cost effective, but estimates of fishing effort are more likely to be biased unless precautions are taken. Precision of fishing effort may increase by keeping count circuits as short as possible. See section 20.5.1 for a discussion of count procedures for estimating fishing effort.

It is suggested that for an on-water roving count, the clerk consider only those anglers fishing on shore, or between shore and the center of the fishing area, as the clerk passes them. Generally, anglers should be actively engaged in fishing activities (line in water, casting, trolling, or changing gear) to be counted as an angler. People away from the shore, moving from one place to another, or not engaged in angling should not be counted.

Phippen and Bergersen (1987, 1991), however, evaluated the effect of various definitions of anglers on estimates of fishing effort, harvest rate, and harvest at Lake Parvin, Colorado. They found the best definition of an angler to be someone with line in the water, changing gear, or walking toward a fishing location; other definitions did not provide more accurate estimates of the true fish harvest (known from the check station). At Lake Parvin, because nearly all visitors to the site were anglers, a broader definition that encompassed anyone who looked like an angler, even if not engaged in angling activities, provided equally accurate estimates of true population values.

20.3.1.2 Access Point Surveys

The access point creel survey represents an on-site intercept method by which the creel clerk is stationed at an access point (boat landing, pier, jetty, or beach) so anglers can be contacted at the ends of their fishing trips. As an on-site method, access point surveys share the same advantages and disadvantages as roving surveys relative to household surveys. Statistical validity requires that access points be randomly chosen; it is common to sample access points based on nonuniform probabilities proportional to use.

The primary advantage of the access point approach relative to the roving method is that information is based on completed fishing trips rather than on uncompleted trips. Thus, trip length biases potentially associated with roving surveys can be avoided. Access points can be so numerous, however, that few anglers use any one point; contact rates are thus low, and clerk time is inefficiently used. Also, it is usually impossible to sample all angler types proportional to their level of effort. This is a particular problem with bank anglers who may be widely dispersed along the shoreline and not associated with well-defined access sites. Shoreline facilities may be quite varied—for example, public and private piers and launch sites, jetties, beaches, parks, and other recreational sections of shoreline—so that a very complete sampling frame (site list) is required. The access point survey is ideal where all anglers must leave from only a small number of points or where anglers must report their catches at a central location (a concession stand, for example). Hayne (1991) provided an overview of access point survey procedures relative to the roving creel method.

The "bus route" method (Robson and Jones 1989; Jones et al. 1990; Jones and Robson 1991) combines the access point and roving intercept methods. Information is collected via access points, but several access points are sampled within any given work period by roving among selected accesses by means of a vehicle. Travel routes are predetermined, and travel time among access points, and time spent at access points, are very precisely scheduled. Time spent at each access point during a sample day is set proportional to expected use. As originally formulated, fishing effort estimates are obtained via car counts, so that for multiuse sites, cars must be distinguished according to the activities of the drivers—anglers, water skiers, picnickers, and so on. Normally, if use estimates must be partitioned among various kinds of activities, then car counts cannot be used, and interviews must be taken by means of a traditional access point survey.

Jones and Robson (1991) compared estimates of fishing effort, derived through computer simulation, for the bus route method and the traditional access point method. Generally, precision was better for the bus route method when car counts were used, but there was no advantage when completed trip interviews were used. It is difficult to obtain interviews when wait times at access points are short; as wait times increase, the bus route method approaches the traditional access point approach in practice. If interview efficiency is not high, then all descriptors obtained via interviews will be difficult to estimate. Generally, the bus route design would be preferred when the number of access points is large (>5) and access points are not too dispersed so that travel time (and cost) can be minimized relative to wait time at accesses (contact opportunity). It is possible to establish more than one travel circuit to accommodate access points that are dispersed over larger geographical areas.

20.3.1.3 Aerial Surveys

Aerial surveys are made from an airplane that is flying low enough and slow enough to accurately count individual anglers in boats and on the bank. Accurate counts are difficult if the shoreline is irregular or heavily wooded, and biases occur if portions of the population cannot be counted. It is also important to establish criteria for deciding if people are engaged in fishing or some other recreational activity.

Aerial surveys basically are a type of roving survey that yield only data on fishing effort, measured as the number of anglers or fishing boats operating over a given time period within a given area (see section 20.5.1 on measuring fishing effort). The primary advantage is that large areas can be covered in relatively short periods of time so that total enumeration is possible. Where sampling areas have been divided into spatial or temporal sampling units, aerial surveys are particularly useful for establishing sampling probabilities (based on angler counts) and for adjusting these probabilities as necessary. Major disadvantages are plane rental costs, which can limit the number of overflights possible; inclement weather, which may deter pilots but not anglers; and observation bias, such as when it is difficult to distinguish fishing from nonfishing boats or when anglers simply are not visible, for example at night or along wooded shoreline areas (Essig and Holliday 1991).

Angler counts from overflights used in conjunction with daily angler activity curves generated from on-site interviews provide an effective way to estimate daily fishing effort from overflight counts taken during a fixed hour of the day (McNeish and Trial 1991). Daily angler activity curves simply show the proportion of daily angling activity expected to occur during each hour of the day for the temporal and spatial strata defined for the survey. Pollock et al. (1994) devoted a chapter to aerial surveys and provided several examples of their use for counting anglers.

20.3.2 Household Survey Methods

Household surveys usually require that the sample be drawn from a list of names (sampling frame), such as fishing license receipts, boat registrations, or a telephone directory. Random sampling is applied directly to the list. Thus, the anglers themselves, or their residences, become the sampling units, instead of fishing days, access points, time periods, and lake sections commonly applied to on-site surveys. Generally, simple random, stratified random, or systematic random sampling are used to choose samples from lists.

Sampling from lists of names can have biases; for example, not all anglers have licenses, not all anglers have telephones, and boat registrations allow only boat anglers to be sampled. These may be acceptable limitations for certain survey objectives. As Deuel (1980b) noted, the primary advantages of household surveys are that data can be related to the entire population, response rates are high for telephone and door-to-door interviews, and cost per interview is low for mail and telephone surveys.

Information from household surveys can be biased because of faulty memory recall over time. There are many kinds of recall biases that can affect data quality. Telescoping, for example, occurs when anglers include events outside the recall period, and omission or memory decay occurs when anglers omit events within the recall period (Deuel 1980a; Fisher et al. 1991). Other recall biases include digit bias (rounding quantitative responses to the nearest 5 or 10 or larger whole numbers), prestige bias (exaggerating good events and underreporting bad events to boost self-esteem), and inflation bias (unintentionally overreporting memorable or pleasurable events).

Fisher et al. (1991) evaluated the effect of several recall periods on the accuracy of certain characteristics of fishing trips, such as the proportion of people who fished, the number of trips taken, the number of days fished, and trip-related fishing expenditures. A 1-month recall period was assumed to provide minimum recall bias, and responses for longer recall periods were compared with the 1-month recall values. In general, the authors found that as recall period increased, estimates of the trip characteristics became larger. For example, the number of fishing days increased 44% when a 3-month recall period was used and average trip expenditures increased 39%. Relative to the 3-month recall estimates, these trip values increased again, by

34% and 42%, respectively, when an annual recall period was used. Carline (1972) found that a postcard follow-up gave an estimate of catch rate for brook trout that was double that obtained from an access point survey of fishing ponds in Wisconsin. Pollock et al. (1994) suggested using no more than a 1-week recall period for mail and telephone surveys if biological measures of harvested fish are desired. See Pollock et al. (1994), Chapters 4–7, for a detailed summary of biases associated with event recall survey methods.

20.3.2.1 Mail Surveys

With due consideration given to sampling from lists of names and to recall biases as discussed above, mail surveys are particularly useful when trying to describe characteristics of anglers relative to an entire population of people. This is a common objective for regional or statewide surveys or surveys of populations living in defined drainages or along specific waterways or sections of coastline. Mail surveys have been used most successfully to develop social and economic profiles of anglers and to determine attitudes and opinions (Brown 1991; Pollock et al. 1994; Chapter 22 of this book).

Generally, it will be inefficient to use mail surveys to obtain estimates of fishery descriptors for a particular site unless a list of anglers who use the site is available from which to draw a sample. It is possible, however, to hand out questionnaires on site to be filled out later and returned by mail, so that information collected on site can be compared with, or extended by, a follow-up mail survey (Brown 1991; Kokel et al. 1991; Pollock et al. 1994).

Brown (1991) reported that mail surveys avoid the bias inherent in face-to-face and telephone surveys due to unpredictable differences in how people respond to interviewers of varied appearances, personalities, and skills (Hudgins and Malvestuto 1985). Also, respondents are not pressured for immediate responses, and mail surveys allow for the presentation of more complex questions.

Nonresponse bias, when a particular portion of the target population does not respond to the questionnaire, is an inherent problem associated with mail surveys in particular. For angler surveys, typically it is the more enthusiastic, avid, or knowledgeable anglers that are most likely to respond. Less serious anglers will be less likely to return the questionnaire, thus increasing the potential for nonresponse bias. Follow-up mailings and phone calls are possible to reduce the proportion of nonrespondents or to determine if nonrespondents differ in some manner relative to those who responded (a test of nonresponse bias). Factors such as saliency of the survey topic, specificity of the survey audience, length of the survey, size of print, month in which the survey was conducted, and amount of space devoted to hypothetical topics all significantly affected response rates of natural resource users (Brown 1991). See Chapter 22 for more information on designing mail surveys.

20.3.2.2 Telephone Surveys

Telephone surveys share some of the positive and negative aspects of mail and on-site surveys. Telephone surveys generally are more expensive than mail surveys but less expensive than on-site surveys. Like mail surveys, telephone surveys generally are not efficient for the collection of site-specific information, and responses are subject to recall biases depending on recall period. Nonresponse bias, however, is not as serious a problem because people are less likely to refuse an interview when called by telephone. Dillman (1978) found, when analyzing characteristics of well-designed mail and telephone surveys, that average response rates increased from 74% with mail surveys to 91% with telephone surveys.

Sampling frames for telephone surveys are based on random-digit dialing, directories, or special registration lists. Special registration lists include fishing license lists, boat registration lists, and lists of angling club members (Pollock et al. 1994). Random-digit dialing allows all telephone numbers, listed and unlisted, to be included in the frame. Directories are telephone company subscriber lists that do not include unlisted numbers. Simple, stratified, systematic, and two-stage random sampling commonly are applied to these frames.

A special role for telephone surveys is as a follow-up to mail surveys, not only to increase response rate but also, and perhaps more importantly, to evaluate mail surveys for nonresponse bias (Pollock et al. 1994). Obviously, phone numbers of people sampled with the mail survey must be obtained so that those people who do not return a questionnaire can be called at a later time. Telephone contacts also can be made prior to a mail survey to elicit the cooperation of individuals to whom questionnaires will be mailed (Dillman 1978).

Weithman (1991) provided an example of an angler telephone survey conducted in Missouri that was designed to estimate statewide angling effort and fishing success for various fish species at certain places in the state. Anglers' names were obtained from fishing license receipts and stratified by type of permit, such as resident fishing, resident fishing and hunting, 3-d trip permits, and 14-d trip permits. Names of anglers were randomly chosen from each permit type to give sample sizes proportional to the actual numbers of each permit type sold in the state. Phone numbers were associated with anglers' names by means of directories; thus, anglers with unlisted numbers or no telephones were excluded from the sample. Anglers sampled were given a screening questionnaire to elicit their participation over a 2-year period; cooperating anglers were sent a letter of confirmation, instructions, data recording forms, and maps in an effort to reinforce the legitimacy of the survey.

Weithman and Haverland (1991) discussed differences in application of angler telephone surveys and on-site roving creel surveys in Missouri to explain contrasts in estimates of effort and fishing success obtained from the two methods. In general, hours of fishing per hectare were twice as high and catch and harvest rates averaged 20% lower for several species when estimated with the telephone surveys. The on-site creel surveys were limited in application (only 9 months each year and no night fishery sampling), probably explaining the differential in effort values. The authors felt that recall bias likely contributed to lower harvest rate estimates based on the telephone survey, though this conclusion contrasts with findings presented earlier in section 20.3.2 concerning recall bias and catch rate estimators.

20.3.3 Complemented Surveys

Complemented surveys are those in which more than one survey method is used. Some of the biases of each individual survey approach can be overcome by using a complemented survey. For example, a night fishery cannot be monitored reasonably using an on-water roving technique but could be surveyed via access points; a roving survey might be used to obtain angler counts over large areas and completed trip interviews could be obtained via access points to avoid length-of-stay biases; or an access point survey may not adequately sample bank anglers but a roving survey would, so that both together would cover all anglers.

In a strict sense, a complemented survey is one that has two sampling populations. A good example of this is the national (United States) Marine Recreational Fisheries Statistics Survey, which has been conducted by the National Marine Fisheries Service since 1960 (Deuel 1980a, 1980b; Essig and Holliday 1991). Based on extensive presurvey evaluation of contact methods, the designers determined that a telephone survey (based on a sampling population of all coastal households with telephones) and an access point survey (based on a sampling population of all fishing sites) together would most effectively provide the information needed to estimate total harvest of marine sport fish. The telephone survey provides estimates of the percentage of anglers in the entire population and of the number of fishing trips by type and location of fishing; the intercept survey provides estimates of harvest per trip by species. Pollock et al. (1994) classified survey methods into a variety of two-survey combinations in which the first survey is used to estimate fishing effort and the second is used to estimate catch rates, for example, telephone–access designs or mail–roving designs. Estimators for effort and catch are given for each combination.

The survey methods discussed here are only a few of many possible types. In many cases, catch is recorded in one form or another, and catch records can be sampled: charter boat captains keep logbooks, anglers may be required to record catch on fishing permits or they may use diaries, and fishing clubs and tournaments typically keep records. These surveys usually cover only a specific segment of the fishery, are subject to the biases of self-reported data, and must be interpreted accordingly. Pollock et al. (1994) devoted a short chapter to logbooks, diaries, and catch cards; Quertermus (1991) reviewed creel data collected from bass tournaments and gave an example from Georgia; and Sztramko et al. (1991) described an angler diary program on Lake Erie.

20.4 THE INTERVIEW PROCESS

The verbal interview is a behavioral interaction between an interviewer (creel clerk) and respondent (angler). The data collection instrument used by the interviewer is a questionnaire, often called an interview schedule or scripted questionnaire, if questions are read aloud by the interviewer. The interview schedule consists of predetermined, exactly worded questions that, ideally, are easily and clearly understood and elicit responses pertinent to survey objectives. Because anglers are clients as well as respondents, interviews should provide a positive social interaction between management agency personnel and resource users.

This section provides a brief overview of the interview process by categorizing the subject into two subtopics: behavioral protocol, and questionnaire design and presentation. Detailed consideration is given to social research methodology in books by Babbie (1973), Miller (1977), Baily (1978), and Dillman (1978). Pollock et al. (1994) devoted a chapter to questionnaire construction specifically for angler surveys and gave examples of questionnaires for mail, telephone, and on-site surveys. Chapter 22 of this book provides a more in-depth treatment of method options for collection of human dimension data.

20.4.1 Behavioral Protocol

On-site intercept surveys require that anglers be contacted during their fishing trips. The interviewer must realize he or she is interrupting the respondent's privacy and leisure time to request information. At the same time, the respondent, a resource user, is likely to judge the interviewer who is representing the agency conducting the survey. The delicacy of this interaction is readily apparent, and there is a behavioral protocol that will help ensure a successful interview.

Establish contact in as courteous a manner as possible. The situation is especially challenging during roving surveys when the interview takes place from boat to boat on the water. There usually will be entries on the interview form that can be answered prior to verbal contact (see top of Figure 20.1), and there is no need to interrupt the respondent before recording this information. Approach anglers by slowing down far enough away to minimize (if not eliminate) boat wake and to avoid tangling the anglers' fishing gear. (A trolling motor is a handy tool for boat-to-boat interviews.) Call to the anglers from a distance that does not interrupt their fishing. Unless the respondent has harvested fish that must be measured, the interview can be conducted with minimum inconvenience to the angler.

Try to gain anglers' trust from the beginning of the interview. Dress in a manner acceptable to the people being interviewed and be officially identifiable (e.g., emblem on shirt or cap or boat label). After greeting the respondent, provide a brief explanation of the purpose of the survey as soon as possible. Anglers are not required by law to answer questions, and their rights should be respected; ask if they are willing to respond to the questionnaire with the understanding that their answers will remain anonymous. Emphasize to the angler that his or her responses are very important because only a small portion of all anglers using the resource will be interviewed and the information is needed to improve the fishery. If they do not want to participate, do not pressure them to respond.

If accurate harvest data are desired, try to check the creel yourself rather than relying on anglers' recollections. Measuring fish usually will interrupt fishing activity, especially for boat-to-boat interviews. Measure fish at the end of the interview (see Figure 20.1). Do not pressure anglers to allow their fish to be measured but emphasize that the information will be important to fishery managers.

Because of the positive social relationship that creel clerks seek to establish with anglers, the objectives of creel clerks and law enforcement officers are not complementary. If an objective is to remind anglers when they are in violation of fishing regulations, a standard method should be established so that the creel survey is not jeopardized. If a violation is noted, for example, point it out to the anglers at the end of the interview, after fish have been measured, and politely let them know that the violation brings a fine and that the game warden will not take the violation lightly. The value of the interview as an information exchange mechanism can be enhanced by providing anglers with written information about the survey, including current survey results if available. Creel clerks thus can act more as extension agents rather than as simply information gatherers, and respondents have a better appreciation of the end result of donating their time and information.

20.4.2 Questionnaire Design and Presentation

Design and presentation of a questionnaire are critical to the collection of high-quality data. Questionnaire design refers to the intent, sequence, and wording of questions; questionnaire presentation refers to the interviewer's demeanor, knowledge of question intent, phrasing of questions, and use of verbal probes and visual prompts.

Only include questions in the interview schedule that are relevant to the objectives of the survey. Baily (1978) suggested that if you cannot decide in advance how the answers will be statistically analyzed and published (or otherwise presented), then you should not ask the questions. Take care to avoid two-part questions, ambiguous questions, negatively phrased questions, and biased terms or phrases. See Babbie (1973) for examples.

The interview schedule should be well organized, not only to enhance the ease with which anglers respond but also to help the interviewer. Place questions in a logical order; for example, items on the interview form shown in Figure 20.1 are ordered so data that can be collected prior to the interview are entered at the top of the form. The actual angler interview begins in the next section of the form with exactly worded questions that progress from asking the place of trip origin to requesting an estimate of the time that the respondent will finish fishing that day (a time sequence). The final section on the form concerns harvest information and provides a space for recording lengths, numbers, and weights of fish by species. The interview schedule ends with an invitation to respondents to ask questions or express opinions.

Interviewers should concentrate on accurately recording the views of respondents (keeping in mind that open-ended responses will have to be categorized for analysis) and refrain from discussion that might destroy an otherwise positive encounter. Some general rules for question order are (1) ask easy-to-answer questions and questions needed for subsequent interviewing first; (2) put sensitive and open-ended questions late in the questionnaire; (3) vary questions in type and length to keep the interest of respondents; and (4) avoid establishing stereotyped responses. See Baily (1978) for details.

It is tempting to increase questionnaire length because the cost of adding questions is small relative to that of survey equipment and operating costs. However, there is a point of diminishing returns because inconvenience to anglers and data processing time increase to accommodate information of marginal value. Additionally, longer interview schedules reduce the number of anglers that can be interviewed during the sampling period. Hudgins and Malvestuto (1985) found that interview length would have to be 6 min or less to sample 20% of the anglers present in 4-h sampling periods when fishing pressure counts exceeded 60 anglers. They also found that creel clerks, in an effort to obtain more interviews and to gain time for angler counts, tended to shorten interview length as the number of anglers increased. Information trade-offs such as more anglers versus more information per angler must be weighed carefully when the survey is designed.

To aid in the consistency of data collection, all questions should be fully expressed on the interview form. Include the introductory remarks as well as other connecting statements (Figure 20.1). The interviewer should play a neutral role but facilitate the interview process. Interviewers should be trained in questionnaire delivery and should understand the intent of all questions. Data comparability depends on consistent use of the instrument; different interviewers can elicit systematically different responses from the same questions, which can decrease the accuracy of the information collected (Hudgins and Malvestuto 1985). Visual prompts, such as holding up a ranking scale (1 = not important, 2 = slightly important, 3 = important, and 4 = very important) can orient the respondent and decrease response error. Chapter 22 provides more details on questionnaire construction and interview techniques.

20.5 OVERVIEW OF QUANTITATIVE PROCEDURES

The primary focus of this section is on collecting information necessary to estimate fishing effort, CPUE or HPUE, and total catch or harvest. Regardless of the sampling design or contact method, catch can be estimated as the product of effort and CPUE. Therefore our primary concern is estimating these two components. Effort can be estimated using angler counts, but CPUE is obtained through the interview process. The basic information that must be recorded for each interview is the amount of time spent fishing (unless the chosen unit of fishing effort is something other than the angler-hour) and the number and weight of each species harvested. Taking length measurements will provide additional information on the length structure of harvestable-sized stocks.

20.5.1 Fishing Effort Estimates

Fishing effort estimates from roving creel surveys are based on angler count data, whereas fishing effort estimates from access point surveys are based on completed trip lengths of anglers as obtained through interviews. Completed trip lengths are simply added over all interviews to obtain an estimate of fishing effort for a given access point during a given sampling period. If there are too many anglers leaving through a given access to interview everyone, then anglers also must be counted; fishing effort would be estimated by multiplying the angler count by the average length of a completed fishing trip (Box 20.6). Also see Hayne (1991) and Pollock et al. (1994).

Whether "instantaneous" counts are made from a vantage point or "progressive" counts are made by roving over a longer period of time, the creel clerk should strive to count all anglers operating within the specified sampling area. Counts are converted to angler-hours by multiplying the number of anglers by the number of hours in the sampling period (Neuhold and Lu 1957; Lambou 1961; Box 20.6: example 3). Alternatively, total angler-hours can be estimated by multiplying the count by the average length of a completed fishing trip (as for the access point example above); however, this estimator is likely to be less precise because an additional variable, trip length, is involved in the calculation. Also, unbiased estimates of completed trip length may be hard to get with the roving survey technique (see discussion in section 20.3.1.1).

Avoiding the use of mean trip length and using the number of hours in the time period as a multiplier instead rests on the assumption that the number of anglers counted is an unbiased estimate of the number of angler-hours in progress at any given instant. Twenty anglers fishing during one instant means that after 1 h, 20 angler-hours have been expended and after 4 h, 80 angler-hours have been expended. The assumption remains the same regardless of how the count is taken, from a vantage point or from a plane or boat progressively circling a sample section.

Field studies show that short counting periods provide better data, and the ideal situation is to take short counts within any given sampling period and average them to obtain an estimate of the true instantaneous count. However, Neuhold and Lu (1957) showed that progressive counts taken over a 1-h sampling period were similar to instantaneous counts taken from a vantage point, and Malvestuto (unpublished) found that progressive counts taken over a 4-h sampling period were similar to counts taken over a 1-h sampling period at West Point Lake, Alabama–Georgia. Lambou (1961) gave a detailed quantitative discussion of angler count data, and Pollock et al. (1994) contrasted instantaneous and progressive counts.

Box 20.6 Estimating Lakewide Fishing Effort

Expansion of secondary sampling unit (SSU) fishing effort data to daily lakewide estimates. Calculations are shown for three on-site survey design examples.

Example 1

Example 1 represents an access point survey with three landings (A, B, and C) where the clerk remains all day long. Landings A, B, and C were assigned sampling probabilities of 0.50, 0.30, and 0.20, respectively, based on percentages of expected use by boat anglers. On each sample day, one landing was chosen randomly to be the work site. The data below show fishing times in hours from six completed trip interviews of anglers taken on a sample day when landing C was chosen as the work site. The six interviews of 11 anglers represent all of the use at landing C that day. The first step is to calculate the number of angler-hours (e_i) for each (ith) interview by multiplying the number of anglers (m_i) by the time spent fishing (t_i). Then the number of anglers and the number of angler-hours are summed (Σ) over the number of interviews (n) to expand the data.

Interview number	Number of anglers (m_i)		Fishing time (t_i)		Angler-hours (e_i)
1	1	×	3.3	=	3.3
2	2	×	7.0	=	14.0
3	3	×	4.6	=	13.8
4	1	×	5.2	=	5.2
5	2	×	4.1	=	8.2
6	2	×	6.5	=	13.0
$n = 6$	$\sum_{i=1}^{n} m_i = 11$				$\sum_{i=1}^{n} e_i = 57.5$

Landing C is expected to exhibit 20% of the total angling activity, and 0.20 was the probability used to choose landing C randomly. Thus, total angler-hours of use at landing C on any given day would be expected to represent about one-fifth of the total fishing effort over the entire lake. The expansion is calculated as

$$\text{Total lakewide daily effort} = \hat{E} = 57.5/0.20 = 287.5 \text{ angler-hours.}$$

Example 2

Example 2 is the same as Example 1, except that the six interviews of 11 anglers now represent only a portion of all anglers who used landing C on the day sampled. If there were 24 anglers who used landing C, and only 11 were included in interviews, then lakewide daily fishing effort can be calculated as follows.

Box 20.6 Continued.

1. Compute mean fishing time per angler ($\bar{t}$) by summing angler-hours over the number of interviews (n) and dividing by total anglers interviewed.

$$\text{Mean fishing time} = \bar{t} = \sum_{i=1}^{n} e_i / \sum_{i=1}^{n} m_i = 57.5 \text{ angler-hours}/11 \text{ anglers} = 5.23 \text{ h.}$$

2. Compute total use at landing C ($\hat{e}$) during the sample day by multiplying mean fishing time per angler ($\bar{t}$) by the total number of anglers (M) who passed through the access point.

$$\text{Angler-hours at access} = \hat{e} = M \cdot \bar{t} = 24 \text{ anglers} \cdot 5.23 \text{ h}$$

$$= 125.5 \text{ angler-hours.}$$

3. Expand total angler-hours at the access that day to a lakewide value ($\hat{E}$) by dividing $\hat{e}$ by the sampling probability (p) associated with landing C.

$$\text{Daily lakewide effort} = \hat{E} = \hat{e}/p = 125.5/0.20 = 627.5 \text{ angler-hours.}$$

Example 3

Example 3 represents a roving creel survey in which sampling was conducted as per Box 20.4. A lake was divided into three sections (1, 2, and 3) that had sampling probabilities of 0.5, 0.25, and 0.25, respectively. Additionally, each sample day was divided into two 6-h time periods (AM and PM) that had sampling probabilities of 0.4 and 0.6, respectively. During a given sample day, only one time period-lake section combination (secondary sampling unit, SSU) was chosen randomly to sample. Calculation of daily fishing effort requires only that counts of anglers be taken. All anglers fishing within the chosen SSU were counted during a randomly chosen hour within a 6-h time period. If on a given sample day, the AM time period was chosen at random along with lake section 3, and 46 anglers were counted within the SSU, then total daily lakewide fishing effort can be calculated as follows.

1. Compute angler-hours for the SSU by multiplying the angler count (C) by the number of hours in the sampling period (H).

$$\text{Angler-hours for SSU} = \hat{e} = C \cdot H = 46 \text{ anglers} \cdot 6 \text{ h} = 276 \text{ angler-hours.}$$

2. Expand angler-hours for the SSU to a lakewide estimate for the entire sample day by dividing $\hat{e}$ by the product of the time period probability (p_t) and the lake section probability (p_s).

$$\text{Daily lakewide fishing effort} = \hat{E} = \hat{e}/p_t \cdot p_s = 276 \text{ angler-hours}/(0.40 \cdot 0.25)$$

$$= 2{,}760 \text{ angler-hours.}$$

If several progressive counts of anglers were taken within an SSU, then a mean count ($\bar{C}$) would be calculated and multiplied by H to give $\hat{e}$ as in step 1 above.

Care must be taken if counts and interviews are conducted concurrently (the count-as-you-go method). When time is spent interviewing anglers, then the probability of intercepting the remaining anglers decreases (Robson 1991). Thus, estimates of fishing effort can be negatively biased if obtained by the roving clerk method (Jones et al. 1990; Wade et al. 1991). If the creel clerk stays on schedule, however, the bias is negligible, always less than 1.4%, even when 15-min interviews are conducted (Wade et al. 1991). To stay on schedule, creel clerks can strive to arrive at checkpoints along the travel route at specified times, or clerks can pace themselves by skipping anglers to interview (in an objective manner), thus maintaining a relatively constant rate of travel around the sample section. The potential for negatively biased effort estimates logically is reduced if sampling periods are short (not greater than the average trip length), given that the probability of anglers leaving over short periods is less than the probability over longer periods.

The sampling unit frequently represents only a portion of the angling population present on any given day, as when the day has been divided into sampling periods, a body of water has been divided into sections, or there are several access points to sample (Box 20.4). In these cases, angler-hours within the sampling unit are expanded to estimate total angler-hours for the entire day by dividing the sampling unit value by the sampling probability associated with the particular unit. Box 20.6 gives examples of these types of expansions. Total angler-hours can be converted to angler-trips by dividing by average time per trip spent fishing. Angler-trips can be converted to boat-trips by knowing the mean number of anglers per boat. It is usually feasible to count anglers separately by fishing type (bank versus boat), and independent effort estimates then can be calculated. Fishing effort over a given time period (usually a year) at a given body of water can be expressed on a per-unit-of-surface-area basis for purposes of comparison with other sites.

In most instances, it is desirable to partition fishing effort according to the species or group of species sought. This requires that anglers be asked to identify the type of fish they intend to catch. Given a statistically valid sampling design, the sample percentages of effort expended for each class of fish can be multiplied by total effort to estimate intended or target effort. Sample percentages of intended effort should be based on completed trip times.

20.5.2 Estimation of Catch per Unit Effort

There are three major reasons to measure CPUE: (1) to estimate total catch over a specified time period; (2) to obtain an index of stock density (see Box 6.1); and (3) to measure fishing quality or fishing success. Creel survey estimates of CPUE or HPUE are obtained by dividing measured (recorded) catch or harvest, respectively, by measured effort (usually angler-hours). This can be done for each fishing party or by summing catch and effort over interviews within some specified time frame (e.g., a day or month). For the roving survey, measured effort is the uncompleted trip time; for access point surveys, it is completed trip time.

Whereas harvest can be directly measured by the creel clerk, catch is a combination of observable harvest plus fish caught and returned to the water; therefore, catch estimates rely, to some degree, on anglers' recall and are subject to recall bias. Given that anglers are asked to recall only fish caught and released during the fishing trip in progress at the time of interview, catch estimates likely are not seriously biased, particularly for sought-after species. When fishing days are subsampled, the calculated CPUE for the subsampling unit is taken to represent CPUE for the entire day.

Catch per unit of effort is a ratio estimate. There are several different ways to estimate CPUE (Crone and Malvestuto 1991); however, two primary methods are the mean-of-ratios and ratio-of-means, or total-ratio, estimators (Box 20.7). For the mean-of-ratios method, separate CPUE values are calculated, for each day or for each fishing party (party values). An average then is taken over these separate values. Based on an empirical analysis of data from four lakes in Alabama, Crone and Malvestuto (1991) showed that mean party values of HPUE were inherently more variable than mean daily values and also that use of party data generally produced higher estimates of HPUE.

Mean-of-ratios estimators place equal weighting on each value of CPUE (or HPUE) used to calculate the mean, regardless of the amount of fishing on which the value was based (Box 20.7). Thus, a daily CPUE based on 60 h of fishing gets the same weight as a daily value based on 6 h of fishing when the mean ratio is calculated. Conceptually, then, mean-of-ratios estimators probably are not the best to use for measurements of stock density or fishing success, cases in which a proportional weighting of fishing effort seems most appropriate. It is appropriate, however, to use mean daily CPUE for calculation of daily catch, which then can be expanded to an estimate of total catch for a specified period of time by multiplying by the number of days in the period (Malvestuto et al. 1978). Pollock et al. (1994) stated that the mean-of-ratios estimator is the most appropriate one to use when data are based on uncompleted fishing trips. Additionally, very short fishing trips (<0.5 h) should not be included in the estimate (Pollock et al. 1994). The variance of mean-of-ratios estimators is calculated as for any set of independent observations (Malvestuto et al. 1978; Crone and Malvestuto 1991; Pollock et al. 1994).

The other approach to estimating CPUE is to calculate a single ratio by dividing the sum of measured catch over all sampling days (or fishing parties) by the sum of measured effort over all days (or parties). This is the ratio-of-means, or total-ratio, estimator (Box 20.7). This estimator is self-weighting, that is, it is influenced by differences in daily (party) fishing effort. (See Snedecor and Cochran 1980, section 21.12, or Pollock et al. 1994, equation 15.9, for variance estimators.) Because individual contributions to the total-ratio estimate are weighted by the amount of fishing effort expended, this estimator seems most appropriate for measuring angling success and indexing stock density. Pollock et al. (1994) stated that it is the appropriate estimator to use for calculation of total catch when completed trip data are used. Crone and Malvestuto (1991) found that precision associated with the total-ratio estimator was similar to that of the mean-ratio estimator when daily values of harvest and effort were used in the computations (rather than party values).

To evaluate angling success, it is most relevant to calculate catch rate according to species sought. It is generally accepted that estimates of CPUE based on total measured effort are not appropriate for evaluating fishing success for particular species unless all species are equally vulnerable to angling or unless the proportional contribution to total effort by anglers fishing for different species stays constant over time. The CPUE for a given species of fish can be calculated by dividing the harvest of that species by the angler-hours directed toward that species (Lambou and Stern 1959; Lambou 1966; Von Geldern 1972; and Von Geldern and Tomlinson 1973). As a measure of fishing success, this form of CPUE assumes either that (1) anglers catch primarily what they seek or (2) fish caught, but not sought, have little bearing on anglers' perceptions of fishing success. From a computational point of view, using mean daily values of CPUE for particular species is not practical because too many

Box 20.7 Estimating Harvest per Unit Effort

Calculation of harvest per unit effort (HPUE) based on the mean-of-ratios ($\hat{R}_1$) and the ratio-of-means, or total-ratio ($\hat{R}_2$), estimators. Calculations are based on five hypothetical interviews of anglers. Each interview provides total number of fish harvested and total number of hours spent fishing.

Sample Data

As raw data, HPUE_i for each (ith) angler can be expressed as a ratio in which the numerator is number of fish harvested (h_i) and the denominator is number of hours fished (e_i).

	Angler				
	1	2	3	4	5
$\text{HPUE}_i = \dfrac{\text{Fish harvested}}{\text{Angler-hours}} = \dfrac{h_i}{e_i} =$	$\frac{3}{2}$	$\frac{0}{1}$	$\frac{15}{12}$	$\frac{6}{8}$	$\frac{15}{10}$

Mean-of-Ratios Estimator ($\hat{R}_1$)

$$\hat{R}_1 = \frac{\sum_{i=1}^{n} (h_i/e_i)}{n}.$$

1. Compute an HPUE value for each (ith) angler by dividing number of fish harvested (h_i) by number of hours fished (e_i).

	Angler				
	1	2	3	4	5
$\text{HPUE}_i = h_i/e_i =$	1.50	0.00	1.25	0.75	1.50

2. Sum HPUE_i values over the anglers interviewed (n).

$$\text{Sum of HPUE}_i = \sum_{i=1}^{n}(h_i/e_i) = 1.50 + 0.00 + 1.25 + 0.75 + 1.50 = 5.00$$

3. Calculate a mean HPUE ($\hat{R}_1$) by dividing the sum of HPUE_i by the number of anglers interviewed (n).

$$\text{Mean-ratio HPUE} = \hat{R}_1 = \frac{\sum_{i=1}^{n} (h_i/e_i)}{n} = \frac{5.00}{5} = 1.00 \text{ fish/h.}$$

Ratio-of-Means, or Total-Ratio, Estimator ($\hat{R}_2$)

$$\hat{R}_2 = \frac{\sum_{i=1}^{n} h_i}{\sum_{i=1}^{n} e_i}.$$

Box 20.7 Continued.

1. Sum fish harvested per angler (h_i) over all anglers interviewed (n) as a numerator for the ratio.

$$\text{Sum of harvests} = \sum_{i=1}^{n} h_i = 3 + 0 + 15 + 6 + 15 = 39 \text{ fish.}$$

2. Sum hours fished per angler (e_i) over all anglers interviewed (n) as a denominator for the ratio.

$$\text{Sum of hours} = \sum_{i=1}^{n} e_i = 2 + 1 + 12 + 8 + 10 = 33 \text{ h.}$$

3. Calculate a mean HPUE ($\hat{R}_2$) by dividing the numerator from step 1 by the denominator from step 2.

$$\text{Total-ratio HPUE} = \hat{R}_2 = \frac{\sum_{i=1}^{n} h_i}{\sum_{i=1}^{n} e_i} = \frac{39 \text{ fish}}{33 \text{ h}} = 1.18 \text{ fish/h.}$$

There is nearly a 20% difference between the two ratio estimates of HPUE.

days are lost from the sample when there is no fishing effort directed toward certain fishes. The total-ratio estimator based on catch and effort data from individual parties (interviews), sorted according to target species, is the most practical way to measure fishing success for individual species or species groups.

20.5.3 Statistical Considerations

Statistically, the quantitative procedures for calculating means, totals, and variances of various fisheries statistics depend on the survey sampling design used to collect the data (see section 20.2). Survey design books such as Cochran (1977) and Jessen (1978) provided specific statistical formulas; statistical design alternatives for angler surveys are succinctly presented by Pollock et al. (1994) and, for small-scale fishery surveys, by Bazigos (1974). Pollock et al. (1994) also provided several numerical examples, with step-by-step computations, for several angler survey designs. Malvestuto et al. (1978) provided a summary of the statistical formulas for calculation of effort, HPUE, and harvest for a roving survey based on nonuniform probability sampling (see section 20.2.3).

Keep in mind that even if the catch and effort information collected is unbiased, it will be of little value for documenting change in the fishery unless it is also precise. A convenient measure of precision is the relative standard error, or RSE. The RSE simply expresses the standard error as a percentage of the estimate with which it is associated (also called the proportional standard error). An RSE in excess of 20% is not desirable—at this level of precision, 95% confidence intervals around estimates would be roughly ±40% or greater. With this level of precision, estimates of important fishery characteristics could not be reported in a concrete fashion, and real changes in the fishery, either temporally or spatially, would be difficult to verify.

Fishing effort appears to be inherently less variable than CPUE (Bayley et al. 1991; Malvestuto 1991; Malvestuto and Knight 1991; Meredith and Malvestuto 1991; Palsson 1991). Relatively precise estimates of fishing effort (RSEs of 15–20% or less) are attainable for most fisheries with moderate levels of monthly sampling effort (6–12 d), stratified by weekdays and weekend days. It can be difficult to obtain the required precision for estimates of CPUE because of within-day variability in catchability among species and because of variability in success from party to party. Sample sizes should be increased to at least 60 to 70% of all days within the survey period to keep RSEs for CPUE less than 20% (Bayley et al. 1991; Malvestuto 1991). It is likely that within-day stratification (morning, afternoon, and evening, for example) and geographical stratification for large water bodies also will be necessary to obtain consistently improved estimates of catch and harvest rates.

20.6 REFERENCES

American Fisheries Society. 1987. Social assessment of fisheries resources. Transactions of the American Fisheries Society 116.

Anderson, R. O. 1975. Optimum sustainable yield in inland recreational fisheries management. American Fisheries Society Special Publication 9:29–38.

Babbie, E. R. 1973. Survey research methods. Wadsworth, Belmont, California.

Bailey C., C. K. Harris, and C. K. Vanderpool, editors. 1986. Proceedings of the workshop of fisheries sociology. Woods Hole Oceanographic Institute Technical Report WHOI-86-34.

Baily, K. D. 1978. Methods of social research. The Free Press, New York.

Bayley, P. B., and M. Petrere, Jr. 1989. Amazon fisheries: assessment methods, current status and management options. Canadian Special Publication of Fisheries and Aquatic Sciences 106:385–398.

Bayley, P. B., S. T. Sobaski, M. H. Halter, and D. J. Austen. 1991. Comparisons of Illinois creel surveys and the precision of their estimates. American Fisheries Society Symposium 12:206–211.

Bazigos, G. P. 1974. The design of fisheries statistical surveys. FAO (Food and Agriculture Organization of the United Nations) Fisheries Technical Paper 133.

Beverton, R. J. H., and S. J. Holt. 1957. On the dynamics of exploited fish populations. Her Majesty's Stationery Service Office, London.

Brown, T. L. 1991. Use and abuse of mail surveys in fisheries management. American Fisheries Society Symposium 12:255–261.

Caddy, J. F., and G. P. Bazigos. 1985. Practical guidelines for statistical monitoring of fisheries in manpower limited situations. FAO (Food and Agriculture Organization of the United Nations) Technical Paper 257.

Carlander, K. D., C. J. DiCostanzo, and R. J. Jessen. 1958. Sampling problems in creel census. Progressive Fish-Culturist 20:73–81.

Carline, R. F. 1972. Biased harvest estimates from a postal survey of a sport fishery. Transactions of the American Fisheries Society 101:262–269.

Cochran, W. G. 1977. Sampling techniques, 3rd edition. Wiley, New York.

Crone, P. R., and S. P. Malvestuto. 1991. A comparison of five estimators of fishing success from creel survey data on three Alabama reservoirs. American Fisheries Society Symposium 12:61–66.

Crow, D. G., and S. P. Malvestuto. 1996. Evaluation of a roving creel-survey design for a large reservoir with an emphasis on subsampling within days. American Fisheries Society Symposium 16.

Dent, R. J., and B. Wagner. 1991. Changes in sampling design to reduce variability in selected estimates from a roving creel survey conducted on Pomme de Terre Lake. American Fisheries Society Symposium 12:88–96.

Deuel, D. G. 1980a. Special surveys related to data needs for recreational fisheries. Pages 77–81 *in* J. G. Grover, editor. Allocation of fishery resources. European Inland Fisheries Advisory Commission and Food and Agriculture Organization of the United Nations, Rome, Italy.

Deuel, D. G. 1980b. Survey methods used in the United States marine recreational fishery statistics program. Pages 82–86 *in* J. G. Grover, editor. Allocation of fishery resources. European Inland Fisheries Advisory Commission and Food and Agriculture Organization of the United Nations, Rome, Italy.

Dillman, D. A. 1978. Mail and telephone surveys: the total design method. Wiley, New York.

Essig, R. J., and M. C. Holliday. 1991. Development of a recreational fishing survey: the marine recreational fishery statistics survey case study. American Fisheries Society Symposium 12:245–254.

Fisher, W. L., A. E. Grambsch, D. E. Eisenhower, and D. R. Morganstein. 1991. Length of recall period and accuracy of estimates from the national survey of fishing, hunting, and wildlife-associated recreation. American Fisheries Society Symposium 12:367–374.

Guthrie, D., and seven coeditors. 1991. Creel and angler surveys in fisheries management. American Fisheries Society Symposium 12.

Hayne, D. W. 1991. The access point creel survey: procedures and comparison with the roving-clerk creel survey. American Fisheries Society Symposium 12:123–138.

Hudgins, M. D., and S. P. Malvestuto. 1985. An evaluation of factors affecting creel clerk performance. Proceedings of the Annual Conference Southeastern Association of Fish and Wildlife Agencies 36(1982):252–263.

Jessen, R. J. 1978. Statistical survey techniques. Wiley, New York.

Jones, C. M., and D. S. Robson. 1991. Improving precision in angler surveys: traditional access design versus bus route design. American Fisheries Society Symposium 12:177–188.

Jones, C. M., D. S. Robson, D. Otis, and S. Gloss. 1990. Use of a computer simulation model to determine the behavior of a new survey estimator of recreational angling. Transactions of the American Fisheries Society 119:41–54.

Knight S. S., and S. P. Malvestuto. 1991. Comparison of three allocations of monthly sampling effort for the roving creel survey on West Point Lake. American Fisheries Society Symposium 12:97–101.

Kokel, R. W., J. S. Stanovick, L. A. Nielsen, and D. J. Orth. 1991. When to ask: angler responses at different times in the fishing trip and year. American Fisheries Society Symposium 12:102–107.

Lambou, V. W. 1961. Determination of fishing pressure from fishermen or party counts with a discussion of sampling problems. Proceedings of the Annual Conference Southeastern Association of Game and Fish Commissioners 15(1961):380–401.

Lambou, V. W. 1966. Recommended method of reporting creel survey data for reservoirs. Oklahoma Fishery Research Laboratory Bulletin 4, Norman.

Lambou, V. W., and H. Stern, Jr. 1959. Creel census methods used on Clear Lake, Richland Parish, Louisiana. Proceedings of the Annual Conference Southeastern Association of Game and Fish Commissioners 12(1958):169–175.

Lester, N. P., M. M. Petzold, and W. I. Dunlop. 1991. Sample size determination in roving creel surveys. American Fisheries Society Symposium 12:25–39.

MacKenzie, C. 1991. Comparison of northern pike catch and harvest rates estimated from uncompleted and completed fishing trips. American Fisheries Society Symposium 12:245–254.

Malvestuto, S. P. 1983. Sampling the recreational fishery. Pages 400–423 *in* L. Nielsen and D. Johnson, editors. Fisheries techniques. American Fisheries Society, Bethesda, Maryland.

Malvestuto, S. P. 1989. Sociological perspectives on large river management: a framework for the application of optimum yield. Canadian Special Publication of Fisheries and Aquatic Sciences 106:589–599.

Malvestuto, S. P. 1991. The customization of recreational fishery surveys for management purposes in the United States. Pages 201–213 *in* I. G. Cowx, editor. Catch effort sampling strategies: their application in freshwater fisheries management. Fishing News Books, Oxford, UK.

Malvestuto, S. P., W. D. Davies, and W. L. Shelton. 1978. An evaluation of the roving creel survey with nonuniform probability sampling. Transactions of the American Fisheries Society 107:255–262.

Malvestuto, S. P., W. D. Davies, and W. L. Shelton. 1979. Predicting the precision of creel survey estimates of fishing effort by use of climatic variables. Transactions of the American Fisheries Society 108:43–45.

Malvestuto, S. P., and S. S. Knight. 1991. Evaluation of components of variance for a stratified two-stage roving creel survey design with implications for sample size allocation. American Fisheries Society Symposium 12:108–115.

Malvestuto, S. P., and E. K. Meredith. 1989. Assessment of the Niger River fishery in Niger, 1983–1985, with implications for management. Canadian Special Publication of Fisheries and Aquatic Sciences 106:533–544.

Malvestuto, S. P., R. J. Scully, and F. F. Garzon. 1980. Catch assessment survey design for the upper Meta River fishery, Colombia, South America. Alabama Agricultural Experiment Station Research and Development Series 27. Auburn University, International Center for Aquaculture, Auburn, Alabama.

McNeish, J. D., and J. G. Trial. 1991. A cost-effective method for estimating angler effort from interval counts. American Fisheries Society Symposium 12:236–243.

Meredith, E. K., and S. P. Malvestuto. 1991. An evaluation of survey designs for the assessment of effort, catch rate and catch for two contrasting river fisheries. Pages 223–232 *in* I. G. Cowx, editor. Catch effort sampling strategies: their application in freshwater fisheries management. Fishing News Books, Oxford, UK.

Miller, D. C. 1977. Handbook of research design and social measurement, 3rd edition. David McKay Company, New York.

Neuhold, J. M., and K. H. Lu. 1957. Creel census method. Utah State Department of Fish and Game Publication 8, Salt Lake City.

Palsson, W. A. 1991. Using creel surveys to evaluate angler success in discrete fisheries. American Fisheries Society Symposium 12:139–154.

Phippen, K. W., and E. P. Bergersen. 1987. Angling definitions and their effects on the accuracy of count-interview creel survey harvest estimates. North American Journal of Fisheries Management 7:488–492.

Phippen, K. W., and E. P. Bergersen. 1991. Accuracy of a roving creel survey's harvest estimate and evaluation of possible sources of bias. American Fisheries Society Symposium 12:51–60.

Pollock, K. H., C. M. Jones, and T. L. Brown. 1994. Angler survey methods and their application in fisheries management. American Fisheries Society Special Publication 25.

Quertermus, C. J. 1991. Use of bass club tournament results to evaluate relative abundance and fishing quality. American Fisheries Society Symposium 12:515–519.

Ricker, W. E. 1975. Computation and interpretation of biological statistics of fish populations. Fisheries Research Board of Canada Bulletin 191.

Robson, D. S. 1961. On the statistical theory of a roving creel census of fishermen. Biometrics 17:415–437.

Robson, D. S. 1991. The roving creel survey. American Fisheries Society Symposium 12:19–24.

Robson, D. S., and C. M. Jones. 1989. The theoretical basis of an access site angler survey design. Biometrics 45:83–98.

Snedecor, G. W., and W. G. Cochran. 1980. Statistical methods, 7th edition. The Iowa State University Press, Ames.

Stanovick, J. S., and L. A. Nielsen. 1991. Assigning nonuniform sampling probabilities by using expert opinion and multiple- use patterns. American Fisheries Society Symposium 12:189–194.

Sztramko, L. K. 1991. Improving precision of roving-creel-survey estimates: implications for fisheries with a closed season. American Fisheries Society Symposium 12:116–121.

Sztramko, L. K., W. I. Dunlop, S. W. Powell, and R. G. Sutherland. 1991. Applications and benefits of an angler diary program on Lake Erie. American Fisheries Society Symposium 12:520–528.

Van Den Avyle, M. J. 1986. Measuring angler effort, success, and harvest. Pages 57–64 *in* G. E. Hall and M. J. Van Den Avyle, editors. Reservoir fisheries management strategies for the 80's. American Fisheries Society, Southern Division, Reservoir Committee, Bethesda, Maryland.

Von Geldern, C. E., Jr. 1972. Angling quality at Folsom Lake, California, as determined by a roving creel census. California Fish and Game 58:75–93.

Von Geldern, C. E., Jr., and P. K. Tomlinson. 1973. On the analysis of angler catch rate data from warmwater reservoirs. California Fish and Game 59:281–292.

Wade, L. W., C. M. Jones, D. S. Robson, and K. H. Pollock. 1991. Computer simulation techniques to assess bias in the roving creel-survey estimator. American Fisheries Society Symposium 12:40–46.

Weithman, A. S. 1991. Telephone survey preferred in collecting angler data statewide. American Fisheries Society Symposium 12:271–280.

Weithman, A. S., and P. Haverland. 1991. Comparability of data collected by telephone and roving creel surveys. American Fisheries Society Symposium 12:67–73.

Chapter 21

Commercial Fisheries Surveys

MARY C. FABRIZIO AND R. ANNE RICHARDS

21.1 INTRODUCTION

In this chapter we describe methods for sampling commercial fisheries and identify factors affecting the design of sampling plans. Commercial fisheries are operations that produce economic benefits by harvesting fish, shellfish, marine plants, or other aquatic resources. The harvest is marketed for consumption (e.g., fresh or frozen fish), industrial processing (e.g., into fish oil, fish meal, or pet food), or other uses. On a global scale, commercial fisheries provided just over 98 million metric tons of fish and shellfish in 1992; the U.S. share of the harvest was slightly less than 6% and was valued at US$4 billion (NMFS 1994).

When sampled properly, commercial fisheries can provide important information on the response of aquatic organisms to exploitation; such information can be used by management agencies to develop regulations for ensuring long-term production of the resource and long-term economic benefit. Fishery statistics are typically used to estimate abundance, mortality, recruitment, growth, and other vital characteristics of populations. Fishery statistics can also be used to study changes in fish community composition resulting from differential exploitation of species. For example, catch statistics from artisanal fisheries in Papua New Guinea have been used to reveal changes in species composition (Medley et al. 1993), and Murawski (1991) used catch statistics to confirm a shift from a community dominated by gadoid and demersal species to one dominated by elasmobranch species on Georges Bank in the northwest Atlantic Ocean.

Commercial fishery sampling generally focuses on two components: the catch and attributes of the fishery itself. The catch is the quantity of fish (or other resource) captured by commercial fishers or fishing gear. The fishery refers to the operation of fishing units (fishers, vessels, gear types, or combinations of these) that harvest the catch and, more importantly, to the amount of effort expended in harvesting the resource. Data from both components can be used to infer the response of populations to removals. Thus, commercial fishery statistics are the building blocks of many important techniques for assessing the condition of exploited fish stocks.

Although fishery statistics are extremely valuable, they have limitations when used to study the response of populations to exploitation. This is because the fishery does not "sample" or collect resources in a random manner; fishers seek concentrations of organisms and expend their effort where and when the likelihood of obtaining large catches is high. In addition, fisheries frequently are selective, preferentially harvesting certain individuals, such as the largest fish in a year-class or only one sex. These practices can produce biases in vital statistics derived from commercial fisheries data.

Because of this, research surveys of commercially valuable species are often used to supplement information gathered from commercial fisheries.

In the sections that follow, we discuss commonly used sampling approaches (section 21.2) and methods for characterizing the commercial catch (section 21.3) and attributes of the fishery (section 21.4). In practice, the catch and the fishery are often sampled simultaneously. We also explore key issues that influence the sampling design and its execution. We conclude with an example that illustrates the design concepts presented in the chapter. We do not discuss commercial fishery surveys for economic indicators or the human dimension aspects of commercial fisheries. However, if the sampling objectives require collection of economic or sociological data, then sampling plans to incorporate appropriate measures must be developed. For further guidance on these topics, the interested reader is referred to Chapter 22 and Anderson's (1986) text on fisheries economics.

21.2 SAMPLING APPROACHES

Sampling commercial fisheries is not a trivial task, even in relatively simple systems. In complex multispecies fisheries that are often prosecuted by a variety of gears and have many landing sites and fishing units, the task becomes even more daunting. A carefully designed approach with clearly stated objectives is essential to success.

Before a sampling program can be developed, objectives must be unambiguously defined. Obvious as this may seem, objectives are often neglected or stated only in broad terms. Sampling programs frequently have multiple objectives, which must be balanced with clear recognition of the trade-offs. A design that is optimal for one purpose may not be for another, and flexibility is desirable because fishery conditions may change.

Once objectives are defined, a sampling approach can be chosen based on the structure of the fishery, target levels of precision, cost considerations, and other factors. Structure refers to such aspects as the temporal and spatial distribution of fishing effort, gear types used, number of vessels or other fishing units, and location of landing sites and market outlets. A fishery's structure must be well understood because this, along with cost, dictates the approaches that are feasible. Within the range of feasible options, target levels of precision and bias are critical considerations in deciding on methods and sampling intensity. Target levels should be determined by examining the effect of errors on the management advice that ultimately emanates from analysis of the data (Gavaris and Gavaris 1983; Pope and Gray 1983; Rivard 1983; Pelletier 1990; Pelletier and Gros 1991). Cost considerations also influence the selection of methods in most cases. Finally, sociological factors may play a considerable role in the success of a sampling design (Powles 1983) and should not be ignored.

Two general approaches exist for determining the catch, effort expended, and other fishery characteristics: censusing and sampling. A census is a complete enumeration of the population of interest (e.g., catch or effort), whereas sampling involves examining only a portion of the population. The sample is used to make inferences about the population as a whole. The advantage of censusing is that the quantity of interest is determined without error because it is counted completely. The disadvantages of censusing are that it is relatively expensive and it may be difficult to accomplish. If a complete census is attempted but not actually achieved, the resulting data have some unknown error that cannot be estimated. Sampling provides estimates that have an associated error, but the error can be estimated if the sampling plan is designed appropriately. A primary advantage of sampling is its cost-effectiveness. Although these two approaches are presented as a dichotomy, they can be used in complementary ways. For instance, to estimate fishing effort, a feasible approach might be to census small ports and sample large ports.

For most problems, sampling is more practical than censusing based on logistical and cost considerations. Sampling techniques are reviewed in Chapters 1 and 2, and Cochran (1977) is an excellent source for more comprehensive information. Here we briefly review sampling techniques in the context of estimating commercial fishery statistics.

Given specific objectives, the first step in designing a sampling plan is deciding exactly what population is to be sampled and how to subdivide it so that it can be easily sampled. In the jargon of sampling methodology, this is called defining the sampling frame and sampling units. Sampling units are nonoverlapping subdivisions of the population, and the sampling frame is the sum of all possible units. In the case of estimating commercial catch or effort, a possible choice of sampling units might be vessels, and the sampling frame would be all vessels landing the species of interest.

Once the sampling frame is defined, a variety of approaches can be used to sample the units. The most basic approach is simple random sampling in which all the sampling units have equal probability of being sampled. For example, to estimate total landings (the amount harvested), a subset of vessels (or other sampling unit) included in the sampling frame is randomly selected, and landings of those vessels enumerated. Landings from vessels that were not selected are not sampled. The estimate of total landings is the mean landings of the sampled vessels multiplied by the number of vessels in the sampling frame (Cochran 1977). This procedure results in an unbiased estimate, but the precision of the estimate may be low if landings vary widely among vessels. To improve precision, stratified random sampling is frequently used.

In stratified random sampling, the population is divided into subpopulations called strata, and independent random samples are then taken in each stratum. A benefit of stratifying is a reduction in the variance of estimates. For example, vessel size-classes (e.g., large and small trawlers) could be defined as strata because catches probably are more uniform within than they are between vessel classes. Strata are defined by one or more variables significantly influencing the quantity of interest. For instance, a combination of fishing area, season, and gear type is commonly used to define strata in temperate fisheries. Subsampling within strata (further partitioning strata) also can improve sampling efficiency by increasing precision for a given amount of sampling effort. For example, to estimate a vessel's landings in weight, a random subsample of boxes of fish is weighed; the average box weight is then multiplied by the total number of boxes to estimate the vessel's total landings. Weighing the entire landings would not only be impractical but also unnecessary because weight per box probably varies relatively little within strata.

In addition to stratified random sampling, many other types of sampling plans are possible, including systematic sampling, cluster sampling, and multistage stratified random sampling (Cochran 1977). The development of a sampling plan is influenced by a suite of considerations and may ultimately include a number of sampling techniques. The important goal in developing a sampling plan is to use an approach that is efficient and leads to appropriate estimation procedures.

After a sampling approach has been chosen, the number of samples to be taken must be decided. Methods for estimating the sampling intensity required for a given level of precision are provided in various texts (Gulland 1966; Cochran 1977; Schweigert and Sibert 1983). Preliminary data are required either from a trial run of the sampling plan or from an existing, similar program. Target levels of precision should take into account not only the desired precision of the basic statistics (e.g., catch, age composition, and effort) but also the precision of the output at higher levels of analysis (e.g., estimates of population size derived from catch-at-age data; Gavaris and Gavaris 1983; Pope and Gray 1983; Rivard 1983; Sampson 1987).

Monitoring data quality is an important component of the fishery sampling process. Unrecorded and misrecorded data can be minimized by careful attention during data collection and recording and by simple checking processes during data compilation (Pope 1988). Intentional misreporting cannot be avoided or corrected, and scientists using commercial fisheries data must be aware of this possibility. Pope and Garrod (1975) discussed the sources and effects of such errors on fisheries management schemes.

Another aspect of data quality concerns potential biases in the information. Bias refers to the systematic difference between an estimated quantity and its true value; for example, the tendency to regularly overestimate or underestimate the true value. Bias may result if sampling programs are poorly designed. Improvements in data quality and consequently in advice to fisheries managers will result from increasing representativeness of sampling for relative abundance, population size structure, and spatial distributions of harvestable stocks (Hilborn and Walters 1992). Such improvements will be possible only through careful design of sampling surveys and collaboration of commercial fishers.

21.3 CHARACTERIZATION OF THE COMMERCIAL CATCH

The most fundamental aspect of sampling a commercial fishery is characterizing its catch. The catch represents removals from the population and thus provides basic information on the productivity of a stock. When combined with information on fishing effort (catch per unit of effort), catch statistics can provide a crude index of relative abundance (see Box 6.1). In addition, the composition of the catch (size, age, and sex) can provide insight on the structure of the population from which it was taken.

The terms catch and landings are not interchangeable. Catch refers to all fishes (or other resources) captured, whereas landings are the portion of the catch retained and brought to market. This distinction arises because part of the catch usually is undesirable (e.g., too small, unmarketable, or nontarget species) and is discarded. Thus, catch is the sum of landings and discards. For some fisheries, discarding is insignificant and landings may be nearly equivalent to catch, whereas for others discarding is extensive and landings are only a fraction of the catch (e.g., Saila 1983). An example of a fishery with extensive discarding is the Gulf of Mexico shrimp fishery. For every kilogram of shrimp caught in Texas and Louisiana, between 1.7 and 137 kg of finfish were discarded during 1980–1981 (Watts and Pellegrin 1982).

Because some discarded fish may survive (e.g., Hill and Wassenberg 1990; Alverson et al. 1994), removals from the population are actually not equal to the catch (landings + discards) but to the sum of the landings and the fraction of the discarded catch that dies (landings + [discards × percent mortality]). Little information is available on discard mortality of commercially captured fish. Such information must be obtained through experimental work rather than sampling and is not further discussed in this chapter.

Catch and landings generally need to be known in terms of both weight and numbers for each species. In addition, most population assessments require information on the composition of the catch in terms of size, age, and sex ratio. The analytic goals for a particular problem will determine the specific data requirements; however, the following basic information is fundamental to a broad range of assessment techniques.

- *Length composition*—the number of fish in the catch or landings that fall into given length intervals. The number of length intervals chosen for measurement is generally somewhat arbitrary but should be related to the growth rate and average or maximum length of a species. Information on length composition can be used to estimate the age composition of the catch or landings by means of age–length keys (Fridriksson 1934; Kimura 1977; Westrheim and Ricker 1978; Chapter 16), to conduct length-based assessments (e.g., for species that cannot be aged; Pauly and Morgan 1987; Rosenberg and Beddington 1988; Gulland and Rosenberg 1992), or to estimate the selectivity of the fishery given fishery-independent estimates of size composition of the population (Gulland 1983; McBride 1991).
- *Age composition*—the number or weight of fish of each age in the catch or landings. Age composition data have many uses, including growth analysis (based on data on size at age; see Chapter 16), mortality estimation by means of following the decline in numbers of a year-class over time (Ricker 1975; Gulland 1983), prediction of yield by combining growth and mortality rate estimates (Gulland 1983), and estimation of absolute population size in numbers or weight by means of catch-at-age analysis or virtual population analysis and its variants (Jones 1964; Gulland 1965; Murphy 1965; Deriso et al. 1985; Hilborn and Walters 1992).
- *Sex ratio*—the proportional representation of males and females in the catch or landings. Sex ratio information is most valuable when combined with length or age data or both, and is used to estimate spawning stock biomass because usually only female fish and their maturity and fecundity are considered in such analyses. Sex-specific data are particularly important for species that show sexual dimorphism in growth, distribution, habitat use, vulnerability to capture, or other behavior. For example, in several species of crustaceans, mature males are more vulnerable to traps than are females (Miller 1990), most likely due to differences in their physiology and behavior. In other fisheries, females are preferentially targeted for their larger size and their roe (e.g., American shad). An extreme form of sexual dimorphism occurs in species that are sequentially hermaphroditic, beginning life as one sex and transforming later to the other (e.g., northern shrimp; Shumway et al. 1985).

Methods for biological sampling of population characteristics (e.g., collection of otoliths for aging [Chapter 16] and inspection of gonads to determine sex or maturity [Chapter 14]) are covered elsewhere in this book. Here we focus on methods for developing statistical sampling programs that allow inferences to be made about the population.

21.3.1 Methods for Data Collection

Given a sampling plan, a number of methods are available for actually collecting data from commercial fisheries. These may be characterized as either direct or indirect (Rowell 1983) and are applicable to either censusing or sampling. Examples of direct methods are onboard sampling, in which an observer is present during the fishing operation to record data (Kulka and Waldron 1983), and port sampling, in which a specialist censuses or samples the harvest at landing sites (Burns et al. 1983; Quinn et al. 1983a; Rowell 1983). In indirect methods, the data are collected secondhand, for example, by compilation of catch data recorded by vessel captains (logbooks), from dealers' purchase slips (market sampling; Rowell 1983), or by verbal reports from the person in charge of the fishing operation (interviews; Rowell 1983). Several methods can be combined in sampling a given fishery; the methods are not mutually exclusive.

Direct methods. Port sampling is perhaps the most commonly used method of collecting commercial fisheries statistics. It involves a direct encounter between the data collection specialist and the fishing unit at the point of landing. Often, each port sampling encounter can be used to meet a number of objectives, such as estimating total catch in weight and numbers, species composition, and length, age, and sex composition, as well as associated fishing effort and economic data. Port sampling is efficient when the number of landing locations is small relative to the number of fishing locations. If landing areas are many and diffuse or if it is not possible to census all ports where landings are made, then a program targeting a subset of ports may be more practical. For example, although artisanal fishers use about 200 landing sites along the Kenyan portion of Lake Victoria, 50% of the 50,000 to 60,000 fishers operate from 65 principal sites (Rabuor 1991). A note of caution, however: when only a portion of the fishery is sampled, inferences can be made about the sampled portion only (Cochran 1977). If the unsampled portion differs in characteristics, then a restricted port sampling plan that targets a subset of the fishery will not represent the entire fishery.

In a typical port sampling approach, the specialist meets the fishing unit at the landing site to collect samples before the landings are unloaded for sale. How samples are collected depends on the fishery, including operational aspects such as how the catch is stored, whether it is processed before landing, and how it is off loaded (Burns et al. 1983). For instance, in some fisheries, a mixed-species catch may be placed in the boat's hold and left unprocessed until sold. In other fisheries, the catch may have been sorted and boxed according to size or quality or both into market categories. Such aspects affect how the sampling will be conducted.

In most fisheries, port sampling can provide direct estimates only of landings, not of catch, because some of the catch is discarded before the vessel returns to the landing site. The only direct approach to estimating catch is by collecting data onboard during fishing operations. It is usually impractical for fishers to do this in a rigorous manner; thus, a data collection specialist is sent as an observer. Onboard sampling provides an opportunity to collect detailed information on a fine scale (e.g., set by set), not only on the amount and composition of the catch but also on fishing effort, area fished, performance of the fishing gear, and other such quantities (Table 21.1). These programs, while costly, provide invaluable information on fishing effort as well as catch, bycatch, and discard rates (Kulka and Waldron 1983; Marasco et al. 1991). One problem with observer programs is the possible effect of the presence of the observer on fishing procedures such as discarding. Another problem with onboard sampling is that it often depends on voluntary cooperation of fishers; thus, vessels cannot be selected randomly from the entire fishery. To the extent that cooperating fishers differ from noncooperators, this will result in biased estimates of fishery characteristics. In addition, cultural barriers may contribute to difficulties in data collection when observers are placed on foreign vessels. Despite these drawbacks, onboard sampling can provide a wealth of information on fishery operations that could not be obtained in any other way.

A typical approach to onboard sampling would be to randomly select a number of sets to be sampled on each day or in each fishing area. Random subsamples would be drawn from the catch of the selected sets to estimate the characteristics of interest. A disadvantage of onboard sampling, aside from its high cost, is that fishing trips may last several days or weeks, yet a relatively small amount of sampling might be adequate to characterize a particular vessel's catch if the vessel fishes on a single concentration of fish. A more effective use of an observer's time might be to sample a larger number of trips for less time each (the number of samples per trip would depend on relative levels of variation within and among fishing units). This might be possible if observers could transfer between fishing units without returning to port.

Optimal allocation of sampling effort in direct methods is affected by nonrandom distribution of fish in the environment (Dickie and Paloheimo 1965; McGlade and Smith 1983). Fish commonly segregate according to size, sex, or life history stage, and this has important implications for how sampling effort should be allocated. If different vessels fish different schools or groups of fish, each of which differ somewhat in characteristics such as age composition, then the catch from any individual vessel is not representative of the population as a whole but of the group from which the catch was taken. Fish collected from a given vessel will be more similar to each other than to fish from different vessels. This result has been demonstrated for American plaice (Baird and Stevenson 1983) and probably is common. Thus the preferred strategy is to obtain relatively few samples from a large number of vessels rather than many samples from relatively few vessels because this produces more accurate estimates. Similarly, it is not desirable to sample a vessel's catch by always taking samples from a standard point, such as the first or last part of the catch. If the vessel fished sequentially on different schools or if fish are nonrandomly mixed within schools, then a biased sample will result. For further discussion of the implications of nonrandom fish distributions on the optimal allocation of sampling effort, see Pennington and Vølstad (1991).

Indirect methods. Commercial fisheries information may be obtained indirectly from logbooks, interviews with fishers, and market sampling. Because the data collection specialist does not directly sample the catch when using indirect methods, extensive biological characterization is not possible. However, the methods can be useful in estimating or censusing catch, landings, discards, effort, area fished, and other quantities of interest. Data collected using indirect methods are likely to be biased for a variety of reasons discussed below; however, estimates from direct and indirect methods can sometimes be cross-checked to estimate bias (e.g., comparing logbook records with a vessel's sampled catch). For example, Berger (1993) compared estimates derived from catcher and processor reports and onboard observers. The estimated proportion of discards was similar between the two methods for most fisheries, but the catch estimated from reports was generally lower.

Table 21.1 Types of data sources and the information provided by each (adapted from Rowell 1983; additional information comes from Low et al. 1985; Smith 1991; Nicholson 1971; Rounsefell 1948).

Source	Information collected	Information that cannot be collected	Comments
Purchase slips (from vessels, fishers, or markets)	Vessel or fisher's name, landings (weight by market category), port landed, and price or value	Effort, extensive biological characterization, discards, and details of catch (depth, location, and daily catch)	Possible misreporting and missing data
Logbooks	Vessel or fisher's name, port landed, trip landings, daily catch, gear type, depth, location fished, number and duration of sets, number of fish per set, bait used (pot or trap fishery), and discards	Biological characterization of catch	Possible misreporting and missing data
Interviews	Vessel or fisher's name, port, landings (weight by market category), trip duration or time spent fishing, depth, number of traps or pots set, soak days, crew size, and vessel characteristics	Biological characterization of catch and discards	Possible misreporting, missing data, and inaccurate recall
Processor records	Landings by vessel and some effort (menhaden fishery)	Biological characterization of catch, discards, and details of catch or effort	Missing data; landings may be mixed species
Port	Port landed, gear type, total landings, length frequency, sex ratio, molt stage (crustaceans), maturity, food habits, individual mean weights, and parasites	Discards and effort	Possible difficulties in sampling large vessels or censusing many ports; combine with interviews to obtain effort
Onboard	Vessel name, location fished, depth, total catch per trip, catch per set, discards by species, number of traps or pots set and hauled, soak time, gear specifications, water temperature, crew size, length frequency, weight, sex ratio, maturity, food habits, parasites, molt stage (crustaceans), number of berried females (lobsters), time of day, cull size, age structure, and hours of gear operation		Presence of observers may influence fishing practices (such as discarding)

Logbooks are report forms provided to captains of fishing vessels for recording data on their fishing trips (see Figure 21.1 for an example). Logbooks can provide detailed information if filled out accurately. However, logbooks may be subject to a variety of drawbacks. Captains may have neither time nor inclination to fill out logbooks properly, they may prefer not to reveal some of the information requested, they may purposely supply inaccurate information, or they may be unable to estimate some quantities accurately. To encourage accurate reporting, logbooks should be designed to be useful to both fishers and scientists. If fishers can use official logbooks for their personal record keeping as well as for reporting, they are more likely to fill logbooks out completely and accurately. Furthermore, summaries of data obtained from logbooks should be provided to fishers in a timely fashion, while protecting the confidentiality of individual fishers' data. Logbook programs designed with built-in incentives for fishers are likely to be the most successful.

Logbooks may be either voluntary or compulsory. In cases in which logbooks are voluntary, participation of only a portion of the fleet makes interpretation and extrapolation to the entire fleet difficult because fishers who volunteer are not necessarily a random sample of all participants. In addition, with voluntary logbooks, some fishers discontinue participation after a few years while others are added; through time, the fleet may be characterized by rather different components, and the continuity of data is compromised (Simpson 1975). Despite these potential drawbacks, voluntary logbooks are invaluable because volunteers are likely to provide more detailed information than could be obtained reliably from all fishery participants. Thus, voluntary logbooks may be used to collect specialized information, whereas mandatory logbooks should be kept relatively simple.

Another indirect method commonly used to obtain fishery statistics is interviews of fishing vessel captains. The interviewer asks questions similar to those on logbooks and records the information on an interview form (Figure 21.2). Interview data are subject to sources of bias similar to those of logbooks. In addition, because interviews take place on land after one or several trips are completed, accuracy depends on the captain's recall ability or personal record keeping. The quality of information obtained from interviews can be affected by sociological factors, such as the personal relationship between the interviewer and the captain. In some areas, language barriers or other cultural differences can prevent successful interviews; thus, important segments of a fishery may not be sampled (Powles 1983). All these factors make conducting interviews a challenge. However, if conducted successfully, interviewing can be a highly cost-effective method for obtaining commercial fishery statistics.

For fisheries in which the catch is sold primarily to dealers, an expedient method of collecting landings data is to obtain copies of dealers' purchase slips. For fisheries in which landings are sold in market categories, size composition of the landings may be approximated from dealer slips. This will not be as accurate as direct sampling for size composition but can be useful, especially if the size of fish in a given market category is stable over time. Market sampling provides information on landings only, not on catch or associated effort data.

Whether commercial landings data are collected from logbooks or purchase slips or collected directly using port sampling, care should be taken to ensure that recorded weights represent quantities of fish or shellfish processed in an identical manner. This is particularly true for fisheries landing a single species in various processed forms. Unprocessed fish are usually recorded as round weight; dressed weight represents weight of fish after they have been gutted or beheaded; and filleted weight is the weight of standard, edible fillets. Some fisheries may land only part of the organism, for example, roe (American shad and sea urchins), livers (goosefish), and fins (some sharks). Shellfish weights may include or exclude the shell; for instance, sea scallop weights are reported as meat weights because only the adductor muscle is landed (Jamieson 1983). Additionally, historical records may include fish products resulting from other types of processing (such as salting or other preservation techniques).

FISHING VESSEL TRIP REPORT

□ DID NOT FISH DURING MONTH/YEAR _ /_

1. Vessel Name	2. USCG Doc. or State Reg. No.	3. Vessel Permit Number	
4. Date/Time Sailed Date (mm/dd/yy) ____/____/ Time (24 hrs) _ :_	5. Trip Type (Check One) □ Commercial □ Party □ Charter	6. No. of Crew	7. No. of Anglers

FILL OUT A NEW PAGE FOR EACH CHART AREA OR GEAR OR MESH/RING SIZE FISHED

8. Gear Fished	9. Mesh/Ring Size	10. Quantity of Gear	11. Size of Gear	
12. Chart Area	14. LATITUDE/LONGITUDE OR LORAN		15. No. of Hauls	16. Average Tow/Soak Time hrs _ mins _
	Latitude	Longitude		
13. Avg Depth ________fa	Station-Bearing #1 ______- _ - _	Station-Bearing #2 ______- _ - _		

17. Species Code Name	18. Kept Pounds (Comm)	Count (Rec)	19. Discarded Pounds (Comm)	Count (Rec)	20. Dealer Permit No.	21. Dealer Name	22. Date Sold (mm/dd/yy)

23. Port and State Landed	24. Date Landed (mm/dd/yy)	Time Landed
	____/____/_	____:_
	____/____/_	____:_

I certify that the information provided on this form is true, complete and correct to the best of my knowledge, and made in good faith. Making a false statement on this form is punishable by law (18 U.S.C. 1001).

25. Operator's Name (printed) and Permit Number (if required)	26. Operator's Signature	Date

Figure 21.1 Trip report form (logbook) for northeastern U.S. fishing vessels participating in fisheries with mandated reporting.

When a species is landed or sold in several processed states, conversion factors are typically used to transform processed weights to a standard (e.g., dressed weight is converted to round weight). Because processing can vary significantly among segments of a fishery and change over time, conversion factors must be developed experimentally for each segment and periodically verified.

CONFIDENTIAL INTERVIEW RECORD

Vessel	Port	Gear	Interview Date
Date sailed: Time:	Date landed: Time:	Days absent	Time lost

Area Fished (1)	Area fished (2)	Area fished (3)
Loran:	Loran:	Loran:
Depth:	Depth:	Depth:
Dates:	Dates:	Dates:
No. tows:	No. tows:	No. tows:
Tow duration:	Tow duration:	Tow duration:

Landings	Total Weight:

Species	Market category	WEIGHT		
		Area 1	Area 2	Area 3

Figure 21.2 Sample form for interviewing vessel captains.

21.3.2 Biological Characterization of the Commercial Catch

Whereas catch and landings statistics are fundamental to describing commercial fisheries, data on biological characteristics of the catch can provide greater insight into the dynamics of exploited stocks. By sampling the commercial catch, inferences can be drawn about abundance, age structure, sex ratios, maturation rates, and stock composition. From these quantities, population dynamics can be inferred.

Censusing catch or landings for biological characteristics is neither practical nor necessary. A well-designed sampling plan can provide unbiased estimates to meet target levels of precision. Here we discuss considerations for developing sampling plans to characterize the catch.

The most basic level at which the catch and landings can be characterized is species composition (the proportion of each species in the catch or landings). Many fishing gears are relatively nonselective (e.g., trawls and gill nets), catching several species as well as a broad range of sizes. If landings are brought to port unsorted (e.g., industrial fisheries) or sampling is being conducted during fishing operations, species composition can be estimated using random sampling of a vessel's landings or catch. Samples of the mixed harvest can be taken at random; sampling at only one point (e.g., the top or bottom of the pile) should be avoided because the species may not be randomly mixed. The samples can then be sorted by species, weighed, and enumerated. The total weight of landings by species equals the proportion of each species in the sample multiplied by the total weight of the landings. When catch is sampled, the catch by species can be derived in a similar manner.

If landings are brought to port sorted by species, species composition can be estimated simply by enumerating the number of containers of each species and estimating the weight per container. In other fisheries, estimating species composition may be complicated by extensive processing before landing (e.g., filleting).

Sampling for characteristics such as age and sex is more complex, and estimation frequently involves a multistage stratified sampling design. Because size is usually closely correlated with age, a measure of size such as length provides a convenient variable for the lowest level of stratification. Length strata should be defined for each species according to growth rates and the range of sizes in the catch. The number of samples per size stratum should be related to variation in age at size for each stratum. For instance, if nearly all fish in the size range 35 to 45 cm are 5 years old, then a small sample size would suffice to characterize the age of fish in that length-class. However, if individual growth rates are highly variable, fish of the same length-class might range in age from 3 to 7 years, and more samples (and perhaps smaller length strata) would be needed to meet target precision levels. Within size strata, samples should be collected randomly.

Length strata provide a convenient structure for collecting other types of biological data as well. Sex ratios, fecundity, maturity, and stock composition may be correlated with age, size, or both, and may be sampled efficiently by size-group. Subsample sizes required for each characteristic may vary depending on how much each characteristic varies within a size stratum. Thus a final sampling plan might call for 20 age samples, 10 sex determinations, 5 fecundity samples, and 15 stock composition samples in a particular length stratum. Another size stratum for the same species might specify a different mix of sample sizes. Because biological parameters such as growth, maturity, and fecundity may vary over time, sampling levels should be reviewed periodically and adjusted as necessary (see Baird 1983; O'Boyle et al. 1983; Quinn et al. 1983a, 1983b).

21.4 CHARACTERIZATION OF A COMMERCIAL FISHERY

From a population dynamics perspective, one of the most important attributes of a fishery is the amount of effort expended in obtaining the catch. This is because fishing effort is proportional to the rate of fishing mortality (Ricker 1975). Catch information alone can be used to monitor only gross production of the fishery, which may not correspond to gross production of the population. For example, increased catch may reflect an actual increase in stock abundance or simply an escalation of fishing effort.

Information on fishing effort and associated factors is needed for calculation of abundance indices that express the catch realized from a standard unit of effort (Box 6.1). Unlike simple catch data, catch per unit of effort (CPUE) data can be used to study changes in stock density (abundance per unit area). In addition, if the area inhabited by the stock is constant, then CPUE will index abundance of the entire population (Gulland 1969). Because landings do not include discards, landings per unit of effort may not reflect abundance.

Effort data are also important for fisheries managed by regulating total effort. Management by effort control is done by first determining the fishing effort corresponding to the optimal fishing mortality rate for a particular stock and then by limiting total effort to keep mortality at the target level. Effort controls are typically used in conjunction with other management measures, such as area closures and mesh size regulations. For example, fishing effort in the Australian spiny lobster fishery is controlled by limiting the number of participants, boats, and traps and by seasonal closures. Additional management measures include minimum size limits, a ban on taking females with eggs, and requirements regarding trap design (Brown and Phillips 1994). Other fisheries partially controlled by effort regulations include those for American shad (Crecco and Savoy 1985) and Pacific halibut (IPHC 1987).

For most fisheries studies, information is needed on both fishing effort and fleet characteristics that may affect the interpretation of effort statistics. With data on fleet characteristics, relative fishing power can be estimated. Fishing power is the relative efficiency of a vessel or gear fishing in a given place and time with respect to an arbitrarily defined standard vessel or gear. Because vessels and gear types vary in efficiency, fishing power estimates are needed to develop standardized units of effort. For example, larger, more powerful trawlers are more efficient harvesters of benthic species (groundfish) than are smaller vessels (O'Brien and Mayo 1987; Mayo et al. 1992). Thus, larger trawlers fishing for 1 d will produce greater harvests than will smaller trawlers fishing for the same amount of time. Fishing power is determined from studies of fishing time, spatial distribution of the fleet, and fleet characteristics such as tonnage class, vessel length, and vessel age (Gulland 1969). Although the choice of a standard vessel or gear type is arbitrary, some standards will result in CPUE data with better statistical properties (e.g., smaller variance; Gavaris 1980).

The spatial and temporal distribution of fishing effort can provide a valuable index of resource distribution when resources are distributed nonrandomly in space and time. Many commercial fisheries are successful because of the concentration of harvestable resources and fishing effort in the same place and time. For example, commercial fishing converges on American shad spawning grounds during the spawning season and along restricted spawning migration routes of the alewife in New England rivers. Thus, spatial patterns of fishing effort can indicate the distribution of exploited stocks. Historically, the distribution of fishing effort was used to discern gross distributions of stocks, particularly for Atlantic cod and haddock in the northwest Atlantic (Rounsefell 1948). Once the concentration of fishing effort was delineated, statistical areas were established to permit efficient collection of data for monitoring purposes (Rounsefell 1948; Halliday and Pinhorn 1990). Additionally, changes in the spatial distribution of fishing effort can indirectly

Table 21.2 Information needed for calculating fishing power and effective effort (from Gulland 1983). For all gears listed, fishing power is affected by vessel size, engine power, gear design and size, skill of crew, and auxiliary equipment (e.g., global positioning system or scanning sonar).

Gear	Units of effort	Additional factors influencing fishing power
Danish seine	Number of sets	
Bottom trawl	Number of days fishing, search time, and tow time	
Midwater trawl	Number of days fishing, search time, and tow time	
Purse seine	Number of sets, number of hours on fishing ground, and search time	Specialized searching equipment (e.g., spotter planes)
Gill net	Number of sets of standard-sized nets, number of days fished, and search time (for drift gill nets)	Duration of fishing
Hooks and lines	Number of hooks (usually units such as baskets or "skates" of standard length or number of hooks), number of lines, soak time, and time on fishing ground	Bait type and use, active searching, and specialized searching methods (e.g., satellite imagery)
Pots and traps	Number of hauls of pots or traps and soak time	Bait type and use and duration of fishing

indicate declines in stock abundance as fisheries move from depleted areas to less depleted ones to maintain high catch rates. In the case of pelagic fisheries that target schooling fish (e.g., clupeoids such as anchovies, sardines, and herrings), CPUE as an index of abundance may not be reliable. As stock abundance declines, CPUE remains high because the effort necessary to locate and capture the remaining schools of fish does not decrease as rapidly as does overall stock size (Hilborn and Walters 1992).

21.4.1 Definition of Effort

Calculation of total effort in a particular fishery requires estimation of individual vessel effort and an analytical summary of the data over the entire fishery. This requires a definition of effort in appropriate measurement units. In a fishery that targets a single species, effort can be defined as investment of time, number of fishing gear units deployed, or some combination of time and number of operations used to harvest fish or invertebrates (Gulland 1964a). Examples of effort units include days fished (e.g., for otter trawl fisheries), number of hooks on a longline (e.g., for Pacific halibut fisheries; Quinn et al. 1982), and number of trap-days, which is the number of traps multiplied by the number of days fished (e.g., for crustacean fisheries; Simpson 1975). Other examples are presented in Table 21.2.

Appropriate units. Effort units should be chosen so that they can be measured reliably and accurately and so that they reflect the key components of the harvesting process. Depending on the gear type, effort units should account for the time the gear is operating and the number of gear units deployed. Different effort units will be appropriate for different fisheries, and sometimes the best unit will be determined from retrospective analyses. For example, Gulland (1964b) showed that CPUE based on days at sea was inappropriate for studying Atlantic cod catch rates from English trawlers operating in the Barents Sea between 1906 and 1960. He recommended using a CPUE index based on hours fished to reduce the bias resulting from changes in fishing power through time (Gulland 1964b). If time spent searching or in transit is variable, days at sea will always be a biased estimator of fishing effort. For example, a vessel that was away from port for 3 d and spent 2 d traveling to and from the fishing ground actually fished 1 d; in this case, days at sea (3) overestimates the true effort (1 d). The bias introduced by using days at sea will be greater with greater variation in transit or search time.

When units of time are used to measure effort, the best measure is the actual time spent fishing. Thus, variable amounts of time lost to foul weather, transit, preparing gear, and handling the catch (sorting, discarding, processing, and storing) will not be included in the estimate (Gulland 1969). Because the fishing activity includes both time spent searching for fish and time the gear is operating, one or both of these components is expected to represent the actual fishing time. For example, in a trawl fishery, the time the trawl is on the bottom is the better measure of fishing time, but in harpoon fisheries for whales, time spent searching may give a better indication of actual fishing time and effort (Gulland 1969). Hours spent searching was also found to be a more accurate measure of effort in the purse-seine fishery for yellowfin tuna in the eastern Pacific (Allen and Punsly 1984). Sometimes when haul duration is fairly uniform among vessels in the fleet, it is possible to convert number of hauls to hours of gear operation, as Yeh (1984) did for the Taiwanese pair trawl fishery off northwestern Australia.

Directed effort in multigear fisheries. Measuring effort in single-species fisheries becomes complicated when multiple gears are used to harvest the resource. For example, the Canadian fishery for Atlantic cod in the Gulf of St. Lawrence uses gill nets and trap nets (Rose and Leggett 1989), and many artisanal fisheries use combinations of nets, handlines, spears, traps, explosive devices, and simple gathering techniques (Alimoso 1991; Rabuor 1991; Medley et al. 1993). Total effort in multigear fisheries cannot be calculated directly by summing effort across gear types; effort measurements for different gears are likely to be expressed in two or more incompatible units. Instead, total catch is rescaled into units of total effort based on total catch over all gear types and catch and effort for a particular fleet or gear type:

$$\text{total effort} = \text{effort}_A \times (\text{total catch}/\text{catch}_A) = \text{total catch}/\text{CPUE}_A,$$

where catch_A is the catch from one segment of the fishery and effort_A is the corresponding effort (Gulland 1969). This approach facilitates estimation of total effort from a series of measures in unrelated units. Alternatively, statistics for each segment of the fishery can be analyzed separately.

Directed effort in multispecies fisheries. In a multispecies fishery, where a single gear is used to harvest many species, it becomes difficult to estimate the effort associated with landings of a single species. It is not unusual for commercial vessels to target several species during the course of a fishing trip or season. For example, 12 key species are landed by northwest Atlantic groundfish fisheries, 15 by Pacific groundfish fisheries (NMFS 1993), and 9 by Japanese groundfish fisheries in the Bering Sea (Low and Berger 1984). Catches from artisanal fisheries are usually characterized by high diversity, with one fishery in Papua New Guinea harvesting 253 species (Medley et al. 1993).

A number of methods have been developed to allocate effort in multispecies fisheries. Gulland (1969) suggested allocating fishing effort to the targeted species or to whichever species predominates the catch. Thus, if the composition of the catch is 80% Atlantic cod and 20% redfish, total effort associated with this catch is allocated to Atlantic cod only. A similar approach is to use qualification levels in which trip effort is allotted to each species that composes an arbitrarily selected percentage of the total catch or landings (see Mayo et al. 1992). Westrheim (1983) described a method that consists of subdividing the landings and all-species effort by area, time, and depth, then allotting effort to each species according to its prevalence in the landings. These approaches are problematical if the species mix is influenced primarily by relative abundance changes (rather than targeting) because they ignore effort if catches or landings are low. This, in turn, artificially inflates CPUE or landings per unit of effort. A better approach is to associate total effort with all species composing the catch or landings (Tyler et al. 1984). Thus, in a multispecies fishery the effort expended is associated with each species. For purposes of estimating CPUE, it is unnecessary to constrain the sum of individual species effort to equal the total multispecies effort. However, total effort should not be calculated as the sum of the individual species effort.

Nominal and effective effort. Not all units of effort are equivalent in a particular fishery because efficiency varies among vessels and over time. Thus, a large vessel with the latest electronic equipment may harvest more fish for a given unit of effort than might a smaller, poorly equipped boat. The larger vessel has greater fishing power and its effective effort is greater. If a commercial fleet consists of vessels having similar fishing power, effort measurement is fairly straightforward. In these cases, the nominal effort (observed or measured effort) is the effective effort. When a fishery is composed of a mixture of vessels of varying fishing power, or when the composition of the fleet changes through time, nominal fishing effort is not a reliable estimate of effective effort. Fishing effort measurements must be adjusted to account for variations in fishing power.

Various factors have been proposed to account for variation in fishing power among vessels. These include vessel characteristics (such as length, gross tonnage, and design), equipment, and skill of fishers. Although some studies have failed to identify strong associations between fishing power and factors such as gross registered tonnage, overall vessel length, engine horse power, and year of vessel construction (Karger 1975; Hilborn and Ledbetter 1985), others have found significant relations between engine power or horsepower class of trawlers and fishing power (Hovart and Michielsen 1975; Westrheim and Foucher 1985). The skill and motivation of the skipper and crew have also been invoked (Rothschild 1972; Karger 1975; Stocker and Fournier 1984; Hilborn and Ledbetter 1985; Low et al. 1985; Hilborn and Walters 1992), but because these factors can be difficult to quantify, they have rarely been studied. Crew size has been shown to be important in the New England groundfish fleet (Carlson 1975).

Another factor that undoubtedly influences the fishing power of a vessel is the use of navigational and fish-finding equipment (Pope 1975). The introduction of aerial assistance (spotter planes) in the yellowfin tuna fishery in the eastern Pacific had a substantial effect on catch rates of vessels using these techniques (Allen and Punsly 1984). In recent years, innovations such as the global positioning system, scanning sonar, satellite imagery, and improved communications have significantly reduced the time necessary to locate fishing grounds or concentrations of fish. Such aids have greatly increased the efficiency of fishing operations and thus influence fishing power.

A number of statistical methods have been developed to standardize effort and estimate effective effort (see Pope 1975; Stocker and Fournier 1984; Westrheim and Foucher 1985). A common approach is to apply general linear models to estimate vessel efficiency relative to an arbitrarily defined standard vessel (Hilborn and Walters 1992). Robson (1966) and Kimura (1981) provide examples of the general linear model approach for standardizing fishing effort among vessels. These types of models permit incorporation of information on gear type and vessel attributes, such as the presence of specialized searching equipment, as well as information on the spatial distribution of fishing vessels and environmental variables (see Allen and Punsly 1984).

21.4.2 Collection of Effort Data

Methods for collecting fishing effort data depend on the intended use of effort data (e.g., fleet performance indicators, abundance indices, or stock area definition), details of fishery operations, and costs of sampling. Where possible, effort data are collected along with catch or landings statistics, and similar survey designs are used (see section 21.2). Like catch and landings statistics, effort data can be collected either by censusing or sampling and frequently by means of a combination of logbooks, purchase slips, interviews, port agents, or onboard observers (see section 21.3.1 and Table 21.1). A well-designed sampling program and subsequent effort standardization protocol can often provide better information from a small sample than will poor or unstandardized data collected from a greater proportion or the entire fleet (Shepherd 1988).

21.5 CATCH PER UNIT EFFORT STATISTICS

As a quantitative field of study, fisheries science developed around theories concerning changes in population biomass. Derivation of biomass indices from information on commercial fisheries allowed fisheries scientists to investigate and model responses of populations to exploitation. The most direct index of population biomass that can be derived from fishery data is CPUE. Stock production has been estimated from such indices (Schaefer 1954; Pella and Tomlinson 1969; Schnute 1977) when other sources of data are lacking. Other methods based on analysis of age-specific catches are also used to study production of fish populations. These methods include cohort analysis, virtual population analysis, and statistical catch-at-age models. (For a review of age-structured models see Megrey 1989; Sparre 1991 reviews multispecies virtual population analysis.) However, CPUE remains a basic component of most descriptions of exploited populations.

Catch per unit effort is a reliable index of stock biomass only when certain assumptions are met concerning fishing gear and the distribution of fish and fishing effort. Specifically, these assumptions are that gear efficiency and catchability are constant through time, all effort units operate independently, and all members of the stock are equally vulnerable to the fishery (Paloheimo and Dickie 1964; Ricker 1975). The first assumption requires gear efficiency and catchability to be independent of stock density. Another way to think of this is that for an equal amount of effort, fishing gear will capture the same proportion of the stock. The last assumption—equal vulnerability—holds only when fish and fishing effort are distributed randomly throughout the area inhabited by the stock. This assumption rarely, if ever, is valid because fish occur in clumps (e.g., schools), and commercial fishing effort is concentrated on aggregations of fish (Paloheimo and Dickie 1964). Hilborn and Walters (1992) provide a detailed discussion of some of the problems associated with using CPUE as an index of abundance.

Catchability. Much attention has been focused on the effect of variation in catchability, q, which is the coefficient of proportionality between CPUE and stock density (Gulland 1966). It is also defined as the proportion of a stock captured by a standardized unit of effort (Ricker 1975). If q varies, the relation between CPUE and abundance is not reliable. Because catchability is defined relative to a standardized effort unit, any change in fishing power or shift in the spatial distribution of fishing effort will affect estimation of q (Gillis et al. 1993). Additional problems in estimating q occur when landings and effort are erroneously attributed to a particular fishing area (Shepherd 1988). Improvements in fishing tactics will also affect estimation of q; these improvements include use of equipment or techniques that result in better catches from a particular ground or increased cooperation among vessels in locating concentrations of fish. Such changes in fishing tactics can be difficult to detect and measure (Gulland 1966). Even though stock density may be declining, CPUE indices may remain stable if apparent catchability is increasing.

Variations in catchability can result directly from changes in the distribution or behavior of fish. Abiotic factors may affect catchability by altering fish distributions. Rose and Leggett (1989) reported that environmental influences (current strength and salinity) were as important as stock abundance in accounting for variability in CPUE of Atlantic cod. Other causes of variation in catchability, including gear saturation, are fully discussed in Gulland (1983).

If changes in stock abundance are associated with changes in the spatial distribution of fish, then catchability of the stock will be affected. This is because catchability may be inversely proportional to the area occupied by the stock (Paloheimo and Dickie 1964). Using information from the Atlantic herring fishery in Canada, Winters and Wheeler (1985) demonstrated that as the abundance of herring decreased, the stock was concentrated in a smaller area. Consequently, a greater proportion of the herring stock was captured by a standard unit of effort, and, by definition, catchability increased. Catch per unit effort remained fairly stable (because stock density was stable), even though stock abundance had decreased. When catchability and stock abundance are inversely related, CPUE is a reliable index of stock density but not abundance (Winters and Wheeler 1985). There is growing evidence to support Paloheimo and Dickie's (1964) postulate that catchability and stock abundance are inversely related. This evidence comes from observations on fisheries for schooling fish, such as anchovies, sardines, and herrings; demersal species such as Atlantic cod and haddock; and anadromous fish such as American shad (Pope and Garrod 1975; Peterman et al. 1985; Angelsen and Olsen 1987; Crecco and Overholtz 1990; Gunderson 1993). In these types of fisheries, as population abundance decreases, the proportion of the population captured by a unit of effort (q) increases; this permits these fisheries to maintain a relatively stable catch rate while stock abundance declines.

Another implication of an inverse relationship between catchability and stock abundance is that fishing mortality rates increase even though fishing effort is held constant (Crecco and Savoy 1985; Crecco and Overholtz 1990). The net result of temporal changes in catchability is that commercial CPUE data can seriously underestimate the rate of decline in stock size (e.g., Crecco and Overholtz 1990).

21.6 EXAMPLE

The following example describes a hypothetical fishery to illustrate the concepts presented in this chapter. The example is simplified compared with most real fisheries; however it demonstrates a sampling program to achieve multiple objectives by means of a variety of techniques.

21.6.1 Description and Objectives

The hypothetical fishery is a multispecies one that uses relatively nonselective gear, discards a portion of the catch at sea, lands in several ports, and includes vessels that vary in fishing power. Many actual fisheries fit this general description, such as groundfish fisheries of the northwest Atlantic, Alaskan, and Pacific coasts of the United States, and gill-net and longline fisheries for migratory pelagic species.

The objectives of the fishery sampling program are to estimate

1. catch and landings (in weight and number) for each species in the fishery;
2. fishing effort (by species); and
3. biological characteristics (length, age, and sex composition) of the catch and landings of each species.

To develop an appropriate sampling program, detailed information on the operation of the fishery is needed. Preliminary information revealed that fishing occurs in two distinct areas during a 2-month season. Trips last 3 to 10 d, vessels usually fish in only one of the two areas, and the catch is sorted by species and market category before vessels return to port. Some of the catch is discarded at sea and probably does not survive. The fishery lands in three ports, and vessels vary in size and fishing power. Landings are sold to wholesalers, who are required by law to maintain purchase records.

21.6.2 Sampling Approaches

Table 21.3 outlines the approaches chosen for estimating fishery characteristics based on preliminary information, target precision levels, and cost considerations. For the quantities to be sampled (discards and biological characteristics), the following approach was adopted. The three ports involved in the fishery were considered subpopulations. Within these subpopulations, the most important factors affecting catch rates and biological composition of the catch are vessel class (fishing power) and area fished. Seasonal effects are probably minor because the fishing season is relatively short. Therefore, combinations of vessel class and area fished were chosen to define strata within subpopulations. The primary sampling unit is a trip for both onboard and port sampling. Based on preliminary data on precision, it was decided that sampling intensity would be proportional to landings in each port and stratum. The greater the landings in a given stratum, the larger the number of samples collected in that stratum. Because the actual number of trips or landings during a fishing season cannot be known in advance, it is not possible to meet the requirements of random sampling in the strictest sense. However, after the first year, information collected in previous years can be used to approximate a random sample. In addition, the allocation of sampling effort should be reviewed periodically and adjusted if the pattern of landings is not stable over time.

A variety of sampling methods are used to carry out the sampling plan, ranging from onboard sampling to market sampling. Some methods address more than one

Table 21.3 Sampling approaches and methods recommended for estimating the characteristics of a hypothetical multispecies fishery that discards a portion of the catch, lands in several ports, and includes vessels that vary in fishing power.

Fishery characteristic	Sampling approach	Sampling method
Discards (quantity and biological characteristics)	Multistage stratified random sampling	Onboard sampling
Effort	Census	Interviews
Landings	Census	Port sampling, interviews, and market sampling
Biological characteristics of catch (length, age, and sex)	Multistage stratified random sampling	Port sampling and onboard sampling

objective. For instance, port sampling is used to obtain some of the landings data and some of the information for biological characterization of the catch; interviews are used to collect a portion of the landings data and all of the effort information. Therefore, a practical way of organizing sampling activities is by method of data collection rather than by type of information sought.

21.6.2.1 Onboard Sampling

Onboard sampling is used to estimate the proportion of the catch that is discarded. The weight and species composition of the discarded catch are estimated so that landings data (obtained by other methods) can be expanded to estimate total catch. Further, the size composition of each discarded species is sampled so that the number of fish discarded can be estimated. There are three levels at which samples must be selected: trips, sets within trips, and samples within sets (for biological characteristics). Ideally, trips are selected randomly within strata (vessel class × area fished). However, this is not strictly possible because the distribution of trips is not known in advance and because all vessels may not cooperate in the program. A practical approach is to devise an ideal sampling plan and execute it as opportunities arise. The ideal plan specifies the desired number and distribution of samples within strata given historical patterns in the fishery. By keeping close track of strata that have been sampled, it may be possible to "fill in the blanks" and approach the ideal plan.

Within a trip, it is somewhat easier to obtain a strict random sample. Random numbers can be used to select the sets that will be sampled. For example, if four sets are usually completed each day, and half of these are to be sampled, then two random numbers from 1 to 4 are drawn and those sets sampled. Within each set, a random subsample is drawn to estimate species composition and weights of the landed and discarded catch. This is further subsampled to estimate biological characteristics of the catch. In selecting subsamples from a tow, the size of the subsample is proportional to the size of the catch. If the catch is sorted into landings and discards before samples are drawn, then a sample is taken from each. The samples are selected randomly within the set, avoiding selection of only the first or last portion of the catch. The samples are examined for species composition, and a sample of each species is measured to estimate length composition. Within length intervals, further subsampling can be conducted to collect structures for age determination (otoliths or scales) or sex data if information on age and sex of the discards is needed.

21.6.2.2 Port Sampling

Port sampling is used to supply some of the landings data and for biological characterization of the landings. Again, because the distribution of trips is not predictable, it is necessary to establish target sampling goals based on the sampling design and attempt to fulfill them. In this fishery, fishes are sorted and boxed by species and market category prior to returning to port. Thus, landings are estimated by enumerating the number of boxes of each species and market category. A random subset is weighed to develop an average weight per box (which may or may not vary by species and market category). The average weight is then applied to the total number of boxes to estimate landings in weight. Landings in number are estimated by selecting a subsample(s) from each combination of species and market category and counting the number of fish per unit weight. The landings in number are the weight of landings multiplied by the number of fish per unit weight, summed over market categories and species. Biological characterization requires further sampling. For each species–market category combination, a random subsample is measured to estimate length composition. Within length-groups (which have been defined for each species), a random subsample is selected for collection of age structures and sex determination.

Interviews of vessel captains are used to supply part of the landings data and all of the data on fishing effort. Ideally, vessel captains are interviewed when each vessel returns to port; however, this may not always be feasible, so some interviews will collect information pertaining to several trips. The interview form (Figure 21.2) includes sections for recording all pertinent data and guides the interviewer. Each trip is recorded on a separate form.

21.6.2.3 Market Sampling

Market sampling supplies part of the landings data. If dockside sampling coverage has been adequate, most landings data may have been obtained through port sampling and interviews. However, there may be cases in which interviews or port sampling were inadequate to document landings and market records can provide the missing data. Market data are also used to verify landings data obtained from interviews and to collect important economic data. Market sampling is done by visiting all wholesalers who purchase the species of interest and requesting copies of dated sales receipts. These generally show the vessel or owner's name and weight of fish purchased by market category and can be matched up with interviews from the corresponding trips. For our purposes, only landings data (by species and market category) are obtained from market sampling.

When combined appropriately, the data obtained from the sampling program outlined here should provide a comprehensive picture of the catch, landings, effort, and biological characteristics of the fishery. The accuracy of the picture will depend on how well the sampling plan was designed and executed and the quality of information obtained through indirect methods (such as interviews). Once established, the sampling plan should be reviewed periodically to evaluate whether its objectives were met. The example demonstrates only one way of approaching fishery sampling; many other possibilities exist, and the exact plan developed must be tailored to the specifications of the fishery and data requirements for population assessment.

21.7 REFERENCES

Alimoso, S. B. 1991. Catch effort data and their use in the management of fisheries in Malawi. Pages 393–403 *in* I. G. Cowx, editor. Catch effort sampling strategies. Blackwell Scientific Publications, Cambridge, Massachusetts.

Allen, R., and R. Punsly. 1984. Catch rates as indices of abundance of yellowfin tuna, *Thunnus albacares*, in the eastern Pacific Ocean. Bulletin of the Inter-American Tropical Tuna Commission 18:303–379.

Alverson, D. L., M. H. Freeberg, S. A. Murawski, and J. G. Pope. 1994. A global assessment of fisheries bycatch and discards. FAO (Food and Agriculture Organization of the United Nations) Fisheries Technical Paper 339.

Anderson, L. G. 1986. The economics of fisheries management, 2nd edition. Johns Hopkins University Press, Baltimore, Maryland.

Angelsen, K., Jr., and S. Olsen. 1987. Impact of fish density and effort level on catching efficiency of fishing gear. Fisheries Research 5:271–278.

Baird, J. W. 1983. A method to select optimum numbers for aging in a stratified approach. Canadian Special Publication of Fisheries and Aquatic Sciences 66:161–164.

Baird, J. W., and S. C. Stevenson. 1983. Levels of precision—sea versus shore sampling. Canadian Special Publication of Fisheries and Aquatic Sciences 66:185–188.

Berger, J. D. 1993. Comparison between observed and reported catches of retained and discarded groundfish in the Bering Sea and Gulf of Alaska, 1990–91. NOAA (National Oceanic and Atmospheric Administration) Technical Memorandum NMFS (National Marine Fisheries Service) AFSC-13.

Brown, R. S., and B. F. Phillips. 1994. The current status of Australia's rock lobster fisheries. Pages 33–63 *in* B. F. Phillips, J. S. Cobb, and J. Kittaka, editors. Spiny lobster management. Blackwell Scientific Publications, Oxford, UK.

Burns, T. S., R. Schultz, and B. E. Brown. 1983. The commercial catch sampling program in the northeastern United States. Canadian Special Publication of Fisheries and Aquatic Sciences 66:82–95.

Carlson, E. W. 1975. The measurement of relative fishing power using cross section production functions. Rapports et Procés-verbaux des Réunions Conseil International pour l'Exploration de la Mer 168:84–98.

Cochran, W. G. 1977. Sampling techniques, 3rd edition. Wiley, New York.

Crecco, V. A., and W. J. Overholtz. 1990. Causes of density-dependent catchability for Georges Bank haddock *Melanogrammus aeglefinus*. Canadian Journal of Fisheries and Aquatic Sciences 47:385–394.

Crecco, V. A., and T. F. Savoy. 1985. Density-dependent catchability and its potential causes and consequences on Connecticut River American shad, *Alosa sapidissima*. Canadian Journal of Fisheries and Aquatic Sciences 42:1649–1657.

Deriso, R. B., T. J. Quinn, II, and P. R. Neal. 1985. Catch-age analysis with auxiliary information. Canadian Journal of Fisheries and Aquatic Sciences 42:815–824.

Dickie, L. M., and J. E. Paloheimo. 1965. Heterogeneity among samples of the length and age compositions of commercial groundfish landings. International Commission for the Northwest Atlantic Fisheries Research Bulletin 2:48–52.

Fridriksson, A. 1934. On the calculation of age-distribution within a stock of cod by means of relatively few age determinations as a key to measurements on a large scale. Rapports et Procés-verbaux des Réunions Conseil International pour l'Exploration de la Mer 86:1–14.

Gavaris, S. 1980. Use of a multiplicative model to estimate catch rate and effort from commercial data. Canadian Journal of Fisheries and Aquatic Sciences 37:2272–2275.

Gavaris, S., and C. A. Gavaris. 1983. Estimation of catch at age and its variance for groundfish stocks in the Newfoundland region. Canadian Special Publication of Fisheries and Aquatic Sciences 66:178–182.

Gillis, D. M., R. M. Peterman, A. V. Tyler. 1993. Movement dynamics in a fishery: application of the ideal free distribution to spatial allocation of effort. Canadian Journal of Fisheries and Aquatic Sciences 50:323–333.

Gulland, J. A. 1964a. Catch per unit effort as a measure of abundance. Rapports et Procés-verbaux des Réunions Conseil International pour l'Exploration de la Mer 155:8–14.

Gulland, J. A. 1964b. The abundance of fish stocks in the Barents Sea. Rapports et Procés-verbaux des Réunions Conseil International pour l'Exploration de la Mer 155:126–137.

Gulland, J. A. 1966. Manual of sampling and statistical methods for fisheries biology. Part 1: sampling methods. FAO (Food and Agriculture Organization of the United Nations) Manuals in Fisheries Science 3.

Gulland, J. A. 1969. Manual of methods for fish stock assessment. Part 1: fish population analysis. FAO (Food and Agriculture Organization of the United Nations) Manuals in Fisheries Science 4.

Gulland, J. A. 1983. Fish stock assessment: a manual of basic methods. Wiley, New York.

Gulland, J. A., and A. A. Rosenberg. 1992. A review of length-based approaches to assessing fish stocks. FAO (Food and Agriculture Organization of the United Nations) Fisheries Technical Paper 323.

Gunderson, D. R. 1993. Surveys of fisheries resources. Wiley, New York.

Halliday, R. G., and A. T. Pinhorn. 1990. The delimitation of fishing areas in the northwest Atlantic. Journal of Northwest Atlantic Fishery Science 10:1–51.

Hilborn, R., and M. Ledbetter. 1985. Determinants of catching power in the British Columbia salmon purse seine fleet. Canadian Journal of Fisheries and Aquatic Sciences 42:51–56.

Hilborn, R., and C. J. Walters. 1992. Quantitative fisheries stock assessment. Chapman and Hall, New York.

Hill, B. J., and T. J. Wassenberg. 1990. Fate of discards from prawn trawlers in Torres Strait. Australian Journal of Marine and Freshwater Research 41:53–64.

Hovart, P., and K. Michielsen. 1975. Relationship between fishing power and vessel characteristics of Belgian beam trawlers. Rapports et Procés-verbaux des Réunions Conseil International pour l'Exploration de la Mer 168:7–10.

IPHC (International Pacific Halibut Commission). 1987. The Pacific halibut: biology, fishery, and management. International Pacific Halibut Commission Technical Report 22.

Jamieson, G. S. 1983. Commercial catch sampling: a review of its usage in the management of contagiously distributed subtidal mollusc species. Canadian Special Publication of Fisheries and Aquatic Sciences 66:240–247.

Jones, R. 1964. Estimating population size from commercial statistics when fishing mortality varies with age. Rapports et Procés-verbaux des Réunions Conseil International pour l'Exploration de la Mer 155:210–214.

Karger, W. 1975. Relationship between fishing power and characteristics of stern trawlers in midwater trawling. Rapports et Procés-verbaux des Réunions Conseil International pour l'Exploration de la Mer 168:27–29.

Kimura, D. K. 1977. Statistical assessment of the age–length key. Journal of the Fisheries Research Board of Canada 34:317–324.

Kimura, D. K. 1981. Standardized measures of relative abundance based on modelling log (c.p.u.e.), and their application to Pacific ocean perch (*Sebastes alutus*). Journal du Conseil International pour l'Exploration de la Mer 39:211–218.

Kulka, D. W., and D. Waldron. 1983. The Atlantic observer programs—a discussion of sampling from commercial catches at sea. Canadian Special Publication of Fisheries and Aquatic Sciences 66:255–262.

Low, L. L., and J. Berger. 1984. Estimation and standardization of fishing effort in data bases of bering sea trawl fisheries. International North Pacific Fisheries Commission Bulletin 42:15–22.

Low, R. A., Jr., G. F. Ulrich, C. A. Barans, and D. A. Oakley. 1985. Analysis of catch per unit of effort and length composition in the South Carolina commercial handline fishery, 1976–1982. North American Journal of Fisheries Management 5:340–363.

Marasco, R., R. Baldwin, N. J. Bax, and T. Landen. 1991. By-catch: a bioeconomic assessment of north Pacific groundfish fisheries. Pages 275–280 *in* N. Daan and M. P. Sissenwine, editors. Multispecies models relevant to management of living resources. ICES (International Council for the Exploration of the Sea) Marine Science Symposia 193.

Mayo, R. K., M. J. Fogarty, and F. M. Serchuk. 1992. Aggregate fish biomass and yield on Georges Bank, 1960–87. Journal of Northwest Atlantic Fishery Science 14:59–78.

McBride, M. M. 1991. Estimation of unreported catch in a commercial trawl fishery. International Council for the Exploration of the Sea, C.M. 1991/D:12, Copenhagen.

McGlade, J. M., and S. J. Smith. 1983. Principal component methods for exploratory data analysis of commercial length-frequency data. Canadian Special Publication of Fisheries and Aquatic Sciences 66:235–239.

Medley, P. A., G. Gaudian, and S. Wells. 1993. Coral reef fisheries stock assessment. Reviews in Fish Biology and Fisheries 3:242–285.

Megrey, B. A. 1989. Review and comparison of age-structured stock assessment models from theoretical and applied points of view. American Fisheries Society Symposium 6:8–48.

Miller, R. J. 1990. Effectiveness of crab and lobster traps. Canadian Journal of Fisheries and Aquatic Sciences 47:1228–1251.

Murawski, S. A. 1991. Can we manage our multispecies fisheries? Fisheries 16(5):5–13.

Murphy, G. I. 1965. A solution of the catch equation. Journal of the Fisheries Research Board of Canada 22:191–202.

Nicholson, W. R. 1971. Changes in catch and effort in the Atlantic menhaden purse-seine fishery 1940–68. U.S. National Marine Fisheries Service Fishery Bulletin 69:765–781.

NMFS (National Marine Fisheries Service). 1993. Our living oceans: report on the status of U.S. living marine resources, 1993. NOAA (National Oceanic and Atmospheric Administration) Technical Memorandum NMFS-F/SPO-15, Washington, DC.

NMFS (National Marine Fisheries Service). 1994. Fisheries of the United States, 1993. Current fishery statistics 9300. NMFS, Silver Spring, Maryland.

O'Boyle, R., L. Cleary, and J. McMillan. 1983. Determination of the size composition of the landed catch of haddock from NAFO division 4X during 1968–81. Canadian Special Publication of Fisheries and Aquatic Sciences 66:217–234.

O'Brien, L., and R. K. Mayo. 1987. Sources of variation in catch per unit effort of yellowtail flounder, *Limanda ferruginea* (Storer), harvested off the coast of New England. U.S. National Marine Fisheries Service Fishery Bulletin 86:91–108.

Paloheimo, J. E., and L. M. Dickie. 1964. Abundance and fishing success. Rapports et Procés-verbaux des Réunions Conseil International pour l'Exploration de la Mer 155:152–163.

Pauly, D., and G. R. Morgan. 1987. Length-based methods in fisheries research. ICLARM (International Center for Living Aquatic Resources Management) Conference Proceedings 13, Manila, Philippines.

Pella, J. J., and P. K. Tomlinson. 1969. A generalized stock production model. Bulletin of the Inter-American Tropical Tuna Commission 13:421–496.

Pelletier, D. 1990. Sensitivity and variance estimators for virtual population analysis and the equilibrium yield per recruit model. Aquatic Living Resources 3:1–12.

Pelletier, D., and P. Gros. 1991. Assessing the impact of sampling error on model-based management advice: comparison of equilibrium yield per recruit variance estimators. Canadian Journal of Fisheries and Aquatic Sciences 48:2129–2139.

Pennington, M., and J. H. Vølstad. 1991. Assessing the effect of intra-haul correlation and variable density on population estimates from marine surveys. International Council for the Exploration of the Sea, C.M. 1991/D:14, Copenhagen.

Peterman, R. M., G. J. Steer, and M. J. Bradford. 1985. Comment on "Density-dependent catchability coefficients." Transactions of the American Fisheries Society 114:438–440.

Pope, J. G., editor. 1975. Measurement of fishing effort. Rapports et Procés-verbaux des Réunions du Conseil international pour l'Exploration de la Mer 168.

Pope, J. G. 1988. Collecting fisheries assessment data. Pages 63–82 *in* J. A. Gulland, editor. Fish population dynamics, 2nd edition. Wiley, New York.

Pope, J. G., and D. J. Garrod. 1975. Sources of error in catch and effort quota regulations with particular reference to variations in the catchability coefficient. International Commission for the Northwest Atlantic Fisheries Research Bulletin 11:17–30.

Pope, J. G., and D. Gray. 1983. An investigation of the relationship between the precision of assessment data and the precision of total allowable catches. Canadian Special Publication of Fisheries and Aquatic Sciences 66:151–157.

Powles, H. 1983. Planning the sampling of commercial catches in the Gulf region. Canadian Special Publication of Fisheries and Aquatic Sciences 66:263–267.

Quinn, T. J., II, E. A. Best, L. Bijsterveld, and I. R. McGregor. 1983a. Port sampling for age composition of Pacific halibut landings. Canadian Special Publication of Fisheries and Aquatic Sciences 66:194–205.

Quinn, T. J., II, E. A. Best, L. Bijsterveld, and I. R. McGregor. 1983b. Sampling Pacific halibut (*Hippoglossus stenolepis*) landings for age composition: history, evaluation and estimation. International Pacific Halibut Commission Scientific Report 68.

Quinn, T. J., II, S. H. Hoag, and G. M. Southward. 1982. Comparison of two methods of combining catch-per-unit-effort data from geographic regions. Canadian Journal of Fisheries and Aquatic Sciences 39:837–846.

Rabuor, C. O. 1991. Catch and effort sampling techniques and their application in freshwater fisheries management: with specific reference to Lake Victoria, Kenyan waters. Pages 373–381 *in* I. G. Cowx, editor. Catch effort sampling strategies. Blackwell Scientific Publications, Cambridge, Massachusetts.

Ricker, W. E. 1975. Computation and interpretation of biological statistics of fish populations. Bulletin of the Fisheries Research Board of Canada 191.

Rivard, D. 1983. Effects of systematic, analytical, and sampling errors on catch estimates: a sensitivity analysis. Canadian Special Publication of Fisheries and Aquatic Sciences 66:114–129.

Robson, D. S. 1966. Estimation of the relative fishing power of individual ships. International Commission for the Northwest Atlantic Fisheries Research Bulletin 3:5–14.

Rose, G. A., and W. C. Leggett. 1989. Predicting variability in catch-per-effort in Atlantic cod, *Gadus morhua*, trap and gillnet fisheries. Journal of Fish Biology 35(Supplement A):155–161.

Rosenberg, A. A., and J. R. Beddington. 1988. Length-based methods of fish stock assessment. Pages 83–103 *in* J. A. Gulland, editor. Fish population dynamics, 2nd edition. Wiley, New York.

Rothschild, B. J. 1972. An exposition on the definition of fishing effort. U.S. National Marine Fisheries Service Fishery Bulletin 70:671–679.

Rounsefell, G. A. 1948. Development of fishery statistics in the north Atlantic. U.S. Fish and Wildlife Service Special Scientific Report 47.

Rowell, T. W. 1983. Sampling of commercial catches of invertebrates and marine plants in the Scotia–Fundy region. Canadian Special Publication of Fisheries and Aquatic Sciences 66:52–60.

Saila, S. B. 1983. Importance and assessment of discards in commercial fisheries. FAO (Food and Agriculture Organization of the United Nations) Fisheries Circular 765.

Sampson, D. B. 1987. Variance estimators for virtual population analysis. Journal du Conseil International pour l'Exploration de la Mer 43:149–158.

Schaefer, M. B. 1954. Some aspects of the dynamics of populations important to the management of commercial marine fisheries. Bulletin of the Inter-American Tropical Tuna Commission 1:25–56.

Schnute, J. 1977. Improved estimates from the Schaefer production model: theoretical considerations. Journal of the Fisheries Research Board of Canada 34:583–603.

Schweigert, J. F., and J. R. Sibert. 1983. Optimizing survey design for determining age structure of fish stocks: an example from British Columbia Pacific herring (*Clupea harengus pallasi*). Canadian Journal of Fisheries and Aquatic Sciences 40:588–597.

Shepherd, J. G. 1988. Fish stock assessments and their data requirements. Pages 35–62 *in* J. A. Gulland, editor. Fish population dynamics, 2nd edition. Wiley, New York.

Shumway, S. E., H. C. Perkins, D. F. Schick, and A. P. Stickney. 1985. Synopsis of biological data on the pink shrimp, *Pandalus borealis* Krøyer, 1838. FAO (Food and Agriculture Organization of the United Nations) Fisheries Synopsis 144 and NOAA (National Oceanic and Atmospheric Administration) Technical Report NMFS (National Marine Fisheries Service) 30.

Simpson, A. C. 1975. Effort measurement in the trap fisheries for crustacea. Rapports et Procés-verbaux des Réunions du Conseil International pour l'Exploration de la Mer 168:50–53.

Smith, J. W. 1991. The Atlantic and Gulf menhaden purse seine fisheries: origins, harvesting technologies, biostatistical monitoring, recent trends in fisheries statistics, and forecasting. U.S. National Marine Fisheries Service Marine Fisheries Review 53:28–41.

Sparre, P. 1991. Introduction to multispecies virtual population analysis. ICES (International Council for the Exploration of the Sea) Marine Science Symposium 193:12–21.

Stocker, M., and D. Fournier. 1984. Estimation of relative fishing power and allocation of effective fishing effort, with catch forecasts, in a multi-species fishery. International North Pacific Fisheries Commission Bulletin 42:3–9.

Tyler, A. V., E. L. Beals, and C. L. Smith. 1984. Analysis of logbooks for recurrent multi-species effort strategies. International North Pacific Fisheries Commission Bulletin 42:39–46.

Watts, N. H., and G. J. Pellegrin, Jr. 1982. Comparison of shrimp and finfish catch rates and ratios for Texas and Louisiana. U.S. National Marine Fisheries Service Marine Fisheries Review 44:44–49.

Westrheim, S. J. 1983. A new method for allotting effort to individual species in a mixed-species trawl fishery. Canadian Journal of Fisheries and Aquatic Sciences 40:352–360.

Westrheim, S. J., and R. P. Foucher. 1985. Relative fishing power for Canadian trawlers landing Pacific cod (*Gadus macrocephalus*) and important shelf cohabitants from major offshore areas of western Canada, 1960–81. Canadian Journal of Fisheries and Aquatic Sciences 42:1614–1626.

Westrheim, S. J., and W. E. Ricker. 1978. Bias in using an age–length key to estimate age-frequency distributions. Journal of the Fisheries Research Board of Canada 35:184–189.

Winters, G. H., and J. P. Wheeler. 1985. Interaction between stock area, stock abundance, and catchability coefficient. Canadian Journal of Fisheries and Aquatic Sciences 42:989–998.

Yeh, S. Y. 1984. Standardization of fishing effort of the Taiwanese pair trawl fishery off northwestern Australia. International North Pacific Fisheries Commission Bulletin 42:23–27.

Chapter 22

Measuring the Human Dimensions of Recreational Fisheries

BARBARA A. KNUTH AND STEVE L. MCMULLIN

22.1 INTRODUCTION

22.1.1 Importance of Human Dimensions Data

The fisheries management triad includes fish populations, their habitats, and people (Chapter 1). Previous chapters presented techniques for measuring aspects of fish populations and their habitats. The techniques featured in this chapter help produce the information needed to make management decisions involving and affecting the people part of the management triad. Management decisions are often allocation decisions that require sociological, economic, legal, and political data. These are the human dimensions of fisheries management.

Human dimensions is the term commonly used to describe the body of theory and techniques that provide information about the human element of fisheries management (Brown 1987). To make informed decisions, managers require a variety of human dimensions data. Information about people's attitudes, beliefs, and values helps managers understand what people think and feel about a fishery resource and its importance. Human motivations and expectations are important to understand because they provide a major impetus for people's involvement with fisheries resources. Studies of attitudes, beliefs, values, motivations, and expectations help provide the overall context in which to understand why people use fisheries resources in certain ways (e.g., why do anglers practice catch-and-release fishing?) and why they choose certain fishing locations (e.g., why do some New Yorkers choose to fish Lake Ontario rather than an Adirondack pond?).

Measuring satisfactions allows the fisheries manager to determine to what extent people's needs and desires were met through the fishery resource. Fisheries management actions may be developed to increase satisfactions, either by manipulating the biota (e.g., stock different species), the physical environment (e.g., provide more observation or access points), or the people (e.g., change unrealistic expectations by providing factual information).

Information about the social world or culture surrounding the human element of the fishery helps provide a rich context for the fisheries manager. This context helps the manager understand the source of human beliefs about the fishery and the importance of the fishery to local or regional communities.

Importance of the fishery to people may also be measured through economic assessments. Economic information is often important for commercial fisheries (e.g., understanding economic demand and supply issues) but also for recreational fisheries (e.g., contributions of recreational fisheries to local economies) and for aquatic biodiversity management (e.g., people's willingness to pay for species preservation initiatives).

Ultimately, fisheries managers are interested in human behaviors. People's actions are what affect the fishery resource. Many fisheries management objectives are targeted toward human behavior (e.g., provide a certain level of harvest or provide target levels of angler-days). A complete understanding of human behaviors, however, often requires understanding many of the other types of information noted above. Human behaviors are prompted by beliefs, cultural concerns, economics, and so on. Fisheries managers seeking to influence people's involvement with fisheries resources must understand not only people's behaviors but also the factors related to behavior.

Human dimensions studies allow managers to measure people's opinions and preferences about alternative management options. Understanding people's opinions about management objectives guides creation of fisheries management programs responsive to human needs. Human dimensions research helps managers assess the likely effects of management decisions on people. Understanding opinions about alternative management strategies and techniques enables managers to judge the probable political and social acceptability of various sets of actions. With this understanding, managers can ultimately choose techniques that have a high degree of probable acceptance, effectiveness, and desirable human outcomes within the affected community.

Identifying socioeconomic characteristics of people permits managers to understand more completely who are the stakeholders in fisheries management. The concept of stakeholders reflects the view that people hold a variety of stakes in fisheries resource management. An individual's stake in a fishery reflects his or her behaviors relative to the fishery, the importance of the fishery in the culture to which the individual belongs, or the economic importance of the fishery to the individual. The human dimension information described above helps characterize the type of stake an individual holds in a fishery. Sociodemographic information characterizes who the individuals are and includes items such as age, income, education, residence, and gender. Understanding the relationship of these sociodemographic characteristics and the other human dimensions variables discussed above helps managers design management programs responsive to the needs and abilities of fisheries stakeholders.

From which stakeholders should fisheries managers gather information? The answer is simple, yet complex: all of them. If fisheries managers consistently gather human dimensions information from only one type of stakeholder (e.g., currently licensed anglers), their view of how people are affected by management decisions will be limited.

Human dimensions information may be gathered from the general public or specially targeted groups. Debates about potential fisheries management goals, funding mechanisms, effects on people, and effects on other resources may be germane to general public audiences. Fisheries managers may require information from targeted audiences (e.g., licensed anglers or youths) when considering how to create management programs responsive to user needs. Managers may require human dimensions information about themselves or other professionals. Fisheries and other environmental managers bring their own sets of beliefs, values, and expectations to their professions. Identifying what these are can help avoid professional conflicts and instill an atmosphere of understanding and open communication to the profession (Magill 1988; Knuth et al. 1995).

22.1.2 Types and Characteristics of Data Collection Techniques

Several types of human dimensions research techniques exist. Selection of which type to use for a particular research study depends on the research objectives and data requirements, the characteristics of the populations to be studied, and the time, staff, and funds available for the study. Major types of techniques, described below, include document review, individual interviews, group interviews, mail surveys, telephone surveys, and direct observation. Each of these techniques require proper training of the research staff. In-depth treatment of many of these techniques applied to recreational fisheries is provided in Pollock et al. (1994).

Document review and content analysis. Documents produced or used by stakeholder groups of interest offer insights about the issues and opinions most germane to those stakeholders. Documents useful for review include local newspapers (especially editorial sections), letters received by the management agency, and brochures or fact sheets prepared by interest or advocacy groups that address issues related to fisheries management. Content analysis is the process of collecting data from documents and organizing those data into major themes and categories (Krippendorff 1980).

Document review provides an accessible source of data about perceptions associated with fisheries management programs. The review process is relatively low cost, although it does require staff trained in use of document review techniques. Information available in documents reflects some of the most salient issues for stakeholders because those stakeholders generated the information themselves rather than submitting it in response to a researcher's request (as in the other methods below). Excerpts taken directly from stakeholder-generated documents can be a rich source of material for illustrating human dimensions concepts. Such anecdotes are often useful to help interpret more quantitative data collected through other methods or to help identify the need for more intensive studies.

Several drawbacks are associated with using document review to provide human dimensions information. Documents produced by members of stakeholder groups may not be representative of the beliefs and perceptions within the entire population. Only those people with the most positive or negative perceptions may have been stimulated to write editorials, for example. Document review does not include direct interaction between researcher and participant, so few opportunities for clarification are available. Information within documents also may not be in a form particularly useful to an agency or for addressing issues on the management agenda.

Individual interviews. Interviewing individuals in person (see Payne 1951) allows the researcher to probe deeply for the reasons behind a person's responses. Direct interaction between researcher and participant allows for an exchange of information rather than a one-way transfer from participant to researcher (as in mail surveys, for example). Both closed- and open-end questions can be asked, but many benefits of conducting in-person interviews will be lost if only closed-end questions are used. Individual interviews make information available to the researcher relatively quickly and can be used with individuals who have limited literacy skills.

Individual interviews can be costly, especially in terms of interviewer staff time. Large numbers of participants cannot be accommodated at one time, unless many trained interviewers are available. Interviews require that the researcher and the participant share the same language.

Central-location intercept interviews are a type of individual interview in which the researcher is stationed at a central location frequented by members of the study population. Such locations may include fishing or boating access points, docks, marinas, or fishing-related exhibits. These contacts ensure the researcher that the study participants are in fact involved in the activities of interest and sometimes allow the researcher to observe behavior and verify the accuracy of some of the responses. Central-location intercept interviews, however, must often be limited in scope because the participant does not want to spend much time with the researcher but would rather be involved in the activity at that site. These interviews also require the researcher to be present at various hours during the day to ensure contacts with early and late arrivers (e.g., late-night anglers).

Group interviews. Focus group interviews (see Morgan 1993) typically involve 8–12 participants. The interview sessions last about 2 h. Participants may receive information in advance to help prepare for the group session. A trained facilitator presents the topic and guides the group discussion, allowing interaction between participants. Focus group interviews are sometimes held in facilities designed specially for videotaping and viewing the discussion through one-way glass by interested parties.

Because focus groups encourage group interaction, this method can provide in-depth insights about people's perceptions, beliefs, and values, leading to better understanding by the researcher or manager of reasons behind people's behavior and beliefs. Information produced by focus groups is available quickly.

The utility of information produced through focus groups depends in large part on the ability of the moderator or facilitator to keep the group on task and probe for more information when needed. The technique is limited in terms of the numbers of people that can be accommodated. Costs in addition to staff time may include incentives for individuals to participate in the group (e.g., dinner prior to meeting or a small cash payment) because this technique is very demanding of participant time. Focus groups do not produce quantitative data that can be generalized to the population of interest. This technique is used for its other benefits, such as producing readily available, in-depth information. Insights gained through focus groups may then be used when designing more quantitative, generalizable research approaches.

Mail surveys. Self-administered mail questionnaires (see Dillman 1978, 1983; Pollock et al. 1994) are a popular means of obtaining information from fisheries stakeholders (Brown 1991). Mail surveys allow the collection of detailed data from large numbers of people and allow the respondents to receive certain resource materials (e.g., maps) to aid in the quality of their responses. Mail surveys also allow respondents time to reflect on their answers, which may be desirable if complicated questions about potential management alternatives are asked. The typical time of involvement for the respondent is between 30 and 60 min.

Mail survey costs will vary with the scope of the research study, such as the number of questionnaires sent, the number of reminders required, and the length of the research instrument. This approach often requires a lengthy implementation time to allow for reminders and responses to be received. Several reminder mailings may be required to achieve a reasonable response rate. Mail surveys, however, are usually less costly than in-person interviews and telephone surveys because they require less staff time.

Mail surveys cannot be used to study people who cannot be reached through mailing addresses, who are illiterate, or who read in a language different from that used in the research instrument. Questions and directions in mail surveys must be extremely clear because participants cannot ask the researcher for clarification. Depending on the commitment of the participants, mail survey questions must sometimes be limited to those requiring short answers only, such as check marks or circling multiple choice answers.

Telephone surveys. Contacting people by telephone provides interaction between researcher and participant but in a limited way compared with in-person interviews. Compared with mail surveys, telephone surveys (see Groves et al. 1988; Pollock et al. 1994) may achieve a better overall participation (or response) rate. The personal contact involved in a telephone survey may stimulate more people to participate and ensures the desired respondent is reached. (It is not unusual for mail surveys to be completed by someone other than the intended person, for example, one spouse answering for the other.) The conversation on the telephone may also help stimulate a person's memory more than might a mail survey. Telephone surveys, however, do not usually allow the respondent the time for personal reflection and thought that is available through a mail survey. Telephone surveys, therefore, may be most appropriate for types of questions that do not require much reflection.

Computer-assisted telephone interviewing may be available, in which the data collected are recorded directly into the computer during the interview. This approach reduces the chances of error produced when data are entered from printed forms onto the computer. Direct data entry also reduces the amount of staff time required. Telephone interviews, computer assisted or not, provide immediate access to data, unlike mail surveys in which the researcher must wait to receive the returned questionnaire.

Telephone surveys may involve certain difficulties. Several calls may be necessary to reach potential participants. Some people important to the study objectives may not have telephones in their residence or may have unlisted telephone numbers. Some types of questions cannot be asked effectively on the telephone. Questions that require much thought, questions that require reaction to printed material, or questions that address sensitive subjects may be asked better through mail surveys or other methods.

Telephone surveys are generally less expensive than personal interviews, primarily because staff time is reduced. They may be more expensive than mail surveys, however, due to the staff time involved in telephone calls.

Direct observation. Researchers may visit sites frequented by fisheries stakeholders to observe directly their behaviors and activities. Observation (see Spradley 1980) also enables the researcher to understand the context in which a fisheries management problem exists. Fisheries observation studies are used most frequently when understanding social and cultural patterns is important, particularly those strongly dependent on particular fisheries, such as in commercial fishing regions. Observation studies may be useful for problems such as assessing the role of subsistence fishing in village political relationships, understanding the role of ethnicity as an influence on why people become involved in a fishing industry, or understanding the ways in which participating in a fishery contributes to mental health, childhood socialization, or other aspects of recreational participation. For example, Bryan (1977) combined direct observation with individual interviews to describe a continuum of specialization among trout anglers. He observed differences in clothing, equipment, and preference for management strategies among occasional anglers, generalists, and specialists.

Observers may also be participants in the community rather than passive observers. Participant observation may increase the ability of the researcher to understand the context of the fisheries management problem and may also result in increased trust within the community studied. Gaining such trust is often a requirement for the researcher to have access to people or events in the community.

Direct observation is time-intensive and requires excellent observational skills. More than one observer is often desirable to increase the reliability of the data collected. Participant observers may find themselves drawn into a community to such an extent that their abilities to observe and record data are compromised. Direct observation is not discussed in detail in this chapter due to its limited applicability for general fisheries management issues.

22.1.3 Characteristics of Data

Qualitative versus quantitative data. Researchers can choose to collect quantitative data, qualitative data, or both. Research objectives will determine what types of data are needed. The type of data sought will influence which research methods are selected. In short, qualitative data provide depth and detail for smaller numbers of people, whereas quantitative data use standardized measures for larger groups of people and often allow generalizations about some broader category of people (Patton 1987).

Closed-end questions are the norm for collecting quantitative data; open-end questions are ideal for collecting qualitative data. Closed-end questions ask respondents to choose their answers from among a list of preselected options. Open-end questions require respondents to answer in their own words. Pollock et al. (1994) discussed the advantages and disadvantages of alternative question structures in detail.

Qualitative data can be extremely useful in gauging the variety of potential reactions to alternative fisheries management strategies and in helping fisheries managers understand the reasons underlying the behavior or beliefs of fisheries stakeholders. Qualitative methods may also be valuable prior to designing a quantitative human dimensions study. Qualitative information helps the human dimensions researcher ensure that the range of variables relevant to the stakeholders of interest have been identified. Summarizing and interpreting qualitative data can be time-consuming. However, qualitative data often complement quantitative human dimensions data.

Gathering quantitative data is appropriate when adequate information exists prior to the study to enable the researcher to develop valid standardized measures. Quantitative methods are characterized by confirmation and categorization rather than by expansion and discovery as with qualitative methods. Quantitative approaches typically allow for data collection from more individuals than is possible with qualitative approaches and often allow the researcher to cover a larger range of content areas than is possible in many qualitative studies. Quantitative data are readily summarized through the use of statistical techniques and allow the researcher to make inferences concerning the population under study.

Although each of the methods discussed in this chapter can be used to collect qualitative and quantitative information, certain methods are more commonly used for each type of data. Focus group interviews, some types of personal interviews, document review, and direct observation are the key methods for gathering qualitative data. Mail and telephone surveys and some types of personal interviews are key methods for gathering quantitative data.

Cross-sectional versus longitudinal data. Comparisons of data collected in different locations, from different populations, or at different times increase understanding of the consequences of management actions and actual or potential responses to fisheries management. Cross-sectional studies compare data collected from different stakeholder populations at one point in time. Longitudinal studies collect data from the same stakeholders, or at least the same type of stakeholders, over time, producing a time series of information about how the human dimensions of a fishery have changed. Any type of comparative research requires close coordination of study design, specifically measurement instruments, to ensure that data can indeed be compared.

22.2 THEORETICAL FRAMEWORK

One step easily overlooked in research related to fisheries management is developing the theoretical framework on which the study design should be based (see Chapter 1). The theoretical framework for the study includes a list of the concepts to be measured and their predicted relationships. When faced with pressing management decisions, it is very easy to rush into efforts that answer only the most immediate of those information needs. Such approaches are often incomplete, and important aspects of the immediate management problem may be overlooked in the rush to answer the most obvious questions. Opportunities to collect human dimensions information that will direct future fisheries management program planning may be lost. Potential contributions to improving our general understanding of human dimensions theory may not be realized. Sufficient planning is a key element of human dimensions research that will help ensure later use of the study results (see Pollock et al. 1994).

Advancing the understanding of human dimensions concepts depends on development of a theoretical model, empirical testing of hypotheses based on the model, and refinement of alternative theories or models in response to the data collected (Wagner 1984; Brown et al. 1989). The utility of human dimensions research for fisheries management problems is increased when it follows this general approach.

Developing the theoretical framework helps define the specific concepts that should be studied and refine the information needs. Two forces guide development of the theoretical framework, the practical application (immediate management information needs) and the application of human dimensions theory to the management problem.

22.2.1 Immediate Information Needs

Much of fisheries research is driven by management problems. Resource allocation decisions must be made. To do so, managers must understand the effects associated with various alternatives. Human dimensions studies can provide insight about the likely response to and compliance with proposed new regulations. Human dimensions research can also help managers assess the effects of various management strategies on the people affected by a fishery with shifting biological characteristics.

The information needs created by the immediate fishery management problem usually include questions of where, who, and how much. What locations will be affected by the management decision? Who (what kinds of people) will be affected? How great will the effect of the decision be? Answering only these questions may help solve the pressing fishery management problem, but it may do little to help managers predict or avoid such problems in the future.

Economic data are a type of specific management information need. Methods discussed in this chapter can be used to conduct economic assessments, which characterize the value of the fishery resource in monetary terms (see Box 22.1). For a more detailed discussion of the socioeconomic aspects of fisheries, see Weithman (1993).

22.2.2 Use and Expansion of Existing Models and Theory

Proactive management means being able to predict future scenarios and respond with management before crises develop, perhaps in the process changing what the future will hold. Being able to predict future scenarios depends on answering the question "why." Why do people react in various ways? Why do they hold different expectations about the fishery? Why do they remain involved or lose interest in fishing activities?

A growing body of human dimensions theory exists to answer the question, "Why"—or to at least identify the questions to ask in research so managers can begin to understand the reasons why people believe and behave as they do. Models that have been developed over time by human dimensions researchers can offer insights about the important variables to explore and relationships to examine in fisheries management research.

Existing theory may not provide the answer for a particular management problem, but it may suggest what data a manager should collect. Management-related research can include tests of the relationships predicted by theory. In this way, research about practical management problems can also contribute to development and expansion of human dimensions theory.

Some of the most important theoretical foundations for human dimensions research in fisheries management issues include the body of literature addressing human behavior. Many different models of human behavior exist, some developed explicitly for natural resource-related behavior ("A Social-Psychological Model for Wildlife Recreation Involvement" in Decker et al. 1987 and "Fishing Specialization Theory" in Ditton et al. 1992) and some developed for general human behavior ("Theory of Reasoned Action" in Ajzen and Fishbein 1980 and "Personal Investment Theory" as synthesized by Maehr and Braskamp 1986).

The theoretical foundations underlying a particular research study can often be depicted graphically, illustrating important concepts and their hypothesized relationships (see Box 22.2). Such illustrations help the researcher identify those variables that are important and will be measured in the current study and those that cannot or will not be addressed. Understanding the limitations of a particular study is important, especially when the data are later analyzed and interpreted. Pollock et al. (1994) provided a brief review of theoretical aspects of recreational angler research,

Box 22.1 Summary of Economic Information Components Relevant to Management Decisions[a]

Three primary economic components related to resource users are relevant for managers to consider: (1) trip expenditures; (2) investments in durable equipment items; and (3) willingness to pay, or consumer surplus. Trip expenditures are variable trip costs for items such as gas, bait, food, and lodging incurred while traveling to, engaging in, and returning from a fishing trip. These items can be determined on a per-angler-hour basis, based on completed trip time (see Chapter 21). These data are then expanded for the entire fishery by multiplying average cost per angler-hour by total angler-hours. Investments in fishing equipment represent fixed costs for items that can be used over several years. Fixed expenditures are difficult to estimate accurately because anglers must recall what they paid for the items, which may have been purchased years previously. The cost of each item must be depreciated over the period of ownership to determine current value. Money spent on trip costs and equipment items goes to businesses that provide these goods and services. These expenditures are more a measure of economic benefit to the business community than to the angler.

Consumer surplus is the amount of money over and above the actual expenditures that an angler would be willing to spend at a particular site. Consumer surplus is the amount of money that theoretically could be collected from anglers for the opportunity to fish. Because most recreational fisheries are offered free of charge, this value cannot be directly estimated because a market economy does not exist. There are two alternative methods for measuring consumer surplus indirectly: travel cost estimation and contingent valuation. The travel cost technique uses the costs of travel as a proxy for price to develop a demand curve, from which consumer surplus can be estimated. The contingent valuation approach measures consumer surplus by asking anglers to state their willingness to pay for the opportunity to use the fishery. Each of these economic assessment methods is described in detail in Chapter 16, "Surveys for Economic Analysis," in Pollock et al. (1994).

[a]Prepared by Stephen P. Malvestuto, Fishery Information Management Systems, Auburn, Alabama.

including attitudes, involvement and commitment, motivations, satisfaction, economic impact assessment, and program evaluation.

Researchers, including those seeking to answer practical management questions, can review such human dimensions literature and theory. Researchers assess the relevance of existing theory to the management problem and then develop models to represent the potentially important variables and those variables' interactions related to the management problem. This approach typically indicates information needs that were not apparent when the immediate management problem was considered in isolation.

All management activities, including research, are limited by time, money, and staff resources. The theoretical development phase of a human dimensions study must reflect this reality. Although many types of data might be "nice to know" in any

Box 22.2 Flow Chart for Analyzing Human Responses to Fish Consumption Advisories (from Knuth 1995)

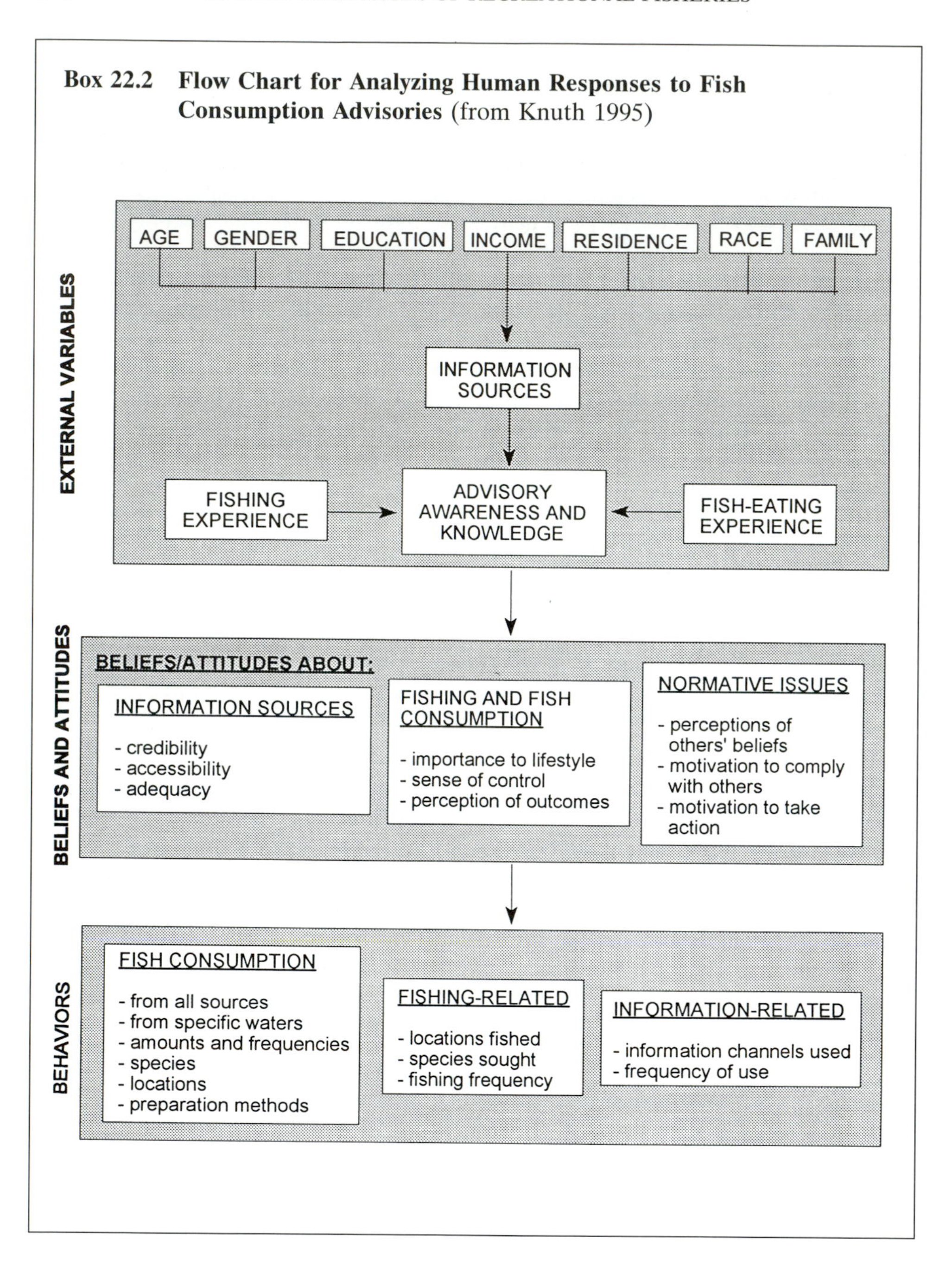

study, the researcher must make careful decisions about what information theory suggests is needed, what information the pending management decision suggests is needed, and the scope of research allowed by the resources available. Questions that will produce data for which the researcher sees no application, either practically or theoretically, should be dropped from the study.

22.3 DESIGN AND IMPLEMENTATION OF THE HUMAN DIMENSIONS STUDY

The theoretical framework provides the foundation on which to design the human dimensions study. Based on the theoretical framework, the researcher should identify specific study objectives (Pollock et al. 1994). The kinds of information to be collected, the specific hypotheses to test or questions to answer, and the population of interest should also be specified. These decisions then influence the choice of the research technique and development of the measurement instrument. A research study need not be limited to one type of research technique. When chosen carefully use of mixed methods may increase the quality of the information and possibly lower the costs of the research study (Dillman and Tarnai 1988).

22.3.1 Selection of the Population of Interest

The theoretical framework should be used to help identify the human population to be studied. Who is likely to be affected by the management decision? Whose values, behaviors, and expectations should be considered when evaluating management alternatives? Whose opinions should be included when evaluating the effectiveness of current fishery management strategies?

The selection of the population to be sampled for the study should be considered for potential biases. For example, angler opinion data collected during a creel survey reflect only the opinions of anglers using that body of water. Anglers who are unhappy with current management of that body of water will not be adequately represented because they are less likely to fish there. Evaluations of management effectiveness will be incomplete if only the current fishery users are consulted. Current fishery users are not the only important fishery stakeholders.

Sampling considerations. The process of sampling, or selecting the members of the study population to participate in the research, depends on the study objectives. Specific considerations for sampling were discussed in Chapter 1. Factors to consider when designing a sampling plan include the following.

The study objectives should help identify if a random or nonrandom sample is required. Random sample selection is required when the study objective is to produce generalizations about the entire population based on data collected from a limited sample. Nonrandom sampling may be used when study objectives include identifying the range of potential responses within a population but not necessarily generalizing about their frequency within the entire population. For example, during early fishery program planning stages, managers may seek to know the potential range of reactions to several management alternatives before proceeding to develop any of them in detail. Purposefully selecting participants who are likely to reflect different values may be the most effective means for collecting such information.

The size of the sample required depends on the desired levels of accuracy and precision for the data (Chapter 2), the desired response rate, the expected variability in the responses, and the resources available for the study. Seeking professional statistical advice will improve the research design (Fowler 1988).

Names, addresses, and telephone numbers of potential study participants can be obtained in several ways. The method selected depends on the population targeted.

Members of specific groups can often be identified via group lists. For example, licensed anglers can be identified through fishing license sales records, boat anglers can be identified through lists of boat permit registrants, and youth involved in fishing can be identified through youth groups such as 4-H or fishing clubs.

Members of the general public can be identified through a variety of means such as telephone books, property owner files, and taxpayer records. None of these methods, however, is a true representation of the entire general public population. Telephone listings, for example, include only those who have a telephone and a listed number.

Most techniques for sampling participants from a study population involve potential bias. The study population should therefore be described carefully and realistically. A study purporting to include all anglers in a given region, for example, may be incomplete if the sample is drawn from license records. Some anglers may be unlicensed (legally or illegally) and would therefore be missed in the sampling process. Anglers may be underage and not required to purchase a license. In many states, senior citizens either are not required to have a license or may fish legally with a general "conservation" license that does not distinguish anglers from hunters or wildlife management area visitors. The number of reported unlicensed anglers fishing in a region has been as high as 65%—40% due to age and 25% due to illegal fishing (Dunning and Hadley 1978).

22.3.2 Development of the Instrument

Any of the methods discussed in this chapter require a data collection instrument. Instruments may be self-administered by the participant (e.g., mail survey) or administered by the researcher (e.g., telephone interview form).

The process of designing a research instrument takes time. Simon and Burstein (1985) suggested beginning with a list of the concepts of interest, based on the theoretical framework, with no regard to specific wording or logical order. After all ideas are recorded, the researcher then begins ordering, clarifying the wording, and simplifying the questions until a useable instrument is produced. Pollock et al. (1994) suggested that all questions, regardless of the type of instrument used, should contribute to important study objectives, be worded clearly and unambiguously, and evoke the most accurate response a participant can provide.

Research instruments should be clear, as concise as possible, and understandable to the research participants. The level of detail should be suitable for meeting the research objectives. Increasing the length and complexity of research instruments may result in limited participation in the study. Even the order of the questions presented, or the sequence of response categories, may influence people's answers (Schuman and Presser 1981). Each question should pertain to a single concept, be relevant to the research objectives, be precise, and be neutral (i.e., the question should not lead the respondent to give a particular answer). Sudman and Bradburn (1983) provided useful advice on questionnaire construction. Pollock et al. (1994: Chapter 4) thoroughly addressed issues of wording and ordering questions for fisheries surveys. These detailed references should be consulted when drafting a survey instrument. Even for the experienced researcher, the reminders included in these references provide a convenient checklist of the do's and don'ts of questionnaire construction.

Careful choice of terminology is critical to ensure the instrument is capable of producing the information needed for the management program. It may be necessary for the researcher to scope out the local community's common terminology before designing a research instrument to ensure the terms used will be commonly understood. For example, common fish names may not be those used in fisheries textbooks but rather those based on local dialect. Asking about satisfaction with a black crappie fishery in many southern states may elicit few responses because anglers do not recognize the fish name. Rather, the researcher should have asked about the "speckled perch" or "silver" fishery. Such scoping or reconnaissance activities to understand the community to be researched will improve the study design. It will also help ensure the range of relevant alternatives will be identified for each closed-end question to be used in the research instrument.

Designing questions that will be good measures of particular concepts includes (1) using complete sentences; (2) carefully defining terms; (3) providing meaningful response categories; and (4) avoiding impressions that the researcher would think negatively about certain answers (Fowler 1988). Pollock et al. (1994) noted that most incorrect answers in human dimensions research are given because (1) the question was vague or misunderstood; (2) the response categories listed were incomplete or not understood by the respondent; (3) the respondent's memory was imperfect regarding the details of the inquiry; or (4) the respondent deliberately provided inaccurate information. Research designs should attempt to avoid these potential problems at the outset.

A variety of question structures are possible. These vary according to the type of response categories used and therefore how the data can eventually be analyzed. Common structures, most of which are discussed in detail in Pollock et al. (1994: Chapter 17), include:

1. checklists, for which the respondent is asked to check off one or more items that apply to his or her preferences, beliefs, attitudes, or behaviors;
2. Likert format agreement scales, in which the respondent is asked to respond to a statement by choosing one of five to seven response options ranging from "strongly disagree" to "strongly agree" (this type of scale may also measure importance: "not at all important" to "extremely important"; frequency: "never" to "always"; or satisfaction: "extremely dissatisfied" to "extremely satisfied");
3. semantic differential lists, in which the respondent considers a specific concept (e.g., flyfishing) and indicates on a continuum where his or her attitude lies (the continuum is anchored on either end by polar opposite adjectives or phrases; for example, "challenging" and "easy");
4. rating questions, in which a respondent is asked to characterize the quality or preferability of a specific item (e.g., 7-point scale ranging from "very poor" to "very good"); and
5. ranking questions, in which a respondent is asked to consider multiple items and rank them in order of importance within the group (e.g., rank five items from 1 to 5 with 5 indicating the most important).

Box 22.3 illustrates examples of commonly used question structures that can be adapted for most of the techniques discussed in this chapter. Other specific examples from fishery studies are included throughout Pollock et al. (1994).

Mail survey questionnaires. The potential effectiveness of mail questionnaires for producing quality data is influenced by several factors. Among these are the population studied and its interest in the research topic and the ease with which members of the population can complete the questionnaire. Many factors may affect

Box 22.3 Example Question Structures for Human Dimensions Research, Adaptable to Mail Surveys, Telephone Surveys, and Interviews

Open-end questions

1. What do you think should be the agency's priority for improving inland fisheries in Ohio?
2. In what year were you born?

Closed-end questions

1. What is your age? (Check one.)
   _____ under 25 years old
   _____ 25 to 44 years old
   _____ 45 to 64 years old
   _____ 65 years or older
2. Some waters can be managed to produce more large (15 inches or greater) largemouth and smallmouth bass, but this usually requires that anglers keep fewer fish. Alternatively, these waters can be managed to provide greater numbers of largemouth and smallmouth bass for anglers to harvest but with fewer large fish. Which option do you prefer? (Check one.)
   _____ More large (15 inches or greater) bass but fewer fish available for harvest
   _____ More bass available for harvest but with fewer large fish
   _____ No preference
3. Think of the type of fishing trip you enjoy the most. How important are the following factors to making the trip a really satisfying experience for you? (Circle one number for each item, a–c.)
   The numbers mean:

   0 = of no concern at all
   1 = not very important
   2 = somewhat important
   3 = important but not essential
   4 = essential for a really satisfying trip

   a. Catching several fish 0 1 2 3 4
   b. Catching a large fish 0 1 2 3 4
   c. Being with friends and family 0 1 2 3 4
4. New York should continue to allow snagging for salmon in at least one section of each major tributary on the Great Lakes. (Check one.)
   _____ Strongly agree
   _____ Agree
   _____ Neither agree nor disagree
   _____ Disagree
   _____ Strongly disagree

Box 22.3 Continued.

Partially closed-end questions

1. Since you learned about fish consumption health advisories, what changes have you made in your fishing habits or in the way you eat the fish you catch? (Check all that apply.)
   _____ I no longer eat any sport-caught fish.
   _____ I eat less fish now than before I learned about the advisories.
   _____ I take fewer fishing trips since I learned about the advisories.
   _____ Other (please specify):________________________

how easy a questionnaire is to complete, including size of the lettering used, length of the survey (number of questions and pages), complexity of individual questions, and the timespan of memory recall required (Brown et al. 1989).

Self-administered questionnaires must be totally self-explanatory. Although a cover letter often accompanies the questionnaire, the letter may be lost. The purpose of the cover letter should be to elicit participation in the study, not provide directions about how to complete the questionnaire.

Mail questionnaires often are designed in a booklet format, the size of a standard sheet of letter paper folded in half (10.8 cm × 14 cm). The booklet format is preferred because is it attractive and less intimidating to the respondent than larger sheets of paper. Larger formats are sometimes required to accommodate diagrams, maps, or matrices that will be completed by the respondent.

The cover of the mail questionnaire should be attractive and enticing to the prospective participant. Cover artwork is often effective in eliciting interest in the content of the questionnaire.

A questionnaire typically begins with brief introductory information explaining the purpose of and general topics addressed in the research instrument. The introduction may include the researcher's name, telephone number, and address in case the respondent has any questions about the study.

The first few questions in a mail questionnaire should be sufficiently captivating to hold the respondent's interest but easy to answer in order to avoid early drop-out due to the complexity of questions. Screening questions may also be placed early in a questionnaire. To help screen out respondents from whom only some information may be necessary. Those respondents are directed to skip forward to another section of the instrument (see Box 22.4). Use of screening questions helps increase response rate because respondents do not become frustrated answering a series of questions not relevant to them. Screening questions also minimize the collection of data that have little utility relative to the study objectives.

Complicated questions may be best placed toward the middle of the instrument. The attention of participants has been captured if they have reached midquestionnaire, and they are not yet as tired of the exercise as they may be by the end of the questionnaire.

Sociodemographic questions, unless used for screening purposes, are typically reserved for the last section of the mail questionnaire for several reasons. First, questions about age, gender, education, income, and residence typically are easy for

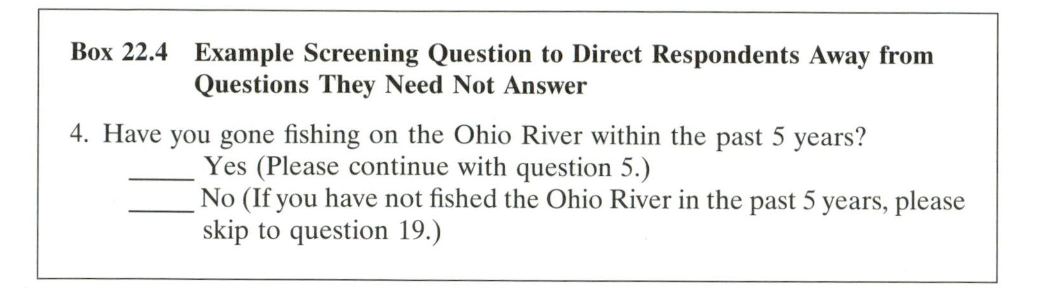

Box 22.4 Example Screening Question to Direct Respondents Away from Questions They Need Not Answer

4. Have you gone fishing on the Ohio River within the past 5 years?
   _____ Yes (Please continue with question 5.)
   _____ No (If you have not fished the Ohio River in the past 5 years, please skip to question 19.)

the respondent to answer. Second, some respondents may be averse to providing such personal data to the researcher. Rather than not return the questionnaire after devoting time to completing the initial portions, a respondent may leave the last section blank but still return the questionnaire. Often, such questionnaires still provide useful data, although one or more sociodemographic characteristics may be unknown. Third, although some respondents may be hesitant to provide personal data, they may be comfortable enough with the research project after having read the other questions in the instrument that they will provide most or all of the demographic data requested.

Mail questionnaires may include space for the respondent to write any additional comments they wish to share. Other required elements are a note of thanks to the respondent for his or her time and effort and directions about how to return the completed questionnaire. An example mail survey instrument is given in Appendix 22.1.

Telephone survey questionnaires. Telephone survey instruments must reflect the needs of the interviewer who will be using them to record data as well as the research objectives. The format of the questionnaire must permit a smooth flow during the interview so participant interest is not lost. Directions to the interviewer may be inserted within the text of the telephone questionnaire but must be set off enough so that the interviewer does not begin reading the text to the participant. Visual cues throughout the questionnaire can be used to direct the interviewer to the next appropriate question depending on the response of the participants.

The writing style used in telephone questionnaires must be conducive to the spoken language because the content will be read by the interviewer. Typically this means shorter sentences and even simpler vocabulary than may be used in written questionnaires.

Some special considerations are required for telephone survey instruments because the respondent cannot see the questions or list of potential responses (Fowler 1988). The number of possible response categories may be limited (e.g., to three or four). The research instrument may instruct the telephone interviewer to read response categories slowly and then read them again and ask the respondent to choose the appropriate category on the second reading. Complex questions may be broken down into sets of simpler questions.

The telephone questionnaire typically begins with an introductory statement by the interviewer, including the interviewer's name, professional affiliation, and the name of the person called (see Box 22.5). Greater use of telephones for telemarketing of products and for solicitation by charitable organizations has made many people wary of accepting telephone calls from strangers. Therefore, introductory statements

Box 22.5 Example Introductory Statement for a Telephone Questionnaire

Hello. May I speak to (Mr. or Ms.) ________________ please? My name is ____________________. I am working for the Nebraska Game and Parks Commission. May I ask you a few questions to learn your opinions about fishing opportunities in Nebraska? (If yes, continue; if no, thank them for their time and hang up.)

should quickly establish the credibility of the interviewer and the research study and stimulate cooperation of the person called (Lavrakas 1987).

Typical questions asked of the interviewer by the participant before agreeing to participate include (1) What is the purpose of the study and how will the information be used?; (2) How did you get my telephone number?; and (3) What group is conducting the study? (Lavrakas 1987). Answers to these and any other anticipated questions should be included on the written research instrument (or a supplementary sheet) and be readily accessible to the interviewer.

Space is provided on the questionnaire to record the telephone number, date, and time of the call. If the person called is unavailable, space is provided to record a set appointment for a call back as well as to list multiple attempts to reach a participant.

Engaging, easy-to-answer questions are usually placed toward the beginning of the telephone instrument, as are screening questions that may allow the interviewer to terminate the session if the participant does not meet the criteria of interest in the study. Difficult, confidential, or highly controversial questions are often left to the end of the telephone survey. These questions may prompt the respondent to terminate the interview. If placed at the end of the session, useful data will have been collected on the less contentious questions before the interview ends.

The telephone questionnaire ends with specific instructions to the interviewer about how to terminate the interview. These include thanking the respondent for participating and describing any rewards or follow-up contacts the respondent may expect.

Individual and group interview protocols. Interview instruments typically begin with a description of the purpose of the study and an explanation of the format of the interview. Standardized descriptions of what types of information will be addressed in the interview and in what form participants will respond to questions should be prepared to ensure consistency among interviewers. Early instructions help form shared expectations between the interviewer and the participant and make the participant feel at ease.

The instrument also includes space for the researcher to record information about the interview itself. Items such as interview number, interviewer identification code, date, time, and site may be included. Other descriptive information that the researcher may check off on a form includes approximate age and gender of each participant, a space for notes regarding the physical setting of the interview location, and any other comments that may be pertinent as the data are analyzed and interpreted.

Interview questions may include a combination of closed- and open-end styles. For open-end questions, lists of key probes may be included on the instrument to remind the interviewer about appropriate prompts to elicit more in-depth responses. If true open-end responses are desired, care must be used to ensure the prompts do not sway the respondent to provide a specific response only because it seems to be desired by the researcher. Sufficient space on the data-recording instrument must be left for the interviewer to record responses and observations during the interview.

In many cases, tape recorders or video cameras are used to record the actual group or individual interview for later transcription and analysis. Permission for recording must be secured from the participants. Even if an electronic record will be made, the interviewer should still have the research instrument from which to read the prepared questions and on which to record observational notes from the interview and reminders about important items to examine in the electronic record and transcript.

Document review protocols. Content analysis instruments must reflect the types of documents that will be reviewed as well as the types of data to be recorded. The instrument typically begins with administrative information such as a descriptor of the document being analyzed (e.g., title, code number, page numbers analyzed, and total pages), an identification code for the document reviewer, time, date, and space for general comments.

The rest of the instrument is formatted to facilitate recording information about the content of the document. Both open-end and closed-end data categories are possible. For example, a reader may list the topics actually encountered in a newspaper editorial (open end) or check off a series of boxes in a predetermined list of topics (closed end). Closed-end lists should typically include an "other" category to allow records of information not specifically listed previously. The instrument may also allow the researcher to record information about the frequency or magnitude (importance) of specific topics in the document or the tone of the text. Characteristics such as reading level, use of tables and graphics, reference to specific information sources, and multiple authorship also may be important to record.

22.3.3 Instrument Quality Considerations

Validity. A valid question is one that measures the concept it is intended to measure. Validity of questions therefore depends on definitions. For example, a researcher may intend that a question asking about the level of satisfaction with a fishing trip encompasses all aspects of the trip: planning, travel to the site, fishing, and travel from the site. The respondent, however, might presume that "fishing trip" means the time on the water actually fishing. Careful, clear wording is necessary to minimize interpretation errors that may reduce validity. Using terminology familiar to the population studied will further enhance validity.

Several approaches are used to assess validity. Pragmatic validation, or predictive validation, involves testing how well a measure enables the researcher to predict behavior or events. In construct validity, the researcher infers validity of the measure by assessing its relationship to other measures of the same concept or its relationship to measures of other concepts to which it theoretically should be related. Discriminant validity infers validity of a measure from the extent to which it is unrelated to measures of theoretically distinct concepts. Face validity is assumed validity based on the intuitively obvious relationship between a measure and the concept to which it is related.

Box 22.6 A Concept-by-Question Matrix

Excerpt from a concept-by-question matrix for a study of resource managers' attitudes toward lake trout rehabilitation in the Great Lakes.

Objective 2. Identify the barriers or constraints managers perceive are most likely to hinder attainment of lake trout rehabilitation goals.

A. Beliefs about the possibility of attaining lake trout rehabilitation goals
 Questions: 10, 11
B. Organismal and species oriented constraints (biological factors)
 Questions: 11, 12, 13
C. Social and society oriented constraints (psychological, sociological, economic, and philosophical factors)
 Questions: 11, 12, 14

Reliability. Reliable measures produce consistent results for the same individual or group each time those measures are employed. A measure may be reliable but not valid if it consistently produces the wrong results. However, a measure cannot be valid if it is not reliable. When a measure is not reliable, it may be due to errors in measurement or because it is not a valid measure of the concept.

Three basic approaches are used to assess the reliability of measures. The test–retest method involves repeated testing of the same sample subjects. If the measure is reliable, the subjects should give consistent answers. A potential problem with the test–retest method is the test effect, that is, subjects may remember how they responded in previous tests. The alternative form method involves developing multiple forms of the same measure, which can then be applied to separate groups of test subjects. Although this avoids problems with the test effect, it requires substantially more work to develop the alternative measures. The third approach to assessing reliability is the subsample method. This involves applying the measure to similar subsamples of a single larger sample group. The subsample method eliminates the test effect and avoids the extra work of developing alternative measures. However, it requires subsample group sizes large enough to yield statistically valid comparisons.

Objective-by-question matrix. An important research tool is the objective-by-question matrix, or the concept-by-question matrix. In such matrices are listed the research objectives or concepts to be studied, accompanied by the individual question numbers from the research instrument indicating the items on the instrument that are designed to address the specific concept area or objective (see Box 22.6).

Researchers review the matrix to identify (1) if all objectives or concept areas have been included in the instrument or if more instrument items should be developed; (2) if certain objectives or concept areas are represented in too many instrument items and some items could be deleted; and (3) if certain questions in the instrument do not correspond to study objectives or concept areas and should be deleted.

Pretesting. The purpose of pretesting is to help ensure that the content and structure of the research instrument is consistent with the research objectives. Is the instrument likely to provide the information sought for the specific management problem the research was designed to answer?

Pretesting may involve several types of people who can each provide different types of information. Professional colleagues may be involved to provide feedback about the readability and clarity of the draft research instrument. Even those unfamiliar with the subject area can indicate which questions are difficult to understand or how wording might be changed to elicit a more precise answer. Those who are familiar with the study objectives can also assess the likely ability of the research instrument to meet the objectives. Is all of the content implied by the objectives addressed in the research instrument? Can any content be dropped from the instrument and still achieve the objectives? The concept-by-question matrix will help the pretest reviewer make these evaluations.

Pretesting also may involve people from among the population to be studied or at least people who share similar characteristics. For example, if the ultimate study population is Ohio River anglers, a subsample of Ohio River license buyers may be selected to participate in a pretest of the research instrument. Sometimes, particularly when study populations are small, the researcher may not want to involve study population members in the pretest. Once individuals have participated in a pretest, they are usually dropped from participation in the full study because they may be predisposed to answering questions in a particular way after having seen the draft research instrument. In such cases of small populations, researchers might choose pretest participants from similar populations. For example, if the study population consists of members of a sporting group in a particular region, the researcher may select members of a group in a neighboring region for the pretest.

It is usually unnecessary to select a large, statistically representative sample of the study population for the pretest. The purpose of the pretest is not to be able to draw statistically significant conclusions about the likely responses of the study population. Rather, the purpose of pretesting is to identify the range of concerns and potential problems that may occur should the draft research instrument be implemented.

Pretesting ideally uses the same format that will be used when the final research instrument is administered, that is, if a mail survey is planned, the pretest is also a mail survey. In some cases, however, direct interaction between researcher and pretest participant will be useful, even if direct interaction is not planned for the final research implementation. For example, the researcher may share a draft mail survey instrument with a select group and allow them to discuss their reaction to the instrument freely in a group setting. The interactions among participants and questions asked of the researcher may produce critical information about the quality of the draft research instrument.

Participation rate. With any human dimensions method, some people contacted are not likely to participate in the research study. Self-selection of participants may result in "nonrespondent bias." This type of bias occurs when the characteristics of the group of respondents are fundamentally different from the group of nonrespondents. For example, people may choose not to participate in a fishery research study because they are less active anglers than those who do respond. Estimations of fishery use based only on information reported by respondents may result in overestimates. Opinions about management activities may reflect only the opinions of current users who are not pleased with management. Those who are pleased may think fishery management will continue as is, so no input is needed from them. In many human dimensions studies, it is important to institute methods to address and avoid potential nonrespondent bias problems.

Definition of adequate response rate varies depending on the diversity of the population studied, the research objectives, and the sample characteristics. Response rates ranging from 65 to 90% have been considered adequate (Dolsen and Machlis 1991), requiring no further need for rigorous follow-up activities with nonrespondents. Response rates below these require collecting information from nonrespondents for comparison with respondent information. Ideally, nonrespondent data would be compared with respondent data to detect differences (Brown and Wilkins 1978; Brown et al. 1981). Where discrepancies are found, weighting techniques can be used to adjust the data for nonresponse bias (Pollock et al. 1994). Researchers can consider stimulating participants to respond by offering an incentive such as cash, maps, keychains, tackle, copies of study results, or other fishery-related information.

Other types of potential bias. Depending on the type of questions asked, the respondent may be required to recall events or activities that occurred over a considerable period of time. Recall bias can reduce the quality of the data. Potential problems with recall include underestimation and overestimation of actual events (Westat, Inc. 1989). Whether underestimation or overestimation is likely to occur depends on the characteristics of the information sought. Some studies have shown that day-to-day activities are frequently underestimated in studies relying on long periods of recall (e.g., 12 months), but unique activities may be overestimated. Asking anglers who frequently fish a local pond to recall how often they fished during the last year may therefore result in underestimates. Asking an angler who fished for Great Lakes salmon a few times in her life how often she fished may result in an overestimate. The researcher should be aware of potential recall bias. Steps to minimize recall bias include using as short a recall period as possible and using prompting questions to stimulate a respondent's memory.

Social desirability bias, or prestige bias, occurs when respondents try to answer in a way that is most desirable, not necessarily most accurate. This type of bias is most likely to occur in controversial situations or with questions related to illegal behavior.

Sampling bias may occur when the population to be sampled does not reflect the population the study was intended to address. For example, a study may be designed to evaluate stakeholder opinions about creating a proposed catch-and-release fishery in a location already supporting a harvest fishery. The study population may be identified as the anglers who currently fish that area. This definition, however, ignores the array of potential anglers who may exist who do not frequent that location because they are seeking a catch-and-release fishery and so must go to other locations. Information collected from current anglers may reflect only those who would feel a loss of opportunity from the proposed change. In fact, the proposed change might create additional use opportunities, but the anglers potentially affected may not have an opportunity to express their opinions and support.

22.3.4 Implementation of the Study

Researchers should remember that any contact with fisheries stakeholders during a research process is part of an ongoing relationship between stakeholder and management agency. Communication during the research process must be clear, professional, and respectful of stakeholders. This includes written directions in mail surveys and the spoken word and body language in personal contact methods.

Legal and procedural concerns. Assuring respondent confidentiality is often a critical component of human dimensions research, particularly when potentially sensitive information is collected. When possible, individual responses should not be reported; if they are, all identifiers should be removed.

Respondent confidentiality during the research process can be enhanced if certain procedures are followed. For example, a code number may be assigned to each participant. As mail surveys are returned, the code number is recorded. Only when producing follow-up mailing lists does the researcher need to link code numbers back to respondent names. Absolute respondent confidentiality may not be possible to ensure because data from a research study conducted by a government agency may be subject to information requests under various freedom-of-information statutes.

Reviews and approvals from within the researcher's own agency are often required for human dimensions research. These usually involve review of the information to be collected and the proposed methods. Reviewers consider the ability to maintain confidentiality; the potential burdens placed on the participant in terms of time, other resources, or physical or emotional stress; and the necessity of including some types of information relative to the research objectives. Concerns focus on the rights of privacy, paperwork and administrative burdens, and any potential risks to participants. Studies conducted with U.S. federal funds are often required to undergo a review by the Office of Management and Budget. State agencies and universities may have their own internal review requirements.

Mail survey. Implementing a mail survey involves multiple contacts with the individuals in the sample. Multiple mailing waves may include several or all of the following:

1. a postcard or letter alerting the potential participant that a questionnaire is on the way;
2. a questionnaire accompanied by a cover letter to elicit interest in participating, and a postage-paid return envelope;
3. 7 to 10 d after the first questionnaire was mailed, a reminder postcard or letter demonstrating the importance of personal involvement in the study;
4. 7 to 10 d after the reminder was mailed, another reminder letter with a second copy of the questionnaire in case the first was lost or misplaced and a postage-paid return envelope;
5. 7 to 10 d after the second copy of the questionnaire was sent, a reminder letter or postcard indicating the importance of receiving responses from everyone initially contacted; and
6. a thank-you for participants that may include a token of appreciation, or a summary of the study results, or both.

The timing of repeat contacts with potential participants allows the researcher to receive responses back before contacting the remaining nonrespondents. As questionnaires are received, they are logged in and the name of the participant is removed from the nonrespondent list. Data are then entered onto computer and analyzed.

The month in which the first mail contact is initiated may influence the ultimate response rate. Brown et al. (1989) suggested that the winter months (January–March) yield the highest response rates, followed by early spring and fall (April and September–November); the summer months and December produce the lowest response rates. The researcher should consider the unique characteristics of the sample population when selecting the time of first contact. One of the authors conducted a survey of state legislators that yielded a poor response rate because the initial mailing was made in October, a time when legislators were preoccupied with election campaigns. An adequate response rate is critical to the quality of a mail survey (Heberlein and Baumgartner 1978).

Telephone survey. Conducting telephone surveys requires contacting potential participants and enlisting their participation. If the researcher anticipates resistance by potential participants or suspicion that the survey is really a telemarketing ploy, the researcher may send a letter to potential participants prior to telephoning that explains the research and advises them of the impending telephone call. Telephone contacts should be scheduled at a time of the day when it is likely to find potential participants at the telephone number available (e.g., usually early evening for residential telephone numbers). Weekends and holidays are usually avoided because individuals may not be home or may not desire to spend their leisure time on the telephone.

Following the survey form, the interviewer requests to speak with the desired individual. If that person is not available, the interviewer attempts to arrange a time to call back. Enlisting cooperation in the study is critical.

Once the interview is underway, the interviewer reads through the questions on the telephone survey form and records the answers provided by the respondent. The interviewer should ask questions consistently from one interview to the next. Depending on the type of study, certain prompting questions or responses to participant inquiries may be allowed. The interviewer must guard against inserting his or her own biases into the telephone interview and thus influencing the answers of the respondent.

Interviews. The researcher conducting personal or group interviews must travel to where the participants are located. Participants for group interviews are identified in advance of the meeting. In-person interviews are often conducted spontaneously, however, and require the interviewer to entice potential participants into being interviewed. Fowler (1988) suggested that successful interviewers seem to be confidently assertive and instantly engaging to potential interviewees.

The physical setup of the interview and the body language of the interviewer are important for successful interviewing. For example, conversation during interviews conducted in an office setting may be enhanced if the interviewer and interviewee are seated on the same side of the table rather than across from each other. The interviewer should maintain an appearance of being interested in what the respondent is saying. An inadvertent frown caused by concentrating on how the interview is going may inhibit the exchange of information. An interviewer who sits with arms crossed during the interview may be perceived as not being open to the respondent's ideas.

At the beginning of the interview, the researcher presents the purpose of the study, providing enough orientation to the project so that the participants will be enticed to participate and will understand (to some degree) why questions are being asked of them. The interviewer should instruct each participant about how responses should be given. In group sessions, it is important to lay the ground rules for the group meeting at the very beginning, such as time limits on individual speakers, protocol for taking the floor (e.g., being called on or taking turns around the room), courtesy to other participants, and the role of the meeting facilitator.

The researcher asks questions during the interview in a standardized manner. Probing questions or deviations from a prepared plan may be necessary and desirable in certain kinds of research, especially in exploratory studies where the purpose is to discover concepts and relationships important to the study participants. In fact, the ability to probe and follow up is one of the strengths of the personal interview technique.

Responses from participants should be recorded accurately. In many cases, a tape recorder is appropriate, assuming permission of the participant is secured. Tape recorded information, however, must be transcribed accurately. If confidential or controversial topics are involved, participants may not agree to be tape recorded. The researcher must have an alternative form of recording data immediately available.

Document review. Reviewing documents requires identifying and securing the appropriate documents that are likely to provide the information desired. The research objectives will help determine the data categories used to design the research instrument. Trained readers review each document by use of the research instrument. Training is required to ensure consistency in data recording if multiple readers are involved.

Typically, two or more people will review separately a subset of the documents to be analyzed, particularly in studies from which quantitative or detailed data are desired. The results of their separate reviews will be compared and discrepancies noted. The researchers discuss further refinements needed in the document review protocol to ensure consistency among readers.

22.3.5 Implementation Considerations

Timing. Timing of a human dimensions study is influenced by the type of information desired and by the characteristics of the fishery and its participants. Information about particular seasonal activities should be collected during or at the close of that season to minimize recall problems. If participants can be contacted at only certain times of the year, implementation must correspond to those limitations. For example, a study of the fish consumption behavior of migrant farmworkers in New York State was limited to the growing season; otherwise, the population of interest (migrant farmworkers) would have already left the state for widely dispersed destinations across the South.

Coding considerations. Coding involves entering the data onto computer for analysis. This process is relatively straightforward for most closed-end responses. Open-end responses must be coded "as is" for text analysis or categorized according to some systematic approach during the coding process. Data collection forms (e.g., returned questionnaires and interview forms) should be checked manually to ensure all answers were entered correctly. Verification techniques such as entering all data twice and comparing the two data sets should be used when possible. Careful analysis of the data should be performed to check for invalid values or answers and those data either corrected or discarded.

Nonrespondent considerations. To increase response rate, a mixed-methods approach may be necessary. Telephone surveys of nonrespondents are often included as the follow-up phase in mail surveys. Data from nonrespondents are compared with data from respondents to judge potential bias. If people refuse to participate in telephone surveys or interviews, they may be willing to at least answer a question about why they do not wish to participate. Such information can help the researcher judge if any pattern exists within the nonrespondent group that is likely to compromise the utility of the data from respondents.

22.4 TOOLS FOR MEASURING THE HUMAN DIMENSION

22.4.1 Computer Software

Implementation software. Many types of computer software may be helpful in the design of survey instruments and implementation of human dimensions studies. Because new software becomes available frequently and existing software is updated regularly, we confine this discussion to general classes of software rather than specific company products.

Attractive design contributes to high return rates of mail surveys. The human dimensions researcher may use word processing software, desktop publishing software, or software designed specifically for use in designing a survey questionnaire. Word processing software has become increasingly sophisticated, allowing the user to design questionnaires in a variety of size and orientation formats, use a variety of fonts, and import graphic images. The primary advantages of word processing software are that almost anyone with a computer has access to a word processing package and printer. The primary disadvantage is that elementary word processing packages that are preloaded on many personal computers occasionally are incapable of performing some of the more sophisticated formatting required to design attractive questionnaires. Desktop publishing software is designed specifically for producing documents that require sophisticated formatting. Although desktop publishing software can easily produce attractive questionnaires, these packages usually require more memory and faster operating speed than word processing software. Neither word processing nor desktop publishing software is capable of assisting the researcher in data analysis.

A few specialized survey software packages are available for personal computers. Because they are designed especially for survey work, they generally contain convenient features for designing a survey and framing questions in commonly used formats. The human dimensions researcher who is minimally computer literate can produce an attractive questionnaire rapidly with a survey software package. These packages are also convenient for data entry and elementary data analysis.

Telephone surveys can make use of more sophisticated computer software. Although survey software packages may be used for telephone surveys, more sophisticated (and expensive) software called computer assisted telephone interviewing (CATI) systems are commonly used by professional telephone survey researchers. In addition to instrument design functions, CATI systems facilitate on-line telephone interviewing at multiple computer work stations. These systems include features such as tracking when telephone calls are made, recording results of each dialing, automatically skipping to appropriate questions based on responses to screening questions, and producing continuous tabulation of survey results. Computer assisted telephone interviewing systems allow many interviewers to conduct different surveys at the same time, and all data entry automatically feeds into the same file. Although the cost of CATI systems (many thousands of dollars) puts them beyond the reach of most individual users, they are valuable aids for agencies or firms that often conduct telephone surveys.

Statistical analysis software. Human dimensions researchers also have many software options available for analysis of survey data. Survey software packages usually include easy-to-use elementary analysis functions such as frequency and cross-tabulation analysis. Generation of tables and figures also is relatively easy to accomplish. For more sophisticated analysis, most survey software packages allow the user to export data files to spreadsheet or statistical analysis software packages. The researcher who produces a survey with word processing or desktop publishing software may wish to enter data directly into spreadsheet or statistical software packages. These applications allow the researcher to perform more complicated sorting procedures, calculations, and statistical tests. In addition, spreadsheet and statistical software packages generally produce higher quality graphics than survey software.

22.4.2 Research Assistance and Organizations

Most fisheries managers have little or no training in human dimensions research methods, although this is changing as newly educated managers enter the field. Most managers, however, quickly discover the importance of human dimensions information in making management decisions. Two kinds of assistance are available to fisheries managers interested in the information human dimensions research can produce. The manager can contract with a social science research organization to do the research or seek advice from other fisheries professionals with human dimensions expertise.

Many diverse types of organizations conduct human dimensions research on a contract basis. Some environmental consulting firms employ social science researchers. The large survey research firms that routinely conduct political polls may be appropriate for some larger projects. A growing number of universities have human dimensions researchers on their faculties.

The fisheries manager seeking advice about human dimensions research should look in-house first. Human dimensions researchers, once extremely rare on agency staffs, are becoming increasingly common. Since 1955, staff of the U.S. Fish and Wildlife Service have coordinated the National Survey of Fishing, Hunting, and Wildlife Associated Recreation. The National Marine Fisheries Service, which has responsibility for most marine fisheries management and must therefore deal routinely with socioeconomic issues, has a small staff of social science researchers and numerous economists. Larger state fish and wildlife agencies, such as those in Missouri and Wisconsin, have had human dimensions researchers on staff for many years. Although these states are the exception among state agencies, the demand for human dimensions researchers in state agencies is growing. Several states have recently hired human dimensions researchers, either in fisheries or wildlife management programs.

Another source of information for fisheries managers who wish to learn more about human dimensions research is professional organizations. The American Fisheries Society offers two subunits of interest, the Socioeconomics Section and the Committee on the Human Dimensions of Recreational Fisheries (under the auspices of the Fisheries Management Section). The Human Dimensions in Wildlife Study Group (Cornell University, Ithaca, New York) publishes a newsletter and the *Journal of the Human Dimensions of Fish and Wildlife Management*.

22.5 REFERENCES

Ajzen, I., and M. Fishbein. 1980. Understanding attitudes and predicting social behavior. Prentice-Hall, Englewood Cliffs, New Jersey.

Brown, T. L. 1987. Typology of human dimensions information needed for Great Lakes sport-fisheries management. Transactions of the American Fisheries Society 116:320–324.

Brown, T. L. 1991. Use and abuse of mail surveys in fisheries management. American Fisheries Society Symposium 12:255–261.

Brown, T. L., C. P. Dawson, D. L. Hastin, and D. J. Decker. 1981. Comments on the importance of late respondent and nonrespondent data from mail surveys. Journal of Leisure Research 13:76–79.

Brown, T. L., D. J. Decker, and N. A. Connelly. 1989. Response to mail surveys on resource-based recreation topics: a behavioral model and an empirical analysis. Leisure Sciences 11:99–110.

Brown, T. L. and B. T. Wilkins. 1978. Clues to reasons for nonresponse and its effect on variable estimates. Journal of Leisure Research 10:226–231.

Bryan, H. 1977. Leisure value systems and recreational specialization: the case of trout fishermen. Journal of Leisure Research 9:174–187.

Connelly, N. A., T. L. Brown, and B. A. Knuth. 1990. New York statewide angler survey, 1988. New York State Department of Environmental Conservation, Albany.

Decker, D. J., T. L. Brown, B. L. Driver, and P. J. Brown. 1987. Theoretical developments in assessing social values of wildlife: toward a comprehensive understanding of wildlife recreation involvement. Pages 76–95 *in* D. J. Decker and G. R. Goff, editors. Valuing wildlife: economic and social perspectives. Westview Press, Boulder, Colorado.

Dillman, D. A. 1978. Mail and telephone surveys: the total design method. Wiley, New York.

Dillman, D. A. 1983. Mail and other self-administered questionnaires. Pages 359–376 *in* P. H. Rossi, J. D. Wright, and A. B. Anderson, editors. Handbook of survey research. Academic Press, New York.

Dillman, D. A., and J. Tarnai. 1988. Administrative issues in mixed mode surveys. Pages 509–528 *in* R. M. Groves, P. P. Biemer, L. E. Lyberg, J. T. Massey, W. L. Nichols, II, and J. Waksberg, editors. Telephone survey methodology. Wiley, New York.

Ditton, R. B., D. K. Loomis, and S. Choi. 1992. Recreation specialization: reconceptualization from social worlds perspective. Journal of Leisure Research 24:33–51.

Dolsen, D. E., and G. E. Machlis. 1991. Response rates and mail recreation survey results: how much is enough? Journal of Leisure Research 23:272–277.

Dunning, D. J., and W. F. Hadley. 1978. Participation of nonlicensed anglers in recreational fisheries, Erie County, New York. Transactions of the American Fisheries Society 107:678–681.

Fowler, F. J., Jr. 1988. Survey research methods. Sage Publications, Beverly Hills, California.

Groves, R. M., P. P. Biemer, L. E. Lyberg, J. T. Massey, W. L. Nichols, II, and J. Waksberg, editors. 1988. Telephone survey methodology. Wiley, New York.

Heberlein, T. A., and R. Baumgartner. 1978. Factors affecting response rates to mailed questionnaires: a quantitative analysis of the published literature. American Sociological Review 43:447–462.

Knuth, B. A. 1995. Guidance for assessing chemical contaminant data for use in fish advisories, volume 4: risk communication. U.S. Environmental Protection Agency, Washington, DC.

Knuth, B. A., S. Lerner, N. A. Connelly, and L. Gigliotti. 1995. Fishery and environmental managers' attitudes about and support for lake trout rehabilitation in the Great Lakes. Journal of Great Lakes Research 21(Supplement 1):185–197.

Krippendorff, K. 1980. Content analysis: an introduction to its methodology. Sage Publications, Beverly Hills, California.

Lavrakas, P. J. 1987. Telephone survey methods. Sage Publications, Newbury Park, California.

Maehr, M. L., and L. A. Braskamp. 1986. The motivation factor: a theory of personal investment. Heath, Lexington, Massachusetts.

Magill, A. W. 1988. Natural resource professionals: the reluctant public servants. The Environmental Professional 10:299–303.

Morgan, D. L., editor. 1993. Successful focus groups: advancing the state of the art. Sage Publications, Newbury Park, California.

Patton, M. Q. 1987. How to use qualitative methods in evaluation. Sage Publications, Newbury Park, California.

Payne, S. L. 1951. The art of asking questions. Princeton University Press, Princeton, New Jersey.

Pollock, K. H., C. M. Jones, and T. L. Brown. 1994. Angler survey methods and their applications in fisheries management. American Fisheries Society Special Publication 25.

Schuman, H., and S. Presser. 1981. Questions and answers in attitude surveys: experiments on question form, wording, and context. Academic Press, New York.

Simon, J., and P. Burstein. 1985. Basic research methods in social science. Random House, New York.

Spradley, J. P. 1980. Participant observation. Holt, Rinehart, and Winston, New York.

Sudman, S., and N. M. Bradburn. 1983. Asking questions: a practical guide to questionnaire construction. Jossey-Bass, Washington, DC.

Wagner, D. G. 1984. The growth of sociological theories. Sage Publications, Beverly Hills, California.

Weithman, A. S. 1993. Socioeconomic benefits of fisheries. Pages 159–177 *in* C. C. Kohler and W. A. Hubert, editors. Inland fisheries management in North America. American Fisheries Society, Bethesda, Maryland.

Westat, Inc. 1989. Investigation of possible recall/reference period bias in national surveys of fishing, hunting, and wildlife-associated recreation. U.S. Fish and Wildlife Service, Washington, DC.

APPENDIX 22.1 1988 NEW YORK STATEWIDE FRESHWATER FISHING SURVEY[1]

Research conducted by the

CORNELL UNIVERSITY COLLEGE OF AGRICULTURE AND LIFE SCIENCES

Department of Natural Resources
in cooperation with

the NEW YORK STATE DEPARTMENT OF ENVIRONMENTAL CONSERVATION

This study concerns your sport fishing in New York State during the 1988 calendar year. We would like you, as the addressee, to fill out the questionnaire, counting only the fishing you personally did or the dollars you personally spent.

Your answers to the following questions will help us draw a composite picture of 1988 New York anglers, their fishing, and their opinions and preferences about a number of fishing topics.

THANK YOU FOR YOUR COOPERATION.

[1]From Connelly et al. 1990.

1988 NEW YORK STATEWIDE FRESHWATER FISHING SURVEY

1. Did you go freshwater fishing in New York between January 1 and December 31, 1988?
   _______ Yes (Please continue with Question 2)
   _______ No (If you did not go freshwater fishing in New York in 1988, please skip to Question 8)

Fishing Preferences and Interests

2. Some waters can be managed to produce more large (15 inches or greater) largemouth and smallmouth bass, but this usually requires that anglers keep fewer fish. Or, these waters can be managed to provide greater numbers of bass for anglers to harvest, but with fewer large fish. Which opinion do you prefer?
   _______ More large (15 inches or greater) bass, but fewer fish available for harvest
   _______ More bass available for harvest, but with fewer large fish
   _______ No preference
3. Higher minimum size limits for trout in streams can produce larger catches, more "recycling" of stocked fish caught and released, and modest gains in the total weight of the take-home catch. But anglers would be allowed to take home fewer of these larger trout. Which option do you prefer?
   _______ Increased use of higher minimum size limits, more larger trout to catch, but fewer to take home
   _______ No change from existing conditions
   _______ Decreased use of higher minimum size limits for trout and allow keeping of smaller trout
   _______ No preference
4. Hook and line caught fish that are not protected by law, such as yellow perch, crappies, bluegill, and bullheads, may be sold legally. Do you think the sale of these species should be:
   _______ Unregulated, as it is now
   _______ Kept legal, but regulated
   _______ Prohibited entirely
   _______ No opinion

5a. Have you ever participated in fishing tournaments such as the ESLO Derby, bass derbies, or other tournaments or derbies in New York?
   _______ Yes _______ No

5b. What is your general attitude about such tournaments or derbies?
   _______ I like the idea of having such tournaments
   _______ I have little interest in tournaments, but do not oppose them
   _______ I am opposed to fishing derbies and tournaments

6. Think of the type of fishing trip you enjoy the most. How important are the following factors to making the trip a really satisfying experience for you? (Circle one number for each item.)

0 = Of no concern at all
1 = Not very important
2 = Somewhat important
3 = Important but not essential
4 = Essential for a really satisfying trip

a. Catching several fish	0 1 2 3 4
b. Catching a large fish	0 1 2 3 4
c. Catching at least one fish	0 1 2 3 4
d. Catching a particular type of fish	0 1 2 3 4
e. Being with friends or family	0 1 2 3 4
f. Being where the scenery is pleasant	0 1 2 3 4
g. Fishing in areas where I know the fish are safe to eat	0 1 2 3 4
h. Trying out new fishing gear	0 1 2 3 4
i. Mastering fishing skills	0 1 2 3 4
j. Catching the most fish of anyone in my group	0 1 2 3 4
k. Catching fish to eat	0 1 2 3 4
l. Fishing where there are few other people	0 1 2 3 4
m. Exploring new fishing areas	0 1 2 3 4

7a. FISHING LOCATION AND EXPENSE TABLE.

Please answer the questions below about all your freshwater fishing in NEW YORK FROM JANUARY 1 to DECEMBER 31, 1988. Since we are only interested in totals for each location, please list each location only once.

WHERE DID YOU FISH IN NEW YORK? (Please indicate below)			HOW MANY TRIPS DID YOU TAKE AND HOW MANY TOTAL DAYS DID YOU FISH AT EACH LOCATION?			ON HOW MANY OF THOSE DAYS WERE YOU FISHING *PRIMARILY* FOR THE FOLLOWING TYPES OF FISH? (Your total for each location should add up to the number of days fished in the preceding column)																HOW MUCH DID YOU SPEND FOR *ALL* TRIPS TO EACH LOCATION? (Gas and oil, food, lodging, rental of boat, and tackle, bait, etc.)	
Location	Name of stream or lake	County or nearest post office	Approximate mileage from your home (one way)	Number of trips	Number of days fished	Yellow perch	Walleye (yellow pike)	Bass	Bluegill/sunfish	Northern pike	Muskie (muskellunge)	Chain pickerel	Lake trout	Rainbow/steelhead trout	Brown trout	Coho or chinook salmon	Atlantic/landlocked salmon	Brook trout	Bullheads/catfish	Other	No Specific Type	Total spent at each location	Total spent while traveling to and from each fishing location
EXAMPLE	Indian Lake	Hamilton	90	4	8			5					1								2	$175	$45
1																							
2																							
3																							
4																							
5																							
6																							
7																							
8																							

7b. If you feel that boat access facilities need improvement at any of the locations you listed in lines 1–8 above, circle the number corresponding to the line(s) in which that location is listed: 1 2 3 4 5 6 7 8

THE FOLLOWING INFORMATION WILL HELP US CATEGORIZE FISHING PARTICIPATION IN NEW YORK AND PREDICT FUTURE INTEREST IN FISHING. ALL INFORMATION IS KEPT STRICTLY CONFIDENTIAL AND IS NEVER ASSOCIATED WITH YOUR NAME.

8. How many years have you been a resident of New York State?
   _______ Years

9. Which of the following best describes the area where most of your first 16 years were spent (Check one):
   _______ Rural, hamlet, or village (under 5,000 population)
   _______ City of 5,000 to 24,999 population
   _______ City of 25,000 to 99,999 population
   _______ Major city of over 100,000 population
   _______ Suburb of major city

10. At what age did you first fish on a fairly regular basis (at least 5 days per year)?
    Age when you first started fishing regularly: _______
    Check here _______ if you have only fished occasionally in the past

11. In approximately what year did you buy your FIRST New York fishing license?
    19_______

12. Prior to last year (the October 1, 1987, through September 30, 1988, license year), in which of the previous 3 years do you think you bought a license that permits fishing in New York (check all that apply):
    _______ 1986–87
    _______ 1985–86
    _______ 1984–85

13. As of today, have you purchased a license that permits fishing in New York in 1989?
    _______ Yes
    _______ No: If not, do you think you will buy a 1988–89 New York license?
    _____ Yes _______ No _______ Uncertain

14. Did you go hunting in New York between January 1 and December 31, 1988?
    _______ Yes _______ No

15. Are you presently a member of a fish and game club or an organized sportsman's group?
    _______ Yes _______ No

16. How many years of school did you complete, counting 12 years for high school graduation and 1 year for each additional year of college, technical, or vocational training?
    _______ years

17. Please circle your approximate TOTAL HOUSEHOLD INCOME before taxes, in thousands of dollars:

    5 6 7 8 9 10 11 12 13 14 15 16 17 18 19 20 22 24 26 28
    30 32 34 36 38 40 45 50 55 60 65 70 75 80 More than 80

Cornell University normally follows a policy of never associating your name with the information you provide. However, it would be extremely valuable to state fisheries managers to be able to contact a sample of anglers who fish a particular waterway at some point in the future. If information is needed in the future that pertains to a waterway that you have fished, may Cornell University or DEC contact you for further information? (Other information such as your education or income would still be kept confidential and not associated with your name.)

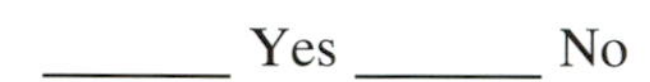

THANK YOU FOR YOUR COOPERATION

TO RETURN THIS QUESTIONNAIRE, simply fold it and place it in the enclosed, self-addressed envelope; postage has been provided.

Symbols and Abbreviations

The following symbols and abbreviations are used without definition in this book. Also undefined are standard mathematical and statistical tests, and symbols of elements in the Periodic Table.

A	ampere
AC	alternating current
°C	degrees Celsius
cm	centimeter
Co.	Company
Corp.	Corporation
cov	covariance
DC	direct current
	District of Columbia
D	dextro (as a prefix)
d	day
df	degrees of freedom
dL	deciliter
E	east
E	expected value
e	base of natural logarithm (2.71828 . . .)
e.g.	(exempli gratia) for example
et al.	(et aliae) and others
etc.	et cetera
F	filial generation
°F	degrees Fahrenheit
g	gram
G	giga (10^9, as a prefix)
ha	hectare (2.47 acres)
Hz	hertz
h	hour
IU	international unit
Inc.	Incorporated
i.e.	(id est) that is
J	joule
K	Kelvin (degrees above absolute zero)
k	kilo (10^3, as a prefix)
kg	kilogram
km	kilometer
L	levo (as a prefix)
L	liter (0.264 gal, 1.06 qt)
log	logarithm
Ltd.	Limited
M	mega (10^6, as a prefix)
	molar (as a suffix or by itself)
m	meter (as a suffix or by itself)
	milli (10^{-3}, as a prefix)
min	minute
mol	mole
N	normal (for chemistry)
	north (for geography)
	newton
N	sample size
NS	not significant
n	ploidy
o	ortho (as a chemical prefix)
Pa	pascal
P	probability
p	para (as a chemical prefix)
pH	negative log of hydrogen ion activity
p	pico (10^{-12}, as a prefix)
n	nanno (10^{-9}, as a prefix)
R	multiple correlation or regression coefficient
r	simple correlation or regression coefficient
S	siemens (for electrical conductance)
	south (for geography)
SD	standard deviation
SE	standard error
s	second
tris	tris(hydroxymethyl)-aminomethane (a buffer)
UK	United Kingdom
U.S.	United States (adjective)
USA	United States of America
V	volt
V	variance
W	watt (for power)
	west (for geography)

α	probability of type I error (false rejection of null hypothesis)
β	probability of type II error (false acceptance of null hypothesis)
Ω	ohm
μ	micro (10^{-6}, as a prefix)
′	minute (angular)
″	second (angular)
°	degree (temperature as a prefix, angular as a suffix)
%	per cent (per hundred)
‰	per mille (per thousand)

Glossary

Terms are defined as they are used in this book. Definitions may be specific to particular contexts and may have legitimately different or broader meanings in other situations.

Abundance: Biomass or numbers of individuals in a population, a portion of the population (such as a year-class), or a sample.

Access point: A specific location where anglers fish from the shore or where they launch and land boats.

Access point survey: A creel or angler survey for which the clerk remains at a defined access point during a sample day or part day and interviews anglers at the ends of their fishing trips.

Acclimation: The process of metabolic compensation in response to change in an environmental factor.

Accuracy: A measure of how close an estimate is to the true value. Compare *precision*.

Acoustic axis: The center axis of the transmitted acoustic beam. Sound intensity is highest along the acoustic axis.

Acoustic pulse or ping: The burst of sound transmitted into the water by a transducer. The pulse has a specific frequency and duration determined by the transducer and echosounder.

Acoustic scatterer: See *acoustic target*.

Acoustic target (aquatic): An object in water, such as a fish, that reflects sound back to a transducer as an echo.

Active capture: Capture of organisms by moving gear. See *passive capture*.

Active metabolism: Maximum aerobic metabolism under a specific set of environmental conditions; often measured while a fasted fish is swimming at its maximum sustainable speed.

Adhesive egg: An egg that adheres to a substrate (e.g., rock, plant, other eggs) upon contact, although adhesiveness may not persist.

Aerial survey (angler): An on-site survey in which anglers fishing in a designated area are counted from an aircraft during a specified time.

Age–length key: A method of assigning ages to fish based on their lengths for purposes of establishing an age-frequency distribution. Fish in a subsample are aged by analysis of their hard parts, and the data are typically organized in a table showing a length range for each age. Fish of unmeasured age then can be assigned to age-groups according to their lengths.

Age validation: Confirmation of aging accuracy (e.g., by monitoring the growth of known-age fish).

Age verification: Measurement of the precision of age determination from hard parts (e.g., by comparing ages assigned by two or more readers).

Alidade: A surveying instrument consisting of a telescope mounted on a rule.

Allele: A form (nucleotide sequence) of a single gene.

Allozyme: A form (amino acid sequence) of an enzyme produced by a specific allele at a given locus (gene). Compare *isozyme*.

Anadromous fishes: Fishes that migrate between marine habitats, where they do most of their growing, and freshwater habitats, where they breed.

Angler contact method: Any method of contacting anglers for a survey. A face-to-face contact at a fishery is often called an "angler intercept."

Angler-hour: One hour of fishing by a single angler.

Angler trip: A visit by an angler to a fishing site or area. Depending on the purposes of the data, an angler trip can be defined as hours fishing at the site in one day, a 24-h day, the number of days in the vicinity of the site, or the number of days away from home.

Annulus: A distinguishable zone on a hard structure that separates successive annual growth zones (plural, annuli).

Anode: The positive electrode or terminal of a direct current circuit.

Anoxic: Devoid of oxygen.

Antimycin: An antibiotic compound produced by species of the fungal genus *Streptomyces*. Antimycin is toxic to fish at low concentrations and is used to sample or eradicate ("reclaim") fish communities. See *reclamation*.

Artisanal fishery: Any small-scale commercial or subsistence fishery, usually associated with developing countries, characterized by one or more traditional gear types such as seines, weirs, gill nets, and scoop nets. The diversity of species landed is often great.

Autoclave: A laboratory device that uses steam at high pressure and temperature to sterilize laboratory materials.

Autocorrelated data: Data whose values are influenced by other data in the set; nonindependent data.

Back-calculation: Use of a consistent relationship between the size of a hard part of a fish (e.g., scale, otolith) and fish length to estimate fish lengths corresponding to previous growth marks on the hard part.

Backwater: A standing-water habitat associated with the floodplain of a stream or river; may be seasonally isolated from the main channel.

Balanced fish population: A population that produces sustainable yields of harvestable-size fish. However, some biologists believe that the concept of balance can also pertain to unexploited fish populations.

Ballast: Heavy material used to sink buoyant objects or to maintain and stabilize towed objects such as nets at specific depths.

Bandwidth: The range of radio, ultrasonic, or other spectral frequencies (unit, hertz) transmitted or received through the narrowest filter in a telemetry, echo-sounding, or other electronic system.

Bar measure (mesh): The distance between adjacent knots in netting, also known as square measure; half the length of stretch measure.

Bayluscide: A nitrosalicylanilide salt, also known as Bayer 73, used to control sea lamprey larvae.

Beam angle (echo sounding): The full angle (in degrees), bisected by the acoustic axis of a transducer, at which the sound intensity is one-half (−3 dB) of that on the acoustic axis.

Beam directivity pattern (echo sounding): The pattern of sound intensities transmitted from or received by a transducer.

Bearing: The horizontal direction of one point with respect to another or to a compass.

Belief: A confidence or conviction held by a person about another person, an object, or a management program.

Benefit (fishery): A return to anglers or society via fishing measured in terms of money or other sociocultural values.

Benthic: On, at, or in the substrate of a fresh- or saltwater body. Compare *demersal*.

Bias: A consistent under- or overestimate of the true population parameter by a statistic.

Biomass: The aggregate weight of a given group of organisms. When biomass is expressed per unit area or volume (e.g., kg/ha or kg/m^3), it is often called standing crop or standing stock.

Biotelemetry: Monitoring an animal via ultrasonic or radio signals transmitted by a device attached to or inserted in the animal. See *radiotelemetry*; *ultrasonic telemetry*.

Buffer (water chemistry): Any substance or mixture of compounds that can maintain the pH of a solution within a narrow range by controlling the concentration of hydrogen ions.

Buoyant egg: An egg that floats or remains suspended in the water column, often because it contains oil globules.

Bus route method: An access point survey in which creel clerks visit several points during a sample day by traveling along a prescribed route and spending a specified amount of time contacting anglers at each point.

Carlin tag: A transbody tag attached by a U-shaped stainless steel wire, the ends of which are passed through the dorsal fish body and crimped together on the far side from the tag.

Cartesian coordinates: The X and Y coordinates that locate a point on a plane or graph with respect to two axes (X and Y) that are perpendicular to each other.

Case: All the attributes measured for one member (e.g., one fish) of a data set.

Catch: The total amount of organisms captured, including organisms kept (harvested) and those returned to the water dead or alive.

Catch and release: Practice of returning angler-caught fish to the water alive.

Catch per unit effort (CPUE): The number or weight of organisms captured with a defined unit of sampling or fishing effort. The most common sampling units are water volume and time. Examples of CPUE are number of larvae caught per cubic meter of water filtered by a net and weight of fish caught per angler-hour. See *catch rate*; *fishing success*.

Catch rate: The number or weight of organisms caught per unit time. See *catch per unit effort*; *fishing success*.

Cathode: The negative electrode or terminal of a direct current circuit.

Check: A zone or ring on a hard body structure used to age fish (e.g., scale, otolith) that represents a period of structural erosion (resorption) associated with energy demands and slowed body growth. Daily checks often can be discerned on the structures of very young fish. In older fish, checks usually form during annual periods of metabolic stress (e.g., winter), but less pronounced checks may result from spawning, brief starvation, temperature shocks, stocking, and other sources of stress.

Circulus: A raised, mineralized ridge-like structure on the surface of a scale that appears as a ring around the focus (plural, circuli).

Clinometer: An instrument for measuring angles of slope.

Coded wire tag: An internal tag made of a small piece of magnetized stainless steel wire that is coded by a system of notches for individual identification.

Community: All populations within a defined geographic area at a given time.

Complemented survey: A survey that comprises more than one contact method or has more than one target population.

Condition factor: A ratio relating fish length to fish weight and measuring the relative plumpness of a fish.

Conductance: A measure of the ability of a circuit to carry an electrical current, measured in siemens (S); the inverse of resistance.

Conductivity: A measure of the ability of a three-dimensional medium (e.g., water or fish) to carry an electrical current, usually measured as conductance per linear distance (μS/cm) in electrofishing; the inverse of resistivity.

Cone hydrophone: A device consisting of a cone surrounding a transducer that receives ultrasonic signals in water.

Continuous data: Data that can assume all real numbers in a given interval.

Continuous signal: An ultrasonic or radio signal that is transmitted continuously, in contrast to a pulsing signal that is intermittently on and off. See *pulsing signal*.

Convex polygon home range method: An animal's home range area measured as the area of a polygon formed by connecting the outermost locations such that all internal angles are less than 180°.

Correlation: A measure of the degree to which two or more variables vary in association with one another.

Count as you go: A roving survey technique in which anglers are both counted and interviewed during the same circuit of a sample section.

Count circuit: A circuit of a sample section made during a roving survey to count all active anglers as a basis for estimating fishing effort.

Cover: An element of habitat that provides shade, concealment, or protection.

Creel: Traditionally, the woven basket used to store harvested fish. Now, the fish harvested by an angler.

Creel census: Complete enumeration of anglers and their harvests in a target population. See *creel survey*.

Creel clerk: A person conducting the field portion of a creel survey.

Creel survey: An on-site survey designed to estimate fishing effort and fish harvest from a sample of anglers. See *creel census*.

Ctenii: Prominent spines on the portion of the scales of ctenoid fishes posterior to the focus.

Current (electric): The flow of electric charge from a region of one potential to a region of another potential, measured in amperes (A). If the flow is constant through time, it is called direct current (DC). If the flow periodically reverses direction, it is called alternating current (AC).

Current density: A spatial expression of current in an electrical field, usually measured as amperes per square centimeter (A/cm^2) in electrofishing.

Daily rings: Zones on a structure (typically an otolith) used to estimate the daily age of fishes.

Dangler tag: Any transbody tag consisting of a message-bearing plate trailing from a (usually U-shaped) stainless steel wire that passes through the fish and is secured on the side opposite the plate.

Dart tag: An external tag made of plastic or thin wire with an arrowhead-like anchor that catches on the dorsal skeleton and a shaft that protrudes from the fish and displays information.

Data logger: A device that can store data (input manually or by other means) in a microprocessor for later transfer to a computer.

Data set: The entire batch of information gathered for analysis.

Day-type stratification: Temporal stratification of survey sampling by the type of day, usually weekdays and weekend days.

Decibel (dB): A dimensionless unit for expressing ratios of sound intensities: 10 times the logarithm of the ratio of a measured sound intensity (I_M) to a reference sound intensity (I_R)—$10 \cdot \log_{10} (I_M/I_R)$.

Demersal: On, at, or near the bottom of the sea. Compare *benthic*.

Densiometer: A spherical mirror divided into grids used to estimate tree canopy density.

Depressor: A weight attached to a net to keep the net at specific depths as it is towed. Compare *diving plane*.

Detritivore: An animal that feeds chiefly on decomposing organic particles and the microbes thereon.

Detritus: Particulate organic or inorganic matter resulting from the decomposition of parent material. Detritus can be inorganic, but most fishery uses of the term refer to organic particles.

Dewar flask: A highly insulated vacuum container designed for holding ultracold liquids, such as liquid nitrogen, at normal atmospheric pressures; commonly used for freezing and storing tissues.

Diel: Pertaining to a 24-h period.

Diel cycle: Daily rhythm or pattern.

Discharge: The volume of water passing a point per unit time; usual metric units are cubic meters per second (m^3/s).

Discontinuous data: Data having no values possible between two points in a series (e.g., integers).

Discrete data: See *discontinuous data*.

Distal: Situated away from the point of origin. Compare *proximal*.

Diurnal: Pertaining to daylight hours. Compare *diel*; *nocturnal*.

Diving plane: A hydrodynamic device (sometimes weighted) attached to a net in such a manner that it exerts a downward force on the net as it is towed. Compare *depressor*.

Doppler frequency shift (Doppler effect): An increase or decrease in the frequency of sound, radio, or other electromagnetic waves caused by movements of the source, a reflective target, a receiver, or all three with respect to one another.

Drift: Movement of organisms due to the force of riverine or oceanic currents.

Drift gill net: An unanchored gill net that floats free with prevailing water currents.

Dual-beam transducer (echosounding): A transducer that has both a wide-angle beam and a narrow-angle beam. Sound is transmitted on the narrow beam and received on both the narrow and wide beams. The ratio of the strengths of the two received echoes allows the position of an acoustic target to be calculated.

Duty cycle: In pulsed direct current or pulsed electronic transmission, the ratio of on-time to total time of one cycle, expressed as a percentage.

Echo: Sound reflected from an acoustic target or scatterer.

Echo counting: A signal-processing technique that counts the number of echoes received from individual targets.

Echogram: A qualitative graphic (visual) representation of echo voltages across time.

Echo sounder: An instrument used to transmit and receive electrical signals via a transducer.

Echo-squared integration: A signal-processing technique used to measure the total amount of acoustic energy reflected back to a transducer. The energy reflected is proportional to total backscattering cross-section.

Effective effort: The amount of fishing effort relative to the proportion of a population caught. Effective effort is affected by daily, seasonal, or other changes in behavior of the target species or age-class.

Efficiency: Any measure of the catch of target organisms in relation to fishing effort.

Effort: See *fishing effort*.

Elastomer tag: A visible implant tag made of a biocompatible polymer.

Electrical circuit: A closed path (e.g., insulated wire) along which electrical charge is carried by electrons. See *electrical field*.

Electrical field: An open path or three-dimensional space (e.g., water) in which electrical charge is carried by ions. See *electrical circuit*.

Electricity: Energy formed by the dissociation of neutral particles into charged particles (i.e., charge carriers such as electrons, protons, and ions).

Electrode: The terminal of an electrical path. In electrofishing, electrodes are the metal elements immersed in water to create an electrical field.

Electrofishing: The use of electricity to capture fish.

Electronic signal detector: An electronic device for detecting a signal from a transmitter.

Electrophoresis: The movement (and separation) of charged particles (usually organic molecules) in response to an electric field.

Electrotaxis: Involuntary movement of fish to orient to an anode and cathode in a direct current or pulsed direct current field.

Embeddedness: The degree that gravel and cobble is infiltrated by sand and silt.

Emergence: Movement of larval fishes out of a substrate such as a buried nest or redd; also hatching of aquatic insects.

Encapsulation (transmitters): Coating the components of a transmitter with epoxy, silicone, or some other weather-resistant and nontoxic material.

Entanglement gear: Devices that capture fish or other aquatic animals by holding them ensnared or tangled in a mesh.

Entrainment: The carrying of aquatic organisms by a natural or artificial current; the term usually is applied with respect to artificial structures such as power plant cooling or irrigation intakes.

Entrapment gear: Devices that capture fish or other aquatic animals when they enter an enclosed area through one or more funnel- or V-shaped openings that deter escape.

Epilimnion: The upper stratum of warm, circulating, and best-oxygenated water in a stratified water body during summer.

Epiphytic: Referring to organisms that live on the surface of plants.

Equal allocation: Assignment of equal numbers of sampling units to each stratum in a stratified random sampling design.

Equal (uniform) probability: Sampling probabilities that are the same for all sampling units.

Error: The difference between a statistic and the true value of a population parameter.

Eutrophic: Having high concentrations of phosphorus, nitrogen, or other nutrients that result in high algal productivity. Water bodies can be naturally eutrophic. Nutrient increase caused by human activities is called eutrophication.

Eutrophy: Trophic state of a water body characterized by high inputs of inorganic or organic nutrients and high biological production rates. See *mesotrophy*; *oligotrophy*.

Expectations: The activities and associated benefits that individuals anticipate they will experience from a fishery resource.

Experimental gill net: A gill net consisting of several panels of different mesh sizes to reduce the overall effects of fish size selectivity of individual mesh sizes.

Experimental unit: Any group of organisms that receives a unique experimental treatment.

Exploitation: Removal of organisms from a population by humans (usually) or other consumers.

External transmitter: An electronic transmitter attached to the exterior of an animal.

Extrusion: Forcing of organisms or other materials through a restraining barrier, such as a net mesh, by the pressure of water or other substances.

False annulus: A zone or ring on an aging structure that has formed at other than an annual interval.

Fecundity: The reproductive capacity of an individual, usually measured as the number of eggs produced by a female in a specified period of time.

Finger throat: An internal funnel-shaped throat of a hoop or fyke net composed of two half-cones of twine on each side of the mouth and secured to a back hoop.

Fishery: A system that includes target organisms, the habitat in which they exist, the community of species in which the target organisms live, and the humans who exploit or affect the target species.

Fishery-dependent data: Data obtained from commercial or recreational fisheries.

Fishery-independent data: Data obtained from scientific sampling programs independently of commercial or recreational fisheries.

Fishing effort: The amount of fishing taking place over a specified period of time in an area or at a particular site. In recreational fisheries, where fishing effort is also called fishing pressure, effort usually is measured in angler-hours or angler-trips. In commercial fisheries, effort usually is expressed in terms of equipment and the number of days it is used: vessel-days, hook-days (longlines), trap-days, etc.

Fishing mortality: See *mortality*.

Fishing power: A measure of the efficiency of a vessel or gear, usually the ratio of the catch taken by the vessel or gear to the catch taken by an arbitrarily defined standard vessel or gear.

Fishing pressure: See *fishing effort*.

Fishing success: Normally the number of fish caught per unit time (catch per unit effort) or harvested per unit time (harvest per unit effort) by an angler or group of anglers.

Fishing trip: A visit to a fishing site or area. Depending on the purposes of the data, a trip can be defined as hours spent fishing in a day or a 24-h period; it can also be total time spent in the area, or total time away from port or home.

Fishing unit: Fishers, vessels, gear types, or combinations of these that harvest commercial resources.

Fixation: The use of chemicals to prevent decomposition of whole bodies, tissues, and cells by stopping enzymatic and microbial functions.

Fixed receiving station: A stationary antenna or hydrophone system for detecting signals from transmitters, often coupled with a device that automatically records data.

Fixed-stratified samples: Samples that are stratified, or segregated, by category, such as by length-group.

Float line: The upper horizontal line of a gill net or trammel net that is either buoyant itself or to which floats are attached so the net will be an upright "wall."

Flowmeter: A device, usually involving a mechanical rotor, that measures the amount of water passing the meter in a given period of time.

Flushing rate: The proportion of a body of water discharged per unit time, calculated as discharge divided by volume.

Focal position: The point location occupied by a fish or other organism. The immediate spatial context is often called the organism's microhabitat.

Focus: The part of a scale, otolith, fin spine, or other circumferentially growing structure that formed first.

Fork length: The length of a fish from the most anterior part of the body to the tip of the medial caudal fin ray.

Formalin: A solution of formaldehyde gas in water, commonly used as a fixative and (in hatcheries) to control parasites and fungus. A saturated solution contains approximately 37% formaldehyde by weight. A 5% formalin preparation, the common dilution for fixatives, thus is approximately 2% formaldehyde.

Frequency (electromagnetic): The number of sound, radio, or other wave oscillations that occur per second, measured in hertz (Hz).

Fulton condition factor (*K* or *C*): An index of fish condition calculated by dividing the weight of a fish by the cubed length of the fish, and multiplying that value by a scaling constant. The condition factors *K* and *C* are used when weight and length are measured in metric or English units, respectively.

Fyke net: A modified (entrapping) hoop net with one or two wings or leaders that guide fish to the net mouth. It is used in areas of little or no water current. See *hoop net*.

Gain (acoustic): Amount of amplification of an acoustic signal (unit, decibels).

Gear selectivity: The bias of a sample obtained with a capture gear.

General linear model (GLM): A statistical model relating a dependent variable to one or more independent variables. The underlying relation is linear such that a unit change in the independent variable(s) produces a unit change in the dependent variable. Examples include simple linear regression, multiple regression, and log-linear models.

Generator: An electromechanical device that converts rotational energy into electrical energy to supply alternating or direct current.

Geographic information system (GIS): A computer database that can produce maps of specialized information overlaying physical maps.

Geomorphology: The science dealing with the nature and origin of the earth's topographic features; the shape of a watershed.

Gilled: Caught in netting that has slipped behind an operculum.

Gill net: A passive capture gear constructed of vertical panels of netting set in a straight line in which fish can become entangled.

Global positioning system: A satellite-based navigational system that provides latitude and longitude to a radio receiver.

Gonadosomatic index (GSI): A ratio of gonad weight to body weight used to assess a fish's reproductive state.

Gradient: The slope or drop in elevation per unit length of stream channel, reach, or watershed.

Grid-cell home range method: A method of determining the area of an animal's home range by summing the areas of all the squares on a gridded map that contain animal locations.

Growth: The addition of length or weight by individuals or the net addition of numbers or biomass to a population. Growth can be negative.

Growth rate: The increase in length, weight, number of individuals, or biomass per unit time.

Habitat: The physical, chemical, and biological features of the environment where an organism lives.

Habitat preference: The proportional use of one habitat type by an organism or species relative to its use of all other available habitats, measured by occurrences or time spent in each habitat.

Habitat use: The number of detected occurrences in, or the time spent in, a particular habitat type by an animal or species.

Halo band: A zone or ring on an age structure that has formed subannually, typically associated with dentary bones.

Harmonic mean: The reciprocal of the arithmetic mean of reciprocal values. The reciprocal values of a characteristic are first averaged, and the reciprocal of that mean is calculated.

Harmonic mean home range: An animal's home range area determined with the harmonic mean of the distances from each animal location to each line intersection (node) on a grid.

Harvest: Fish permanently removed from the water by recreational or commercial fishers.

Harvest per unit effort (HPUE): The number or weight of fish harvested (permanently removed from the water) per unit time (usually per angler-hour). See *fishing success*; *harvest rate*.

Harvest rate: The number or weight of fish harvested per unit time (e.g., angler-hour). See *harvest per unit effort*; *fishing success*.

Herbivore: An animal that feeds chiefly on plants.

Heterogeneity: Having a nonuniform structure (e.g., patches of animals such as zooplankton and larval fish in the water column).

Histology: The branch of science dealing with tissue structure and function.

Home range: The area over which an animal travels in its normal activities, exclusive of migrations.

Homologous (body structures): Similar in position, histology, and embryonic development between different organisms.

Hoop net: A cylindrical or conical entrapment net supported by a series of internal hoops or frames. The net has one or more internal funnel-shaped throats with tapered ends directed inward from the mouth, through which fish enter the enclosure. It is most often used in flowing waters. See *fyke net*.

Household survey: A mail, telephone, or door-to-door survey in which the sampling units are households.

Human dimensions: The body of theory and techniques that provides insights and information about the human element in various endeavors, including fisheries management.

Hyaline: Optically transparent or translucent. See *opaque*.

Hydrograph: A plot of water discharge or water stage through time.

Hydrophone: An underwater microphone that receives signals from ultrasonic transmitters. It contains a transducer that converts vibrations from ultrasonic signals in water to electrical impulses that are sent to a receiver.

Hyperthermia: Stress from high body temperature.

Hypolimnion: The deep, cool, undisturbed, and increasingly oxygen-deficient stratum of water in a stratified water body during summer.

Hypothermia: Stress from low body temperature.

Hypothesis: An as yet unsupported or poorly supported statement that something is true. Null hypotheses (that two or more states, conditions, or experiment outcomes do or will not differ) and research (alternative) hypotheses (that conditions or results do or will differ) are used in statistical analyses.

Hypoxia: Below-normal concentration of oxygen in the blood.

Ichthyoplankton: Fish eggs or larvae drifting in the plankton.

Implanted transmitter: An ultrasonic or radio transmitter surgically placed in the body cavity of an animal.

Increment: An increase; often refers to annual growth that has occurred on hard body parts used to age fish.

Inference: A statistical conclusion about a population based on samples drawn from that population.

Innoculating loop: A wire with a loop end attached to a handle used to move bacterial cultures from one medium to another.

Instantaneous count: A quick count of anglers in a sample area, often taken from a high vantage point.

Instantaneous mortality: See *mortality*.

Intended effort: See *target effort*.

Intensity (sound): The power of a sound wave per unit area (unit, watts per square meter, W/m^2).

Internal transmitter: An ultrasonic or radio transmitter that is placed in the stomach or surgically implanted in the body cavity of an animal.

Interspecific: Between two or more distinct species, referring to comparisons or interactions.

Interval data: Numerical data with a fixed interval between observations.

Interview schedule: See *scripted interview*.

Isotonic: Equal osmotic pressure. When ambient water is isotonic with fish fluids, no net water exchange occurs between the fish and the ambient medium.

Isozyme: One of a set of structurally different (in amino acid sequence) but functionally similar enzymes produced by alternative alleles at the same locus or gene (in which case they are called allozymes) or by different loci.

Karyotype: The number, shape, and appearance of chromosomes within a nucleus; often used to compare or distinguish species.

Kernel: The center of an otolith (also called the nucleus).

Lacustrine: Pertaining to lakes.

Landings: The portion of a commercial catch retained and brought to market.

Lead: A wall or panel of netting extending from the mouth of an entrapment device to guide fish into the enclosure.

Lead line: The horizontal weighted line at the bottom of a gill or trammel net.

Length-of-stay bias: Bias associated with roving angler surveys that arises because anglers who fish for long periods of a day are more likely to be counted and interviewed than are anglers who fish for short periods.

Limnetic: In the open area of a lake away from the shore. Compare *littoral*.

Linear array hydrophone: A series of spatially separated but linked transducers that can sequentially record the passage of ultrasonically tagged fish.

Littoral: The aquatic zone extending from the shoreline of lakes and oceans out to depths where light is insufficient for growth of rooted macrophytes. Compare *limnetic*.

Littoral organisms: Organisms associated with nearshore habitats.

Logbook (fishing): A book of standardized forms used by vessel captains or anglers to record their fishing localities, catches, and landings or harvests.

Longline: A passive angling gear, chiefly used in marine systems, composed of a main or ground line, deployed horizontally and often supported off the bottom by floats, to which is attached short vertical or drop lines, each with a baited hook.

Loran C: A navigational system in which signals from any two (of many) land-based transmitters allow a vessel's position to be calculated accurately by triangulation. The calculation is performed by an on-board receiving unit based on delays in the arrival of the two signals.

Lyophilization: Freezing and drying of tissue; also freeze-drying.

Macrophytes (aquatic): Vascular plants and macroscopic algae that are rooted or attached below the water level and that may be completely submerged, partly floating, or emergent.

Mail survey: A survey in which respondents are mailed a questionnaire to be filled out and mailed back to the survey agency.

Mark: Any induced external, internal, or integumental modification that can be used for recognition of an organism.

Market sampling: Sampling of commercial fishery landings based on inspection of dated sales receipts issued by fish dealers; the receipts or purchase slips record the amount of fish purchased by species and sometimes by market category.

Maximum sustained (sustainable) yield (MSY): The maximum harvest that can be sustained year after year, derived from population models. See *optimum sustained yield*.

Mean-of-ratios estimator (of CPUE): An overall estimate of catch (or harvest) per unit effort for a fishery (CPUE or HPUE, the ratio) calculated by averaging CPUE or HPUE values for individual fishers or sampling days. See *ratio-of-means estimator*.

Measured effort: The fishing effort expended by recreational or commercial fishers who were interviewed.

Melanophore: A cell containing melanin, a black pigment.

Memory recall bias: Bias caused by the inability of survey respondents to remember exactly what occurred at a previous time. This bias is most acute for surveys done well after the experience(s) queried.

Meristics: Counting of serial or segmental structures (e.g., scales, fin rays, myomeres).

Mesenteric fat: Adipose (fatty) tissue in mesenteries of the body cavity.

Mesotrophy: Trophic state of a water body characterized by intermediate inputs of inorganic and organic nutrients and intermediate biological production rates. See *eutrophy*; *oligotrophy*.

Messenger: A device used to open or close sampling gears at depth. Messengers may be mechanical or electronic, but simple messengers are typically weights that are released down a support or tow line to trigger gear closure.

Metalimnion: The layer of a stratified water body in which temperature decreases (top to bottom) most rapidly. It encompasses the thermocline.

Microhabitat: Specific combination of habitat elements in the place occupied by an organism for a particular purpose (feeding, reproduction, etc.).

Migration: An extended movement by an animal from one place to another, typically followed by a return trip to the area previously occupied.

Modified fyke net: A fyke net modified with one or two rectangular frames that improve stability.

Morphology: The form and structure of an organism.

Morphometric: A morphological measurement, or referring to one.

Mortality: The rate of death, expressed as percentage loss, as loss per unit time (per day, per year), or as both (e.g., percent loss per day). When loss is expressed logarithmically, the rate is called instantaneous mortality. If no time units are given for mortality, an annual rate can be assumed. In fisheries contexts, total mortality is often divided into fishing mortality (mortality caused by human exploitation) and natural mortality (mortality caused by all other factors). By convention, instantaneous rates are denoted Z for total mortality, F for fishing mortality, and M for natural mortality ($Z = F + M$). In nonlogarithmic units, total annual mortality is denoted A.

Motivation: The set of beliefs, attitudes, and other factors that stimulate a person to be involved with an issue or process.

Multistage probability sampling: A sampling design in which both primary and secondary sampling units are randomly selected.

Multivariate: More than one variable. Used in reference to statistical techniques that analyze more than one dependent or independent variable simultaneously, including analysis of variance, multiple regression, principal components analysis, and discriminant function analysis.

Narcosis: Muscle relaxation accompanied by stupor and loss of equilibrium.

Natural mortality: See *mortality*.

Necropsy: The postmortem external and internal examination of an animal; analogous to "autopsy" in human medicine.

Neuston: Organisms that live at the air–water interface of a water body.

Neutral buoyancy: Weightlessness, neither sinking nor rising in a liquid or gaseous medium. Fish and their eggs achieve neutral buoyancy by structural (e.g., oil globules) or physiological (e.g., secretion or absorption of swim bladder gas) means.

Nocturnal: Pertaining to nighttime hours. Compare *diel*; *diurnal*.

Noise: Any electrical, electromagnetic frequency, or physical (e.g., wind) interference that hinders the ability of a receiver or listener to detect an artificial or natural signal.

Nominal data: Data that may be separated according to categories but do not have any numerical value or relationship.

Nonresponse bias: The bias in survey results that arises when some people do not or cannot respond to a questionnaire or scripted interview.

Normal distribution: A frequency distribution following a particular mathematical relationship or, more generally, having a symmetrical "bell shape."

Notochord: A cartilaginous longitudinal rod that supports the axis of fish larvae and that later (in bony fishes) is subsumed by the axial skeleton.

Nucleus: The center of a circumferentially growing structure such as an otolith.

Null signal: No or least signal; used in connection with the orientation of an antenna.

Nutrient agar: A gel-like medium used to culture bacteria.

Ohm's law: The resistance (ohms) of any linear electrical circuit is equal to the voltage (volts) divided by the current (amperes).

Oil globule: A discrete sphere of low-density lipids found in both fish eggs and yolk, which aids in buoyancy.

Oligotrophy: Trophic state of a water body characterized by low inputs of inorganic and organic nutrients and low biological production rates. See *eutrophy*; *mesotrophy*.

Omnivore: An animal that eats both plant and animal foods.

Ontogeny: The developmental history of an individual organism from zygote to maturity.

Opaque: Optically dense; used to denote zones on hard body parts representing periods of active growth. See *hyaline*.

Optimum allocation (statistical): Allocation of randomly chosen sampling units in proportion to the size of the stratum, the variability within the stratum, and the cost of sampling the stratum for a specified characteristic of the target population.

Optimum sustained (sustainable) yield (OSY): The maximum sustained yield modified to obtain a sustained mix of biological, social, economic, and political benefits. See *maximum sustained yield*.

Ordinal data: Data that can be arranged from smallest to largest (also known as rank scale). The numbers assigned to items within a variable have a categorical significance such as greater than, less than, or equal to.

Oscillotaxis: Forced movement of fish, a thrashing motion, without orientation to the electrodes in an alternating current field.

Osmotic: Pertaining to the diffusion of water through a semipermeable membrane from a region of lower to higher solute concentration.

Osteological: Pertaining to the structure and development of bone.

Osteological characters: Size, shape, number, and position of bony structures.

Otolith: One of three (paired) structures in the inner ears of fishes that are formed from alternating layers of high and low-density calcium carbonate.

Parameter: A characteristic of a population. See *variable*.

Passive capture: Capture of organisms by stationary gear. See *active capture*.

Passive integrated transponder (PIT): A small transmitter attached to an animal that transmits an identification signal only when it is stimulated to do so by an external electronic "query."

Patchy distribution: Nonrandom aggregations (and voids) of organisms. Patchy distributions can result from physical forces (currents, temperatures, etc.) or biological factors (availability of food or spawning sites, etc.).

Pelagic: In or of the open water column of a lake or ocean.

Pericardial cavity: The body cavity surrounding the heart, lined by the pericardial membrane. See *peritoneal cavity*.

Periphyton: Sessile organisms, such as algae, attached to underwater surfaces.

Peritoneal cavity: The abdominal body cavity, lined by the peritoneal membrane. See *pericardial cavity*.

Peterson disc tag: A transbody tag consisting of two discs attached by a wire that passes through the dorsal muscle of a fish.

Photomicrography: Photography through microscopes.

Photophores: Organs (modified mucus glands in fishes) that produce light neurally or from symbiotic phosphorescent bacteria.

Phototaxis: Movement by an organism toward light (positive phototaxis) or away from light (negative phototaxis).

Phylogenetic: Relating to evolutionary descent.

Physicochemical: Referring to physical and chemical properties. Physicochemical properties of aquatic habitats include dissolved oxygen concentration, temperature, and substrate composition.

Pigmentation: Coloration due to pigments. In fishes, pigments are contained in chromatophores (pigment cells).

Piscicide: Any naturally produced or synthetic compound that, applied to water, kills fish.

PIT tag: See *passive integrated transponder*.

Planimeter: A device consisting of a tracing arm and a measuring scale used to determine the area of polygons (e.g., lake contours or home range maps).

Plotboard: A map with a Cartesian coordinate system (*X–Y* axes) used to plot animal locations and determine the coordinate values of each location.

Population: All individuals of the same species within a defined geographic location at a given time.

Population dynamics: The interactions of recruitment, growth, and mortality that determine the abundance, age structure, and sizes of individuals in a population.

Population structure: The proportional distribution of sizes, ages, or genders in a population resulting from processes of recruitment, growth, and mortality.

Potassium permanganate: An oxidant widely used to detoxify piscicides such as rotenone and antimycin.

Pot gear: Entrapment devices that are portable, rigid traps with small openings through which aquatic animals enter.

Power: Energy per unit time, measured in watts (W); calculated as the product of voltage and current.

Power density: A volumetric expression of power in an electrical field, usually measured as milliwatts per cubic centimeter (mW/cm^3) in electrofishing.

Power rating: Maximum recommended horsepower for boats.

Power transfer theory: A developing concept stating that electrofishing is a power-based phenomenon in which the efficiency of fish capture depends on the transfer of power from water into fish.

Precise estimate: An estimate with a small standard error.

Precision: A measure of repeatability, or of how close repeated measurements are to one another. Compare *accuracy*.

Preservation: Long-term storage of organisms, usually in aldehyde-based or alcohol-based solutions, to prevent enzymatic or bacterial breakdown of tissues and structures. Compare *fixation*.

Pressure (physics): The force per unit area (units, pascals [Pa] or newtons per square meter [N/m^2]).

Primary sampling units: In two-stage probability sampling, the sampling units first subjected to random selection. See *secondary sampling units*.

Proactive management: An approach to managing fishery resources that emphasizes the ability to predict future scenarios and develop appropriate management programs before crises develop; opposite of reactive management.

Probability home range: Home range of an organism calculated as an ellipse or circle based on the variance of the *X* and *Y* spatial values for an animal's locations and enclosing a specified percentage of the animal's locations.

Production: Formation of new biomass by a given group of organisms in a unit of time.

Productivity: The capacity of an ecosystem to produce new biomass in some particular form. It is a relative term without strict qualification.

Progressive count: A count of anglers taken over a protracted period of time as a creel clerk travels through a survey area.

Prophylactic treatment: A treatment used to prevent infections or disease.

Proportional standard error: See *relative standard error*.

Proportional stock density: The percentage of a sample of "stock-length" fish that also are greater than or equal to "quality length." Stock and quality lengths are species-specific. See *relative stock density*.

Proximal: Situated toward the point of origin. Compare *distal*.

Pseudobranch: Small, well vascularized area of tissue in the buccal cavity (mouth) of many fish. The function of pseudobranchs is uncertain; they may supply extra oxygen to the eyes.

Pseudoreplication: The use of inferential statistics to test for treatment effects on data from experiments in which the treatments have not been replicated or the replicates are not statistically independent.

Pulsed direct current: Unidirectional current with periodic interruptions.

Pulse duration: The on-time of one pulse of pulsed direct current, usually measured in milliseconds.

Pulse interval: The time between pulses ("beeps") in a signal from a radio or ultrasonic transmitter that is often used to transmit (code) environmental measurements.

Pulse length: The amount of time for one pulse ("beep") in a signal from a transmitter. Alternatively, the length of an acoustic pulse (units, meters).

Pulse rate: The number of pulses that a signal from a transmitter makes in a given time interval.

Pulse transmission rate: The number of acoustic pulses transmitted per unit time (unit, number of pulses per second).

Pulse width: See *pulse length*.

Pulsing signal: An ultrasonic or radio signal that is alternately on (giving a "beep") and off. See *continuous signal*.

Purse line: A line surrounding an opening of a net (top of a plankton net, bottom of a purse seine) that is drawn tight after sampling to close the net.

Qualitative: Not measured.

Quantitative: Measured.

Radio receiver: A device that is fed radio frequency signals from an antenna, filters unwanted frequencies from the signals, amplifies the signal, and converts the signal to a form that is audible to an investigator or is processed by an electronic signal detector.

Radio signal: An electromagnetic wave frequency that is intermediate between audio frequencies and infrared frequencies (15 to 10^9 megahertz).

Radiotelemetry: A method to monitor the location, behavior, and physiology of free-ranging animals by attaching radio transmitters to the animals and monitoring the signals. Signal frequencies usually are 27–30 megahertz. See *biotelemetry*, *ultrasonic telemetry*.

Radio-tracking: Locating an animal equipped with a radio transmitter and determining its position and movement patterns.

Radio transmitter: A device that transmits radio frequency signals.

Radius: A groove-like depression radiating from the focus to the edge on some scales.

Randomization: Assuring that each element in a population has some probability of selection.

Random sample: A sample from a population in which each element had some defined probability of selection. If it is a simple random sample, each element had an equal probability of selection. See *simple random sampling*, *stratified random sampling*, *systematic random sampling*.

Ratio estimate: A survey estimate calculated as a ratio (e.g., number of fish caught per angler-hour).

Ratio-of-means estimator (of CPUE): An overall estimate of catch (or harvest) per unit effort for a fishery (CPUE or HPUE) calculated by dividing mean catch (harvest) per fisher or sampling day by mean effort (hours, days, gear units, etc.). See *mean-of-ratios estimator*.

Reach: A continuous, uninterrupted stretch of stream or river.

Recall period: The length of time after which respondents are asked to recall prior events. For example, a 2-month recall period implies that respondents are asked to recall events that took place 2 months in the past.

Receiver: A device that is fed electromagnetic signals from an antenna or hydrophone, filters unwanted frequencies from the signals, amplifies the signal, and converts the signal to a form that is audible to an investigator or is processed by an electronic signal detector.

Receiving antenna: A device consisting of rods or wire that is used to receive radio waves.

Reclamation (fish populations): Elimination of all fish in a system, usually by draining the system or treating it with a poison such as rotenone or antimycin, and restocking of the system with selected species. The system can be closed (a pond, lake, or reservoir) or open (a river or stream section). Synonymous with *renovation*.

Recreational fishery: A fishery in which fish are caught for pleasure, not for sale.

Recruitment: The number of fish surviving to a defined size or age. Commonly defined recruitment size- or age-classes are those first vulnerable to the predominant fishing gear, those fully vulnerable to the gear, and those that are reproductively mature.

Redd: A covered gravel nest constructed by a trout or a salmon.

Reference collection: A collection of preserved organisms to be used for comparisons with future collections. Accompanying each organism should be information on capture location, capture date, collector, common and scientific name, etc.

Relative abundance: The proportional (percentage) numerical abundance of a species within a collection of species.

Relative condition factor (K_n): An index of condition calculated by dividing the weight of a fish by a length-specific predicted weight defined from a single population or limited geographic range; contrast with *relative weight*.

Relative species composition: The proportional (percentage) numerical abundance of a species within a collection of species.

Relative standard error (RSE): The standard error expressed as a percentage of the estimate with which it is associated.

Relative stock density (RSD): The percentage of "stock-length" fish that also are in a defined length interval of larger fish. Stock lengths and larger length-classes ("quality," "preferred," "memorable," and "trophy") are species-specific. See *proportional stock density*.

Relative weight (W_r): An index of condition calculated by dividing the weight of a fish by a length-specific standard weight for that species. See *standard weight*.

Renovation: See *reclamation*.

Replicate samples: Two or more samples drawn from the same field or laboratory conditions or treatments.

Resistance (electrical): Opposition to the flow of an electrical current, measured in ohms (Ω); the inverse of conductance.

Resistivity: A volumetric expression of resistance in an electrical field, usually measured in ohm-centimeters (Ω-cm); the ratio of voltage gradient to current density.

Response rate: The percentage of a human sample that respond to an interview or questionnaire.

RNA:DNA ratio: An index of recent growth. The quantity of RNA varies with protein synthesis whereas the quantity of DNA is relatively stable.

Roe: Eggs; caviar is salted and sometimes smoked fish roe. Roe is harvested by removing the ripe ovaries of female fishes or invertebrates.

Rotenone: An organic compound that is toxic to fish and other aquatic organisms; used routinely to reclaim and sample fish communities. Rotenone inhibits cellular respiration.

Roving survey: An on-site survey in which a creel clerk moves through a sample section during a specified length of time to count and contact anglers for interview while they are fishing.

Salinity: The amount of ions in water, expressed as parts per thousand (‰; g/kg water or g/L water).

Sample: A set of items or measurements drawn from a population.

Sample size: The number of items or measurements in a sample.

Sampling bias: A systematic error in sampling inherent in a given sampling technique (e.g., size selectivity of a gill net).

Sampling design: A statistical method of choosing a random sample to estimate specified characteristics of a given population.

Sampling effort: The amount of time (e.g., number of days) and other resources spent to sample a specified population of individuals.

Sampling frame: A list of sampling units (e.g., fishing license receipts or telephone numbers) from which a random sample is drawn.

Sampling probability: The chance of choosing a particular sampling unit at random. Sampling units can be assigned equal or unequal probabilities.

Sampling regime: A definable sampling scheme.

Sampling unit: The basic unit of sampling: a habitat, an angler, a particular combination of space and time, etc.

Sampling variance: A statistic derived by measuring the difference between each sampling unit and the mean of the sample for a given characteristic. Variance is inversely related to precision.

Satisfaction: A measure of the extent to which people's needs and expectations are met by a fishery resource.

Saturation (gear): The point at which an individual gear has captured so many animals that it is no longer effective. In practice, saturation is rarely reached, but capture effectiveness declines as saturation is approached.

Scanning receiver: An ultrasonic or radio receiver that automatically scans specific frequencies or channels at specific time intervals; useful in recording data from transmitters set to various identifying frequencies.

Scope for activity: The maximum potential metabolic energy available for activity, including swimming, growth, and reproduction, under a specific set of environmental conditions; also the difference between standard and active metabolism.

Scripted interview: A survey form that is read to the respondent during a face-to-face or telephone interview. The script is sometimes called a schedule.

Secondary sampling units: In two-stage probability sampling, the sampling units secondarily subjected to random selection after primary units have been selected. Secondary units are "nested" within primary units. See *primary sampling units*.

Seine: A length of netting (usually 1–3 m in depth) with weights at the bottom and floats at the top that is pulled from the ends through the water to sample fishes.

Selectivity: The ability of a gear to catch a certain size or kind of fish relative to its ability to catch other sizes or kinds. For example, the selectivity of a large-meshed trawl for small fish is lower by some factor than its selectivity for larger fish.

Sensing transmitter: An ultrasonic or radio transmitter attached to an animal that also can measure and relay environmental or physiological variables affecting the animal.

Sensitivity (instrumental): The ability of a receiver, hydrophone, or other electronic device to detect a weak signal.

Sextant: A hand-held navigational device used to measure angles between landmarks or altitudes of celestial bodies.

Signal decoder: An electronic device that measures or counts characteristics of ultrasonic or radio signals.

Signal-to-noise ratio: The ratio of the strength of a signal to the background noise. The higher the ratio, the better the signal.

Simple random sampling: Sampling of a nonstratified population such that all members of the population have an equal chance (probability) of being selected.

Sinuosity: A measure of the curvature of a stream; a descriptor of the general meander pattern of a river.

Size selectivity: Relative over- or underrepresentation of specific sizes (lengths) of fish or other animals in a sample taken with a particular gear.

Slurp gun: A hand-operated vacuum tube used by a diver to collect fish by suction.

Soak time: The time that passive sampling gear is left fishing in the water.

Spaghetti tag: A transbody tag of thin vinyl tubing tied off as a loop.

Species selectivity: Over- or underrepresentation of a species, relative to other co-occurring species, in samples taken with a particular gear.

Specific conductance: The movement of electrons through water, which is affected by temperature and the concentration of electrolytes in the water.

Square measure (mesh): Same as *bar measure*.

Square throat: An internal funnel of a hoop or fyke net with a square or circular opening and no twine extension in the constricted end of the funnel; contrast with *finger throat*.

Stadia rod: A graduated, upright surveying rod used in measuring distance and elevation; it also can be used to measure water depth.

Stakeholders: Individuals (or groups) who are affected by, or perceive they are affected by, a fishery resource and its management.

Standardized sampling: Sampling conducted in a prescribed manner that defines conditions such as specific gear, methods of operation, timing, and location.

Standard length: The length of a fish from the tip of the upper jaw to the end of the hypural plate, an array of altered vertebral elements that support the rays of the caudal fin. Maximum standard length is measured from the most anterior portion of the fish, which may or may not be the tip of the upper jaw.

Standard metabolism: Metabolism of an inactive organism.

Standard target (echosounding): An acoustic target with a known target strength that is used to calibrate acoustic hardware.

Standard weight (W_s): A weight established by a standardized regression of weight on length for a particular species. "Standard weight" equations usually embrace fish throughout a species' range and are based on a 75th percentile weight rather than average weight in a length-class.

Standing crop: The total weight of a fish species, or fish community, per unit area at a given time.

Standing stock: Same as *standing crop*.

Statistic: An estimator of a population parameter.

Stock: A group of fishes, frequently a population, believed to constitute a unique genetic resource in a fishery.

Storage ratio: The ratio of the average annual volume of a water body to its average annual discharge volume.

Stratification (statistics): Division of a target population into subpopulations (strata), which are then subjected to simple random sampling. Strata may be based on different elements of the population, on discrete areas, or on time. Ideally, subpopulations are relatively homogenous within strata but differ substantially between strata. The purpose of stratification is to reduce the overall variance of the population estimate; the separate estimates of the subpopulations also may have value.

Stratification (water): The vertical segregation of water masses in a body of water into distinct layers (strata) as a result of differences in water density. Density differences may be induced by dissolved substances or differential heating of the water. For thermal strata, see *epilimnion*; *hypolimnion*; *metalimnion*.

Stratified random sampling: Random sampling of defined subpopulations (strata) for which separate estimates are desired.

Stratified two-stage probability sampling: Stratified random sampling in which primary sampling units (PSUs) are chosen within each stratum and secondary sampling units (SSUs) are randomly sampled within each PSU. Sampling probabilities associated with PSUs and SSUs can be equal (uniform) or unequal (nonuniform).

Stratum weights: The proportions assigned to individual strata that determine how sampling effort will be allocated (partitioned) among the strata.

Stream order: A ranking of the relative sizes of streams within a watershed based on the nature of their tributaries. A first-order stream receives no defined tributaries, a second-order stream results from the confluence of two first-order streams, and so on.

Stress: A physiological response caused by an external stimulus (stressor) that results in an energetic cost to the organism.

Stressor: Any stimulus that causes a stress response.

Stretch measure (mesh): The length of a single mesh of netting, between diagonal knots, when the net is stretched taut; twice the length of bar or square measure.

Substrate: The bottom material of a lake, stream, or ocean (e.g., mud, gravel, bedrock). Also, any biological or physical element to which something is or could be attached.

Sulcus acusticus: The central groove that extends from the anterior to the posterior end of the inner surface of an otolith.

Survey design: The random sampling plan used to choose a sample of elements from a specified population.

Systematic random sampling: Identifying sampling units by randomly choosing the initial unit and then using a fixed interval to choose the remaining units.

Systematics: Hierarchical classification of organisms into a series of groups according to perceived phylogenetic relationships. Compare *taxonomy*.

Tag: An object attached externally or internally to an animal that identifies that animal individually or as part of a defined group.

Target effort: Fishing effort (e.g., angler-hours) directed at particular species of fish.

Target population: The population of individuals to which sampling will be applied. The population of individuals about which inferences will be made based on a sample.

Target strength (echosounding): A measure of the proportion of sound that is reflected from an acoustic scatterer back to the transducer. It is expressed in decibels and is equivalent to 10 times the $\log_{10}$ of the backscattering cross-section.

Taxon: A taxonomic group of any rank, including all subordinate groups. A taxon could be a single species, such as Atlantic herring, or it could be a higher grouping, such as clupeids, which would include all herring species.

Taxonomic key: A branching set of paired statements used for identification of unknown specimens. A specimen is identified by progressing through a series of "yes–no" decisions until a unique decision is made.

Taxonomy: The theory and practice of describing, naming, and classifying organisms. Compare *systematics*.

T-bar anchor tag: An external tag made of plastic with a T-shaped anchor that catches on the dorsal skeleton and a shaft that protrudes from the fish and displays information.

Telemetry: Measurement and transmission of information via a radio or ultrasonic signal to a remote receiving station. See *biotelemetry*; *radiotelemetry*; *ultrasonic telemetry*.

Telephone survey: A survey in which randomly selected respondents (sometimes randomly selected by telephone number) are interviewed over the phone.

Tempering: The gradual change of a fish's thermal environment from one temperature to another when the fish is to be transferred between environments.

Tetany: Muscle contraction and rigidity accompanied by loss of equilibrium, a characteristic response of fish in an electrical field.

TFM: Abbreviation for 3-trifluoromethyl-4-nitrophenol, a synthetic compound that is used to control sea lamprey larvae in tributaries of the Great Lakes and elsewhere.

Thalweg: Path of a stream's or river's deepest water.

Theodolite: A surveying instrument used to measure vertical and horizontal angles.

Thermistor: An electrical resistor whose resistance varies in relation to temperature.

Thermocline: A water stratum with a sharp vertical drop in temperature, sometimes more than 1°C per meter. The thermocline (included in the metalimnion) separates a warmer epilimnion from a cooler hypolimnion, which typically do not exchange waters during the period of stratification.

Threshold (electrical): The minimum level of input (e.g., electrical power) required to achieve a specified output (e.g., narcosis in a fish).

Time blocks: Temporal strata, usually months or seasons.

Time delay tracking system: A system that locates an animal by the differences in time that it takes a signal from a tag transmitter to reach each hydrophone (or sometimes antenna) in an array of three or more fixed hydrophones (or antennas).

Time period probabilities: Sampling probabilities assigned to time periods within a fishing day. Time periods generally are defined as secondary sampling units.

Time-varied gain: An amplification of an acoustical signal that increases with time and is used to compensate for sound spreading and absorption in the water.

Tissue homogenate: A suspension of ground tissue in a buffer.

Total length: The length from the most anterior part of a fish to the tip of the longest caudal fin ray when the lobes are compressed dorsoventrally; also referred to as maximum total length. Natural total length is the length without the lobes compressed.

Total mortality: See *mortality*.

Total ratio estimator: See *ratio-of-means estimator*.

Towing bridle: Rope looped from one side of a towing boat to the other; the attached towing rope is allowed to center itself during a tow.

Trammel net: A passive capture gear generally consisting of two large-mesh panels and a sandwiched small-mesh panel; all three panels are suspended vertically between a float line and a lead line. Fish swim through one large mesh, push the small-mesh panel through the opposite large mesh, and are entrapped in the pocket so formed.

Transbody tag: Any tag affixed by penetrating a fish body from one side to the other.

Transducer (echosounding): A pressure-sensitive device that converts electrical energy into sound energy for transmission and sound energy into electrical energy when reflected sound is received.

Transect: A spatial line or strip along which samples are taken.

Transintestinal expulsion: Movement of a transmitter implanted in a fish's body cavity through the intestinal wall; once in the intestine, the transmitter is expelled through the anus.

Transmission loss (echosounding): The attenuation of sound intensity as sound travels through water. The reduction in sound intensity is caused by the spherical spreading of the sound wave and by the absorption of sound by the water and objects therein.

Transmitter: An electronic device that sends radio or ultrasonic signals. Transmitters can be attached to animals to allow them to be tracked.

Transponder: A transmitter that returns a signal in response to one sent to it.

Trap net: An entrapment device similar to a fyke net in which some frames are square or rectangular and one to four leads guide fish into the enclosure.

Trawl: A triangular pocket-shaped net that is towed through the water to catch aquatic organisms. The net opening has a weighted footrope (bottom) and a buoyed headrope (top) that help keep the net open as it is towed through the water.

Triangulation: A method of determining a location from the bearings of two or more known sites.

Trip expenditures: The amount of money spent on items such as gas, bait, food, lodging, and equipment rental associated with a fishing trip.

Trip length: The length of a fishing trip, generally measured in hours for a day trip, or days for an extended trip.

Trophic state: A relative descriptor of nutrient and organic content of a water body. Lakes are typically classified as oligotrophic, mesotrophic, or eutrophic (lower to higher organic contents) on this basis.

Trot line: Passive angling gear used in warmwater inland fisheries, composed of a main line or ground line strung horizontally to which is attached short vertical "drop" lines, each with a baited hook.

Turnover time: The number of days required to discharge a volume equivalent to the volume of a water body.

Ultrasonic receiver: A device that is fed ultrasonic frequency signals from a hydrophone, filters unwanted frequencies from the signals, amplifies the signal, and converts the signal to a form that is audible to an investigator or is processed by an electronic signal detector.

Ultrasonic signal: An electromagnetic wave or vibration that has a frequency that is just above the upper limit of human hearing, which is about 20,000 cycles per second, or 20 hertz.

Ultrasonic tag: See *ultrasonic transmitter*.

Ultrasonic telemetry: A method to monitor the locations, behavior, and physiology of free-ranging animals by using tags that generate electromagnetic waves with frequencies usually of 20–300 kilohertz. See *biotelemetry*; *radiotelemetry*.

Ultrasonic transmitter: A device that transmits ultrasonic frequency signals.

Unbiased estimates: Estimates from samples that are representative of true population values.

Uncompleted fishing trips: Trips in which anglers are interviewed in the act of fishing before they have completed their fishing trips.

Unequal (nonuniform) probabilities: Sampling probabilities that are assigned in proportion to the expected magnitude of a specified characteristic (e.g., fishing effort).

Values: Ideals, often reflecting social customs, that reflect ethical convictions about the relationship of objects or qualities.

Variable: A characteristic of a sample or a characteristic that varies without reference to a defined population. See *parameter*.

Vertical migration: Repeated patterns of movement (e.g., on a diel cycle) by fishes and invertebrates up and down through the water column in response to light, predators, etc.

Visible implant tag: An internal tag that is injected into transparent tissue so that it can be discovered with the naked eye.

Voltage: The energy per charge carrier in a path (e.g., circuit), measured in volts (V).

Voltage gradient: A linear expression of voltage in an electrical field, usually measured as volts per centimeter (V/cm) in electrofishing.

Voucher specimens: Specimens archived in permanent collections to serve as physical evidence that documents the existence and physical presence of species.

Wading staff: A pole used for support when streams are waded.

Watershed: An area of land from which water drains into a stream or lake; also referred to as drainage basin, catchment, or detached water closet.

Wavelength: The length of one cycle of sound and other electromagnetic phenomena (unit, meters).

Wegener ring: A cylindrical net constructed around hoops, used to sample small fishes in shallow, vegetated habitats; the net is typically thrown from shore or a boat.

Weir: A barrier built across a stream to divert fish into a trap.

Wings: Walls or panels of netting extending at an angle from the mouth of an entrapment device to guide fish into the enclosure.

Within-day sampling: Choosing secondary sampling units (e.g., time periods) within randomly chosen fishing days (primary sampling units).

Zoonotic diseases: Diseases in humans that can be contracted from animals.

Zooplankton: Animals in the plankton, usually dominated by rotifers, copepods, and cladocerans.

Index